Mushrooms of the Georgia Piedmont and Southern Appalachians

Mushrooms

of the Georgia Piedmont and Southern Appalachians

A Reference

by Mary L. Woehrel
and William H. Light

THE UNIVERSITY OF GEORGIA PRESS ● ATHENS

A Wormsloe
FOUNDATION
nature book

Athens, Georgia 30602
www.ugapress.org

Designed by Louise OFarrell
Set in 10/13 Minion by Graphic Composition, Inc.
Printed and bound by Pacom
The paper in this book meets the guidelines for permanence and durability of the Committee on Production Guidelines for Book Longevity of the Council on Library Resources.

Most University of Georgia Press titles are available from popular e-book vendors.

Printed in Korea
21 20 19 18 17 c 5 4 3 2 1

Names: Woehrel, Mary L. | Light, William J. Haugen.
Title: Mushrooms of the Georgia piedmont and southern Appalachians : a reference / by Mary L. Woehrel and William H. Light.
Description: Athens : The University of Georgia Press, [2017] | Series: A Wormsloe Foundation nature book | Includes bibliographical references and indexes.
Identifiers: LCCN 2016049003 | ISBN 9780820350035 (hard bound : alk. paper)
Subjects: LCSH: Mushrooms—Georgia—Identification. | Mushrooms—Appalachian Region, Southern—Identification. | Mycology—Georgia. | Mycology—Appalachian Region, Southern.
Classification: LCC QK605.5.G46 W64 2017 | DDC 579.609758—dc23
LC record available at https://lccn.loc.gov/2016049003

Contents

Preface *vii*

Scientific Synonyms *viii*

The Geographic Area of This Book *ix*

Other Useful Guides *x*

Acknowledgments *xiii*

Introduction *1*

Three Domains and Six Kingdoms: Is That All? *1*

So What Are Fungi Anyway? *6*

Who's Who among the Fungi *8*

Fungi: An Overview *10*

Mating Types and Sexuality *10*

Division/Phylum Basidiomycota (Club Fungi and Mushrooms) *12*

Division Ascomycota (Yeasts and Sac Fungi) *14*

Zygomycetous Fungi (Pin Molds, Bread Molds, and Other Zygospore-Forming Fungi) *16*

Division Chytridiomycota (Chytrids) *17*

Mycorrhizae *17*

- Division Glomeromycota (arbuscular mycorrhizae) *17*
- Ectomycorrhizal Fungi *18*
- Mycoheterotrophic Plants *19*

Carnivorous Fungi *22*

Endoparasitic Fungi *23*

Fungus-Gardening Beetles and Ants *25*

Mycetozoa/Myxomycota (Slime Molds) *27*

Lichens *28*

Mushroom Toxins and Mushroom Poisoning *29*

Medicinal Properties of Some Southeastern Mushrooms *34*

Collecting and Identifying Mushrooms *40*

- What's Needed *40*
- Chemical Staining *41*
- Microscopy *42*

Scientific Nomenclature *45*

References Cited *49*

Mushroom Morphology: A Brief Pictorial Primer *51*

Dichotomous and Pictorial Keys to the Major Mushroom Forms *54*

SPECIES ACCOUNTS

Morels, False Morels, and Elfin Saddles *61*

The Carbonaceous Fungi (Xylariales) *70*

Black Knot *97*

Jelly Fungi 99

Earth Tongues, Club Fungi, and Coral Fungi 115

Parasitic Fungi 147

- Arthropod-Parasitizing Fungi 149
- The Insect Destroyers (Order Entomophthorales) 152
- Mycoparasitic Fungi 156
- Molds, *Hypomyces* 158

Cup and Disk Fungi 168

Class Agaricomycetes: An Overview 185

Puffballs, Earthstars, Truffles, False Truffles, and Deer Truffles 186

Stinkhorns and Stinky Squids (Order Phallales, Family Phallaceae) 212

The Boletes 222

- Old Man of the Woods 224
- Other Boletes 228

Inky Caps (Coprinoid Mushrooms) 265

The Amanitas (Order Agaricales, Family Amanitaceae) 280

The Lepiotas (Order Agariales, Family Agaricaceae) 310

Stropharia-Like Mushrooms (Order Agaricales, Family Strophariaceae) 318

The Corts (Webcaps) (Order Agaricales, Family Cortinariaceae, Genus *Cortinarius*) 324

The Rooted Collybia Complex 336

Oyster-Like (Pleurotoid) Mushrooms 339

Miscellaneous Euagarics 362

- The Magic Mushrooms (*Psilocybe* and *Panaeolus*) 371
- Brown Look-Alikes—Some Edible, Some Deadly Poisonous 382

The Russulas and Milkcaps 418

- Russulas 420
- Milkcaps 430

Chanterelles 445

Corticioid Fungi (Crusts, Spreads, and Parchment Fungi) 467

Polypores (Shelf and Bracket Fungi) 493

Hydnoid (Toothed) Fungi 573

Bird's Nest Fungi and Artillery Fungi 591

Slime Molds (Mycetozoa/Myxomycota) 597

Glossary of Some Mycological Terms 617

Index of Common Names 633

Index of Scientific Names 643

Preface

THIS BOOK is intended to be a field-durable guide to the wild mushrooms of Georgia in the Piedmont region. The public's enthusiastic response to the establishment of a wild mushroom club in Atlanta in 2004 (the Mushroom Club of Georgia) demonstrates the wide appeal of this growing and unique pastime in this area. The timing is right because of the green movement and the emphasis on local food sources, organic produce, "growing your own," and enjoying nature.

We had several reasons for writing this book:

- To extend the continuity of species accounts provided in other books for the Northern and Central Appalachians to the Southern Appalachians and Georgia Piedmont
- To provide accounts of the unusually wide range of mushrooms in our area from northerly species found in North Georgia/North Carolina to subtropical species found in the Piedmont
- To bring rigorous taxonomic analysis to bear on the many species having confused nomenclature
- To provide accounts of both seldom-seen species and those that are most familiar. The Southeast is rich with unusual and rarely found varieties.
- To provide a resource specific to the area for the growing number of mycophiles in the Georgia Piedmont and the Southern Appalachians

Nomenclatural problems abound in most of these works because of the many synonyms and very confused taxonomic history of most of the species treated. This is discussed in detail in the introductory section, and a complete list of synonyms of all taxa treated will be available online in Appendices A (fungi) and B (slime molds).

Specialists often tend to throw their keys together as an afterthought, casting about for features they can fit into a dichotomous-style key without any consideration given to the features' relevance or usefulness. They do not place keys high on their priority list for rigorous attention and treatment. In fairness, we note that some online keys and species accounts are excellent, such as those presented by the amateur (but very professional) mycologist Dr. Michael Kuo (MushroomExpert.Com); Dr. Gary Emberger of Messiah College in Grantham, Pennsylvania; and several others.

Clear-cut and accurate descriptions and illustrations of mushrooms are critical to mycophagists who enjoy harvesting and consuming wild mushrooms, a pastime that is fast approaching the popularity of birdwatching in this country. Many of the photographs in current guides are "Hollywood" depictions of these fungi that do not show the various forms and color morphs that fungi assume in nature. This is especially critical for individuals who immigrate to the United States from Europe and Asia, where wild mushrooms are regularly harvested and eaten. Some of our dangerously poisonous mushrooms closely resemble those routinely consumed abroad, and most cases of mushroom poisoning involve immigrants who are naive about our native fungi.

We aimed to produce a useful, well-organized, completely illustrated, and workable guide to many of the common and not-so-common mushrooms of the Georgia Piedmont and southern Appalachian region that is scientifically and nomenclaturally accurate and easy to use for both beginning mushroom enthusiasts and more advanced mycophiles. Photographs and graphics are presented together with the verbal descriptions of the species they illustrate rather than in a separate section. The user will thus be able to directly compare the verbal descriptions by looking at the referenced photograph or image. When appropriate, specific features will be indicated by arrows.

An organizational problem in a book such as this is deciding whether to group several species according to their taxonomic groups (i.e., families and genera) or on the basis of their superficial appearance. In other words, should an effuse, crust-like, charcoal-like spread be grouped in a section with other somewhat similar-looking crust-like fungi, or should it be placed together with the erect members of carbonaceous mushrooms to which it is closely related? We have made this and similar determinations on a case-by-case basis that we think will best serve our (and our readers') interests. For some groups we have provided a brief introductory section, as well as dichotomous verbal, and in some cases also tabular, or matrix keys to the genera and species before the individual species accounts.

We have made every effort to ensure that all verbal and dichotomous keys in our field guide contain *exactly matching* couplets so that direct comparisons can be made between matching suites of characters. And when needed, tabular keys will permit the identification of fungi in various growth and reproductive stages based on a suite of morphological and other features not easily presented in dichotomous keys.

We have attempted to provide specific and useful characters that distinguish the species presented here from similar-looking taxa with which they might be confused. And since we are biologists, we have also included detailed (and occasionally extensive) comments about the natural history of these fascinating organisms.

The introduction is intended to

- give a brief discussion on collecting and identifying mushrooms;
- give an overview of the format and structure of the book;
- give a brief overview of edible and medicinal mushrooms;
- discuss the different types of mushroom toxins and their medical significance;
- discuss how to distinguish and avoid consuming poisonous species;
- discuss mycorrhizal relationships; and
- give an overview of mushroom biology, natural history, and ecology.

The introduction also presents an overview of

- the currently recognized three-domain, six-kingdom classification of living organisms, in which the Fungi are one of the four kingdoms currently comprising the Domain (Supraregnum) Eucarya;
- the basic biology and reproductive strategies of the Fungi, including sections breaking down the divisions of fungi that have the tongue-twisting names of Ascomycota, Basidiomycota, "Zygomycota,"[1] and Chytridiomycota: what they are and the differences between them;
- the taxonomy and scientific nomenclature as applied to the Fungi, including the distinction between teleomorphic and anamorphic nomenclatures; and
- mushroom toxins, their mode of action, and the medical management of mushroom intoxication, plus simple caveats on avoiding dangerous mushrooms (several of which are *extremely* lethal).

Scientific Synonyms

Most mushroom taxa have a tortuous and highly confused nomenclatural history, which has often resulted in many different scientific names being applied to the same species. The reasons for this are many and will be discussed in the section dealing with scientific nomenclature. For example, one of the species given as an example in the introduction has nearly two dozen different scientific names—synonyms—and many widespread species have many more synonyms than that. We provide these extensive synonymies for the benefit of professional mycologists and taxonomically inclined mycophiles. Since including these often extensive lists is not suitable for a field guide such as this, we have made the synonyms of each species discussed available as two online appendices, which will be hosted by the University of Georgia and will be readily accessible to the reader(s). Appendix A is for the Fungi, and Appendix B is reserved for the Mycetozoa (slime molds).

Lichens and slime molds have traditionally been treated

1. The Zygomycota no longer exists as a division or phylum. It included a wide array of quite unrelated organisms and has been split up into six evolutionary lineages, which are discussed in the introduction. We use the term "Zygomycota" in the interim in its former sense.

within the domain of mycology, and they are briefly discussed below in the introduction. Because slime molds are of special interest to most professional and amateur mycologists, some of the more common species are treated here. But time and space considerations prevent us from including a selection of lichens in this work, even though they involve a symbiotic fungal partner.

The Geographic Area of This Book

The geographic area covered by this book includes the Georgia/South Carolina Piedmont and Southern Appalachian Mountains.

The Piedmont is bounded on the south and east by the Fall Line, which separates it from the Coastal Plain. The Fall Line separates the hard metamorphosed crystalline rocks of the Piedmont from the softer sedimentary rocks of the Coastal Plain. Rivers arising in the Piedmont flow primarily southeast and typically form waterfalls at the Fall Line where the Piedmont and Coastal Plain meet. To the northwest the Piedmont is bordered primarily by the Blue Ridge Mountains, the easternmost range of the southern Appalachians, which are especially prominent in the Carolinas and that portion of Georgia where the Brevard Fault lies. The Blue Ridge Mountains consist of very highly metamorphosed and crystalline rocks to form the Crystalline Appalachian Province. Northwest of the Blue Ridge Mountains lie the Valley and Ridge Province and the Cumberland Plateau, respectively. These last two provinces comprise the Sedimentary Appalachian Province.

The dissected plains and rolling hills of the southern Piedmont form a transitional upland region between the flat Coastal Plain and the mountains, with elevations ranging from about 180 to 455 meters (600 to 1,500 feet). Less diverse than either the Coastal Plain to the southeast or the Blue Ridge Mountains to the northwest, the Southern Piedmont nevertheless supports mixed oak-hickory-pine forests, with American Beech, Tuliptree (*Liriodendron tulipifera*, often called "Yellow Poplar"), Sweetgum, and three species of magnolias, and with River Birch, Black Willow, Sycamore, and both River and Giant Cane along waterways. The dominant oaks are Southern Red Oak,

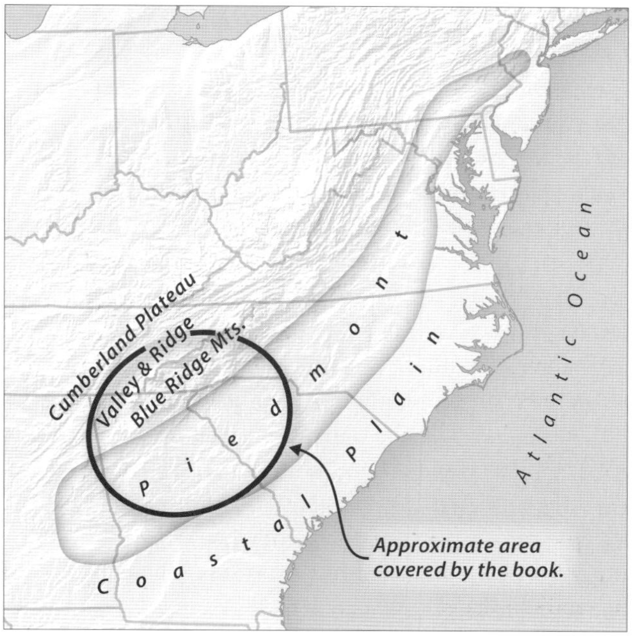

Approximate Area Covered by the Book

Northern Red Oak, Chestnut Oak, Swamp Chestnut Oak, Black Oak, Scarlet Oak, Post Oak, and Blackjack Oak, with White Oak in the northern part of our area. Several species of hickories are also abundant. Conifers include Loblolly Pine, Shortleaf Pine, and Longleaf Pine along the Fall Line. A wide variety of smaller trees and shrubs, some of them invasive exotics like Privet, make up a dense understory.

Elevations in the Blue Ridge Mountains range from 600 meters (2,000 feet) or so near Asheville, North Carolina, to 2,025 meters (6,684 feet) at the top of Mount Mitchell less than 20 miles away. This is accompanied by an average temperature change of about 14°C (25°F). This is one of the richest and most biodiverse regions in North America and boasts some 2,000 species of fungi. Trees include Red Spruce and Fraser Fir, which comprise an ecosystem at the highest elevations, Eastern and Carolina Hemlock, Eastern White Pine, Virginia Pine, Table Mountain Pine, Pitch Pine, many of the same oaks and other hardwood trees that occur in the Piedmont, in addition to Eastern White Oak, Yellow Birch and Sweet Birch, Yellow Buckeye, and others. Many of the same understory trees and shrubs found in the Southern Inner Piedmont (closer to the mountains) also occur here, plus various species of blueberries and the spectacular rhododendrons that characterize this ecosystem. The American Chestnut, which once dominated this region, was eliminated as a canopy species by Chestnut Blight (http://en.wikipedia.org/wiki/Chestnut_blightCryphonectaria parasitica), an introduced fungal parasite that is discussed in the species accounts section.

We clearly cannot come close to presenting all of the several thousand species of mushrooms found in our area, so this cannot be considered a comprehensive guide. But we have tried to give a good representation of the common species, as well as a number of the less often and rarely encountered taxa. Some of these will be widespread cosmopolitan species treated in most other field guides. We have attempted to provide additional taxonomic and natural history information about these taxa and to tie them into related species with which they might be confused. But many other species will be more or less restricted to the Georgia Piedmont and southern Appalachians or, in some cases, are Neotropical taxa that range up into our area. A number of small, nondescript mushrooms will be encountered that can only be referred to the catch-all category known as LBMs—Little Brown Mushrooms. Certain other groups will have a number of species that can be identified but a great many others that are nearly impossible to distinguish from each other by macroscopic—and, often enough, also not even by microscopic—features. Among such groups are species of *Russula* and *Lactarius* and many of the boletes.

Other Useful Guides

Print Sources

This book should nicely complement William C. Roody's excellent field guide: 2003. *Mushrooms of West Virginia and the Central Appalachians*. Lexington: University of Kentucky Press. 520 pp.

Other useful books and field guides, each with its own strengths and weaknesses, include the following:

Arora, David. 1986. *Mushrooms Demystified*. 2nd ed. Berkeley: Ten Speed Press. 958 pp. An excellent guide with dichotomous keys and keen insights about many North American mushrooms. However, the black-and-white photographs are somewhat limiting, and the book tends to emphasize western species.

Bessette, Alan, David B. Harris, & Arlene R. Bessette. 2009. *Milk Mushrooms of North America: A Field Identification Guide to the Genus* Lactarius. Syracuse, NY: Syracuse University Press. xii + 297 pp.

Bessette, Alan, William C. Roody, & Arlene R. Bessette. 2000. *North American Boletes: A Color Guide to the Fleshy Pored Mushrooms*. Syracuse, NY: Syracuse University Press. xiii + 396 pp.

Bessette, Alan, William C. Roody, Arlene R. Bessette, & Dail L. Dunaway. 2007. *Mushrooms of the Southeastern United States*. Syracuse, NY: Syracuse University Press. xiii + 373 pp.

Beug, Michael W., Alan E. Bessette, & Arlene R. Bessette. 2014. *Ascomycete Fungi of North America*. Austin: University of Texas Press. xii + 488 pp. An excellent book with a unique set of combined dichotomous verbal and pictorial keys.

Binion, D.E., S.L. Stephenson, W.C. Roody, H.H. Burdsall Jr., L.N. Vasilyeva, & O.K. Miller Jr. 2008. *Macrofungi Associated with Oaks of Eastern North America*. Morgantown: West Virginia University Press. xv + 466 pp. A very useful book that discusses species not treated in other works.

Horn, Bruce, Richard Kay, & Dean Abel. 1993. *A Guide to Kansas Mushrooms*. Lawrence: University Press of Kansas. xv + 297 pp.

Kuo, Michael. 2005. *Morels*. Ann Arbor: University of Michigan Press. 205 pp.

Lincoff, Gary H. 1997. *National Audubon Society Field Guide to North American Mushrooms*. New York: Alfred A. Knopf. 926 pp.

Stamets, Paul. 1996. *Psilocybin Mushrooms of the World: An Identification Guide*. Berkeley: Ten Speed Press. xx + 245 pp.

Stuntz, Daniel E., David L. Largent, & Roy Watling. 1977. *How to Identify Mushrooms to Genus IV: Keys to Families and Genera*. Eureka: Mad River Press. 94 pp.

Trappe, Matt, Frank Evans, & James M. Trappe. 2007. *Field Guide to North American Truffles*. Berkeley: Ten Speed Press. 136 pp.

Weber, Nancy S., & Alexander H. Smith. 1985. *A Field Guide to Southern Mushrooms*. Ann Arbor: University of Michigan Press. vi + 280 pp. This book contains descriptions and photographs of southern species not treated in other works and is a useful adjunct resource for anyone in our area. Unfortunately, some of the species described and illustrated have never been properly described or named, and the scientific names given were qualified as "in press" and still have no taxonomic standing.

Online Sources

A number of good websites also provide a great deal of useful information and are very helpful in identifying difficult taxa. These sites are primarily concerned with more northern species, but most of the taxa treated also occur here in the Southeast.

MushroomExpert.Com. Among the very best of these are the numerous sites of MushroomExpert.Com, established and authored by Dr. Michael Kuo. Dr. Kuo is an English instructor at Eastern Illinois University and, as he is quick to acknowledge, an "amateur" (but oh so very professional) mycologist. He is universally recognized as an expert in the field. He has also authored and coauthored several books on the subject. Typing in the scientific name of almost any North American mushroom (including rare and/or unusual ones) will very likely bring up one of his excellent websites, which will be loaded with details to help you distinguish it from look-alikes. He is more likely than most sources to provide useful tricks of the trade about microscopy and chemical staining for the species he discusses.

Mycoquébec.org (*Les champignons du Québec*). This is another excellent site for identifying many widely distributed North American mushrooms. Alphabetically arranged links will take you to the genus and species of interest. The site is, however, completely in French, but an online translator (e.g., Google Translate) will get you through most of the descriptions with not too much difficulty. This site is available at http://www.mycoquebec.org/bas.php?l=l&trie=A.

Studies in the Amanitaceae. Edited and maintained by Dr. Rodham E. Tuloss (Herbarium Rooseveltensis Amanitarum) and Dr. Zhu-liang Yang (Chinese Academy of Sciences). An excellent site with detailed descriptions and illustrations of amanitas, including those found in Great Smoky Mountains National Park and environs, by two leading authorities on the subject. It should be noted that many of the names are provisional ones for new species that have never been validly published. These are thus *nomina nuda* (naked names) and have no taxonomic standing. When forced to use such names, we place them within quotation marks in roman typeface (not italicized) to indicate that they are not valid scientific names. This site is available at http://whos.amung.us/stats/2ldynzwmgwng/http://www.amanitaceae.org/.

Mushroom Observer. A collaborative amateur mycology website started by Nathan Wilson in 2006. Available at http://mushroomobserver.org/observer/how_to_use.

Tom Volk's Fungus of the Month Pages. Established and authored by Dr. Tom Volk, professor of biology at the University of Wisconsin–La Crosse. His monthly (more or less) features offer a wide variety of natural history and other information about the featured fungi. The site naturally focuses on mushrooms found in the north-central United States, but many of them are also found in our area. Available at http://botit.botany.wisc.edu/toms_fungi/fotm.html.

George Barron's Website on Fungi. Established and maintained by Dr. George Barron. This site is a particularly good resource of images and information for those interested in nematophagous and other carnivorous and endoparasitic fungi by the leading authority on the subject. Available at http://www.uoguelph.ca/~gbarron/.

Acknowledgments

WE ARE INDEBTED to more people than we can list who directly or indirectly contributed to the completion of this work. We especially thank the following individuals for their friendly support and advice during our initial proposal of this project to the University of Georgia Press, for generously providing us with their photographs and their kind permission to use them in this work, and/or for critically reviewing the entire manuscript, both informally as a personal favor to us and as official outside reviewers:

Charles Seabrook ("Wild Georgia with Charles Seabrook," *Atlanta Journal Constitution*), who initially encouraged us to contact the University of Georgia Press and who strongly recommended us to the editorial staff there.

Dr. David Porter, professor emeritus of plant biology, University of Georgia, who encouraged us to write the book, gave us a number of valuable pointers along the way, and exhaustively reviewed both our original proposal and the submitted manuscript. His invaluable and detailed comments and suggestions to include additional material greatly improved the final manuscript.

Dr. Britt A. Bunyard, publisher and editor in chief, *FUNGI* magazine, Richfield, Wisconsin, who has supported our efforts throughout this project. His critical comments and detailed review of the submitted manuscript likewise contributed enormously to its improvement.

We also thank the following colleagues and friends:

John Plischke III, mycologist, photographer, and past president of the Western Pennsylvania Mushroom Club. He generously provided both his professional photos and numerous PDFs of difficult-to-obtain literature. John also critically reviewed the final manuscript and offered many helpful comments and suggestions.

Dr. Jack D. Rogers, professor emeritus of plant pathology, Washington State University, for his friendly correspondence and many helpful insights and suggestions about the carbonaceous fungi of the order Xylariales.

Dr. Steven L. Stephenson, Department of Biological Sciences, University of Arkansas, who kindly provided us with some of the late Emily Johnson's wonderful mushroom photographs.

Alan Rockefeller, a noted mushroom taxonomist, photographer, and lecturer specializing in the fungi of Mexico and California and a frequent contributor to Mushroom Observer. We are grateful for his generous permission to use his photographs and micrographs, and especially for permission to use some of his unpublished information about *Psilocybe caerulescens*.

Dan Molter, Department of Philosophy, University of Utah, for permission to use a number of his outstanding photographs and micrographs.

Ron Pastorino, a current regional trustee for the North American Mycological Association and a well-known mycologist and award-winning mushroom photographer, for generously providing a number of his photographs and micrographs.

Dr. Timothy B. Brenneman, professor of plant pathology, University of Georgia, Tifton, for providing us with the wonderful photos of the ascomata of the Pecan Truffle, *Tuber lyonii*. He first discovered the Pecan Truffle in Georgia at the bases of Pecan trees in 1989.

Dr. Matthew Smith, Department of Plant Pathology, University of Florida, kindly provided the excellent micrograph of the spiny ascospores of the Pecan Truffle and gave us permission to use it.

Robert Searfoss, Agency for Toxic Substances and Disease Registry at the Centers for Disease Control and Prevention, Atlanta, Georgia, for his unflagging support and for providing us with a cloud backup of our files totaling nearly 90 gigabytes.

A number of individuals have released their photos to the public domain or made them available under a Creative Commons or GNU General Public License. They include the following (some identified only by their pseudonyms):

Amanita77
Andreas Kunze
André-Philippe Drapeau Picard
Ann F. Berger
Antonio Abatiello
Berndt Gliwa
Bob Zimmer
Boleslaw Kuznik
Borch3kawki
Botond Balogh
Captainpixel, Wikipedia
Carl Mueller
Christine Braaten
D. F. Farr & E. B. McCray
Damon Brunette
Danny Steaven
Dr. Lorne Stobbs
Drahkrub
Drew Parker
Eddee
Ehecatl (Alonso)
Elizabeth Bush, USDA Forest Service
Emily Johnson
Erich G. Vallery, USDA Forest Service
Erich Witter
Franco Folini
George Chernilevsky
gsharpnolack
Hans Hillewaert
Heather Hallan-Adams
Helen Ginger
Herbert Baker
Holger Krisp, Ulm, Germany
hríb
Ieponumos
James Solomon, USDA Forest Service
Jerzy Opiola
John Carl Jacobs
John S. Harper
Jonathan Mack
Jörg Hempel
Joseph O'Brien, USDA Forest Service
Josh Milburn
K. Põldmaa
Kaldari
Karen (Oldmanofthewoods)
Kristine (JanusGD)
Leblanc
Marko Eesti
Miika Silverberg
Nathan Wilson
Norbert Nagel
Paffka
River Bissonnette
Rob Hille
Robert Sasata (Sasata)
Rosemary Winnall
Severine Meißner
Stanislaw Szydlo
Steven Russell
Strobilomyces
Thomas Pruß
W. Siegmund
Walt Sturgeon
Walter Siegmund
weiliiiiiii
William Tanneberger
Ylem
Zonda Grattus

We also gratefully acknowledge the following individuals who generously allowed us to use their photographs and other materials in this work:

Richard Bell, Chattahoochee Nature Club, Roswell, Georgia

Bill Blanton

Mike Christison, Georgia Botanical Society

Jim Edwards

Teresa Fortenberry

Stephanie Hannagud

Dr. Linas Kudzma

Stan Machosky, who arranged for the acquisition of Bill Blanton's marvelous photos

Dr. Michael Moribaldi, who collected the wonderful specimen of Dryad's Saddle which he donated to us

Debbie Viess

Ken Zinkand

Photos not specifically otherwise credited were taken by the authors.

The following individuals invited us to lead mushroom walks and/or give PowerPoint presentations for their respective organizations:

Jerry Hightower, ranger naturalist, Chattahoochee National Recreation Area, US National Park Service, Atlanta, Georgia

Richard and Carla Bell, Chattahoochee Nature Club, Roswell, Georgia

Nancy Jones, executive director, Blue Heron Nature Preserve, Buckhead (Atlanta), Georgia

David Dunagan and Patricia Thernell, John Ripley Forbes Big Trees Forest Preserve, Sandy Springs, Georgia

And finally, we thank our former and current acquisitions editors at the University of Georgia Press: Laura Sutton, Regan Huff, and James Patrick Allen, as well as our production editor, Jon Davies. They have all gone out of their way to accommodate our needs and provide us with technical and other support.

Mushrooms of the Georgia Piedmont and Southern Appalachians

Introduction

Three Domains and Six Kingdoms: Is That All?

Scientific Classification

The business of science is twofold: (1) to describe the Universe as best we can and (2) to create models of how we think it works. To do this, scientists classify the things and phenomena they observe into various categories. However, it must always be remembered that *all* scientific categories are artificial devices constructed by the human mind. Mother Nature does not always (or even usually) accommodate herself so readily to our classification systems.

How Many Kingdoms?

Even so elementary a concept as a biological species cannot be precisely or even adequately defined, and many biological entities simply do not fit into our conceptions of where we think they belong. And when we are dealing with categories higher than the genus level, all bets are off. Well, perhaps this is overly harsh—we do have a fair understanding of the evolutionary relationships of many groups; others, however, are still quite problematic. We simply do the best we can with what we have and acknowledge the inherent inadequacies in any system we may devise. We need to classify things in order to think about and discuss natural phenomena, but our systems are only models of our view about things; they are not the things themselves. As Alfred Korzybski noted in 1931, "*A map is not the territory* it represents, but if correct, it has a similar structure to the territory, which accounts for its usefulness" (italics added). Keep these caveats in mind when going through this section.

A Little Background

In the 4th century BCE the Greek scientist-philosopher Aristotle first classified all living things as either plants or animals and divided these two large groups into smaller subgroups. He categorized plants as trees, shrubs, or herbs and animals on the basis of how they moved. Flying animals were put into one subgroup; those that swam were put into another; and those that walked, crawled, or ran upon the ground were placed into a third category.

Some two thousand years later, the great 18th-century Swedish biologist Carolus Linnaeus also categorized all living creatures as either plants or animals. He called these two groups "kingdoms" (inanimate things were placed in a third kingdom, the Mineralia). But unlike Aristotle, Linnaeus grouped animals (Kingdom Animalia) and plants (Kingdom Plantae) according to similarities in their structures rather than where they lived or how they moved. Thus he placed animals having a backbone into the class Vertebrata in the kingdom Animalia. Plants were classified on the basis of the number and arrangement of their reproductive organs: a plant's class was determined by its stamens (male organs) and its order by its pistils (female organs). This, of course, produced some unnatural groupings, which even Linnaeus himself acknowledged were necessarily "artificial" because of inadequate knowledge at the time.

Kingdom Animalia contained mainly **motile** and **heterotrophic** (feeding on other organisms) organisms, whereas **Kingdom Plantae** contained mainly **sessile** (remaining more or less fixed in one place) and **autotrophic** organisms that obtained nourishment by photo- or chemosynthesis. Although they are heterotrophs and incapable of photosynthesis, the fungi were included in the plant kingdom since they have a cell wall resembling that of plants and are nonmotile. However, certain ambiguous organisms included within the "Protozoa" could be classified as either animals or plants: some were clearly animal-like and heterotrophic and either moved about or remained stationary; others were simultaneously both animal- and plant-like (being both autotrophic *and* heterotrophic) and were motile.

A Few Changes to the System

This simple but rather inadequate system prevailed until the mid-1860s, when Ernst Haeckel and J. Hogg proposed a third kingdom: **Protista**. Haeckel's kingdom contained all microscopic organisms, including the bacteria, protozoa, and fungi. Haeckel's proposal did not really catch on. In his time the distinction between prokaryotes (organisms lacking a nucleus) and eukaryotes (organisms having a membrane-bound nucleus) was not understood; however, Haeckel did consider the "Monera" (bacterial forms) to be a unique group within the Protista. Even as recently as the middle of the last century most taxonomists still crammed everything that was not clearly animal into Kingdom Plantae.

In 1925 the French biologist Édouard Chatton distinguished between the eukaryotic and prokaryotic grades of cellular organization. In 1937 he divided all living things into two "empires" based on their cell structures: (1) the **Prokaryota** (organisms lacking a nucleus and membrane-bound organelles), the bacteria, and (2) the **Eukaryota** (organisms whose cells have nuclei and membrane-enclosed organelles), which included the animals, plants, fungi, and protists.

In the mid-1950s Herbert Copeland grouped all prokaryotic organisms into the kingdom Mychota (later changed to Monera), placing all the remaining eukaryotic protists into the kingdom Protoctista. Copeland's five-kingdom system thus included the Monera, Protoctista (= Protista), Fungi, Animalia, and Plantae. Despite this improvement, Copeland's system did not take into account the major distinctive features of organisms' modes of nutrition: photosynthesizing autotrophs (photoautotrophs), ingesting heterotrophs (phagotrophs), and/or absorbing heterotrophs (osmotrophs). And in many forms, the type of nutrition depends upon environmental conditions: some species can alternately photosynthesize in the manner of plants or capture and ingest prey the way animals do. The multiple modes of nutrition among the protoctistan eukaryotes thus still posed a major problem.

In 1969 an American biologist, Robert H. Whittaker, recognized that the fungi are quite different from other eukaryotes in many important respects, and he assigned them to a new kingdom, the Fungi. In Whittaker's five-kingdom system, which is still favored by many biologists, three kingdoms of more complex eukaryotic organisms are classified according to the three forms of nutrition: photosynthesis (plants), food ingestion (animals), and food absorption (fungi). The other two kingdoms (Monera and Protista) are distinguished according to differences in cellular structure and physiology.

In this approach, Whittaker tried to place organisms into kingdoms that more closely reflected their supposed evolutionary relationships. This approach to classifying organisms was an important step in trying to form **monophyletic** (as contrasted with **polyphyletic**) groups, or **clades**, that include a common ancestor and all its descendants and to create a system where similarities and evolutionary relationships are apparent. It was based on the obvious premise that a natural classification system should mirror evolutionary descent as closely as possible.

Here it is important to bring up the concept of **paraphyly**, which has the cladistic school of phylogenetics so riled up. Then we will dispense with it and ignore it altogether from this point forward. A **paraphyletic** group is a monophyletic clade from which one or more of its descendant lines has been excluded. For some reason, most cladists find this intolerable,[1] but it is regularly practiced by most practical biologists. Examples of paraphyly in the animal kingdom include the reptiles (Class Reptilia) and the branch of theropod dinosaurs that gave rise to the

1. No doubt because Willi Hennig said so. But, in fact, paraphyly is a normal part of arranging living things into any hierarchical classification scheme based on evolution. At some point crossopterygian fishes became land-dwelling tetrapods and turned into amphibians, some of which gave rise to the reptiles. The reptiles, in turn, gave rise to a number of diverse groups, including (1) the dinosaurs, one group of which turned into birds, and (2) the therapsids, some of which gave rise to diverse groups of mammals, three of which—the egg-laying monotremes, the pouched marsupials, and the placentals—persist today.

birds (Class Aves). By removing the birds from the reptiles and placing them in their own class, the Reptilia are now defined as paraphyletic, with the implication that this somehow violates some vague natural "law" that, to our knowledge, has never been elucidated. To this we say no, the reptile-bird clade is indeed monophyletic, but at some point in the Late Mesozoic era, birds became sufficiently different from their theropod ancestors that we are justified in placing them in their own class. A botanical example is the exclusion of the monocotyledons as a distinct group from the dicotyledons from which they arose, thus defining the dicots as a paraphyletic group.

Still More Tinkering

But not being content with this, in 1977 Carl Woese, a microbiologist at the University of Illinois, and his research associate, George Fox, proposed dividing Kingdom Monera into two distinct "urkingdoms" (from the German prefix *ur-*, meaning "original" or "primeval") based on previously unrecognized but important differences: the Archaebacteria and Eubacteria. He further attempted to establish a three-urkingdom system in which plants, animals, protists, and fungi were all grouped together into one superkingdom containing all eukaryotes: the Eukaryota. In 1990 Woese introduced the concept of three domains for his three major supraregnum (a group above the level of kingdom) branches of living things and dropped the name Archaebacteria in favor of Archaea. He also advocated discarding the term *prokaryote* altogether.

The initial use of a six-kingdom system blends Whittaker's classical five-kingdom system and Woese's original three-urkingdom system. In an attempt to provide a workable model incorporating some of these current concepts, one authority proposed another model with five kingdoms: Protozoa, Chromista, Plantae, Fungi, and Animalia, each of which contains some unicellular protistan representatives (Corlis, 1998).

Variations of such six-kingdom systems have become standard in most biological works because they at least approximate a broad-brush overview of multiple domains/empires and kingdoms, retain the classical organizing concepts of hierarchical taxonomic ranks, and are of practical use for most biological purposes. In a separate publication, Woese and Ralph Wolff showed that the Archaea are actually more closely related to the Eucarya than to the true bacteria and yet are distinct from both of those major groups. In 1990 Woese proposed the now generally accepted division of life into three major domains above the level of kingdom: the Eubacteria (now usually referred to as the Bacteria), Archaebacteria (= Archaea of current usage), and Eucaryota (also spelled Eukaryota). In current terminology, animals, plants, fungi, and protists (the eukaryotes) comprise **Domain Eucarya**, with organisms at a prokaryotic grade of cellular organization being allocated to either the **Archaea** or the **Bacteria**.

Where Will It All End?

Despite the apparent acceptance of the three-domain, six-kingdom system by most biologists, much dissension remains, and the issue is anything but settled. Thomas Cavalier-Smith of the University of British Columbia and his colleagues have recombined the Archaea and the Eubacteria back into a single prokaryotic category, the Monera. They also regard the Chromista—formerly in the kingdom Protista—as a distinct kingdom. Thus, they still wind up with six kingdoms, but they acknowledge only two domains.

There is also a two-empire, eight-kingdom scheme, with Archaebacteria and Eubacteria as kingdoms in Empire Bacteria and six kingdoms in Empire Eukaryota: Archezoa (very primitive eukaryotes without chloroplasts or mitochondria); Protozoa (unicellular eukaryotes with mitochondria but lacking cell walls); Chromista (photosynthetic organisms containing chloroplasts in the lumen of the rough endoplasmic reticulum, e.g., diatoms, brown algae, cryptomonads, and oomycetes); Plantae (photosynthetic organisms with chloroplasts within the cytoplasm, including all land plants and many green algae); Fungi; and Animalia. "[E]very archezoan examined bears vestigial mitochondria, or genes inherited from mitochondria. Thus, all modern eukaryotes evolved from a mitochondrion-bearing ancestor" (A. Poole and D. Penny, 2007, *Nature* 447, p. 917).

To further muddy the waters, James Lake and his

colleagues recently proposed splitting the Archaea into the Halobacteria, Methanogens, and Eocytes, thus yielding four major groupings of prokaryotes (Eubacteria + Halobacteria, Methanogens, and Eocytes). All of these could constitute kingdoms on their own, since each is as unrelated to the others as any are to the Eucarya. This gives yet another two-empire, eight-kingdom scheme. And one microbiologist, Jack R. Holt of Susquehanna University, has proposed splitting Domain Bacteria into nine separate kingdoms!

Are you confused? I'll let you in on a big open secret. So are we scientists.

The enormously complex emerging phylogenetic diagrams based upon strict monophyletic lineages have made it almost impossible to erect meaningful high-level, hierarchically ranked taxonomic categories or to sustain many of those already in place. This has increasingly resulted in the replacement of the classical system of hierarchical megasystematics—which in recent decades has produced descending-order monstrosities like domain/empire > unikonta/opisthokonta/bikonta (unranked categories) > grade > superkingdom > subkingdom > branch > infrakingdom > superphylum > subphylum > infraphylum,[2] followed by the more traditional ranks of class on down (and their supra- and infra- categories)—by unranked phylogenetic clades. These higher taxa (ranked or not) are all still very speculative despite all the molecular evidence claimed in their support. And different authorities have proposed vastly different arrangements for the same evidence. Indeed, the same authors frequently revise their own evolutionary diagrams as often as every other year.

However, while phylogenetic schemes based upon unranked monophyletic clades may more accurately reflect biological reality—or at least our current model of what we think it is—a vast amount of information about relationships is lost.

2. These higher taxonomic categories are very different from the former classical categories of domain > kingdom > division/phylum > class > order > family > genus > species. These new megacategories are recent consequences of all the genetic, ultrastructural, and biochemical studies that have produced these cladeograms. The ranks and their bizarre-sounding names are simply attempts to impose a neoclassical sense of order on it all. Alas, we fear this has become an impossible goal.

Arranging closely related species into genera, related genera into families, families into orders, and so on up the hierarchy of ranks certainly involves making subjective decisions. Although having a subjective element, such decisions are for the most part very well educated guesses that, even if later shown to be wrong, provide a deeper understanding of the relationships and their magnitude and significance of the taxa "nested" within the hierarchy.

The debate over the number of kingdoms of all living and fossil taxa still is far from over, and things are still in flux. Domains and kingdoms really represent the several trunks and major branches of the evolutionary tree of living things—in recent years commonly referred to as the universal tree of life. Like taxonomy itself, splitting the tree into kingdoms is arbitrary and depends just where you make your slices along the trunk and branches. The higher you slice, the more kingdoms you will produce. Some recently proposed classification schemes have more than *15 kingdoms!* This would seem to render the entire concept of a kingdom utterly meaningless. Nevertheless and despite all the debate, on the basis of molecular sequencing and other criteria, a system of three domains and six kingdoms now seems a reasonable interim approach and has been adopted by most biologists. This will most assuredly change, however, as new information becomes available. Table 1 shows 3 domains containing 11 kingdoms. The number of kingdoms will surely expand as new relationships are inferred from a variety of chemical and genetic data.

Table 1. 3 domains, 11 kingdoms . . . and counting

Domain Bacteria prokaryotic grade	**Domain Archaea** prokaryotic grade	**Domain Eukarya** eukaryotic grade
Division Proteobacteria	Kingdom Euryarchaeota	Kingdom Protista
Division Cyanobacteria	Kingdom Crenarchaeota	Kingdom Plantae
Division Eubacteria	Kingdom Koryarchaeota	Kingdom Fungi
Division Chlamydia		Kingdom Animalia

Table 2. Domain Eucarya with eight kingdoms. The following cladogram shows the presumed evolutionary relationships among the Eucarya.

DOMAIN EUCARYA		
BIKONTA	**UNIKONTA**	
KINGDOM EXCAVATA	KINGDOM OPISTHOKONTA	KINGDOM AMOEBOZOA
KINGDOM RHIZARIA	KINGDOM FUNGI	**Phylum Mycetozoa**
KINGDOM CHROMALVEOLATA	SUBKINGDOM DIKARYA	Class Myxogastria
KINGDOM PLANTAE	**Phylum Ascomycota**	Class Dictyostelia
	Phylum Basidiomycota	Class Protostelia
	Phylum Glomeromycota	**Phylum Lobosa**
	"Zygomycetous Fungi"	**Phylum Archamoebae**
	KINGDOM METAZOA	
	KINGDOM CHOANOFLAGELLATA	

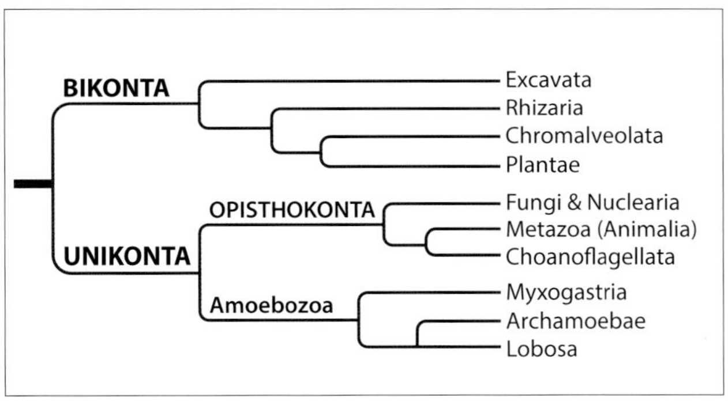

FIGURE 1. Cladogram showing the presumed evolutionary relationships among the Eucarya.

So, having been through all this, we will adopt a classical approach in this book and will pretend that the arrangement of taxa in the hierarchical system of descending rank—namely, domain > kingdom > division/phylum > class > order > family > genus > species and their numerous supra- and infra- categories—is a valid one.

Terms like *prokaryote*, *protist*, and *zygomycete* are useful descriptors of a grade of evolution, whereas Kingdom Protista or Division Zygomycota would represent a polyphyletic and too heterogeneous a group to be useful or to reflect biological reality (Cavalier-Smith, 1998). This is a good place to remind you of Korzybski's caveat: "A map is not the territory."

The Prokaryotic Grade of Organization

The former kingdom Monera contained most organisms that lack a cell nucleus and other membrane-bound organelles, including the bacteria, the Cyanobacteria (formerly called the "blue-green algae"), the Actinomycetes (Actinobacteria), and what are now known as Archaea (= Archaebacteria of some authors). However, recent DNA and RNA sequence analysis and other evidence have indicated that two major groups of prokaryotes exist, the Bacteria (often called the Eubacteria) and the Archaea, the last group being more closely related to the eukaryotes than they are to the Bacteria. The Monera has thus been divided into the domains Archaea and Bacteria, which together with the four eukaryotic kingdoms comprise the three-domain, six-kingdom systems. Most current schemes now recognize three domains: Archaea (Archaebacteria), Bacteria (Eubacteria), and Eucarya (Eukaryota). But dissenters abound.

Monerans, the simplest of all living organisms, are at a prokaryotic (prenuclear) grade of evolution. This means that they lack a true cell nucleus and other membrane-bound organelles such as chloroplasts (where photosynthesis occurs in plants and photosynthesizing protists), mitochondria (where aerobic respiration occurs inside cells), Golgi bodies, and the complex internal membrane

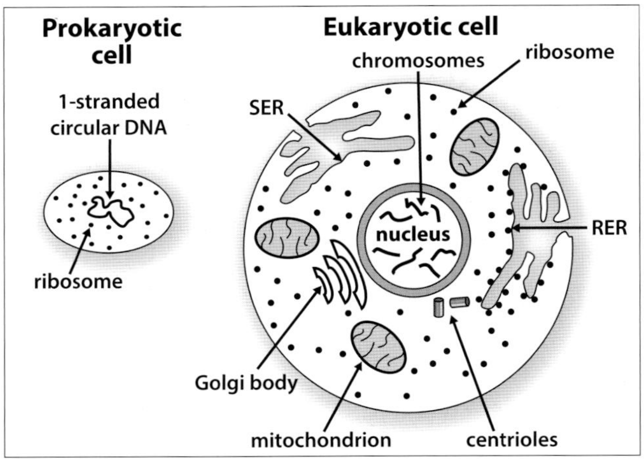

FIGURE 2. Prokaryotic and eukaryotic grade of cellular organization. RER = rough endoplasmic reticulum; SER = smooth endoplasmic reticulum.

system known as the endoplasmic reticulum. All other organisms do possess such membrane-bound cellular organelles and are known collectively as eukaryotes.

Eukaryotic Organization

Prokaryotes (Fig. 2) differ from members of the eukaryotic kingdoms in being single-celled organisms or colonies of cells with little organization to their cell structure. It is important to emphasize that, as used here, the adjectives "moneran" and "prokaryotic" refer to a grade of cellular organization and do not imply any phylogenetic relationships between the organisms so described. Monerans have no membrane-bound structure to protect their genetic material. Their DNA consists of a single circular strand of DNA with no or only very few introns contained within an irregularly shaped region of the cytoplasm known as the nucleoid. Lacking chromosomes, they do not undergo mitosis or meiosis and reproduce by simply dividing into two separate cells, a process known as binary fission. Many members of Kingdom Protista also consist of single cells and often also reproduce by binary fission, but there are many other important differences, as we shall see.

Eukaryotic cells contain a nucleus bounded by a double membrane inside which the double-stranded DNA is complexed to special proteins known as **histones** and is contained within linear **chromosomes** in a diffuse chromatin network. When present, the cell wall is composed of **cellulose** (some algae and all plants), **chitin** and other polysaccharides (fungi), β-1,3-linked xylan (the structural polysaccharide of many marine algae that lack cellulose entirely), silicon (diatoms), and calcium carbonate calcite (some groups of red algae, certain green algae, and brown and red algae). Eukaryotic cells also contain a variety of membrane-bound organelles such as mitochondria, Golgi bodies, and both smooth and rough endoplasmic reticulum, an internal membrane-bound labyrinth that respectively lacks or is lined with ribosomes. Centrioles are present in animal cells and participate in organizing the microtubule system during cell division; they are absent in conifers, flowering plants, and most fungi.

The Eucarya are now often divided into four kingdoms: **Protista** (= Protoctista), **Fungi**, **Plantae**, and **Animalia**. However, the protists do not form a monophyletic group but include many phylogenetically unrelated groups, including the slime molds, several distinct and unrelated groups of algae, and many polyphyletic groups of protozoa. Although the higher taxonomic relationships of organisms are still in dispute, a simplified and generally accepted theme has begun to emerge that we will summarize briefly.

So What Are Fungi Anyway?

Fungi are eukaryotic **heterotrophic** organisms having a cell wall composed of varying amounts of **chitin** (pronounced "*kite*-in"). This distinguishes them from plants, which have cell walls made up of **cellulose** (see Fig. 3). In

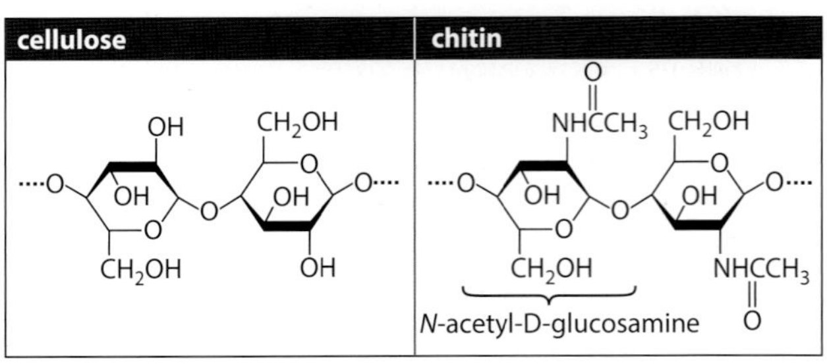

FIGURE 3. Cellulose and chitin are both polysaccharides. Cellulose is made up of a linear chain of linked D-glucose molecules. Chitin is a similar polymer of *N*–acetyl-D-glucosamine—a glucose derivative.

many fungi, chitin is the major cell wall constituent; in others, it is present only during cell division or reproduction and is otherwise absent. Animals have no cell walls at all; their cells are enclosed in the same type of **plasma membrane** that surrounds fungal and plant cells beneath the cell wall. Most fungi grow as multicellular filamentous **hyphae** (sing. hypha) that mass together to form an entangled tissue known as a **mycelium** (my-*see*-lee-uhm). However, many yeasts exist as single cells. Both sexual and asexual reproduction typically occur through spores and commonly involve specialized **sporocarps**, or fruiting bodies, such as the familiar mushrooms you see sprouting up. However, some fungi propagate solely by vegetative growth. Fungi are more closely related to animals than to plants. Nevertheless, for historical reasons, mycology, the scientific study of fungi, usually falls under a branch of botany, and the rules by which fungi are named are specified by the *International Code of Nomenclature for algae, fungi, and plants* (Melbourne Code), or simply the *ICN* (previously known as the *International Code of Botanical Nomenclature* [Vienna Code], or *ICBN*).

Chitin $(C_8H_{13}O_5N)_n$ is made up of several hundred to many thousands of polymers of *N*–acetyl-D-glucosamine, a derivative of glucose (see Fig. 3). Chitin is essentially cellulose: a hydroxyl (–OH) group on each glucose monomer has been exchanged for an acetylamine ($-NHCOCH_3$) group. Chitin is the main component in the cell wall of many fungi, as well as in the exoskeletons of nearly all arthropods. It also occurs in cnidarians, in brachiopods, and in the mouthparts of many mollusks (Tasch, 1980; Palaeos, 2003).

The chitin found in fungi is essentially the same as that occurring in other organisms, but fungal chitin is associated with other polysaccharides not found in arthropod exoskeletons (Peter, 2002). Chitin is widely distributed in varying amounts among the fungi, but it is not universally present. Other polysaccharides such as glycoproteins and β-glucans (polymers of glucose) are found in fungi, and chitin may be nearly or even completely lacking in some cases. In animal glycoproteins, the sugar that is bound O-glycosidically to the corresponding hydroxy amino acid is *N*-acetylgalactosamine. In fungi, these substituents are mannose or mannose oligomers.

Fungi also produce glycogen, which, like chitin, is

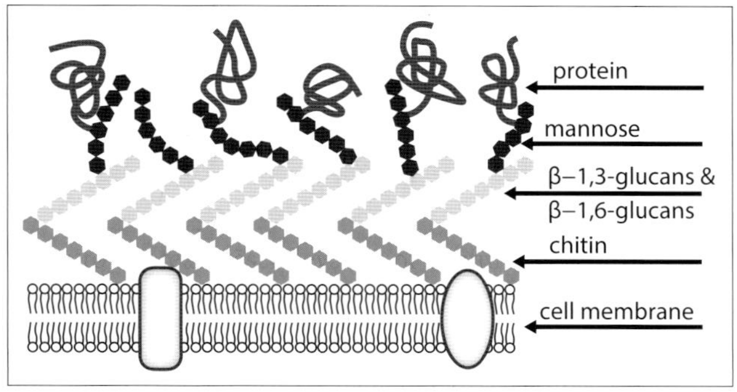

FIGURE 4. Typical fungal cell wall.

otherwise only found in animals. Like animals, fungi are also obligate heterotrophs, although they absorb nutrients and organic compounds into their hyphae instead of ingesting them. Fungi must withstand extreme environmental conditions; the mature cell wall is thus extremely rigid and resistant to enzymatic degradation and osmotic stress. Characteristically, the fungal cell wall (Fig. 4) consists of an inner chitin layer just outside the plasma membrane surrounding the cell, an intermediate layer of β–1,3- and β–1,6-glycans, and an outer layer of mannose-containing glycoproteins. True fungi lack cellulose entirely; those cellulose-containing fungus-like organisms previously included within the Fungi are now placed in the kingdom Protista (about which more below).

Characteristics of Fungi

- Fungi are **heterotrophic**, that is, they must derive their energy by feeding upon other organisms or assimilating nourishment from their microenvironment, which they digest by secreting extracellular enzymes.
- Some fungi are **saprobic**,[3] deriving their nourishment from decomposing dead or decaying organic matter. Others are **parasitic** on living plants or animals. Most grow by producing slender filaments known as **hyphae**; an entangled mass of hyphae constitutes a **mycelium**.

3. The noun *saprophyte* and its adjective, *saprophytic*, are often used for fungi, but these terms really apply to plants. Because decomposing fungi and bacteria are no longer considered plants, *saprobe* and *saprobic* are now the appropriate terms.

- Many fungi form symbiotic **mycorrhizal associations** with the roots of plants. The plant benefits from the minerals absorbed by the fungal mycelium, and the fungus is nourished by the photosynthetic sugars produced by the plant and translocated to the roots.
- **Lichenicolous fungi** have formed symbiotic (mutualistic) associations with algae, cyanobacteria ("blue-green algae"), or both to produce composite superorganisms known as **lichens**. Lichens have morphological and physiological features that are quite distinct from those of their symbiotic components.
- Some fungi are **carnivorous**, in addition to being saprobic, and have developed elaborate weapons to trap and kill their animal prey as a source of nitrogen.
- A few fungi have formed complex symbiotic associations with insects and other organisms that have evolved over tens or even hundreds of millions of years.
- The hyphal walls are usually strengthened with **chitin**, a derivative of N-acetyl glucosamine, instead of the cellulose found in plants. Chitin is otherwise found only in certain animals.
- Fungi exhibit **apical growth** in which the hyphae extend their tips; this contrasts with the **intercalary growth** of filamentous algae, which undergo repeated cell divisions within a chain of cells from formative tissue (meristem) located below the tip of the growing structure.
- Fungi do not possess chloroplasts or chlorophyll, do not engage in photosynthesis, and are **obligate heterotrophs**, that is, they must obtain their nourishment from consuming other organisms. In most cases, they accomplish this by parasitizing and consuming other living or decomposing dead organisms. But many species form symbiotic associations with photosynthetic plants or cyanobacteria.

Evolutionary History

Probable fungi have been reported from the Proterozoic eon in North America 2 billion years ago (Tyler and Berghoorn, 1954)[4] and 1.43 billion years ago (Butterfield, 2005);[5] in South Australia ~1.5 billion years ago (Javaux, Knoll, & Walter, 2001) and 1,050–900 million years ago (Sherwood-Pike & Gray, 1985);[6] and in China from poorly dated "Mesoproterozoic" deposits (Yin, 1997). It is believed that saprobic decomposers and parasites were well established by the end of the Proterozoic Eon and that the main cell wall constituent of these forms would have been cellulose, not chitin (Butterfield, 2005).

Higher chitin-containing fungi (Ascomycota and Basidiomycota) were present from the late Silurian period (443–419 million years ago), in which an association of terrestrial fungi with green algae resembling modern lichens was reported (Sherwood-Pike & Gray, 1985). A definite 400-million-year-old fossil lichen, *Winfrenatia reticulata*, was described from the early Devonian period of Scotland (Taylor, Hass, Remy, & Kerp, 1995). And a fossil marine lichen-like association between a fungus and coccoidal cyanobacteria was recently reported from South China (Yuan, Xiao, & Taylor, 2005), indicating that fungus-photoautotroph partnerships developed before the appearance of vascular plants.

A fossil endomycorrhizal symbiosis between a fungus (*Glomites*; see Taylor, Remy, Hass, & Kerp, 1995) and the aerial stems and rhizomes of *Aglaophyton major* (Kidston & Lang, 1920) D.S. Edwards 1986—a prevascular plant intermediate between the bryophytes and the higher vascular plants—is known from the early Devonian period (400 million years ago). These arbuscules are morphologically identical to living arbuscular mycorrhizae (Taylor, Remy, Hass, & Kerp, 1995).

Who's Who among the Fungi

Basidiomycota, Ascomycota,[7] Zygomycota, Chytridiomycota, Glomeromycota—what in the world do these

4. However, these may represent the sheaths of cyanobacteria instead. See W.N. Stewart & G.W. Rothwell, 1993, *Paleobotany and the Evolution of Plants*, 2nd ed. (Cambridge: Cambridge University Press), 521 pp.; A.H. Knoll, 1985, Patterns of evolution in the Archaean and Proterozoic eons, *Paleobiol.* 11:53–64.

5. "Reliably, if not conclusively," identified as *Tappania*—a fungus and definitely a eukaryote, probably a sister group to the higher fungi and more derived than the Zygomycota (see below).

6. Described as sporangia of an oomycete (see the following section for a breakdown of the higher fungal taxa).

7. The Basidiomycota and Ascomycota were formerly referred to as the classes Basidiomycetes and Ascomycetes, respectively, when

bizarre terms refer to? We know that many of you have encountered and been very puzzled by these names. Well, these are just the five divisions (phyla) into which the kingdom Fungi has been divided on the basis of their modes of sexual reproduction. In common parlance, these are the club fungi (mushrooms, jelly fungi, smuts, and rusts), the sac fungi and yeasts, the bread molds, the chytrids,[8] and the arbuscular mycorrhizae, respectively. However, things are by no means settled. The Zycomycota no longer exists as a taxonomic category; both it and the Chytridiomycota have been broken up into several new categories (about which see more below).

So what is a division? In botany and mycology, a division is a major biological category equivalent to an animal phylum. Both terms—*division* and *phylum*—are permitted in the scientific literature by the *ICN* and its predecessor, the *ICBN*. Examples of plant divisions include the liverworts, mosses, ferns, cycads, conifers, and angiosperms (flowering plants). Equivalent animal phyla would include (among others) the mollusks (clams, snails, octopuses, and their relatives), the annelid worms (earthworms, leeches, and polychaetes), the echinoderms (sea stars, sea urchins, crinoids, sea cucumbers, and their relatives), and the huge phylum Chordata: certain primitive lancets, the sac-like tunicates and the vertebrates—lampreys and hagfishes, cartilaginous fishes (sharks, rays, skates, and the curious chimaeras, or ratfishes), bony fishes, amphibians, reptiles, birds, and mammals. Thus, the five divisions of fungi discussed here represent groups as distinct from each other as clams are from raccoons.

The divisions **Basidiomycota** and **Ascomycota** are evolutionarily related sister groups that comprise the subkingdom **Dikarya**, loosely known as the "higher fungi." Members of both divisions have wall-like septa that compartmentalize the fungal hyphae into segments. These compartments contain pairs of unfused haploid (see note 12) nuclei (**dikaryons**) following mating and before nuclear fusion (see below). The structures that coordinate simultaneous mitosis of the two dikaryotic nuclei—the clamp connections of the Basidiomycota (Fig. 4) and the hook-like croziers of the Ascomycota (Fig. 5)—also appear to be homologous, that is, they evolved from a common ancestral form (Taylor et al., 2004).

The now defunct division/phylum **Zygomycota** ("conjugation fungi") is a former polyphyletic (i.e., comprising several evolutionarily unrelated groups that did not arise from a common ancestor) and artificial catch-all category characterized by the formation of a **zygospore** during sexual reproduction and the complete absence of hyphal cell walls except in the **gametangia** (sexual reproductive structures).[9] The former Zygomycota are now understood to represent a polyphyletic assemblage of portions of some four different lineages, including a clade that does not even belong to the kingdom Fungi: *Amoebidium*, a trichomycete living in arthropod guts that is now assigned to the Mesomycetozoea/Ichthyosporea, which is closely related to the choanoflagellates, near the base of the animal evolutionary line.

Fungal groups formerly included within the Zygomycota are now separated into the following phyla/subphyla: Glomeromycota (arbuscular mycorrhizae), Mucoromycotina (pin molds and related groups), Mortierellomycotina (a large group of saprobic basal fungi), Entomophthoromycotina (a group of specialized pathogens in insects and other invertebrates), Kickxellomycotina (four zygomycete orders having disciform septal pores containing lenticular plugs), and Zoopagomycotina (specialized predatory and/or parasitic fungi on amoebae and microscopic invertebrate animals) (Hibbett et al., 2007; Hoffmann, Voigt, & Kirk, 2011). Only

the fungi were still thought to represent a phylum within Kingdom Plantae. Basidiomycetes is currently a class within the division Basidiomycota.

8. In older classifications, the chytrids, or Mastigomycota, were included in the slime molds and water molds—now classified as Phylum Mycetozoa (Division Myxomycota of the mycologists) and Division Oomycota, respectively, first placed in the "kingdom Protista"—another artificial and polyphyletic catch-all category that has likewise been broken up into varying numbers (depending on the authority) of quite unrelated clades *incertae sedis* (evolutionary lines of uncertain status). The Mycetozoa are in Kingdom Amoebozoa (Unikonta), while the Oomycota are in Kingdom Chromalveolata (Bikonta).

9. Zygospores are thick-walled spores in certain fungi and some algae formed by the fusion of two gametangia, followed by the fusion of the two haploid nuclei to produce a diploid resting spore (the zygospore). Zygospores also occur in certain protists, another polyphyletic catch-all "kingdom" that is being broken up in various ways by different authorities.

the Glomeromycota (arbuscular mycorrhizae) and the Entomophthoromycotina (insect pathogens) will be discussed further in this section and in a later section dealing with the Entomophthorales. All these ancient terrestrial fungi share ancestral evolutionary characters, including a coenocytic (having multinucleated cells) mycelium and exhibiting gametangiogamy (a sexual process resulting in the formation of zygospores).

The division **Chytridiomycota** represents a very ancient lineage of fungi, with fossils found in the Vendian period of Russia, 630–542 million years ago (Mya). Chytrids are unique among the fungi in having flagellated gametes; they are saprobic, degrading cellulose, chitin, and keratin, or parasitic on prokaryotes, algae, and amphibians in aquatic and terrestrial environments. The Chitrydiomycota may not be monophyletic; the Blastocladiomycota and Neocallimastigomycota are now segregated into two phylum-level sister clades of flagellated fungi (James et al., 2006; Hibbett et al., 2007).

And finally, the **Glomeromycota** is a similarly ancient (460 million years ago) and basal group of symbiotic fungi that form intracellular (living inside host cells) arbuscular mycorrhizae (branching filamentous hyphae, about which more below) within the roots of the vast majority of higher plants.

Two groups of fungi without formal taxonomic standing are the former and now defunct Phylum **Deuteromycota** ("**Fungi Imperfecti**") and the lichens. Fungi Imperfecti constitute an informal wastebasket category for those fungi that can no longer reproduce sexually and that are thus *incertae sedis* (of uncertain taxonomic status). They include asexually reproducing members of both the Basidiomycota and the Ascomycota.[10] The lichens actually represent a symbiotic association between a thallus-producing fungus (the mycobiont) and a photosynthesizing alga and/or a cyanobacter (the photobiont).[11] The taxonomy of lichens is based on mycobiont nomenclature and, like that of the fungi in general, is still governed by the nomenclatural rules governing botanical names.

Fungi: An Overview

Fungi exist in two basic morphological forms: **hyphae** and **yeasts**. Hyphae form filamentous **mycelia** (sing. mycelium) with or without cross-walls (**septa**) that partition the hyphae into cell-like compartments. Nonseptate hyphae form **coenocytic** (multinucleated), noncompartmentalized mycelia. Hyphae grow terminally and branch laterally. In contrast, yeasts are unicellular fungi that reproduce asexually by means of budding or binary fission.

Most fungi exhibit a life cycle with both a **haploid** (n) phase and a **diploid** ($2n$) phase.[12] In the dominant haploid phase, the hyphae contain two nuclei of opposite mating strains, (+) and (−), in certain compartments—the **dikaryotic** stage. The two nuclei then undergo fusion (**karyogamy**) to form a $2n$ **zygote**. This is followed by a series of reduction divisions (meiosis) to form four haploid nuclei—two each of the (+) and (−) mating types. The details vary in the various fungi, with yeasts exhibiting the simplest pattern. Some fungi live for extended periods as haploid organisms; others live as dikaryotic hyphae. The specific patterns of nuclear movements and interactions are consistent within the major groups of fungi and form the basis for their major classification.

Mating Types and Sexuality

Fungi commonly reproduce asexually through mycelial fragmentation or by asexual spores (**conidia**),[13] which permits rapid dispersal under favorable environmental circumstances. In sexually reproducing fungi, geneti-

10. Some Basidiomycota and Ascomycota and all but one of the Glomeromycota have lost the capacity to form sexual stages. These fungi form asexual reproductive units via blastic (when the spore is evident before it separates from the conidiogenic hypha) or thallic (when a septum first appears, creating a cell that becomes the spore) development.

11. The Cyanobacteria—formerly known as "blue-green algae"—were the earliest photosynthesizing organisms known to have appeared on Earth (fossils date from 2.8 billion years ago). Cyanobacters are now classified as a phylum in the domain Bacteria.

12. Haploid nuclei contain only one member of each pair of homologous chromosomes. Diploid nuclei have both copies (one from each parent) of each homologous pair of chromosomes.

13. Conidiospore is an outdated term often encountered in older literature.

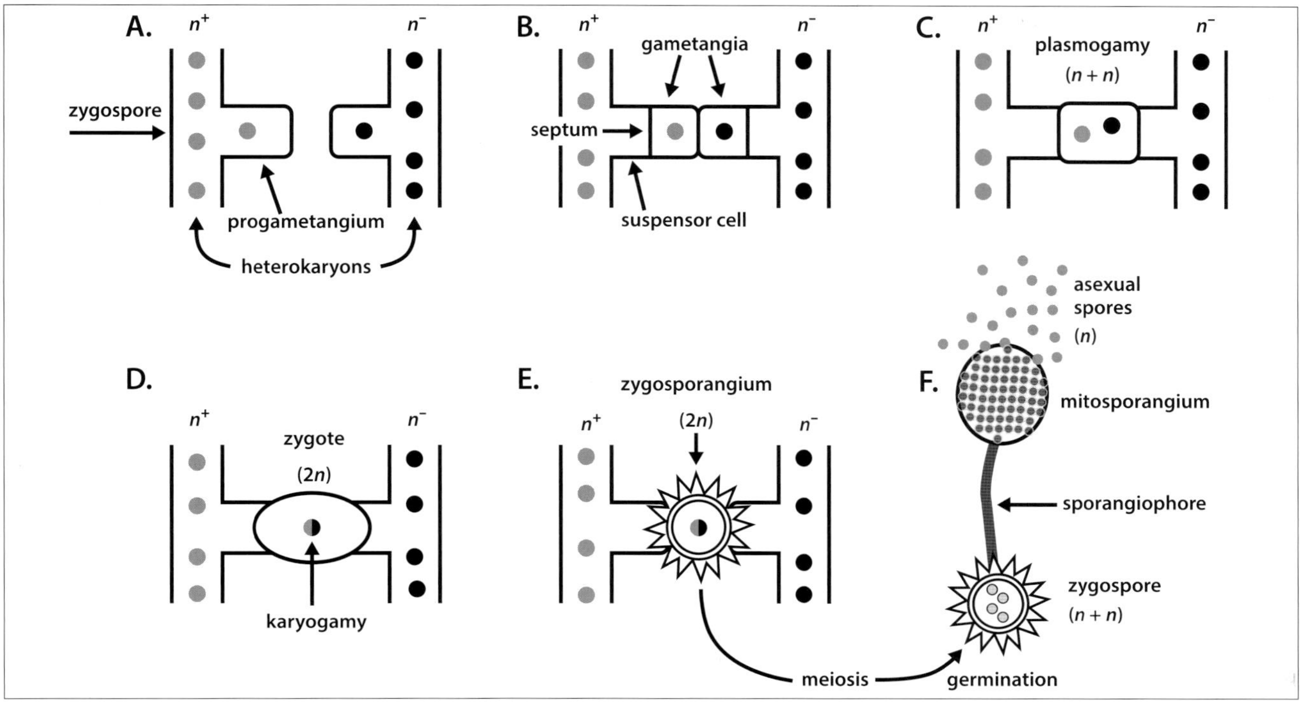

FIGURE 5. Zygospore formation. A, heterokaryotic hyphae of opposite but compatible mating types (n+ & n–) send out lateral extensions (progametangia) toward each other. B, when these make contact, a septum forms and isolates a nucleus from each mating type to form a gametangium. C, plasmogamy, the fusion of the two gametangia, follows to produce an $n + n$ dikaryon. D, following plasmogamy, the two nuclei fuse (karyogamy) to form a diploid ($2n$) zygote. E, a thick, roughened wall forms on the outside to produce a resistant zygosporangium that can survive harsh conditions. F, the diploid nucleus undergoes meiosis and genetic recombination to produce a zygospore containing four haploid nuclei. When environmental conditions warrant, the zygospore germinates to produce either a new mycelium or a sporangiophore bearing a mitosporangium, which releases haploid asexual spores. A number of zygosporous fungi produce azygospores in which karyogamy followed by subsequent meiosis has not been observed.

cally determined differences in mating types (so-called sexes) exist among individual fungi. In general, individual hyphae of the same mating type cannot mate with one another. However, some fungi with genetically compatible hyphae are **homothallic** and produce **zygospores** (diploid spores resulting from the fusion of two haploid nuclei) from different branches of the same mycelium. This is seen, with variations, more often in the polyphyletic groups of zygomycetous fungi and in the Ascomycota and is relatively uncommon among the Basidiomycota. **Heterothallic** fungi require mating between mycelia containing two different but compatible types of nuclei, generally denoted as + and – strains. All heterothallic ascomycetes have a single-locus, two-allele mating system. Basidiomycetes, on the other hand, have a two-locus, multiple-allele mating system that allows for an enormous number of different mating types. For example, *Schizophyllum commune* (the Common Splitgill) is estimated to have more than 28,000 different mating strains, the fungal equivalent of sexes, making it the most widespread and successful mushroom in the world.[14]

14. For those readers with a genetics background, this number is calculated from the number of possible combining permutations of multiple alleles of four gene loci in this "tetrapolar heterothallic" mating species: 9 alleles at *AA*, 32 at $A\alpha$, 9 at $A\beta$, $\beta\alpha$, and 9 at $\beta\beta$. The original calculations yielded >20,000 possible mating types (J.R. Raper, 1966, *Genetics of sexuality in higher fungi* [New York: Ronald Press]); that number has now been increased to >28,000 (J. Whitfield, 2004, Everything you always wanted to know about sexes, *PLoS Biol.* 2[6], e183, DOI:10.1371/journal.pbio.0020183). Because the wild populations contain many different *A* (>300) and β (>90) alleles, the outbreeding potential is >98.9%.

Division/Phylum Basidiomycota (Club Fungi and Mushrooms)

With a few exceptions, these are the critters you go hunting for on your forays into the woods to gather mushrooms for dinner. The fruiting bodies known as mushrooms and jelly fungi such as wood ears are your usual quarry. (As you shall see below, the much-sought-after morels and truffles are an exception belonging to another division, the Ascomycota.)

The division Basidiomycota contains some 30,000 described species—about 37% of the described species of true Fungi (Kirk et al., 2001). Included here are the mushrooms, puffballs, stinkhorns, polypores, jelly fungi, boletes, earthstars, smuts, rusts, and several miscellaneous groups. These fungi are important agents of wood decay and decomposers of leaf litter, animal dung, and other organic matter. Others form mycorrhizal symbioses with plants or are plant parasites.

The Basidiomycota were formerly known as the class Basidiomycetes, a sister group to the class Ascomycetes (now the division Ascomycota) before the Fungi were elevated to the rank of kingdom and the two classes were promoted to the rank of phyla (divisions). The Basidiomycota are the most evolutionarily advanced fungi, and their septate hyphae have a distinctly "cellular" organization. An undisputed fungal clamp connection (see Fig. 8)—diagnostic of the division Basidiomycota—was reported in the wood of a mid-Pennsylvanian coenopterid fern, *Zygopteris illinoiensis* (Andrews, 1942) Baxter, 1952; this establishes a date of 290 million years ago for this group (Dennis, 1970).

The Basidiomycota are filamentous, hyphae-forming fungi (except for certain unicellular yeasts) that reproduce sexually by forming specialized terminal cells known as **basidia** (sing. basidium), which usually produce four external, haploid **basidiospores** (Fig. 6). Some Basidiomycota also reproduce asexually. In the Basidiomycota, a long-lasting dikaryotic mycelium periodically produces club-shaped end cells (basidia) in which two sexually compatible nuclei fuse to form a diploid nucleus (Fig. 7). Meiosis follows shortly, generally resulting in four haploid basidiospores. Basidiospores are typically ballistic and are forcefully ejected to produce new monokaryotic hyphae.

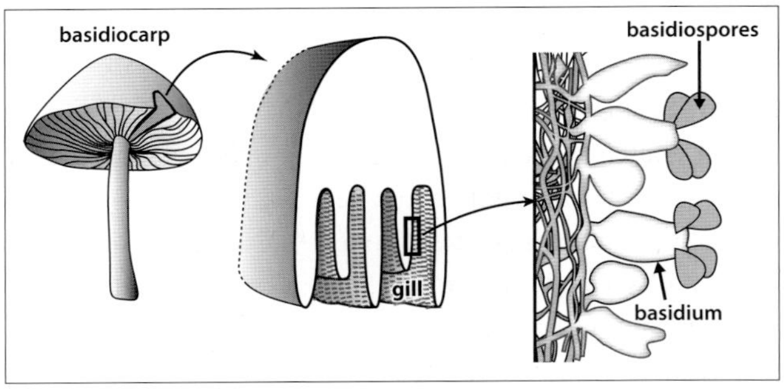

FIGURE 6. Basidiocarp: the diploid reproductive structure of a basidiomycete showing the fruiting body, hymenium (gills), and basidia. (Modified from an image by Debivort.)

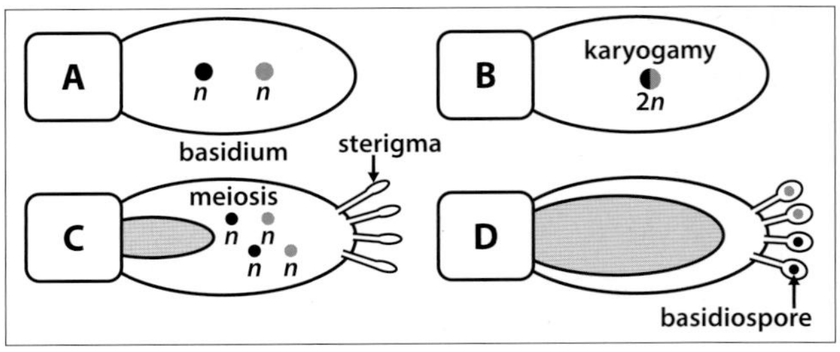

FIGURE 7. Formation of basidiospores within a basidium. Two haploid (n) nuclei in an apical dikaryotic hyphal compartment (A) undergo karyogamy (fuse) within a basidiocarp to form a diploid ($2n$) nucleus (B). This diploid nucleus undergoes meiosis and genetic recombination to form four haploid nuclei (C). Four sterigmata form, and a growing fluid-filled vacuole at the base of the compartment (basidium) forces protoplasm into the sterigmata, inflating their tips (D). One haploid nucleus passes into each sterigma and matures into a basidiospore. The black and gray circles represent nuclei of sexually opposite poles. The grey oval represents the growing vacuole. (Redrawn and modified after figures by Vicki Tariq.)

The diploid basidia are produced on basidiocarps—the prominent fruiting bodies (Fig. 6) that emerge aboveground and are so sought after by mycophiles. Plasmogamy (the fusion of two parent cells without the fusion of their nuclei) between compatible individuals through cross-clamps and delayed karyogamy produces a long-lasting dikaryon. Eventually, the paired nuclei in the dikaryotic basidium fuse, and the resulting diploid basidium begins the cycle anew.

In the Basidiomycetes, a lateral connection forms between two adjoining cells of a filament and arches over the septum dividing them. To maintain the dikaryon, a short lateral branch of a terminal cell forms a **clamp connection** with another compatible hyphal compartment (Fig. 8). One of the nuclei migrates through the connection to form a dikaryon of two opposite nuclear types, (+) and (−). This process is repeated to form new dikaryotic cells asexually as the hypha grows.

Two major spore-releasing adaptations exist among the Basidiomycota: **lamellae** (gills) and **pores**. In the gilled

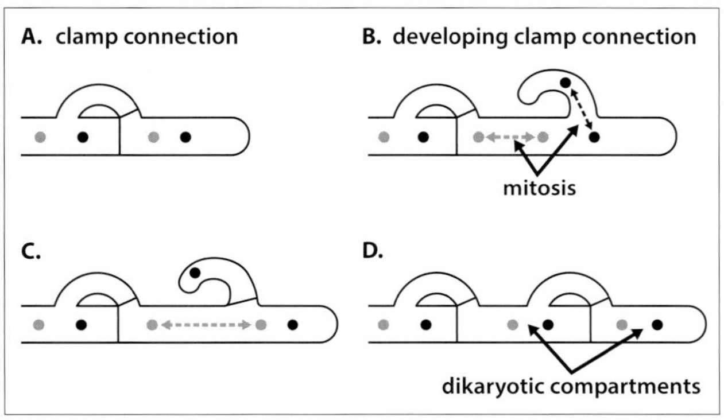

FIGURE 8. Formation of a clamp connection. A, apical (terminal) compartment of hypha where growth occurs. B, elongation of hyphal tip and beginning of clamp connection; synchronous mitotic division of (+) and (−) nuclei, with one nucleus migrating into the developing clamp. C, septum forms at the base of the clamp, trapping the nucleus, while daughter nuclei from mitotic divisions migrate into hyphal tip. D, septum forms below the clamp, creating a new apical compartment; clamp fuses to the adjacent cell and releases nucleus into the subterminal compartment. Both the terminal and subterminal cells are now dikaryotic, with (+) and (−) nuclei.

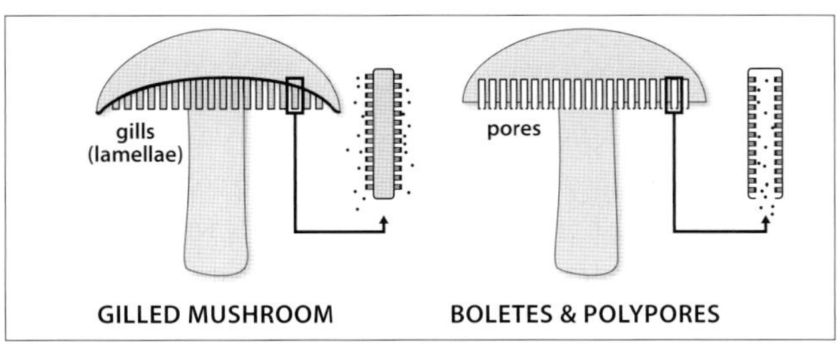

FIGURE 9. In gilled mushrooms, the spores are borne on the outer surfaces of the lamellae (gills). In boletes and polypores, the spores are borne on the inside walls of the tubes.

mushrooms, the spore-carrying basidia are borne on outer surfaces of numerous curtain-like gills suspended from the fruitbody (see Fig. 9). But many basidiomycetes—most notably the boletes and polypores—have developed a system of closely packed tubes in place of gills. In these mushrooms, the basidia are borne on the inside wall of the tubes, whence they are released into the air. In both cases, the spores are primarily carried away by air currents or in some cases by water and/or the activities of animals.

Hydnoid basidiomycetes bear cylindrical rounded to pointed teeth instead of gills or pores; these teeth are covered by the hymenium with outward-facing basidia, as in the gilled mushrooms. In some groups, the teeth may be awl shaped, peg-like and stubby, or variously flattened; in some cases, they form a maze-like network. And parchment fungi and crust-like spreads have a smooth, wrinkled, or ridged hymenium, with no special basidia-bearing structures. Some of the latter are resupinate (i.e., the hymenium faces upward or outward instead of downward).

Ballistospory

Most types of basidiospores are forcibly discharged from the tips of their sterigmata; such spores are known as **ballistospores** to distinguish them from passively released spores (**statismospores**). Some 30,000 species of mushrooms and other fungi discharge ballistospores. The ballistospores of the Wood Ear, *Auricularia auricula-judae*, were found to discharge at speeds of 0.87–1.57 meters per second (2.85–12.56 feet per second). The viscosity of air acting against such a small body traveling at an initial speed of 1.25 meters per second (m/s) causes a very rapid deceleration, and the spore would be propelled for a distance of just 0.4 millimeter (mm) after a flight time of only 2 milliseconds (ms). The flight of the spore is abruptly braked, and gravity exerts control over its trajectory after a flight time of 2 ms or less.[15] To avoid hitting the surfaces of adjacent gills or the opposite walls of narrow tubes (e.g.,

15. A. Pringle et al., 2005, *Mycologia* 97(4):866–871, note that it takes 10 microseconds for *A. auricula-judae* to launch a ballistospore, implying an acceleration greater than 12,000 g to reach an initial velocity of 1.20 m/s.

in polypores and boletes), the spores travel no more than about 0.1 mm before falling downward to be dispersed by the turbulent air below the cap.

The mechanism behind the discharge of ballistospores in the Basidiomycota is explained in Figure 10 and the accompanying legend. Ballistospore discharge in the Ascomycota (see the following section) is brought about by hydrostatic turgor pressure within the ascus. Ballistosporic ascospores are blasted away from the sporangiophore. In many apothecial and perithecial fungi, the spores can be ejected through the ostiole for some distance.

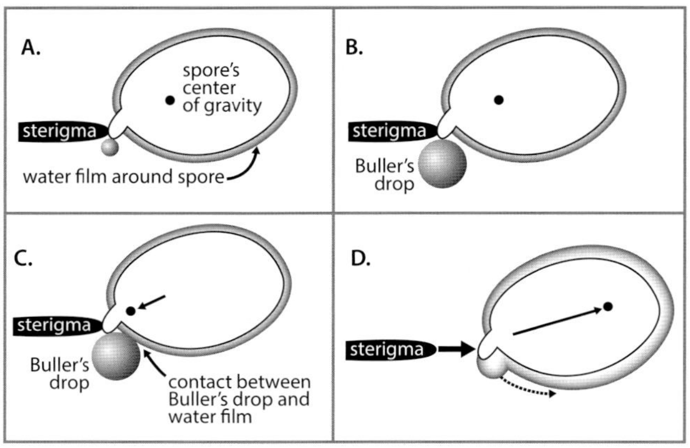

FIGURE 10. Mechanism underlying forcible ballistospore discharge. A, sugars secreted at the apiculus by the spore cause a drop of water (Buller's drop) to condense, and the high humidity between the gills (or within a pore) causes a film of water to form around the spore. B, Buller's drop expands, thereby increasing the mass of the spore toward the apiculus. C, the continued growth of Buller's drop shifts the spore's center of gravity toward the apiculus (black dot and arrow). D, the forces in the surface tension give energy to Buller's drop; when Buller's drop contacts the water film on the side of the spore, it instantly collapses and flows into the surface film, expanding it toward the distal end of the spore (dashed curved arrow). This shifts the spore's center of gravity rapidly back toward the distal end of the spore (black dot and arrow), imparting great momentum in that direction to the spore. Newton's Third Law of Motion mandates that every force exerts an equal and opposite force; thus, an equivalent force is exerted back against the sterigma. The contact between the sterigma and the apiculus is simultaneously broken, and the spore is catapulted away, as shown by the heavy arrow. (Redrawn and modified from figures by Pringle et al., 2005 [see note 15] and the Australian National Botanic Gardens at https://www.anbg.gov.au/fungi/spore-discharge-mushrooms.html.)

Statismospory

Gasteroid fungi like the puffballs, earthstars, truffles, false truffles, stinkhorns, and bird's nest fungi produce their spores (basidiospores) within the **gleba** (a fleshy or gelatinous tissue consisting of a **capillitium** of densely interwoven sterile threads intermingled with spores) of an enclosed **angiocarp** or **periderm** and are unable to forcibly discharge them. Their **statismospores** are passively released. Mushrooms that are intermediate in form between gasteroid and the agarics or boletes are termed **secotioid** (see Fig. 11). The term **sequestrate** is sometimes used to refer to either gasteroid or secotioid mushrooms. Many gasteroid fungi are also **hypogeous** (maturing completely or partially underground), such as the truffles and false truffles; others, like the puffballs, stinkhorns, and certain club fungi (e.g., *Clavariadelphus*), are **epigeous**, with their sporocarps (fruitbodies) emerging aboveground. Bird's nest fungi produce their spores inside small, disk-shaped **peridioles**, which contain the gleba, basidia, and basidiospores enclosed in a hardened wall.

Gasteroid and secotioid fungi have evolved independently in many unrelated agaricoid (gill-bearing mushrooms) groups. All of them depend on the force of falling raindrops (bird's nest peridioles), wind (puffballs), or insects and other animals (truffles, stinkhorns) for spore dispersal.

Division Ascomycota (Yeasts and Sac Fungi)

The division Ascomycota (sac fungi) is a monophyletic group containing some 75% of all described fungi. It includes most of the mycobionts that form lichen associations with algae.

In contrast to the basidium of the Basidiomycota, a single, modified reproductive cell called an **ascus** (pl. asci) typically contains eight ascospores. These are produced by a meiotic reduction division followed by mitosis. In most Ascomycota, the asci form in groups within fruiting bodies known as ascocarps (Fig. 12).

The apical cell of a dikaryotic ascogenic hypha elongates

FIGURE 11. Gasteroid and secotioid fruitbodies. A, a puffball (*Calvatia*), showing undifferentiated gleba enclosed in a peridium; B, a secotioid mushroom resembling an unopened mushroom (*Nivatogastrium*) with a rudimentary stipe and a gleba consisting of irregularly anastomosing gills; C, a stinkhorn (*Phallus*) egg with a developing receptacle and gelatinous gleba enclosed by a peridium. (Photos: A, Ron Pastorino; B, Alan Rockefeller; C, Stanislaw Szydlo.)

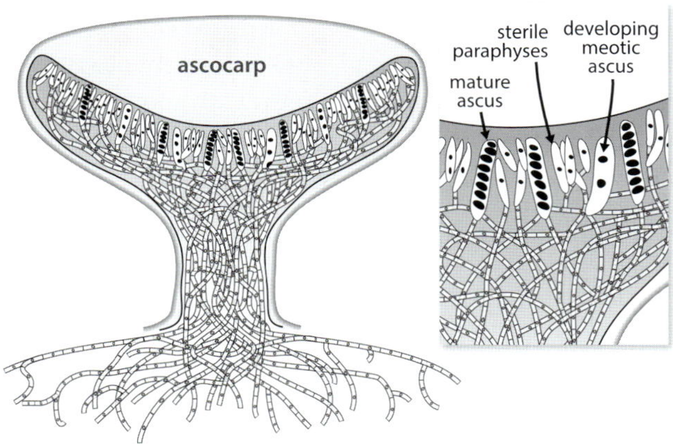

FIGURE 12. Typical ascocarp—the fruiting body of an ascomycete. (Modified from a figure by Debivort.)

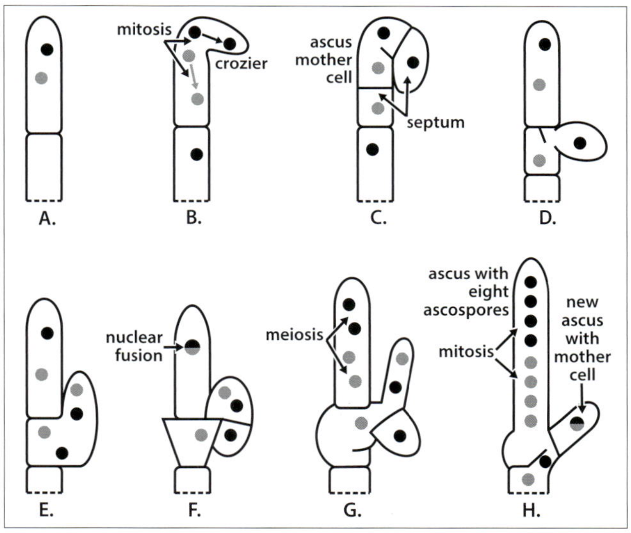

FIGURE 13. Formation of a crozier, an ascus, and eight ascospores. The black and gray circles represent sexually opposite (and thus compatible) nuclei. See text for discussion. (Modified after figures by Vicki Tariq.)

and forms a hook-like **crozier**—the counterpart to the clamp connection of the Basidiomycota (Fig. 13). Two sexually compatible nuclei in the compartment simultaneously divide by mitosis, and two septa develop that divide the crozier into three compartments.

The tip and basal sections are uninucleate, but the middle compartment—the **ascus mother cell**—contains

A

B

FIGURE 14. Representative Ascomycota. A, Devil's Urn, *Urnula craterium*; B, North American Yellow Morel, *Morchella esculentoides*. (Photos: A, W. Light; B, Ken Zinkand.)

two sexually compatible haploid nuclei. The nuclei fuse into a diploid nucleus, which then undergoes meiosis to form four haploid nuclei. The four nuclei then divide by mitosis to form eight haploid nuclei, four of each sexual type. Each nucleus is then surrounded by protoplasm and develops a cell wall to become an **ascospore**.

The sac fungi include morels (*Morchella* spp.), truffles (*Tuber* spp.), brewer's and baker's yeasts (*Saccharomyces* spp.), Dead Man's Fingers (*Xylaria* spp.), the cup fungi, and the great majority of lichen associations. A number of plant pathogens also belong in this division: ergot (*Claviceps* spp.), apple scab (*Malus* spp.), and the powdery mildews.

Zygomycetous Fungi (Pin Molds, Bread Molds, and Other Zygospore-Forming Fungi)

Primarily terrestrial, these fungi may occasionally be found in water. All produce sexual zygospores, and their coenocytic (multinucleate) hyphae are typically aseptate and lack the cross-walls that divide the hyphae into compartments; however, they may produce septa when forming zygospores. Reproduction is primarily asexual by means of bulbous sporangia containing haploid spores that develop at the tips of upright hyphae known as sporangiophores. Germination of these spores produces the haploid hyphae of a new mycelium.

Sexual reproduction involves conjugation (the fusion of two hyphae containing sexually compatible gametes to

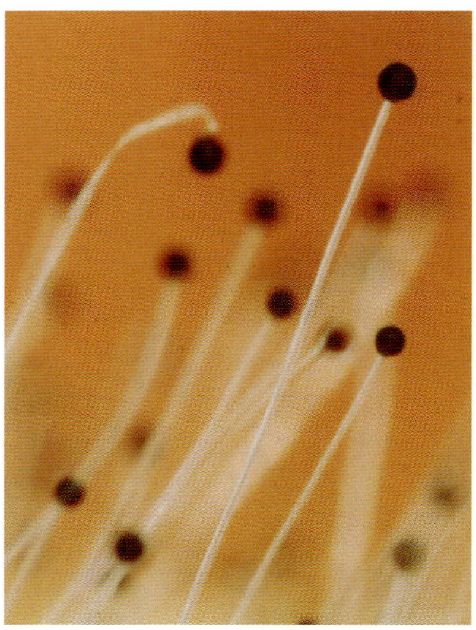

A

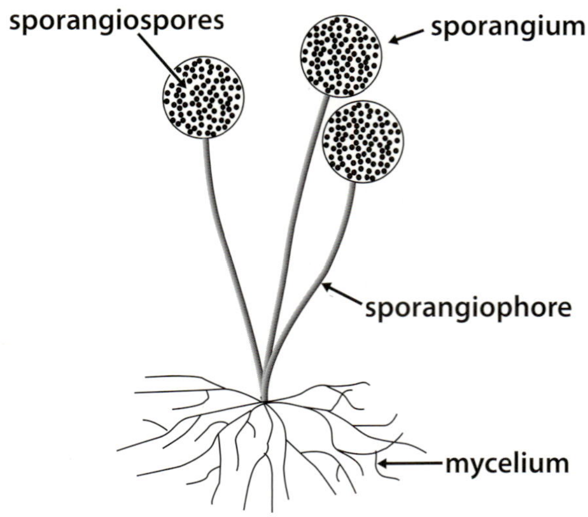

B

FIGURE 15. Zygomycetous fungi. A, pin mold on a peach (micrograph by Zephyris); B, graphic of a filamentous zygomycete showing major structures.

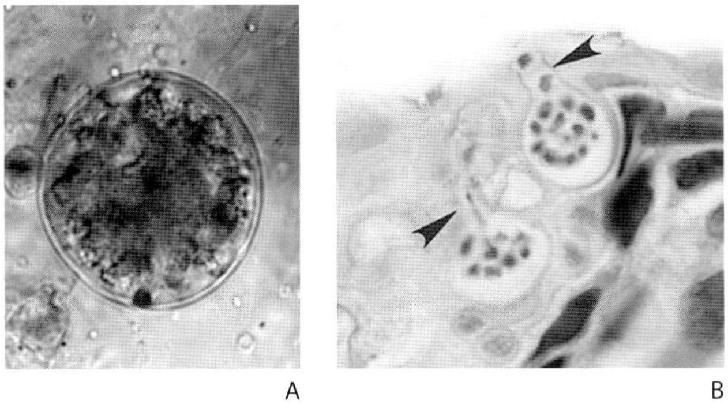

FIGURE 16. Chitridiomycota. A, a chytrid (photo, Dr. David Midgley); B, chytridiomycosis. Skin of upper hind limb of a Central American Smooth-skinned Toad, *Bufo haematiticus* (Cope, 1862). Two sporangia (arrows) of *Batrachochytrium* cf. *dendrobatidis* with many zoospores visible. (Source, Daszak et al., 1999, http://wwwnc.cdc.gov/eid/article/5/6/99-0601_article.)

form zygosporangia). Single nuclei from each mating type fuse to form the diploid nucleus; a zygospore then forms around the diploid nuclei, which later undergo meiosis, usually at germination. The hyphae thus produced are again haploid. There are no fleshy fruiting bodies. Zygomycetes form endomycorrhizal associations with a great many green plants. Bread mold—*Rhizopus stolonifer* (Ehrenberg 1918 *ex* Fries) Vuillemin 1902—is a classic zygomycete that covers the surface of bread and other foods, while its inwardly penetrating hyphae absorb nutrients.

Division Chytridiomycota (Chytrids)

The Chytridiomycota resemble some protists in having flagellated zoospores and gametes (reproductive cells) that enable them to swim. All other fungi lack flagella, suggesting that they lost this trait very early in their evolutionary history.[16] However, they resemble fungi in having absorptive extracellular nutrition, typical fungal metabolism, and cell walls composed of chitin and, in a few, in forming hyphae.

Batrachochytrium dendrobatidis Longcore, Pessier & Nichols 1999 (Fig. 16B) is a parasitic, nonhyphal chytrid fungus that has caused catastrophic amphibian population crashes in many parts of the world. It causes an often lethal fungal skin infection—**chytridiomycosis**—in frogs and salamanders. First discovered and described in 1998, it is the only chytrid known to parasitize vertebrates and is the sole species in its genus. *Batrachochytrium dendrobatidis* infects other animals besides amphibians, for example, crayfish and other freshwater arthropods, which can maintain the infection in aquatic ecosystems for many weeks and transmit it to amphibians.

Mycorrhizae

Division Glomeromycota (arbuscular mycorrhizae)

Arbuscular mycorrhizae (AM) are symbiotic **endomycorrhizal associations** between fungi of the division Glomeromycota and the root cortical cells, or thalli, of plants. AM fungi produce highly branched processes known as **arbuscules** (Fig. 17) between the cell wall and the

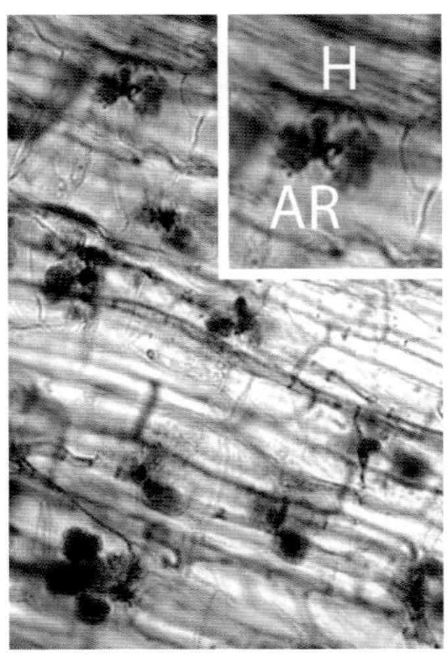

FIGURE 17 at upper right. Glomeromycota. Micrograph of flax root cortical cells containing arbuscular mycorrhizae (AR) and fungal hyphae (H). (Photo: M.S. Turnell, University of Manitoba.)

16. The 500-odd species of water molds and downy mildews (Phylum Oömycota), once classified as fungi, are now placed in the kingdom Protista. They were formerly considered to be fungi because of their hyphae-like, filamentous form and external absorptive style of nutrition. However, their cell walls are composed of cellulose instead of chitin, as in the fungi, and the nuclei within the filaments are diploid rather than haploid, as in the fungi.

plasma (cell) membrane of the root cells of a host plant. These mycorrhizae provide a large surface area for nutrient exchange between the plant and the fungus. The fungus thus obtains a reliable source of carbohydrates from its photosynthesizing host; in turn, the fungus provides the plant with water and minerals absorbed from the soil by its mycelium, which has a vastly larger surface area than the roots of the plant (Harrison, 2005). These arbuscules are continually formed and digested by the plant cells. In this symbiosis, the fungal hyphae penetrate into the cortical cells of the host plant but without piercing the cell membrane. Arbuscular mycorrhizae are by far the more common and are associated with 85%–90% of all plant species.

Sexual reproduction is unknown in these fungi (Kuhn, Hijri, & Sanders, 2001), except for the genus *Endogone*, which forms sexual zygospores (and may actually belong in the division Zygomycota). Plant-fungus interactions are very widespread today, and 80%–90% of all green plants form associations with these mycorrhizal fungi. This is an ancient symbiosis, with evidence dating to about 460 million years ago during the Ordovician period (Redecker, Kodner, & Graham, 2000; Redecker, Morton, & Bruns, 2000), and mycorrhizal nutrient transfer mutualism may have been in existence when plants first invaded the land (Remy et al., 1994). The major fungal lineages—Basidiomycota, Ascomycota, and Glomeromycota—may have diverged from each other around 500 million years ago (Berbee & Taylor, 2001, 2007).

Ectomycorrhizal Fungi

As in the case of AM, in exchange for the plant's photosynthetic carbon and other compounds, ectomycorrhizal fungi benefit their plant hosts by greatly increasing the effective surface area through which the plant roots can absorb water, minerals, and other nutrients from the soil. Ectomycorrhizal hyphae form a fungal sheath or mantle around the short lateral roots of their host plant and an intricate hyphal network called a **Hartig net** that penetrates between the outer cortical cells of the root (see Fig. 18). In the Northern Hemisphere, ectomycorrhizal fungi are mainly associated with trees in the families Fagaceae (beeches and oaks), Betulaceae (birches), and Pinaceae (pines, spruces, firs, hemlocks, and the like). These fungi usually belong to the division Basidiomycota, but a few are in the Ascomycota. Prominent mushroom genera

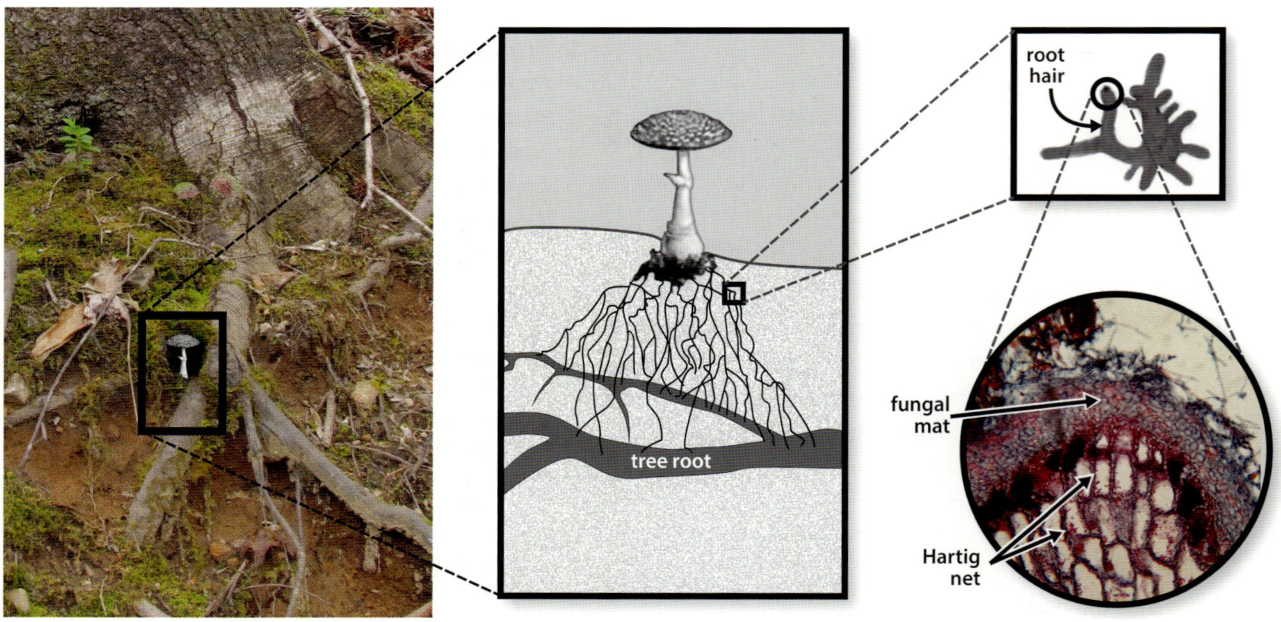

FIGURE 18. Ectomycorrhizal association between a Fly Agaric, *Amanita muscaria*, and its host tree. The ectomycorrhizal cross-section incorporated into this figure is a micrograph we made from a slide prepared by Carolina Biological Supply Company. (© Carolina Biological Supply Company, used with permission. Reproduction of this image without written permission from the copyright holder is unlawful.)

FIGURE 19. A, *Monotropa uniflora*; B, *Monotropa hypopitys*; C, *Corallorhiza maculata*. (Photos: A, Anonymous; B–C, Walter Siegmund.)

that form ectomycorrhizal associations include *Amanita*, *Russula*, *Lactarius*, *Laccaria*, *Boletus*, *Suillus*, *Strobilomyces*, *Cantharellus*, *Hygrophorus*, *Cortinarius*, *Entoloma*, and *Tricholoma*, among a great many others far too numerous to list here. Many are highly specific as to host species.

Mycoheterotrophic Plants (Fig. 19)

More than 400 species of achlorophyllous plants (plants that lack chlorophyll) obtain their carbon compounds by forming heterotrophic associations with fungal ectomycorrhizae—a form of nutrition known as **mycoheterotrophy** (Min et al., 2012). These plants are directly parasitic on the fungal ectomycorrhizae and steal nutrients and carbon compounds from them, apparently without the fungi being aware of their loss. Because the fungi obtain their carbon compounds from the plants with which they are mutually beneficially associated, this direct theft from the fungi is sometimes referred to as a form of "indirect parasitism" on the host plants. Perhaps a more appropriate term is **epiparasitic cheating** (Bitartondo & Bruns, 2011). Epiparasitic cheating by *Monotropa* and related genera involves one of the most pervasive and ubiquitous mutualistic associations known: ectomycorrhizal associations of fungi with their host plants. Because the host and epiparasitic plant never come into direct contact, the host cannot mount a defense against the parasite without also selecting against its needed mycorrhizal fungus (Bitartondo & Bruns, 2011).

Epiparasitic cheating is especially well developed in the species of the cosmopolitan genus *Monotropa* Linnaeus in the heath family (Ericaceae). In our area, Indian Pipe (*Monotropa uniflora* Linnaeus) parasitizes various ectomycorrhizal species of *Russula* and *Lactarius*, while Pinesap (*M. hypopitys* Linnaeus)[17] is associated with the genus *Tricholoma* (Bitartondo & Bruns, 2011).[18] Related monotropoid mycorrhizal plants such as Pinedrops (*Pterospora andomedea* Nuttall) and Snow Plant (*Sarcodes sanguinea*

17. The epithet *hypopitys* (below pines) is based on the Greek words *hypo* (below) and Pitys, an Oreiad nymph who was transformed into a fir tree or a pine to escape the persistent and unwanted amorous pursuits of Pan. Linnaeus misspelled the epithet as *hypopithys*, an orthographic error that is properly correctable to *hypopitys*. However, this epithet continues to be spelled as both *hypopitys* and *hypopithys* by various authors in both the botanical and mycological literatures.

18. Beech Drops (*Epifagus americana* Nuttall) is widespread and common in our area. It superficially resembles species of *Monotropa* but belongs to the Broomrape family (Orobanchaceae), which is not associated with fungi. Beech Drops is directly parasitic on the roots of American Beech trees (*Fagus grandifolia* Ehrhart) and does not involve mycorrhizal fungi.

Torrey) are widely distributed in western North America. *Pterospora* and *Sarcodes* both form mycorrhizae with false truffles of the genus *Rhizopogon* Fries. All are members of the subfamily Monotropoideae.

Achlorophyllous orchids of the genus *Corallorhiza* Gagnebin (Family Orchidaceae), several species of which occur in our area, are also obligately mycoheterotrophic on mycorrhizal fungi.[19]

19. *Corallorhiza trifida* Châtelain, a widespread species in eastern North America, has some chlorophyll and is only partly dependent upon fungi for its nutrition.

Monotropoid species form **monotropid mycorrhizae** with their fungal hosts, whom they trick into accepting them as their normal mutualist host tree. A typical monotropoid mycorrhiza involves a fungal mantle and a Hartig net, much as one sees in a typical ectomycorrhiza, but with the additional presence of a **fungal peg**. A fungal peg is produced when a hypha from the fungal mantle or the Hartig net invades an epidermal cell of *Monotropa*, *Pterospora*, or *Sarcodes*. The plant responds by invaginating its cell wall to accommodate the invading hypha and in laying down additional wall material to form finger-like projections around each fungal peg.

To Summarize

SUBKINGDOM DIKARYA (THE "HIGHER FUNGI")

DIVISION/PHYLUM ASCOMYCOTA

Subphylum Pezizomycotina: Filamentous ascomycetes that reproduce by fission rather than budding; includes almost all ascus-bearing fungi that have fruiting bodies visible to the unaided eye. Most members of this subphylum form asci following the formation of croziers at their bases. The members of the obsolete class Discomycetes are now included in the classes Pezizomycetes, Leotiomycetes, Sordariomycetes, and Lecanoromycetes (lichenized fungi). The subphylum includes 90% of all Ascomycota and contains 13 classes, of which the following include taxa we discuss in this work:

Class Pezizomycetes: Fungi with operculate asci (with a lid-like cover over the exit pore). Morels and their allies, the "true" truffles, and most of the cup-shaped and disk-shaped fungi are included in this class.

Class Leotiomycetes: Generally small fruitbodies with an exposed hymenium in a cup-like apothecium and inoperculate (lacking a lid) asci with an apical perforation pore for releasing ascospores. Commonly encountered members include the green elf cups and jelly babies.

Class Geoglossomycetes: The earth tongues, *Geoglossum*, *Trichoglossum*, and related genera.

Class Sordariomycetes (Pyrenomycetes): Includes the insect-parasitizing families Cordycipitaceae and Ophiocordicipitaceae (Order Hypocreales); the family Hypomyceaceae, with the mycoparasitic genus *Hypomyces* (Order Hypocreales); the carbonaceous fungi (Order Xylariales); and the family Cryphonetriaceae (Order Diapothales), containing the serious forest pathogen *Cryphonectria parasitica*, which causes Chestnut Blight.

Subphylum Taphrinomycotina: Dimorphic plant parasites having both yeast and hyphal stages, which infect leaves, catkins, and branches. Members of this subphylum do not form croziers.

Subphylum Saccharomycotina: The budding yeasts, including *Saccharomyces cerevisiae* Meyen *ex* E.C. Hansen, which is important for baking, wine making, and beer brewing; and the yeast genus *Candida* Berkhout, which can cause fungal infections in humans and other mammals. Members of this subphylum do not form croziers.

DIVISION/PHYLUM BASIDIOMYCOTA

Subphylum Agaricomycotina (Hymenomycetes): The hymenium (the fertile spore-bearing layer) is exposed before the spores develop; includes most mushrooms, the polypores, jelly fungi, "wood ears" and their allies, crusts and spreads, gasteroid fungi (puffballs and earthstars, false truffles, bird's nest fungi), and the hydnoid (toothed) fungi. Although the systematics of this subphylum are still in flux, three classes are currently recognized: the **Agaricomycetes**, the **Tremellomycetes**, and the **Dacryomycetes**.

Class Agaricomycetes: By far the largest class in this subphylum, it contains the following orders (several minor orders are omitted):

Agaricales: The gilled mushrooms (excluding the russulas, which are now in their own order)—the eucarics or "agarics" (*sensu lato*); contains 33 families and some 413 genera.

Boletales: The boletes, as well as some gilled mushrooms, the puffballs, earthstars, earthballs (Sclerodermataceae), and false truffles (Rhizopogonaceae).

Russulales: The russulas and milkcaps, certain polypores (*Bondarzewia*), some toothed fungi (e.g., Lion's Mane), and many clavarioid fungi (club and coral fungi).

Corticiales: Corticioid fungi (spreads and crusts).

Polyporales: The polypores and shelf fungi, some spreads, and a few gilled forms (e.g., *Lentinus*).

Thelephorales: Most hydnoid (toothed) mushrooms, a few polypores, and certain club fungi; all are ectomycorrhizal.

Hymenochaetales: Certain imbricating (overlapping) spreads; saprobic and parasitic wood-decaying fungi.

Gomphales: Certain club fungi (e.g., *Clavariadelphis* and *Gomphus*).

Phallales: The stinkhorns and their allies.

Geastrales: The earthstars.

Auriculariales: The wood ears (*Auricularia*) and certain related fungi (e.g., Crystal Brain Fungus and Black Witches' Butter), all formerly included in the order Tremellales.

Cantharellales: The chanterelles, some toothed fungi, and some club and coral fungi.

Sebacinales: Certain encrusting and enveloping jelly-like fungi (e.g., *Sebacina* and *Tremellodendron*) that form ectomycorrhizal or endomycorrhizal associations with plants.

The gasteroid fungi were formerly grouped together in the artificial and now defunct class Gasteromycetes ("stomach fungi") because they produce their basidiospores enclosed within an outer cortex or peridium. These gasteroid fungi included the puffballs, earthstars, stinkhorns, and false truffles, which are now divided among several orders.

Class Tremellomycetes: The classic jelly fungi, typified by Witches' Butter and its relatives that are parasitic on other fungi. This group is characterized by having tremelloid basidia. The class also includes a number of anamorphic yeasts, including *Cryptococcus*, some of which are dangerous human pathogens.

To Summarize (*Continued*)

Class Dacrymycetes: Consists of a single order (Dacrymycetales) containing the sole family Dacrymycetaceae, characterized by branching basidia. It includes such jelly-like fungi as *Calocera*, *Dacrymyces*, and *Dacryopinax*, all of which are treated in this book.

Subphylum Ustilaginomycotina: Faultative parasites of higher plants with simple life cycles that cause leaf spots, galls, and other growths; contains three classes, including the **smuts (Class Ustilaginomycetes)** and a small order *incertae sedis* of fungi that cause opportunistic skin infections in animals, including humans.

Subphylum Pucciniomycotina: Obligate plant parasites (primarily angiosperms, or flowering plants) with no fruiting body and complex life cycles that involve a haploid yeast phase and a parasitic dikaryotic hyphal phase; contains eight classes, including the **rusts (Class Urediniomycetes)**.

Clade (unranked) Wallemiomycetes: Three species of molds in the genus *Wallemia*; found worldwide in hypersaline saltern water (salt evaporation ponds) that contaminates low-moisture foods such as flour, cake, and fried fish. This clade may include the Entorrhizomycetidae, commonly considered to be basal (ancestral) to the Ustilaginomycotina.[20]

The cladogram to the right shows the taxonomic relationships between the major groups of fungi as they are presently understood.

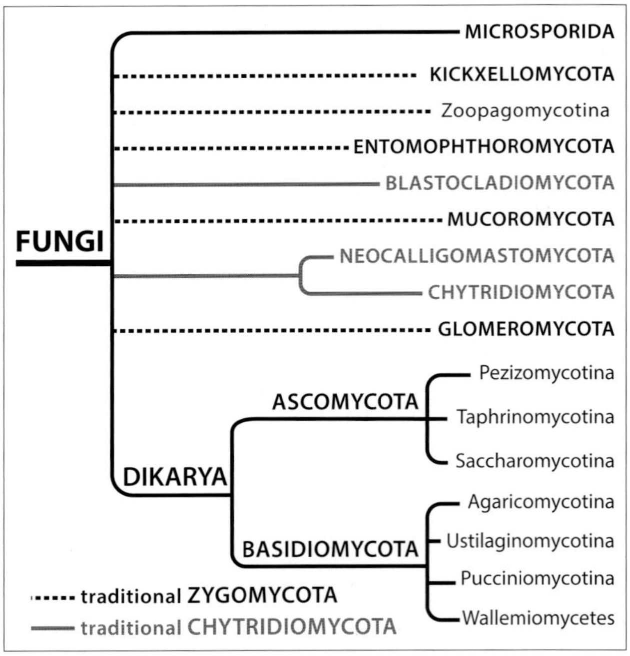

FIGURE 20. Cladogram of the kingdom Fungi as currently conceived. The divisions (phyla) and one subdivision (subphylum) into which the outdated concepts of the Zygomycota and Chytridiomycota, respectively, have been divided are indicated by dotted and grey lines, respectively.

Carnivorous Fungi

Approximately 150 species of fungi attack, kill, and digest small animals and protists, including protozoans, rotifers, small arthropods (copepods and crustacean nauplii), and nematodes. Most of these carnivorous fungi are anamorphic (asexual) forms, and for more than a century they have been given their own special scientific names involving "form-genera" and "form-species" (see the section below dealing with the scientific nomenclature of fungi). Many of these fungi are only known as anamorphs, while others have been putatively or definitively identified with specific teleomorphic (sexual) stages. Thus, species of the teleomorphic mushroom genus *Orbilia* commonly have one or more form-species of the form-genus *Arthrobotrys* that attack and kill nematode worms. (The teleomorph, together with its anamorph[s], constitute the entire fungus, known as the holomorph.)

20. P.B. Matheny et al., 2006, *Can. J. Bot.* 84:1794–1805.

Fungi and nematodes have coexisted in soil for a very long time. A nematophagous fungus and its prey preserved in amber are known from the Late Cretaceous period some 100 million years ago. Some nematophagous fungi are species-specific and eat only one kind of nematode. Others will attack several species but often are more virulent against a particular species.

Why do fungi do this? For the same reason that certain green plants do. Nitrogen is essential for synthesizing proteins and nucleic acids and requires a dietary carbon-to-nitrogen (C:N) ratio no more than ~30:1. However, the wood and soil inhabited by most fungi are very carbon rich and nitrogen poor, with C:N ratios of ~300:1. Wood-rotting fungi are very efficient nitrogen recyclers and can grow in very low nitrogen levels. But many species trap, kill, and consume small invertebrate animals, protists, and bacteria as an added source of nitrogen, just as do Venus Flytrap, sundews, and pitcher plants. Microscopic nematodes (roundworms of the phylum Nematoda) are a favorite prey of many carnivorous fungi, including a number of choice edibles, like oyster mushrooms (*Pleurotis* and some related genera). Some nematophagous fungi are species-specific and eat only one kind of nematode. Others are more eclectic in their diets and will attack several species but often are more virulent against a particular species.

Nematophagous fungi employ one or more of the following weapons, often in combination, to immobilize their prey (Fig. 21):

- adhesive networks (AN), the most widespread device
- adhesive knobs (AK) and columns (AC)
- nonconstricting (NCR, always together with AK) and constricting rings (CR)
- lacerating spikes or spiny balls (SB)
- toxin-secreting cells (TSC)

Most nematode-trapping fungi are ascomycetes in the order Orbiliales (Orbiliomycetes)—in particular the genera *Orbilia* Fries and *Hyalorbilia* Baral & G. Marson—and many exist primarily in their anamorphic (asexual) form. But a number of basidiomycetes trap and consume

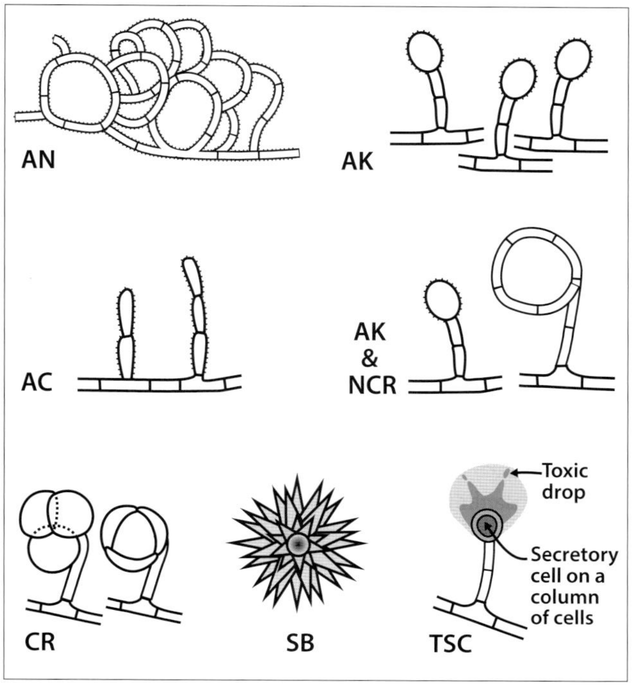

Figure 21

nematodes in their teleomorphic form as sexual fruit-bodies. Many gilled fungi in the genera *Hohenbuehelia*, *Pleurotus*, *Resupinatus*, and *Peniophorella* kill nematodes with adhesive traps and toxins. Others, like *Coprinus*, mechanically damage the worm's cuticle with awl-shaped spines and/or spiny balls. Nematodes have high internal turgor (100–200 mm Hg), and multiple cuts and tears in the cuticle cause the internal body contents to leak out, killing the victim. A potent toxic cocktail is added to the mix by some species.

Endoparasitic Fungi

Arthropod Pathogenic (AP) Fungi

Macrofungi that parasitize arthropods infect at least nine orders of insects and at least one order of arachnids (spiders, Araneae), but the host range of a given species is generally quite restricted. Insect orders affected include

FIGURE 22. A, *Ophiocordyceps sphenocephala* stromata sprouting from a dead paper wasp (*Polistes* sp.); B, *Cordyceps militaris* on dead lepidopteran pupa; C, *Ophiocordyceps melolonthae* on dead beetle larva. (Photos: A, Erich G. Vallery, USDA Forest Service—SRS-4552, Bugwood.org; B & C, John Plischke III.)

- Odonata (dragonflies and damselflies)
- Blattaria (cockroaches)
- Hemiptera (true bugs)
- Coleoptera (beetles)
- Phasmida (walkingsticks)
- Isoptera (termites)
- Hymenoptera (ants, bees, hornets, and wasps)
- Lepidoptera (butterflies and moths)
- Diptera (true flies)

Species of the well-known ascomycetous genus *Cordyceps* (*sensu lato*) appear to emerge from the ground as a club-like fungus (Fig. 22) but are in fact attached to a host organism (a dead arthropod or another fungus) buried in leaf litter, in moss, or in the ground. The genus *Cordyceps* Fries (*sensu lato*) was divided into four genera by Sung et al.:[21]

21. G.-H. Sung et al., 2007, *Studies Mycol.* 57:5–59.

CORDYCEPS (*SENSU STRICTO*): Almost all species have pallid to brightly colored and soft, fleshy stromata and parasitize the larvae and pupae of Lepidoptera and Coleoptera buried in the soil, leaf litter, or moss.

ELAPHOCORDYCEPS: Characterized by dark-colored stromata; parasitize "false truffles" or deer truffles of the genus *Elaphomyces* and cicada nymphs.

METACORDYCEPS: Four species, all but one restricted to East Asia.

OPHIOCORDYCEPS: The largest genus of arthropod-pathogenic fungi; stromata mostly dark-colored; parasitize Hymenoptera, Lepidoptera, Coleoptera, Hemiptera, Isoptera, Odonata, Diptera, and Aranaea (spiders).[22]

22. *Torrubiella arachnophila* (J.R. Johnston) Mains, originally assigned to the genus *Cordyceps* (to which it is very closely related), parasitizes spiders in eastern North America. It has two anamorphs (which are typically encountered), characterized by the form of their synnemata (conidial forms): *Gibellula pulchra* (Saccardo) Cavara and *G. leiopus* (Vuillemin) Main. This taxon is described with a photograph in M.W. Beug, A.E. Bessette, & A.R. Bessette,

In the case of *Cordyceps* or *Ophiocordyceps*, one or more spores of the endoparasitic fungus attach to the exoskeleton of an arthropod or peridium of a fungal host. These spores germinate and develop a short germ tube with a flattened terminal appresorium that produces an infection peg that penetrates the exoskeleton of its victim. The fungus quickly grows inside the host's body cavity, and hyphae spread throughout the host's body, eventually killing the victim. The fungal endosclerotium develops one or more stromata that break through the victim's exoskeleton to produce perithecia with meiotic sexual spores and/or asexual conidiophores bearing mitotic conidia. These spores are then spread through the air or carried by insects to infect new hosts.

Entomophthorales

The fungal order Entomophthorales is a monophyletic group of obligately pathogenic fungi that repetitively discharge their conidia—a unique feature that distinguishes this group—and that reproduce in nature only in living arthropod hosts. Because these fungi produce zygospores they were formerly placed in the polyphyletic and now defunct phylum/division Zygomycota. It is currently placed in the phylum/division Entomophthoromycota and the class Entomophthoromycetes (Humber, 2012). However, it is not yet known whether the zygospores of the Entomophthoromycota all actually undergo karyogamy and meiosis. Some species do produce thick-walled azygospores in which no gametangial conjugations have been observed but that might still occur after spore formation.[23] This group of parasitic fungi is discussed in much more detail later in this book, along with accounts of some representative species.

2014, *Ascomycete fungi of North America* (Austin: University of Texas Press), 347.

23. A.P. Gryganskyi et al., 2013, *Persoonia* 30:94–105.

Fungus-Gardening Beetles and Ants

Attine Ants

Certain New World leaf-cutting ants of the subfamily Myrmicinae, Tribe Attini, have developed a mutualistic relationship with an agaric mushroom, *Leucocoprinus gongylophorus* (A. Möller) Singer. In this obligate relationship, the ants assiduously cultivate underground gardens of this mushroom, which they carefully groom and weed to eliminate a ubiquitous parasitic fungus of the genus *Escovopsis* that contains five known species that attack the ants' fungal gardens. The ants also provide the fungus garden with nourishment in the form of leaves and other plant material they cut and gather and bring back to the nest. Neither the ants nor the fungus could survive without the other, and *L. gongylophorus* occurs only inside attine ant nests. It has almost lost the ability to reproduce sexually and is solely propagated into new colonies by dispersing ant queens. The asexual form of the fungus is also known by its monotypic form-genus and form-species: *Attamyces bromatificus* Kreisel. The ants also harbor filamentous bacteria (*Pseudonocaria*) that produce *Escovopsis*-inhibiting antibiotics, which the ants apply to the garden. In turn, the fungus provides the ants with food in the form of spherical or swollen hyphae called gongylidia, which are consumed by the ants.

The only other animals known to grow their own food are humans, certain African fungus-gardening termites, and ambrosia beetles (some of which occur in our area). This relationship is most highly developed in the American tropics, with 13 genera and a number of species of attines participating in gardening this one species of fungus. In the southeastern United States, it involves the ant *Trachymyrmex septentrionalis* (McCook) Creighton, which establishes locally abundant underground fungus gardens in the Coastal Plain (south of the Fall Line in Georgia) in a wide variety of oak-and-pine-dominated habitats with sandy soils and sparse shade. It also ranges as far north as Long Island, New York, as far west as Texas, and sporadically to Arkansas and Illinois. Unlike the other higher attine ants, all of which cultivate *Attamyes bromatificus* (≡ *Leucocoprinus gongylophorus*), *T. septentrionalis* cultivates the anamorphs of several unidentified species of *Leucocoprinus*. Because this fascinating interaction between four different organisms apparently does not occur in the Piedmont itself, we will not discuss it further here.

Ambrosia Beetles

In North America certain "true weevils" from three different tribes of the subfamily Scotylinae (Family Curculionidae) that form mutualistic associations with fungi are loosely referred to as "ambrosia beetles."[24] These mutualistic associations form a polyphyletic ecological guild that independently arose from scotyline bark beetle ancestors at least seven times by convergent evolution; some 3,400 species of these highly derived weevils are involved in these relationships worldwide.[25] They have the following characteristics in common: (1) the fungi, which are parasitic on plants, are the main source of nourishment for both the larval and adult beetles; (2) the fungi are dependent on the beetles for transportation from one host to another and for inoculation into new plant hosts; and (3) neither partner can independently survive without the other.

Adult female ambrosia beetles create tunnels and galleries in the nutrient-poor sapwood (outer part of the xylem) of dying or dead trees (figure 23). They inoculate their fungal symbionts and lay their eggs along these galleries; both the larvae and adults feed on fungal mycelia exposed in the tunnel walls, as well as on tight clusters (**sporodochia**) of conidiophores produced by the mycobionts. The related bark beetles (Subfamily Scotylinae), on the other hand, deposit their eggs along galleries in the nutrient-rich living tissue (phloem) of the inner bark on which they also feed.[26] As the female bores through the sapwood, a distinctive toothpick-like sawdust strand is pushed out of the tunnel, as shown in the images at right taken by Jim Edwards (Fig. 23), which probably represent the frass of one of the introduced Asian ambrosia beetles of the genus *Xylosandrus*. Emerging adult ambrosia beetles collect fungal spores into specialized pockets in their mandibles (or elsewhere on their bodies, depending on the species) known as **mycetangia** and can fly 2–3 kilometers (1¼–1¾ miles) to locate a new host tree.

FIGURE 23.
Photos: Jim Edwards.

The beetle larvae are exclusively mycophagous, and many adults also feed primarily or exclusively on the fungus; by way of payback, the nitrogenous wastes from the beetles also provide needed nutrients for the fungus. In their tunneling, the beetles damage the wood, thus facilitating the ability of the fungus to penetrate into the wood. In turn, the fungus weakens the wood and facilitates the beetles' excavation activities. Ambrosia beetles generally attack shrubs and trees that are weakened by stress or dis-

24. Tribes Cortylini, Xyleborini, and Xylotrerini; other tribes participate in this association in other parts of the world.

25. T.C. Harrington et al., 2010, *Mycotaxon* 111:337–361.

26. See note 22.

ease or are dying or already dead. As they tunnel through the xylem (either the outer sapwood or the inner heartwood or both), they inoculate their host plants with these spores. Most fungal partners are polymorphic anamorphs of two ascomycete form-genera, *Ambrosiella* and *Raffaelea*; the form-genera *Dryadomyces* and *Fusarium* are also sometimes involved in these associations. Some species of *Ambrosiella* are associated with the teleomorphic genera *Ophiostoma* Sydow & P. Sydow and *Grosmannia* Goid (Sordariomycetes, Ophiostomatales).

Mycetozoa/Myxomycota (Slime Molds)

Slime molds were long considered to be fungi because of a superficial resemblance and the fact that they produce spores. However, slime molds differ from fungi in the following characteristics:

- They are enclosed in a cell or plasma membrane instead of a chitin-based cell wall.
- The walls of their spores are composed of cellulose rather than chitin.
- They engulf their bacterial and protozoan prey by phagocytosis through amoeboid pseudopodia instead of releasing exoenzymes that break down plant and organic material outside the cell.
- They are motile and able to move about as myxamoebae and plasmodia.

The unranked clade "Myxomycota" refers to the former fungal classification of these organisms, while the term *Mycetozoa* reflects their protozoa-like characteristics within the Amoebozoa. Although the protozoa are named according to the rules of the *International Code of Zoological Nomenclature* (*ICZN*), the slime molds are also still named according to the botanical code, or *ICN*/*ICBN*. This has given rise to dual nomenclatural systems for these organisms, which is guaranteed to cause nomenclatural confusion and continuing instability for decades to come. Major problems will arise as various botanically and zoologically oriented workers describe families, genera, and species under the two different systems. These vagaries are due to historical accident and obscure the relationships between the slime molds and other amoeboid forms. Once thought to be fungi, the slime molds are now classified in the catch-all "kingdom" Protista, an unnatural group of 80 or so unrelated phylum-level groups with diverse evolutionary histories. A more realistic taxonomic grouping includes these organisms in the kingdom Amoebozoa, but much controversy exists over how to rank these groups. The higher taxonomic categories of these organisms are still very much in flux, and we won't pursue it further here. Of the two major but unrelated groups of slime molds, only the plasmodial, or "acellular," slime molds are large enough to be seen with the naked eye, and all of those given species accounts in this book are in this category.

Myxogastria/Myxomycota (plasmodial, or "acellular," slime molds)

Plasmodial slime molds form a supercell, or **syncytium** (pl. syncytia), containing thousands of individual diploid nuclei within a cytoplasmic matrix enclosed inside a single cell membrane. Such a syncytium is known as a **plasmodium** (pl. plasmodia), or fruiting body, which can range from a few millimeters in size to quite large—up to a few meters across. The plasmodium moves across the forest floor or other substratum, engulfing bacteria, fungal spores, and other microorganisms by phagocytosis through pseudopodia in classic amoeboid style. The plasmodium eventually produces a fruiting body that produces haploid spores by meiosis; these spores in turn give rise to uninucleate **myxamoebae**, or flagellated swarm cells. These haploid amoebae or swarm cells then fuse, their nuclei fuse (syngamy), and the resulting diploid nuclei repeatedly divide mitotically to produce a massive, multinucleate plasmodium enclosed within a plasma (cell) membrane.

Plasmodial slime molds produce three main types of fruiting bodies, shown in Figure 24:

FIGURE 24. Three main types of fruiting bodies for plasmodial slime molds. (Photos: A & B, W. Light; C: John Carl Jacobs.)

A. **SPORANGIUM** (pl. sporangia): small fruiting bodies enclosed by a peridium (Fig. 24a);

B. **AETHALLIUM** (pl. aethallia): a large, cushion-shaped, or amorphous fruiting body covered by a peridium (Fig. 24b);

C. **PLASMODIOCARP**: the plasmodium is concentrated inside main channels or arteries that are covered by a common peridium (Fig. 24c).

Because of their dramatic appearance, often striking beauty, and resemblance to fungi, slime molds have been traditionally included in mushroom field guides. We will likewise include a few of the more commonly encountered forms in our area, but they really deserve a book of their own.

Lichens

Lichens are composite, symbiotic organisms composed of members of as many as three kingdoms. The dominant partner is a fungus (the **mycobiont**) that determines the form of the lichen. Lichenized fungi are all ascomycetes. They were formerly placed in the obsolete class Discomyetes but are now assigned to the subphylum Pezizomycotina and distributed in two classes: the **Lecanoromycetes** (most species), with a small number in the **Lichinomycetes**. These fungi form symbiotic associations with a photosynthesizing partner (the **photobiont**), usually either a green alga (Kingdom Protista), a cyanobacterium (Kingdom Eubacteria), or, in some cases, both.[27] The fungus and the photobiont have their own scientific names, but the lichen always takes the Latin name of the dominant fungus. The mycobiont provides a safe living structure (the **thallus**) and absorbs minerals from the substratum for its photobiont partner(s). In return, the photobionts provide the carbon products of photosynthesis to the fungus. Cyanobacteria also fix atmospheric nitrogen into biologically available compounds. The lichen thallus produced by the fungus in response to chemical signals from its photobiont partner(s) is a very different structure from what is produced by either partner acting alone. The carbon-containing products of photosynthesis are converted by the fungus into specialized carbohydrates called **lichen acids**, which, like the thallus, cannot be produced by any of the partners acting alone. These lichen acids are deposited externally on the surface of the thallus and play a major role in the primary chemical weathering of rocks that leads to the formation of gravels, sands, soils, and clays.

The typical foliose or crustose lichen thallus (shown in cross-section in Fig. 25) consists of 95% fungal hyphae, which provide moisture and shelter. It consists of an upper-

27. Cyanobacteria were formerly known as "blue-green algae."

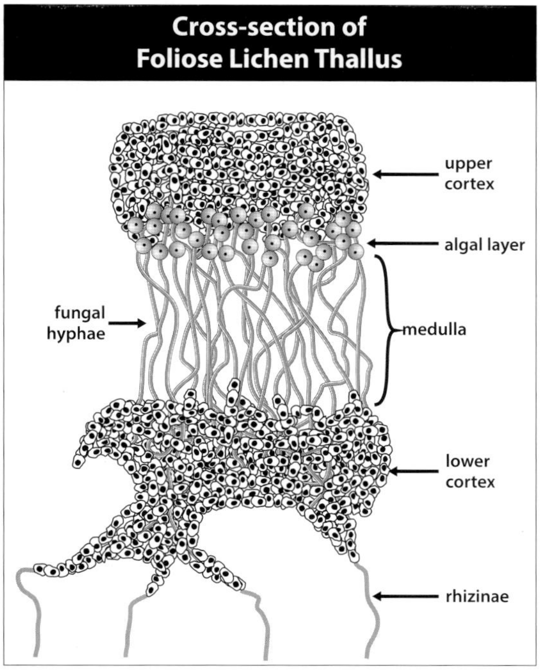

FIGURE 25

most protective layer, the **upper cortex**, formed by densely agglutinated fungal hyphae. Embedded below this is the **algal layer**, which makes up 5% of the lichen community and provides food through photosynthesis. Beneath the algal layer is a third layer of loosely interwoven fungal hyphae without algal cells called the **medulla**. The **lower cortex** resembles the upper surface and consists of densely packed fungal hyphae. It often bears root-like fungal rhizines, which attach the thallus to the substratum and which absorb and conduct water to the thallus.

Lichens reproduce asexually by means of **soredia** (clusters of algal cells in a meshwork of hyphae) and **isidia** (simple symbiotic outgrowths of the medulla with a cortical covering). These soredia and isidia break off as symbiotic fragments and are disseminated by wind and rain. The fungal partner can also reproduce itself sexually by forming fruitbodies in the same way as its nonlichenized counterpart. Lichenized ascomycetes produce ascospores inside asci within prominent, open, cup-like **apothecia** or small, rounded, flask-shaped **perithecia** on the surface of the thallus. Lichenized basidiomycetes grow their usual mushroom fruitbodies from the lichen thallus and produce basidiospores at the tips of basidia. In order for a sexually reproduced lichenized fungus to develop into a new lichen it must encounter and incorporate an appropriate algal partner, which is not guaranteed.

Lichens are found in most terrestrial and aquatic environments (including the marine intertidal and spray zones). They range from 86° N latitude in the Arctic to 80° S in Antarctica and from sea level to over 5,000 meters (16,400 feet) in elevation. However, they are largely absent from urban and other areas with high levels of air pollution due to their sensitivity to sulfur dioxide and fluorine.

Mushroom Toxins and Mushroom Poisoning

Sve gljive su jestive, ali neke samo jednom.
(All mushrooms are edible, but some only once.)
—Croatian proverb

Commercial button and Portabella mushrooms (*Agaricus bisporus*) are commonly served raw in salads in restaurants and are routinely eaten raw by many people at home. However, all *Agaricus* species contain various hydrazines, including gyromitrin and agaritine, which are both potent toxins and carcinogenic. Hydrazines are readily driven off by sautéing or cooking, and *all mushrooms should be cooked before consumption!*

Mushrooms also contain a number of compounds that affect some individuals adversely but not others. This is why most discussions about collecting and eating wild mushrooms contain various caveats: (1) be absolutely sure of your identifications, since many poisonous species closely resemble edible ones, and (2) eat only small amounts when first consuming a species new to your dinner table. We will limit this brief discussion to the truly toxic species that cause significant and sometimes fatal intoxications in humans who eat them.

Mushroom poisons fall into seven major categories: (1) protoplasmic toxins causing organ failure—the cyclopeptides (amatoxins) and hydrazine derivatives (gyromitrin); (2) neurotoxins (muscarine, ibotenic acid/muscimol, and psilocybin); (3) toxins that act like disulfiram (Antabuse®) (coprine); (4) orellanine; (5) allenic norleucine; (6) gastric irritants; and (7) myotoxins.

Cyclopeptide Toxins Causing Organ Failure

Amatoxins. These are the deadliest mushroom toxins known and are responsible for most fatal mushroom intoxications (i.e., poisonings) worldwide.[28] They are the lethal toxins found in some (but not all) species of *Amanita*, *Galerina*, *Lepiota*, and *Conocybe*. Amatoxins are highly stable and are ***not neutralized*** by cooking or leaching. Amatoxin-containing mushrooms in our area include *Amanita bisporigera*, *Lepiota clypeolaria*, and *Galerina marginata*.[29] Some *Amanita* species also contain allenic norleucine (2-amino-4,5-hexadienoic acid), which can cause gastrointestinal distress, disorientation, and kidney failure.

α-Amanitin, the major mycotoxin in all species of Destroying Angels, Death Caps, and the Deadly Galerina, is a cyclopeptide composed of eight amino acids. It decreases levels of reduced glutathione (GSH), an important free radical scavenger, and specifically and irreversibly binds to and blocks the enzyme RNA polymerase II in the liver. This inhibits protein synthesis and eventually causes cytolysis in the hepatocytes (liver cells), resulting in cell death, necrosis, and liver failure. Severe cramps and diarrhea typically arise between 6 and 24 hours after ingesting the mushroom. A temporary and false remission typically occurs on day 3, followed by a relapse on days 4–5 that often leads to catastrophic liver and kidney failure, massive hemorrhaging, and, all too often, coma and death within a week to 10 days unless heroic measures, including liver transplantation, are carried out.

The deadly amanita toxins have an oral LD_{50} (the amount that will kill half of the animals tested) of 0.1 mg/kg for humans.[30] The toxin is present in all tissues of the fungus. The tissue concentration of α-amanitin in *Amanita bisporigera* is 1.70 ± 0.68 milligrams per gram of dry weight of mushroom tissue and 1.70 ± 0.45 mg/g in the cap and the stipe, respectively. However, it contains a whopping 2.2–5.0 mg/g of dry weight of tissue of both α- and β-amanitin, and this does not appear to vary geographically.[31] Just 5 grams of dried mushroom—the weight of two post-1982 American pennies—is a potentially lethal dose for an adult human. The dry-weight concentration is much higher than in the fresh weight, perhaps by a factor of 10, due to water loss in desiccation. Thus, the consumption of just 10 or 20 grams of freshly collected Eastern Destroying Angel—just ⅓ to ⅔ ounce, less than 1/25 of a pound, or the amount of meat two-thirds the size of a matchbox (one modest mouthful)—is sufficient to kill an adult human being weighing 154 pounds!

In cases where an early diagnosis is achieved, massive doses of penicillin[32] and intravenous infusions of silymarin/silibinin/silybin from a mixture of flavonolignans extracted from the Blessed Milk Thistle, *Silybum marianum* (L.) Gaertner, can prevent the most severe liver injury *if administered within 48 hours of ingesting the poison.*[33] However, in most cases of severe amanitin poi-

28. Unlike most field guides, we do not use the term *poisonous* or *poisoning* in reference to the psychedelic properties or experience resulting from the ingestion of certain psilocin-containing mushrooms. "Poisonous" is here restricted to toxic mushrooms that can cause moderately serious to severe or even life-threatening medical (not psychiatric) emergencies.

29. The Death Cap, *Amanita phalloides*, was introduced into North America from Europe and is now found along the West and East Coasts from California to British Columbia and from Maryland to Maine, respectively. It has not yet been reported in our area (B.E. Wolfe et al., 2009, *New phytologist*, 14 pp., DOI:10.1111/j.1469-8137.2009.03097.x.).

30. The lower the LD_{50}, the more lethal the poison.

31. β-Amanitin also inhibits eukaryotic RNA polymerase II and III and directly damages the heart, liver, and kidneys.

32. Recent studies have indicated that monotherapy with silibinin alone offers better protection against liver damage than combination therapy with silibinin and penicillin or penicillin alone, with mortalities less than 10% for silibinin monotherapy compared with 20% (U. Mengs, R.-T. Pohl, & T. Mitchell, 2012, *Current Pharmaceutical Biotechnology* 13:1964–1970).

33. Silibinin alters the membrane permeability of the hepatocytes (liver cells), preventing entry of the toxin into the cells, and increases RNA protein synthesis, which is impaired by the poison's inhibition of RNA polymerase; this promotes cellular regeneration. It also competitively prevents binding of the toxin to its receptor and counteracts amanitin-induced glutathione depletion, which promotes the scavenging of superoxides and other free radicals that

soning, diagnosis is typically delayed because of false remission, and liver transplantation becomes the only hope of survival. Silibinin dihydrogen disuccinate disodium (Legalon® SIL, Madaus GmbH) is now available in this country for all patients with amatoxin poisoning. Prompt medical treatment can reduce the mortality in amanitin poisoning from over 50% to about 10% (Beug, 2008, 42). Any physician suspecting a case of amanitin poisoning should *immediately* call the 24-hour Legalon® SIL study hotline at 866-520-4412. Once the principal investigator for the study determines that the situation qualifies, a supply of the antidote will be shipped overnight and free of charge to the physician's hospital pharmacy.

More than 90% of all fatal mushroom poisonings are due to certain species of *Amanita* in Section Phalloideae, such as the Death Cap (*A. phalloides*), and the Destroying Angel species complex (*A. bisporigera, A. verna, A. virosa*, and *A. ocreata*).

Phallotoxins. Phallotoxins are bicyclic heptapeptides found in the Death Cap, *Amanita phalloides*, and several other *Amanita* species. Phallotoxins are limited poisons because, unlike amatoxins, they are easily degraded by heat or digestion and are not absorbed from the gut. They thus do not contribute to the toxicity of the Death Caps.

Gyromitrin and other hydrazines. Gyromitrin (*N*-methyl-*N*-formylhydrazone), an organic derivative of hydrazine,[34] is found in certain species of *Gyromitra*, in particular the False Morel, *G. esculenta*, and possibly also *G. infula*. Both species occur in our area. And all species of *Agaricus*, including the commercially sold "button" and Portabella mushrooms (*A. bisporus*), and many species of *Leucoagraricus* and *Macrolepiota* contain gyromitrin and agaritine, another hydrazine compound. Hydrazines are both directly toxic and known carcinogens, but because of their volatility (low boiling points), they are readily eliminated by cooking.

Gyromitrin is hydrolyzed in the body into monomethylhydrazine (MMH), which is both a carcinogen and a highly toxic metabolite that affects the liver, central nervous system, and kidneys.[35] MMH is further metabolized into *N*-methyl-*N*-formylhydrazine, which in turn produces nitrosamides via the cytochrome P_{450} microsomal enzymes. This generates methyl radicals, leading to liver and renal necrosis. MMH also interferes with the actions of pyridoxine (vitamin B_6), which results in decreased activities of gamma-aminobutyric acid (GABA) and glutamate in the brain.

Besides irritation of the eyes, nose, and throat, headache, and nausea, MMH and its metabolites can result in pulmonary edema, seizures, and coma, as well as severe injury to the liver, kidneys, and central nervous system. GI symptoms (stomach cramps, vomiting, and diarrhea) begin a few hours after consuming *Gyromitra*, followed by lethargy and delirium. Severe poisoning can lead to enlarged liver and spleen, jaundice, destruction of red blood cells, and methemoglobinemia, leading to renal failure and even coma, circulatory collapse, and death within a week. The LD_{50} for MMH in humans is 30–50 mg/kg (Andary & Privat, 1985). Gyromitrin poisonings account for only 1% of all mushroom poisonings in the United States but are more common in Europe. Poisonings by *G. esculenta* appear to be much more severe and dangerous in the Midwest than in western North America. The mushrooms also appear to vary in toxicity with both season and altitude.

Gyromitrin poisoning can also result from inhalation of the vapors during cooking.

Neurotoxins

Muscarine. Symptoms of muscarine poisoning appear within half an hour to two hours after ingestion and include excessive and involuntary salivation, sweating, urination, tears, severe vomiting, and diarrhea and may also involve visual disturbances, pulse irregularities, hypotension, and difficulty breathing. Most cases recover within

---

tear apart cell membranes. The extract also offers some protection against liver damage from acetaminophen (Tylenol®) poisoning, presumably by these same mechanisms, especially by maintaining GSH reserves in the hepatocytes (Campos, Garrido, & Valenzuela, 1989; Threlkeld, 1997).

34. Even the most popular edible "button" mushroom, *Agaricus bisporus*, produces organic hydrazine derivatives, including agaritine, a hydrazine derivative of an amino acid, and gyromitrin.

35. MMH is also used as a rocket fuel by NASA.

a few hours without complications, but severe poisoning can lead to convulsions and death from respiratory failure. Atropine is an antidote to muscarine poisoning but is very rarely required.

Muscarine activates the postsynaptic acetylcholine receptors in cells innervated by the parasympathetic nervous system. It is found in significant and potentially dangerous amounts in some species of the genera *Inocybe* and *Clitocybe* and in fairly harmless trace amounts in some *Russula*, *Lactarius*, *Entoloma*, certain red-pored *Boletus*, and *Omphalotus*. Muscarine is present in minuscule amounts in *Amanita pantherina* and the Fly Agaric, *A. muscaria*, from which it was first isolated and named.

Ibotenic acid/muscimol. The major toxins present in *Amanita muscaria*, *A. pantherina*, *A. gemmata*, and certain other members of the genus are the isoxazole derivatives muscimol and ibotenic acid and not muscarine, which is present in only trace amounts, if at all. Muscimol is a potent agonist (activator) of gamma-aminobutyric acid (GABA)—the main inhibitory neurotransmitter in the central nervous system—at $GABA_A$ receptors. Ibotenic acid is a potent neurotoxin used to induce lesions in the brains of experimental animals. It also serves as a prodrug (a preconverted form) for muscimol, is structurally similar to the stimulatory neurotransmitter glutamic acid, and is known to induce exictotoxicity. Excessive excitation of excitatory receptors can lead to neuronal dysfunction, resulting in neurodegenerative disease.[36]

Most intoxications arise in individuals deliberately seeking a psychotropic experience who consume these mushrooms. Symptoms of intoxication usually occur within 30 minutes following ingestion and include drowsiness, dizziness, and nausea, followed by confusion, euphoria or dysphoria, dissociative sensations of floating and/or out-of-body experiences, visual distortions and occasionally frank hallucinations, delusions, and, in severe cases, a comatose state that can last for 24 hours. Children manifest more excitatory signs, such as ataxia, muscle jerks, and fasciculations (muscle twitching), mental clouding, lethargy or euphoria, and occasionally seizures.

Deaths from *Amanita muscaria* intoxication in North America are almost nonexistent and are very rare from *A. pantherina* (Buck, 1963). Despite the common medical recommendation to administer atropine to victims of severe *A. muscaria* or *A. pantherina* poisoning, this is definitely contraindicated! Muscarine is not involved in these intoxications (widespread belief to the contrary notwithstanding); atropine mimics and can actually worsen the effects of muscimol and ibotenic acid.

Though ibotenic acid and muscimol are not heat labile, blanching these mushrooms and discarding the water significantly reduces their toxicity.

Hallucinogens: psilocybin and psilocin. Mushrooms containing the psychedelic compounds psilocybin and psilocin include species of the following genera: *Psilocybe* (116 spp.); *Gymnopilus* (14 spp.); *Copelandia*; *Panaeolus* (13 spp.); *Hypholoma* and *Pluteus* (6 spp. each); *Panaeolina* and *Conocybe* (4 spp. each); *Gerronema* (2 spp.); and *Agrocybe*, *Galerina*, *Inocybe*, and *Mycena* (1 sp. each) (Guzmán, Allen, & Gartz, 2000). Psilocybin is the phosphorylated (having an attached phosphate group)[37] prodrug of psilocin present in most hallucinogenic mushrooms. Upon ingestion, this prodrug is converted into its active form, psilocin (4-hydroxy-*N*, *N*-dimethyltryptamine)—a tryptamine-based indole that acts upon serotonin (5-hydroxytryptamine, or 5-HT) receptors in the brain (Fig. 26). In this respect, psilocin is very similar to the psychedelic effects of LSD and mescaline.

The effects of psilocin include vivid visual and auditory perceptions and even hallucinations, altered mental states and perception of time, dissociative psychological states and out-of-body experiences, and profound spiritual experiences lasting up to six hours. Some individuals may experience nausea early on, which usually quickly abates, and a few may suffer a panic attack. The toxicity of these substances is extremely low, and fatalities are essentially unheard of. These agents have no physically addicting properties, although tolerance to their effects does occur with frequent use, and psychological dependence may develop in some individuals. Psilocin-containing mush-

36. W.C. Zincand et al., 1992, *Mol. Chem. Neuropathol.* 16(1–2): 1–10.

37. Phosphate groups (PO_4^{3-}) form high-energy bonds, which are vital in most metabolic processes.

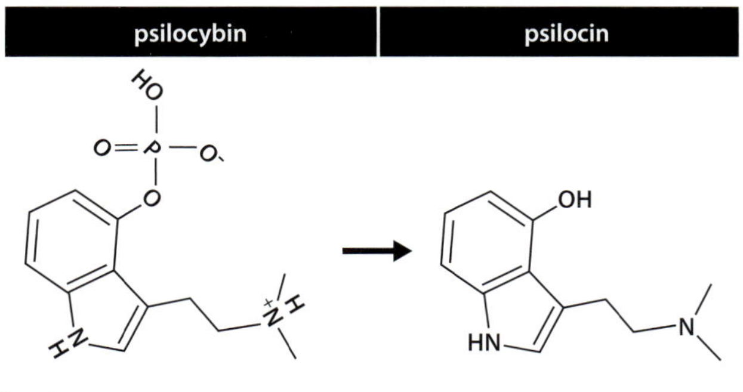

FIGURE 26

rooms have been used in religious rituals by shamans and others in native societies for thousands of years, particularly in Central and South America.

Disulfiram (Antabuse®)–Like Toxins

Coprine. Certain otherwise edible species in the genus *Coprinopsis* contain coprine, which is metabolized into a compound that causes the disulfiram-ethanol (Antabuse®) reaction when consumed within a few days of drinking beverage alcohol. Drinking even one alcoholic beverage three hours *before* or up to five days *after* eating coprine-containing mushrooms can result in severe headache, flushing, nausea, vomiting, and heart palpitations lasting from a few hours to several days. The symptoms typically arise within 15 minutes and usually subside without complications. But several severe coprine-ethanol poisonings required monitoring for cardiac arrhythmias, and one victim suffered a fatal rupture of the esophagus after extremely severe vomiting (Mayer et al., 1971).

Coprine and disulfiram both inhibit aldehyde dehydrogenase, the enzyme that converts acetaldehyde—the first metabolite of ethanol, which is responsible for "hangover" symptoms—into harmless acetic acid (vinegar). This causes acetaldehyde to accumulate to toxic levels in the blood and produces acute acetaldehyde poisoning:

- vasodilation, resulting in sweating, flushing of the face and upper chest, and hypotension, which can lead to shock;
- hyperventilation and dyspnea, which can lead to respiratory depression;
- nausea and vomiting;
- chest pains, palpitations, and tachycardia, which can lead to arrhythmias, cardiovascular collapse, myocardial infarction (in compromised persons), and acute congestive heart failure;
- vertigo, syncope, and confusion, which can lead to seizures, unconsciousness, and death.

Coprinoid mushrooms found in the Southeast that have been *reported* to contain coprine include (1) genus *Coprinus*: *C. comatus* and *C. sterquilinus*; (2) genus *Coprinopsis*: *C. atramentaria* and *C. variegatus*; (3) genus *Coprinellus*: *C. disseminatus*, *C. domesticus*, *C. micaceus*, *C. radians*, and *C. truncorum*; and (4) *Parasola plicatilis*. However, despite widespread reports to the contrary, coprine does *not* occur in either *Coprinus comatus* or *Coprinellus micaceus* (Antowiak & Antowiak, 1991).

Other nondisulfiram alcohol-induced syndromes occur in some persons who have consumed *Coprinus comatus*, *Clitocybe clavipes*, *Boletus luridus*, morels, *Pholiota squarrosa*, *Armillaria mellea*, *Pleurotus ostreatus*, and *Boletus edulis*, among a number of other species.

Orellanine

Orellanine is a potent nephrotoxin (kidney poison) found in certain species of *Cortinarius*. It accumulates in the renal tubules, where it causes severe tubular necrosis with ensuing kidney failure that can lead to death. Severe oxidative stress and a shutdown of important antioxidant enzymes contribute to the pathogenesis of renal failure, which may also involve liver disease and severe intestinal inflammation. Orellanine poisoning was unknown in North America until July 2008, when the first case surfaced from eating an undescribed species of *Cortinarius*.[38] The main danger of poisoning comes from confusing certain *Cortinarius* species with the edible (when cooked) Wood Blewit, *Clitocybe nuda*.

Orellanine poisoning is characterized by a long latency period before the first symptoms appear. This can range from two to three days to up to three weeks or more after

38. North American Mycological Association, "Mushroom poisoning syndromes," http://www.namyco.org/toxicology/poison_syndromes.html.

ingesting the toxic mushrooms. Mild nausea, vomiting, and diarrhea may occur one to two days after ingestion, followed by a prolonged period of apparent remission, after which the symptoms of renal failure suddenly appear: intense thirst, frequent urination, kidney pain, headaches, myalgias, and muscle cramps, followed by loss of consciousness, convulsions, and death in 15% of cases. Recovery may take weeks or months, and victims may require indefinite hemodialysis or kidney transplant.

Allenic Norleucine

Allenic norleucine is an intoxicant found in *Amanita smithiana* (in the Pacific Northwest), *A. pseudoporphyria* and *A. sphaerobulbosa* (in Japan and Southeast Asia), *A. proxima* (in the Mediterranean region of Europe), and *A. abrupta* (widespread in eastern North America, including our area). It produces renal tubulopathy similar to that seen in orellanine poisoning. More than a dozen cases are known from the Pacific Northwest in individuals who mistook *Amanita smithii* for the highly desirable "American Matsutake," a species of *Tricholoma* usually identified in field guides as *Tricholoma magnivelare* (but which may actually represent an undescribed species).

Early symptoms include nausea, stomach cramps, vomiting, diarrhea, excessive sweating, weakness, generalized discomfort, and dizziness, typically appearing after a latent period of between half an hour to 12 hours. This is followed by acute renal failure four to six days after consuming the mushrooms, characterized by low urinary output or complete inability to urinate and marked elevation of blood urea nitrogen and creatinine—all signs of acute renal failure. Alanine aminotransferase (ALT) and lactate dehydrogenase, nonspecific indicators of tissue damage, are also often elevated. The renal failure resembles that seen in orellanine poisoning, but it appears earlier, is generally less severe, and has a better prognosis, with most patients gaining full renal function after a period of hemodialysis. Nevertheless, fatalities have occurred.

Gastrointestinal Irritants

A number of mushrooms can cause nonspecific and ill-defined gastrointestinal poisonings in some individuals but not in others. Among these are *Chlorophyllum molybdites*, *Omphalotus illudens*, *Amanita flavorubescens* and *A. brunnescens*, *Tricholoma* spp., *Russula* spp., *Lactarius* spp., and *Boletus* spp. Nausea, vomiting, abdominal cramps, and diarrhea typically begin within one to two hours after ingesting the suspect mushrooms. Most symptoms subside without further complications within 8–12 hours, and only supportive care is needed.

Myotoxins

Tricholoma equestre (Linnaeus) P. Kummer was formerly widely consumed as a choice edible in Europe. However, the mushroom is now known to contain one or more uncharacterized myotoxins that attack muscle cells and can cause severe skeletal and cardiac muscle injury when consumed in large amounts. Since 1992 a dozen cases of severe myopathy following the consumption of large amounts of this mushroom produced rhabdomyolysis (muscle cell damage with the release of myoglobin into the urine), acute myocarditis, arrhythmias, and cardiac failure. Three of the 12 known cases died despite intensive treatment. All the cases occurred in southwestern France, and all the victims had eaten three or more meals of *T. equestre* within a 2-week period (Bedry et al., 2001).

The Yellow Knight Fungus, *Tricholoma flavovirens* (Persoon *ex* Fries) Lundell, which is widely distributed in the southeastern United States, is considered conspecific with *T. equestre* by many mycologists. Thus, this species should not be eaten until its taxonomic status and toxicity have been better defined.

Medicinal Properties of Some Southeastern Mushrooms

Besides being fun to discover, photograph, and learn about, why else would anyone study mushrooms? One compelling reason is an abundance of good evidence that certain mushrooms act as superfoods, effectively improving the immune system and even curing disease.

It has long been known that some mushrooms contain anywhere from 10% to 60% protein, making them especially valuable as a protein source in a strict vegetarian diet. They also contain vitamin D, potassium, selenium, riboflavin, and niacin, among other nutrients.

But perhaps their greatest value is their ability to rapidly boost the immune system without apparent adverse side effects. In an era when even last-ditch antibiotics are rapidly losing their effectiveness and cancer is still a leading cause of death despite decades of new types of chemotherapy, it makes sense to see what mushrooms can offer in the way of safe, medicinal effects.[39] About 700 species of mushrooms exhibit significant pharmacological activity.[40] Some of the supposed beneficial effects of mushrooms are based on anecdotal evidence and hearsay, but many of the medicinal effects are very real and have been scientifically verified. Indeed, one author noted, "It is . . . remarkable how reliable the facts collected by traditional eastern medicine are in the study of medicinal mushrooms."[41] And many of our pharmaceutical drugs, like the statins, were initially derived from mushrooms and other fungi.

The mushroom compounds responsible for these effects include a variety of polysaccharides, glycoproteins, ergosterols, triterpenes, and antibiotics, some of which are extracted from the fruiting body of the mushroom and others from the mycelium. Those that have drawn the most attention for their immunoenhancing effects are various types of triterpenoids (cyclic derivatives of squalene) and polysaccharides (complex chains of simple sugar molecules). The following anticancer polysaccharides from mushrooms in our area have undergone clinical trials in Asia, and all are currently produced by Asian pharmaceutical agencies: PSK (polysaccharide-K, commercially sold as Krestin®) and PSP (polysaccharopeptide, a polysaccharide linked to a small protein or polypeptide) from Turkey Tail, *Trametes versicolor*; and Grifron™-D from Hen-of-the-woods (Maitake), *Grifola frondosa*.[42]

The obvious medical potential of mushrooms has precipitated a great deal of medical and pharmacological research throughout the world. But most of the clinical studies on human patients thus far have been undertaken in Asia, and it is often difficult to interpret or extrapolate the results of these studies. As noted by Sullivan et al. in their abstract, "[T]ranslating traditional Eastern practices into acceptable evidence-based Western therapies is difficult. Different manufacturing standards, criteria of purity, and under-powered clinical trials make assessment of efficacy and toxicity by Western standards of clinical evidence difficult."[43]

Because these mushrooms have potent bioactive compounds, some caution is advised about taking them as nutritional and/or medicinal supplements. Many of these pharmacologically active compounds could have unwanted side effects or deleterious interactions with certain over-the-counter and prescribed medications. For example, the lovastatin found in oyster mushrooms could potentially seriously affect muscle, liver, and brain function in patients taking statin medications. Likewise, certain β-glucans lower blood glucose and can exacerbate the hypoglycemic effects of drugs used by diabetic patients. Reishi can inhibit platelet aggregation and may cause excessive bleeding in patients taking blood thinners. And interactions with drug-metabolizing pathways can variously exacerbate or diminish the bioavailability and efficacy of medications and mushroom supplements. Readers are encouraged to discuss the supplemental and medicinal use of mushrooms with their physicians and/or pharmacists.

Although many mushrooms, both wild and cultivated, produce immunoenhancing effects, we will discuss just a few of the important species occurring in our area with the longest history of confirmed medicinal use. Except for Chaga, all of the species discussed in this section are

39. Each year at least 2 million people in the United States become infected with antibiotic-resistant bacteria, and at least 23,000 people die as a direct result of these infections. Many more die from other conditions that were complicated by antibiotic-resistant infections (CDC, 2013, *Antibiotic resistance threats in the United States, 2013* [Atlanta: Centers for Disease Control and Prevention, September 16], 114 pp., http://www.cdc.gov/drugresistance/threat-report-2013/).

40. S.P. Wasser, 2002, *Appl. Microbiol. Microtechnol.* 60:258–274.

41. S.P. Wasser, 2002, *Appl. Microbiol. Microtechnol.* 60:258.

42. P.H. Kidd, 2000, *Alt. Med. Rev.* 5:4–27; M. Mukhopadhyay, 2008, *J. Mycopatholog. Res.* 46(1):13–21.

43. R. Sullivan et al., 2006, Medicinal mushrooms and cancer therapy: Translating a traditional practice into Western medicine, *Perspect. Biol.* 49(2):159–170.

described in detail in the species accounts in this book. Many wild varieties can be identified and harvested by those willing to spend the time and effort to learn about them. Others are readily available in stores such as Whole Foods and GNC, herbal shops, Oriental shops, or online.

- **Birch Polypore (*Piptoporus betulinus*)**: Though primarily a northern species, we have found the Birch Polypore in northern Georgia in the southern Appalachians, where it is apparently not uncommon. Tea brewed from this polypore has anti-inflammatory, antifatiguing, immunoenhancing, and antibiotic properties against mycobacteria. It contains resins and agaric acid, which are potent purgatives, as well as oils toxic to the intestinal roundworm *Trichuris trichiura*. Ötzi, the Tyrolean "Iceman," who died about 5,300 years ago, was found to be infected with *T. trichiura*; he also carried Birch Polypore conks with him, which he is thought to have used to treat his trichuriasis infection.[44,45,46,47]

  Like Chaga (see the following account), Birch Polypore extracts betulin and betulinic acid from the birch trees (*Betula* spp.) on which it exclusively grows. These pentacyclic triterpenes are potently cytotoxic to melanoma cells with no adverse effects to the experimental animals. A novel antibiotic, piptamine, extracted from this mushroom exhibited antibiotic activity against a variety of Gram-positive bacteria, yeasts, and fungi and was particularly potent against *Staphylococcus aureus* and *Enterococcus faecalis* (MIC values of 0.78 and 1.56 μg/mL, respectively).[48]

- **Hen-of-the-woods, Maitake (*Grifola frondosa*)**: The Japanese name, *maitake* (舞茸), means "dancing mushroom." This choice edible is infrequently found in the southern Appalachians and the Piedmont on stumps or at the base of dead or dying hardwood trees. It is used in traditional Chinese and Japanese medicine to enhance the immune system. Scientific studies have identified a protein-bound β-glucan in this mushroom with immunoenhancing properties and also found that Maitake can stimulate both the innate and acquired (adaptive) immune responses, as well as regulate blood pressure, glucose, and lipids in the blood and liver.[49,50] Scientific studies of maitake D-fraction, which contains various β-glucans, have reported antitumor effects in mice, but definitive studies with human patients have not been completed.

  Grifron™-D (Maitake Products), a highly purified form of maitake D-fraction, was highly effective in inducing apoptosis (95% cell death within 24 hours) in human prostatic cancer cells *in vitro*.[51] And a Japanese study reported improvements in patients with various cancers who were given a combination maitake D-fraction and whole maitake powder along with conventional chemotherapy. But no control group was included in the study, and it was not possible to distinguish between any effects produced by the maitake and the conventional treatments. In another study, these same authors reported that maitake D-fraction given without other treatment hindered metastatic progress, lessened the expression of tumor markers, and increased NK (natural killer) cell activity in human cancer patients.[52]

- **Lion's Mane Mushroom, Yamabushitake (*Hericium erinaceus*)**: Used in traditional Chinese medicine, this mushroom is rich in β-glucan polysaccharides with potent anticancer, immunoenhancing, lipid- and glucose-lowering, antioxidant, and neuroprotective properties. This mushroom also has antimicrobial and wound-healing effects.[53] Atrophic gastritis (chronic

44. L. Capasso, 1998, *Lancet* 352(9,143):1864.
45. B. Schlegel et al., 2000, *J. Antibiot.* (Tokyo) 53(9):973–974.
46. T. Kamo et al., 2003, *J. Nat. Prod.* 66(8):1104–1106.
47. H.V. Wangung et al., 2004, *J. Antibiot.* (Tokyo) 57(11): 755–758.
48. J.-K. Liu, 2005, *Chem. Rev.* 105(7):2723–2744. MIC = minimum inhibitory concentration.
49. G. Deng et al., 2009, *J. Cancer Res. Clin. Oncol.* 135(9):1215–1221, DOI:10.1007/s00432-009-0562-z, PMID 19253021.
50. N. Kodama et al., 2004, *J. Med. Food* 7(2):141–145, DOI:10.1089/1096620041224012, PMID 15298759.
51. S.A. Fullerton et al., 2000, *Mol. Urol.* 4(1):7–13.
52. N. Kodama et al., 2002, *Altern. Med. Rev.* 7:451; 2003, *J. Med. Food* 6(4):371–377.
53. M.A. Khan et al., 2013, *J. Complem. Integr. Med.* May 24:10, pii: /j/jcim.2013.10.issue-1/jcim-2013-0001/jcim-2013-0001.xml, DOI:10.1515/jcim-2013-0001.

inflammation of the stomach mucosa) was also successfully treated with oral *Hericium* extract granules.[54] See the Lion's Mane species account in the "Hydnoid (Toothed) Fungi" chapter of this book.

- **Oyster Mushroom (*Pleurotus ostreatus*)**: This mushroom is a source of lovastatin, the first statin HMG co-A reductase inhibitor used to lower cholesterol levels; lovastatin is also found in Red Yeast Rice (i.e., rice cultured with a mold, *Monascus purpureus* Went [Ascomycota: Monascaceae]). Of 20 species of edible mushrooms tested for antitumor potential, a water-soluble extract from *P. ostreatus* exhibited the strongest cytotoxicity (inducing apoptosis) to human androgen-independent PC-3 cancer cells.[55]

- **Turkey Tail (*Trametes versicolor*)**: One of the most common yet most powerful of the medicinal mushrooms, it is boiled to make tea. It would be hard *not* to find this mushroom during a walk in the woods in our area or anywhere in North America. Turkey Tail is a small shelf mushroom that grows on downed logs, often in rosettes. It has white pores underneath, distinguishing it from a very similar species, False Turkey Tail, *Stereum ostrea*, which is completely smooth and lacks pores underneath the cap. See the detailed accounts of both these species in this book.

  Trametes versicolor is known to be antibacterial, antiviral, and antitumoral and to reduce serum cholesterol. It is also a powerful antioxidant and is probably the best-known and best-documented medicinal mushroom. Protein-bound polysaccharide K (PSK; Krestin®, Sankyo), isolated from the fruitbody of Turkey Tail, appears to be a promising adjuvant with chemotherapy for treating gastric, esophageal, colorectal, breast, and lung cancers.[56,57] PSK (Krestin®) is an approved drug in Japan and is given in oral doses up to 3 grams per day as an adjuvant with other chemotherapy and radiologic agents. It provides improved quality of life and extended 5- and 10-year survival rates for stomach and colorectal cancer patients.[58] PSK exerts its antitumor effects by activating toll-like receptor 2 (TLR2) and stimulating both innate and acquired immunity.[59] PSK has also been found to have a protective effect on healthy cells while making cancerous cells more sensitive to chemotherapy. It is used in Europe, Asia, and Japan in cancer treatment for gastric, esophageal, breast, and some lung cancers.

  PSP (polysaccharopeptide, a polysaccharide linked to a small protein or polypeptide) obtained from Turkey Tail has also shown *in vitro* antitumor activities and inhibits the growth of induced tumors in animal models. It also significantly extends survival times in dogs with hemangiosarcoma.[60] However, evidence of clinical benefits in cancer patients is lacking.

  As potent as it is, however, *Trametes versicolor* is not as effective when taken by mouth as some other mushrooms, notably Reishi, *Ganoderma lucidum*.

- **Reishi, Lingzhi (*Ganoderma lucidum*)**: This powerful medicinal mushroom is widely distributed in eastern North America, Europe, and Asia and is very common in our area.[61] Reishi, also known as Varnish Shelf and a variety of other names, is a polypore that grows on deciduous woods, usually near the soil surface on stumps. (See the account of this *Ganoderma* species complex in the "Polypores" chapter of this book.) Named *ling zhī* (Divine Mushroom of Longevity) by the Chinese, it has been used as a medicinal throughout East Asia for

54. H.E. Lei, 2009, *Chinese J. Hospit. Pharm.* 14:1199–1201.
55. Y.-H. Gu and G. Sivam, 2006, *J. Med. Food.* 9(2):196–204.
56. M. Fisher & L.X. Yang, 2002, *Anticancer Res.* 22(3):1737–1754.
57. Y. Maehara et al., 2012, *Surgery Today* 42(1):8–28.
58. C. Hobbs, 2004, *Int. J. Med. Mushrooms* 6(3):195–218.
59. H. Lu et al., 2011, *Clin. Cancer Res.* 17(1):67–76. Toll-like receptors are a class of proteins that play a major role in the innate immune system. They function as receptors that recognize foreign antibodies in macrophages and dendritic cells and send signals to activate cells in the innate immune system.
60. D.C. Brown & J. Reetz, 2012, *Evidence-Based Compl. Altern. Med.*, article ID 384301, 8 pp., http://dx.doi.org/10.1155/2012/384301. Some of the dogs in the Brown and Reetz study survived this highly aggressive canine cancer for more than a year, compared with the longest median survival time of 86 days for untreated dogs.
61. It is not yet known whether these properties are equally present in all populations of *G. lucidum* or if they are also present in the closely related *G. curtisii* (also called Reishi) and *G. tsugae* (Hemlock Varnish Shelf).

more than two millennia[62] and is typically prepared as a bitter-tasting tea.

A number of different commercial products made from this mushroom are sold as powders, dietary supplements, and teas, and in 1995 their total market value was more than US$1.6 billion.[63] These mushrooms have demonstrated anti-inflammatory and antioxidant properties and enhance the human immune response both *in vitro* and *in vivo* by activating macrophages and cytotoxic NK (natural killer) cells and T-cells. Studies on the anticancer effects of Reishi have not been convincing, but it is likely that it may prove beneficial as an adjuvant to chemotherapy.[64] Earlier reports that antioxidants interfere with cancer chemotherapy proved to be unfounded.[65] Like many other mushrooms, Reishi contains a wide variety of bioactive molecules (terpenoids, steroids, phenols, nucleotides and their derivatives, glycoproteins, and polysaccharides) plus all the essential amino acids. They are high in polyunsaturated fatty acids, which adds to their overall health value.[66] Several compounds in Reishi were shown to inhibit a variety of human cancers, including leukemia, liver cancer, lung cancers, prostate cancer, colon cancer, and cervical cancer.[67] A number of studies have demonstrated the immunoenhancing effects of Reishi in stimulating T and B lymphocytes, NK cells, and dendritic cells both *in vitro* and *in vivo*, as well as enhancing both the innate and adaptive immunity pathways.[68] Reishi also lowers serum glucose, reduces blood pressure, and may help reduce the pain of rheumatoid arthritis. To date, however, well-designed human trials are lacking.[69]

The triterpenoids are more concentrated in the fruitbodies than in the mycelium. Those looking for relief from bronchitis, asthma, and allergies can thus prepare an effective medicinal tea from the fruitbody itself rather than buying an extract of the mycelium.[70]

- **Artist's Conk (*Ganoderma applanatum*)**: This mushroom is widespread and common throughout our area. Like several other *Ganoderma* species, Artist's Conk produces a number of bioactive chemical compounds with potent medicinal properties. Exopolymers produced by *G. applanatum* reduced plasma glucose by up to 22% in streptozocin-induced diabetic rats, lowered total cholesterol and triglyceride levels by 20.3% and 22.5% and alanine transaminase (ALT) and aspartate transaminase (AST) by 23.2% and 20.7%, respectively.[71] Several highly branched β-1,3-D-glucan polysaccharides in the cell walls of *G. applanatum* and other *Ganoderma* species have shown potent antitumor activities.[72] Ethanolic (ethyl alcohol) extracts from *G. applanatum* showed high OH^- radical scavenging activity and inhibited lipid peroxidation, suggesting a likely antioxidant therapeutic potential for this species.[73] And polysaccharide preparations from *G. applanatum* showed marked antitumor activity against transplanted sarcoma 180 (connective tissue cancer) in

62. S.P. Wasser et al., 2005, Reishi or Lingzhi (*Ganoderma lucidum*), *in* P.M. Coates, M.R. Blackman, G.M. Cragg, M. Levine, J. Moss, & J.D. White (eds.), *Encyclopedia of dietary supplements* (New York: Marcel Dekker), 680–690.

63. S.T. Chang & J.A. Buswell, 1999, *Int. J. Med. Mushrooms* 1:139–146.

64. X. Jin et al., 2012, *Cochrane Database Syst. Rev.* 13:6, CD007731, DOI:10.1002/14651858.CD007731.pub2.

65. C.B. Simone et al., 2007, *Alt. Ther. Health Med.* 13(2):40–47.

66. See note 19.

67. S. Wachtel-Galor et al., 2011, *Ganoderma lucidum* (Lingzhi or Reishi): A medicinal mushroom, *in* I.F.F. Benzie & S. Wachtel-Galor (eds.), *Herbal medicine: Biomolecular and clinical aspects*, 2nd ed. (Boca Raton: CRC Press), 175–197.

68. P. B. Matheny et al., 2006, *Can. J. Bot.* 84:1794–1805.

69. C. Ulbricht et al., 2010, *J. Soc. Integrat. Oncol.* 8(4):148–159.

70. M. Zhu et al., 1999, *Phytother. Res.* 13:529–531.

71. B.-K. Yang et al., 2007, *Phytotherapy Res.* 21(11):1066–1069. Elevated ALT is a specific indicator of liver damage, while elevated AST can indicate liver disease, as well as heart attack, acute pancreatitis, acute hemolytic anemia, and other serious medical problems.

72. S.C. Jong & J.M. Birmingham, 1992, Medicinal benefits of the mushroom *Ganoderma*, *Advances Appl. Microbiology* 37:101–134. β-D-glucans are monomers of D-glucose linked by β-glycosidic bonds and are basic constituents of the cellulose in plant walls. β-glucans derived from yeasts and medicinal mushrooms are potent modulators of the body's immune system. They appear to activate the body's natural killer cells, T-cells, B-cells, neutrophils, and macrophages, as well as stimulating inflammatory cytokines. See L. Ren et al., 2012, Antitumor activity of mushroom polysaccharides: A review, *Food Funct.* 3(11):1118–1130.

73. K. Acharya et al., 2005, *Indian J. Exp. Biol.* 43:926–929.

mice, with complete regression of tumors in more than half the animals and with no evidence of toxicity.[74]

- **Royal Sun Agaricus, Almond Mushroom, "Agaricus blazei" (*Agaricus subrufescens*):** This mushroom has been widely cultivated for many decades in Brazil, Japan, Korea, Denmark, and the Netherlands, where it is sold as a gourmet food and medicinal specialty. Although rare in our area, this species is described in detail in the "Miscellaneous Euagarics" chapter of this book.

One of the most effective and potent of all medicinal mushrooms, *Agaricus subrufescens* is widespread but rare and very sporadic in eastern North America. It is common in the Atlantic region of Brazil, where it is widely cultivated, and is also native to portions of western Europe. It was extensively cultivated in the northeastern United States from the late 19th to the early part of the 20th century but was soon replaced by *A. bisporus* (Lange) Imbach—the widely sold edible mushroom known as the Button Mushroom when young and Portobello when fully mature. It is also widely cultivated in Denmark and the Netherlands, as well as in Japan (from strains obtained in Brazil by Japanese mycologists) and in China and Korea. A few farmers in the United States are again growing this mushroom indoors under controlled conditions for the alternative/integrative health market.

The Almond Mushroom was traditionally used to treat diabetes, hyperlipidemia, atherosclerosis, heart and circulatory disease, digestive disorders, and cancer and to prevent osteoporosis and peptic ulcer.[75] Due to its widespread popularity in alternative medicine circles, this mushroom has in recent decades undergone extensive scientific scrutiny, primarily by Asian but increasingly also by Western investigators.

The main bioactive compounds in *A. subrufescens* are polysaccharides such as riogluccans, β-glucans, and glucomannans. Ergesterol, an analog of mammalian cholesterol with antitumor activity, is present in the cell membranes of this species, as well as in other known medicinal mushrooms.[76] Many *in vitro* studies have shown that both the mycelium and fruitbodies of *A. subrufescens* possess immunostimulant and immunomodulating effects, but the results have sometimes been contradictory. Water and ethanol extracts from this mushroom act on many different cytokine receptors and pathways, which can both stimulate and suppress immune responses. Similarly conflicting results have been obtained in studies on the cancer-suppressing properties of *A. subrufescens*.[77] Different mushroom strains and extraction methods can affect the anticancer properties, and β-glucans, the most important bioactive components, are distributed differentially in the soluble and the insoluble dietary fraction.[78]

Nevertheless, *Agaricus subrufescens* has shown strong antitumor and immunoenhancing properties both *in vitro* and *in vivo*. It contains compounds that inhibit the enzyme aromatase, which is associated with tumor growth. It has potential for prevention and treatment of breast cancer, but much remains to be learned about the mechanisms. In many of the studies done, it was found that certain of its compounds, although not effective against several cancers *in vitro* (in culture), have pronounced activity against these same cancers *in vivo* (in living animals) and vice versa. Human trials have not yet been completed.

However, one problem has yet to be resolved: *Agaricus subrufescens*, like many other mushrooms, also contains carcinogenic hydrazines. Heating the mushrooms can remove these hydrazines, but consumers should be aware that many of the products sold as "Agaricus blazei" may not have been heat treated and may contain these potentially dangerous compounds. This species also concentrates highly toxic heavy metals such as cadmium, mercury, and lead (as do many mushrooms); such mushroom products coming from China contain dangerous concentrations of these elements.[79]

74. S. Takuma et al., 1971, *Chem. Pharm. Bull.* 19(4):821–826.
75. T.K. Mizuno, 1995, *Food Rev. Int.* 11:167–172.
76. K. Wisitrassameewong et al., 2012, *Chiang Mai J. Sci.* 39(2):281–291.
77. F. Firenzuoli et al., 2008, *Evidence-based Complement. Alternat. Med.* 5(1):3–15.
78. P. Manzi & L. Pizzoferrato, 2000, *Food. Chem.* 68:315–318.
79. G.M. Halpern, 2007, *Healing Mushrooms* (Garden City Park: Square One Publishers), 90.

This mushroom has long been referred to as *Agaricus blazei* Murrill (a very different species), stemming from a misidentification by a Belgian mycologist to whom the Brazilian specimens were sent by the Japanese collector who brought samples of this species to Japan.[80,81] This identification error was later recognized, and the species was described as a new species, *Agaricus brasiliensis* Wasser, Didukh & Stamets (2002).[82] Alas, this turned out to be a homonym of yet a different taxon, *Agaricus brasiliensis* Fries (1830) and is therefore an illegitimate name. Despite this, it is sometimes still erroneously called *A. brasiliensis*. Molecular genetic and breeding studies finally revealed that the North American, Brazilian, European, Hawaiian, and Asian populations are all later synonyms of a North American species, *Agaricus subrufescens* Peck (1893).[83] Nevertheless, this species and the medicinal extracts and pharmaceutical products prepared from it continue to be widely and incorrectly called *A. blazei* (and to a lesser extent also *A. brasiliensis*) in both the biomedical and popular literature and in commercial packaging. Bioactive extracts from this mushroom are still typically referred to as "ABM [for *Agaricus blazei* Murrill] glucans" in the biomedical literature, and a steroid recently isolated from this mushroom in Japan (where it is popularly called *himematsutake*) was recently named "blazein" on the basis of this continuing misidentification.[84]

Collecting and Identifying Mushrooms

What's Needed

Hand lens. A good hand lens is an indispensable tool while mushrooming in the field. We find that a good geologist's hand lens is a much better optical device than a jeweler's loupe for this purpose. The hand lens is ruggedly made and can be attached to a lanyard and worn around the neck. A good hand lens should be free of chromatic aberration. This is accomplished by combining two or preferably three separate pieces of glass in a doublet or triplet, respectively. Coddington lenses achieve the same effect with deep cuts inside a single piece of solid glass to create air gaps. A 10× hand lens is the maximum magnification that is practical in the field. Recommended triplet hand lenses include the Bausch & Lomb 10× Hastings Magnifier, Edmund Optics 10× Hastings Triplet Magnifier, and the Belomo 10× Loupe. These can be bought for $35–$45 at Amazon.com, about $10 more at Edmund, or even more cheaply at a good rock and mineral show.

Knife. A rugged field knife in a scabbard or a pocket knife is useful for digging up mushrooms or prying them loose from tree bark. This is especially important when you need to dig up an amanita intact to check out the volva.

Binoculars. This may seem like a counterintuitive item for a field mycologist or mycophile, but a decent pair of binoculars can save you much time and wasted energy. More times than we can recall we have avoided an unnecessary climb up a steep hill or wading across a fast-running cold mountain stream when the distant large and beckoning "agaric" or "polypore" turned out to be a fallen leaf, a fragment of broken bark, lichen, or even a piece of discarded picnic trash when viewed through binoculars. Good binoculars are also indispensable for checking out the inevitable birds and mammals you will encounter in your forays in the field—that is, assuming you are also a dedicated general field naturalist, like most of us.

Thermometer. Dedicated morel hunters often carry a soil or meat probe thermometer in the spring to check for optimum soil temperatures for these much-sought-after mushrooms. Optimum conditions exist for morels when temperatures 10 centimeters (4 inches) below the ground are between about 13°C and 17°C (55°F and 62°F). These individuals tend to be very secretive and proprietary about when and where they seek (and find) their quarry.

Basket or other collecting bag.

80. P. Heinemann, 1993, *Bull. Jard. Bot. Belgium* 62:365–368.
81. R.W. Kerrigan, 2005, *Mycologia* 97(1):12–24.
82. In A.F. da Eira, M.Y. Didukh, P. Stamets, S.P. Wasser, & M.A.L. de Amazonas, 2002, *Internat. J. Med. Mushrooms* 4(4):274.
83. *Agaricus subrufescens* Ellis & Everhart (1894) is an illegitimate homonym of *A. subrufescens* Peck (1893) and is properly referred to as *Tricholoma subrufescens* Saccardo (1895).
84. H. Ito et al., 2009, *Oncol. Rep.* 20(6):1359–1361.

Chemical Staining

A variety of chemical stains are used to help distinguish certain microscopic features (e.g., the different types of hyphae and spore structures) and to assist in identifying the different species of certain groups of fungi having specific staining characteristics. These may include immediate or quick responses (color "flashes") that then resolve into one or more different hues that may themselves be stable or fade. A negative response can also be diagnostically important. Aqueous solutions should be made with distilled water, as tap water contains chemical contaminants that can alter the staining characteristics of your samples. Some of these reagents are quite toxic and corrosive and must be handled with care. Eye protection is strongly recommended.

Potassium hydroxide (KOH). A 3%–5% or 10% solution (we use 10% reagent grade KOH) is generally used. Potassium hydroxide is an important stain for identifying corts (*Cortinarius* spp.) and boletes. Species of the genus *Agaricus* may turn yellow or green or not react with KOH. Other mushrooms may turn red, magenta, olive, or black. This stain is also useful in distinguishing between species of *Amanita*, *Russula*, and *Lactarius*, as well as many polypores. KOH tends to crystallize after a year or so. It is useful for restoring dried material (3%–5%) and for softening hard bracket fungi but forms a cloudy precipitate when used together with Melzer's reagent. Sodium hydroxide (NaOH) can be used in place of KOH.

Ammonium hydroxide (NH_4OH). In contrast to KOH, ammonium hydroxide lasts for several years. Use a 10% solution for microscopic staining and 40%–75% for macroscopic staining of tissues. Useful for boletes and for staining the marginal cystidia of *Pholiota* and *Stropharia*, *Psathyrella*, *Hypholoma*, *Collybia*, *Cystoderma*, and *Xeromphalina*.

Ammonia (NH_3). Ordinary household ammonia is suitable for this purpose. A 10% aqueous solution can be used to reconstitute dried material and produce specific color changes in certain mushrooms, particularly boletes of the genera *Boletus* and *Suillus* and milkcaps (*Lactarius* spp.). Also useful for *Crinipellis* and some species of *Mycena* and for staining spores for microscopy.

Melzer's reagent. This solution is an aqueous solution of chloral hydrate, potassium iodide (KI), and elemental iodine (I_2). Since it contains chloral hydrate,[85] a Schedule IV controlled substance, it requires a medical prescription and must be prepared by a licensed compounding pharmacist according to the following formula: 1.5 grams I_2, 5.0–7.5 gm KI, 100 gm chloral hydrate, and 100 mL distilled water.[86] Because of the chloral hydrate and elemental iodine, this dangerously toxic solution must be labeled as poisonous and not fit for human consumption. Melzer's reagent forms a cloudy precipitate with alkaline chemicals like KOH. It is useful for restoring dried material, but not if KOH was previously used.

Spores and tissues that stain bluish-grey, violet, violet-black, or bluish-black with Melzer's reagent are **amyloid** and indicate the presence of amylose (a component of starch); those that stain brown to reddish-brown are **dextrinoid**, presumably due to a glycine-betaine-IKI complex. A negative reaction or if the spores/tissues remain yellowish with Melzer's reagent is termed **nonamyloid (inamyloid)**. It may take up to a half hour to obtain a reaction.

Lugol's solution (IKI). Due to the difficulty of obtaining chloral hydrate, Lugol's solution containing 1%, 2%, or 5% I_2 and 10% KI in distilled water can be substituted for Melzer's reagent. However, this reagent does not stain as effectively or consistently as Melzer's reagent because of the lack of chloral hydrate.[87] Because of the elemental iodine, Lugol's solution is also toxic to humans. Lugol's solution is the preferred stain to test the tip and outer wall of the asci of ascomycetes for **hemiamyloidity** (a reddish reaction in the presence of IKI and the absence of KOH),

85. Chloral hydrate ($C_2H_3Cl_3O_2$) is a powerful sedative-hypnotic agent that activates GABA (gamma aminobutyric acid) receptors in the brain to induce sedation and unconsciousness. It was formerly called "knockout drops." It can be added to an alcoholic drink to create a Mickey Finn and was once used as a date-rape drug.

86. L.M. Leonard, 2006, *McIlvanea* 16(1):43–51.

87. Chloral hydrate has a high refractive index (1.4280) and can deeply penetrate and clear a variety of tissues.

as the presence of chloral hydrate masks this hemiamyloid reaction.[53]

Iron salts, ferrous sulphate (Fe_2SO_4). A 10% aqueous solution is very useful for identifying russulas and boletes. These mushrooms may turn olive, green or blackish-green, or reddish-pink, or they may not react. This solution is also useful for identifying certain monotypic genera (e.g., *Rhodotus palmatus*) or distinguishing between certain species of *Cantharellus*.

Cotton Blue (aniline blue) or lactophenol Cotton Blue. A solution of lactic acid (lactate) and/or lactate and phenol used to study spores and the gill cystidia (pleurocystidia) of some mushrooms. Made by dissolving 1 gm of Cotton Blue in 99 mL of lactate. Spore walls and/or ornamentation staining blue are **cyanophilous**; nonreactive spores are **acyanophilous**.

Cresyl Blue. A 0.5%–1.0% solution is made by dissolving 0.5–1.0 gm of Cresyl Blue in 99.5–99.0 mL of water. Spore walls and/or hyphae staining magenta-red to violet are **metachromic**; those staining dark blue are **orthochromic**.

Congo Red. Best used in a solution of 1% Congo Red in concentrated ammonia, although a 1% aqueous solution can be used. This stain is useful for seeing hyphal structures and clamp connections.

Wieland-Meixner test for amatoxins. This qualitative test for the presence of deadly amatoxins in the genera *Amanita*, *Lepiota*, and *Galerina* uses concentrated hydrochloric acid (HCl) and low-quality paper containing lignin (i.e., newspaper). The test must be done away from sunlight or high heat.[88] A small amount of juice from the mushroom to be tested is squeezed onto the surface of a newspaper or other low-grade paper and allowed to dry at room temperature. A single drop of concentrated HCl is then added. If one or more amatoxins are present, a greyed bluish-green spot will appear, usually within 1–3 minutes, but one should wait for up to 20 minutes. Hospital laboratories have often obtained false negatives with this test when using dilute HCl or higher quality paper that is low in lignin content (e.g., filter paper). Psilocin-containing mushrooms of the hallucinogenic genera *Psilocybe* and *Panaeolus* will also give a 100% false positive reaction with the Meixner test.[55]

CAUTION: HCl is a strong and extremely corrosive acid. **Never add water to a strong acid!** It will generate excessive heat and can cause the acid to violently boil over and splatter. **Always add the acid to the water** to avoid this dangerous reaction.

Schäffer's test for the genus *Agaricus*. This test uses a mixture of aniline ($C_6H_5NH_2$) and nitric acid (HNO_3) on the surface of species of the genus *Agaricus*. A positive reaction (orange to red) indicates the agaric belongs to Section *Flavescentes*.

Microscopy

Careful attention to the macroscopic features of mushrooms will allow you to accurately identify many of them to species and even to varieties. However, and alas, this will only get you so far. Many genera will require detailed examination of the size, shape, and ornamentation of the spores; structural details of the hyphae, cystidia, basidia, or asci; presence of clamp connections; and other microscopic features to pin specimens down further. And in all too many cases, even this may not be definitive. Nevertheless, the serious amateur mycologist will sooner or later find it necessary to invest in a decent microscope.

This is not the appropriate place, nor do we have sufficient space, to discuss the details of how best to use stereo (dissecting) and compound microscopes. The details of illumination, focusing, and adjustment of the diaphragm and condenser and other tricks of the trade are readily available online. We refer you to Michael Kuo's MushroomExpert.Com[89] and several other excellent web-

88. When exposed to direct sunlight or heat, for example, 29.4°C–35°C (85°F–95°F), concentrated HCl will turn bluish within 20–30 minutes (T.J. Duffy, 2008, Toxic fungi of western North America, *Mykoweb*, http://www.mykoweb.com/TFWNA/P-17.html).

89. M. Kuo, February 2006, Using a microscope to study mushrooms, retrieved from the MushroomExpert.Com website, http://www.mushroomexpert.com/microscope.html, and linked websites.

sites on this subject: First Nature;[90] an online portion of a book by J. F. Ammirati, J. A. Traquair, and P. A. Horgen;[91] Move Humanity Forward;[92] and elsewhere.

What you need in a microscope. To examine spores, basidia, asci, cystidia, hyphae, and other microscopic structures you will need a good-quality compound microscope with at least a triple (quadruple is preferred) nosepiece to accommodate objectives of various magnifications. While a monocular microscope (one eyepiece) will work, you have to acquire the skill to keep both eyes open while peering through the ocular with your dominant eye and allowing the weak eye to remain unfocused. This can also be quite tiring, and we strongly recommend investing in a binocular microscope. The modest extra cost will be well worth it. It is not necessary to buy an expensive research-quality Nikon, Olympus, Leica, or Zeiss. (The former high quality of Leica and Zeiss upright compound microscopes has deteriorated in recent years, as has the companies' customer service.) Excellent binocular compound microscopes are available for between US$200 and $300 with free shipping from AmScope in Irving, California.[93] These microscopes are also sold at reduced prices on Amazon.com, with free shipping with an Amazon Prime account. They also come with a 5-year warranty. We use AmScope's 40X-1600X Lab Clinic Vet Trinocular T490A brightfield microscope, which cost less than $280 three years ago (it still costs less than $300) and which has been most satisfactory. The third "simul-focal" ocular (photo-port) permits mounting a camera for taking photographic micrographs, which are in focus with the two viewing oculars.[94]

90. http://www.first-nature.com/fungi/facts/microscopy.php.

91. J.F. Ammirati, J.A. Traquair, and P.A. Horgen, 1985, *Poisonous mushrooms of the northern United States and Canada* (Minneapolis: University of Minnesota Press), 38–40, available at http://books.google.com.

92. Available at http://www.movehumanityforward.org/forums/phpbb/viewtopic.php?f=7&t=138400.

93. AmScope's microscopes and prices can be viewed at http://www.amscope.com/?gclid=CPjrs6PKsL8CFcnm7AodSEQA9A.

94. Decent used student and high-quality research microscopes are also often available from Amazon.com, Fisher Scientific, and other biological supply sources, as well as various colleges and universities.

You will need 10× oculars, or eyepieces (we have both 10× and 16×), and the objectives should include at least 10× and 40× objectives and a 100× oil immersion lens with a spring (ours are 40×, 64×, and 100×). The total magnification is that of the objective multiplied by that of the oculars. Optical magnifications greater than 1,500×–1,600× are impractical because although the magnification (the size of the image relative to its actual size) can increase indefinitely, the resolution (the smallest distance at which two discrete objects can be distinguished as separate entities) begins to fail at around 1,500×–1,600×. Because of inherent limitations in the wavelengths of visible light, the maximum resolution is with the shortest wavelength of visible light ($\lambda \approx 400$ nanometers), and the maximum realizable numerical aperture (NA = 1.40) in the best apochromatic lenses and substage condensers is about 200 nanometers (0.20 micrometer)—about the size of the smallest cocci (spherical bacteria). Nothing smaller than this can be discerned no matter how much you magnify it. In fact, however, most light microscopy is done with white light (midrange wavelength of about 550 nm) from an incandescent bulb and objectives and condensers with much lower NAs—typically around 1.25 for a 100× plan achromat objective (plan apochromat lenses with NA = 1.40 cost quite a bit more). With such a system (like ours) the maximum resolution will be about 0.268 µm, which is still more than adequate for fungal microscopy.

We are concerned here only with standard brightfield illumination involving the absorbance of ordinary white light by the sample. We will not discuss other approaches to obtain increased contrast such as darkfield, phase contrast, differential interference contrast, and cross-polarized light.

Your microscope should have a rack-and-pinion substage condenser and diaphragm with a numerical aperture of at least 1.25 and adjustable internal brightfield illumination. Köhler illumination is by far the best; it produces very even illumination of the entire field and ensures that an image of the illumination source (i.e., a tungsten filament) does not appear. Cheaper microscopes will likely have critical illumination for either transmitted or reflected light.[95] These can create problems when an image

95. Avoid cheap "kiddie" or very old microscopes that use a mirror to reflect light from an external source.

of the light sources clutters up the field of view, which then must be compensated for by adjustments to the light intensity and the condenser. You should also have a mechanical stage with coaxial movement control knobs.[96]

Optical reticles and stage micrometers. To measure the size of spores and other structures you will need a 10× ocular with a micrometer reticle (around $35 new on sale) and a stage micrometer with a 1 mm scale marked off in 0.01 mm increments. This provides measurement increments of 1 µm at 1,000× magnification. Stage micrometers are more expensive than ocular micrometers, but AmScope and Amazon.com have them on sale for as little as $35–$75. And we have seen them advertised for as little as $20. The details of how to calibrate these micrometers can readily be found online, for example, at the websites of Rice University (http://www.ruf.rice.edu/~bioslabs/methods/microscopy/measuring.html) and MushroomExpert.Com (http://www.mushroomexpert.com/microscope_spores.html).

Slides and coverslips. Standard glass microscope slides are inexpensive and available from a number of sources, including Fisher Scientific and Amazon.com. Glass coverslips are likewise inexpensive and come in a variety of sizes and thicknesses. Number 1 coverslips (0.13 to 0.16 mm thick) are best for most biological work. We prefer dimensions of 22 × 22 mm, but smaller (and longer) sizes are also available. Permanent mounts can be made by sealing the edges of the coverslip over the specimen in its mounting medium with clear nail polish.

Immersion oil. The use of immersion oil of the proper refractive index can enhance resolution at higher magnification; it is commonly used with 40× and higher objectives (our 40× and 64× objectives are "high-dry" lenses, which do not use oil), and oil is essential with the 100× objective. Immersion oils with an index of refraction of 1.515 prevent scattering of the light rays at the coverslip-air interface and allow all of the light to enter the objective. This is shown in Figure 27.

96. Trying to move a slide by hand in the reverse direction of the apparent motion in the field under higher magnification is very tedious and is a difficult skill to master.

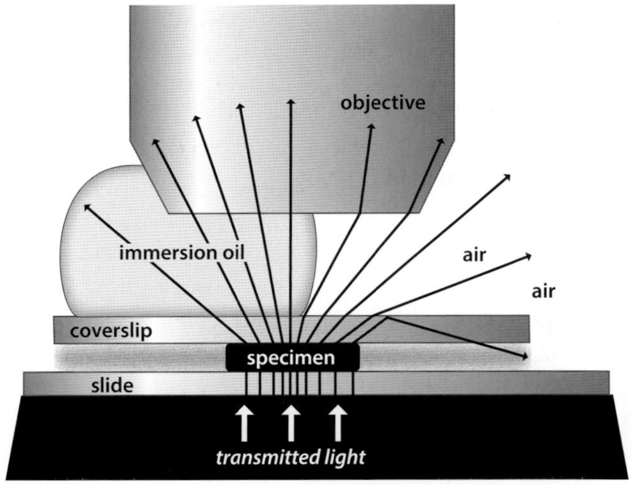

FIGURE 27

Inexpensive immersion oils of the right clarity, viscosity, and refractive index are available from the same sources where you can obtain slides and coverslips.

Glycerol, glycerin ($C_3H_8O_3$). This clear, viscous liquid is water soluble and can be used as a mounting medium for microscope slides when viewing the natural color of a specimen is desired. Various stains can also be added to elicit specific chemical staining reactions or to make microscopic structures more visible. Glycerol is cheap and available at most pharmacies.

Lens paper. Lens-cleaning paper or Kimwipes will be needed to wipe the oil off the oil immersion lenses. These are very cheap and readily available at Amazon.com, Fisher Scientific, and similar outlets.

Other useful tools include needlenose forceps (tweezers), a good dissecting needle, and a package of single-blade razors for cutting thin sections from mushroom fruitbodies for microscopic examination.

Additional information and instructions about microscopes and microscopy are available at the following sites:

Molecular Expressions: Optical Microscopy Primer: http://micro.magnet.fsu.edu/primer/index.html

Microscope: Basics and Beyond, by Mortimer Abramowitz (renowned introduction to optical microscopy in full color): http://micro.magnet.fsu.edu/primer/pdfs/basicsandbeyond.pdf

Optical Microscopy, by Michael W. Davidson and Mortimer Abramowit: http://micro.magnet.fsu.edu/primer/pdfs/microscopy.pdf

Dissecting and mounting techniques. Space does not permit us to give detailed instructions on calibrating optical and stage micrometers; making hand sections; preparing slides; teasing out tissue from the cap, gills, and stipe; obtaining and examining spores, basidia, cystidia, hyphal tissues, and clamp connections; and the like. However, several excellent online sources are available, especially those created by the amazing Dr. Michael Kuo at his MushroomExpert.Com website.

Using a Microscope to Study Mushrooms: http://www.mushroomexpert.com/microscope.html#references

Basic Mycological Techniques by R. T. Moore, University of Ulster, UK. Discusses stains, slide preparation, permanent mounting techniques, and spore print preservation: http://www.science.ulst.ac.uk/rtm/Techniques.html#I

Mycorrhizal Associations: The Web Resource, Section 10, Methods for Identifying Mycorrhizas. This website describes how to collect and prepare specimens of mycorrhizal associations for microscopy, including clearing and staining methods, the chemicals involved, and safety precautions concerning caustic and toxic reagents: http://mycorrhizas.info/method.html

Microscope.com. Instructions for making wet mounts, dry mounts, and smears and staining techniques. Detailed information on all aspects of microscopy are also available here: http://www.microscope.com/education-center/how-to-guides/mount-slides/

Notes on Microscopy & How to Get Started—Move Humanity Forward. An excellent website presenting a wealth of illustrated information on basic microscopy and slide-making techniques, mushroom anatomy, spore structure, and many other tricks of the trade. It also presents a great deal of information about basic mycology and an extensive, partially illustrated glossary. It can be accessed at http://www.movehumanityforward.org/forums/phpbb/viewtopic.php?f=7&t=138400&p=297276

Scientific Nomenclature

The following is a brief discussion of scientific nomenclature as it pertains to organisms covered by the *International Code of Nomenclature for algae, fungi, and plants* (*ICN*, Melbourne Code), which covers the Fungi. A botanical name is a formal scientific name that conforms to the *ICN* or its predecessors. A similar code, the *International Code of Zoological Nomenclature* (*ICZN*), exists for the scientific names of animals. The *ICN* and the *ICZN* are similar in many respects, but there are also a number of important differences.

You all know the basic reason behind scientific names: to provide for a single, unique, and universally understood name for every species we know about, living and extinct, on the planet. Fat chance! The total number of species estimated to exist on Earth ranges from 5 million to 30 million, of which about 2 million have been formally described and named.

As an example let's take the common Turkey Tail Fungus, *Trametes versicolor*. This cosmopolitan polypore has numerous common names, including but not limited to the following:

English: Turkey Tail, many-zoned polypore, many-colored polypore, varicolored bracket, rainbow fungus, white-rot fungus

German: *die Schmetterlingstramete* (the butterfly trametes)

French: *tramète versicolore* (varicolored tramete)

Spanish: *el yesquero multicolor, yesquero de varios colores* (many-colored fire starter)[97]

Dutch: *gewoon elfenbankje* (just an elves' bench)

Danish: *broget læderporesvamp* (many-zoned polypore)

97. Dried polypores are commonly used in Latin America as tinder to light fires with sparks struck from flint.

Slovak: *trúdnikovec pestrý* (variegated trametes)

Polish: *wrośniak różnobarwny* (turkey tail)

Chinese: *yun zhi* (cloud mushroom)

Japanese: *kawaratake* (mushroom by the river)

You can see the problem immediately. Not only does this one species have many common names, but some of these names also apply to other species.

Well, what about scientific names? One species, one name, right? That was the whole idea behind Linnaeus' system. Well, that was the idea all right, but it doesn't always work out that way. For example, our Turkey Tail Fungus has the following validly published scientific names, *all of which* have been used in the scientific literature (and several of which continue to be used by various authors):

Trametes versicolor* (Linnaeus 1753 *ex* Fries 1821) Lloyd 1920[98]

Synonymy (list of other validly published scientific names that refer to this species):

Boletus versicolor Linnaeus 1753

Agaricus versicolor (Linnaeus 1753) Lamarck 1783

Poria versicolor (L.) Scopoli 1783

Agaricus versicolor (L.) Paulet 1793

Agarico-suber versicolor (L.) Paulet 1793

Polyporus fuscatus Fries 1818

Polyporus versicolor (L.) Fries 1821

Sistotrema versicolor (L.) Trattinnick 1830

Polystictus versicolor (L.) Fries 1851

Polyporus nigricans Lasch 1859

Polyporus versicolor var. *nigricans* (Lasch 1859) Fries 1874

Hansenia versicolor (L.) Karst 1880

Bjerkandera versicolor (L.) Karst 1881

Coriolus versicolor (L.) Quélet 1886

Microporus fuscatus (Fries 1818) Cooke 1886[99]

Microporus nigricans (Lasch 1859) Cooke 1886

Microporus versicolor (L.) Kuntze 1898

Polyporus versicolor var. *fuscatus* (Fries 1818) Rea 1922

Polyporus versicolor var. *nigricans* (Lasch 1859, *non* Fries 1874) Rea 1922

Xerocomus versicolor (Kuntze 1898) Gilbert 1931

Trametes versicolor (Linnaeus 1753) Lloyd 1920

Trametes versicolor f. *fuscata* (Fries 1818) Domański, Orloś & Skirg 1967

Ochroporus nigricans (Lasch 1859) Fiasson & Niemelä 1984

Pretty nightmarish, isn't it? It's quite a list. "How can this be?" you may ask. "The whole point of the Linnaean system is to achieve nomenclatural stability. Why do I still see names like *Coriolus versicolor* and *Polyporus versicolor* appearing in the current scientific and medical literature when they are talking about the same species? And the common Turkey Tail at that? A species that was described by Linnaeus himself back in 1753!"

One issue is that not every author agrees with a specific combination of genus and species. Thus, while most authorities currently accept the combination of *Trametes versicolor* as the valid name for the ubiquitous Turkey Tail, some holdouts insist that the species is properly referred to the genus *Polyporus* Micheli *ex* Fries 1821 or *Coriolus* Quélet 1886. (Don't panic! We will explain the use of personal names and dates shortly.) But a major problem stems from some geneticists and molecular biologists with no understanding of systematic biology or taxonomy who

98. The combination of Pilát 1939 is superfluous, as this was already published by Lloyd in 1920; however, Pilát 1939 is commonly but erroneously cited as the authority for this genus and species combination by those workers who have not done their homework.

99. The combinations of Kuntze 1898 commonly cited are superfluous, as these names were previously published by Cooke in 1886.

insist on willy-nilly publishing their own names to fit their imagined (and often imaginary) evolutionary relationships. Mitochondrial DNA nucleotide and certain protein amino acid sequences can greatly aid our understanding of the phylogenetic relationships between organisms, but they are by no means the whole story. And their practitioners often fail to recognize that mutation rates are not constant between various groups of organisms.

Some of these Johnnies-come-lately are responsible for a great deal of confusion,[100] and no one is sure anymore just what is meant by a given scientific name. It is complicated enough when nomenclatural experts are involved. The problem is guaranteed to become even more severe now that universities have disallowed systematic biology and taxonomy as a valid academic discipline since the 1980s. It is considered to be "intellectually uninteresting" and suitable only for amateur naturalists and "fuzzy caterpillar lovers," as one particularly arrogant and narrow cell biologist once put it to one of us (WL). How quickly they condemn that which they do not understand!

The purpose of scientific names is indeed to provide a single and unique name for each species. Binomial nomenclature consists of assigning each species a genus name and a specific epithet. But nomenclature is also intended to reflect our understanding of biological relationships. A genus is a group of related species. But one investigator's concept of a genus may be wide and that of another narrow. Thus certain species will be included within a given genus by one worker but excluded by another. Our Turkey Tail is placed in the genus *Trametes* by most authorities today, and it has the specific epithet, or species name, of *versicolor*. But some disagree, so we still see *Polyporus versicolor* and *Coriolus versicolor* appearing in the literature. A species name can never be used alone (except in the most informal situations when experts are talking casually) but must always be accompanied by its generic name. However, a genus name can stand alone. Once a generic name has been introduced, it may thereafter be abbreviated as long as no possible confusion exists (like another genus with the same abbreviation). Thus, once you have presented *Trametes versicolor*, it may thereafter be referred to as *T. versicolor*, except that *an abbreviated genus name cannot begin a sentence!* (I know, some experts who should know better apparently don't.)

The genus name or its abbreviation is always capitalized *and italicized* (or underlined if in roman typeface). Many journals adamantly refuse to adhere to this requirement (italics of all nomenclatural codes) but will italicize anything else that suits their fancy. Incidentally, if a scientific name appears in a passage that is itself in italics, then the scientific name should be set in roman typeface to set it off. In nearly all cases (this is *always* true for animal names), a botanical species name is written in lowercase, even in titles, but a very few botanical specific epithets may be capitalized under certain circumstances.

Now for the names of persons in a scientific name. The name of the person(s) who first validly publishes a name, particularly at the species-, genus-, or family-rank level, is part of the organism's full, formal scientific name, along with the date of the valid publication. This is to establish priority in case of identical names being applied to more than one taxonomic category (homonymy) or when two or more names (synonyms) are found to refer to the same taxon (i.e., a species, genus, or family). The oldest validly published name takes precedence over the "later synonyms," which are said to be "sunk" into synonymy with the oldest valid name. You can see this in the synonymy list provided above for *Trametes versicolor*.

However, the principle of nomenclatural priority has the following limitations: Article 13.1 of the *ICN* considers that the valid publication of the scientific names for fungi begins in May 1753 with the publication of *Species Plantarum* (1st ed.) by Carolus Linnaeus. However, fungal names in the orders Uredinales and Ustilaginales and the class Gasteromycetes adopted by Christiaan Hendrik Persoon (*Synopsis Methodica Fungorum*, December 31, 1801) and those of other fungi adopted by Elias Magnus Fries (*Systema Mycologicum*, vols. 1 [January 1, 1821] to 3, with an additional *Index* [1832] and *Elenchus Fungorum*)

100. As a needless complication, J.L. Fiasson and T. Niemelä resurrected *Polyporus nigricans* Lasch and placed it in the genus *Ochroporus* J. Schröter 1888 (see J.L. Fiasson & T. Niemelä, 1984, The Hymenochaetales: A revision of the European poriod taxa, *Karstenia* 24:14–28). However, *Ochroporus* is a junior synonym of *Phellinus* Quélet 1886. The genus *Trametes* belongs to an entirely different order, the Polyporales, from the genus *Phellinus*, which is in the order Hymenochaetales. This is like saying that a cat and an antelope should be placed together in the same genus!

are sanctioned and are to be conserved over all other earlier validly published synonyms and homonyms. In other words, if a later synonym or homonym was sanctioned by either Persoon or Fries, the sanctioned name is conserved as the basionym for that taxon. Earlier versions of the *ICBN* had similar provisions.

Let's look at this briefly to see what it means. Carl Linnaeus originally described this fungus as *Boletus versicolor* in 1753.[101] Lamarck transferred it to the genus *Agaricus*, and it became known as *Agaricus versicolor* (Linnaeus 1753) Lamarck 1783. Linnaeus's name was placed inside parentheses to indicate that he had originally described the species in a genus other than the one it was now in. Lamarck's name and the date of publication follow Linnaeus's name inside parentheses to indicate that he, Lamarck, was the authority for this nomenclatural change. That same year, another reviser (Scopoli) placed Linnaeus's species into the genus *Poria*, and it became known as *Poria versicolor* (L.) Scopoli 1783. You get the picture. Notice that we are now abbreviating Linnaeus's name as "L." This is routinely done in botanical nomenclature with well-known botanists who have standard abbreviations assigned to them. A list of fungal authors' names and their official abbreviations has been published online by *Index Fungorum*. This list is current as of 2000 and is available in several formats at http://www.indexfungorum.org/authorsoffungalnames.htm.

Now let's examine the complete scientific name of our fungus: *Trametes versicolor* (Linnaeus 1753 *ex* Fries 1821) Lloyd 1920. The Latin preposition *ex* (of, from, out of, after) is often replaced by a colon. What on earth is all this? It means that the name of *Trametes versicolor* (as *Boletus versicolor*) comes to us from Linnaeus through the authority of a defining publication by Fries in 1821 and that the last reviser who transferred the specific epithet to its last and current genus, *Trametes* (after a whole slew of such transfers, as you can see from the synonymy), was Lloyd in 1920 (see the comment in note 101).

On a final note, when a subsequent author creates a new infraspecific taxon (a subspecies, variety, or form) for a species, this automatically creates an **autonym**; that is, the basionym becomes expanded to include the new category. Let's take our Turkey Tail as our nomenclatural example once again. *Trametes versicolor* (Linnaeus) Lloyd was originally described by Linnaeus as *Boletus versicolor*. In 1821 Fries erected the taxon *Polyporus versicolor* var. *fuscatus* (Fries) Fries; he had deprecated *Polyporus fuscatus* Fries (1818), which had described earlier a variety of *Boletus versicolor* (which he had transferred to the genus *Polyporus*). He thus created the taxon *Polyporus versicolor* var. *fuscatus* (Fries) Fries (1821). This automatically created the autonym *Boletus versicolor* Linnaeus (1753) var. *versicolor*. The fact that Linnaeus's name precedes the name of the variety indicates that this is an autonym and not a taxon originally designated by Linnaeus.

Although homonymy (two different taxa having identical scientific names) is prohibited *within* a given nomenclatural code, it is permissible *between* codes. One of many such examples is the butterfly genus *Pieris* Schrank, 1801 and the plant genus *Pieris* D. Don 1834 (heath family).[102] This apparent homonymy is permissible; neither name is a homonym of the other, and both are valid. Notice that the naming authority is followed by a comma in the case of the butterfly but not in the case of the plant. The year of publication is often placed inside parentheses in botanical but not in zoological nomenclature. These are some of the minor differences between the two codes.

Anamorphic Form-Genera and Form-Species

Now just when you are beginning to grasp all this (more or less), we have to muddy the waters even further.

For more than a century Article 59 of the various revisions of the *International Code of Botanical Nomenclature* permitted mycologists to give asexually reproducing fungi

101. Linnaeus was his original family name—it was not Latinized from Linné, as is commonly stated; rather, Carl's father, Nils Ingemarsson (Son of Ingemar), upon entering university took on a family name based on the Latin word for the linden tree in his yard: Linnaeus. In those days, Scandinavians did not normally have last names. Carl would normally have been called Carl Nilsson (Son of Nils) but automatically became Carl Linnaeus because his father had assumed a surname. Carl later assumed the name Carl von Linné following his ennoblement.

102. A very minor difference between the *ICBN/ICN* and the *ICZN* is that in the former there is no comma between the naming authority's name and date of publication; a comma is required under the rules of the *ICZN*. There are other, much greater differences.

(anamorphs) separate names from their sexual states (teleomorphs). And since some fungi have more than one anamorph, a single biological species of fungus can have any number of concurrently valid scientific names that may include a genus and species plus a number of different anamorphic "form-genera" and "form-species." When names are available for both anamorph and teleomorph states of the same fungus, the holomorph usually takes the teleomorph name; under some circumstances it can take the anamorph name if it is subsequently epitypified with a teleomorph. Thus, the teleomorphic (sexual) stage of the Cuban Club was named *Xylaria cubensis* (Montagne) Fries (1851), while its distinctive anamorph, the Bubblegum Fungus, was called *Xylocoremium flabelliforme* (Schweinitz) J.D. Rogers (1984). The basionyms of these two taxa are *Hypoxylon cubense* Montagne (1840) and *Sphaeria flabelliformis* Schweinitz *ex* Fries (1828), respectively (this little detail will become important momentarily).

At least that's the way it was up until recently. The renamed *International Code of Nomenclature for algae, fungi, and plants* (Melbourne Code), adopted by the Eighteenth International Botanical Congress in Melbourne, Australia, in July 2011, has all superseded that: now the rule is one fungus, only one valid scientific name. Now the different names applied to the sexual and asexual forms of the same species of fungus must compete for priority on the basis of their dates of publication.

In our above example of the Cuban Club/Bubblegum Fungus, the holomorph clearly belongs to the genus *Xylaria*. However, the basionym *Sphaeria flabelliformis* was published before *Hypoxylon cubense*. Strict application of this rule would seem to mandate that the teleomorph widely known as *Xylaria cubensis* and the equally well-known anamorph (*Xylocoremium flabelliforme*) must now be called *Xylaria flabelliformis*. This is exacerbated by the fact that *Sphaeria flabelliformis* Schweinitz was sanctioned by Fries in 1828, while *Hypoxylon cubense* was not sanctioned. This creates a nomenclatural nightmare, and we are unclear just how to proceed. When we queried Dr. Jack D. Rogers, a leading authority on the order Xylariales to which this fungus belongs, about this specific case, he strongly urged us to retain both names in our account. And we have done just that in all cases of form-taxa that we discuss until the dust settles.

References Cited

Andary C. & Privat G. 1985. Variations of monomethylhydrazine content in *Gyromitra esculenta*. *Mycologia* 77(2):259–264.

Antowiak R. & Antowiak W.Z. 1991. Chapter 2, Alkaloids from mushrooms. *In* Brossi A. (ed.), *The alkaloids: Chemistry and pharmacology*. San Diego: Academic Press, 298 pp.

Bedry R. et al. 2001. Wild-mushroom intoxication as a cause of rhabdomyolysis. *N. Engl. J. Med.* 345(11):798–802.

Berbee M.L. & Taylor J.W. 2001. Systematics and evolution. Fungal molecular evolution: Gene trees and geologic time. *In* Hock B. (ed.), *The Mycota: A comprehensive treatise on fungi as experimental systems for basic and applied research*. Vol. 7: *Systematics and evolution*, Part B. Berlin: Springer Verlag, pp. 229–245.

Berbee M.L. & Taylor J.W. 2007. Rhynie chert: A window into a lost world of complex plant-fungus interactions. *New Phytologist* 174(3):475–479.

Beug M.W. 2008. Toxicology: Reflections on mushroom poisoning in North America. *Fungi* 1(2):42–44.

Bitartondo M.I. & Bruns T.D. 2011. *Molecular Ecology* 10:2285–2295.

Buck R.W. 1963. Toxicity of *Amanita muscaria*. *J. Am. Med. Assoc.* 185:663.

Butterfield N.J. 2005. Probable Proterozoic fungi. *Paleobiol.* 31:165–182.

Campos D.D., Garrido A., & Valenzuela A. 1989. Silybin dihemisuccinate protects against glutathione depletion and lipid peroxidation induced by acetaminophen on rat liver. *Planta Med.* 55:417–419.

Cavalier-Smith T. 1998. A revised six-kingdom system of life. *Biol. Rev.* 73:203–266.

Corlis J.O. 1998. Haeckel's kingdom Protista and current concepts in systematic protistology. *Stapfia 56, zugleich Kataloge des OÖ. Landesmuseums*, Neue Folge Nr. 131:85–104.

Daszak P., Berger L., Cunningham A.A., Hyatt A.D., Green D.E., & Speare R. 1999. Emerging infectious diseases and amphibian population declines. *EID* 5:735–748.

Dennis R.L. 1970. A Middle Pennsylvanian basidiomycete mycelium with clamp connections. *Mycologia* 62:578–584.

Guzmán G., Allen J.W., & Gartz J. 2000. A worldwide geographical distribution of the neurotropic fungi, analysis and discussion. *Ann. Mus. Civ. Rovereto* 14(1,998):189–280.

Harrison M.J. 2005. Signaling in the arbuscular mycorrhizal symbiosis. *Ann. Rev. Microbiol.* 59:9–42.

Hibbett D.S. et al. 2007. *Mycol. Res.* 111:509–547.

Hoffmann K., Voigt K., & Kirk P.M. 2011. *Mycotaxon* 115: 353–363.

Humber, R.A. 2012. Entomophthoromycota: A new phylum and reclassification for entomophthoroid fungi. *Mycotaxon* 120:477–492.

James T.Y. et al. 2006. A molecular phylogeny of the flagellated fungi (Chytridiomycota) and description of a new phylum (Blastocladiomycota). *Mycologia* 8(6):860–871.

Javaux E.J., Knoll A.H., & Walter M.R. 2001. Morphological and ecological complexity in early eukaryotic ecosystems. *Nature* 412:66–69.

Kirk P.M., Cannon P.F., David J.C., & Stalpers J. 2001. *Ainsworth and Bisby's dictionary of the fungi.* 9th ed. Wallingford, UK: CAB International.

Kluge M., Mollenhauer D., Wolf E., & Schüßler A. 2002. The Nostoc-Geosiphon endosymbiosis. *In* Rai A.N., Bergman B., & Rasmussen U. (eds.), *Cyanobacteria in symbiosis.* Dordrecht, the Netherlands: Kluwer Academic Publishers, pp. 19–30.

Korzybski A. 1931. A non-Aristotelian system and its necessity for rigour in mathematics and physics. Address given at the American Association for the Advancement of Science annual meeting in New Orleans, Louisiana, December 28, 1931. Reprinted in *Science and Sanity*, 1933, International Non-Aristotelian Library Publishing Company, pp. 747–761.

Kuhn G., Hijri M., & Sanders I.R. 2001. Evidence for the evolution of multiple genomes in arbuscular mycorrhizal fungi. *Nature* 414:745–748.

Longcore J.E., Pessier A.P., & Nichols D.K. 1999. *Batrachochytrium dendrobatidis* gen. et sp. nov., a chytrid pathogenic to amphibians. *Mycologia* 91:219–227.

Mayer J.H. et al. 1971. Esophageal rupture after mushroom and alcohol ingestion. *N. Engl. J. Med.* 285:1323.

Min S. et al. 2012. *Symbiosis* 57:1–13.

Palaeos. 2003. Eukarya: Glossary C–E. Palaeos.com/Eukarya/Lists/EuGlossaryC.htm.

Peter M.G. 2002. Chapter 5, Chitin and chitosan in fungi. *In* Steibüchel A. (ed.), *Biopolymers, polysaccharides from Eukaryotes.* Weinheim, Germany: Wiley-VCH, pp. 123–132.

Redecker D., Kodner R., & Graham L.E. 2000. Glomalean fungi from the Ordovician. *Science* 289:1920–1921.

Redecker D., Morton J.B., & Bruns T.D. 2000. Ancestral lineages of arbuscular mycorrhizal fungi (Glomales). *Mol. Phylogen. Evolut.* 14:276–284.

Remy W., Taylor T.N., Hass H., & Kerp H. 1994. Four-hundred-million-year-old vesicular arbuscular mycorrhizae. *Proc. Natl. Acad. Sci.* 91:11841–11843.

Sherwood-Pike M.A. & Gray J. 1985. Silurian fungal remains: Probable records of the class Ascomycetes. *Lethaia* 18:1–20.

Tasch P. 1980. *Paleobiology of the invertebrates.* New York: John Wiley and Sons. 975 pp.

Taylor J.W., Spatafora J., & O'Donnell K. 2004. Chapter 12, The fungi. *In* Cracraft J. & Donoghue M.J. (eds.), *Assembling the tree of life.* Oxford: Oxford University Press, pp. 171–179.

Taylor T.N., Hass H., Remy W., & Kerp H. 1995. The oldest fossil lichen. *Nature* 378:244.

Taylor T.N., Remy W., Hass H., & Kerp H. 1995. Fossil arbuscular mycorrhizae from the Early Devonian. *Mycologia* 87:560–573.

Threlkeld D.S. (ed.). March 1997. Central nervous system drugs, acetaminophen. *In Facts and comparisons drug information.* St. Louis, MO: Facts and Comparisons, pp. 247–f.

Threlkeld D.S. (ed.). April 1997. Central nervous system drugs, antidepressants, selective serotonin reuptake inhibitors. *In Facts and comparisons drug information.* St. Louis, MO: Facts and Comparisons, pp. 264q–4r.

Tyler S.A. & Berghoorn E.S. 1954. Occurrence of structurally preserved plants in Pre-Cambrian rocks of the Canadian Shield. *Science* 119:606–608.

Yin L. 1997. Acanthomorphic acritarchs from Meso-Neoproterozoic shales of the Ruyang Group, Shanxi, China. *Rev. Palaeobot. Palynol.* 98:15–25.

Yuan X., Xiao S., & Taylor T.N. 2005. Lichen-like symbiosis 600 million years ago. *Science* 308:1017–1020.

Mushroom Morphology

A Brief Pictorial Primer

Gill Distance Terminology

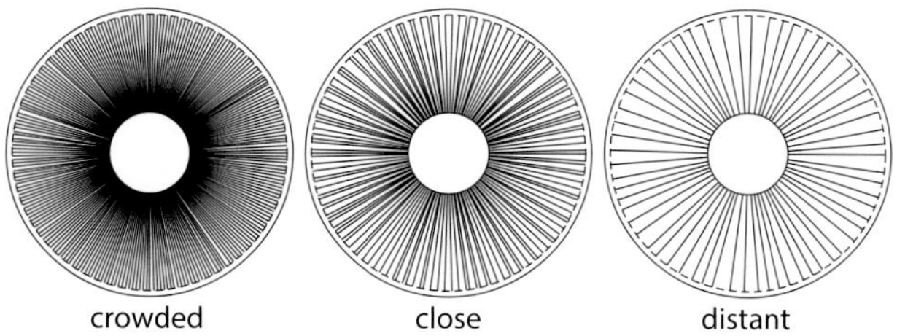

Types of Gill Attachments

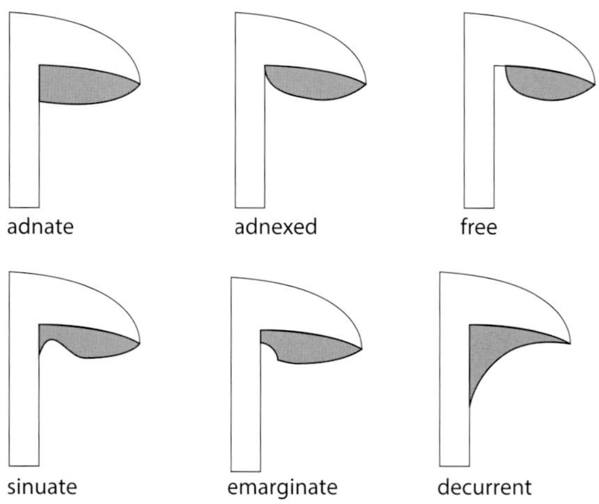

Basic Mushroom Shapes

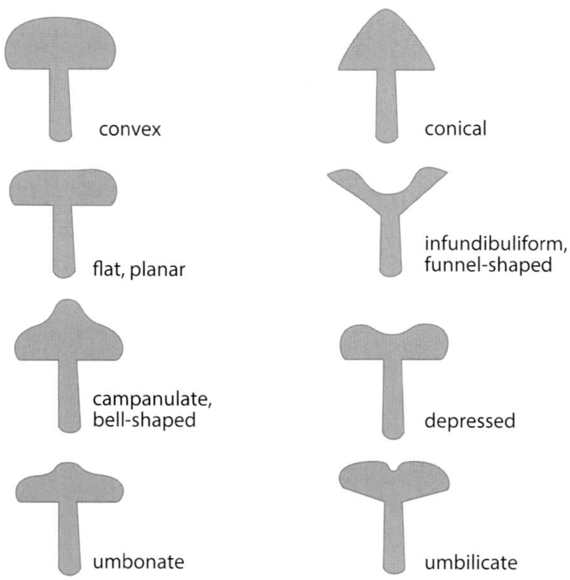

Modified from drawings by Debivort.

Some Spore Types

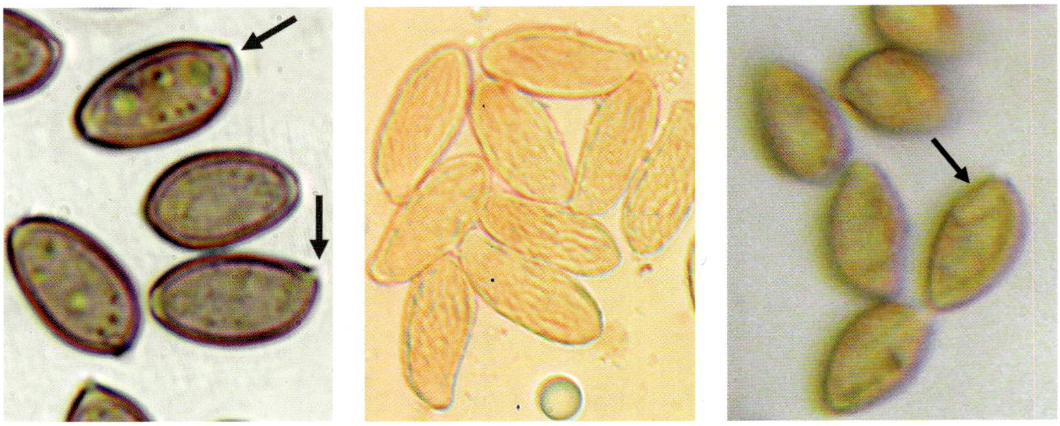

1 2 3

Figures: 1, elliptical, smooth, thick-walled, with apical germ pore (arrows), uniguttulate (with a single oil droplet, greenish circular areas); 2, oblong, inequilateral, longitudinally ridged or striate; 3, ovoid, slightly inequilateral, with subhilar plage (lighter flattened area, arrow). (Micrographs: 1, Alan Rockefeller; 2, Ron Pastorino; 3, Dan Molter.)

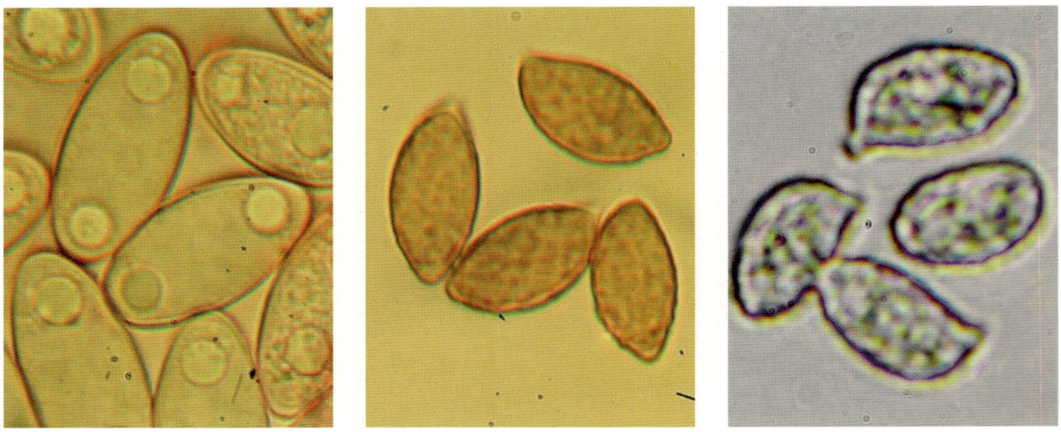

4 5 6

Figures: 4, ellipsoid to subfusiform, smooth, biguttulate (two oil droplets); 5, broadly spindle-shaped (football-shaped) to broadly elliptical, verrucose; 6, elliptical to drop-shaped, with offset apiculus, with irregularly lobed ridges. (Micrographs: 4 & 5, Ron Pastorino; 6, Alan Rockefeller.)

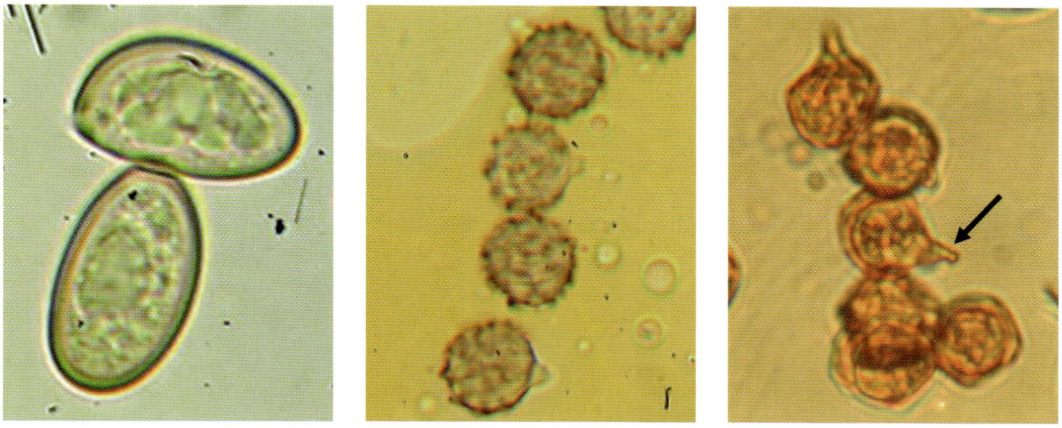

7 8 9

Figures: 7, elliptical, smooth; 8, globose, prominently verrucose (warty); 9, angular, with prominent apiculus or hilar appendage (arrow), uniguttulate (with oil droplet).

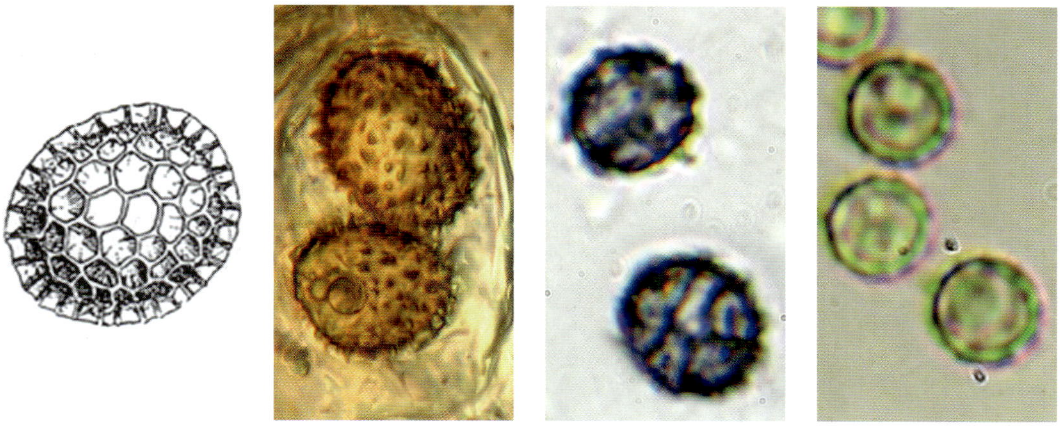

10 11 12 13

Figures: 10, subglobose, alveolate (honeycombed); 11, ovoid-elliptical, echinate (strongly spiny); 12, subglobose, warts and ridges forming a partial reticulum; 13, globose, echinulate (minutely spiny). (Images: 10, Drawing by C.L. Shear; 11, Matthew Smith, University of Florida; 12, Alan Rockefeller; 13, Ron Pastorino.)

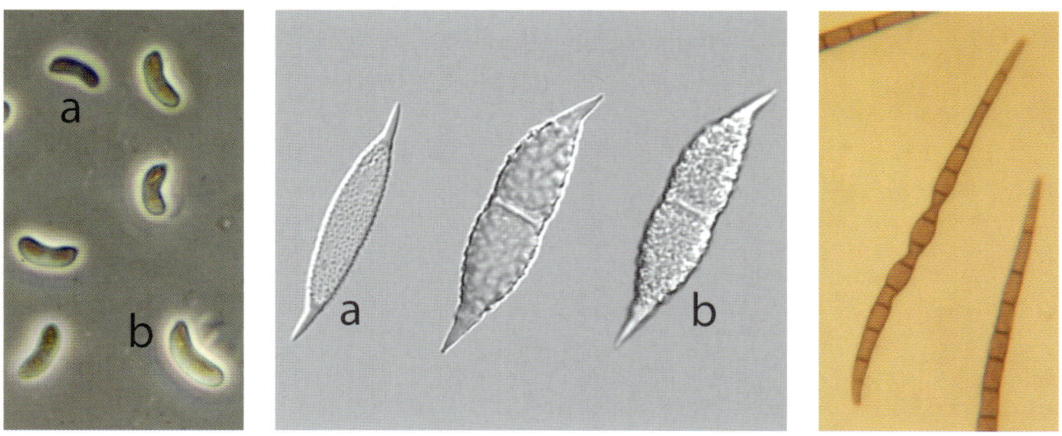

14 15 16

Figures: 14, allantoid (sausage-shaped and curved); 15a, fusiform (spindle-shaped), with sharp apiculi, aseptate, finely verrucose, (warted) aseptate; 15b, fusiform, with long, sharp apiculi, uniseptate, heavily verrucose (warted); 16, needlelike, multiseptate. (Micrographs: 14, unknown; 15, K. Põldmaa, D.F. Farr, & E.B. McCray, Hypomyces Online, Systematic Mycology and Microbiology, Laboratory, ARS, USDA; 16, W. Light.)

Dichotomous and Pictorial Keys to the Major Mushroom Forms

This key is not exhaustive and is intended only as a general guide to the basic types and categories of mushrooms for those new to mushrooming. Many atypical mushrooms do not fit the general shape and fruitbody form of their group (e.g., the Black-footed Polypore and the Yellow-gilled Bolete) and will not key out here. A number of polypores, for example, are more typically mushroom-shaped than bracket-like or shelf-like. Similarly, many gilled mushrooms have sessile shelf-like fruitbodies rather than the typical "umbrella on a stem" one normally expects. Alas, the exceptions to the rules comprise a significant proportion of most mushroom groups. This includes both the natural (biologically related) taxonomic groups (e.g., the amanitas, the russulas and milkcaps, and the boletes) and the artificial categories to which many biologically unrelated mushrooms have been assigned on the basis of superficial resemblances (e.g., the jelly fungi, the cup and disk fungi, and the club and coral fungi). The figure numbers in the verbal dichotomous key refer to the collection of images that immediately follow.

1\. Fruitbody a pileus (cap) with thin gills (Fig. 1), thick gill-like ridges (Fig. 2), pores (Fig. 3) or maze-like walls (Figs. 4 & 5), or teeth (Figs. 6 & 7), with or without a stipe 3

1′. Fruitbody otherwise 2

2\. Fruitbody: a puffball (Fig. 8), earthstar (Fig. 9), phallus-shaped (Fig. 10), morel-like (Figs. 11 & 12), cup-like (Fig. 13) or nest-shaped with "eggs" (Fig. 14), club-shaped (Fig. 15), periscope-shaped (Fig. 16) to spoon-shaped (Fig. 17), or branching and coral-like (Figs. 18 & 19) 8

2′. Fruitbody otherwise 11

3\. Fruitbody: shaped like a classical "umbrella-on-a-stick" mushroom, pileus with gills (Fig. 20), stipe present or absent . . . 4

3′. Fruitbody as above or variously shaped (e.g., shelf-like), pileus with fleshy ridges, pores, maze-like networks, or teeth, stipe present or absent 5

4\. Stipe brittle, snaps cleanly in two like chalk . . . . *Russula*/*Lactarius* (p. 418)

4′. Stipe flexible, does not snap cleanly in two . . . . most agarics (p. 265)

5\. Fruitbody highly variable, vase-like or deeply depressed (Fig. 21), trumpet-like (Fig. 22) to mushroom-shaped (e.g., Figs. 20, 23), with thick, blunt, decurrent (running down stipe) ridges . . . . chanterelle-type mushrooms (p. 445)

5′. Fruitbody with pores, daedeloid (maze-like), or teeth; stipe present or absent 6

6\. Cap with teeth . . . . hydnoid fungi (p. 573)

6′. Cap with pores, daedeloid, or daedeloid mixed with pores and/or teeth 7

7\. Cap fleshy, mushroom-like, with central stipe and pores beneath in place of gills (Fig. 23) . . . . boletes (p. 222)

7′. Cap typically shelf-like (Fig. 24), fleshy, leathery, or woody, but may be mushroom-shaped (Fig. 25); with eccentric stipe, fused basal stipes, or stipe absent; typically with pores (Fig. 25), but pores may form a maze-like web or elongate to resemble gills (Fig. 5) or teeth (Fig. 6) 16

8. Fruitbody round, oval, or pear-shaped (Fig. 8), soft when young, papery and brittle when older and releasing dark, powdery spores, with or without a star-shaped exoperidium (outer layer) at base (Fig. 9) . . . . . puffballs and earthstars (p. 186)

8′. Fruitbody otherwise 9

9. Fruitbody blackish-brown to blackish, hard, charcoal-like round (Fig. 26), cushion-shaped (Fig. 27), or finger-like (Fig. 28) . . . . carbonaceous fungi (Xylariales) (p. 70)

9′. Fruitbody otherwise 10

10. Fruitbody vase- or cup-shaped (Fig. 13), or urn-shaped (Fig. 29), or resembling tiny bird's nest (Fig. 14) . . . . cup and disk fungi (p. 168)

10′. Fruitbody otherwise 11

11. Fruitbody phallus-shaped (Fig. 10), with slimy tip, basal volva (cup), often having a very disagreeable smell . . . . stinkhorns (p. 212)

11′. Fruitbody otherwise 12

12. Fruitbody roughly conical, pinecone-like or saddle-shaped, with prominent ridges and pits (Fig. 11), wrinkled or smooth; with central stalk . . . . morels and false morels (p. 61)

12′. Fruitbody otherwise 13

13. Fruitbody amorphous, jelly-like, white, yellow, purplish, or blackish (Fig. 30) . . . . jelly fungi (p. 99)

13′. Fruitbody otherwise 14

14. Fruitbody club- or coral-shaped, with one or more erect finger-like projections (Figs. 18 & 19), may be united into a fused base 15

14′. Fruitbody more or less flat and crust-like (Fig. 31), with (Fig. 32) or without uplifted margins, or shoehorn-shaped (Fig. 33) . . . . crusts, spreads, and parchment fungi (p. 467)

15. One or more long, club-like, finger-like projections emerging from a buried false truffle or dead arthropod (Figs. 34 & 35) . . . . *Cordyceps* (*sensu lato*) (p. 149)

15′. Fruitbody club-like or coral-like, not emerging from buried false truffle or dead arthropod . . . . club and coral fungi (p. 115)

16. Fruitbody shelf-like (Fig. 24) or hoof-shaped (Fig. 36), stipe lateral and typically lacking or rudimentary, fertile surface typically smooth (Fig. 37), with pores (Fig. 3), daedeloid channels (Fig. 5), parallel lamellae (Fig. 38), or flattened teeth (Fig. 39) . . . . typical polypores (p. 493)

16′. Fruitbody otherwise 17

17. Fruitbody compound, cabbage-like (Fig. 40) . . . . cauliflower mushrooms, black-staining and umbrella polypores (p. 523)

17′. Fruitbody typically with a well-developed stipe (Fig. 41) 18

18. Lower surface of pileus smooth or with pores . . . . reishis and other stalked polypores (p. 501)

18′. Lower surface with teeth (Fig. 42) . . . . stipitate toothed mushrooms (p. 575)

Pictures for the Dichotomous Key

1

2

3

4

5

6

7

8

9

10

11

12

13

14

15

16

17

18

19

20

21

22

23

24

25

26

27

28

29

30

31

32

33

34

35

36

37

38

39

40

41

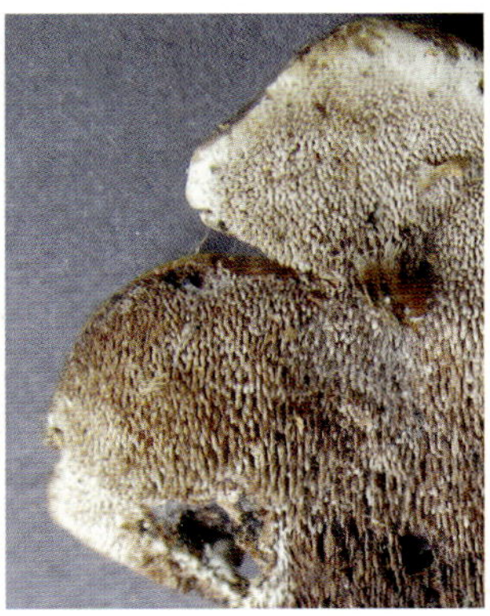
42

Species Accounts

Morels, False Morels, and Elfin Saddles

Morels, false morels, and elfin saddles are superficially similar but are not closely related. They belong to three separate families in the order Pezizales. Although morels and certain false morels are widely sought and consumed as choice edibles, all three groups have caused severe poisonings in Europe and North America. This is discussed a little further below under each group separately.

The family Mochellaceae includes the true morels (*Morchella* spp.), the verpas (*Verpa* spp.), and certain cup fungi in the genus *Disciotis* (not found in our area). The members of this family are characterized by fruitbodies that are completely hollow when sliced in half, caps with a system of deep pits and ridges, and smooth, multinucleate (20–60 nuclei), aguttulate (not containing oil droplets) ascospores.

The North American species of *Morchella* have recently undergone a major taxonomic revision,[1] which distinguished three major clades (evolutionary lineages): (1) the Rufobrunnea Clade, containing the sole species *M. rufobrunnea*, restricted to the West Coast; (2) the Esculenta Clade, corresponding to the traditional group of "yellow morels"; and (3) the Elata Clade, corresponding to the traditional "black morels." None of the European species of *Morchella* whose names have traditionally been applied to North American morels actually occurs on this continent. This is discussed in detail in the species account of the North American Yellow Morel (*M. esculentoides*).

Morels contain a heat-labile hemolysin—a toxin that destroys red blood cells—that must be eliminated by thorough cooking before they can be safely eaten. And even properly cooked morels cause severe but non-life-threatening gastrointestinal distress for some people, especially when consumed with alcohol. The consumption of large amounts of morels in several consecutive meals can also cause a poorly understood neurological syndrome. This is discussed in more detail in the species account of the North American Yellow Morel (*Morchella esculentoides*).

The family Discinaceae includes five genera, of which the best known are the false morels, or lorchels, of the genus *Gyromitra*. Unlike *Morchella*, the stipe of *Gyromitra* species is not hollow but is partially to completely filled with a chambered flesh or a cottony mycelium, the cap is convoluted and brain-like rather than pitted and ridged, and the ascospores are uni- to multiguttulate (contain one or more large or small oil droplets). Also unlike most *Morchella*, the cap is attached to the stipe only at its apex.

The elfin saddles, or helvellas, are now in the family Helvellaceae, which also includes five other minor genera and is most closely related to the true truffles (Family Tuberaceae). Species of the genus *Helvella* have highly irregular caps that are typically bilobately saddle-shaped (or even trilobate) at some point in their growth cycle but may be cup- or saucer-shaped. Like the false morels, their caps are attached to the stipe only at the top, and their spores are guttulate and contain oil droplets.

False morels and many elfin saddles contain dangerous levels of gyromitrin or related hydrazine-based toxins associated with both acute toxicity and cancer. Severe poisonings have occurred with all three groups of mushrooms, including undercooked morels or morels that have been consumed in large quantities over three or more consecutive meals. This is discussed in the account of the North American Yellow Morel. False morel poisonings are frequently fatal and account for 2%–4% of all fatal mushroom poisonings in Europe. Gyromitrin is discussed in detail in the introduction under the heading "Mushroom Toxins and Mushroom Poisoning."

1. M. Kuo et al., 2012, *Mycologia* 104(5):1159–1177.

Key to the Southeastern Genera of Morels, False Morels, and Elfin Saddles

1. Stipe and cap hollow throughout, never filled with cottony hyphae, cap attached by one-half or all its length, ornamented with deep pits and ridges . . . . *Morchella*

1′. Stipe and fruitbody not hollow throughout, cap attached only at its apex, without deep pits and ridges; typically (1) bell-shaped, (2) deeply and irregularly lobed, convoluted, brain-like, (3) saddle-shaped to irregularly lobed or trilobate, (4) cup-shaped, or (5) thimble- to bell-shaped 2

2. Cap thimble- or bell-shaped, smooth or finely wrinkled . . . . *Verpa conica*

2′. Cap deeply lobed and brain-like, saddle-shaped, trilobate, or cup-shaped 3

3. Cap deeply irregularly lobed, convoluted, brain-like . . . . *Gyromitra* (in part)

3′. Cap saddle-shaped, like a double bishop's cap, trilobate, or cup-shaped 4

4. Cap wide, cup- or saucer-shaped . . . . *Helvella macropus*

4′. Cap saddle-shaped, like a double bishop's hat, or trilobate 5

5. Cap uniformly reddish-brown to dark brown . . . . *Gyromitra infula*

5′. Upper surface of cap pale cream, tan, pale greyish-tan, greyish-brown, or brown; lower surface creamy white, white, to pale tan or buff . . . . *Helvella* spp.

North American Yellow Morel

Morchella esculentoides

Morchella esculentoides. (Photos: A & D, Teresa Fortenberry; B, Ken Zinkand; C, W. Light.)

Morchella esculentoides **Kuo, Dewsbury, Moncalvo & S.L. Stevenson**, in Kuo et al., *Mycologia* 104(5):1163 (2012) [Ascomycota, Pezizomycotina, Pezizomycetes, Pezizomycetidae, Pezizales, Morchellaceae]

Misapplied name: *Morchella esculenta* [*non Phallus esculenta* Linnaeus (1753)]

Derivation: *Morchella*, origin uncertain, possibly from the German word for morel, *Morchel*; *esculentoides*, Latin, from *esculentus*, "edible," plus the Latin suffix -*oideus*, from the Greek word *eidos*, meaning "resembling," "having the form of"; epithet refers to the similarity with the European *Morchella esculenta* (Linnaeus) Persoon (1801), with which it was long confused.

Cap: 2.3–11 (up to 22) cm high by 1.5–4.2 (up to 12.6) cm wide at widest point (1–4½[–9] × ⅔–1⅔[–3⅔] in); cap hollow, typically ovoid or egg-shaped with bluntly conical apex (Figs. A & C), occasionally subcylindrical with convex tip (Fig. B), young specimens sometimes subglobose; hymenophore covered with deep, randomly arranged pits and ridges, with whitish or pale yellowish, bluntly rounded ridges and greyish- to dark-brown or blackish pits when young (Fig. C), becoming sharper and yellowish to brownish-yellow, with similar-colored pits in maturity (Figs. A, B, & D); base of cap attached directly to stipe with no groove.

Flesh: Whitish, 1–3 mm thick in the cap (hymenophore), thicker in stipe and often chambered and/or pitted at the base.

Stipe: 2–12 cm tall by 1.5–10 cm wide (⅔–4 in), occasionally larger; base typically expanded (Figs. A & B), whitish (Fig. B) to pale yellowish (Figs. A & D), smooth (Fig. A) or covered with fine granules (Fig. D).

Microscopic: Spores (ascopores) (17–)18–22(–24) × 11–13 (–15) µm, smooth, elliptical; asci 8-spored.

Occurrence: Variously saprobic and mycorrhizal depending on stage in life cycle; solitary, scattered, or gregarious in various mixed broadleaf forest ecosystems; beneath various hardwoods, especially Green Ash (*Fraxinus pennsylvanica*), dead American Elm (*Ulmus americana*), apple orchards, and occasionally conifers; widespread and fairly common in Southeast, March–June, when soil temperatures reach 12.7°C (55°F) for "grey morels" (actually immature Yellow Morels) and 15.5°C (60°F) for larger, mature Yellow Morels.

Edibility: Edible, choice; highly desired and sought after, but requires thorough cooking (see comments below).

Comments: Morels contain a heat-labile hemolysin, a toxin that destroys red blood cells, which must be eliminated by thorough cooking before they can be safely eaten. Even well-cooked morels have caused severe gastrointestinal distress (nausea, vomiting, diarrhea, cramps, and other symptoms) for some people, especially when consumed with alcohol. This is quite distinct from the Antabuse®-like reaction that occurs with alcohol and coprine-containing mushrooms (certain species of coprinoids in the genera *Coprinus* and *Coprinopsis*). Furthermore, only very fresh, unspoiled mushrooms should be consumed. When consumed in large quantities morels can also cause a poorly understood neurological syndrome characterized by tremor, dizziness/inebriation, drowsiness/confusion, unsteadiness, and/or visual disturbances.[1] Despite being considered one of the choicest of edible mushrooms, non-life-threatening morel poisonings have been reported in the medical literature of Europe and North America for many decades. Morels found in old apple orchards may contain toxic levels of lead arsenate and should not be consumed.[2]

This highly prized and much-sought-after edible mushroom has long been known as *Morchella esculenta* (Linnaeus) Persoon and is still described as such in most field guides. However, despite a strong superficial resemblance, the European *M. esculenta* does not occur in North America. In fact, none of the European species of *Morchella*—*M. conica*, *M. crassipes*, *M. deliciosa*, *M. esculenta*, *M. elata*, or *M. semilibera*—whose names have long been applied to North American morels actually occurs here. Only 2 of the 41 species of *Morchella* occur naturally in both Europe and Asia, and none are shared between North America and Asia.

North American species of *Morchella* have recently undergone a major taxonomic revision by Kuo et al.,[3] who described five new species of yellow morels, including *M. esculentoides* (Esculenta Clade), and nine new species of black morels (Elata Clade). The Yellow Morel of North America was formally described and named as a new species, *Morchella esculentoides*, by Kuo et al. in 2012. DNA studies have shown that the eastern grey morels with pale ridges and randomly oriented blackish pits (Fig. C) formerly thought to be a distinct species are, in fact, immature *M. esculentoides*.[4]

Four other *Morchella* species occur in our area and are described below: two yellow morels (Esculenta Clade)—*M. diminutiva* Kuo, Dewsbury, Moncalvo & S.L. Stephenson (2012) and *M. virginiana* O'Donnell & S.A. Rehner (2012); and two black morels (Elata Clade)—*M. angusticeps* Peck (1887) (previously called *M. elata* Fries) and *M. punctipes* Peck (1903) (previously called *M. semilibera* De Chambre *ex* Fries). All have pits and ridges arranged more or less vertically (Fig. E) instead of randomly, as in *M. esculentoides*.

Morchella diminutiva is much smaller than *M. esculentoides*, its fruitbody only 9.5 cm (3¾ in) tall or less. Its conical cap is only 2–4 cm high by 1–2.7 cm wide (¾–1⅔ × ⅖–1 in) at its widest point, and when young it often has a slight sinus (gap) where it attaches to the stipe. It occurs in eastern North American hardwood forests under White Ash (*Fraxinus americana*), rather than Green Ash, and Tulip Tree (*Liriodendron tulipifera*) but is also found under hickories (*Carya* spp.) and other hardwoods.

Morchella virginiana is another small-capped (cap 3–7 cm high by 2.2–3.5 cm wide [⅘–1½ in]) yellow morel restricted to the Southeast and associated exclusively with Tulip Trees in riparian and upland ecosystems in the northern regions of our area.

Morchella angusticeps—"the classic North American Black Morel"—is slightly smaller than *M. esculentoides*; its distinctively conical to bluntly conical cap is 3–8 cm tall by 2.5–5 cm wide (1¼–3¼ × 1–2 in) and is attached to the stipe with a distinct sinus about 2–5 mm deep by 2–5 mm wide, which may become obscured at maturity. The vertically oriented ridges are tan to dark brown, becoming black at maturity, and the pits are pale tan to brownish yellow, sometimes tinged with olive. It is associated with several hardwood trees but particularly White Ash and Tulip Tree.

The Half-free Morel, *M. punctipes* (Fig. E), cannot be mistaken for anything else. Only the top half of the tiny

E *Morchella punctipes.* (Photo: Jason Hollinger.)

2–4.5 cm tall and wide (¾–1¾ in) cap is attached to the stipe, thus creating a prominent skirt or overhang. The mature cap exhibits vertically arranged greyish-tan to yellowish-brown shallow pits bordered by anastomosing dark-brown to brownish-black ridges. When mature the fragile stipe is disproportionately tall for its diminutive cap: 1.5–15 cm (⅔–6 in). In the Southeast it occurs in northern and montane hardwood forests containing White Ash, Tulip Tree, and American Elm.

The dangerously poisonous false morels (*Gyromitra* spp.) can be mistaken for morels, which they superficially resemble.

1. S. Berndt, 2010, Neurologisches Syndrom nach Morchelgenuss, *Zeitschrift für Mykologie* 76(1):7–12; P. Saviuc et al., 2010, *Clinical Toxicology* 48(4):365–372.

2. E. Shavit & E. Shavit, 2010, *Fungi* 3(2): 11–18.

3. M. Kuo et al., 2012, *Mycologia* 104(5):1159–1177.

4. Molecular studies (K. O'Donnell et al., 2001, *Fungal Genet. Biol.* 48:252–265) indicate that the family Morchellaceae originated in the Middle Triassic period approximately 243 million years ago (Mya), with the genus *Morchella* diverging in the early Cretaceous period about 130 Mya. Both radiations occurred in western North America, with dispersal to Asia and Europe (then still part of the united paleocontinent Laurasia) taking place across the Bering and North Atlantic land bridges, respectively. The morels as a group exhibit a high degree of continental endemism, with very little or no sharing of species between continents.

Key to the Southeastern Species of *Morchella*

1. Cap attached to stipe by only its upper half, leaving a substantial free-hanging skirt . . . . *M. punctipes*

1′. Cap attached throughout its length to stipe, skirt lacking 2

2. Pits and ridges disorganized, randomly arranged . . . . *M. esculentoides*

2′. Pits and ridges mostly arranged vertically 3

3. Cap attached to stipe by distinct sinus or space up to 5 mm deep and wide (sometimes obscured with age); mature ridges dark brown or black . . . . *M. angusticeps*

3′. Cap adnate where attached or with slight sinus when very young; mature ridges brownish-yellow to yellowish-brown 4

4. Fruitbody no more than 9.4 cm (3¾ in) tall; cap conical to subconical, no more than 4 cm (1⅔ in) tall; associated with ashes, tulip trees, and other hardwoods . . . . *M. diminutiva*

4′. Fruitbody 5–12.5 cm (2–5 in) tall; cap ovoid (egg-shaped), with bluntly conic apex, 3–7 cm (1¼–2¾ in) tall; associated exclusively with tulip trees . . . . *M. virginiana*

False Morels and Lorchels

Gyromitra caroliniana, G. esculenta, G. infula

A B C

A, *Gyromitra caroliniana*; B, *G. esculenta*; C, *G. infula*. (Photos: A, Ron Pastorino; B, Severine Meißner; C, Robert Sasata.)

***Gyromitra caroliniana* (Bosc) Fries**, *Öfvers. K. Svensk. Vetensk.-Akad. Förhandl.* 28:173 (1872)

Basionym: *Morchella caroliniana* Bosc, *Mag. Gesell. Naturf. Freunde, Berlin* 5:86, t. 5:6 (1811)

***Gyromitra esculenta* (Persoon) Fries**, *Summa Veg. Scand.*, Section Post. (Stockholm):346 (1849)

Basionym: *Helvella esculenta* Pers., *Comm. Schaeff. Icon. Pict.*:64 (1800)

***Gyromitra infula* (Schaeffer) Quélet**, *Enchir. Fung.* (Paris):272 (1886)

Basionym: *Helvella infula* Schaeffer, *Fung. Bavar. Palat.* 4:105 (1774)
[Ascomycota, Pezizomycotina, Pezizomycetes, Pezizomycetidae, Pezizales, Discinaceae]

Derivation: *Gyromitra*, Greek, *gyros*, "round," and *mitra*, "headband"; *caroliniana*, Latin, "from Carolina"; *esculenta*, Latin, "edible"; *infula*, Latin, "fillet," "headband," "badge of honor."

Cap: ***G. caroliniana***—more or less irregularly round, convoluted, and brain-like, usually 5–18 cm (2–7¼ in) wide and tall (but may grow much larger), reddish-brown; undersurface whitish to greyish, hidden beneath fertile outer layer. ***G. esculenta***—convoluted and brain-like, 4–8 cm tall by 4–12 cm wide (1⅔–3¼ × 1⅔–4¾ in), pinkish-tan to reddish-brown, aging darker to blackish; undersurface whitish to tan, not usually exposed. ***G. infula***—prominently bilobate, saddle-shaped, occasionally trilobate, smooth to wrinkled but not convoluted or brain-like, 2–13 cm tall by 2–10 cm wide (¾–5¼ × ¾–4 in), yellowish-brown, reddish-brown to dark brown; undersurface whitish to brownish.

Flesh: ***G. caroliniana***—whitish to greyish, brittle, densely packed throughout cap and stipe, with many pockets and chambers. ***G. esculenta***—brittle, whitish to tan, sometimes chambered. ***G. infula***—very thin, brittle, whitish to brownish.

Stipe: ***G. caroliniana***—stipe thick, surface vertically ridged, massive in proportion to cap, flesh dense, with many cavities. ***G. esculenta***—stipe smooth to slightly granular, hollow or filled with cottony hyphae, not ridged, sometimes with chambers. ***G. infula***—hollow, flesh very thin (1–2 mm), surface covered with fine hairs.

Microscopic: Spores (ascospores): ***G. caroliniana***—22–35 × 10–16 µm, ornamented with ridges and projections, containing one large and two to three smaller oil droplets.

G. esculenta—19–28 × 10–13 µm, smooth, usually with two small oil droplets. ***G. infula***—17–24 × 7–11 µm, smooth, two large (sometimes one or three) oil droplets.

Occurrence: Saprobic, but likely also mycorrhizal; solitary, scattered, or in groups; occasional. ***G. caroliniana***—on ground under hardwoods and mixed woods, spring (March–June). ***G. esculenta***—on ground under conifers, especially White Pine, spring (April–June). ***G. infula***—saprobic on decaying wood or humus, in mountains, fall (July–October).

Edibility: All are dangerously poisonous and potentially deadly! *Gyromitra esculenta* is commonly consumed despite the danger, especially in Europe, sometimes with disastrous and even fatal results. See comments.

Comments: *G. infula* is unique for *Gyromitra* species in usually fruiting in fall (July–October) rather than spring (March–June).

Gyromitra species contain a potentially deadly toxin called gyromitrin (*N*-methyl-*N*-formylhydrazone). Gyromitrin is hydrolyzed in the body into monomethylhydrazine (MMH), which was and is used as a rocket fuel in orbital control systems and space shuttles and to maneuver lunar landing modules. MMH in even small amounts is both a carcinogen and a highly toxic metabolite that affects the liver, the central nervous system, and the kidneys.[1] False morel poisoning can result in pulmonary edema, seizures, and coma. Stomach cramps, vomiting, and diarrhea typically begin a few hours after consumption, followed by lethargy and delirium. Severe poisoning can lead to enlarged liver and spleen, jaundice, destruction of red blood cells leading to renal failure, coma, circulatory collapse, and even death within a week. Gyromitrin poisonings are relatively infrequent in the United States but are more common in Europe. *Gyromitra esculenta* is the main culprit in both regions. The mushrooms also appear to vary in toxicity with both season and altitude, and poisonings by *G. esculenta* appear to be much more severe and dangerous in the Midwest than in western North America.

False morels are generally parboiled twice in an attempt to drive off the volatile gyromitrin and related hydrazine derivatives, with the water being discarded each time. However, the volatile fumes given off by this process are readily inhaled and can also cause poisoning. Furthermore, sufficient toxins to cause poisoning from ingestion can also remain in the mushrooms after the double boiling. Unless you have an overt or latent death wish and are prone to such life-threatening games as Russian roulette, these mushrooms should never be consumed! Despite the obvious risk, false morels are often consumed by knowledgeable mycophiles in Europe and the Great Lakes region of America who consider them choice edibles. They are also inadvertently consumed by amateurs who mistake them for their true morel look-alikes.

1. MMH is further metabolized into *N*-methyl-*N*-formylhydrazine, which in turn produces nitrosamides via cytochrome P_{450} microsomal enzymes. These nitrosamides generate methyl radicals. leading to liver and renal necrosis.

Key to the Southeastern Species of *Gyromitra*

1. Cap irregularly lobed, convoluted, brain-like; fruiting in spring 2

1′. Cap saddle-shaped, like a bishop's cap, or trilobate; fruiting in summer and fall . . . . . *G. infula*

2. On ground in deciduous and mixed woods; stipe thick, massive in proportion to cap, surface vertically ridged, flesh dense, with many cavities . . . . . *G. caroliniana*

2′. On ground under conifers; stipe not massive in proportion to cap, stipe smooth, not ridged, hollow or filled with cottony hyphae . . . . . *G. esculenta*

Elfin Saddles and Helvellas

Helvella albella, H. crispa, H. elastica, H. macropus

A B C

A, *Helvella crispa*; B, *H. elastica*; C, *H. macropus*. (Photos: A, W. Light; B, John Plischke III; C, Holger Krisp, Ulm, Germany.)

Helvella albella **Quélet** [as "Elvella"], *Compt. Rend. Assoc. Franç. Avancem. Sci.* 24(2):621 (1896)

Helvella crispa **(Scopoli) Fries**, *Syst. Mycol.* (Lundae) 2(1):14 (1822)

Basionym: *Phallus crispus* Scopoli, *Fl. Carniol.*, 2nd ed. (Wien) 2:475 (1772)

Helvella elastica **Bulliard**, *Herb. Fr.* 6:tab. 242 (1785)

Helvella macropus **(Persoon) P. Karsten**, *Bidr. Känn. Finl. Nat. Folk* 19:37 (1871)

Basionym: *Peziza macropus* Persoon, *Observ. Mycol.* (Lipsiae) 1:26 (1796)
[Ascomycota, Pezizomycotina, Pezizomycetes, Pezizomycetidae, Pezizales, Helvellaceae]

Derivation: *Helvella*, Latin, "potherb"; *albella*, Latin, diminutive of *alba*, "little white one"; *crispa*, Latin, "wrinkled"; *elastica*, Latin, "elastic"; *macropus*, Greek, "large-footed."

Cap: ***H. albella***—1–2.5 cm across (~½–1 in), bilobately saddle-shaped, margin rolled over upper surface when young, expanding and flaring when mature, free from stipe; upper (fertile) surface greyish-brown to dark brown; lower (sterile) surface finely pubescent at first, becoming smooth, greyish-white to very pale brown or beige. ***H. crispa***—2–6 cm (¾–1⅔ in) across, saddle-shaped to irregularly lobed and convoluted (Fig. A); margin initially inrolled, later expanding and curling upward; upper surface smooth to slightly wrinkled, whitish to pale cream or pale buff; undersurface finely hairy, concolorous with upper surface. ***H. elastica***—2–6 cm (¾–1⅔ in) across, loosely saddle-shaped to irregular, lobes deeply convex, margins strongly infolded over lower surface, often fusing in maturity (Fig. B, arrow); upper surface smooth, tan, brown or greyish-brown; lower (sterile) surface smooth, white to pale tan. ***H. macropus***—1.5–4 cm across, cup- or disk-shaped to sometimes almost flat; upper surface medium grey to dark greyish-brown, smooth (Fig. C); undersurface lighter greyish or greyish-brown, hairy and/or covered with scurfy scales, especially toward margin.

Stipe: ***H. albella***—3–5.4 cm tall by 4–8 mm thick (1¼–2¼ × 1/10–⅓ in), hollow (ribbed inside), often slightly enlarged toward base, slightly chambered or pitted, smooth or with fine powdery coating; whitish. ***H. crispa***—3–10 cm tall by 0.5–3 cm thick (1¼–4 × ¼–1¼ in), tapering toward base and apex, deeply pitted and vertically ribbed with anastomosing crossribs (Fig. A); whitish to pale or pinkish buff. ***H. elastica***—3–11 cm tall by 0.3–1 cm thick (1¼–4½ × up to ⅖ in), equal, smooth, hollow, cream-colored (Fig. B). ***H. macropus***—2–5 cm tall by 1.5–5 mm thick (¾–2 × ~¼ in), solid, slightly expanded downward, pubescent to hairy (Fig. C); grey to grey-brown.

Microscopic: Spores (ascospores): ***H. albella***—20–24 × 12–15 µm, elliptical, smooth, clear, containing one large oil droplet and a variable number of small droplets at each end. ***H. crispa***—(14–)16–21(–23) × 10–14 µm, elliptical, smooth, containing one to five oil droplets. ***H. elastica***—18–24 × (10–)11–14(–15) µm, elliptical to oblong, smooth, with one large, central and zero to five smaller oil droplets. ***H. macropus***—19–28 × 10–13 µm, fusiform-elliptical, smooth to finely punctate, clear, with one large oil droplet and a smaller one at each end.

Occurrence: Both saprobic and likely mycorrhizal at different stages in life cycle; solitary and gregarious; late summer–fall (July/August–November/December). ***H. albella***—on ground in woods; occasional. ***H. crispa***—on ground in hardwood and conifer forests; occasional. ***H. elastica***—on ground under conifers or hardwoods, rarely from wood; fairly common. ***H. macropus***—on ground, often in moss, and on decaying wood in broadleaf and sometimes conifer woods; occasional.

Edibility: Not recommended. All species are suspect and potentially toxic, although *H. crispa* is eaten in Europe.

Comments: Many (and perhaps all) species of *Helvella* contain varying amounts of gyromitrin. They can and have caused severe poisonings in people who consume them, although others appear to eat them with impunity. As in the case of the false morels, gyromitrin intoxication is not an idiosyncratic or allergic reaction but is elicited in a direct dose-dependent manner.

Key to the Southeastern Species of *Helvella*

1. Cap weakly to strongly saddle-shaped, irregularly lobed, or trilobate 2

1′. Cap cup- or saucer-shaped . . . . . *H. macropus*

2. Cap and stipe uniformly pale cream to pale buff . . . . . *H. crispa*

2′. Upper surface of cap tan, greyish-tan, grey-brown to brown; lower surface creamy white, buff, or pale tan 3

3. Lower (sterile) surface of cap pubescent when young, margin initially rolled upward and inward over upper (fertile) surface . . . . . *H. albella*

3′. Lower surface never pubescent, margin never rolled over upper surface . . . . . *H. elastica*

The Carbonaceous Fungi

(Xylariales)

Many Xylariales (and all of those treated here) are saprobes or weak parasites on the wood of living or dead angiosperms. Anamorphs are prominent stages in the life cycle of these fungi and often form coremia on the stromata (fruitbodies).[1] In the case of the Cuban Club (*Xylaria cubensis*), the anamorphic Bubblegum Fungus (form-genus and form-species *Xylocoremium flabelliforme*) is the dominant life stage and is the form most often found in Nature.

For you die-hard fungal taxonomists, the order Xylariales is a highly diverse but monophyletic group of ascomycetous fungi with the following characteristics:

- well-developed superficial stromata;
- dark, leathery, woody, or carbonaceous, thick-walled perithecial (occasionally cleistothecial)[2] ascomata that are superficial or embedded in the stroma;[3]
- a persistent peripheral or basal layer of cylindrical, subglobose, or club-shaped, 8-spored (rarely 16-spored) asci usually bearing an amyloid (sometimes nonamyloid) apical ring;[4]
- typically brown to dark-brown (occasionally clear or lightly pigmented) ascospores with a germ slit or pore (lacking in one family) that may be unicellular, unequally bicellular, or transversely septate;
- asci that within a given perithecium mature at different times over a long period (when present, the anamorphs are typically hyphomycetous[5] and holoblastic).[6]

This order includes seven families:[7] the Amphisphaeriaceae, Apiosporaceae, Clypeosphaeriaceae, Diatrypaceae, Graphostomataceae, Hyponectriaceae, and Xylariaceae.

We are concerned with just two of them: the Xylariaceae and the Diatrypaceae. Of the taxa treated in this book, the Xylariaceae includes the following genera: *Xylaria* Hill *ex* Schrank; *Daldinina* Cesati & De Notaris; *Entonaema* A. Möller; *Hypoxylon* Bulliard and *Annulohypoxylon* Y.-M. Ju, J.D. Rogers & H.-M. Hsieh;[8] *Kretzschmaria* Fries; and the crust-like *Biscogniauxia* Kuntze. The Diatrypaceae contains the crust-like genus *Diatrype* Fries, as well as a number of other genera not discussed here.

Hypoxylon (*sensu lato*), *Daldinia*, and *Biscogniauxia* are typical members of the family Xylariaceae. They generally exhibit well-developed, superficial[9] stromata containing numerous embedded perithecia in a single layer (Fig. 1B, C, & E); asci bearing a discoid apical ring that usually stains blue with Melzer's reagent or other iodine solutions; and dark-colored, unicellular (occasionally unequally bicellular) ascospores with a germination slit of various lengths and configurations. Except for the genus *Biscogniauxia*, their stromata lack an outer dehiscing layer, that is, a layer that splits apart and falls away. *Hypoxylon* and *Daldinia* both form rounded, cushion-shaped or subglobose stromata, whereas *Biscogniauxia* forms a flattened whitish-grey or blackened crust. *Hypoxylon* differs from *Daldinia* in having homogeneous stromatal tissue below the perithecial layer; that of *Daldinia* consists of concentric zones of alternating carbonaceous perithecial and noncarbonaceous stromatal tissues (Fig. 1B). *Hypoxylon* and *Daldinia* both differ from *Biscogniauxia* in having pigments that are extractable with KOH.

Annulohypoxylon differs from *Hypoxylon* (*sensu stricto*) in having the perithecia enclosed in a carbonaceous layer of stromatal tissue (Fig. 1E); the ostioles (perithecial openings to the exterior) protrude above the stromatal surface and are encircled by a flattened disk-like ring. The perithecia of *Hypoxylon* are not encased in a carbonaceous layer, and the ostioles generally open below the stromatal surface (Fig. 1D).

Biscogniauxia species develop within (rather than upon the surface of) the bark of their host plants and possess a bipartite stroma with an outer juvenile layer that eventually dehisces along with the covering bark of the host plant to expose the inner layer of developing perithecia.

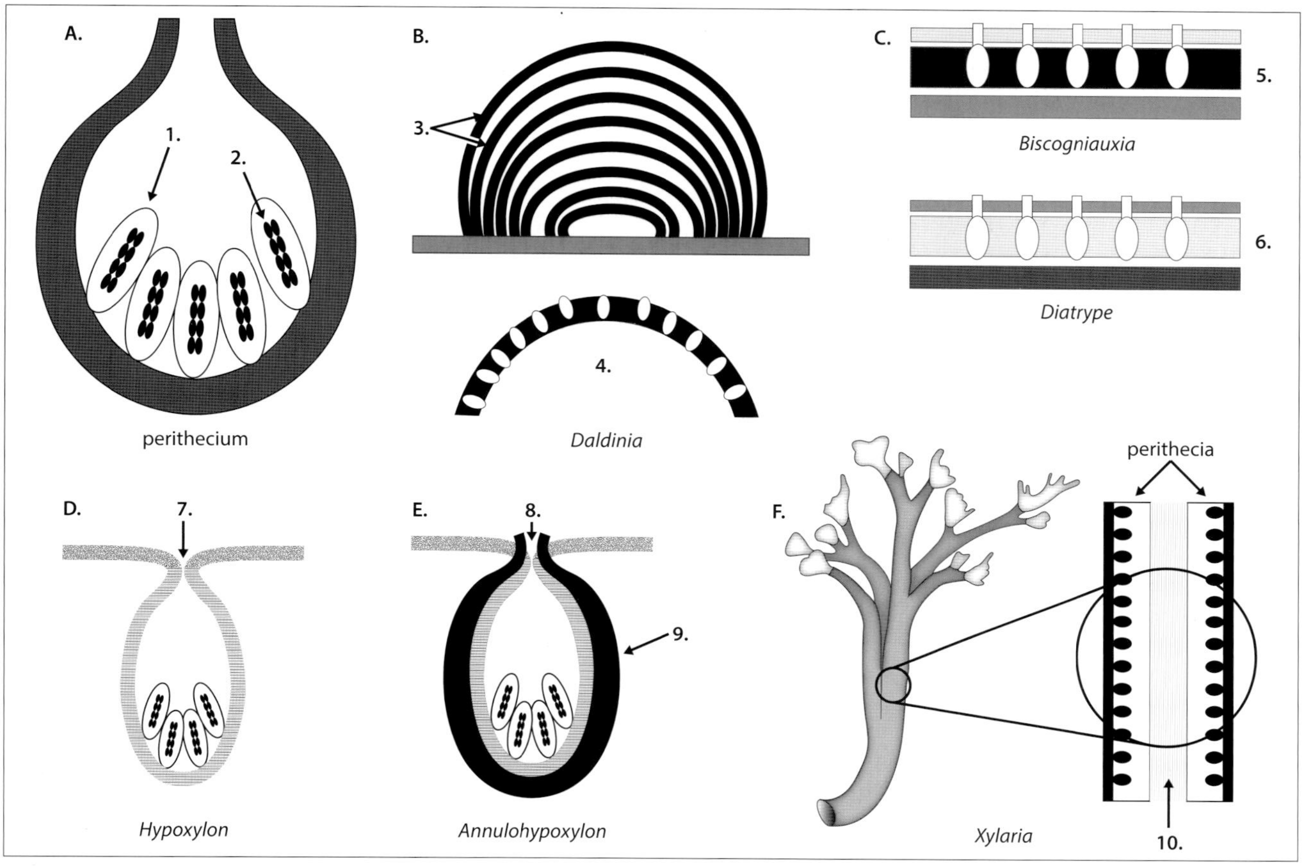

FIGURE 1. Distinguishing perithecial features of various xylarian fungi. A, generic xylarian perithecium showing basal development of asci; B, cross-section of *Daldinia* stroma showing concentric layers of perithecia (3 & 4) and intervening stromatal tissue; C, sections through stromata showing the black carbonaceous layer (5) of *Biscogniauxia* (Xylariaceae) and the tough white flesh (6) of *Diatrype* (Diatrypaceae); D & E, showing the unencased perithecia and embedded ostiole (7) of *Hypoxylon* (*sensu stricto*) and the emergent ostiole (8) and carbonaceous covering enclosing the perithecial layer of *Annulohypoxylon* (9); F, *Xylaria*, showing a cross-section through a typical cylindrical stroma with the perithecia surrounded by whitish matrix just beneath the outer carbonaceous layer and a central woody core (10).

Hypoxylon and *Daldinia* lack this outer sloughing layer. *Biscogniauxia* is distinguished from the superficially similar genus *Diatrype* (Family Diatrypaceae) by the perithecia being embedded in a black carbonaceous stromatal layer (Fig. 1C:5), by dark brown cylindrical ascospores with a prominent germ slit, and by an apical ring that stains blue in Melzer's reagent. The perithecia of *Diatrype* are embedded in a layer of tough whitish flesh (Fig. 1C:6), the ascospores are yellowish and sausage-shaped and lack a germ slit, and the apical rings do not react with Melzer's reagent.

Xylaria species form erect and elongate club-like, subcylindrical or flattened, long, thin filiform or variously branching antler-like stromata with a blackened, carbonaceous surface covering a layer of perithecia embedded in a whitish stromatal matrix (Fig. 1F:10), this underlain by a central, woody core.

The mature, coalesced, subspherical carbonaceous stromata of *Kretzschmaria* differ from those of *Daldinia* in having thin, carbonaceous tissue immediately below the surface, with more or less homogeneous (not concentrically zonate) leathery to woody and white to grey tissue between and beneath the perithecia that becomes dark brown and disintegrates. Alternating concentric layers of perithecial and stromatal tissue that characterize *Daldinia* are lacking in *Kretzschmaria*, and no pigments are released with KOH.

The curious genus *Entonaema* A. Moeller has cushion-like, hemispherical to globose, hollow, liquid-filled stromata that are turgid and under pressure when fresh and hollow when dry. In contrast to most xylariaceans, the ectostroma (the thin, membranous outer pellicle) is brightly colored in most species, the underlying stroma surrounding and beneath the perithecia and surrounding the cavity is gelatinous when fresh and horny and hard when dry. The asci are stalked with an apical ring that stains blue in Melzer's reagent, and the unicellular ascospores are dark brown, elliptical to subrectangular, with a longitudinal germ slit.[10]

1. Erect thread-like asexual reproductive structures formed by fused conidiophores and giving rise to the conidia (asexual spores).

2. Perithecia are subglobular or flask-shaped ascocarps (ascomycetous fruitbodies) that release their spores through a pore or ostiole; cleistothecia (cleistocarps) are completely enclosed spore-bearing fruitbodies that release their spores by disintegration.

3. An ascoma is an ascomycetous fruitbody.

4. As used here amyloid refers to the apical structure of an ascus that stains blue with Melzer's reagent.

5. Producing conidia (asexual spores) on exposed conidiophores (simple or branched hyphae that produce bead-like chains of condia.

6. A type of blastic production of cell walls in which, following completion of any developmental stage, the fungus in a new stage lays down wall layers which are continuous with all of the wall layers used in the previous stage.

7. The number of families included in the Xylariales varies from 3 to 11, depending on the authority; most authors recognize 6 or 7 (by excluding or including the Apiosporaceae), depending on the authority. The taxonomic delineation of this group is still in flux, and the boundaries are not yet generally agreed upon.

8. *Hypoxylon* (*sensu stricto*) and *Annulohypoxylon* are considered as two different sections (*Hypoxylon* and *Annulata*, respectively) of the genus *Hypoxylon* (*sensu lato*) by some authors.

9. Stromata that develop on the surface of (instead of within) their host plants.

10. *Fide* J.D. Rogers, 1981, *Mycologia* 73(1):28–61.

Keys to the Taxa Treated Here

Key to the Genera

1. Stroma bright irregularly cushion-like to subglobose, bright yellow to olivaceous, pellicle thin, pliable, with hollow, liquid-filled core . . . . *Entonaema*[a]

1′. Stroma variably shaped, surface hard, rusty-red, blue-grey, greyish-white, and/or black carbonaceous when mature 2

2. Mature stroma black, broadly effuse, forming a hard, more or less flat, crust-like spread 3

2′. Mature stroma rusty-red, blue-grey or black, may be coated with white conidia (asexual spores) when young; rounded, cushion-shaped to erect and columnar to spade-like or whip-like to branched and antler-like 4

3. Mature surface chocolate-brown to black, smooth, glazed; perithecial ostioles ring-like; perithecia embedded in tough whitish tissue . . . . *Diatrype*[b]

3′. Mature surface whitish to silvery-grey, evenly dotted with black, papillate perithecial ostioles, eventually becoming entirely black; perithecia embedded in black carbonaceous tissue . . . . *Biscogniauxia*[c]

4. Mature stromata erect, stubby to columnar, whip-like or branching and antler-like . . . . *Xylaria*

4′. Stromata rounded, cushion-like to subglobose, may coalesce into larger masses 5

5. Stroma in cross-section consisting of alternating black and silverish concentric rings . . . . *Daldinia*

5′. Stroma not consisting of alternating concentric rings of tissue 6

6. Immature stage inconspicuous and thread-like when present; mature stroma rusty-red to blackish; orange or greenish extractable pigments with KOH 7

6′. Immature asexual stage flat, pale grey, soft, powdery, with a white border; mature sexual stage consisting of dull copper-brown to blackish, irregular, coal-like fused mounds; nonreactive with KOH . . . . *Kretzschmaria*[d]

7. Perithecia enclosed in a carbonaceous layer, ostioles projecting above the stromatal surface. . . . . *Annulohypoxylon*

7′. Perithecia not encased in a carbonaceous layer, ostioles opening beneath the stromatal surface . . *Hypoxylon* (*sensu stricto*)

a. Represented in our area by the sole species *E. liquescens*.
b. Represented in our area by the sole species *D. stigma*.
c. Represented in our area by the sole species *B. atropunctata*.
d. Represented in our area by the sole species *K. deusta*.

Key to *Xylaria*

1. Coremium (immature asexual fruitbody) pinkish, resembling a cushion-shaped coral head or chewed piece of bubblegum . . . . *Xylecoremium flabelliforme*[a]

1′. Mature stroma blue-grey or black, may be dusted with white asexual spores; thick and columnar or long, thin, and whip-like 2

2. Stroma columnar with a prominent radial crown of eight or more thin, tentacle-like branches up to 4 cm long by ¾ mm thick . . . . *Xylaria tentaculata*

2′. Stroma without a radial crown of thin, tentacle-like branches. 3

3. Stromata unbranched, as well as columnar and branched, with flattened, strap-like tips . . . . *X. hypoxylon*

3′. Stromata unbranched or mostly unbranched, thick and columnar or long, thin, and whip-like 4

4. Stroma thick, cylindrical to club-shaped, growing on decaying, injured, or buried hardwood. 5

4′. Stromata thin, whip-like, black, growing on fallen fruits or nuts. 6

5. Stroma with bluntly rounded tip, blackish with distinct bronze or copper tint; ascospores lacking germ slit or germ slit obscure . . . . *X. cubensis*

5′. Stromata digitiform, pointed and blue-grey with white tips when young, club-shaped and blackish with no copper tint when mature; ascospores with germ slit . . . . *X. polymorpha*

6. Growing from buried hickory nuts. . . . . *X. oxyacanthae*[b]

6′. Growing on fallen magnolia cones or sweetgum fruits on the ground 7

Keys to the Taxa Treated Here (*Continued*)

7. Growing on fallen Southern Magnolia cones . . . . *X. magnoliae*

7′. Growing on fallen Eastern Sweetgum fruits . . . . *X. liquidambar*

a. This is the anamorphic (asexual) form of *Xylaria cubensis*; the anamorph is more often encountered, and the anamorph and teleomorph are rarely found together.

b. It grows from buried *Carya* nuts in the Southeast; elsewhere this species appears to be restricted to *Crataegus* (hawthorn) fruits.

Key to *Hypoxylon* and *Annulohypoxylon*

1. Stromata shiny blackish-brown to black; perithecial ostioles ringed by a flat disk . . . . *A. truncatum*

1′. Stromata dull brick- to rusty-red or dull black and carbonaceous; ostioles papillate, not ringed by a flat disk 2

2. Stromata rusty dark brick-red to orange-red; perithecia not enclosed in a carbonaceous layer (*Hypoxylon*) 3

2′. Stromata dull blackish-brown to black; perithecia enclosed in a carbonaceous layer (*Annulohypoxylon*) 4

3. Perithecial mounds distinctly papillate or beaded; restricted to bark of dead beech trees . . . . *H. fragiforme*

3′. Perithecial mounds inconspicuous to scaly, not strongly beaded or papillate; found on various hardwood trees . . . . *H. howeianum*

4. KOH-extractable pigments sienna or orangish, rarely with greenish tint; found on birches, alder, and hazelnuts . . . . *A. multiforme*

4′. KOH-extractable pigments greenish-olivaceous to pale yellowish, or pigments lacking; restricted to dead branches and exposed roots of beech trees . . . . *A. cohaerens*

Key to *Daldinia*

1. Internal flesh with whitish or cream-colored zones that are gelatinous when fresh and loculate (disintegrating and developing cavities) when old and dry, associated with fire-burned woody substrata . . . . *D. vernicosa*

1′. Internal flesh without whitish or cream-colored gelatinous or loculate zones, not associated with fire-burned woody substrata 2

2. Always releases pale grey-yellow, honey, amber, or cinnamon pigments in KOH . . . . *D. childiae*

2′. Initially releases dark purple pigments, often changing to yellowish or olivaceous after several minutes . . . . . . *D. petriniae*

Hypoxylon Canker

Biscogniauxia atropunctata

A B C

***Biscogniauxia atropunctata* (Schweinitz) Pouzar**, *Česká Mykol.* 33(4):216 (1979)

Basionym: *Sphaeria atropunctata* Schweinitz, *Schr. Naturf. Ges. Leipzig* 1:31 (1822)
[Ascomycota, Pezizomycotina, Sordariomycetes, Xylariomycetidae, Xylariales, Xylariaceae]

Derivation: *Biscogniauxia*, etymology unknown; *atropunctata*, Latin, from *atro*, "black-cloaked," and *punctatum*, Latin, "dotted," "black-dotted," referring to the blackish perithecia.

Fruitbody (stroma): Young surface whitish to concrete- or silvery-grey, evenly dotted with black, papillate perithecial ostioles (Fig. A right, Fig. B), this eventually falling away, leaving an entirely black surface (Fig. A left, Fig. C); broadly effuse, forming an extensive (several feet), flat to somewhat undulating crust-like spread; flesh black and carbonaceous immediately below the surface and between the perithecia (see Fig. 1C in the introductory remarks to this section); tissue below the perithecia inconspicuous. No KOH-extractable pigments.

Microscopic: Ascospores (23–)26–30(–33) × (11–)12–14(–16) µm, unicellular, elliptical, smooth; brown to dark brown; with straight, spore-length germ slit; asci stalked, cylindrical, spore-bearing portion 140–160 × 16–20 µm, apical ring discoid, bluing in Melzer's reagent.

Occurrence: Saprobic, causing a white rot in living and dead hardwood branches and trunks, especially in oaks (*Quercus* spp.) and beeches (*Fagus grandifolia*), but also hickories (*Carya* spp.); widespread and common in central and eastern North America; year-round.

Edibility: Inedible.

Comments: The fully mature, black stromata of *B. atropunctata* can be mistaken for the Common Tarcrust, *Diatrype stigma* (Hoffman) Fries. However, the latter is smooth, with a glazed reddish-black to black surface resembling burnt wood; the ostial openings are discoid (instead of papillate) and do not turn blue in Melzer's reagent; and the perithecia are embedded in tough, white (rather than black carbonaceous) flesh (see Fig. 1C in the introductory remarks to this section).

The Marginate Hypoxylon Canker, *Biscogniauxia marginata* (Fries) Pouzar, is reported from Great Smoky Mountains National Park.[1] The discrete stromata are discoid or cup-like (rather than effuse and spreading), considerably smaller—6–8 cm (2½–3¼ in) across—and the subglobose (not elliptical) ascospores have an S-shaped (rather than straight) germ slit. It appears to be restricted to beeches.

1. L.N. Vasilyeva et al., 2007, *Fungal Diversity* 25:219–231.

Common Tarcrust

Diatrype stigma

A

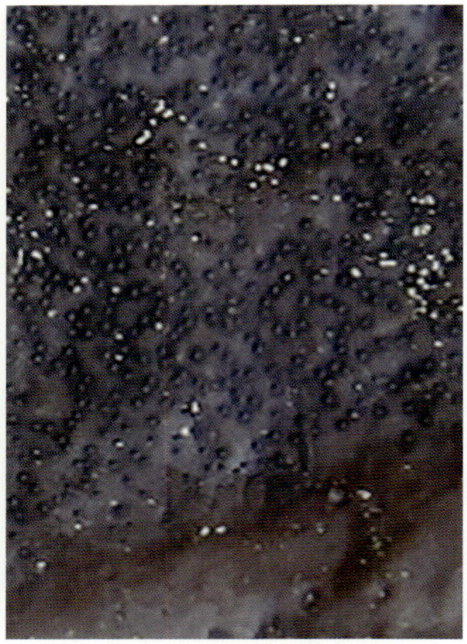

B

***Diatrype stigma* (Hoffman) Fries**, *Summa Veg. Scand.*, Section Post. (Stockholm):385 (1849)

Basionym: *Sphaeria stigma* Hoffman, *Veg. Crypt.* 1:7 (1787) [Ascomycota, Pezizomycotina, Sordariomycetes, Xylariomycetidae, Xylariales, Diatrypaceae]

Derivation: *Diatrype*, etymology unknown; *stigma*, Latin, "a small spot or minute hole," referring to the perithecia.

Fruitbody (stroma): Stromata reddish purple-brown when young, becoming chocolate-brown to black when mature (Fig. A); smooth to finely roughed, sometimes minutely cracked, broadly effused, forming an interrupted or continuous, often extensive, glazed, crust-like, undulating spread of uneven thickness (0.5–1 mm); perithecia ovoid, with short necks, evenly distributed in a single layer within a layer of tough, whitish flesh (see Fig. 1C in the introductory remarks to this section), with disk-shaped or ring-like ostioles opening at the surface (Fig. B).

Microscopic: Spores (6–)7–9(–10) × 1–1.5(–2) µm, allantoid (sausage-shaped), smooth pale yellow, lacking germ slit; asci club-shaped, spore-containing portion (30–)35–40(–50) × 4–8 µm; apical ring not bluing in Melzer's reagent.

Occurrence: Saprobic on a variety of hardwoods, especially beech (*Fagus grandifolia*) and oaks (*Quercus* spp.); widespread and common in eastern North America; year-round.

Edibility: Inedible.

Comments: The smooth, glazed, reddish-black to blackish effusive stromata of this species resembles burnt wood. It can be mistaken for the fully mature black stromata of Hypoxylon Canker, *Biscogniauxia atropunctata* (Schweinitz) Pouzar, which infects the same hardwood trees. However, the surface of *B. atropunctata* is rough due to the numerous papillate (not discoid) ostial openings, and the older black stromata are almost always associated with younger greyish spreads peppered with black perithecial ostia. In contrast, the surface of *Diatrype stigma* is smooth, with discoid ostial openings. In addition, the perithecia [of *B. atropunctata*] are embedded in black carbonaceous (rather than tough, whitish) flesh, and the ascospores are elliptical with a straight, spore-length germ slit (in contrast to the sausage-like spores of *D. stigma*, which lack a germ slit).

Yellow Sack Fungus

Entonaema liquescens

A

B

Photos: A, Mary Woehrel; B, Drew Parker.

Entonaema liquescens Möller, *Phycomyc. Ascomyc. Bras.* 9:306, Figs. 24–26, 47–49 (1901)
[Acomycota, Sordariomycetes, Xylariomycetidae, Xylariales, Xylariaceae]

Derivation: *Entonaema*, etymology unknown; *liquescens*, Latin, "melting."

Anamorph: Form-genus *Nodulisporium* Preuss.[1]

Fruitbody (stroma): 1–13 cm across by 1–6 cm high (½–5¼ × ½–2½ in), stalkless, narrowly connected to substratum, cushion-shaped, hemispherical, or globose, often irregularly folded and convoluted (Figs. A & B); outer pellicle thin, dull to bright yellow (Fig. A) or olivaceous, sometimes with orange tints (Fig. B), surface readily rubbed off to expose darker underlying layer; stroma hollow, filled with watery liquid under pressure, this released and fruitbody collapsing when punctured.

Flesh: Underlying peridium and surrounding liquid-filled cavity, 2–5 mm thick and gelatinous when wet, containing perithecia; <1 mm thick, hard and reddish to blackish when dry.

Microscopic: Spores (ascospores) (8–)9.5–12(–13) × (4–)5–6 (–7.3) µm, elliptical to rectangular, smooth, with inconspicuous to conspicuous longitudinal germ slit shorter than spore length, usually with two oil droplets, light to dark brown; perithecia carbonaceous, apical ring of asci staining blue with Melzer's reagent.

Occurrence: Saprobic on decaying and decorticated hardwood logs; summer–fall; this tropical species is widespread and rare to locally common in North America, Central and South America (first described from Brazil), Africa, and Asia.

Edibility: Presumably inedible.

Comments: This appears to be the only species of the largely tropical genus *Entonaema* reliably reported from North America.[2] This mushroom cannot be confused with any other species in North America. The yellow pellicle, gelatinous flesh, and liquid-filled hollow cavity are diagnostic.

The genus *Entonaema* Möller appears to be most closely related to the xylariacean genera *Hypoxylon* Bulliard and *Daldinia* Cesati & De Notaris. The exact relationship between the anamorph and teleomorph is not known but is likely similar to that seen in *Hypoxylon* (sect. *Hypoxylon*) and *Daldinia*.[1]

1. J.D. Rogers, 1982, *Mycotaxon* 15:500–506.
2. *Fide* J.D. Rogers, 1981, *Mycologia* 73:28–61.

Brittle Cinder

Kretzschmaria deusta

A B C

Kretzschmaria deusta **(Hoffmann) P.M.D. Martin,** *J. S. Afr. Bot.* 36(2):80 (1970)

Basionym: *Sphaeria deusta* Hoffmann, *Veg. Crypt.* 1:3 (1787) [Ascomycota, Pezizomycotina, Sordariomycetes, Xylariomycetidae, Xylariales, Xylariaceae]

Derivation: *Kretzschmaria*, etymology unknown; *deusta*, Latin, "burned up," referring to the charcoal-like appearance of the mature stroma.

Deprecated synonym: Often still referred to as *Ustulina deusta* (Hoffmann) Lind (1913)[1]

Fruitbody (stroma): 4–10 cm wide by about 0.5–3 mm thick (1⅔–4 × ¼–1¼ in), forming irregular cushion-like stromata or coalesced cushion-like spreads up to 50 cm (20 in) long, these loosely attached to the substratum by narrow connections. **Anamorph (asexual stage)**—young asexual stage flat, pale grey, soft, powdery (Fig. B), bordered by a white growing edge, with numerous rounded lobes throughout (Fig. A). **Teleomorph (sexual stage)**—transforming into the mature sexual stage consisting of hard, dull copper-brown to blackish, irregular, coal-like fused mounds (Fig. A, arrow) studded with large (1.5 mm diameter) pimple-like perithecia, each bearing a papilla-like ostium, or pore (Fig. C), these carbonaceous stromata becoming brittle, cracked, and hollow with age.

Flesh: Thin, carbonaceous immediately below the surface, the tissue between and beneath the perithecia leathery to woody, white to grey, becoming dark brown and disintegrating.

Chemical staining: No reaction with KOH.

Microscopic: Spores (ascospores) (27–)28–35(–36) × 7–10 µm, unicellular, spindle-shaped, inequilateral (one side flattened), smooth, with a straight germ slit half the spore length on the flattened side, brown to dark brown; apical ring of asci bluing in Melzer's reagent.

Occurrence: Saprobic, causing white rot. Appearing on lower trunks, stumps, and roots of decaying hardwood trees; widespread in temperate regions of Northern Hemisphere; common in the Southeast, summer–midwinter, overwintering.

Edibility: Inedible.

Comments: The mature sexual stromata look for all the world like fused masses of charcoal briquettes or lumps of coal. They superficially resemble the stromata of Carbon

Cushion, *Annulohypoxylon multiforme* (Fries) Y.-M. Yu, J.D. Rogers & H.-M. Hsieh, whose adjacent fruitbodies also coalesce into larger masses. However, *A. multiforme* occurs on the bark and decorticated branches of birches (*Betula* spp.), the flesh surrounding and beneath the perithecia is hard, black, and carbonaceous, the ascospores are much smaller (8–12 × 3.5–5 µm), and the stromata release sienna or orangish (rarely with greenish tint) pigments in KOH. *Kretzschmaria deusta* is typically found on beech (*Fagus* spp.), oaks (*Quercus* spp.), and maples (*Acer* spp.); the perithecia are embedded in leathery white to greyish (not black or carbonaceous) flesh; and the stromata release no pigments when challenged with KOH.

The effusive light-grey asexual stage of this fungus superficially resembles the resupinate pore surface of the Smoky Polypore, *Bjerkandera adusta* (Willdenow) P. Karsten. However, though the pores in *B. adusta* are tiny, they are distinct (pores are completely lacking in the asexual stage of *K. deusta*), and the effused-reflexed fruitbodies typically form overlapping, imbricated shelves that are never seen in *K. deusta*.

Carbon Balls, *Daldinia childiae* J.D. Rogers & Y.-M. Ju, also resembles the sexual stromata of *K. deusta*, but the solid (not hollow) carbonaceous stromata consist of alternating silverish and black concentric rings and release bright purple to dark purple pigments in KOH.

1. The genus name *Ustulina* Tulasne & C. Tulasne (1863) is a deprecated synonym of *Kretzschmaria* Fries (1849) (*fide* J.D. Rogers & Y.-M. Ju, 1998, *Mycotaxon* 68:345–393).

Carbon Balls, Cramp Balls

Daldinia childiae, *D. petriniae*, and *D. vernicosa*

Daldinia childiae. (Photos: A & C, W. Light; B, D & E, Teresa Fortenberry.)

***Daldinia childiae* J.D. Rogers & Y.-M. Ju,** in Rogers, Ju, Watling & Whallery et al., *Mycotaxon* 72:512 (1999)

Misapplied name: *Daldinia concentrica: sensu auct. mult.; non Sphaeria concentrica* Bolton, *Hist. Fung.* Halifax III (Huddersfield):plate 180 (1789)

***Daldinia petriniae* Y.-M. Ju, J.D. Rogers & F. San Martín,** *Mycotaxon* 61:275 (1997)

***Daldinia vernicosa* Cesati & De Notaris,** *Comment. Soc. Crittog. Ital.* 1:198 (1863) [*nom. nov.* for *Sphaeria vernicosa* Schweinitz (1825); *non S. vernicosa* De Candolle & Lamarck 1815 (Figs. 7A, 14B & C, 47, 48)]

Deprecated synonyms: *Daldinia fissa* Lloyd, *Mycol. Writ.* 7:1313 (1922)

***Daldinia simulans* Child,** *Ann. Missouri Bot. Gard.* 19:453 (1932) [Ascomycota, Sordariomycetes, Xylariomycetidae, Xylariales, Xylariaceae]

Derivation: *Daldinia*, Latin, a genus named in honor of Agostino Daldini, a Swiss Catholic monk; *childiae*, Latinized epithet in honor of the American mycologist Marion Child; *petriniae*, Latinized epithet in honor of the Swiss mycologist Liliane E. Petrini; *vernicosa*, Latin, "varnished."

Fruitbody (stroma): ***D. childiae***—sessile or with short, stout stipe (Fig. D, arrow), globose or hemispheric to top-shaped, may be irregular; 0.5–5 cm in diameter by 0.8–4 cm high (¼–2 × ⅓–1⅔ in); surface hard, friable, brownish-wine, greyish-white, pinkish-brown to chestnut during young asexual stage (Fig. A), becoming greyish-brown to black but dull and unvarnished when sexually mature (Figs. B & D), surface smooth or (usually) covered with indistinct to distinct perithecial mounds with slightly papillate ostioles (Figs. D & E). ***D. petriniae***—sessile or nearly so, initially hemispherical to nearly club-shaped, often becoming cushion-shaped or disk-shaped; 1–5.8 by 0.9–3.4 cm (½–2⅓ × ⅓–1⅓ in); with indistinct to conspicuous perithecial mounds with papillate or faintly discoid ostioles; surface dark brick-red to fuscous, becoming blackened and varnished with age. ***D. vernicosa***—turbinate (shaped like a top), usually with a stipe; surface smooth (rarely wrinkled in rapidly dried specimens), without or with faintly visible perithecial outlines, 0.5–5 by 0.5–5 cm (¼–2 × ¼–2 in), but mostly only up to 2.5 cm (1 in) high; surface brown-vinaceous, blackened, and typically varnished in age.

Flesh: The genus is characterized by purplish-brown to coal-black flesh consisting of alternating concentric zones of woody/pithy or gel-like and waxy/carbonaceous texture and brown, greyish-brown to (less often) whitish and dark brown color (Fig. C), each representing a season's growth. ***D. childiae***—dark brown woody or pithy zones (0.2–0.6 mm thick) alternating with lighter brown woody to pithy zones (0.6–1.0 mm) that persist; ratio of darker to lighter zones 1:1–3. ***D. petriniae***—dark brown woody or pithy zones (0.3–0.7 mm thick) alternating with lighter brown woody to pithy zones (0.5–1.5 mm) that persist; ratio of darker to lighter zones 1:1–3. ***D. vernicosa***—dark brown pithy or woody zones (0.1–0.2 mm thick) alternating with white gelatinous zones (0.6–1.3 mm) that disintegrate and develop cavities when dry; ratio of darker to lighter zones 1:5–10.

Chemical staining: *D. childiae*—always releases pale grey-yellow, honey, amber, or cinnamon pigments in KOH; ***D. petriniae***—initially releases dark purple pigments, often changing to yellowish or olivaceous after several minutes; some specimens release only olivaceous pigments, which may depend on their host species. ***D. vernicosa***—releases dark livid purple or violet pigments with KOH.

Spore print: All species—asexual conidia (conidiospores) colorless or whitish; sexual ascospores dark brown to blackish-brown, often deposited on and around the fruitbodies.

Microscopic: Apical ring bluing in Melzer's reagent; perispore indehiscent in 10% KOH. **Spores (ascospores): All species**—elliptical, often inequilateral (flattened on one side), smooth (under light microscopy),[1] nonseptate, with a straight germ slit running the length of the spore on the convex side of inequilateral spores; perispore either dehiscent or indehiscent.[2] ***D. childiae***—spores (ascospores) 12–16 × 5.5–7.5 µm, with narrowly rounded ends; perispores dehiscent in 10% KOH. ***D. petriniae***—12.5–16.5 × (6–)6.5–7.5 µm, with narrowly rounded ends; perispores dehiscent in 10% KOH. ***D. vernicosa***—11.5–15 × 6.5–8(–9) µm, ends narrowly to broadly rounded; perispores indehiscent in 10% KOH.

Occurrence: All species—saprobic, causing a soft rot, found year-round, alone or infrequently in clusters on decaying hardwoods. This genus is likely weakly and facultatively parasitic and continues to decay its host trees after their death. ***D. childiae***—little host preference; widespread and common throughout North America; cosmopolitan: Europe, Asia, South America, Africa, Oceania, Australia. ***D. petriniae***—found almost exclusively on members of the birch family (Betulaceae); throughout temperate Northern Hemisphere.[3] ***D. vernicosa***—no strong host preference, but commonly found on Betulaceae and fire-damaged wood; widespread in temperate and subtropical Northern Hemisphere,[4] rarely in tropics and only at high altitudes.

Edibility: Inedible.

Comments: The familiar and cosmopolitan Carbon Balls long known as *Daldinia concentrica* has been found not to be that species at all. In fact, *Daldinia concentrica* (Bolton) Cestati & De Notaris (*sensu stricto*)—the type species of the genus—does not naturally occur outside of the warmer portions of western, central, and southern Europe, including the British Isles.

Most early *Daldinia* collections were identified as *D. concentrica* regardless of their true identities, and this concept actually included several different species. This occurred because the type specimens used by Bolton to describe and name *Sphaeria concentrica* (the basionym, or Bolton's original published genus and species combination) were missing and thought to be lost. Bolton's original illustrations were therefore declared to be the holotype,[5] and this mushroom was interpreted according to the concepts then prevalent,[6] which included the long-held assumption that *D. concentrica* released various shades of yellowish, greenish, olivaceous, or amber pigments with KOH. Bolton's original type material was later found, and an epitype (see note 5) was selected.[7] When tested with KOH, the type material of *D. concentrica* was found to release deep purple rather than yellowish or greenish pigments (as previously believed). The cosmopolitan taxon widely thought to represent *D. concentrica* turned out to be an undescribed species, which was formally described and named as *D. childiae* by Rogers and Ju in 1999 (see note 7). Several other species were also included in the earlier and broader concept of *D. concentrica*. Three of these occur in our area and are discussed below and in the above taxonomic section. This issue is by no means fully resolved, and additional new taxa will likely be discovered that were previously erroneously reported as *D. concentrica*.

Daldinia childiae is cosmopolitan and is the species most often encountered in North America, as well as on other continents. It also thrives under the widest ecological conditions, ranging from the tropics to high latitudes and high altitudes. It shows little host preference and is found on a wide variety of angiosperm species. Its immediate release of pale grey-yellow, honey, amber, or cinnamon pigments in response to KOH and its wide range of hardwood host species distinguish this species from the other two members of the genus found in our area.[8] These and other species in the genus *Daldinia* are discussed in great detail by Stadler et al. (2014), and much of this material was taken from that source.[9]

Daldinia petriniae is only found on members of the family Betulaceae, specifically birches (*Betula* spp.), alders (*Alnus* spp.), and hornbeams (*Carpinus* spp.). It initially releases dark purple pigments, which often change to yellowish or olivaceous after several minutes.

Daldinia vernicosa always releases dark livid purple or violet pigments with KOH. Like *D. childiae*, it has no strong preference for host species but is commonly found on members of the Betulaceae and on fire-burned woody substrata. The presence of dark brown pithy or woody zones alternating with distinctive white gelatinous zones (the latter disintegrating and developing cavities when dry) having a width ratio of darker to lighter zones of 1:5–10 contrasts markedly with the persistent (nondisintegrating) zones of darker and lighter brown woody to pithy zones of the other two species, both of which have darker-to-lighter zone width ratios of 1:1–3.

The concentric gel and carbonaceous zones in the flesh characterize the genus and appear to be an adaptation for dry environments. The gel portion of the rings is thought to retain water, while the encasing waxy to carbonaceous crust inhibits evaporation (see Fig. 1B-3 in the introduction to this section). Each ring represents a growth season much as do the annual growth rings in trees.

A blackish dust-like film is often observed surrounding Carbon Balls that serves as a diagnostic feature. This results from the forcible ejection of the ascospores during both wet and dry weather. This forcible ejection is due to turgor pressure as water accumulates within the perithecia housing the asci. However, this forcible nocturnal discharge has been observed to continue for 20–30 days during subsequent dry spells and results from the water stored in the stromatal tissues. This water and the weight of the stroma decrease gradually over this period of continual spore discharge.[10]

Legend has it that King Alfred hid in a homestead to avoid capture during a war and was told to keep an eye on and remove cakes baking in the oven. He fell asleep, and the cakes burned to a crisp. Because Carbon Balls are thought to resemble these burned cakes, they are sometimes called "King Alfred's cakes." They are also known as Cramp Balls because it was once thought that carrying them as a charm would ward off cramps; this was apparently practiced until quite recently in parts of England.[11] They are also used in traditional Chinese medicine to treat cramps and other spasmodic conditions. When fully mature, carbon black and dry, Carbon Balls make a good tinder that will take a spark from a steel-struck flint.

See the distinguishing remarks in the species account of the similar sexual stage of the Brittle Cinder, *Kretzschmaria deusta* (Hoffmann) P.M.D. Martin.

1. Under scanning electron microscopy (10,000×), *D. vernicosa* still exhibits smooth spores; *D. petriniae* reveals faint transverse striations at 5,000×; and *D. vernicosa* shows conspicuous transverse striations at 2,500–5,000×.
2. The perispore is an outer sheath covering the ascospore. It may or may not dehisce in 10% KOH—i.e., split apart and fall away, leaving a transparent empty shell. This feature has taxonomic significance.
3. Documented in Great Smoky Mountains National Park, southeast of Gatlinburg, Tennessee (*fide* Stadler et al., 2014, *Studies in Mycology* 77[1]:105).
4. Reported in our area from Hickorynut Mountain and Tallulah Falls, Georgia (Stadler et al., 2014, *Studies in Mycology* 77[1]:87).
5. The holotype is the single specimen in a type series chosen by the original naming author to which the name of a species is permanently attached. This is necessary in case the specimens constituting the type series turn out to represent more than the one taxon being described and named. This is not a rare occurrence in biology. An epitype is a specimen or illustration selected to serve as an interpretative type when the holotype, lectotype, or previously designated neotype, or all original material associated with a validly published name, is demonstrably ambiguous and cannot be critically identified for purposes of the precise application of the name of a taxon.
6. Ju et al., 1997, *Mycotaxon* 61:262.
7. J.D. Rogers & Y.-M. Ju, *in* Rogers, Ju, Watling, & Whalley, 1999, *Mycotaxon* 72:512.
8. Two or three other species occur in the Coastal Plain region but are not reported from the Piedmont or southern Appalachians.
9. M. Stadler et al., 2014, A polyphasic taxonomy of *Daldinia* (*Xylariaceae*), *Studies in Mycology* 77(1):1–143.
10. C.T. Ingold, 1946, *Trans. Brit. Mycol Soc.* 29:43–51.
11. B.A. Oso, 1977, *Economic Botany* 31(3):367–371.

Red Carbon Cushion

Hypoxylon fragiforme and *H. howeianum*

A

B

C

D

Hypoxylon fragiforme **(Persoon : Fries) Jean Kickx**, *Fl. Crypt. Louvain* (Bruxelles):116 (1835)

Basionym: *Sphaeria fragiformis* Persoon, *Ann. Bot.* (Usteri) 11:21 (1794)

Hypoxylon howeianum **Peck**, *Ann. Rep. N.Y. State Mus.* 24:98 (1872)
[Ascomycota, Sordariomycetes Xylariomycetidae, Xylariales, Xylariaceae]

Derivation: *Hypoxylon*, Greek, "beneath wood"; *fragiforme*, Latin, "strawberry-shaped"; *howeianum*, Latinized epithet in honor of Elliot C. Howe.

Fruitbody: Both species are minutely bumpy (pimple-dotted) at maturity. The pimples contain the openings (ostioles) of spore-forming structures (perithecia) embedded just below the surface. Both have a layer of white granules between the perithecia when young. The interior flesh is black and hard. KOH-extractable pigments orange or rust. ***H. fragiforme***—perithecial mounds distinctly papillate (pimple-like) or beaded, giving fruitbody a berry-like appearance (Figs. A & B). ***H. howeianum***—hemispherical to nearly spherical, occasionally coalescing, perithecial mounds inconspicuous to scaly, not strongly beaded or papillate (Figs. C & D); surface rusty to dark brick-red, with orange-red perithecia covered by a white to greyish tissue when young; tissue beneath perithecial layer black; KOH-extractable pigments orange to strongly red-brown or rust; young *fruitbodies covered with distinctive thread-like anamorphic spore-bearing structures when young*.

Spore print: Dark brown to blackish.

Microscopic: Asci with discoid apical ring bluing in Melzer's reagent; spores (ascospores) dark brown, ellipsoid-inequilateral, germ slit running length of spore; perispore opening in 10% KOH. ***Hypoxylon fragiforme***—asci 80–90 × 6–8 µm, stipes to 90 µm; ascospores 10.5–15 × 5–7 µm. ***Hypoxylon howeianum***—asci 50–70 × 5–6 µm, stipes 40–60 µm; ascospores 6–9 × 3–3.5 µm.

Occurrence: ***Hypoxylon fragiforme***—Europe and eastern North America, restricted to bark of dead beech (*Fagus grandiflora* in North America and *F. sylvatica* in Europe), in clusters July–November. ***Hypoxylon howeianum***—North America and Europe on recently dead (or, more rarely, rotten) wood of various hardwood trees, including occasionally beech.

Edibility: Inedible.

Comments: *Hypoxylon fragiforme* is primarily restricted to the bark of dead beech trees and is rarely reported on other deciduous host species; *H. howeianum* occurs on a wider variety of hardwood species. The spores of *H. fragiforme* (10.5–15 × 5–7 µm) are larger than those of *H. howeianum* (6–9 × 3–3.5 µm). The perithecia of *H. fragiforme* are significantly more pronounced and pimple-like (Figs. A & B), and the brick-red fruitbody resembles a beaded berry that turns black with age. The perithecia of *H. howeianum* are less distinct and resemble fuzzy scales (Figs. C & D), the fruitbodies do not change color, and young fruitbodies commonly sprout from distinctive radiating hyphal cords (sennemata).

Hypoxylon howeianum is sometimes spelled *H. howeanum*; this latter version is technically the correct spelling. However, for reasons discussed in detail in online Appendix A, we concur with Dr. Jack Rogers and his colleagues that *H. howeianum* is the preferred spelling.

Carbon Cushion and Woodwort

Annulohypoxylon multiforme and *A. cohaerens*

A

B

C

Annulohypoxylon multiforme (Fries) Y.-M. Yu, J.D. Rogers & H.-M. Hsieh, *Mycologia* 97(4):859 (2005)

Basionym: *Sphaeria multiformis* Fries, *Observ. Mycol.* (Havniae) 1:169 (1815); Fries *ex* Fries, *Observ. Mycol.* 1:169 (1815)

Annulohypoxylon cohaerens (Persoon) Y.-M. Ju, J.D. Rogers & H.-M. Hsieh, *Mycologia* 97(4):857 (2005)

Basionym: *Sphaeria cohaerens* Persoon, *Neues Mag. Bot.* 1:82 (1794)
[Ascomycota, Sordariomyecetes, Xylariomycetidae, Xylariales, Xylariaceae]

Derivation: *Annulo*, Latin, "little ring"; *hypoxylon*, Greek, "beneath wood," the name referring to the stromata (fruitbodies), which resemble those of *Hypoxylon* but with ostioles (openings) surrounded by an annulate disk; *multiforme*, Latin, "with many shapes"; *cohaerens*, Latin, "sticking together, adhering."

Fruitbody (stroma): ***A. multiforme***—cushion-shaped, semispherical, 0.5–4 cm in diameter by 1.5–7 mm thick (¼–1⅔ × ¼ in); typically coalescing into lumpy (Fig. A) or elongate (Fig. B) fused masses up to 5 cm (4 in) long or wide (but usually less); rusty or red-brown when young, becoming black and carbonaceous with age; with large perithecial mounds 4–25 mm in diameter by 2–25 mm thick and bearing papillate ostioles (nipple-like openings) (Fig. C); perithecia enclosed in black carbonaceous layer; spherical to egg-shaped, 0.5–0.6 mm in diameter by 0.6–0.7 mm high. ***A. cohaerens***—subspherical, flattened above, often constricted below, 1.5–4 (up to 8) cm in diameter by 1.5–2.7 mm thick (⅔–1⅔ × ⅔–1 in), in clusters or confluent, brown when young, purplish-black with age, with papillate ostioles.

Flesh: Dark-brown to black below perithecial layer, up to 4 mm thick in both species.

Macrochemical staining: KOH—***A. multiforme***—extractable pigments sienna or orangish, rarely with greenish tint; ***A. cohaerens***—extractable pigments greenish-olivaceous to pale yellowish, or pigments lacking.

Microscopic: Spores (ascospores) of both species light brown to brown, elliptical to crescent-shaped, one side flattened, with acute ends, smooth. ***A. multiforme***—8–12 × 3.5–5 µm, with faint straight germ slit shorter than spore length; asci 70–80 × 5–6 µm, stipes of equal length; apical ring bluing in Melzer's reagent. ***A. cohaerens***—9–12 × 4–5 µm, with prominent germ slit equal to the spore length; asci 75–85 × 5–7 µm, stipes 50–70 µm, apical ring usually bluing in Melzer's reagent.

Occurrence: ***A. multiforme***—common, year-round in temperate regions of North America, Europe, Asia; on bark and decorticated branches, trunks, and stumps of Betulaceae: birches (*Betula*), alder (*Alnus*), and hazelnuts (*Corylus*). ***A. cohaerens***—common on dead branches and exposed roots of beeches (*Fagus*), highly host specific; North America and Europe.

Edibility: Both inedible.

Comments: *Annulohypoxylon multiforme* is found on birches; the very similar *A. cohaerens* occurs on dead beeches, often together with Red Cushion Hypoxylon (*Hypoxylon fragiforme*). *Annulohypoxylon multiforme* could also be confused with *A. annulatum* (Schweinitz) Y.-M. Ju, J.D. Rogers & H.-M. Hsieh (2005), which occurs on dead oak branches, but the papillate ostioles are encircled by a faint convex disk that is completely lacking in *A. multiforme* and *A. cohaerens*. The stromata of the Black Raspberry Hypoxylon, *A. truncatum* (Schweinitz) Y.-M. Ju, J.D. Rogers & H.-M. Hsieh—another species found on oak branches—form tight, conglomerate, raspberry-like masses, and the papillate ostioles are encircled by a sharply defined flat disk 0.2–0.4 mm in diameter that resembles a washer.

Annulohypoxylon multiforme differs from aged and blackened Red Carbon Cushion (*Hypoxylon fragiforme*) in its tendency for adjacent fruitbodies to coalesce into larger masses, the larger size of its perithecial mounds, the black layer around the perithecia, and the length of the germ slit on the spores. *Annulohypoxylon cohaerens* differs from *H. fragiforme* in its greenish or yellowish KOH-extractable pigments; those of *H. fragiforme* are orange or rusty. The only other fungus these might be confused with is Carbon Balls, *Daldinia concentrica*. However, that fungus does not coalesce into irregular clumps or elongate masses, and the black flesh is concentrically ringed with alternating silver-grey zones. *Daldinia concentrica* has yellowish, greenish, or olivaceous extractable pigments in KOH; this distinguishes it from *Annulohypoxylon multiforme* (but not *A. cohaerens*), which releases sienna or orangish pigments in KOH. These fungi were previously placed in the genus *Hypoxylon* but are now in the genus *Annulohypoxylon* because of a black carbonaceous layer that surrounds each of the perithecia. This black layer is lacking in *Hypoxylon* (*sensu stricto*).

Black Raspberry Hypoxylon

Annulohypoxylon truncatum

A

B

***Annulohypoxylon truncatum* (Schweinitz) Y.-M. Ju, J.D. Rogers & H.-M. Hsieh**, *Mycologia* 97(4):861 (2005) [Ascomycota, Sordariomyecetes, Xylariomycetidae, Xylariales, Xylariaceae]

Basionym: *Sphaeria truncata* Schweinitz, *Schr. Naturf. Ges. Leipzig* 1:44 (1822)

Derivation: *Annulo*, Latin, "little ring"; *hypoxylon*, Greek, "beneath wood," the name referring to the stromata (fruitbodies), which resemble those of *Hypoxylon* but with ostioles (openings) surrounded by a ring-like disk; *truncatum*, Latin, "cut off," "truncated."

Fruitbody (stroma): Stromata tightly clustered, rounded, conglomerate, cushion-shaped, raspberry-like masses with prominent globular perithecial mounds (Fig. A), 2–12 mm (up to ~½ in) in diameter by 0.6–2 mm high; with papillate ostioles (nipple-like openings) encircled by a flat disk 0.2–0.4 mm in diameter (Fig. B); each perithecium surrounded by a carbonaceous layer; stromata dark reddish-brown when young, blackish-brown when mature (Fig. A).

Flesh: Tissue beneath the perithecial layer dark brown to blackish, persistent.

Macrochemical staining: Extractable olive-green or dull olive-green pigments in KOH.

Microscopic: Spores (ascospores) brown to dark brown, 8–10.5 × 4.5–6 μm, ellipsoid-inequilateral, with narrowly rounded ends, broadly spindle-shaped; straight germ slit running the length of the spore. Stipitate, cylindrical asci (spore-bearing sacs within the perithecia) 70–90 μm long × 4.5–6 μm wide, with stipes 70–90 μm long; apical ring bluing or not in Melzer's reagent.

Occurrence: On dead branches of oaks (*Quercus* spp.); North and Central America, Caribbean; reported from Korea.[1]

Edibility: Inedible.

Comments: This is the type species for the genus *Annulohypoxylon* Y.-M. Ju, J.D. Rogers & H.-M. Hsieh (2005).[2] A similar species, *Annulohypoxylon annulatum* (Schweinitz) Y.-M. Ju, J.D. Rogers & H.-M. Hsieh (2005), also occurs on dead oak branches, but the stromata are larger (up to 20 mm wide × 4.5 mm high), diffusely cushion-shaped to semiglobose (never forming a tight, conglomerate, raspberry-like mass), and the perithecia are inconspicuous and much less prominent. The papillate ostioles are encircled by a faint convex (rather than a sharply defined flat) disk, and the asci also have shorter stipes (45–55 μm). The spores of these two species are indistinguishable. See also the distinguishing remarks in the account of the Carbon Cushion, *A. multiforme* and *A. cohaerens*.

1. J. Cha et al., 2012, *Mycobiol.* 40(1):79–81, http://www.ncbi.nlm.nih.gov/pmc/articles/PMC3385149/.
2. Y.-M. Ju et al., 2005, *Mycologia* 97(4):844–865.

Carbon Antlers, Candlesnuff Fungus

Xylaria hypoxylon

A B C

Photos: A & B, W. Light; C, Norbert Nagel.

***Xylaria hypoxylon* (Linnaeus : Fries) Greville**, *Fl. Edin.*:355 (1824) [Ascomycota, Sordariomyecetes, Xylariomycetidae, Xylariales, Xylariaceae]

Basionym: *Clavaria hypoxylon* L., *Sp. Pl.* 2:1182 (1753)

Derivation: *Xylaria*, Greek, pertaining to wood, on which this genus grows; *hypoxylon*, Greek, "under or beneath wood."

Fruitbody: Up to 8 cm tall by 2–7 mm wide (3¼ × ⅘–⅓ in), cylindrical to irregular, typically black with branched, flattened, strap-like white tips, these later becoming entirely black and minutely pimpled; woody to carbonaceous, surface smooth or roughened by perithecia (sexual spore-bearing structures).

Flesh: White, surrounded by black line of perithecia containing stalked spore-sacs (nonsessile asci) having umbilicate to slightly nipple-like openings that stain blue (amyloid reaction) in Melzer's reagent.

Stalk: Long or short, black.

Microscopic: Spores (ascospores) 12–15 × 5–6 µm, brown, ellipsoid, flattened on one side, smooth, with straight germ slit slightly shorter than length of spore.

Occurrence: Saprobic on decaying hardwoods, June–August; widespread and common in eastern North America.

Edibility: Inedible.

Comments: The white tips on the fruitbodies are due to a coating of conidia (asexual spores). This normally slender species occasionally occurs as a more robust form, which can make identification problematic. This species is generally more highly branched than *X. polymorpha* (which is usually unbranched), and the branches are flattened and strap-like; the spores are much larger (20–30 × 5–9 µm), and the germ slit is only ½–⅔ the length of the spore. Keys to the species of *Xylaria* and other Pyrenomycetes are provided by Rogers (1986) and Rogers et al. (2008).[1]

1. J.D. Rogers, 1986, *Mycotaxon* 26:85–97; J.D. Rogers et al., 2008, Pyrenomycetes of the Great Smoky Mountains National Park. VI. *Kretzschmaria*, *Nemania*, *Rosellinia*, and *Xylaria* (*Xylariaceae*), *Fungal Diversity* 29:107–116, http://fungaldiversity.org/fdp/sfdp/29-9.pdf.

Dead Man's Fingers

Xylaria polymorpha (*sensu lato*)

A B C

***Xylaria polymorpha* (Persoon) Greville**, *Fl. Edin.*:355 (1824) [Acomycota, Sordariomycetes, Xylariomycetidae, Xylariales, Xylariaceae]

Basionym: *Sphaeria polymorpha* (Pers.), *Comm. Fung. Clav.* (Lipsiae):17 (1791)

Derivation: *Xylaria*, Greek, *xylos*, "wood," on which this genus grows; *polymorpha*, Greek, "many forms," referring to the varied shapes the fruitbody can assume.

Fruitbody (stroma): 2–10 cm (0.8–4 in) tall by 1–4 cm (⅖–1⅔ in) thick; narrowly digitiform, cylindrical, bluntly pointed to rounded when young (Fig. A). Distally expanding when mature to knobby, broadly club-shaped forms (Fig. C), or flattened, spatulate, or tongue-shaped, occasionally branched (Fig. B); blue-grey with white tips when young (Fig. A), later (when sexually mature) becoming dark grey to black (Fig. C), hard, carbonaceous, studded with tiny pimple-like perithecia (spore-bearing structures [Fig. C, inset]), surface finely wrinkled or cracked; in spring covered with dust-like coat of white asexual spores (Fig. B).

Stalk: Short to medium length, irregularly columnar, merging imperceptibly with fruitbody.

Flesh: White, fibrous, tough, with single layer of black perithecia just below the surface.

Microscopic: Spores dark brown, 20–30 × 5–9 µm, spindle-shaped, long-elliptical, one side flattened with germ slit running ½–⅔ the length of the spore; surface smooth.

Occurrence: Saprobic, causing soft rot; solitary, several, or gregarious (Fig. A) from base of rotting or injured hardwood trunks, stumps, or buried roots, often appearing to sprout from the ground; June–November, often overwintering; widespread in North America.

Edibility: Inedible.

Comments: *Xylaria polymorpha* (*sensu lato*) actually comprises a group of several morphologically indistinguishable and poorly understood cryptic species. The stromata (fruitbodies) usually arise in the spring and early summer and may persist for many months or several years, during which time they continue to release spores. The mature stromata are morphologically indistinguishable from those of *X. longipes* and *X. cubensis* but can be distinguished microscopically by their spores. Those of *X. cubensis* are short-ellipsoid (8.5–11 × 4–5 µm) and lack an obvious germ slit; those of *X. longipes* do not exceed 17 µm in length and have spiral germ slits. The spores of *X. hypoxylon* (Carbon Antlers) do not exceed 16 µm, and the straight germ slit runs nearly the length of the spore. Keys to the species of *Xylaria* and other Pyrenomycetes are provided by Rogers (1986) and Rogers et al. (2008).[1]

1. J.D. Rogers, 1986, *Mycotaxon* 26:85–97; Rogers et al., 2008, Pyrenomycetes of the Great Smoky Mountains National Park. VI. *Kretzschmaria*, *Nemania*, *Rosellinia*, and *Xylaria* (Xylariaceae), *Fungal Diversity* 29:107–116, http://fungaldiversity.org/fdp/sfdp/29-9.pdf.

Cuban Club/Bubblegum Fungus

Xylaria cubensis/Xylocoremium flabelliforme

A B C D

A, *Xylaria cubensis* and *Xylocoremium flabelliforme*; B, *Xylaria cubensis*; C & D, *Xylocoremium flabelliforme*. (Photos: A, John Plischke III; B–D, W. Light.)

Teleomorph: ***Xylaria cubensis*** **(Montagne) Fries**, *Nova Acta R. Soc. Scient. Upsal.*, ser. 3, 1(1):126 (1851)

Basionym: *Hypoxylon cubense* Montagne, *Annls Sci. Nat.*, Bot., sér. 2, 13:345 (1840)
[Ascomycota, Pezizomycotina, Sordariomycetes, Xylariomycetidae, Xylariales, Xylariaceae]

Derivation: *Xylaria*, Greek, *xylos*, from "wood"; *cubensis*, Latin, "from Cuba" (the type locality).

Fruitbody (stroma): Solitary, usually unbranched, mature stroma up to (1.5–)3–7.5(–10) cm tall by (0.2–)0.5–1(–2.5) cm thick (1¼–3 [up to 4] × ~¼–½ [up to 1] in), cylindrical to club-shaped, with bluntly rounded tip; *bronze or copper-colored when young*, becoming blackish-brown at maturity, but faint copper tinge always present (Figs. A & B); surface developing minute pimples (perithecia 0.5 mm across), with papillate ostioles surrounded by network of cracks visible with hand lens and with texture like fine sandpaper to the touch; deep fissures or cracks lacking.

Stipe: Short (Fig. B) or lacking (Fig. A), when present often arising from reddish felt-like base.

Flesh: White at first, becoming carbonaceous or hollow.

Microscopic: Spores (ascospores) 9–10.5 × 4.5–5 µm, ellipsoid-inequilateral, smooth, brown to dark brown, germ slit usually lacking or obscure; asci stalked, apical ring bluing in Melzer's reagent.

Occurrence: Widespread and common on decaying hardwood throughout the Southeast as far north as New Jersey; tropical, subtropical, and temperate regions of the world; August–November in our area.

Edibility: Inedible.

Anamorph: ***Xylocoremium flabelliforme*** **(Schweinitz) J.D. Rogers**, *Mycologia* 76(5):914 (1984)

Anamorphic form-genus and form-species: See comments below and in online Appendix A
[Ascomycota, Sordariomycetes, Hypocreomycetidae, Hypocreales, Cordycipitaceae]

Basionym: *Sphaeria flabelliformis* Schweinitz : Fries, *Elench. Fung.* 2:58 (1828) [see online Appendix A for anamorph and teleomorph synonymies]

Derivation: *Xylocoremium*, Greek, *xylos*, "wood"; *coremium*, Greek, "broom" (referring to the shape of the fungus); *flabelliforme*, Latin, "fan- or leaf-shaped."

Fruitbody: (coremium)—this separate, distinctive anamorphic form is erect, with a total height of 5–30 mm (¼–1¼ in) and a width of 2–15 mm (0.08–⅔ in), bases swollen, black (Fig. A, arrow), with enlarged whitish, yellowish, or pinkish leaf-like or coral-shaped heads (Figs. A, C, & D). Typically, a coremium is not connected with the teleomorph, although it is sometimes initiated from the base of a deteriorating teleomorph (Fig. A).

Microscopic: Conidiophores mostly unbranched, forming dense palisades on the lamellae; conidia (3)4–6(7) × (1.5)2–3(4) µm, 1-celled, lacking germ slits, clear, obovate to ellipsoidal, with flattened bases, produced sympodially and forming pinkish masses beneath the coremia.

Occurrence: Coremia widespread but occasional throughout Southeast (as far north as New Jersey) and tropics, absent from the West; solitary or scattered on decaying hardwood; March–June, into later summer months in northern regions. Teleomorphs maturing August–November.

Edibility: Inedible.

Comments: The anamorphic (asexual) form of this fungus, *Xylocoremium flabelliforme*, is separate and distinct from the teleomorphic (sexual) stage, *Xylaria cubensis*. In this respect, it differs from most other *Xylaria* species, which bear their anamorphic stages on immature sexual stromata. Figure A, photographed by John Plischke III, shows a rare example of the anamorph and teleomorph occurring together. For this reason, a distinct form-genus, *Xylocoremia*, was proposed for it by Rogers.[1]

This pink little asexual fungus looks like well-chewed bubblegum stuck on a log or branch and can hardly be mistaken for anything else. *Xylocoremium flabelliforme* is the anamorphic (asexual) form of *Xylaria cubensis*—the teleomorphic, or sexual, form. Some works continue to refer this fungus to the form-genus *Isaria*. See online Appendix A for the synonymies of both the anamorphic and teleomorphic forms of this species.

The sexual form, *Xylaria cubensis*, is erect, club-shaped, and up to 10 cm (4 in) tall and 2.5 cm (1 in) thick. The finger-like coppery-brown to brownish-black fruitbody has a white interior, which may become hollow and carbonaceous. Unlike many *Xylaria* species, it is never covered with white conidiospores (asexual spores). The surface is smooth except for tiny cracks around the ostioles (tiny holes through which the ascospores are released), which feel like fine sandpaper. *Xylaria cubensis* can be difficult to distinguish from other very similar mature *Xylaria* species in our area that have shed their initial powdery white dusting of asexual spores. These include *X. corniformis* (Fr.) Fries, *X. longipes* Nitschke (Dead Moll), and *X. polymorpha* (Persoon) Greville (Dead Man's Fingers). See the discussion under Dead Man's Fingers. *Xylaria hypoxylon* (Linnaeus *ex* Fries) Greville (Carbon Antlers) is also similar, but the stromata are thinner and less robust and are usually branched and flattened near the tips.

1. J.D. Rogers, 1984, *Mycologia* 76(5):914. The holomorph clearly belongs to the genus *Xylaria*. However, the basionym *Sphaeria flabelliformis* was published before *Hypoxylon cubense*. Strict application of the new "one species, one name" rule required by the *ICN* (Melbourne Code) would seem to mandate that the teleomorph widely known as *Xylaria cubensis* and the equally well-known anamorph (*Xylocoremium flabelliforme*) must now be called *Xylaria flabelliformis*. This is exacerbated by the fact that *Sphaeria flabelliformis* Schweinitz was sanctioned by Fries in 1828, while *Hypoxylon cubense* was not sanctioned. In a personal communication to us, Dr. Jack D. Rogers, a leading authority on the order Xylariales and the naming authority for the anamorph's form-species, strongly urged us to retain both names, a decision we had already made and for which we had sought his support.

Fairy Sparklers

Xylaria tentaculata

A B C

Photos: Jim Edwards.

Xylaria tentaculata **Ravenel** ***ex*** **Berkeley & M.A. Curtis**, *J. Linn. Soc., Bot.* 10(46):381 (1868) [1869][1]
[Ascomycota, Sordariomycetes, Xylariomycetidae, Xylariales, Xylariaceae]

Derivation: *Xylaria*, Greek, *xylos*, "wood"; *tentaculata*, from Neo-Latin *tentaculum*, "feeler," referring to a slender, flexible process or filament.

Fruitbody: 1.5–4 cm (⅔–1⅔ in) tall, consisting of a stalk topped by a fertile head (Fig. A, arrow and letter H) with a radiating crown of 8–20 asexual tentacles (Figs. A & B); teleomorphic head (Fig. B) subcylindrical, up to 1 cm high by up to ⅔ cm thick (½ × ¼ in), prominently studded with grey to dull blackish-brown perithecial mounds (Fig. B, letter P and arrows) up to 1 mm in diameter with a papillate ostiole; with a crown of 8–20 pale grey tentacle-like branches up to 4 cm (1⅔ in) long by ¾ mm thick, these initially erect, later spreading out and tapering toward the tips, occasionally branched; sterile crown and fertile head covered with whitish powder at maturity (Figs. A & B).

Flesh: Soft, white, fragile.

Stalk: 1.5–2.5 cm tall by up to 3 mm thick (⅔–1 × 1⁄12 in), erect, subequal, twisted, often curved near base (Figs. A & C); surface granular, grey to blackish.

Microscopic: Spores (ascospores) 14–19(–22) × 6–8.5(–9) µm, elliptical-inequilateral to sickle-shaped, smooth, biguttulate (containing two large oil droplets), germ slit slightly shorter than spore length, dark brown; apical ring of ascus bluing in Melzer's reagent; hyaline (clear) conidia (asexual spores) produced on thin, hair-like conidiophores on the branches.

Occurrence: Saprobic on organic debris, in leaf litter, on decaying wood, or in moss in forests; solitary, scattered, or gregarious, summer–fall; widespread and locally rare to common in southeastern North America as far north as Quebec; Mexico, Cuba, Sri Lanka.

Edibility: Inedible.

Comments: The distinctive crown of thin tentacle-like branches makes it highly unlikely that this species will be confused with any other. However, because of its tiny size and fragility it is often overlooked. It appears to be more common in the Southeast.

Xylaria tentaculata develops as an erect club-shaped stroma that produces anamorphic tentacle-like branches when it attains a height of 1½ to 2 cm; thin lateral hair-like conidiophores form on these branches, which produce conidia sympodially on thin, hair-like conidiophores.[2] A swollen subcylindrical teleomorphic (sexual) structure studded with mound-like perithecia later forms at the top of the stalk and just beneath the tentacular apical branches.[3] Once the teleomorphic head has formed, the production of conidia on the branches ceases.[4]

Xylaria tentaculata belongs to a small section containing just three species that bear their conidia on a crown of sterile (anamorphic) branches on a teleomorphic stroma that subsequently develops a club-shaped head just below the branches in which the sexual asci develop. In addition to *Xylaria tentaculata*, this section includes *X. comosa* (Montagne) Fries (Central and South America) and *X. fimbriata* Lloyd (Puerto Rico).

1. Berkeley later (1875, *Grevillea* 4[30]:48) listed this species as "*Xylaria tentaculata* B. & Br. [Berkeley & Broome]." This appears to have been a *lapsus calami*, which has caused considerable confusion about the correct authorship of this taxon. Despite what you will find in many mushroom guides and online sources, this taxon is correctly referred to as *Xylaria tentaculata* Berkeley & M.A. Curtis, not Berkeley & Broome.

2. Sympodial conidia form sequentially at the base of an apical blastic cell. The main axis of this cell continues to grow and produce new conidia beyond and to one side of the original conidium.

3. M.A. Vincent & K. Metcalf, 2006, *Mich. Bot.* 45:207–209.

4. J.D. Rogers, 1985, *Sydowia, Annales Mycologici*, Ser. II, 38:255–262.

Sweetgum Fruit Fungus

Xylaria liquidambar

A

B

C

***Xylaria liquidambar* J.D. Rogers, Y.-M. Ju & F. San Martín,** *Sydowia* 54(1):92 (2002)[1]
[Ascomycota, Sordariomycetes, Xylariomycetidae, Xylariales, Xylariaceae]

Misapplied name: *Xylaria persicaria* (*ex auctore*) [*non Sphaeria persicaria* Schweinitz], *Schr. Naturf. Ges. Leipzig* 1:28 (1822)

Derivation: *Xylaria*, Greek, *xylos*, "wood," on which this genus typically grows; *liquidambar*, from the Latin *liquidus*, "fluid," and the Arabic *ambar*, "amber"—a noun in apposition standing for the sweetgum genus (*Liquidambar*), with which this fungus is exclusively associated.

Fruitbody (stroma): Asexual (conidial) stage—erect, slender, wiry, twisted and crooked, thickening toward base, with sharply pointed, needle-like tip (Fig. C), longitudinally striate (Fig. C, inset), tough; 2–5 cm tall by 1–2 mm thick at base; upper ⅔–¾ coated with white conidiospores, lower portion blackish; sexual stage similar to asexual stage but entirely blackish, covered by bumpy perithecia (Figs. A & B).

Microscopic: Conidiospores 7.5–10 × 3–4 µm,[2] broadly spindle-shaped to elliptical, smooth, clear; **ascospores** 10–16 × 4–6 µm, elliptical to bean-shaped, flattened on one side, with long spiral germ slit, smooth, clear when young, becoming dark brown; apical pore bluing in Melzer's reagent.

Occurrence: Found singly or in clusters exclusively on the fruits of *Liquidambar* spp., in our area only on the fallen fruits of American Sweetgum, *L. styraciflua* Linnaeus; widespread in the Southeast; reported from Mexico,[3] China,[4] and Korea (on *Liquidambar styraciflua* and *L. formosana* Hance).[5]

Edibility: Inedible.

Comments: The thin, crooked, wiry stromata and its occurrence on the fallen fruits of Eastern Sweetgum immediately identify this species. The correct spelling of the scientific name of this species is *Xylaria liquidambar* and not *X. liquidambaris*, as cited by most authors (see note 1).

1. The emendation of the original spelling of the specific epithet by *Index Fungorum*, *Species Fungorum*, and *Mycobank*, which has been followed by most subsequent authors, is a misapplied, incorrect, and invalid orthographic correction. We adhere to the original spelling by Rogers and his colleagues, who insist that the epithet is "a noun in apposition" and not an adjective, as incorrectly emended (J.D. Rogers, personal communication, June 19, 2013). A detailed explanation of our reasons is available in online Appendix A.
2. Ascospore dimensions are given as 15–17(–19) × 4–6 µm by R. Medel et al. (2008) for samples on *Liquidambar styraciflua* from montane cloud forests in the state of Veracruz, Mexico.
3. J.D. Rogers, 1979, *Canadian J. Bot.* 57:941–945; F. San Martín and J.D. Rogers, 1989, *Mycotaxon* 34:283–373.
4. J.D. Rogers, 1979, *Canadian J. Bot.* 57:941–945.
5. J.-G. Han & H.-D. Shin, 2007, *Mycobiology* 35(4):171–173 [as *Xylaria persicaria* on *Liquidambar styraciflua* & *L. formosana* Hance].

Magnolia Cone Fungus

Xylaria magnoliae

A

B

***Xylaria magnoliae* J. D. Rogers**, *Can. J. Bot.* 57(8):941 (1979)

[Ascomycota, Sordariomycetes, Xylariomycetidae, Xylariales, Xylariaceae]

No synonyms

Derivation: *Xylaria*, Greek, *xylos*, "wood"; *magnoliae*, Latin, "found on magnolias."

Fruitbody: Long, tapering, tentacle-like stromata with acute tip (Fig. A), branched or unbranched, arising from finely hairy to felt-like base; up to 12 cm long by 5 mm (4 × ~¼ in) wide at base; smooth to somewhat tomentose, brown to dull black, woody, densely studded with knob-like perithecia (Fig. C), these 0.3 mm in diameter, with small nipple-like openings (ostioles).

Microscopic: Spores (ascospores) pale yellow, 11–15(–17) × 3–5(–6) µm, irregularly spindle-shaped to crescent-like, smooth, germ pore inconspicuous; asci stalked, 90–120 × 6–8 µm, spore-bearing region 50–70 µm, apical ring blue in Melzer's reagent.

Occurrence: Common, widespread, May–November; in North America restricted to magnolia cones.

Edibility: Inedible.

Comments: The small white gilled mushrooms sprouting from the same magnolia cone as *X. magnoliae* in Figure B are Magnolia Cone Mushroom (Magnolia Coincaps), *Strobilurus conigenoides* (Ellis) Singer. This widespread species is also restricted to magnolia cones. This makes identification of both species very easy and straightforward, as no other fungi occur on this substratum.

Two other species in our area that may be confused with *X. magnoliae* are *Xylaria liquidambar* J.D. Rogers, Y.-M. Ju & F. San Martín, and *X. oxyacanthae* Tulasne. *Xylaria liquidambar* grows on the fallen fruits of Tulip Magnolia (*Liriodendron tulipifera*) and Sweetgum (*Liquidambar styraciflua*), not magnolia cones; in our area *X. oxyacanthae* occurs only on hickory and pecan nut shells (*Carya* spp.). *Xylaria liquidambar* previously has been misidentified as *X. persicaria* (Schweinitz : Fries) Berkeley & M.A. Curtis. This is discussed in detail by J.D. Rogers, Y.-M. Ju, & F. San Martín, 2002, *Sydowia* 54:91–97.

Hickorynut Club

Xylaria oxyacanthae

A

B

Photos: Teresa Fortenberry.

Xylaria oxyacanthae **Tulasne & C. Tulasne**, *Selecta Fungorum Carpologia: Nectriei-Phacidiei-Pezizei* (Paris) 2:15 (1863) [Acomycota, Sordariomyecetes, Xylariomycetidae, Xylariales, Xylariaceae]

Derivation: *Xylaria*, Greek, *xylos*, "wood," on which this genus typically grows; *oxyacanthae*, Greek, combining form of *oxýs*, "sharp," and *akanthos*, "thorn," referring to the form of the fruitbody.

Fruitbody (stroma): Conidial (asexual) stage 5–10 cm tall by 1–3 mm thick at base (2–4 × ¹⁄₁₆–⅛ in), erect, tough, thin, cylindrical to slightly laterally compressed, simple or occasionally branched, with acutely pointed yellowish to pink tips (Fig. B), upper portion covered with white layer of conidiospores (Figs. A & B), lower black portion buried within the soil; sexual stage black, brighter tips usually lost, slightly thicker (3–4 mm), with visible embedded globose to subglobose perithecia.

Microscopic: Conidiospores (asexual spores) 3.1–4.3 × 2.4–3.8 µm, globose to ovoid, clear, smooth; spores (ascospores) (8–)10–13(–14) × 4.5–6.0 µm, smooth, dark brown, elliptical, flattened on one side, with straight germ slit running the length of the spore; apical ring shaped like an inverted hat, bluing in Melzer's reagent.

Occurrence: Solitary, in pairs, or in dense stands sprouting in litter from the buried husks of hickory nuts (*Carya* spp.) or the seeds of rotted hawthorn fruits (*Crataegus* spp.); conidial stage sprouting in spring/early summer (May–June), sexual stage in summer/fall (July–October); reportedly fairly common.

Edibility: Inedible.

Comments: This fungus is immediately identified by its substratum. It only grows from the buried husks of hickory nuts or hawthorn seeds after the fleshy fruit has rotted away.[1] Its asexual form (shown above) is also distinguished from other *Xylaria* species by its tall, thin, cylindrical shape and acute yellowish to pinkish tip, which contrasts with the white coat covering most of the column. The mature, dark brownish to blackish, club-like sexual form is macroscopically indistinguishable from several other members of this genus. When infecting *Crataegus* seeds in the Pacific Northwest, the immature conidia-bearing stromata of *X. oxyacanthae* appear from the seeds in the spring or second spring following fruitfall after the fruit pulp has rotted away. In at least some cases, the fungus appears to infect the seeds while they are still in the flower on the tree. This may also be true of other *Xylaria* species with narrow host specificities.[2] It is not known if this also applies to *Carya* infections.

1. Previous reports have insisted that *X. oxyacanthae* is restricted to *Crataegus* seeds (e.g., A.J.S. Whalley, 1985, *Sydowia, Annales Mycologici*, Ser. II, 38:369–382, 1987, *Agarica* 8[16]:68–72). However, it has also been at least occasionally reported growing from the husks of *Carya* nuts (E.A. Stowell & J.D. Rogers, 1983, *Mycotaxon* 17:433–444). Samples collected in the Georgia Piedmont (Atlanta area) by Teresa Fortenberry growing on buried hickory nut husks that we had identified as *X. oxyacanthae* were sent to Dr. Jack D. Rogers at Washington State University. Molecular tests undertaken by his colleagues Yu-Ming Ju and Monica Ju confirmed that this species is indeed *X. oxyacanthae* and that its hosts definitely include both *Crataegus* fruits and *Carya* nuts (J.D. Rogers, personal communication, November 12, 2013).

2. J.D. Rogers et al., 2008, *North American Fungi* 3(8):1–5.

Black Knot

Apiosporina morbosa

A

B

C

Black Knot, *Apiosporina morbosa*, on Wild Black Cherry.

Apiosporina morbosa (Schweinitz) von Arx, *Acta Bot. Nederl.* 3:86 (1954)

Basionym: *Sphaeria morbosa* Schweinitz, *Schr. Naturf. Ges. Leipzig* 1:40 (1822)
[Ascomycota, Pezizomycotina, Dothideomycetes, Pleosporomycetidae, Venturiales, Venturiaceae]

Derivation: *Apiosporina*, Latin, from *apium*, "celery," "parsley," and *sporarum*, "spored"; *morbosa*, Latin, from *morbidus*, "sick," "diseased."

Description: Fruitbody 3–30 cm and up to 5 cm in circumference (1¼–12 × 2 in), forming an elongated irregularly club-shaped or spindle-shaped, black, nodular, very hard, carbonaceous mass resembling animal scat wrapped around a twig or branch (Figs. A & C). Often forming massive, irregular sclerotia on the trunks of badly infected trees (Fig. B)—the sclerotium in Figure B was approximately 81 cm tall by 41 cm wide (32 × 16 in) growing at a height of 3½ m (~12 ft) on the trunk of a tall, mature Wild Black Cherry.

Microscopic: Spores (ascospores) (13–)16–18(–22) × (4.5–)5–6.5(–7.5) µm, smooth, club-shaped, pale green, becoming olivaceous; septate near base, divided into two unequal cells.

Occurrence: Parasitic and highly pathogenic on various trees of the genus *Prunus*, most commonly on cultivated and wild cherry and plum trees. In our area most commonly found on Wild Black Cherry, *Prunus serotina* var. *serotina* Ehrhart. Widespread and common throughout North America, including Alaska and Mexico; year-round.

Edibility: Inedible.

Comments: Some two dozen species of the genus *Prunus* are known to be infected by Black Knot, but this disease most commonly appears on wild and cultivated plums and cherries. We have found very large Black Knot cankers (Fig. B) on the main trunks of Wild Black Cherry trees in the Atlanta area and in North Georgia. Descriptions in the literature and online photos typically show the much smaller earlier-stage stromata forming on and around twigs and branches. We have also found these on nearby trees and high in the crown branches of the trees with large cankers (Figs. A & C), but have more frequently encountered the larger, more dramatic, and advanced lesions shown in Figure B. These lesions are tumor-like responses of the host to the fungal infection. This serious pathogen is found throughout North America and appears to be restricted to this continent.

The first signs of infection—light-brown spindle-shaped swellings (Fig. A, inside the white oval)—appear in late summer or in the following spring or fall. Fungal mycelia overwinter in the infected branches. These swellings enlarge over the next growing season to form olive-green velvety pseudothecia, which produce spindle-shaped conidia (asexual spores). After two winters these enlarge in the spring into hardened, black, spherical or elongate, canker-like pseudothecia (Fig. 1C) that produce eight ascospores (sexual spores) consisting of two unequal-sized cells contained within asci. Young branches and twigs of both wild and cultivated *Prunus* spp. are typically infected in the spring as the ascospores are forcibly released during rainy weather between the period of bud break and the setting of fruit. The ascospores are the primary inoculum and are dispersed by wind and rain.

Figure A shows a typical spindle-shaped knot on a Black Cherry twig. The U-shaped bend in the twig is due to asymmetrical cellular growth caused by growth factors released by the fungus. The white oval indicates the swelling of the twig produced by new fungal growth. The dead portions of old knots are often tinged whitish or pinkish, indicating infections by a secondary parasitic fungus, *Trichothecium roseum* (Fig. E, below right).

Black Knot sclerotia look very much like some of the Crown Gall tumors on the trunks and branches of many hardwoods and conifers. These tumors are caused by a common soil-living bacterium, *Agrobacterium tumefaciens*, which transfers a segment of its DNA into the nucleus of the host tree. This transforms the host's DNA, causing it to produce the tumorous cankers.

Figure D shows a Crown Gall canker high up on the trunk of a Swamp Chestnut Oak, *Quercus michauxii* Nuttall, in one of the units of the Chattahoochee National Recreation Area in Atlanta. The tumors produced by Black Knot and Crown Gall can appear virtually identical, but Black Knot is restricted to the genus *Prunus*, while Crown Gall infects a wide range of plant species. Crown Gall can also infect *Prunus*, but the canker is not associated with the characteristic spindle-shaped sclerotia wrapped around the branches and twigs.

D

Crown Gall, *Agrobacterium tumefaciens*, on a Swamp Chestnut Oak. Photo: W. Light.

E

Photo: John Plischke III.

Jelly Fungi

The species in this section are not arranged in any particular taxonomic sequence, but according to their superficial resemblance to facilitate distinguishing them. *Sebacina* and *Tremellodendron* are discussed here even though they superficially more closely resemble the club fungi. This illustrates once again the logistical arrangement problem posed by fungi in a book of this nature. Completely unrelated groups all too often closely resemble each other, while related taxa often look very different.

The jelly fungi include a number of quite unrelated but morphologically very similar fungi representing three different classes of Basidiomycota: Tremellomycetes, Agaricomycetes, and Dacrymycetes. These were all originally included in the family Tremellaceae, which was named and described by Fries in 1821 in the class Hymenomycetes and the order Tremellini. Besides *Tremella*, Fries included the genera *Agyrium* Fries,[1] *Dacrymyces* Nees, *Naematelia* Fries,[2] *Exidia* Fries, and *Hygromitra* (Nees).[3] The family Tremellaceae was subsequently (1900) restricted by Patouillard to include only those jelly fungi with **tremelloid basidia**—basidia that are globose to ellipsoid and are longitudinally or diagonally septate and cruciate (i.e., the septa forming a cross when viewed from above; shown at right in Fig. C). Tremelloid basidia were long considered to be the *condicio sine qua non* for the order Tremellales (Class Tremellomycetes). The family Tremellaceae thus included the genera *Tremella* and *Exidia*, in addition to *Sebacina* and *Tremellodendron*, but *Dacrymyces* and *Auricularia* were excluded.

In 1897 Gustav Lindau erected the family Auriculariaceae for the genus *Auricularia* and its allies with **auricularioid basidia**—cylindrical and transversely multiseptate basidia with a single elongate and sinuous sterigma arising from each segment (Fig. B). These basidia also characterize the order Auriculariales, which is in the class Agaricomycetes. A variant of this is the **myxarioid basidium**, which is divided into two segments by a single transverse septum. However, in *Myxarium* species it is the hypobasidium (basal portion of the basidium) that bears the sterigmata and their spores and the distal segment that is stem-like. A third group of jelly-like fungi have bifurcate, or Y-shaped, **"tuning fork" basidia** (Fig. D), which distinguish all members of the class Dacrymycetes, which contains the sole family Dacrymycetaceae. Nearly all other Basidiomycota have aseptate **holobasidia** (Fig. A).

However, R.J. Bandoni's investigations of the ultrastructure of the septal pores showed that genera with tremelloid basidia such as *Tremella* and *Exidia* and their respective relatives are not closely related. Similarly, he found that genera with auriculoid basidia like *Auricularia* also represented several unrelated clades.[4] *Tremella* thus remains the major teleomorphic genus defining the order Tremellales and family Tremellaceae (itself an artificial collection of several ill-defined clades), which also includes a number of anamorphic yeast genera. All teleomorphic members of the Tremellaceae are parasitic on other fungi (including lichens), but the anamorphic yeasts infect a variety of hosts, and some (e.g., *Cryptococcus*) are pathogenic to humans. *Exidia*, despite having tremelloid basidia, is now

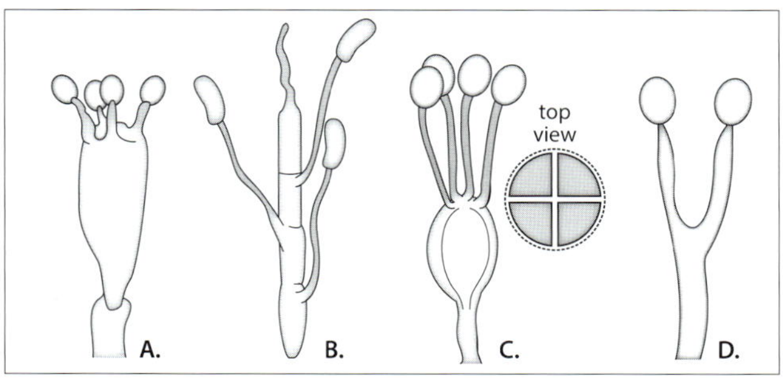

Types of Basidia. A, aseptate holobasidium, characteristic of most Basidiomycota; B, laterally septate auricularioid basidium, found in the order Auriculariales; C, longitudinally septate and cruciate tremelloid basidium of the class Tremellomycetes (top view showing the cruciate septa); D, Y-shaped "tuning fork" basidium found in all members of the class Dacrymycetes.

grouped together with *Auricularia* and *Myxaria* in the order Auriculariales. The first two genera are in the family Auriculariaceae, while *Myxaria* is placed in the obscure family Hyaloriaceae; all are believed to be saprobic on dead wood or plant material.

Dacrymyces, *Dacryopinax*, and *Calocera* all have Y-shaped tuning fork basidia, a feature that unequivocally places them in the class Dacrymycetes and the order Dacrymycetales containing the sole family Dacrymycetaceae. Moreover, *Dacrymyces* is also parasitized by *Stereum* species rather than the other way around, as is the case with *Tremella*. Because *Dacrymyces* has a tremelloid fruitbody that looks for all the world like a species of *Tremella* (one of the "witches' butter" fungi), it is discussed in detail in this section. But because of their more structured fruitbodies, which resemble tiny club fungi, *Dacryopinax* and *Calocera* are treated in the following section along with the earth tongues, club fungi, and coral fungi.

The spreading, white, encrusting jelly fungi of the genera *Sebacina* and *Tremellodendron* belong to the class Agaricomycetes, the order Sebacinales, and the family Sebacinaceae. Both were previously assigned to the Tremellaceae. The genus *Sebacina* Tulasne & C. Tulasne was erected for highly polymorphic, effuse (spreading) to coral-like, cartilaginous to rubbery-gelatinous and encrusting fungi with tremelloid basidia. Both genera are ectomycorrhizal with forest trees and orchids, and the distinction between them is problematic. *Sebacina* is cosmopolitan; *Tremellodendron* is restricted to the Americas.

1. E. Fries, 1821, Systema Mycologicum I:55 (LV). Fries erroneously noted that the type of *Agyrium* (*A. rufum*, actually a lichen in the division Ascomycota) was synonymous with the *Tremella cinnabarina*.

2. *Naematelia* was based on a type species (*Tremella encephala* Persoon) whose distinguishing feature (its solid core) turned out to be the remnants of the host *Stereum* species parasitized by *T. encephala*.

3. The two known species of *Hygromitra* belong to the division Ascomycota.

4. R.J. Bandoni, 1984, *Trans. Mycol. Soc. Japan* 25:489–530.

Yellow Witches' Butter

Tremella mesenterica

A

B

***Tremella mesenterica* Retzius**, *K. Svenska Ventenskaps-Akad. Handl.* 30:249 (1769)
[Basidiomycota, Tremellomycetes, *incertae sedis*, Tremellales, Tremellaceae]

Derivation: *Tremella*, Latin, from *tremere*, "to tremble," and the suffix *-ella*, "little," referring to the gelatinous mass; *mesenterica*, Latin, "middle intestine."

Fruitbody: Gelatinous and firm when wet, irregularly lobed, brain-like; collapsing to a thin hard film when dry; translucently deep yellow to bright yellow-orange, fading to pale yellow; rarely colorless; up to 7.5 cm (3 in) wide by 2.5–5 cm (1–2 in) high.

Spores in mass: whitish to pale yellow.

Microscopic: Basidia sessile, 20–31 × 16–22 µm, bearing four spores, separated into longitudinal compartments by two or four vertical or diagonal septa; basidiospores elliptical to cylindrical, nonseptate, smooth, 10–16 × 6–12 µm; asexual conidia 2.0–3.0 by 2.0–2.5 µm, spherical, ovoid, or ellipsoid, sometimes coating young fruitbodies in a bright yellow slime. Clamp connections throughout.

Occurrence: Parasitic on the hidden hyphae of *Peniophora* spp., but rarely seen with host fruitbodies. Cosmopolitan; common on hardwoods in broadleaf and mixed temperate and tropical forests: North and South America, Europe, Africa, Asia, Australia.

Edibility: Edible but bland; used for texture in soups and other dishes.

Comments: This species superficially resembles *Tremella aurantia* Schweinitz (1822),[1] which grows on hardwoods and parasitizes *Stereum hirsutum* (Willdenow) Persoon (1800), with which it is typically associated. Unlike *T. mesenterica*, *T. aurantia* is usually not greasy or slimy but matte, and its lobes are thicker. *Tremella aurantia* often has a whitish fleshy internal layer formed by its *Stereum* host, which is often visible in a vertical section through the fruitbody; this layer may form cottony masses, and the lobes may be hollow. Because of the internal host hyphae, it does not collapse into a film but retains its shape when dried. Its spores are also smaller, 8.5–10 × 7–8.5 µm.

Tremella mesenterica also resembles *Dacrymyces palmatus*, which is somewhat smaller, is more orange in color, grows with a white mycelial attachment on conifer (not deciduous) wood, and, like all members of its family (Dacrymycetaceae), has Y-shaped basidia bearing two transversely septate spores.

All members of the genus *Tremella* alternate between a yeast-like asexual phase that produces conidia and a sexual reproductive stage that generates basidiospores. The fruitbodies arise from the mycelia of *Peniophora* located beneath the bark.

1. *Tremella aurantia* is not a synonym of *T. mesenterica*, as is often erroneously stated.

Orange Jelly, Orange Witches' Butter

"Dacrymyces palmatus"

A

B

Dacrymyces palmatus **(Schweinitz) Bresàdola**, *Höhn. Oesterr. Bot. Zeitschr.* 54:425 (1904)[1]
[Basidiomycota, Dacrymycetes, *incertae sedis*, Dacrymycetales, Dacrymycetaceae]

Basionym: *Tremella palmata* Schweinitz, *Trans. Am. Phil. Soc.*, ser. 2, 4(2):186 (1832)[1]

Derivation: *Dacrymyces*, Greek, *dákry*, combining form meaning "tear," "teardrop," and *mykēs*, Greek, "fungus"; *palmatus*, Latin, "palmate," "hand-shaped."

Fruitbody: 1–6 cm (0.4–2.4 in) wide by up to 2.5 cm (1 in) high, irregularly brain-like or multilobed, rubbery or soft gelatinous mass, abruptly narrowed and tough at attachment to substratum; yellowish-orange to orange (Fig. B), drying to reddish-orange, *with white mycelial mass at point of attachment* (Fig. A, B inset), *this feature diagnostic.* Flesh colored like fruitbody, liquefying with age, no distinctive odor or taste.

Spores in mass: Pale yellow.

Microscopic: Spores 17–25 × 6–8 µm, cylindrical, slightly curved, sausage-shaped, thin-walled, smooth, clear, with six to seven transverse septa; contents granular; basidia Y-shaped (like a tuning fork), bearing two spores. Ovoid, clear asexual conidia may also be present. Hyphae septate, lacking clamp connections.

Occurrence: Parasitic, causing brown rot. Single, scattered or clustered on dead conifer branches and logs; year-round; widespread and common in Southeast, especially after rains.

Edibility: Edible, used in soups.

Comments: "Dacrymyces palmatus" is commonly described as growing on decorticated conifers, but, as shown in the photos above, the host's bark (in this case *Pinus taeda*) is commonly intact. Similar-looking but unrelated *Tremella* species such as *T. mesenterica* and *T. aurantia* typically grow on dead hardwood branches and logs rather than on conifers. The basidia of *Tremella* species are also longitudinally septate (not Y-shaped as in *Dacrymyces*), and the spores are nonseptate and smaller (<17 µm long). See also the distinguishing comments under Yellow Witches' Butter (*Tremella mesenterica*). "Dacrymyces palmatus" liquefies into an amorphous mass upon aging. It is often parasitized internally by *Tremella* spp., especially when the edge of an older fruitbody is liquefying.

1. The species is also called *Dacromyces chrysospermus* Berkeley & M.A. Curtis (1873), but their original description is too vague to identify it with any certainty, and no type specimen was designated. Schweinitz's name, *Tremella palmata*, was earlier used for another species of fungus (*T. palmata* Schumacher 1803); it is thus an illegitimate homonym and cannot be used. However, because it has been widely known as *Tremella palmata*/*Dacrymyces palmatus* for nearly two centuries, we see no other option but to likewise provisionally call it "Dacrymyces palmatus." See online Appendix A for a full discussion of this issue.

Golden Ear

Tremella aurantia

A

B

Tremella aurantia parasitizing *Stereum hirsutum*. (Photos: A, W. Light; B, Franco Folini.)

***Tremella aurantia* Schweinitz**, *Schrift. Naturforsch. Gesell. Leipzig* 1:114 (1822)
[Basidiomycota, Tremellomycetes, *incertae sedis*, Tremellales, Tremellaceae]

Derivation: *Tremella*, Latin, from *tremere*, "to tremble," and the suffix *-ella*, "little," referring to the gelatinous mass; *aurantia*, Latin, from *aurantiaco*, "orange."

Fruitbody: Gelatinous, initially irregularly lobed, brain-like, initially compact and yellowish-orange (Fig. B), expanding into a larger, yellowish, more leaf-like mass (Fig. A) 10–15 cm (4–6 in) across, becoming pallid with age or when swollen with rainwater (Fig. A, arrow); surface shiny when wet, otherwise dull, relatively lusterless, sometimes covered with a white pruina (powdery coat); fruitbody retaining its shape when dry, reconstituting itself when wet. Odor and taste not distinctive.

Spores in mass: Presumably white.

Microscopic: Spores 8.5–10 × 7–8.5 μm, subglobose to ovoid, smooth, thin-walled, with hilar appendage; basidia stalked, diagonally septate, two to four spores 9–13 μm in diameter.

Occurrence: Parasitic on and typically associated with *Stereum hirsutum* (Willdenow) Persoon, growing on hardwoods, less often on conifers. Widespread in North and Central America, Europe, Asia, Australia.

Edibility: Edible, but hardly worth the effort.

Comments: This species superficially resembles Yellow Witches' Butter, *Tremella mesenterica* Retzius,[1] which grows on hardwoods and parasitizes the hidden hyphae of *Peniophora* species, while *T. aurantia* is usually found associated with its *Stereum* host (Figs. A & B). Unlike *T. mesenterica*, *Tremella aurantia* has a dull matte surface (except when wet) instead of the greasy or slimy fruitbody typical of *T. mesenterica*, and its lobes are thicker. The spores (8.5–10 × 7–8.5 μm vs. 10–16 × 6–12 μm) and the basidia (9–13 μm vs. 16–22 μm) are also smaller in diameter than those of *T. mesenterica*, and they are stalked rather than sessile. *Tremella aurantia* often has a whitish fleshy internal layer formed by its *Stereum* host, which is often visible in a vertical section through the fruitbody; this layer may form cottony masses, and the lobes may be hollow. Because of the internal host hyphae, unlike *T. mesenterica*, it does not collapse into a film but retains its shape when dried. Its spores are also smaller.

Tremella mesenterica also resembles "Dacrymyces palmatus," which is somewhat smaller, is more orange in color, grows with a white mycelial attachment on conifer (not deciduous) wood, and, like all members of its family (Dacrymycetaceae), has Y-shaped basidia bearing two transversely septate spores.

1. *Tremella aurantia* and *T. mesenterica* are not synonymous, as is often erroneously stated.

Brown Witches' Butter

Tremella foliacea

A

B

C

Tremella foliacea **Persoon**, *Observ. Mycol.* (Lipsiae) 2:98 (1800) *ex* Fries, *Syst. Mycol.* 2(1):212 (1822)
[Basidiomycota, Tremellomycetes, *incertae sedis*, Tremellales, Tremellaceae]

Derivation: *Tremella*, Latin, from *tremere*, "to tremble," and the suffix *-ella*, "little," referring to the gelatinous mass; *foliacea*, Latin, *folium*, "leaf," plus *-aceus*, "having the nature of," "leaf-like."

Fruitbody: Dense clusters of many sessile, gelatinous, closely overlapping leaf-like lobes resembling seaweed (Figs. A–C), 3–20 cm across and 5–10 cm high (1¼–8 × 2–4 in); pale to dark brown, purplish-brown, reddish-brown to nearly black.

Spores in mass: Pale yellowish.

Microscopic: Spores 7–12 × 6–10 µm; broadly elliptical to nearly spherical, thin-walled, smooth, clear, nonreactive in Melzer's reagent; basidia tremelloid (i.e., globose to ellipsoid), with cruciate longitudinal or oblique septa.

Occurrence: Parasitic on the mycelia (not the fruitbodies) of *Stereum* spp. (visible in Figs. A & C); found on dead or fallen hardwood branches, especially oaks; summer through winter in our area. Widespread but occasional in the Southeast; cosmopolitan: North and South America, Europe, northern Asia, North Africa, Australia, New Zealand.

Edibility: Edible but bland.

Comments: This species is most likely to be mistaken for Wood Ear, *Auricularia americana* Parmasto & I. Parmasto *ex* Audet, Boulet & Sirard, which it resembles. Wood Ear belongs to a different class (Agaricomycetes) and is distinguished by its white (not yellowish) spore print, larger (12–19 × 4–8 µm) sausage-shaped basidiospores, and transversely (not longitudinally) septate basidia. See other distinguishing features in the Wood Ear species account. *Tremella foliacea* is parasitic on several species of the poreless shelf fungus genus *Stereum*. It is quite variable and may comprise a complex of similar species throughout its range.

American Wood Ear

Auricularia americana (Deciduous Unit)

A

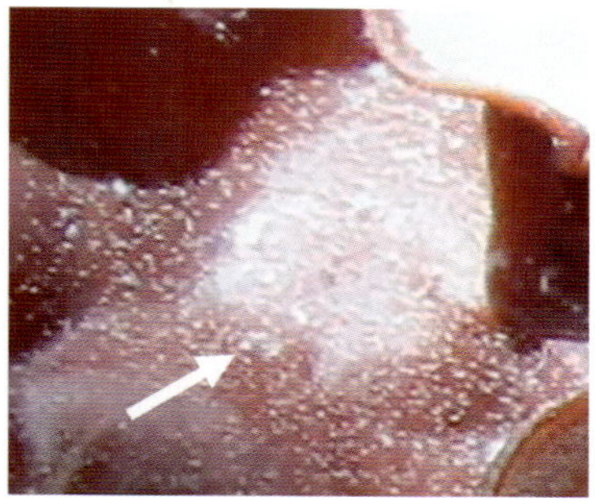

B

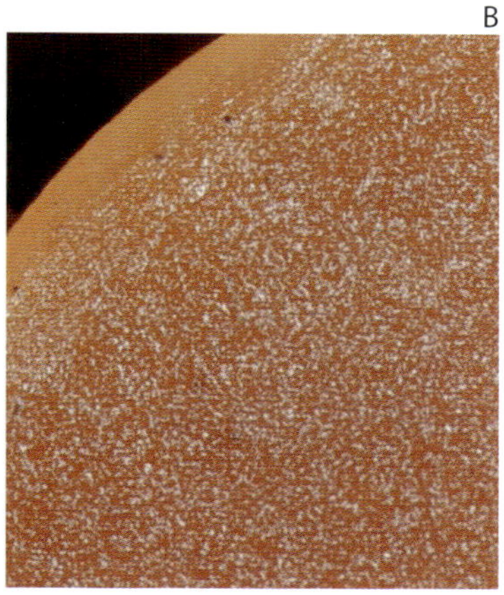

C

D

Photos: A, B & D, W. Light; C, Jim Edwards.

***Auricularia americana* Parmasto & I. Parmasto *ex* Audet, Boulet & Sirard,** in Boulet, *Les champignons des arbres de l'est de l'Amérique du Nord*:287 (2003) [*Auriularia americana* "Deciduous Unit" *sensu* Duncan & Macdonald (1967)[1] and Looney et al. (2013)][2]
[Basidiomycota, Agaricomycotina, Agaricomycetes, *incertae sedis*, Auriculariales, Auriculariaceae]

Derivation: *Auricularia*, Latin, from *auris*, "ear," and *-aris*, "pertaining to," referring to its ear-like shape; *americana*, "American."

Misapplied name (many authors): *Auricularia auricula-judae* (Bulliard) Quélet, *Enchir. Fung*. (Paris):207 (1886)

Fruitbody: Rubbery-gelatinous when fresh, usually 2–10 cm (0.8–4 in), but up to 12 cm (4¾ in) across, in cespitose to overlapping clusters (Fig. D), auricular (shaped like floppy ears) or cup-like (Figs. A & D); abhymenial (sterile upper) surface convex, often covered with minute, downy or silky, greyish hairs (Fig. C); laterally to centrally attached, sessile or with short, stubby stipe; context (flesh) more or less homogeneous, *medulla lacking* (Type B); hymenium (fertile lower surface) concave, moist, smooth; smooth when young, both surfaces later becoming deeply wrinkled or folded (Fig. A) and even more ear-like, tough; reddish-brown with purplish tint to cinnamon, yellowish-brown, or blackish-brown.

Spore print: White (Fig. B, arrow).

Microscopic: Spores 13.0–15.0 × 4.6–5.9 µm, allantoid (sausage-shaped), smooth; basidia up to 80 × 8 µm, divided into four compartments by three transverse septa, with one terminal and three lateral epibasidia; sterigmata rarely present.

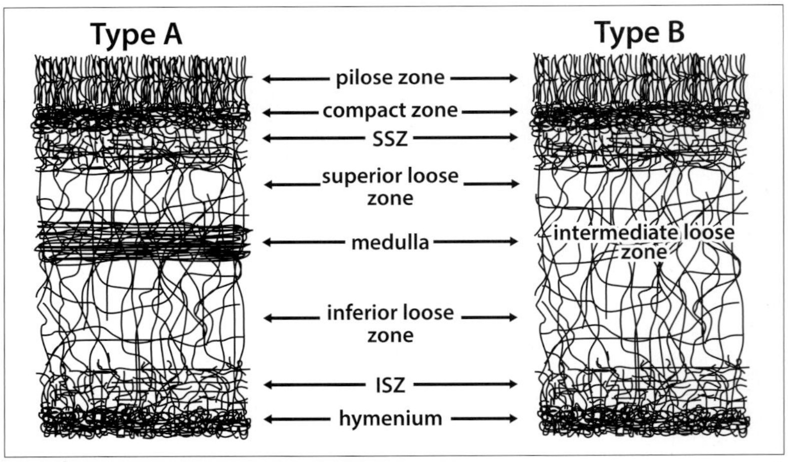

Transverse sections through the two types of *Auricularia basidiocarps.* Type A, with distinct medullary layer consisting of hyphae oriented parallel to the abhymenial and hymenial surfaces, characteristic of *A. fuscosuccinea.* Type B, medulla lacking or very indistinct, characteristic of *A. americana.* SSZ = superior subcompact zone, ISZ = inferior subcompact zone. (Modified after figures by Lowy [1951] and Looney et al. [2013].)

Occurrence: Saprobic; solitary, in clusters, or cespitose on decaying hardwood logs, branches, and sticks, sometimes forming large colonies; year-round but most common in fall; widespread in the Southeast and Midwest.

Edibility: Edible, but not very flavorful; used in soups as a filler.

Comments: Long misidentified as the European species *Auricularia auricula-judae* (Bulliard) Quélet (1886) by nearly all field guides and online sources, the American Wood Ear was recognized as distinct and described as a new species in 2003. This taxon remains confused, with two distinct clades designated by Looney et al. (2013) as follows: *Auricularia americana* (Coniferous Unit), associated with conifers and ranging from eastern Canada to western North America; and *Auricularia americana* (Deciduous Unit), found in the southeastern and midwestern United States south into Mexico.[2]

Auricularia fuscosuccinea (Montagene) Henning is continuously distributed from western Tennessee south to Argentina. It is only known in our area of interest by an apparently isolated population in the vicinity of Knoxville in eastern Tennessee, but it may actually be more widespread.[2] This species is very similar to *A. americana*, but it tends to be a deeper rosy-vinaceous brown to dark reddish-brown. It is distinguished by its prominent and often pigmented Type A medula. The allantoid spores are similar to those of *A. americana*, but they are smaller (11.0–13.6 × 6.5–8.5 µm) and contain one to two large oil droplets.[3]

Wood Ear is most likely to be confused with Brown Witches' Butter, *Tremella foliacea* Persoon, a similar-looking but quite unrelated species.[4] It is pale brown, dark brown, or purplish-brown to almost black, with a typically large seaweed-like gelatinous mass of up to 20 cm (8 in) ruffled, leaf-like blades or lobes with a common base. It is never cup-shaped or ear-shaped like Wood Ear. The spores are smaller, 7–12 x 6–10 µm, subglobose to oval, the spore deposit is pale yellowish, and the basidia are tremelloid (i.e., vertically or obliquely, not transversely, septate). *Tremella foliacea* is parasitic on and usually associated with *Stereum* spp. It grows on *Stereum*'s hyphae within the wood rather than on the host's fruitbody. It is highly variable and likely comprises a complex of species.

1. E.G. Duncan & J.A. Macdonald, 1967, *Mycologia* 59:803–818.
2. B.P. Looney et al., 2013, *North American Fungi* 8(6):1–25.
3. B. Lowy, 1951, *Mycologia* 43(3):351–358.
4. *Tremella* and *Auricularia* belong to two different classes: Tremellomycetes and Agaricomycetes, respectively. They are no more closely related than are cats and sparrows.

Black Witches' Butter

Exidia glandulosa

A

B

Photos: A, W. Light; B, Jim Edwards.

***Exidia glandulosa* (Bulliard) Fries**, *Syst. Mycol.* (Lundae) 2(1):224 (1822)
[Basidiomycota, Agaricomycotina, Agaricomycetes, *incertae sedis*, Auriculariales, Auriculariaceae]

Basionym: *Tremella glandulosa* Bulliard, *Herb. Fr.* 9:tab. 420, fig. 1 (1789)

Derivation: *Exidia*, Latin, from *exidium*, "destruction"; *glandulosa*, "very glandular."

Fruitbody: Individual sporocarps 1–3 cm (0.4–1.2 in) across and 2 cm (0.8 in) high, usually merging into large patches 50 cm (20 in) or more in extent, but individual fruitbodies remaining discrete (Fig. B); attached to wood substratum at base; shiny black fruitbodies amorphously lobed, brain-like; fertile upper surface smooth (Fig. A) or with scattered pimple-like pegs (Fig. B inset); sterile undersurface at first smooth and matte, later densely covered with small gelatinous spines.

Spores in mass: White.

Microscopic: Spores sausage-shaped, 14–19 × 4.5–5.5 µm; basidia tremelloid, longitudinally septate; clamp connections present.

Occurrence: Saprobic on dead hardwood sticks and branches, especially oaks; spring and fall, or in summer during cold spells and in warm winter periods.

Edibility: Inedible.

Comments: *Exidia glandulosa* is commonly confused with the very similar *Exidia nigricans* (Withering 1776) P. Roberts 2009, and the two were long considered to be conspecific. They are microscopically indistinguishable but can be distinguished macroscopically. The fruitbodies of *E. glandulosa* nearly always maintain their discrete identity even though they may be closely crowded together. *Exidia nigricans* fruitbodies coalesce into ill-defined, effused, highly ridged and wrinkled, black masses up to 10 cm (4 in) in extent. Molecular studies also indicate that they represent two distinct taxa. Both species are often called simply "Witches' Butter," which causes confusion with the common name for the yellow jelly fungus, *Tremella mesenterica*. See Donk (1966) and Roberts (2009) for discussions of the nomenclatural issues concerning these species.[1]

Exidia recisa (Ditmar) Fries is morphologically very similar to *E. glandulosa* but is yellow-brown, cinnamon-brown, or purplish-brown, and the surface is covered with minute blackish, peg-like projections visible with a land lens. It is most commonly found on dead willow branches and twigs.

1. M.A. Donk, 1966, *Persoonia* 4:228; P. Roberts, 2009, *Mycotaxon* 109:219–220.

Snow Fungus, Silver Ear Fungus

Tremella fuciformis

A

B

A & B, *Tremella fuciformis*; C, *Ophiostoma epigloeum*. (Photos: A, W. Light; B, Teresa Fortenberry; C, University of Georgia: http://naturalhistory.uga.edu/~GMNH/Mycoherb_Site/GAascos.htm#ophiepi.)

***Tremella fuciformis* Berkeley**, *Hooker's J. Bot. Kew Gard. Misc.* 8:277 (1856)
[Basidiomycota, Tremellomycetes, *incertae sedis*, Tremellales, Tremellaceae]

Deprecated synonym: *Nakaiomyces nipponicus* Kobayasi (1939)

Derivation: *Tremella*, Latin, from *tremere*, "to tremble," and the suffix *-ella*, "little," referring to the gelatinous mass; *fuciformis*, "spindle-shaped," referring to the lobes.

Fruitbody: Sessile, up to 7–7.5 cm (2¾–3 in) across and 3 cm (1⅛ in) high, firm-gelatinous (rubbery) watery, foliose, with numerous well-developed, erect, irregular, crenulate (frilly), lettuce-like flaps or lobes; translucently white when fresh, drying to pale yellow. Surface smooth, shiny; entire surface bearing spores.

Spores in mass: White.

Microscopic: Spores 7–9.5(–14) × 5–6(–8.5) µm, ovoid to elliptical, flattened on one side, clear, smooth, thin-walled; nonreactive in Melzer's reagent; basidia with four vertical septa, bearing four spores, sometimes with a stalk; gloeocystidia present; yeast-like subglobose conidia 4–5 × 3 µm present in context near base, sometimes also near or at hymenial surface; context hyphae with clamp connections and producing haustoria that penetrate the woody substratum, as well as the hyphae of their *Annulohypoxylon* or *Hypoxylon* hosts, but exact relationship unknown (see comments).

Occurrence: On various hardwoods, especially oaks; mainly tropical and subtropical, in southeastern North America; southern and eastern Asia; throughout Central and South America, Caribbean, sub-Saharan Africa, Australia, New Zealand, Oceania. Fruitbodies always associated with *Hypoxylon* species in nature.

Edibility: Edible, highly regarded for its medicinal properties.

Comments: This mushroom starts out as a nondescript slimy mucilaginous mass until it encounters a symbiotic partner: a species of *Hypoxylon* (*sensu lato*, many of which are now in the genus *Annulohypoxylon*). It is unclear whether this association is parasitic (and if so, just who is parasitizing whom) or mutualistic. When it does encounter a suitable partner, it is stimulated to form these beautifully distinctive fruitbodies. Its preferred mycobiont is *Annulohypoxylon archeri* (Berkeley) Y.-M. Ju, J.D. Rogers & H.-M. Hsieh (2005), but other species are also involved.

The somewhat similar Dead Man's Rubber Gloves, *Tremella reticulata* (Berkeley) Farlow (1908), is larger, up to 15 cm across by 8 cm high (6 × 3¼ in), and usually grows on the ground. The finger-shaped, distinctly coral-like lobes are opaque white (not clear or translucent) and develop from a central mass, becoming hollow with age. The spores are slightly longer (9–11 × 5–7 µm) and more elliptical. *Ductifera pululahuana* (Patouillard) Wells (1958) forms dingy whitish to yellowish brain-like masses on well-decayed, decorticated hardwoods. The spores are somewhat longer (9–12 × 4.5–7 µm), more elliptical, and yellowish in mass.

A tiny, often overlooked parasitic fungus, *Ophiostoma epigloeum* (Guerrero) de Hoog [Ascomycota, Sordariomycetidae], occasionally infects *Tremella fusiformis* (its only host).[1] Its minuscule fruitbodies (0.17–0.58 mm long × 0.04–0.05 mm wide at the base) appear as many tiny dark spines on the Snow Fungus's surface (Fig. C at right). It has only been recorded once in

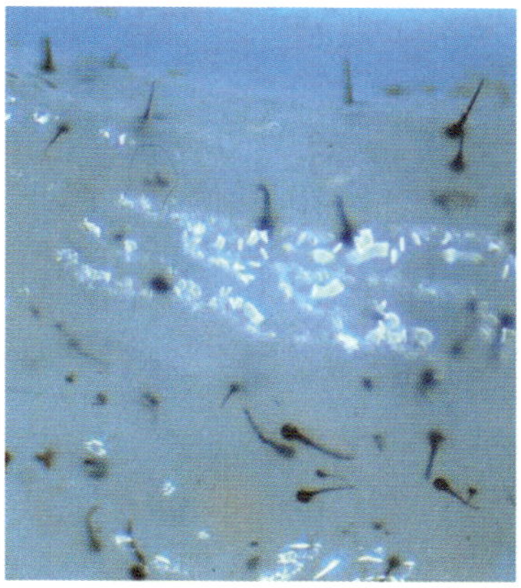

C

Georgia but is probably usually overlooked because of its tiny size. These tiny spine-like fruitbodies in its combined fungal association with *Tremella fusiformis* resulted in the proposal of *Nakaiomyces nipponicus* Kobayasi (1939) as a distinct species in Japan. *Nakaiomyces nipponicus* is otherwise indistinguishable from and is thus a synonym of *Tremella fuciformis*.

1. *Ophiostoma epigloeum* is a hyperparasite, since it parasitizes a fungus that is itself parasitic or mutualistic with another fungus. It appears to be a common parasite of *T. fuciformis* imported from China for restaurants.

Crystal Brain Fungus, Dead Man's Rubber Gloves

Myxarium nucleatum

Photo: Jim Edwards.

***Myxarium nucleatum* Wallroth**, *Fl. Crypt. Germ.* (Norimbergae) 2:260 (1833)
[Basidiomycota, Agaricomycetes, *incertae sedis*, Auriculariales, Hyaloriaceae]

Basionym: *Tremella nucleata* Schwein., *Schr. Naturf. Ges. Leipzig* 1:115 (1822)

Derivation: *Myxa*, Greek, "slime," plus *-arium*, Latin, a place or device associated with a specific feature or function; *nucleatum*, Latin, "kernel," referring to the white granular inclusions.

Fruitbody: Initially subglobose, becoming deeply irregularly convoluted and brain-like, often fused into gelatinous masses up to 20 cm long by 5 cm wide by 8 mm thick (8 × 2 × ⅓ in); gelatinous, wet, surface smooth to warty, translucent, dirty white to pale greyish-brown, pinkish, to reddish- or olive-brown.

Flesh: Gelatinous, clear, translucent, often containing small (⅕–⅓ mm) whitish granules of calcium oxalate; *this feature diagnostic*.

Spores in mass: White.

Microscopic: Spores 9–12 × 4–5.5 µm, sausage-shaped, smooth, thin-walled, containing granules; basidia myxarioid (with a stem-like distal part separated by a transverse septum from the tremelloid spherical basal part).

Occurrence: Saprobic and occasional on lower surfaces of decaying hardwood logs; scattered or fused into broad rows or sheets. Fruiting in winter following rainy periods and in autumn. Widespread, but uncommon in North America, Central America, South America, Europe, New Zealand.

Edibility: Unknown; however, most jelly fungi are edible.

Comments: This species is commonly referred to as *Exidia nucleata*. However, DNA studies suggest that this taxon properly belongs in the genus *Myxarium*. The white granular calcium oxalate inclusions in the context (visible in the photo above) immediately distinguish this species from its look-alikes: *Ductifera pululahuana* (Patouillard) Wells and *Sebacina sparassoidea* (Lloyd) P. Roberts, formerly called *Tremella reticulata* (Berkeley) Farlow. In addition, *D. pululahuana* spores are yellowish in mass, and *Sebacina sparassoidea* grows from the ground. The Snow Fungus, *Tremella fuciformis* Berkeley, exhibits clear, translucent white, prominent, lettuce-like, frilly flaps or lobes that are pristine white, quite unlike the dirty-white to grey-brown amorphous, brain-like masses of *Myxarium nucleatum*.

Coalescing White Jellyfungi

Sebacina concrescens and *S. sparassoidea*

A

B

A, *Sebacina concrescens*; B, *S. sparassoidea*. (Photos: A, W. Light; B, Ron Pastorino.)

***Sebacina concrescens* (Schweinitz) P. Roberts**, *Sydowia* 55(2):349 (2003)

Basionym: *Peziza concrescens* Schwein., *Schr. Naturf. Ges. Leipzig* 1:118 (1822)

***Sebacina sparassoidea* (Lloyd) P. Roberts**, *Sydowia* 55(2):351 (2003)

Basionym: *Tremella sparassoidea* Lloyd, *Mycol. Notes* (Cincinnati) 61:894, plate 135 (1919)
[Basidiomycota, Agaricomycotina, Agaricomycetes, *incertae sedis*, Sebacinales, Sebacinaceae]

Derivation: *Sebacina*, Latin, *sebum*, "fat," "tallow"; *concrescens*, Latin, "growing together of parts," referring to its manner of growing around and encrusting plant stems and debris; *sparassoidea*, Greek, *sparássein*, "to tear."

Fruitbody: *S. concrescens*—a soft, spreading, translucent to opaque, whitish to greyish, irregular cartilaginous or rubbery-gelatinous, irregular, lobate mass (Fig. A) growing from the ground; typically encrusting around and supported by the bases and stems of herbaceous plants; up to 14 and occasionally 25 cm (5½–10 in) across; ***S. sparassoidea***—firm-gelatinous, with multiple erect, wide, *very blunt, hollow, self-supporting* branches, often open at tips (Fig. B), arising from a common base; branches coalescing into a compound, often intricate, coralloid fruitbody up to 10 cm (4 in) tall and wide; fruitbody whitish to cream when fresh, drying to reddish-brown.

Spores in mass: White.

Microscopic: Both species similar and indistinguishable: spores overlapping in size, 9–14 × 5–8 µm, variously subglobose, oval, ellipsoid to allantoid (sausage-shaped), smooth, clear; basidia subglobose, 12–14 × 10–12 µm, tremelloid (i.e., divided longitudinally into four compartments), each cell developing a long, sinuous sterigma (epibasidium) bearing a basidiospore; clamp connections lacking.

Occurrence: *S. concrescens*—nonspecifically ectomycorrhizal with the families Betulaceae (birches) and Fagaceae (beeches and oaks); growing on the ground, fruitbody typically enveloping the stems and leaves of living herbaceous plants, leaves, and debris; widespread but uncommon in north-central, northeastern, and southeastern North America, July–October. ***S. sparassoidea***—growing from humus under hardwoods; widespread but uncommon (more abundant during wet seasons) in north-central and northeastern North America; also reported from the "Upper" Southeast and southern Appalachians.[1]

Edibility: Reported in most works as inedible; however, *S. concrescens* is collected from forests in Guatemala for personal consumption in that country; not sold in markets and of little commercial value.[2]

Comments: Species in the genus *Sebacina* are not readily separable by microscopic features, most of which broadly overlap. *Sebacina sparassoidea* is distinguished from *S. concrescens* by a more complex, less robust, less amorphous, coral-like fruitbody, with erect, wide, bluntly rounded, hollow, self-supporting branches. *Sebacina concrescens* is more ill-defined, shapeless, and typically envelops and is supported by the bases, stems, and twigs of herbaceous plants. Their habit of growing from the ground is characteristic and distinguishes these species from other jelly-like fungi in our area, all of which grow from woody substrata.

Sebacina incrustans (Persoon) Tulasne & C. Tulasne forms a thin, tough, leathery sheet-like, irregular, smooth to bumpy, white, greyish, yellowish, or pale tan spreading mass up to 15.5 cm (6¼ in) or more in extent and 1 mm thick that encrusts and envelops living plant stems, bases, and branches, leaves, cones, and other debris (Fig. C). The spores are 11.5–15 × 6–8 µm, ovoid to ellipsoid, slightly flattened on one side, smooth, and clear. The basidia are subglobose, 15–20 × 10–14 µm, longitudinally divided into two or four compartments, each bearing a single sterigma up to 70 µm long.

Sebacina sparassoidea was formerly called *Tremella reticulata* (Berkeley) Farlow and is still listed as such in many books and online sources.[3] See online Appendix A for a discussion of the nomenclatural issues surrounding this taxon. *Sebacina concrescens* is likewise called *Tremella concrescens* (Schweinitz) Burt in most field guides that discuss it.[4]

C *Sebacina incrustans.* (Photo: Debbie Viess.)

1. N.S. Weber & A.H. Smith, 1985, *A Field Guide to Southern Mushrooms* (Ann Arbor: University of Michigan Press), 260.

2. M.C. Bran et al., 2003, *Hongos comestibles de Guatemala: Diversidad, cultivo y nomenclatura vernácula* (fase III), Informe Técnico Final, Dirección General de Investigación–DIGI, Universidad de San Carlos de Guatemala, Guatemala City; R.A. Cáceres Staackmann, 2011, *Contribución al conocimiento de los hongos comestibles de la comunidad de Xetonox, San Juan Comalapa, Chimaltenango*, thesis presented to the Facultad de Ciencias Químicas y Farmacia, Universidad de San Carlos de Guatemala, for the title of chemical biologist, 59 pp.

3. In 2003 both species were transferred from the genus *Tremella* to the genus *Sebacina* and from the order Tremellales to the order Sebacinales. The two genera are thus not closely related. Nevertheless, some guides published since 2003 continue to refer these species to *Tremella*. However, all *Tremella* species are parasitic on other wood-inhabiting fungi (basidiomycetes or ascomycetes) or lichens. In contrast, *Sebacina* species are ectomycorrhizal with hardwood trees. *Sebacina* species also grow from the ground, while *Tremella* species grow directly upon or from woody substrata.

4. In its listing for *Sebacina sparassoides*, *Index Fungorum* makes the following nomenclatural comment: "Nom. illegit., Art. 52.1" with an accompanying editorial comment asserting that Roberts' "[c]ircumscription includes the type of a name (*Corticium tremellinum* var. *reticulatum* Berk., 1873) which ought to have been adopted." However, Roberts (ibid., 353) clearly states that *Corticium reticulatum* (Berk.) Cooke) "is doubly illegitimate as a later homonym of *C. reticulatum* Berk. & Broome, 1873, and *C. reticulatum* (Fr.) Fr., 1974." Roberts also noted that transferring the epithet *reticulatum* to the genus *Sebacina* would create a new homonym with *Sebacina reticulata* Patouillard (1895) [= *Thelephora recticulata* (Patouillard) Saccardo & Sydow].

False Coral Fungus

Tremellodendron pallidum

Photos: A, B, D & E, W. Light; C, hrib; F, Teresa Fortenberry.

***Tremellodendron pallidum* (Schweinitz) Burt**, *Ann. Missouri Bot. Gard.* 2:734 (1915)[1]

Basionym: *Thelephora pallida* Schweinitz, *Trans. Am. Phil. Soc.*, ser. 2, 4(2):166 (1834)
[Basidiomycota, Agaricomycotina, Agaricomycetes, *incertae sedis*, Sebacinales, Sebacinaceae]

Derivation: *Tremellodendron*, based on both Latin (genus *Tremella*) and Greek, *dendron*, "tree," to include branching thelephoracean fungi with tremelloid (vertically septate) basidia; *pallidum*, Latin, "pale."

Deprecated synonyms: *Thelephora schweinitzii* Peck, *Ann. Rep. N.Y. St. Mus. Nat. Hist.* 29:67 (1878) [1876]
Tremellodendron schweinitzii (Peck) G.F. Atkinson, *J. Mycol.* 8(3):106 (1902)

Fruitbody: 2.5–10 cm tall by 5–15 cm wide (1–4 × 2–6 in), when mature forming a rounded, coral-like cluster of erect, leathery to rubbery-gelatinous, spreading branches arising from a central base, these subcylindrical to flattened and often partially fused at their apices; initially white, creamy white to dingy white, becoming buff or beige or orangish-brown with age; often tinged greenish with age from invading algae.

Flesh: Whitish, cartilaginous, leathery to rubbery-gelatinous, tough, flexible, woody near the base; odorless, taste rancid.

Spores in mass: White.

Microscopic: Spores 7.7–12.5 × 4–6.5 µm, elliptical to sausage-shaped, smooth, clear, nonreactive in Melzer's reagent; basidia ovoid, tremelloid (longitudinally cruciate-septate [cross-shaped], partitioned into four chambers), without basal clamp connections; hyphae lacking clamp connections.

Occurrence: Ectomycorrhizal with hardwoods, especially Fagaceae (beeches, chestnuts, and oaks); solitary or in groups on the ground; fairly common in mixed woods in eastern North America south to North and South Carolina, Tennessee, and Georgia; summer–late fall.

Edibility: Variously reported as both edible and inedible.

Comments: *Tremellodendropsis semivestita* (Berkeley & M.A. Curtis) R.H. Petersen[2] could be mistaken for *Tremellodendron pallidum*, particularly in its younger growth stages. *Tremellodendropsis semivestita* is smaller than *T. pallidum*, 3–4 cm tall by 1 cm across (1¼–1⅔ × ½ in), erect, leathery, repeatedly forking into fairly straight, tapering branches that rise close together, initially covered with whitish down, drying pale brownish-olive to buffy-brown; spores 12–15 × 5–6 µm. The spores are larger than those of *T. pallidum* (12–18 × 6–7 µm), and only the apices of the basidia are cruciately subseptate (incompletely longitudinally septate and forming a cross). Unlike *T. pallidum*, the hyphae also have clamp connections.

1. *Index Fungorum* (*IF*) erroneously states that *Thelephora pallida* Schweinitz is a later homonym of *Thelephora pallida* (Persoon) Persoon and is therefore an illegitimate name under Article 53.1. *IF* also incorrectly states that *Tremellodendron pallidum* is illegitimate under Article 52.1 because "the circumscription of the genus includes the type of a name (*Thelephora schweinitzii* Peck. [*sic*] 1878), the epithet of which ought to have been adopted." *IF* therefore substituted *Tremellodendron schweinitzii* (Peck) G.F. Atkinson as the correct scientific name for this taxon. This has been followed by several authors. However, this is incorrect. In fact, *Thelephora pallida* Schweinitz 1878 was only a subjective and temporary homonym of *Thelephora pallida* (Persoon) Persoon 1800, which was originally described as *Craterella pallida* Persoon 1794 and subsequently transferred to several other genera, thus eliminating the homonymy. *Craterella pallida* Persoon is currently considered a deprecated synonym of *Cotylidia pannosa* (Sowerby) D.A. Reid 1965. See online Appendix A for more details.

2. Orthographic correction. *Tremellodendropsis semivestita* (Berkeley & M.A. Curtis) R.H. Petersen [as "*Tremellodendropsis semivestitum*"], 1987, *Mycotaxon* 29:64. The Greek suffix *-opsis*, "resemblance," "appearance," is always feminine third declension.

Earth Tongues, Club Fungi, and Coral Fungi

This artificial grouping of fungi based on sometimes vague superficial appearances represents both divisions (phyla) of the unranked clade Dikarya (the so-called higher fungi): the Ascomycota and Basidiomycota. They range from very small (<1 cm, or ¾ in, in height), simple awl-shaped growths to very large (>30 cm, or 12 in) and intricately branching coral-like growths. In color they range from black and various drab shades of brown or grey to brilliant violets, living vibrant yellows, oranges, and pristine whites. The few examples presented here represent two subdivisions and five classes of Ascomycota and two subclasses and five orders of Basidiomycota.

Club fungi not treated in this section. The carbonaceous "club fungi" of the genus *Xylaria* (Ascomycota, Class Sordariomycetes, Order Xylariales) could also have been included here but are presented in the first section of this book dealing with the Carbonaceous Fungi. Their dark, blue-grey to blackish or black color[1] and their hard and rock-like or tough and leathery ascocarps (fruitbodies) studded with pimple-like perithecia distinguish them from the clubs treated in this section. The amorphous, semigelatinous, more massive, and indeterminately growing fruitbodies of the Coalescing White Jellyfungi (Genus *Sebacina*) and the False Coral Club (Genus *Tremellodendron*, Class Agaricomycetes, Order Sebacinales), which encrust over and envelop adjacent twigs and plant debris and tremelloid basidia, distinguish them from the club fungi discussed here. *Sebacina* and *Tremellodendron* are treated in the preceding section on the jelly fungi.

Fruitbody: Morphologies of the Club-like and Coral-like Fungi

A B C D

A, *Xylaria polymorpha*; B, *X. hypoxylon*; C, *Geoglossum* sp.; D, *Trichoglossum* sp. (Photos: A, W. Light; B, Norbert Nagel; C & D, Jim Edwards.)

E

F

G

H

E, *Spathularia velutipes*; F, *Calocera cornea*; G, *Dacryopinax elegans*; H, *Clavariadelphus pistillaris*. (Photos: E, Teresa Fortenberry; F, John Plischke III; G, W. Light; H, Holger Krisp.)

I

J

K

I, *Leotia lubrica*; J, *Ramaria formosa*; K, *Thelephora vialis*. (Photos: I & K, W. Light; J, Teresa Fortenberry.)

L

M

N

L, *Tremellodendron pallidum*; M, *Geoglossum* sp.; N, *Trichoglossum* sp. (Photos: L, hrib; M & N, Jim Edwards.)

The graphic below shows the general arrangement of the hymenia (tan color) in various ascomycetous and basidiomycetous club fungi. *Leotia* is essentially a cup fungus whose apothecium (hymenial layer) has been pushed upward and outward and then folded back at the margin. The asci are inoperculate (lacking a lid). *Trichoglossum* and *Geoglossum* bear their apothecia on the outside of stipitate (stalked) club-, spoon-, or paddle-shaped stromata. The asci are inoperculate, and the spores are forcibly discharged one at a time by the buildup of fluid pressure within the ascus. The perithecia of *Xylaria* are embedded beneath a hard, carbonaceous surface layer and discharge their ascospores through an ostiole at the tip of the ascus. The mature stroma often has a hollow center. Basidiomycetous club fungi like *Clavariadelphus* and coral fungi such as *Clavaria* bear their basidia in a hymenial layer covering the outside of the stroma.

As stated above, several of the taxa keyed out below are actually treated in other sections of this book. They are included in the key here because of their superficial similarities. The club-like species of *Cordyceps* (*sensu lato*) are not included in the following general key, although they resemble *Geoglossum* and *Trichoglossum* (Figs. C & D). These parasitic fungi of arthropods and deer truffles are attached underground to their buried hosts or emerge directly from their corpses that are prominently attached to vegetation.[2]

1. *Xylaria* species often exhibit a white dusting of conidia, as shown in Figs. A & B.

2. Before killing it, certain of these insect parasitizers alter the behavior of their host and cause it to climb up to a prominent position best suited for either the dispersal of the fungal spores or to attract and infect other host insects.

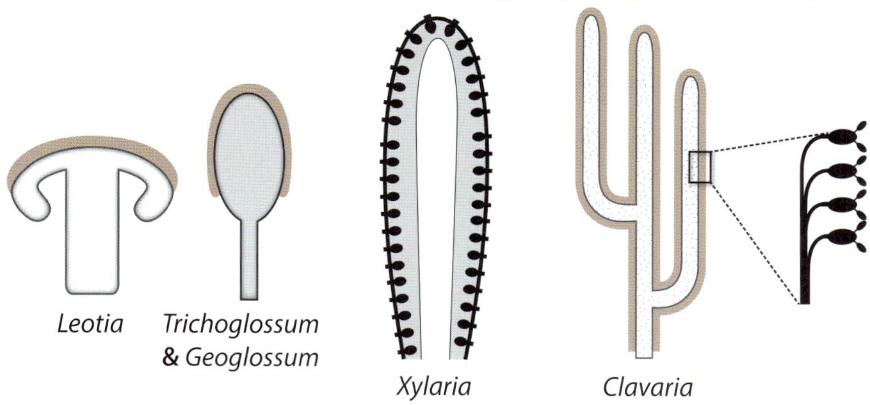

General arrangement of the hymenia (tan color) in various ascomycetous and basidiomycetous club fungi.

General Key to Club-, Tongue-, and Coral-like Fungi Treated in This Book

1. Fresh stromata (fruitbodies) generally blackish (Figs. A, C, & D), may have powdery coating of white conidia (Figs. A & B) 2

1′. Fresh stromata not blackish 4

2. Stromata simple, erect cylindrical columns (Fig. A), with or without a circle of tentacles at the tip, or simple and/or branching and antler-like (Fig. B) or twisted, wiry, tentacle-like; often coated with white conidia .... *Xylaria*

2′. Stromata clavate or paddle-shaped, not coated with white conidia (Figs. C & D) 3

General Key to Club-, Tongue-, and Coral-like Fungi Treated in This Book (*Continued*)

3. Hymenium and stipe with spines or setae visible with hand lens (Fig. N) . . . . *Trichoglossum*

3′. Hymenium lacking spines or setae (Fig. M) . . . . *Geoglossum*

4. Fruitbody cartilaginous to rubbery-gelatinous, forming either (1) an irregular amorphous or lobate mass often encrusting over and enveloping adjacent twigs and plants or (2) radiating coral-like branches often flattened, fusing at the tips and arising from a central base (Fig. L) . . . . *Sebacina* and *Tremellodendron*

4′. Fruitbody different from above 5

5. Fruitbody spoon-shaped, with deep depression or longitudinal furrow; stipe rough, scaly . . . . *Microglossum rufum*

5′. Fruitbody shaped differently 6

6. Fruitbody spathulate (fan-shaped) or irregularly lobed, more or less laterally compressed 7

6′. Fruitbody shaped differently 9

7. Hymenium with Y-shaped basidia . . . . *Dacryopinax spathularia*

7′. Basidia lacking, hymenium with spore-bearing asci 8

8. Fruitbody cream to pale yellow; mature spores needle-like, multiseptate, croziers and paraphyses abundant (Fig. E) . . . . *Spathularia*

8′. Fruitbody bright yellow, irregularly shaped, mature spores elliptical, aseptate, croziers and paraphyses lacking . . . . *Neolecta irregularis*[a]

9. Fruitbody tough-gelatinous, typically an erect, awl-shaped column (Fig. F), tip may be slightly forked . . . . *Calocera*

9′. Fruitbody shaped differently 10

10. Fruitbody stipitate and pileate, cap smooth, irregularly convex, lobed and furrowed, somewhat brain-like (Fig. I), margin strongly inrolled 11

10′. Fruitbody different from above 12

11. Fruitbody gelatinous, spores elliptical to spindle-shaped . . . . *Leotia*

11′. Fruitbody fleshy-leathery, spores needle-shaped . . . . *Cudonia*

12. Fruitbody stipitate, with deep hood-like cup at an angle to the stipe, shaped like a ship's deck air intake (Fig. G) . . . . *Dacryopinax elegans*

12′. Fruitbody different from above 13

13. Fruitbody club-like, expanded and rounded (Fig. H) or flattened at tip. . . . . *Clavariadelphus*

13′. Fruitbody different from above. 14

14. Fruitbody consisting of erect fan-shaped, often fused lobes arising from a common central stipe and forming a more or less vase-like composite structure (Fig. K) . . . . *Thelephora*

14′. Fruitbody (1) cylindrical and worm-like or somewhat flattened, unbranched or branched or (2) intricately branching, complex, coral-like 15

15. Fruitbodies cylindrical and worm-like or somewhat flattened and strap-like . . . . *Clavaria*, *Clavulinopsis*, and *Ramariopsis*

15′. Fruitbodies coral-like (Fig. J), with vertically oriented branches typically arising from a common base . . . . *Artomyces*, *Clavulina*, and *Ramaria*

a. *Neolecta irregularis* is distinguished from the genus *Spathularia* in the account of *S. velutipes*.

Black Hairy Earth Tongue

Trichoglossum hirsutum

A–E, *Trichoglossum hirsutum*; C, setae (arrow); D, seta and paraphyses (arrow); E, 15-septate spore. (Photos & micrographs: W. Light.)

***Trichoglossum hirsutum* (Persoon) Boudier**, *Hist. Class. Discom. Eur.* (Paris):86 (1907)

Basionym: *Geoglossum hirsutum* Persoon, *Comm. Fung. Clav.* (Lipsiae):37 (1797)
[Ascomycota, Pezizomycotina, Leotiomycetes, Leotiomycetidae, Geoglossales, Geoglossaceae]

Derivation: *Trichoglossum*, Greek, *trícha*, "hair," plus *glóssa*, "tongue"; *hirsutum*, Latin, "hairy."

Fruitbody: Small, 3–8 cm (1¼–3¼ in) tall; a dark-brown to black, viscid, club-, spade-, or paddle-shaped, flattened, often grooved head (Figs. A & B) atop a slender, blackish stalk; both head and stalk velvety, covered by minute, hair-like spines (setae) visible with a hand lens (Fig. C, arrow, & Figs. D & F).

Stipe: Up to 6 cm (2½ in) tall by 2–3 mm thick, cylindrical, subequal, pubescent.

Flesh: Thin, tough, brownish-black.

Spore print: Brown.

Microscopic: Spores very long, cylindrical, tapering toward each end, mostly 15-septate (Fig. E), but with considerable variability (11- to 18-septate); 95–160 × 5–7 µm; eight spores per ascus; paraphyses brown, cylindrical, usually with slightly expanded and strongly curved or coiled tips (Fig. D, arrow).

Occurrence: Saprobic, solitary to clustered in soil, on well-decayed wood, in moss, or in leaf litter; widespread in North America and the most frequently encountered species of *Trichoglossum*.

Edibility: Not generally considered edible.

Comments: This is the most widespread and most frequently encountered species in the genus. The shape of the head varies considerably in *T. hirsutum*. It is typically club-like or paddle-shaped, laterally compressed, and often grooved (Fig. B) but is occasionally irregularly lobed and capitate or head-shaped (Fig. A).

The next most prevalent species, *T. velutipes* (Peck)

Durand (1908), is distinguished from *T. hirsutum* by asci containing four (rather than eight) ascospores[1] that are mostly 7- to 11-septate (rather than mostly 15-septate).

Other species that could be encountered in our area include (in decreasing order of prevalence) *T. farlowii* (Cooke) Durand (1908), *T. walteri* (Berkeley) Durand (1908), *T. variabile* (Durand) Nannfeldt (1942), *T. octopartitum* Mains (1940), and *T. confusum* Durand (1921). The spores of *T. farlowii* are 0- to 5-septate (mostly 3-septate) and shorter than in *T. hirsutum*: (45–)57–75(–90) × 5–7 µm; those of *T. variabile* mostly 10- to 13-septate; those of *T. walteri*, *T. confusum*, and *T. octopartitum* are mostly 7-septate (rarely more). The spores of *T. octopartitum* are mostly longer, while those of *T. walteri* and *T. confusum* are mostly shorter than 100 µm. *Trichoglossum walteri* is the most common of the 7-septate species. Its ascospores are typically 72–100 µm; those of *T. confusum* are considerably shorter, generally 55–65 µm.[2]

The genus *Trichoglossum* is distinguished from the very similar genus *Geoglossum* by the small, awl-like, dark-brown setae emerging from both the head and stipe that are visible through a hand lens (Fig. F). The smooth, nonvelvety heads of *Geoglossum* species are also generally more elongate and fusiform than the paddle-shaped heads of *Trichoglossum*, but this feature is variable.

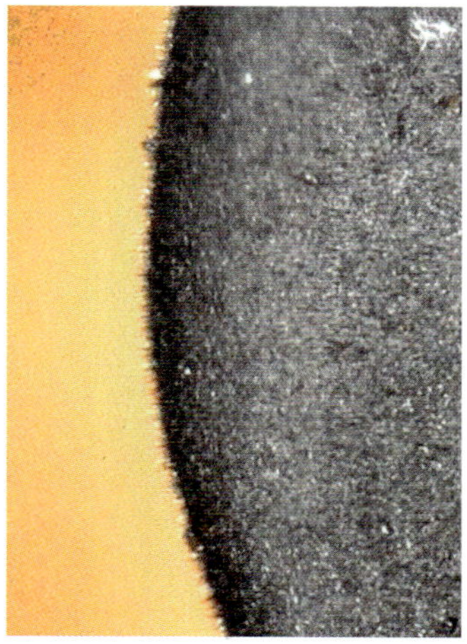

F *Trichoglossum* sp., showing setae. (Photo: Jim Edwards.)

1. *Trichoglossum tetrasporium* Sinden & Fitzpatrick (1930) also has only four spores per ascus, but the spores are mostly 15-septate. It is known only from four widely scattered collections in New York, the UK, and China; its geographic distribution is unclear.

2. See E.B. Mains, 1954, *Mycologia* 46(5):586–631 for a full discussion of these species.

Velvety Fairy Fan

Spathularia velutipes

A

B

A & B, *Spathularia velutipes*. (Photos: A, Teresa Fortenberry; B, Jim Edwards.)

Spathularia velutipes **Cooke & Farlow *ex* Cooke**, *Grevillea* 12(61):37 (1883)[1]
[Ascomycota, Pezizomycotina, Leotiomycetes, Leotiomycetidae, Rhytismatales, Cudoniaceae]

Basionym: *Spathularia velutipes* Cooke & Farlow, in Cooke, *Grevillea* 12(61):37 (1883)

Derivation: *Spathularia*, Latin, "spade-like"; *velutipes*, Latin, *velutinae*, "velvety," plus *pes*, "foot."

Fruitbody: Up to 7 cm (2¾ in) tall with expanded and flattened, hood-like or lobed, wrinkled or smooth head 1–3 cm wide (⅖–1¼ in); cream-colored or yellowish margins of head usually partially decurrent down either side of the supporting stalk (Figs. A & B). Button stage with a veil-like covering, which is very rarely seen.

Flesh: Whitish, odorless.

Stipe: Up to 3–6 cm tall by 1 cm thick (1¼–2½ × ½ in); cylindrical, hollow, finely velvety or fuzzy, reddish-brown, with orange mycelium at base.

Spore print: White, cream, or yellowish.

Microscopic: Spores clear, smooth, translucent, long, needle-like, 33–43 × 1.5–3.0 µm, aseptate at first, becoming multiseptate, enclosed in gelatinous sheath; asci club-shaped, lacking operculum (covering lid), pore nonreactive in Melzer's reagent.

Occurrence: Solitary or gregarious (occasionally cespitose) on decayed conifer logs or on the ground among moss or forest litter in conifer woods; August–September. Widespread but occasional east of the Great Plains.

Edibility: Unknown.

Comments: The very similar Yellow Fairy Fan, or Yellow Earth Tongue, *Spathularia flavida* Persoon (Fig.C), is distinguished from *Spathularia velutipes* by its smooth, pale yellow to light-tan (rather than velvety and reddish-brown) stipe and the dense whitish or yellowish (rather than orange) mycelium at the base. The spore print is buff instead of whitish or yellowish, and the spores are considerably longer (30–95 × 1.5–3 µm). The button stage also lacks the veil-like covering of *S. velutipes*, which is why Maas Geesteranus placed it in its own monotypic genus, *Spathulariopsis*.[1]

C

D

E

C, *Spathularia flavida*; D, *Neolecta irregularis*; E, *Microglossum rufum*. (Photos: C, Paffka; D, Walt Sturgeon; E, Nathan Wilson.)

The Irregular Earth Tongue, *Neolecta irregularis* (Peck) Korf & Rogers (Fig. D), has an irregularly shaped fruitbody, is yellow to orangish or yellow-green at the head with a white stipe, and has much smaller elliptical spores (5.5–8.5 × 3–4 µm). The Orange Earth Tongue, *Microglossum rufum* (Schweinitz) Underwood (Fig. E), has a yellow-orange to orange head and a rough, scaly yellow to orange-yellow stipe, and the mature spores are shorter, thicker, sausage- to spindle-shaped (18–38 × 4–6 µm) and 5- to 10-septate (or more) with numerous oil droplets; the apex (pore) of the asci stains blue in Melzer's reagent.

1. This species is usually referred to as *Spathulariopsis velutipes* (Cooke & Farlow) Geesteranus. This combination was based on a faulty interpretation. For a discussion of this issue, see online Appendix A.

Pestle-shaped and Flat-topped Club Corals

Clavariadelphus spp.

A B

A, *Clavariadelphus pistillaris*; B, *C. truncatus*. (Photos: A, Holger Krisp; B, Damon Brunette.)

***Clavariadelphus pistillaris* (Linnaeus) Donk**, *Meded. Bot. Mus. Herb. Rijks Univ. Utrecht* 9:72 (1933)

Basionym: *Clavaria pistillaris* Linnaeus, *Sp. Pl.* 2:1182 (1753)

***Clavariadelphus americanus* (Corner) Methven**, *Mycotaxon* 34(1):154 (1989)

Basionym: *Clavariadelphus pistillaris* var. *americanus* Corner, *Ann. Bot. Mem.* 1:692 (1950)

Misapplied name: *Clavariadelphus pistillaris* (*auct. mult. amer.*); *non Clavaria pistillaris* Linnaeus (1753)

***Clavariadelphus truncatus* Donk**,[1] *Meded. Bot. Mus. Herb. Rijks Univ. Utrecht* 9:73 (1933)
[Basidiomycota, Agaricomycotina, Agaricomycetes, Phallomycetidae, Gomphales, Clavariadelphaceae]

Derivation: *Clavariadelphus*, etymology uncertain, from *clava*, Latin, "club," and *delph-*, "hollow," or *delphus*, "womb";[2] Latin epithets: *pistillaris*, "pestle," *americanus*, "American," *truncatus*, "truncated."

Fruitbody: *C. pistillaris* and *C. americanus*—initially cylindrical, club-shaped when mature, broadly expanded and rounded at top, constricted toward base (Fig. A); dry, smooth to vertically rugose near base; becoming powdery at apex as spores mature; bruising brownish when injured. ***C. pistillaris***—15–30 cm tall by 2–6 cm across at the top (6–12 × ¾–2½ in), yellowish, orangish-yellow, pinkish-beige, becoming brownish to pale purplish. ***C. americanus***—7–12 cm tall by 3 cm across at the top (2¾–4¾ × 1¼ in), whitish to pale yellowish-white, orange-white, pale pink to pale buffy ochraceous. ***C. truncatus***—5–15 (up to 18) cm tall by 2.5–5(–8) cm across at the top (2–6[–7¼] × 1–2[–3¼] in), club- to top-shaped, tapering downward, top flattened, disk-like to somewhat depressed, dry, smooth near base, vertically wrinkled toward apex (Fig. B); golden- to orange-yellow, buff-orange to pale brownish-orange.

Flesh: All three species—white, thick, firm when young, soft and spongy when mature, darkening when injured; odor not distinctive, taste bland to bitter.

Chemical staining: *C. pistillaris* and *C. americanus*—surface negative with KOH and NH_4OH, greenish with $FeCl_3$. ***C. truncatus***—surface cherry red to light red with KOH; flesh dark green with $FeCl_3$.

Spore print: *C. pistillaris* and *C. americanus*—white to cream; ***C. truncatus***—pale brownish-yellow.

Microscopic: Spores: All three species—broadly ovoid to ellipsoid or almond-shaped, with oblique hilar appendage; smooth, clear, with yellow oil droplets or oil droplets lacking; nonreactive in Melzer's reagent. ***C. pistillaris***—9–16 × 6–9 µm. ***C. americanus***—8–12 × 4.5–6.5 µm. ***C. truncatus***—9–12(–13) × 5–8 µm.

Occurrence: All three species—scattered or clustered on the ground, rare to occasional, widespread in eastern North America. ***C. pistillaris***—mycorrhizal with beeches (*Fagus* spp.); July–November; widespread in eastern North America and Europe. ***C. americanus***—mycorrhizal with oaks (*Quercus* spp.) and pines (*Pinus* spp.); restricted to eastern North America. ***C. truncatus***—mycorrhizal with conifers; North America, Europe, Asia Minor, August–November.

Edibility: All three species edible, but can be bitter-tasting.

Comments: These distinctive, robust, club-shaped fungi can scarcely be mistaken for anything else. Though similar in appearance, *Clavariadelphus pistillaris* is associated with beech trees (*Fagus grandifolia* in eastern North America) and is two and a half to three times larger than *C. americanus*. The latter species is also mycorrhizal with oaks and pines. *Clavariadelphus truncatus* is readily distinguished by its flattened apex and association with conifers. It also stains bright red with KOH, in contrast to the green response of *C. pistillaris* and *C. americanus*. The subtle microscopic differences are difficult to assess and are of limited value. These fungi are not often encountered. Other species of the genus are found farther north and out west.

1. Replacement name for *Clavaria truncata* Quélet, *Enchir. Fung.* (Paris):221 (1886); *nom. illegit.*, Art. 53.1 [homonym]; *non Clavaria truncata* Lumnitzer, *Fl. Poson.*:533 (1791).

2. *Adelphoi* means "siblings born from the same womb" (mother). All these terms may refer to the uterus-shaped fruitbody of the type species of the genus: *C. pistillatus*.

Club-like Tuning Fork

Calocera cornea

Inset, tuning fork basidium and spores. (Drawings based on a figure in L. Trierveiler-Pereira et al., 2009, *Mycotaxon* 109:107–110; photo: John Plischke III.)

***Calocera cornea* (Batsch) Fries**, *Stirp. Agri. Femison.* 5:67 (1827) [Basidiomycota, Dacrymycetes, Dacrymycetales, Dacrymycetaceae]

Basionym: *Clavaria cornea* Batsch, *Elench. Fung.* (Halle):139 (1783)

Derivation: *Calocera*, Greek, "beautiful horn"; *cornea*, Latin, "horny."

Fruitbody: Erect cylindrical columns, awl-shaped or bluntly rounded tips, rarely forked at tip, tough-gelatinous, up to 2 cm (¾ in) tall by 3 mm in diameter, yellow to orange-yellow; hymenium covering fruitbody.

Spore print: White to yellowish.

Microscopic: Spores 7–11 × 3–4.5 µm, cylindrical to slightly curved, with offset hilar appendage, smooth aseptate or with one septum at maturity (inset, b); basidia forked, 25 × 3 µm, each producing two basidiospores (inset, a); clamp connections lacking.

Occurrence: Widespread in North America and Europe; reported from Chile and Brazil. Saprobic on well-decayed barkless hardwood, gregarious or in scattered groups. Summer–late fall following heavy rains.

Edibility: Inedible.

Comments: Though *C. cornea* superficially resembles some club fungi, it is a true jelly fungus and is readily distinguished by its bifurcate basidia. *Calocera viscosa* (Persoon) Fries is larger (3–10 cm in height), branched, and more viscous and grows on decayed conifer wood. Its spores are 3-septate. Dried *Calocera cornea*, like many jelly fungi, readily reconstitutes after rehydration. The hymenium covers the fruitbody, and each basidium forcibly expels its two basidiospores.

Little Periscope

Dacryopinax elegans

A B C

Dacryopinax elegans.

Dacryopinax elegans **(Berkeley & M. A. Curtis) G. W. Martin,** *Lloydia* 11:116 (1948)
[Basidiomycota, Dacrymycetes, Dacrymycetales, Dacrymycetaceae]

Basionym: *Pistillaria elegans* Berkeley & M. A. Curtis, *Grevillea* 2(14):18 (1873)

Derivation: *Dacryopinax*, Greek, *dacry*, "tear," plus *pinax*, "dish," referring to the teardrop- and dish-shaped form of the cap of the type species of the genus (*D. elegans*); *elegans*, Latin, "elegant."

Fruitbody: 0.5–2.5 cm tall, with expanded, cup-shaped, hood-like cap 0.5–1.5 cm (¼–⅔ in) wide, seamlessly mounted atop a columnar stipe (Figs. A & B), cap expanding laterally and flaring with age (Fig. C), at times becoming flattened and palmate. Outer sterile surface finely tomentose, light beige or light brown (Figs. A & B), becoming reddish-brown then hard and darker brown with age (Fig. C). Fertile inner surface flesh-colored to light reddish-brown (Figs. A & B), darker and concolorous with sterile surface with age.

Flesh: Soft, gelatinous, watery, enclosed in tough cuticle.

Spore print: Light yellow to yellow-orange.

Microscopic: Spores 7–11.5 × 3.5–4.5 µm (specimens from Mexico 13.6–15.6 × 5.6–6.4 µm), elongate to cylindrical, curved, with one to three transverse septa; basidia bifurcate, resembling a tuning fork, clavate to cylindrical; sterile cuticle formed of thick-walled, septate, smooth hyphae.

Occurrence: Uncommon, but widespread in North, Central, and South America, Europe, Asia. Saprobic, forming small clusters or lines of fruitbodies on well-decayed hardwood, fruitbodies often persisting for several weeks.

Edibility: Inedible.

Comments: When fresh, this tiny mushroom can hardly be mistaken for anything else. Its equally widespread and similar-sized but edible congener, *Dacryopinax spathularia* (Schweinitz) G. W. Martin (Figs. D & E below), is apically flattened, petal- or spatula-shaped or palmate, and yellow to orange when fresh (Fig. D), darkening to orange-red with age; the sporeprint is white, the spores (7–11 × 3.5–4.5 µm) are ellipsoid and slightly curved and have a single septum when mature. The forked basidia are characteristic of the class Dacrymycetales, which contains the single family Dacrymycetaceae.

D E

Dacryopinax spathularia. (Photos: Jim Edwards.)

Jelly Clubs and Jelly Babies

Leotia lubrica, *L. viscosa*, and *L. atrovirens*

A, *Leotia lubrica*; B, *L. viscosa*; C, *L. atrovirens*. (Photos: A, W. Light; B, Mueller; C, Bob Zimmer.)

***Leotia lubrica* (Scopoli) Persoon**, *Comm. Fung. Clav*. (Lipsiae):31 (1797)

Basionym: *Helvella lubrica* Scopoli, *Fl. Carniol*., 2nd ed. (Wien) 2:477 (1772)

***Leotia viscosa* Fries**, *Syst. Mycol*. 2:30 1821 (1822)

***Leotia atrovirens* Persoon *ex* Fries**, *Mycol. Eur*. (Erlanga) 1:202 (1822) *ex* Fries, *Syst. Mycol*. 2:30 (1822)
[Ascomycota, Pezizomycotina, Leotiomycetes, Leotiomycetidae, Leotiales, Leotiaceae]

Derivation: *Leotia*, uncertain etymology, possibly from *leiótis*, Greek, "smoothness," "slickness";[1] *lubrica*, Latin, "slippery," "smooth"; *viscosa*, Latin, "viscid," "slimy"; *atrovirens*, Latin, "blackish green."

Cap: All taxa—½–4 cm (¼–1⅔ in) across, bald, irregularly convex, lobed and furrowed, somewhat brain-like (Figs. A & B), margin strongly inrolled, viscid when wet; ***L. lubrica***—pale yellow, yellow-ochre (Fig. A), to orangish-yellow, sometimes with tints of olive-green, undersurface paler. ***L. viscosa***—similar to *L. lubrica*, but color olive-green, dark green (Fig. B), to greenish-black. ***L. atrovirens***—pea-green to bluish-green (Fig. C).

Flesh: All taxa—gelatinous.

Stipe: All taxa—1–6 cm tall by 0.5–1 cm thick (¾–3¼ × up to ½ in), subequal, sometimes expanded near base, smooth, but often with tiny flakes or granules, slimy; hollow or, except *L. atrovirens*, with gelatinous core. ***L. lubrica* and *L. viscosa***—pale yellow (Fig. B) to yellowish-orange (Fig. A) or greenish-yellow (Fig. C).

Spore print: All taxa—white to cream.

Microscopic: All taxa—spores 18–25 × 4–6.5 µm, subcylindrical to spindle-shaped, smooth, clear, somewhat curved, with bluntly rounded ends, multiseptate when mature; asci lacking opercula, containing eight spores.

Occurrence: All taxa—saprobic; solitary or in groups under hardwoods or conifers in leaf litter, soil, or moss or on rotten wood; summer–fall. ***L. lubrica***—widespread in North America, Europe, East Asia, Australia, New Zealand. ***L. viscosa***—throughout North America, also Central America, Europe. ***L. atrovirens***—occasional, North America, Europe.

Edibility: Unknown.

Comments: These three closely related species have traditionally been separated on the basis of the color of their caps and stipes, since the microscopic features are indistinguishable. *Leotia lubrica*, the Ochre Jelly Club, has a yellow to brownish-yellow cap and a yellowish to yellow stipe, as shown in Fig. A. *Leotia viscosa*, the Green-capped Jelly Club, has a green to greenish-black cap and yellow stipe, and *L. atrovirens* has both a greenish cap and a greenish stipe.

However, these species exhibit considerable overlap with one another, and it has been suggested that they represent a single highly variable species whose variable phenotype (physical appearance) is produced by an anamorphic hyphomycetous parasite. In particular, *L. atrovirens* is suggested to represent *L. lubrica* parasitized by *Hypomyces leotiicola* Rogerson & Samuels (1985), whose blue-green pigment is produced by the parasitized host.[2] Anecdotal evidence seems to support this notion.[3]

Cudonia circinans (Persoon) Fries and *C. lutea* (Peck) Saccardo [Helotiales, Cudoniaceae] also resemble these *Leotia* species. They can be distinguished from *Leotia* by their generally drabber coloration; smaller size (cap ≤2 cm across); dry, nonviscid cap and stipe; tough, thin, nongelatinous flesh; and longer and much thinner, needle-like spores (30–50+ × 2 μm).

A phylogenetic analysis of DNA sequences indicated that *L. viscosa*, *L. lubrica*, and *L. atrovirens* are all polyphyletic species comprising four distinct clades that do not correspond to the three generally recognized species discussed here. The color of the fruitbody is not sufficient to identify these mushrooms to species.[4]

1. *Fide* J.E. Smith, W.J. Hooker, & M.J. Berkeley, 1854, *English Flora: Class 24, Cryptogamia, Pt. 2, Comprising the Fungi* 5(2):185.
2. J. Breitenbach & F. Kränzlin, 1984, *Fungi of Switzerland*, vol. 1, *Ascomycetes* (Lucerne: Verlag Mykologia).
3. *Index Fungorum*, *Species Fungorum*, and *MycoBank* muddied the nomenclatural waters even further by mistakenly transferring the teleomorph *L. atrovirens* to the hyphomycetous anamorphic form-genus and form-species *Coryne atrovirens* (Persoon 1801) Saccardo (1889), [teleomorph = *Claussenomyces atrovirens* (Persoon) Korf & Abawi (1971)]. They confused the name of the stipitate *Leotia atrovirens* Persoon (1822) [Order Leotiales] with the sessile disk-like *Peziza atrovirens* Persoon (1801) [Helotiales, Helotiaceae].
4. Z. Zhong & D.H. Pfister, 2004, *Mycol. Progr.* 3(3):237–246.

Magenta Coral

Clavaria zollingeri

A

B

Photos: A, Teresa Fortenberry; B, Dan Molter.

***Clavaria zollingeri* Léveillé,** *Annls Sci. Nat.*, Bot., sér. 3, 5:155 (1846)
[Basidiomycota, Agaricomycotina, Agaricomycetes, Agaricomycetidae, Agaricales, Clavariaceae]

Derivation: *Clavaria*, Latin, from *clava*, "club"; *zollingeri*, epithet in honor of the German mycologist Heinrich Zollinger.

Synonym: *Clavaria lavandula* Peck, *Bull. N.Y. St. Mus.* 139:47 (1910)

Fruitbody: Branches 1–10 cm tall by 2–6 mm thick (½–4 × ¼ in), cylindrical to worm-like, tips rounded to bluntly truncated, becoming hollow with maturity; surface smooth, purple, dark amethyst or violet to reddish- or pinkish-purple; basidome 7 cm (2¾ in) or more in width.

Flesh: Reddish-purple, thin, brittle; odor not distinctive, taste faintly radish-like or cucumber-like.

Spores in mass: White.

Chemical staining: Surface negative with $FeSO_4$.

Microscopic: Spores 4(–5)–7.5 × 3–3.5(–4.5) µm, subglobose, ovoid to broadly elliptical, smooth, clear, thin-walled, with a hilar appendage; nonreactive in Melzer's reagent; basidia 4-sterigmate, clamp connections lacking.

Occurrence: Saprotrophic. Solitary, scattered, or clustered on the ground in mixed woods; occasional in eastern North America; South America, Europe (rare and threatened), Asia, Australia, New Zealand; summer–fall.

Edibility: Edible.

Comments: This fungus is often described as limited to northeastern North America but in fact is widespread, if not common, in the Southeast. In our area it could be confused with the Amethyst Coral, *Clavulina amethystina* (Bulliard) Donk.

Clavulina amethystina is macroscopically nearly identical. It differs from *Clavaria zollingeri* as follows: it is much rarer, less branched and longitudinally finely striated with more or less acute purplish-brown tips, the flesh is white rather than purple, the basidia are 2-sterigmate (rather than 4-sterigmate), clamp connections are present instead of absent, and the spores are larger, 7–10(–12) × 6–8 µm. It is also ectomycorrhizal rather than saprotrophic.

White Worm Coral, White Spindles

Clavaria fragilis

A B C D

Photos: A, Teresa Fortenberry; B–D, W. Light.

***Clavaria fragilis* Holmskjold**, *Beata Ruris Otia Fungis Danicis* 1:7 (1790)
[Basidiomycota, Agaricomycotina, Agaricomycetes, Agaricomycetidae, Agaricales, Clavariaceae]

Derivation: *Clavaria*, Latin, from *clavus*, "club"; *fragilis*, "fragile," referring to the easily broken fruitbodies.

Deprecated synonym: *Clavaria vermicularis* Svartz (1811) is still used by many authors.[1]

Fruitbody: Cylindrical, worm-like, unbranched, smooth or slightly furrowed, up to 15 cm tall by 5 mm thick (6 × ¼ in), with no clearly demarcated base, often in dense cespitose clusters (Figs. A, C, & D); opaque white when dry, nearly translucent when wet, tapering to an acute tip that often bends and yellows with age (Fig. B), entire fruitbody yellowing with age (Fig. D).

Flesh: White, very fragile.

Spores in mass: White.

Microscopic: Spores 4.5–7 × 2.5–4 µm, ellipsoid, smooth, clear, containing small oil droplets; basidia lacking clamps; context hyphae thick, up to 12 µm wide, lacking clamp connections.

Occurrence: Saprotrophic (feeding on dead organic matter). Typically in cespitose clusters on soil, moss, or litter in mixed hardwood and conifer woods; late fall–midwinter; widespread throughout the Northern Hemisphere; Africa, Australia.

Edibility: Edible, but tasteless and not worth the effort.

Comments: *Clavaria fragilis* has several close look-alikes. *Clavaria acuta* Sowerby is smaller, seldom exceeding 5 cm (2 in) in height, and the fertile distal one-third to one-half of the fruitbody is expanded to at least twice the diameter of the thin basal stalk, from which it is readily distinguished. Unlike those of *C. fragilis*, the bases of the basidia in *C. acuta* have clamp connections.

The Yellow and Golden Spindle Clubs (*Clavulinopsis fusiformis* and *C. aurantiocinnabarina*) are bright yellow and pale to reddish-orange, respectively, and the bases of their basidia have clamp connections (see the accounts of those species).

Multiclavula mucida (Fries) Petersen is a *tiny*—15 mm tall (⅔ in) by 1–1.5 mm wide—lichenized worm coral *always found growing on wet, rotting, decorticated hardwood logs covered by a green algal scum.*[2] Although in a separate order (Cantharellales), it looks just like a minuscule *Clavaria fragilis*, although the white fruitbodies are sometimes forked near the base.

1. *Clavaria vermicularis* Svarts would normally be an automatic *nomen illegitimum* under Article 53.1 of the *ICN* as a later homonym of *Clavaria vermicularis* J.F. Gmelin (1792). However, in 1821 Fries sanctioned both *Clavaria fragilis* Holmskjold and *Clavaria vermicularis* Svarts, thus rendering both *Clavaria vermicularis* Svarts (1811) and *Clavaria vermicularis* Batsch (1783) [also a later homonym] unpriorizable. See online Appendix A for details.

2. Members of the genus *Multiclavula* enclose a green algal symbiont (*Coccomyxa*) within an unstructured mantle of mycelial tissue that produces a green scum. This symbiotic association has traditionally been considered to be a basidiolichen known as *Botrydina* but might more accurately be described as an ectomycorrhizal association with a unicellular algal photobiont (see the introductory section on lichens for a discussion of these terms and concepts). The fact that all other cantharelloid genera are ectomycorrhizal fungi strengthens this idea. See A. Voitk & E. Ohenoja, 2011, Genus *Multiclavula* in Newfoundland and Labrador, *Fungi* 4(2):26 and 27, 30, and 31.

White Coral

Ramariopsis kunzei

A

B

Photos: A, Kaldari; B, Jim Edwards.

***Ramariopsis kunzei* (Fries) Corner**, *Monograph of Clavaria and Allied Genera (Annals of Botany Memoirs No. 1)*:640 (1950)

Basionym: *Clavaria kunzei* Fries, *Syst. Mycol.* (Lundae) 1:474 (1821)

[Basidiomycota, Agaricomycotina, Agaricomycetes, Agaricomycetidae, Agaricales, Clavariaceae]

Derivation: *Ramaria*, Latin, from *rāmus*, "branched," and *-opsis*, "similar to"; *kunzei*, epithet in honor of the German mycologist Gustav Kunze (1793–1851).

Fruitbody: 2–12 cm tall by up to 10 cm wide (1–4¾ × 2 in); erect, coral-like, with repeatedly forked, cylindrical to somewhat flattened branches with bluntly pointed tips arising from a short cylindrical stipe; smooth, white to ivory-cream, sometimes with pinkish tinge. Base ½–3 cm tall by 1 cm thick (¼–1¼ × ½ in), sometimes yellowish to pinkish.

Flesh: White, fragile or pliable; odor and taste not distinctive.

Spores in mass: White.

Bruising: Not bruising.

Chemical staining: No reaction with $FeSO_4$; dry fruitbody black with KOH.

Microscopic: Spores 3–5.5 × (2.5–)3–4.5 µm, subglobose to broadly ellipsoid; clear, thin-walled, minutely spiny, with apiculus and one oil droplet; nonreactive in Melzer's reagent; clamp connections present.

Occurrence: Saprotrophic. Solitary, gregarious, or cespitose on the ground or on decayed wood in forests; widespread in North America, summer–fall; Europe, Asia, Australia, Solomon Islands.

Edibility: Edible, but without flavor.

Comments: *Ramariopsis lentofragilis* (Atkinson) Corner can be confused with *R. kunzei*. It is distinguished from *R. kunzei* by its more robust fruitbody, branches that do not develop a pinkish tinge, its pale tan (not white) spore print, and by gradually staining (within two minutes) green or blackish-green with $FeSO_4$. *Ramariopsis lentofragilis* is poisonous and causes severe abdominal pain and general weakness persisting for up to 10 days.

Crested Coral Fungus

Clavulina coralloides

A

B

Photos: A & B, W. Light; C, Herbert Baker.

Clavulina coralloides **(Linnaeus) J. Schröter,**[1] in Cohn, *Krypt.-Fl. Schlesien* (Breslau) 3.1(25–32):443 (1888) [1889]

Basionym: *Clavaria coralloides* Linnaeus, *Sp. Pl.* 2:1182 (1753) [Basidiomycota, Agaricomycotina, Agaricomycetes, *incertae sedis*, Cantharellales, Clavulinaceae]

Derivation: *Clavulina*, Latin, from *clava*, "club," and *-ina*, a diminutive suffix; *coralloides*, "coral-like."

Fruitbody: Up to 10 cm (4 in) tall by 10 or more cm wide; sparingly to (more often) repeatedly branched, especially near the tips; branches up to 5 mm (¼ in) thick, smooth, wrinkled, or longitudinally furrowed, tips bluntly pointed (Figs. A & B) to more commonly cristate (Figs. B & C)—flattened with a crest formed of jagged, tooth-like, pointed, thorny projections, these variously white, yellow, or brownish depending on age; branches white when young, becoming various shades of pinkish to pale greyish-ochre; base sometimes stained greyish to blackish by a parasitic fungus, *Helminthosphaeria clavariarum* (Desmazières) Fuckel.

Flesh: White, soft, fragile, odor and taste not distinctive.

Spores in mass: White.

Bruising: Not bruising.

Chemical staining: Nonreactive with $FeSO_4$.

Microscopic: Spores 7–10 × 6–10 µm, subglobose, smooth, clear, thick-walled, with apiculus; basidia 2-sterigmate; clamp connections present.

Occurrence: Solitary, scattered, or aggregated on the ground or in moss in forests, associated primarily with conifers but also with hardwoods; widespread and fairly common in North America; summer–late fall.

Edibility: Edible.

Comments: The Grey Coral, *Clavulina cinerea* (Bulliard) J. Schröter, is marginally distinguished from *C. coralloides* by its lilac to lilac-grey or purplish-grey fruitbody, whose branch tips can be either bluntly rounded or cristate with dark horny points, as in *C. coralloides. Clavulina cinerea* is also occasionally parasitized by *Helminthosphaeria clavariarum*, which can blacken the base. Distinguishing between *Helminthosphaeria*-parasitized *C. coralloides* and *C. cinerea* can thus be problematic, and it is unclear whether these represent two separate taxa or a continuum between differing ecotypes of a single polymorphic species, which may also include *Clavulina rugosa* (Bulliard) J. Schröter.

1. This fungus is almost always referred to as *Clavulina cristata* (Holmskjold) J. Schröter (1888); however, the name *Clavaria coralloides* Linnaeus 1753 predates *Ramaria cristata* 1790 by nearly four decades and thus has nomenclatural priority.

Grey Coral, Ashy Coral

Clavulina cinerea

A B C

Photos: A & C, Teresa Fortenberry; B, Robert Sasata.

***Clavulina cinerea* (Bulliard) J. Schröter**, in Cohn, *Krypt.-Fl. Schlesien* (Breslau) 3.1(25–32):443 (1888) [1889]

Basionym: *Clavaria cineria* Bulliard, *Herb. Fr.* 8:tab. 354 (1788) [Basidiomycota, Agaricomycotina, Agaricomycetes, *incertae sedis*, Cantharellales, Clavulinaceae]

Derivation: *Clavulina*, Latin, from *clava*, "club," and *-ina*, a diminutive suffix; *cinerea*, "grey," "ash-grey."

Fruitbody: Up to 11 cm (4½ in) tall, erect, coral-like; lilac-grey, becoming grey to greyish-brown with age, branches often sinuous (Fig. A), with somewhat flattened and bluntly rounded (Fig. B) or crest-like, with tooth-like, pointed, thorny projections (Fig. C); surface smooth to slightly wrinkled; stipe short, rudimentary or lacking; base sometimes stained greyish to blackish by a parasitic fungus, *Helminthosphaeria clavariarum* (Desmazières) Fuckel.

Flesh: Greyish-white, brittle; odor and taste not distinctive.

Spores in mass: White.

Bruising: Not bruising.

Chemical staining: Nonreactive with $FeSO_4$.

Microscopic: Spores (6.5–)7–10(–11) × (5.5–)6–9(–10) µm; subglobose to broadly ellipsoid, smooth; nonreactive in Melzer's reagent; 2-sterigmate; clamp connections present.

Occurrence: Solitary, scattered, or aggregated on the ground or in moss in forests, associated primarily with conifers, especially pines (*Pinus* spp.); widespread and fairly common in the Southeast and North America, summer–late fall; Europe, Asia.

Edibility: Edible.

Comments: This species is distinguished by its lilac-grey (when young) to dull grey or greyish-brown fruitbody and flattened branch tips, which can be either round or cristate (crested) with concolorous or dark pointed projections.

The Crested Coral Fungus, *C. coralloides* (Linnaeus) J. Schröter,[1] is marginally distinguished from *C. cinerea* by its white fruitbody. Its branch tips can be either bluntly rounded or cristate with dark horny points, as in *C. coralloides*. However, morphological intergrades are not rare. *Clavulina cinerea* is also occasionally parasitized by *Helminthosphaeria clavariarum*, which can blacken the base. Distinguishing between *Helminthosphaeria*-parasitized *C. coralloides* and *C. cinerea* can thus be problematic, and it is unclear whether these represent two separate taxa or a continuum between differing ecotypes of a single polymorphic species, which may also include *Clavulina rugosa* (Bulliard) J. Schröter. Recent genetic studies indicate that European populations of *C. coralloides* and *C. rugosa* are well-delimited species, whereas *C. cinerea* appears to represent more than a single taxon.[2]

1. The Crested Coral Fungus is almost always referred to as *Clavulina cristata* (Holmskjold) J. Schröter (1888); however, the name *Clavaria coralloides* Linnaeus 1753 predates *Ramaria cristata* 1790 by nearly four decades and thus has nomenclatural priority.

2. I. Olariaga et al., 2009, *Mycol. Res.* 113(11):1261–1270.

Yellow and Golden Spindle Club Fungi

Clavulinopsis fusiformis and *C. aurantiocinnabarina*

A B C

A, B, & inset, *Clavulinopsis fusiformis*; C, *C. aurantiocinnabarina*. (Photos: A, B, & inset, Teresa Fortenberry; C, Jim Edwards.)

***Clavulinopsis fusiformis* (Sowerby) Corner**, *Ann. Bot. Mem.* 1:367 (1950)

Basionym: *Clavaria fusiformis* Sowerby, *Col. Fig. Engl. Fung. Mushr.* 2:98 (1799) [1798–1799]

***Clavulinopsis aurantiocinnabarina* (Schweinitz) Corner**, *Ann. Bot. Mem.* 1:358 (1950)

Basionym: *Clavaria aurantio-cinnabarina* Schwein., *Trans. Am. Phil. Soc.*, ser. 2, 4(2):183 (1832)
[Basidiomycota, Agaricomycotina, Agaricomycetes, Agaricomycetidae, Agaricales, Clavariaceae]

Derivation: *Clavulinopsis*, Latin, from *clava*, "club"; and *-opsis*, Greek, "resembling," "resembles a club"; *aurantiocinnabarina*, Latin, "gold-red"; *fusiformis*, Latin, "spindle-shaped."

Fruitbody: Both species—cylindrical, worm-like, unbranched, often in dense cespitose clusters. ***C. fusiformis***—5–15 cm high by up to 1.5 cm wide (2–6 × ⅔ in), cylindrical, often flattening, sometimes grooved (Fig. A); dry; bright or pale yellow, white at the extreme base (Fig. B); usually with a somewhat pointed, reddish-brown tip (Fig. B inset). ***C. aurantiocinnabarina***—unbranched, cylindrical, worm-like, sometimes twisted, deep reddish-orange to pale orange, yellowish to whitish at base, color fading to pale apricot with age; 1.5–8 (up to 15) cm tall by 2–6 mm thick (3¼–6 × ¼ in), smooth, with conspicuous longitudinal groove; tip pointed when young, becoming rounded or blunt at maturity.

Flesh: *C. fusiformis*—yellow, thin, pliable; ***C. aurantiocinnabarina***—reddish-orange, thin, brittle.

Chemical staining: *C. fusiformis*—flesh and surface dull green in $FeSO_4$, flesh negative and surface orange in KOH; ***C. aurantiocinnabarina***—negative in $FeSO_4$ or KOH.

Spore print: White to yellowish.

Microscopic: Basidia 4-sterigmate (both species). ***C. fusiformis***—spores 5–9 × 4.5–9 µm, subglobose to broadly ellipsoid, smooth, with a prominent apiculus 1–2 µm long; clamp connections present. ***C. aurantiocinnabarina***—spores 5.6–7 × 5.2–7 µm; globose to subglobose, smooth, thin-walled, with a weak apiculus <1 µm long, containing a single oil droplet; context with scattered, often inconspicuous clamp connections.

Occurrence: Both species—solitary, scattered, or gregarious to cespitose or joined at the base. ***C. fusiformis***—in woods under hardwoods or conifers, sometimes in grassy areas; ***C. aurantiocinnabarina***—common on soil in grassy areas or under hardwoods in low moist areas, summer–fall in eastern North America.

Edibility: *C. fusiformis*—taste is bitter; ***C. aurantiocinnabarina***—reported to be edible, taste is mild.

Comments: *Clavulinopsis laeticolor* (Berkeley & M.A. Curtis 1868) R.H. Petersen (1965) is distinguished from *C. aurantiocinnabarina* by its smaller fruitbodies, which are typically only 2–5 cm (¾–2 in) tall by 1.5–3 mm thick. The spores are smooth, round to broadly elliptical (4.5–7 [up to 9] × 3.5–5 [up to 6] µm) or pear-shaped and bear a prominent apiculus (1.5–2 µm long), in contrast to the weak apiculus of *C. aurantiocinnabarina*. The surface and flesh *stain dingy greenish in* $FeSO_4$ *and yellow-green or green in 10% KOH*. The flesh is also yellowish and pliable compared to the reddish-orange and brittle flesh of *C. aurantiocinnabarina*.

Clavulinopsis fusiformis (Sowerby 1799) Corner (1950) is dull to bright yellow rather than orange, like *C. aurantiocinnabarina* and *C. laeticolor*. The spores of *C. fusiformis* are similar to those of *C. laeticolor*, but the flesh is yellowish (not reddish-orange) and has a bitter taste. The fruitbody surface is negative in KOH but stains grey-green in $FeSO_4$, which distinguishes this taxon from both *C. aurantiocinnabarina* and *C. laeticolor*. *Clavulinopsis helveola* (Persoon 1797) Corner (1950), a look-alike to *C. fusiformis*, is the size of *C. aurantiocinnabarina* but has thinner yellow to orange-yellow fruitbodies and angular spores with a warty surface.

Yellow Antler Coral

Clavulinopsis corniculata

A

B

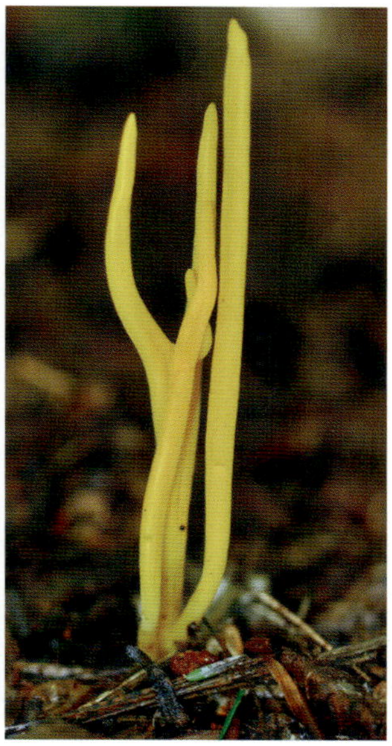
C

Photos: Jim Edwards.

Clavulinopsis corniculata **(Schaeffer) Corner**, *Monograph of Clavaria and Allied Genera (Annals of Botany Memoirs No. 1)*:362 (1950)

Basionym: *Clavularia corniculata* Schaeffer, *Fung. Bavar. Palat. Nasc.* (Ratisbonae) 4:117 (1774)
[Basidiomycota, Agaricomycotina, Agaricomycetes, Agaricomycetidae, Agaricales, Clavariaceae]

Derivation: *Clavulinopsis*, Latin, from *clava*, "club," and *-opsis*, "similar to"; *corniculata*, "crescent-shaped."

Fruitbody: 1–9 cm (more typically 2–5 cm) tall (1–2, up to 3⅔ in); delicate, unbranched (Fig. C) to slightly (Fig. B) or moderately (Fig. A) dichotomously branched one to four times; egg-yellow to orangish ochraceous-yellow, apices typically U-shaped or crescent-shaped (Fig. A), with rounded or bluntly pointed tips. Base ½–6 cm tall by 2–5 mm thick (¼–2½ × ~¼ in), often rudimentary or lacking, sometimes with white, downy basal mycelium.

Flesh: Whitish to pale yellow, thin, firm to tough; odor floury to oily-starchy, taste sweet to starchy.

Spores in mass: White.

Chemical staining: Branches grey-green to olive-green with $FeSO_4$; orange with KOH.

Microscopic: Spores 4–7 × 4.5–7 µm; subglobose to globose, smooth, walls somewhat thick, with prominent apiculus, with or without oil droplets; clamp connections present but often obscure.

Occurrence: Saprotrophic. Solitary, scattered, or gregarious on ground or dead wood under hardwoods in forests or in grassy areas; widespread in North America; Europe; uncommon, early summer–late fall.

Edibility: Reported to be edible.

Comments: *Ramariopsis crocea* (Persoon) Corner is very similar but is generally smaller (4 cm [1⅔ in] or less in height), has a distinct base occupying ¼–½ of the total height, is associated with conifers, and has a negative reaction with $FeSO_4$.

The Yellow Staghorn Coral, *Calocera viscosa* (Persoon) Fries, is also similar but is very greasy and viscid (not dry) to the touch; it grows on the roots and stumps of conifers (not from the ground or on hardwoods), and its cylindrical to sausage-shaped (not subglobose) spores are 3-septate.

Yellow-tipped Coral, Beautiful Coral

Ramaria formosa

A

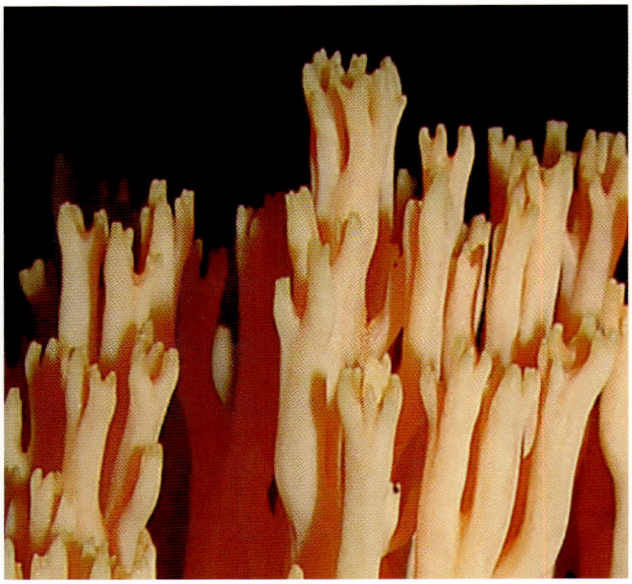

B

Photos: Teresa Fortenberry.

Ramaria formosa **(Persoon) Quélet**, *Fl. Mycol. France* (Paris):466 (1888)

Basionym: *Clavaria formosa* Persoon, *Comm. Fung. Clav.* (Lipsiae):41 (1797)
[Basidiomycota, Agaricomycotina, Agaricomycetes, Phallomycetidae, Gomphales, Gomphaceae]

Derivation: *Ramaria*, Latin, from *rāmus*, "branched"; *formōsa*, "beautifully formed."

Fruitbody: 8–20(–30) cm tall by 5–12 cm wide (3¼–8[–12] × 2–4¾ in); coral-like, with vertically oriented, separated branches arising from a modest to (typically) massive base; branches peach to salmon-pink to light coral-red, with light yellow tips when young, becoming concolorous with the branches to yellowish-brown with age, base whitish; surface smooth or wrinkled. Branching up to six times from a robust base; terminal branches finely divided, <5 mm (¼ in) thick, with acute to rounded tips;

Flesh: Whitish to pinkish, soft and spongy at the base, fleshy-fibrous above, brittle and crumbly when dried; odor and taste not distinctive in North America (taste reported as bitter in Europe).

Spores in mass: Golden-yellow.

Bruising: Flesh bruising brown when handled.

Chemical staining: Surface of branches green to bluish-green with $FeSO_4$; no reaction to KOH.

Microscopic: Spores 8–13 × 3–6 µm, elliptical to drop-shaped, with offset apiculus, distinctly ornamented with irregularly lobed ridges; clear, nonreactive in Melzer's reagent; clamp connections present.

Occurrence: Mycorrhizal with hardwoods; solitary, scattered, or gregarious on the ground under oaks (*Quercus* spp.) and American Beech (*Fagus grandifolia*); under conifers in western North America. Widespread in North America; Europe; August–September.

Edibility: Inedible, toxic; causes acute gastrointestinal upset, nausea, vomiting, and diarrhea.

Comments: This beautiful coral is distinguished by its typically massive size, salmon-pink to coral-pink branches with finely divided yellow tips, tendency to bruise brown when handled, and prominently ridged spores, and in possessing clamp connections.

The question as to whether or not this species has a bitter or unremarkable taste and if it is actually poisonous has also been challenged.[1] *Ramaria formosa* in North America has been incorrectly reported as bitter and poisonous by several field guides and online sources. This may be due to its being confused with *R. pallida* Maire,[2] which does have a bitter, acrid taste and is indeed poisonous, causing gastrointestinal upset.

Ramaria conjunctipes (Coker) Corner differs from *R. formosa* in the following respects: (1) it grows on the ground under conifers (not hardwoods); (2) the branches arise in dense fascicles with downwardly tapering, cylindrical stems with acute tips and white rhizomorphs at an indistinct (not robust) base; (3) the color is intense yellow-salmon when young; (4) the flesh is sooty (not whitish to pinkish); (5) the spores are smaller (7.9–9.4 × 4.7–5.4 µm), broadly ellipsoid to comma-shaped, and sparsely to delicately (not prominently) ridged; and (6) clamp connections are lacking.

1. J.F. Ammirati, J.A. Traquair, & P.A. Horgen, 1985, *Poisonous Mushrooms of the Northern United States and Canada* (Minneapolis: University of Minnesota Press), 307–308.

2. *Ramaria pallida* does not occur in our area.

Straight-branched Coral

Ramaria stricta

This photograph shows multiple tightly aggregated/cespitose fruitbodies. (Photo: Jim Edwards.)

***Ramaria stricta* (Persoon) Quélet**, *Fl. Mycol. France* (Paris):464 (1888)

Basionym: *Clavaria stricta* Persoon, *Comm. Fung. Clav.* (Lipsiae):33 (1797)
[Basidiomycota, Agaricomycotina, Agaricomycetes, Phallomycetidae, Gomphales, Gomphaceae]

Derivation: *Ramaria*, Latin, from *rāmus*, "branched"; *stricta*, "narrow," "tight."

Fruitbody: 5–12.5 cm tall × 5–8 cm wide (2–5 × 2–3¼ in), with many multiply branching, erect, slender and cylindrical, more or less parallel branches arising from a short, stipe-like base; pale yellow to pinkish-buff or pinkish-orange, with slender and pointed yellow tips, these becoming concolorous with the branches with age; base up to 2 cm (¾ in) thick, white, often with white rhizomorphs embedded in the wood.

Flesh: Whitish, tough, thin, brittle; odor of anise, garlic, or radish; taste bitter, astringent.

Spores in mass: Yellow-brown.

Bruising: Branches and flesh bruising brown.

Chemical staining: Branches green with $FeSO_4$; copper to dull orange or brownish with KOH.

Microscopic: Spores (7–)7.5–10(–10.5) × 3.5–5.5 µm, narrowly ovoid to long-elliptical; clear, thin-walled, with prominent, eccentric hilar appendage, finely roughened with low, short, scattered, anastomosing ridges; with or without several oil droplets; clamp connections present; nonreactive in Melzer's reagent.

Occurrence: Saprotrophic on decayed wood; solitary, gregarious or cespitose on dead and decayed or buried conifer or hardwood stumps, trunks, or branches; widespread and occasional to common in North America; occurring in Southern Piedmont and southern Appalachians but more common in western coniferous forests; summer–fall.

Edibility: Inedible, nontoxic but bitter.

Comments: The similar-looking Crown-tipped Coral, *Artomyces pyxidatus* (Persoon) Jülich,[1] is widespread in the Southeast and also grows on decaying hardwood logs and stumps. However, its odor is not distinctive, the taste is mild to peppery, and its branches are crowned with cup-like tips encircled by three to six prominent, finger-like teeth. The branches are negative or stain greyish or brownish (not green) with $FeSO_4$.

1. Referred to as *Clavicorona pyxidata* (Persoon) Doty by many authors.

Crown-tipped Coral Fungus

Artomyces pyxidatus

A

B

Photos: Teresa Fortenberry.

***Artomyces pyxidatus* (Persoon *ex* Fries) Jülich**, *Biblthca. Mycol.* 85:399 (1982) [1981]

Basionym: *Clavaria pyxidata* Persoon : Fries, *Neues Mag. Bot.* 1:117 (1794)
[Basidiomycota, Agaricomycotina, Agaricomycetes, *incertae sedis*, Russulales, Auriscalpiaceae]

Derivation: *Artomyces*, Greek, from *artos*, "loaf of leavened bread," and *myces*, "fungus"; *pyxidatus*, Latin, from *pyxis*, "box," "box-like," referring to the tips of the branches.

Deprecated synonym: *Clavicorona pyxidata* (Persoon) Doty, *Lloydia* 10:43 (1947)

Fruitbody: 5–10 (up to 12) cm tall by 2–8 (up to 10) cm wide (2–4[–4¾] × 1–3¼[–4] in); coral-like, repeatedly branched (Fig. A); whitish to yellowish-cream, becoming pale ochre to tan or pinkish with age; branches 1–5 mm thick, tips shallowly cup-like and rimmed by three to six finger-like projections (Fig. B); base lacking or rudimentary up to 3 mm (~1⁄12 in) tall, finely felt-like, whitish, pinkish, or brownish.

Flesh: Whitish, fibrous, pliable; odor indistinct or starchy (like potatoes); taste mild to delayed peppery.

Spores in mass: White.

Bruising: Not bruising.

Chemical staining: Negative to greyish or brownish with $FeSO_4$.[1]

Microscopic: Spores 4–5 × 2–3 µm, ellipsoidal, clear, smooth (very minutely pitted), with tiny apiculus; amyloid; basidia 4-sterigmate; clamp connections present.

Occurrence: Saprotrophic on rotten hardwood. Solitary, scattered, or densely clustered on dead and decayed hardwood logs; widespread and common in North America east of the Rocky Mountains, with high-altitude extensions into subtropical Mexico and Costa Rica, spring–fall; Europe; Asia.

Edibility: Edible.

Comments: This beautiful and distinctive coral is immediately identified by the crown-like tips of its branches and its growth on rotten hardwood substrata. Several individual *A. pyxidatus* in Great Smoky Mountains National Park were reported to occupy a single log, with mycelia attaining a measured size of 1.9 m (6¼ ft).[2]

Although *A. pyxidatus* populations from all continents

are interfertile, ITS (internal spacer sequences) reveal the existence of several distinct clades that have been genetically isolated for an unknown period, including northeastern and southwestern clades in North America. Southeastern specimens (e.g., from Georgia) appear to represent hybridizing intergrades between these two clades.[3]

1. *Fide* M. Kuo, 2007, *Artomyces pyxidatus*, retrieved from http://www.mushroomexpert.com/artomyces_pyxidatus.html, accessed March 25, 2014.

2. S.M. Tieken, 2002, *Artomyces pyxidatus* (Auriscalpiaceae, Fungi) in the Great Smoky Mountains National Park: Individuals and Populations, master's thesis, University of Tennessee, vii + 57 pp., http://trace.tennessee.edu/utk_gradthes/2188.

3. E.B. Lickey et al., 2002, *Mycologia* 94(3):461–471.

Vase Thelephore

Thelephora vialis

Thelephora vialis.

Thelephora vialis **Schweinitz**, *Trans. Am. Phil. Soc.*, ser. 2, 4(2):165 (1832)
[Basidiomycota, Agaricomycotina, Agaricomycetes, *incertae sedis*, Thelephorales, Thelephoraceae]

Derivation: *Thelephora*, Greek, from *thele*, "nipple," and *phóros*, "bearer"; *vialis*, Latin, "of the roads."

Synonyms: *Phylacteria vialis* (Schweinitz) Patouillard, *Essai Tax. Hyménomyc.* (Lons-le-Saunier):119 (1900)
Thelephora tephroleuca Berkeley & M.A. Curtis, *Grevillea* 1(10):149 (1873)

Fruitbody: 2.5–15 cm wide by 2.5–10 cm tall (1–6 × 1–4 in); highly variable, consisting of erect spoon-like lobes or fan-shaped, often fused flabellae arising from a common central stipe and forming a more or less vase-like composite structure (Figs. A & C) or an irregular, flower-like cluster (Fig. B), margins more or less entire (Figs. A & C) to deeply scalloped (Fig. B); upper surface striated or scaly, dirty off-white (Figs. A & B), outer surface smooth, light greyish-purple, with whitish margin (Fig. C).

Flesh: Thick, leathery, white, emitting a disagreeable fetid smell when dry.

Stipe: 1–5 cm tall by up to 40 mm thick (½–2 × 1⅔ in), leathery, solid whitish to greyish (Fig. C, arrow), finely hairy.

Spore print: Brown.

Chemical staining: KOH—upper surface faint yellow-brown; lower (outer) surface strongly reddish yellow-brown.

Microscopic: Spores 4.5–8 × 4.5–6.5 µm, angular-ellipsoid, ornamented with warts bearing minute spines.

Occurrence: Solitary or in scattered groups on the ground in deciduous woods, associated particularly with oaks (*Quercus* spp.); widespread in the Southeast; summer–midwinter.

Edibility: Inedible.

Comments: *Thelephora palmata* Scopoli *ex* Fries and *T. terrestris* Ehrhart also occur in our area. *Thelephora palmata* (Fig. D) forms highly branching, dark lilac- to purplish-brown, coral-like fruitbodies arising from a short, stubby,

D E

D, *Thelephora palmata*; E, *Thelephora terrestris*. (Photos: Miika Silverberg.)

concolorous stalk no more than 1.5 cm (⅔ in) tall and occurs on the ground in conifer and mixed woods. Its flesh stains deep blue with KOH, and it has larger (8–12 × 6–9 µm) and spinier purplish-brown spores. *Thelephora terrestris* (Fig. E) forms single to overlapping circular to vase-like clusters of dark brown to blackish-brown fruitbodies arising from a very rudimentary stalk. The upper surface is covered with matted fibrous or hairy scales, which immediately distinguish this species from *T. vialis*. The pores are similar to those of *T. palmata*.

Parasitic Fungi

Separating out several unrelated groups of higher fungi under the heading "Parasitic Fungi" is in a sense quite misleading. Many species of polypores and other mushrooms are, in fact, parasitic on living trees, shrubs, and other plants, where they cause various types of wood rot. Some of them also become saprobic after their host has died. And segregating out unrelated mycoparasitic genera like *Elaphocordyceps*, *Hypomyces*, *Phaeocalicium*, and the Aborted Entoloma (*Entoloma abortivum*)[1] does a disservice to other mushroom-parasitizing fungi, such as the jelly fungus genus *Tremella*, as well as hypermycoparasites (fungal parasites on other fungal parasites) like *Helminthosphaeria clavariarum* and *Ophiostoma epigloeum* that are treated in other sections of this book. A number of the anamorphic forms of well-known mushrooms, such as the edible oyster mushroom *Hohenbuehelia petaloides* (Bulliard *ex* Fries) Schulzer,[2] and many other nematode-killing, "carnivorous" fungi also function in precisely the same way as the arthropod-parasitizing cordyceps and insect destroyers (Emtomophthorales) discussed in this section. However, space does not permit us to head down that fascinating but rocky road in this book.

Nevertheless, the fungi discussed in this section are particularly noteworthy and distinctive in their appearance, and they have fascinating natural histories. Three of them are edible (two are considered choice), and one also has potent and widely appreciated medicinal properties.

In general, mitosporic (asexual) conidia released by these fungi typically attach to the exoskeletons of their insect host, germinate, and penetrate the cuticle into the insect's body cavity. Once inside, the fungus sends hyphae and/or protoplasts throughout the victim's body that digest its organs and tissues, eventually killing it. Many of these fungi first send hyphae to the host's brain, where they affect the insect's behavior, causing it to climb to a prominent position on a plant or other structure and adhere to it by its mouthparts before dying. This provides the fungus with a good vantage point from which to produce its fruitbody and shed its infectious spores.

The entomopathogenic (insect-parasitizing) fungi discussed in this book represent two divisions (Ascomycota and Entomophthoromycota), two orders (Hypocreales and Entomophthorales), and five families. This is but a tiny sample of the known arthropod-parasitizing fungi, which involve a number of other classes. In the Ascomycota, the entomopathogenic teleomorphic genus *Cordyceps* (*sensu lato*) has a number of anamorphic form-genera that parasitize various arthropods (e.g., *Beauveria*, *Isaria*, *Hirsutella*, and *Metarhizium*). The Entomophthoromycota includes *Entomophthora*, *Furia*, *Erynia* (these three discussed here), as well as *Pandora*, *Zoophthora*, and others.

The splitting of the genus *Cordyceps* (*sensu lato*) into *Cordyceps* (*sensu stricto*), *Elaphocordyceps*, *Ophiocordyceps*, and *Metacordyceps* is discussed under "Endoparasitic Fungi" in the introduction.

The mycoparasitic fungi discussed here include two species of *Elaphocordyceps*, three of *Hypomyces*, and one each of *Phaeocalicium* and *Entoloma*.

Elaphocordyceps includes nearly two dozen described species that parasitize deer truffles of the genus *Elaphomyces*. *Elaphocordyceps* that parasitize deer truffles appear to have evolved from an *Elaphocordyceps* clade that parasitizes developing underground cicada nymphs.[3] *Elaphocordyceps ophioglossoides* and *E. capitata* (both treated in this section), for example, may have evolved from an Asian cicada-infecting species very close to *E. paradoxa* Kobayasi.[4]

Quandt et al. recently sunk the genus *Elaphocordyceps* into synonymy with *Tolypocladium* (2014, "Phylogenetic-based nomenclatural proposals for *Ophiocordycipitaceae* [Hypocreales] with new combinations in *Tolypocladium*," *IMA Fungus* 5[1]:121–134). The latter is the form-genus of anamorphic fungi of which *Elaphocordyceps* was later determined to be the teleomorphic form. Since

Tolypocladium W. Gams (1971) predates *Elaphocordyceps* G.-H. Sung & Spatafora (2007), these authors recombined *Elaphocordyceps capitatis, E. longisegmentis*, and *E. ophioglossoides* as *Tolypocladium capitatum, T. longisegmentum*, and *T. ophioglossoides*. We do not accept this or similar combinations because of the great nomenclatural instability they produce.[5]

Hypomyces species infect the fruitbodies of various mushrooms to produce a light to vividly colored perithecia embedded in a thin, concolorous, cottony or crust-like layer called the **subiculum**. This subiculum eventually completely covers the infected host fruitbody, typically transforming it into an unrecognizable and amorphous mass. Most *Hypomyces* infect euagarics, boletes, russulas and milkcaps, and polypores assigned to several orders. Most *Hypomyces* are inedible, and some are poisonous. But the Lobster Mushroom (*H. lactifluorum*) infects several species of russulas, as well as *Lactifluus piperatus* (formerly in the genus *Lactarius*), turning the acrid-tasting host mushrooms into choice and much-sought-after edibles.

1. These genera plus the Aborted Entoloma represent three different divisions (phyla) and four separate classes.

2. The teleomorphic fruitbodies are saprobic on dead wood.

3. N. Nikoh & T. Fukatsu, 2000, *Mol. Biol. Evol.* 17:629–638.

4. T. Fukatsu & N. Nikoh, 2003, Interkingdom host shift in the *Cordyceps* fungi, *in* J.F. White et al. (eds.), *Clavipitalean Fungi: Evolutionary Biology, Chemistry, Biocontrol, and Cultural Impacts* (New York: Marcel Dekker), 290–305. Though primarily parasites of deer truffles (*Elaphomyces*), at least three species infect cicadas: *E. inegoensis, E. paradoxa*, and *E. toriharamontana*. See C.A. Quandt et al., 2014, *IMA Fungus* 5(1):121–134.

5. Under the rule of "one fungus, one name" adopted by the International Botanical Congress in Melbourne in 2011, the *ICN* (*International Code of Nomenclature for algae, fungi, and plants*), after January 1, 2013, all names of anamorphs and teleomorphs will compete equally for priority, and the dual-naming system will no longer be allowed. However, the Congress also decided that anamorph-typified names should not displace widely used teleomorph names until the specific case has been considered by the General Committee established by the Congress (D.L. Hawksworth, 2011, A new dawn for the naming of fungi: Impacts of decisions made in Melbourne in July 2011 on the future publication and regulation of fungal names, *Mycokeys* 1:7–20).

Arthropod-Parasitizing Fungi

Beetle Cordyceps

Ophiocordyceps melolonthae

A B

Photos: A, John Plischke III; B, Stefanie Hanagud.

Ophiocordyceps melolonthae **(Tulasne & C. Tulasne) G.-H. Sung, J.-M. Sung, Hywel-Jones & Spatafora**, in G.-H. Sung et al., *Stud. Mycol.* 57:44 (2007)

Basionym: *Torrubia melolonthae* Tulasne & C. Tulasne, *Selecta Fungorum Carpologia: Nectriei-Phacidiei-Pezizei* (Paris) 3:12 (1865)
[Ascomycota, Pezizomycotina, Sordariomycetes, Hypocreomycetidae, Hypocreales, Ophiocordycipitaceae]

Derivation: *Ophiocordyceps*, Latin, from *ophidia* (New Latin, from the Greek *ófis*), "serpent," plus *cord*, "club," and *-ceps*, "head"; *melolonthae*, "cankerworm," from *Melolontha*, "cockchafers," a genus of European scarab beetles: "snake-like clubhead associated with beetles."

Fruitbody (stroma): Mature stromata consisting of a yellowish to yellow-orange stalk capped by an ovoid white to yellow head.

Head: 2–3 cm tall by 1–1.5 cm thick (¾–1¼ × ½–⅔ in); initially cream-colored (Fig. B), becoming yellowish (Fig. A), head indistinct in developing fruitbody (Figs. A & B); surface finely roughened by minute, scarcely visible perithecia, distinctly set off from stipe in mature fungus and becoming ovoid to fusiform.

Flesh: Whitish to pale yellowish.

Stipe: 5–7 cm tall by 3–10 cm thick (2–2¾ × up to ½ in), subequal, slender to stout, smooth, pale buff to yellowish.

Microscopic: Spores (ascospores) initially thread-like and multiseptate, later fragmenting into unicellular part-spores 4–10 × 1–2.5 µm, these elliptical, smooth, clear.

Occurrence: Parasitic on buried scarabid beetle larvae (and occasionally adults), one to several emerging from soil or decayed wood; widespread and fairly common in eastern North America; summer–midwinter.

Edibility: Inedible.

Comments: This fungus is most often encountered before the fertile heads have developed. The similar-looking Military Cordyceps or Trooping Club, *Cordyceps militaris* (Linnaeus) Fries, has a distinctly reddish-orange tint due to its prominent reddish-orange perithecia, and it grows from the buried pupae of lepidopterans (butterflies and moths) rather than beetle larvae. The thread-like, multiseptate ascospores fragment into unicellular part-spores 2–4.5 × 1–1.3 µm in size. *Ophiocordyceps ravenelii* (Berkeley & M.A. Curtis) G.-H. Sung, J.-M. Sung, Hywel-Jones & Spatafora, which also parasitizes beetle larvae and occurs in the southern Appalachians, has a dark purplish-brown to chocolate-brown fruitbody and superficially resembles the Snake-tongued Cordyceps, *Elaphocordyceps ophioglossoides* (see our account of that species).

Ophiocordyceps sphecocephala (Klotzsch *ex* Berkeley) G.-H. Sung, J.-M. Sung, Hywel-Jones & Spatafora grows from dead wasps, hornets, and bees. It also has multiseptate spores that fragment into long, spindle-shaped segments 8–12(–14) × 1.5–2 µm. A close relative with similarly elongated fragmented ascospores, *O. myrmecophila* Cesati, parasitizes ants.

Military Cordyceps, Trooping Club

Cordyceps militaris

A B

Photos: John Plischke III.

***Cordyceps militaris* (Linnaeus) Fries**, *Observ. Mycol.* (Havniae) 2:317 (1818)

Basionym: *Clavaria militaris* Linnaeus, *Sp. Pl.* 2:1182 (1753) [Ascomycota, Pezizomycotina, Sordariomycetes, Hypocreomycetidae, Hypocreales, Cordycipitaceae]

Derivation: *Cordyceps*, Latin, from *cord*, "club," and *-ceps*, "head"; *militaris*, "military."

Fruitbody (stroma): Up to 8–10 cm tall by 5 mm thick (3¼–4 × ¼ in), with expanded club-shaped (Fig. B) to spindle-shaped (Fig. A) head; head orange, dotted with prominent, pimple-like, bright reddish-orange perithecia (Fig. A, inset); stalk smooth, hollow, pale orange, often curved (Fig. B), arising from buried lepidopteran (butterfly or moth) pupa or larva; flesh pale orange.

Microscopic: Spores (ascospores) long, smooth, thread-like, segmented, 300–500 × 4 µm, fragmenting into elliptical part-spores 2–5–6 × 1–1.5 µm.

Occurrence: Parasitic. Solitary to clustered on the buried corpses of butterfly and moth pupae; North America, Europe, Asia; more common east of the Rocky Mountains; fall.

Edibility: Edible; widely used as a medicinal.

Comments: The orange to orangish-red head with its prominent, pimple-like perithecia and lepidopteran host identify this cordyceps, which is the species most often seen in our area. *Cordyceps cardinalis* G.-H. Sung & J.W. Spatafora is similar, but the stroma is much smaller and thinner—0.4–5 cm × 0.5–1.5 mm (~¼–2 in tall), the terminal fertile region is orangish-red to red, the spores (ascospores) are much shorter and thinner (160–320 × 1 µm) than in *C. militaris*, and they do not fragment into part-spores. The difference in the color of the stroma is subtle but distinct, at least in the southern Appalachians.[1] *Cordyceps cardinalis* also emerges from lepidopteran larvae rather than from pupae.

The Beetle Cordyceps, *Ophiocordyceps melolonthae*, arises from beetle larvae (and occasionally adults) and has a cream-colored to yellow head, and the minute perithecia, which are concolorous with the head, are barely visible.

1. *Cordyceps cardinalis* has a disjunct distribution in the southern Appalachian Mountains and East Asia (G.-H. Sung et al., 2004, *Mycologia* 96[3]:658–666).

The Insect Destroyers

(Order Entomophthorales)

The etymology of the order Entomophthorales derives from the Greek roots *entomo-*, "insect," and *-phthora*, the feminine form of *phthoros*, meaning "destroyer." In 2007 this order and its subcategories—previously placed in the polyphyletic and now defunct Zygomycota—were assigned to a new subdivision (subphylum), the Entomophthoromycotina, by David Hibbett and his colleagues.[1] In 2012 Richard Humber elevated the Entomophthoromycotina to the rank of division (phylum),[2] with the following current systematic breakdown: division/phylum Entomophthoromycota, Class Entomophthoromycetes, and a single order, Entomophthorales, with three families: Ancylistaceae J. Schröter, Completoriaceae Humber, and Entomophthoraceae Warming. The Ancylistaceae are either saprobes or parasites of desmids (a group of unicellular green algae) or animals (tardigrades, nematodes, termites, aphids, and a few vertebrates). The Completoriaceae is a monotypic family containing the sole species *Completoria complens* Lohde, known only as an obligate parasite of fern gametophytes in greenhouse cultures. The Entomophthoraceae are all obligate parasites of arthropods.

Two new classes were also erected for this phylum: (1) the Basidiobolomycetes, containing the sole genus *Basidiobolus* of filamentous fungi found in the dung of amphibians, reptiles,[3] and insectivorous bats and in plant debris and soil; and (2) the Neozygitomycetes, containing the sole genus *Neozygites* (and possibly some species currently assigned to the form-genus *Tarichium*), the species of which parasitize aphids and leafhoppers (Class Insecta, Order Hemiptera), thrips (Class Insecta, Order Thysanoptera), springtails (Class Entognatha, Order Collembola), and mites (Class Arachnida, Order Acarina). The Entomophthoromycotina includes more than 250 species, consisting mostly of arthropod pathogens or soil- and litter-borne saprobes.[4]

Phylum Entomophthoromycota, Class Entomophthoromycetes

Order Entomophthorales Schröter

Saprobic on decaying organic matter or animal dung or obligately parasitic in plants or animals; consisting of a well-defined mycelium—which may fragment and form hyphal bodies—or naked protoplasts; sporophores unbranched or branched; spores uninucleate, pleurinucleate (all nuclei alike), or multinucleate, spores forcibly discharged or passively released; resting spores typically zygospores or azygospores; three types of nuclei, each type typical of at least one of the five recognized families.

Family Entomophthoraceae Nowakowski

This is the largest family in the order Entomophthorales. It includes the following 12 genera: *Batkoa*, *Entomophaga*, *Entomophthora*, *Erynia*, *Eryniopsis*, *Furia*, *Massospora*, *Orthomyces*, *Pandora*, *Strongwellsea*, *Tarichium* (in part; some species will likely wind up in the family Neozygitaceae), and *Zoophthora*.[4]

We discuss three representative members of the family Entomophthoraceae—all obligate invertebrate parasites (primarily on arthropods), which are occasionally encountered and are shown in the following photos. This is the largest family in the order, and it contains some 12 genera.[5] It is distinguished from the other families in the order by having relatively large (5–15 μm long) nuclei during interphase with much conspicuous heterochromatin, which readily stains with aceto-orcein and Bismark Brown, the absence of a prominent nucleolus during interphase, and the nucleus remaining visible during mitosis.[6] The condiophore is simple or dichotomously or digitally branched. The primary condia are forcibly discharged, and the resting spores are zygospores or azygospores (haploid or parthenogenetic zygospores formed by mitosis) formed as lateral or terminal buds connected to the parent cell by a narrow isthmus.[7]

We present here a brief discussion of three of the insect destroyers belonging to the order Entomophthorales and the family Entomophthoraceae. Most of our description and comments focuses on *Entomophthora muscae*, because it is the best known.

***Furia ithacensis* (J.P. Kramer) Humber**, *Mycotaxon* 34(2):451 (1989) [Subfamily Erynioideae]

Basionym: *Erynia ithacensis* J.P. Kramer, *Mycopathologia* 75:160 (1981)

A, *Furia ithacensis* on a snipe fly; B, *Entomophthora muscae* on a muscid fly; C, sporangia of *E. muscae* on a muscid fly. (Photos: A & B, John Plischke III; C, Hans Hillewaert, Ostende, Belgium.)

Derivation: *Furia*, Latin, "fury"; *ithacensis*, from (*in sensu* "native of") Ithaca, NY.

Entomophthora muscae **(Cohn) Fresenius (*sensu lato*)**, *Bot. Ztg.* 14:883 (1856) [Subfamily Entomophthorideae]

Basionym: *Empusa muscae* Cohn, *Hedwigia* 1(10):61 (1855)

Derivation: *Entomophthora*, Greek, *entomo-*, "insect," *-phthora*, feminine form of *phthoros*, "destroyer"; *muscae*, Latin, "flies," referring to its major hosts.

Erynia sepulchralis **(Thaxter) Remaudière & Hennebert**, *Mycotaxon* 11(1):302 (1980) [Subfamily Erynioideae]

Basionym: *Empusa sepulchralis* Thaxter, *Mem. Boston Soc. Nat. Hist.* 4(6):133–201 (1888)

Derivation: *Erynia*, Latin, from *Erynnis*, "the Fury"; *sepulchralis*, "of the grave."

Synonym: *Entomophthora sepulchralis* (Thaxter) Saccardo & Traverso, *Syll. Fung.* (Abellini) 19:657 (1910)

The genus ***Furia*** (Bakto) Humber infects only snipe flies of the family Rhagionidae. This fungus secures its victims to a leaf substratum by numerous, whitish, web-like rhizoids (Fig. A, arrows). The rhizoids and cystidia are thicker than the vegetative hyphae, and the conidia are not noticeably elongated and have conical papillae. *Furia ithacensis* forms resting spores that are incised with very irregular ridges, which separates it from other species of *Erynia* that are pathogenic to flies.

Erynia: Produces sexual resting zygospores, which remain dormant for one to two years before germinating. The rhizoids and cystidia are much thicker than the vegetative hyphae, and the conidia have conical papillae.

Entomophthora: The vegetative cells are short and rod-shaped; the conidiophores are simple, and the conidiogenous cells are club-shaped. The species in this genus all have primary conidia with an apical projection and a broad, relatively flat basal papilla. The primary conidia contain 2–12 (up to about 40) nuclei and are forcibly discharged and attach to the substratum in a droplet of discharged cytoplasm; they quickly form and discharge secondary conidia. The unique shape of the conidia and the characteristic halo of conidia about the corpse of the dead fly (Fig. B) immediately diagnose this genus. If present, the rhizoids are about the same diameter as the hyphae.

***Entomophthora muscae* (*sensu lato*):** This is actually a complex of cryptic but genetically distinct species that are very difficult to distinguish from each other. This group of morphological species infects flies from the following families: Anthomyiidae (root-maggot flies), Calliphoridae (blowflies, carrion flies, bluebottle flies, and greenbottle flies), Drosophilidae (fruitflies), Empididae (dagger flies and balloon flies), Muscidae (houseflies and their relatives, the typical hosts), Sarcophagidae (flesh flies), and Syrphidae (flower flies and hoverflies).

After attaching to a fly, a conidium (sporangium) of *E. muscae* germinates to produce a germ tube that penetrates through the insect's cuticle via enzymatic action. It then grows into the victim's hemocoel and sends mycelia into its brain, which affect the host's behavior. Meanwhile, small subspherical fungal bodies proliferate and spread throughout the fly's body, digesting its internal organs and causing the abdomen to become swollen. (Certain other entomophthoran genera [e.g., *Furia*] produce thread-like mycelia instead of fungal bodies.) Flies infected by *E. muscae* typically die on the fifth day or so following infection. During the critical stage of its illness, an infected fly's behavior is altered by the pathogen. The fly typically climbs to an elevated position, attaches to the substratum by its proboscis, straightens its legs, raises its abdomen, and opens its wings before dying to facilitate the wide dispersal of the fungal spores.[8] Great numbers of unbranched fungal conidiophores emerge through cracks in the dead fly's intersegmental membranes, each of which produces an asexual multinucleated (15–20 nuclei with diameters of 3.5–5.5 μm) terminal primary conidium (sporangium), which is then forcibly ejected by the accumulating turgor pressure within the sporangiophore. These sticky conidia (sporangia) adhere to whatever surface they land upon, leaving a white, powder-like, infectious halo all around the fly's corpse (Fig. B; several conidia are shown in Fig. B, inset). These primary conidia are produced soon after the fly dies (typically within two to three hours); smaller secondary conidia with fewer nuclei may later develop from the primary conidia.

Houseflies (*Musca domestica*) often seek elevated temperatures (>40°C, or 104°F) in sunny locations during the first few days of infection. This induced "behavioral fever" appears to inhibit the fungus and increase the flies' survival time up to six to eight days,[9] whereas infected flies in cool locations die within two days. The optimal temperature for Entomophthorales is 15–25°C (59–77°F). After initially moving to higher temperatures, infected flies then move to cooler locations after several days.[10] On the other hand, *E. muscae* killed adult Onion Flies (*Delia antiqua*) in 7 days at 21°C (~70°F) but took more than 17 days at 10°C (50°F).[11]

Time to kill is strongly dependent on incubation temperature. At 21°C, *E. muscae* took 7 days to kill adults, and this time was extended to >17 days at 10°C.

These conidia are highly infectious and will infect any fly they contact. Moreover, male houseflies (*Musca domestica*) are strongly attracted to dead females with distended abdomens infected by *Entomophthora muscae*, and they attempt to copulate with the corpses.[11,12] These behavioral alterations in housefly behavior are induced by the fungal pathogen to promote the pathogen's reproduction and dissemination, in this case to conspecific hosts, which can contribute to epizootic outbreaks—sharp increases in the rates of disease above usual expectations in an animal population—among these flies. The normal forcible discharge range (up to 1.5 cm) of the conidia is insuffi-

D

Erynia sepulchralis on a crane fly (cf. *Tipula* sp.). (Photo: John Plischke III.)

cient to infect other healthy flies. The behavioral changes induced by the fungus that cause infected flies to climb up to prominent and highly visible positions and die in an abnormal but sexually attractive posture that attracts healthy males ensures the propagation of the fungal parasite. Most epizootic outbreaks occur in the spring and fall. Resistant resting spores capable of surviving harsh environmental conditions or host fly scarcity develop as asexual haploid azygospores inside the dead host's abdomen.

What was formerly thought to be a single species that parasitized a wide range of dipteran hosts was shown to be a complex of very similar species with narrower host specificities. Closely related species have different numbers of nuclei in the primary conidia: *E. schizophorae* Keller & Wilding has fewer and larger nuclei (4–8 nuclei per conidium, with diameters of 4.9–5.0 µm);[13] *E. scatophagae* Girard has 15–18 nuclei, with diameters of 3.8–4.2 µm;[14] and *E. syrphi* Giard has 19–22 nuclei, with diameters of 2.8–3.4 µm.[15] *Entomophthora muscae* (*sensu stricto*) typically exhibits 15–20 nuclei per conidium (range of 10–27) and diameters of 3.9–4.4 µm.[14,15]

Entomophthora muscae (*sensu stricto*) infects Muscidae and Anthomyiidae; *E. schizophorae* attacks houseflies, blowflies and carrion flies, fruitflies, dagger flies, and *Platycheirus clypeatus* (Syrphidae); *E. syrphi* attacks flower flies and hoverflies; and *E. scatophagae* infects flesh flies.[14] However, molecular evidence indicates that no genetic variation exists between *E. muscae* isolates from the same host species regardless of sampling times and locations, whereas great variation exists between isolates from different host species; thus no genetic exchange takes place between the fungal pathogens from different species, and even the resting spores develop as asexual azygospores. This strongly suggests that only the fungal isolates from a specific host should be considered the true species, which in the case of *E. muscae* would be *Musca domestica*.[15] Nevertheless, this issue of speciation and host specificity among the Entomophthorales is far from resolved.

Entomopathogenic fungi are important regulators of many insect and mite species. Because they often cause dramatic epizootic outbreaks that result in rapid declines in host populations, much interest has been generated about the possibility of using them to control noxious insect pests. However, accurate information about the ecology and host-parasite interactions has prevented greater utilization of these potent insect pathogens as biological control agents.

1. D.S. Hibbett et al., 2007, A higher-level phylogenetic classification of the fungi, *Mycol. Res.* 111:509–547. The other entomoparasites we treat here all belong to the division Ascomycota.
2. R.A. Humber, 2012, Entomophthoromycota: A new phylum and reclassification for entomophthoroid fungi, *Mycotaxon* 120:477–492.
3. One undescribed genus is known only as a pathogen in snakes (A.P. Gryganskyi et al., 2013, *Persoonia* 30:94–105).
4. A.P. Gryganskyi et al., 2013, *Persoonia* 30:94–105.
5. *Batkoa* Humber, *Entomophaga* A. Batko, *Entomophthora* Fresenius, *Erynia* (Nowakowski *ex* A. Batko) Remaud. & Hennebert, *Eryniopsis* Humber, *Furia* (Batko) Humber, *Massospora* Peck, *Orthomyces* Steinkraus, *Pandora* Humber, *Strongwellsea* A. Batko & Weiser, *Tarichium* Cohn [a form-genus of anamorphic species known only from their resting spores that belong to the families Entomophthoraceae and Neozygitaceae], and *Zoophthora* A. Batko.
6. A.B. Jensen et al., 1998, *Fungal Genet. Biol.* 24:325–334.
7. S. Keller & R. Wegensteiner, 2007, Introduction, *in* S. Keller (ed.), *Arthropod-Pathogenic Entomophthorales: Biology, Ecology, Identification*, COST Action 842 (Luxembourg: COST Office), 4, file:///C:/Users/New%20User/Downloads/QSNA22829ENC_002%20(2).pdf.
8. S.B. Krasnoff et al., 1995, Behavioral effects of the entomopathogenic fungus *Entomophthora muscae* (Cohn) Fresenius, on its host *Musca domestica* L.: Postural control of dying hosts and gated pattern of mortality, *J. Insect Physiol.* 41(10):895–903.
9. H.T. Gul et al., 2014, Entomopathogenic fungi as effective insect pest management tactic: A review, *Appl. Sci. Bus. Econ.* 1(1):10–18.
10. H.E. Roy et al., 2006, Bizarre interactions and end games: Entomopathogenic fungi and their arthropod hosts, *Ann. Rev. Entomol.* 51:331–357.
11. J. Eilenberg & J.K. Pell, 2007, Ecology, *in* S. Keller (ed.), *Arthropod-Pathogenic Entomophthorales: Biology, Ecology, Identification* (Luxembourg: COST).
12. L. Zurek et al., 2002, *J. Invert. Pathol.* 80(3):171–176.
13. S. Keller, 1984, Entomophthora muscae *als Artenkomplex*, *Mitt. Schweiz.—Ent. Ges.* 57:131–132.
14. S. Keller et al., 1999, Redescription of *Entomophthora muscae* (Cohn) Fresenius, *Sydowia* 51(2):197–209.
15. S. Keller, 2002, The genus *Entomophthora* (Zygomycetes, Entomophthorales) with a description of five new species, *Sydowia* 54(2):157–197.

Mycoparasitic Fungi

Head-like Cordyceps

Elaphocordyceps capitata and *E. longisegmentis*

A B C

A & B, *E. capitata*; C, *E. longisegmentis*. (Photos: A & C, John Plischke III; B, Jim Edwards.)

Elaphocordyceps capitata **(Holmskjold) G.-H. Sung, J.-M. Sung & Spatafora**, in G.-H. Sung et al., *Stud. Mycol.* 57:37 (2007)

Basionym: *Clavaria capitata* Holmsk., *Beata Ruris Otia Fungis Danicis* 1:38, table 15 (1790)

Elaphocordyceps longisegmentis **(Ginns) G.-H. Sung, J.-M. Sung & Spatafora,** in G.-H. Sung et al., *Stud. Mycol.* 57:37 (2007)

Basionym: *Cordyceps longisegmentis* Ginns, *Mycologia* 80(2):219 (1988)
[Ascomycota, Pezizomycotina, Sordariomycetes, Hypocreomycetidae, Hypocreales, Ophiocordycipitaceae]

Derivation: *Elaphocordyceps*, Greek, from *eláfi*, "deer," plus *cordyceps*, "deer cordyceps," referring to the "deer truffle" genus, *Elaphomyces*, it parasitizes; *capitata*, Latin, from *capitatis*, "capitate," "head-like."

Fruitbody (stroma): A dark, round-headed mushroom atop a yellow stalk growing from underground deer truffles of the genus *Elaphomyces*.

Stipe: head ratio: ***E. capitata***, 1–4:1; ***E. longisegmentis***, 10:1.

Head: 1–2 cm wide and high (¼–¾ in), knob-like, dark reddish-brown to blackish, finely roughened by protruding perithecia.

Stalk: 2–8 cm tall by up to 1.5 cm thick (¾–3¼ × ⅔ in), at first yellow, becoming olive-brown, smooth to somewhat furrowed.

Microscopic: Spores (ascospores) long, thread-like, smooth, clear, fragmenting into shorter part-spores; ***E. capitata***, 16–28 × 2.5–3.5 µm; ***E. longisegmentis***, 40–65 × 4–5 µm.

Occurrence: Parasitic. Solitary or several on buried deer truffles of the genus *Elaphomyces*, especially *E. granulatus* Fries. Widespread in North America and Europe; occasional, early fall–midwinter.

Edibility: Unknown.

Comments: The Long-spored Cordyceps, *Elaphocordyceps longisegmentis* (Ginns) G.-H. Sung, J.-M. Sung & Spatafora, is very similar and parasitizes the same deer truffles. It differs in its longer stipe (stipe-to-head ratio up to 10:1, compared to 1–4:1 for *E. captitata*) and considerably longer part-spores: 40–65 × 4–5 µm.

Goldenthread Cordyceps, Snake-tongued Cordyceps

Elaphocordyceps ophioglossoides

Photo: John Plischke III.

Elaphocordyceps ophioglossoides **(Ehrhart) G.-H. Sung, J.M. Sung & Spatafora**, in G.-H. Sung et al., *Stud. Mycol.* 57:37 (2007)

Basionym: *Sphaeria ophioglossoides* J.F. Gmelin, *Systema Naturae*, 13th ed., 2(2):1474 (1792) *ex* Fries, *Syst. Mycol.* 2:324 (1823)
[Ascomycota, Pezizomycotina, Sordariomycetes, Hypocreomycetidae, Hypocreales, Ophiocordycipitaceae]

Derivation: *Elaphocordyceps*, Latin, from *cord*, "club," and *-ceps*, "head"; *ophioglossoides*, *ophidia* (New Latin, from the Greek *ófis*), "serpent," and *glossoides*, "tongue-shaped."

Synonym: *Cordyceps ophioglossoides* (Ehrhart) Link.

Fruitbody (stroma): 2–8 cm tall by 1.5 cm thick (¾–3¼ × ⅔ in), cylindrical to club-shaped, attached by golden-yellow cords to underground deer truffles of the genus *Elaphomyces*.

Head: Cylindric to spindle-shaped, 2–2.5 cm high (¾–1 in), at first reddish-brown and smooth, becoming blackish and roughened with pimple-like perithecia.

Stalk: Up to 7.5 cm tall by 1 cm thick (2¾ × ½ in), yellow to smoky brown or blackish-grey; attached to underground deer truffle by stout golden-yellow rhizomorphs.

Microscopic: Spores (ascospores) long, thread-like, segmented, fragmenting into elliptical, smooth part-spores 2–5 × 1.5–2 µm.

Occurrence: Parasitic on deer truffles of the genus *Elaphomyces* Nees. Widespread in forests in eastern North America; late summer–fall.

Edibility: Unknown.

Comments: The dark reddish-brown to blackish heads on a yellow stalk and the golden-yellow rhizomorphs that attach this mushroom to the deer truffle are diagnostic. The part-spores are tiny compared to those of *Elaphocordyceps capitata* (see our account of that species). *Ophiocordyceps ravenelii* (Berkeley & M.A. Curtis) G.H. Sung, J.M. Sung, Hywel-Jones & Spatafora resembles *E. ophioglossoides* but parasitizes beetle larvae rather than deer truffles.

Molds, *Hypomyces*

Bolete-eating Hypomyces

Hypomyces chrysospermus

Hypomyces chrysospermus infecting Yellow-gilled Bolete, *Phylloporus rhodoxanthus*. A, uninfected Yellow-gilled Bolete; B–D, progressive stages of infection; E–G, complete takeover of the host mushroom; E, subiculum entirely white (stage 1), F, some yellow appearing on the cap, G, cap bright golden-yellow (stage 2); stage 3 (sexual stage) very rare, not shown.

***Hypomyces chrysospermus* Tulasne & C. Tulasne**, *Annls Sci. Nat.*, Bot., sér. 4, 13:16 (1860) [teleomorph] [Ascomycota, Pezizomycotina, Sordariomycetes, Hypocreomycetidae, Hypocreales, Hypocreaceae]

Derivation: *Hypomyces*, Greek, *hypo-*, "under, beneath," and *myces*, "fungus," meaning "hidden fungus," referring to its cryptic infectious form; *chrysospermus*, from *khrūso*, "gold," and *spermus*, "seed," referring to the yellow to golden-yellow color of the aleuriospores (second-stage asexual spores).

Anamorph: *Sepidonium chrysosperma* (Bulliard) Link : *ex* Fries, *Syst. Mycol.* 3:438 (1832)

Basionym: *Reticularia chrysosperma* Bull., *Herb. Fr.* 10:tab. 476 (1790)

Fruitbody (subiculum): Anamorph—(asexual stage) appearing first as a powdery, moldy coating covering part or all of the host bolete's fruitbody (including the cap), pore surface (filling the tubes), and stipe (Figs. B–D); subiculum initially typically white (Figs. B & E), becoming yellow (Fig. F) to bright golden-yellow (Figs. B & G), gradually covering and obliterating the host's fungal structure (Fig. F). **Teleomorph**—(sexual stage) rarely seen; perithecia spherical to flask-shaped, orange-yellow, brick-red, or reddish-brown, nonreactive in KOH; most often found on host's tube layer.

Microscopic: Anamorph forming two types of asexual spores: (1) in white stage, **conidiospores** 10–25 × 5–10 µm, elliptical, thin-walled, smooth; (2) in yellow stage, **aleuriospores** 25–30 µm, globose, thick-walled (2–3 µm), yellow, golden-yellow, or yellow-brown, ornamented with prominent warts 1.5–2 µm high; **teleomorph** rare, **ascospores** (15–)20–25(–30) × 4–5(–6) µm, spindle-shaped to lanceolate, septate, divided into two unequal cells, densely covered with prominent warts.

Occurrence: Parasitic on various bolete genera, June–November, widespread in North America, common; Europe, Asia, New Zealand; of worldwide occurrence.

Edibility: Inedible, may be poisonous.

Comments: This is the most frequently reported species of bolete-parasitizing *Hypomyces* in the world and is the one most often encountered in our area.[1] It is usually only found in its anamorph stage, *Sepidonium chrysosperma*. In the uncommon teleomorph, the ascospores are contained in a sac-like ascus with an opening to the outside called an ostiole. The asci are enclosed within a spherical to flask-shaped perithecium generally visible on the subiculum surface with a hand lens.

The closely related *H. microspermus* Rogerson & Samuels (1989) differs from *H. chrysospermus* by its initially white subiculum becoming pale yellow rather than bright golden-yellow and its yellow, buff, deep peach, to brick- to orange-red perithecia, which turn purplish-red in KOH. Its aleuriospores are also smaller, 12–15 µm in diameter. *Hypomyces microspermus* and *H. chrysospermus* are very closely related but distinct species that uniquely arise from the same anamorphic form-genus and form-species: *Sepedonium chrysospermum*.

Hypomyces tulasneanus Plowright (1882) [anamorph: *Sepedonium tulasneanum* Saccardo (1886)] exhibits a yellow-green subiculum that becomes brownish and effusive and only covers the hymenium of the host. The subglobose to globose perithecia are yellow to brown, sometimes green at the apex, and turn pale yellow-brown in KOH. The aleuriospores are narrowly ovoid to oblong, 18–27(–32) × 7–10(–12) µm.

1. Hosts include the following bolete genera: *Boletus*, *Gomphidius*, *Gyroporus*, *Leccinum*, *Paxillus*, *Phylloporus*, *Pisolithus*, *Porphyrellus*, *Strobilomyces*, *Suillus*, *Tylopilus*, and *Xerocomus*.

Lobster Mushroom

Hypomyces lactifluorum

A

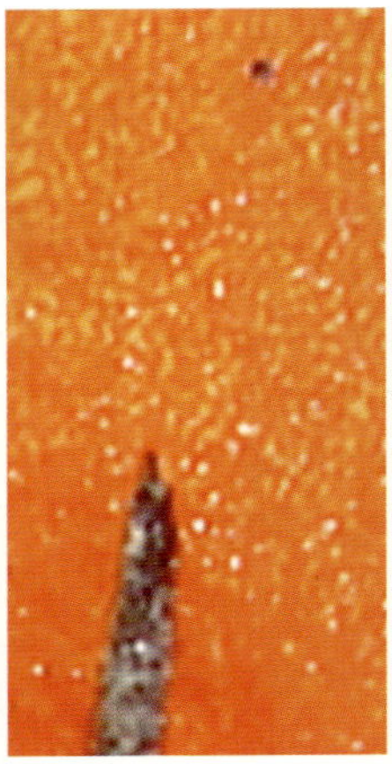

B

C

D

Photos: A, Herbert Baker; B & D, Teresa Fortenberry; C, K. Põldmaa, D.F. Farr, & E.B. McCray, *Hypomyces*—Systematic Botany and Mycology Laboratory, http://nt.ars-grin.gov/taxadescriptions/keys/HypomycesIndex.cfm

***Hypomyces lactifluorum* (Schweinitz *ex* Fries) Tulasne & C. Tulasne**, *Annls Sci. Nat.*, Bot., sér. 4, 13:11 (1860)

Basionym: *Sphaeria lactifluorum* Schweinitz [as "*Lactifluorum*"], *Schr. Naturf. Ges. Leipzig* 1:30 (1822)
[Ascomycota, Pezizomycotina, Sordariomycetes, Hypocreomycetidae, Hypocreales, Hypocreaceae]

Derivation: *Hypomyces*, Greek, *hypo-*, "under, beneath," and *myces*, "fungus," "hidden fungus," referring to its cryptic infectious form; *lactifluorum*, Latin, from *lacte*, "milk," and *fluere*, "to flow," "flowing milk," presumably referring to *Lactarius* (one of its two preferred host genera), whose species typically ooze a milky fluid when injured.

Fruitbody (stroma): Forming a characteristic tough, hard, yellowish-orange to bright orange, skin-like crust (subiculum) completely covering the white fruitbody of the parasitized host and obliterating the host's gills and other structures (Figs. A & D), this orange subiculum becoming deep reddish-purple to dark purple with age;[1] subiculum densely studded with pimple-like, teardrop-shaped perithecia (Fig. B) that release the parasite's ascospores, those often covering the subiculum with a fine, white powder.

Flesh: Consisting of the whitish tissue of the host mushroom underlying the parasite's subiculum; odor fishy.

Spores in deposit: White. Typically oozing from perithecia and covering stroma with a white powder.

Chemical staining: Subiculum stains vivid reddish-purple with KOH.

Microscopic: Spores (ascospores) 35–43(–50) × 4.5–7.5 µm, spindle-shaped, with a sharply pointed apiculus at each end; clear, thin-walled, heavily warted; uniseptate, divided into two equal cells (Fig. C).

Occurrence: Parasitic on *Russula brevipes* Peck in mixed hardwood-conifer woods, possibly also on other *Russula* species; also associated with *Lactifluus piperatus*; widespread and occasional throughout and restricted to North America, midsummer–fall.

Edibility: Edible and considered choice, despite relatively unpalatable taste of unparasitized host mushrooms. Many authors urge caution, since the parasitized host species is rarely identifiable, and some allergic reactions have been reported. Nevertheless, no host of this parasite is known to be poisonous.

Comments: Readily recognized on sight. This parasitic fungus changes the surface of the host's whitish fruitbody to bright orange. It also obliterates the host's gills and causes them to stop forming their own basidiospores; instead, the perithecia of the *Hypomyces* covering the host's stroma produce only the ascospores of the parasitic fungus. The common name refers to the resemblance of the host mushroom parasitized by *H. lactifluorum* to the reddish-orange exoskeleton and underlying white flesh of cooked lobster. The blunt, chanterelle-like ridges to which the *Russula* and *Lactarius* host gills are reduced by *H. lactifluorum* can superficially resemble the orange fruitbodies of certain *Cantharellus* species. However, the dense pimple-like perithecia distinguish *H. lactifluorum* from *Cantharellus*.

Hypomyces lactifluorum is generally believed to attack only edible (if not the most palatable) white russulas and milk mushrooms. Its main hosts appear to be *Russula brevipes* Peck, but mushrooms in the *Lactifluus* (formerly *Lactarius*) *piperatus* species complex are also a commonly associated nonparasitized host.[2] It is thus considered safe to eat even when the host can no longer be identified, as is usually the case. The parasite transforms the bland taste of *R. brevipes* and the hot, fiery, peppery taste of *L. piperatus* into a delicious gourmet feast.

Hypomyces lactifluorum is the type species for the genus *Hypomyces* (Fries) Tulasne. No anamorphic (asexual) stage is known.

Hypomyces luteovirens (Fries) Tulasne is a related species found throughout the Southeast. It forms a yellowish-green, mold-like covering over the gills and stem (but never over the cap itself) of infected *Russula* species.

1. "The change of color from orange to purple is a noteworthy feature" (J. Seaver, 1910, The Hypocreales of North America—III, *Mycologia* 2:72).

2. C.T. Rogerson & G.J. Samuels, 1994, *Mycologia* 86:839–866. This nonparasitic characterization is apocryphal, since *H. lactifluorus* is known to change the sharp, hot, and peppery taste of *L. piperatus* and convert it into a choice edible.

Yellow-green Russula Mold

Hypomyces luteovirens

Photos: A, K. Põldmaa, D.F. Farr, & E.B. McCray, *Hypomyces*—Systematic Botany and Mycology Laboratory, http://nt.ars-grin.gov/taxa descriptions/keys/HypomycesIndex.cfm; B & D, gsharpnolack; C, John Plischke III.

***Hypomyces luteovirens* (Fries *ex* Fries) Tulasne & C. Tulasne**, *Annls Sci. Nat.*, Bot., sér. 4, 13:12 (1860)

Basionym: *Sphaeria luteovirens* Fries, *K. Svenska Vetensk-Akad. Handl.* 38:251 (1817)
[Ascomycota, Pezizomycotina, Sordariomycetes, Hypocreomycetidae, Hypocreales, Hypocreaceae]

Derivation: *Hypomyces*, Greek, *hypo-*, "under, beneath," and *myces*, "fungus," "hidden fungus," referring to its cryptic infectious form; *luteovirens*, Latin, "yellow-green."

Fruitbody: (pseudostroma)—consisting of a pimple-studded, mold-like subiculum or pseudostroma composed of the tissues of both parasite and host covering the gills and stipe of various infected *Russula* species, initially white (Fig. C), becoming yellow to yellowish-green (Fig. D). This parasitic fungus obliterates the structure of the host's gills and in their place produces densely crowded, pimple-like, dark green to blackish (when dry), ascospore-releasing perithecia (Fig. B). No anamorph known.

Spores in deposit: White, cream, or yellow.

Chemical staining: Negative in KOH.

Microscopic: Spores (ascospores) (27–)30–35(–36) × 4.5–5.5 µm, spindle-shaped, with sharp apiculus at each end, clear, nearly smooth to covered with fine warts; unicellular, aseptate (Fig. A); perithecial apex with distinctive moniliform (bead-like) chains of cells extending from surface.

Occurrence: Parasitic on *Russula* spp.; widespread and occasional throughout North America; Europe; Japan; summer–late fall.

Edibility: Unknown, but consumption not advised due to its wide and possibly toxic host specificity.

Comments: The distinctive yellow-green subiculum formed by this fungus on its *Russula* host is not likely to be confused with any other taxon. The subiculum of another mold-like fungus, *Hypomyces lateritius*, forms a hard, reddish-brown to ochre crust (rather than yellow-green), containing recessed (not pimple-like) perithecia on parasitized *Lactarius* species.

Golden Hypomyces

Hypomyces aurantius

A & B, *Hypomyces aurantius* covering pore surface of *Trametes versicolor*. (Photos: A, John Plischke III; B & E, W. Light; C & D, K. Põldmaa, D.F. Farr, & E.B. McCray, *Hypomyces* Online, Systematic Mycology and Microbiology Laboratory, ARS, USDA. Retrieved December 14, 2013, from /taxadescriptions/keys/HypomycesIndex.cfm.)

***Hypomyces aurantius* (Persoon *ex* Fries) Tulasne**, *Selecta Fungorum Carpologia: Nectriei-Phacidiei-Pezizei* (Paris) 3:48, plate 10, fig. 3 (1865)

Basionym: *Sphaeria aurantia* Persoon, *Icon. Desc. Fung. Min. Cognit*. (Leipzig) 2:45 (1800)
[Ascomycota, Pezizomycotina, Sordariomycetes, Hypocreomycetidae, Hypocreales, Hypocreaceae]

Anamorph: *Cladobotryum varium* Nees, *Syst. Pilze* (Würzburg):56 (1816) [1816–1817]

Derivations: *Hypomyces*, Greek, *hypo-*, "under, beneath," and *myces*, "fungus," meaning "hidden fungus," referring to its cryptic infectious form; *aurantius*, Latin, "orange," referring to the orange layer of perithecia on the host; *Cladobotryum*, Greek, from *kládos-*, combining form of "branched," plus *bótry*, "a bunch of grapes," referring to the form of the conidia; *varium*, Latin, "variable."

Fruitbody (subiculum): Teleomorph—subiculum a crust-like or skin-like layer composed of loosely interwoven hyphae (Fig. D), typically covering the pore surfaces of very senescent, decayed polypores (Figs. A & B); initially whitish, gradually becoming typically pale yellowish-orange (Fig. A) to bright orange or rusty-red; with pale pinkish-salmon to flesh-colored pimple-like perithecial papillae partially embedded throughout the subiculum or in only some portions thereof (Fig. B, arrows; Fig. D). **Anamorph**—*Cladobotryum varium* Nees (1816) forms a dense, white to pale bluish-grey, floccose (cottony to powdery) mat (Fig. B) created by the abundant long chains of conidial spores borne on aerial hyphae (Fig. E); often spreading onto adjacent substratum.

Chemical staining: Perithecia staining purple with KOH.

Microscopic: Spores (ascospores) (14–)20–25(–27) × (3–)4–6(–7.5) µm, spindle-shaped, with two long, acute apiculi 2–4.5 µm in length, one slightly constricted median septum

(Fig. C, arrow), 2-celled; clear, surface uniformly covered with warts 0.5 µm high.

Occurrence: Parasitizes fruitbodies of many different fungi, wood, and bark. Hosts include various lignicolous (wood-decaying) gilled mushrooms, but especially polypores and particularly *Trametes* spp.; may cover the entire host. Widely distributed and fairly common in temperate zones, rare or absent in the tropics; North and South America, Europe, northern and eastern Asia, Australia.

Edibility: Inedible and potentially toxic (see comments).

Comments: This eurytopic (generalist) fungus is both facultatively (optionally) saprobic on wood and parasitic on a wide range of gilled mushrooms and polypores. *Hypomyces aurantius* is known to contain the following ergot mycotoxins: ergocryptine, ergocryptimine, argoclavine, elymoclavine, clymoclavine, and chamoclavine.[1]

1. S. Yamatodani & I. Yamamoto, 1983, Peptide-type ergot alkaloids produced by *Hypomyces aurantius*, *Nippon Nogeikagaku Kaishi* 57:453–456 (in Japanese).

Amanita Mold, Amanita Eater

Hypomyces hyalinus

A B C

A & B, *Hypomyces hyalinus* parasitizing *Amanita* sp.; C, *Amanita rubescens.* (Photos: A & B, W. Light; C, Teresa Fortenberry.)

Hypomyces hyalinus **(Schweinitz *ex* Fries) Tulasne**, *Schr. Naturf. Ges. Leipzig* 1:30 (1822)

Basionym: *Sphaeria hyalina* Schweinitz, *Schr. Naturf. Ges. Leipzig* 1:30 (1822)
[Ascomycota, Pezizomycotina, Sordariomycetes, Hypocreomycetidae, Hypocreales, Hypocreaceae]

Derivation: *Hypomyces*, Greek, *hypo-*, "under, beneath," and *myces*, "fungus," meaning "hidden fungus," referring to its cryptic infectious form; *hyalinus*, Latin, from the Greek *hyalinos* (from *hyalos*), "transparent," "translucent."

Fruitbody (subiculum): Teleomorph—a profuse whitish (Fig. A), buff (Fig. B), or pale pinkish (Fig. C) covering over host's pore surface, cap, and stipe, grossly disfiguring and turning the fruitbody firm; not reactive to KOH; minute scattered, gregarious, or cespitose buff to pale brown perithecia embedded throughout subiculum. No anamorph (asexual stage) is known.

Microscopic: ascospores 15–20 × 4–5.6 µm, spindle-shaped, with warts, divided by a submedian septum; each end apiculate (with short, acute point) 1–2.5 µm long.

Occurrence: Parasitic on *Amanita* spp., primarily on *A. rubescens* but also on *A. frostiana*, *A. muscaria*, and other amanitas. Also reported from unidentified agarics and from *Lactarius* sp. (China).[1] Common in eastern, central, and northwestern North America in hardwood and conifer woods, summer–fall; East Asia, Japan.

Edibility: Possibly edible if parasitizing *Amanita rubescens*, but likely poisonous on other hosts and not recommended.

Comments: Identification of the host species is generally problematic because of distortion and degradation of the fruitbody. Most *Hypomyces* species have an anamorphic or asexual stage that in some cases is much more commonly encountered than the teleomorphic (sexual) form. However, no anamorph is known for *H. hyalinus*.

1. K. Põldmaa, D.F. Farr, & E.B. McCray, *Hypomyces* Online, Systematic Mycology and Microbiology Laboratory, ARS, USDA, retrieved July 21, 2016, from /taxadescriptions/keys/Hypomyces Index.cfm.

Fairy Pins

Phaeocalicium polyporaeum

A

B

***Phaeocalicium polyporaeum* (W. Nylander) Tibell**, *Thunbergia* (*Publications from the Herbarium*, University of Uppsala) 4:7 (1979)

Basionym: *Calicium polyporaeum* Nylander, *Flora*, Regensburg 58:7 (1875)
[Ascomycota, Pezizomycotina, Eurotiomycetes, Mycocaliciomycetidae, Mycocaliciales, Mycocaliciaceae]

Derivation: *Phaeocalicium*, Greek, from *phaios*, "dark," "dusky," and *Calicium*, a genus of lichenized fungi, "dark *Calicium*"; *polyporaeum*, Latin, refers to the polypore hosts of this fungus.

Fruitbody (ascoma/ascocarp): Resembles a tiny dark olive-brown to blackish matchstick (Figs. A & B) consisting of a minute, black, stalked, round, ovoid, or cup-like apothecium (head) atop a thin, black, hair-like stalk 1.5–1.75 mm tall (about 1⁄16 in), but can be up to 2.5 mm (≤1⁄10 in) tall (including the heads, which are 0.2–0.6 mm wide and high).

Stalk: 1–2 mm tall by up to 0.3 mm thick (1⁄16 × <1⁄64 in); stalk subequal or expanding toward the base.

Microscopic: Ascospores (9–)11–14(–18) × 3–4 µm, highly variable in shape, narrowly ellipsoid to cylindrical, smooth; initially unicellular, becoming uniseptate when mature; pale brown.

Secondary metabolites: None known.

Occurrence: Parasitic, alone or in scattered groups on the upper surfaces of old sporocarps of Violet-toothed Polypore, *Trichaptum biformis* (Linnaeus *ex* Fries) Lloyd, and Turkey Tail, *Trametes versicolor* (Fries) Ryvarden, often when these polypores are also colonized by green algae. Widespread but sporadic in eastern North America and Arizona, summer–fall; also reported from Europe and northern Asia.

Edibility: Inedible and far too tiny to even consider.

Comments: Although commonly referred to as "lichenicolous fungi" (one of the common names is "Pin Lichen") and studied by lichenologists, *Phaeocalicium* species are not lichens and are not associated with any photobiont (algal symbiont). The genus contains some 10 species, 5 of which occur in temperate–boreal North America. *Phaeocalicium polyporaeum* is the only one occurring in our area. Other species of *Phaeocalicium* are saprobes or weak parasites on thin decaying branches of specific hardwood trees or shrubs. In the northern regions of North America, *P. populineum* (Duby) A. Schmidt occurs on the twigs of poplars (*Populus* spp.), *P. betulinum* (Nylander) Tibell on birches (*Betula* spp.), and *P. compressulum* (Vainio) A. Schmidt on thin decaying branches of alders (*Alnus* spp.).

Aborted Entoloma

Entoloma abortivum

A

B

***Entoloma abortivum* (Berkeley & M.A. Curtis) Donk**, *Bull. Bot. Gardens Buitenz* 18:157 (1949)
[Basidiomycota, Agaricomycetes, Agaricomycetidae, Agaricales, Entolomataceae]

Basionym: *Agaricus abortivus* Berk. & M.A. Curtis, *Ann. Mag. Nat. Hist.*, ser. 3, 4:289 (1859)

Derivation: *Entoloma*, Greek, from *entos*, "inner," and *loma*, "fringe," "inner fringe," referring to the inrolled margin of most species in the genus; *abortivum*, Latin, "abortive," meaning imperfectly formed.

Cap: Normal (unaborted) form 5–10 cm (2–5 in) wide, dry, convex, round to kidney-shaped, with rolled-in margin becoming flat or upturned with age; greyish to grey-tan, sometimes with slight pinkish tinge.

Gills: Usually somewhat decurrent, close, whitish when young, becoming pinkish with age.

Flesh: White, thick, with mealy odor.

Stipe: 2.5–10 cm tall by 5–15 mm (¼ × ⅔ in) thick, often swollen at base; dry, whitish or greyish.

Aborted forms: Highly distorted, amorphous, lumpy chalky white masses (Fig. A, left) with spongy white- and pinkish-marbled flesh, 2.5-cm high by 2.5–10 cm (1–4 in) wide; often (but not always) associated with normal forms together with various *Armillaria* species.

Spore print: Salmon pink.

Microscopic: Spores 8–10 × 4.5–6 µm, elliptical, angular, hexagonal, translucent, nonreactive in Melzer's reagent or iodine.

Occurrence: Saprobic, solitary or clustered on or near decaying hardwood; August–November; usually associated with or near Honey Mushrooms, *Armillaria* spp. (see comments below).

Edibility: Both normal and aborted fruitbodies are edible, but not choice (see comments below).

Comments: The aborted form of this mushroom is due to a parasitic interaction between it and at least three species of Honey Mushrooms: *Armillaria mellea*, *A. tabescens*, and *A. gallica*. It was long thought that the aborted morphs of *E. abortivum* resulted from parasitization by *Armillaria* spp., but recent studies suggest that it is *E. abortivum* that parasitizes *Armillaria*. Michael Kuo has suggested that this relationship might actually go either way, depending on the "strength" of the particular mycelia involved. Examples we have found here in the South incline us to agree with Kuo's notion.

While both forms of *Entoloma abortivum* are edible, the normal form is best identified when found in close proximity to the aborted form and near an *Armillaria* species. Otherwise, *Entoloma* species identification can be problematic, and many are poisonous. See also nomenclatural comments under the Yellow Unicorn Mushroom, *Entoloma murraii*.

Cup and Disk Fungi

Most of these fungi are readily identified by their cup-shaped or disk-like ascocarps, but some hide their identities by pretending to be regular cap-and-stem mushrooms until you examine them more closely. The ascocarp can be sessile or stipitate (stalked), and the exposed hymenium is typically either flat and disk-like or concave and ranging from shallowly depressed to deeply concave like a cup, urn, or goblet.

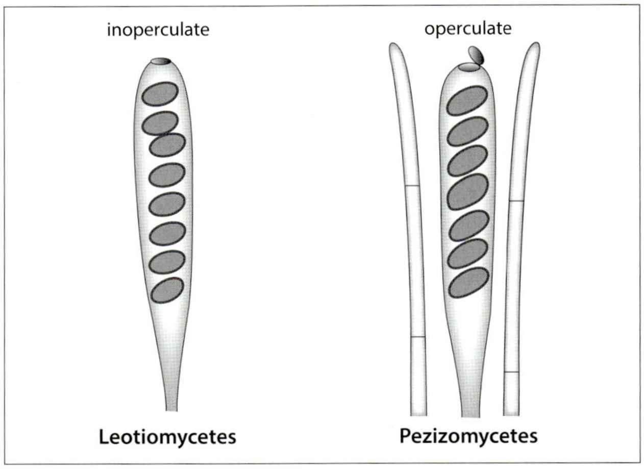

Cup and disk fungi all belong to the division Ascomycota. The taxa presented in this section belong to either the class Pezizomycetes, containing the sole order Pezizales, or the class Leotiomycetes, here represented by the order Helotiales.[1] Fungi in the class Pezizomycetes are **operculate**; that is, the opening to the ascus is secured by a flap-like lid that looks for all the world like a toilet seat cover. The Leotiomycetes are **inoperculate**; the opening to the ascus lacks a cover, but it may have a small, temporary plug. These two types of asci are shown above.

The **apothecium** consists of asci with or without an **operculum** covering the apical pore, usually accompanied by sterile hyphae with simple to modified tips (e.g., often club-like or branched and bent at various angles).[2] The apothecium of a typical cup fungus (genus *Peziza*) is shown at the left of the following figure. The multiple apothecia of a morel (Genus *Morchella*), each occupying a separate cup-shaped depression, is shown at right. Although technically a cup fungus, the morels, related false morels, and elfin saddles are treated separately in their own section.

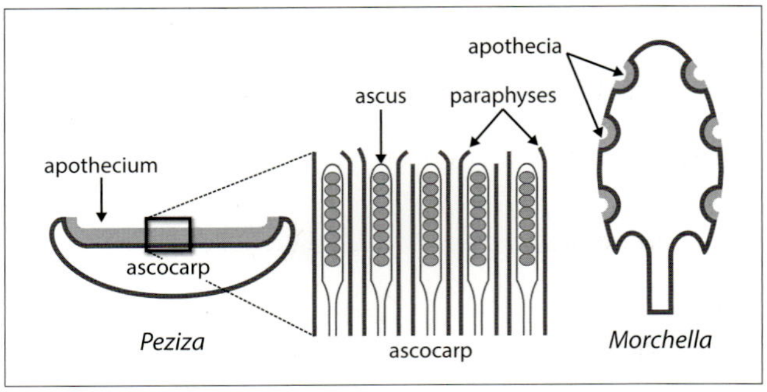

As is the case throughout this book, a number of the genera and species discussed in other sections (e.g., the earth tongues, the jelly fungi, the club fungi, and the morels and their allies) could also have been included here. Indeed, some authors group them all together under the general category of cup and disk mushrooms.

Many cup and disk fungi can be readily identified in the field by their macroscopic features and ecological associations, but others require microscopic examination. Iodine-based stains such as Melzer's reagent and Lugol's solution stain the ascus near the pore blue in certain groups or in certain species within given groups but not in others. This can be an important diagnostic feature, but it is sometimes difficult to discern and in some cases can be quite variable, depending on the age of the specimen and other factors.

1. Members of the order Leotiales are discussed in the section "Earth Tongues, Club Fungi, and Coral Fungi."

2. The term *apothecium* refers to a hymenium covering an exposed flat, concave, or convex surface of an ascocarp in which the asci are directly exposed to the environment.

Key to Cup and Disk Fungi

1. Ascocarp goblet- to urn-shaped, 1–6 inches across, sterile outer surface tough, dark brown to blackish-brown. 2

1′. Ascocarp colored differently 3

2. Inner (upper) fertile surface of cup concolorous with outer surface . . . .*Urnula craterium*

2′. Inner (upper) fertile surface of cup pale orange, pinkish-beige, or reddish-brown . . . . *Galiella rufa*

3. Mature ascocarp disk-like or cushion-like, reddish-purple, greyish-purple, to purplish-pink; immature anamorph club-like or brain-like . . . . *Ascocoryne* spp.

3′. Mature ascocarp colored differently. 4

4. Mature ascocarp deeply and irregularly cup-shaped, up to 4 or more inches across 5

4′. Mature ascocarp disk-like or shallowly cup-shaped, ¼ inch or less across. 6

5. Inner (upper) fertile surface bright yellow, orange, or orange-red; outer sterile surface paler yellow to yellowish-orange . . . . *Aleuria aurantia*

5′. Inner (upper) surface light to chestnut-brown, sometimes tinged with red; outer surface whitish to ash-grey or ochraceous, covered with pimple-like scales. . . . . *Peziza* spp.

6. Ascocarp distinctively blue-green. . . . .*Chlorociboria* spp.

6′. Ascocarp not colored blue-green 7

7. Ascocarp uniformly pale yellow, lemon-yellow, or golden-yellow, disk-like or very shallowly cup-shaped; inner and outer surfaces smooth, margin low, entire. . . . . *Bisporella* spp.

7′. Ascocarp colored differently, or if outer surface yellow, inner (upper) surface sage to pea-greenish. 8

8. Upper surface of mature ascocarp pale sage to pea-green, outer surface yellow . . . . *Chlorosplenium* spp.

8′. Entire mature ascocarp white to cream or pale orange-yellow to pale orange. 9

9. Ascocarp stipitate, white to cream-colored throughout, shallowly cup-shaped, maturing to a flattened, rimmed disk, margin entire, stipe ¼ inch tall. . . . . *Dasyscyphus virgineus*

9′. Ascocarp dull orange, orange-red, orange-yellow, to olive brown 10

10. Ascocarp blob-like, cushion-shaped; dull orange when young, becoming pale dull grey-brown to olive-brown; on conifer wood . . . . *Dacrymyces stillatus*

10′. Ascocarp dull orange to orange-red; disk-like or very shallowly cup-like, margin lined with eyelash-like hairs up to 1 mm long; on hardwood . . . .*Scutellinia* spp.

Devil's Urn

Urnula craterium

A

B

C

***Urnula craterium* (Schweinitz) Fries**, *Nova Acta R. Soc. Scient. Upsal.*, ser. 3, 1:122 (1851)
[Ascomycota, Pezizomycotina, Pezizomycetes, Pezizomycetidae, Pezizales, Sarcosomataceae]

Basionym: *Peziza craterium* Schwein., *Schr. Berl. Ges. Naturf. Fr. Leipzig* 1:117 (1822)

Derivation: *Urnula*, Latin, "little urn"; *craterium*, from *cratera*, a type of bowl.

Fruitbody: Deeply urn- or goblet-shaped, 2–7 cm wide by 4.5–15 cm tall (¾–2¾ × 1¾–6 in), initially narrow opening soon expanding, margin becoming tattered and bluntly dentate; inner fertile and outer sterile surfaces dark brown to blackish, darkening with age; outer surface velvety, initially rough, gritty, leathery, becoming smoother with age.

Flesh: Blackish, tough, initially gelatinous, becoming leathery.

Stipe: 2–4 cm tall by 0.5–1.0 cm thick (¾–1⅔ × ¼–⅖ in), continuous with and same color as fruitbody, tapering downward, with black mycelia at the base.

Spore print: Cream.

Microscopic: Spores 24–36 × 10–15 µm, smooth, broadly elliptical to spindle-shaped, clear, nonamyloid; two distinctive types of paraphyses: (1) filiform, with few septa, and colorless; (2) thicker with brown walls, multiseptate and bead-like, branching at the tip.

Occurrence: Saprobic, solitary, in scattered groups or dense clusters on hardwood sticks and small logs, especially oaks, which may be buried. Early spring (March–May), occasionally even appearing under snow; perennial. Widespread in eastern North America, Europe, Japan, China.

Edibility: Probably edible but not very palatable.

D Photo: Jim Edwards.

Comments: *Urnula craterium* begins in late autumn as a dense black mycelial mass covering decaying branches lying on the ground or partially buried in forest litter. This soon produces cylindrical black stromata (fruitbodies) a few centimeters long and 3 mm thick. These slowly enlarge over the winter and emerge with the onset of warm weather in the early spring as the goblet-like fruitbodies of the Devil's Urn. Figure D shows a developing young fruitbody shortly after the apical cavity has opened. The hymenium (spore-bearing surface) develops from this apical region and grows downward into the expanding cavity, which it will eventually completely cover. The oldest part of the hymenium is thus at the apex of the fruitbody near the pore opening, while the youngest portion lies at the bottom of the cavity.

The paraphyses (erect sterile filaments) at the apex of the fruitbody become swollen and bead-like in response to warming temperatures. This induces the conversion of stored glycogen into glucose, and the increased pressure causes the paraphysis to swell. This, in turn, both creates and expands the apical mouth shown in Figure D and forces adjacent asci to expel their spores through the opening at the top of the fruitbody. This is discussed in detail by F.A. Wolf, *Mycologia* 50(6):837–843 (1958).

Hairy Rubber Cup, Peanut Butter Cup

Galiella rufa

Galiella rufa **(Schweinitz) Nannfeldt & Korf**, *Mycologia* 49(1):108 (1957)
[Ascomycota, Pezizomycotina, Pezizomycetes, Pezizomycetidae, Pezizales, Sarcosomataceae]

Basionym: *Bulgaria rufa* Schweinitz, *Trans. Am. Phil. Soc.*, ser. 2, 4(2):178 (1832) [1834]

Derivation: *Galiella*, named in honor of the French mycologist Marcelle Le Gal; *rufa*, Latin, "reddish-brown," referring to the color of the fertile upper surface.

Fruitbody: Goblet- to cup-shaped, 1.2–4 cm wide by 1.2–3 cm high (½–1⅔ × ½–1¼ in), at first closed, later opening; mature upper fertile surface smooth, shallowly to moderately concave, pale to dull orange, pinkish-beige, reddish-orange, or reddish-brown; sterile outer surface tough, dark brown to blackish-brown, densely covered with hairs 7–8 µm long, giving it a felty texture; margin incurved, typically finely toothed or fringed; entire fruitbody rubbery-gelatinous.

Flesh: Thick, tough, rubbery-gelatinous throughout, greyish, translucent.

Stipe: Lacking or rudimentary, 1–2 cm tall by 3–5 mm thick, connected to black basal mycelium.

Spores in mass: Reddish-brown.

Microscopic: Spores 10–22 × 8–10 µm, elliptical to subfusiform with narrow tips, with thick (1 µm) walls; clear, surface covered with fine warts;[1] nonreactive in Melzer's reagent.

Occurrence: Saprobic, solitary or in loose clusters on decaying hardwood sticks and logs; late summer and fall. Widespread in North America east of the Rocky Mountains; Malaysia.

Edibility: Considered inedible, but reported to be widely eaten in Malaysia.

Comments: Bulgaria iniquinans (Persoon) Fries is similar in both size and shape to G. rufa and is also associated with hardwoods. However, the mature upper fertile surface is smooth, shiny, and dark greenish-black to black; the outer sterile surface is dark brown to blackish and finely or noticeably hairy or scaly. Young fruitbodies have a dark-red wine-colored upper surface and a yellowish-tan outer surface. The mature spores are smooth (which distinguishes the genus Bulgaria Fries from the genus Galiella Nannfeldt & Korf),[1] kidney-shaped, and dark brown or black (rather than reddish-brown) in mass.[2]

Wolfina aurantiopsis (Ellis) Seaver closely resembles *G. rufa* but has a pale yellowish upper fertile surface and becomes tough and corky on drying, in contrast to *G. rufa*'s rubbery-gelatinous texture. *Wolfina aurantiopsis* has much larger (27–33 × 16–18 µm) clear or yellowish spores and is associated with pines and hemlock (instead of hardwoods). The spores are clear or slightly yellowish.

Moose Antlers or Rabbit Ears, *Wynnea americana* Thaxter, superficially resembles *G. rufa*, but emerges from the ground as dense clusters of tall, erect, rabbit- or mule-ear-shaped (rather than goblet-shaped or cup-like) fruitbodies, which can reach a height of 14 cm and a width of 6 cm (5⅔–½ in). The spores are also considerably larger (35–38 × 12–14 µm), boat-shaped or broadly fusiform, and longitudinally grooved (not finely warty) throughout their lengths.

1. The genus *Galiella* is characterized by ascospores with fine to prominent warted surfaces. These ornamentations are composed of callose-pectin, which reacts strongly to cotton dyes such as methyl blue.

2. Only the top four spores in each ascus appear brown under the microscope; the lowermost four remain clear.

Orange Peel Fungus

Aleuria aurantia

Photo: Bill Blanton

***Aleuria aurantia* (Persoon) Fuckel**, *Jahrb. Nassau. Ver. Naturk.* 23–24:325 (1870) [1869–1870]

Basionym: *Peziza aurantia* Persoon, *Obs. Mycol.* (Lipsiae) 2:76 (1799); *ex* Fries, *Syst. Mycol.* 2(1):49 (1822)
[Ascomycota, Pezizomycotina, Pezizomycetes, Pezizomycetidae, Pezizales, Pyronemataceae]

Derivation: *Aleuria*, etymology not known; *aurantia*, Latin, "orange," "golden."

Fruitbody: 1–10 cm (½–4 in) across, irregularly cup-shaped, margin wavy and deeply indented, sometimes torn at maturity; hymenophore (inner or upper, fertile surface) bright yellow, orange, or orange-red, smooth; outer, sterile surface paler yellow to yellowish-orange, finely powdery when young, becoming smooth with age.

Flesh: Thin, brittle, orange to white; no particular odor.

Stipe: Absent or rudimentary.

Microscopic: Spores (ascospores) (13–)17–24 × (7.5–)9–11 (–12) μm, elliptical, coarsely and evenly (forming a regular pattern) reticulate when mature, typically biguttulate (containing two oil droplets), with prominent apicular projection at each end, clear; paraphyses clavate, with expanded ends, filled with orange granules.

Occurrence: Saprobic. Solitary or in clusters in grassy areas on disturbed ground, in gardens, and along roadsides; widespread and fairly common in North America; summer–late fall.

Edibility: Edible.

Comments: This distinctive mushroom is not likely to be confused with any other species in our area. Most similar orange-colored cup-like mushrooms are much smaller (e.g., tiny, <5 mm) and usually have prominent hairs lining the margin, for example, the Orange Eyelash Cup, *Scutellinia erinaceus* (Schweinitz) Kuntze (see the account of this species). The Peanut Butter Cup or Hairy Rubber Cup, *Galiella rufa* (Schweinitz) Nannfeldt & Korf, can also have a reddish-orange fertile (upper) surface, but the outer surface is dark brown to blackish-brown, and the margin is lined with small, irregular teeth (see the account of this species).

Palamino Cup, Recurved Cup

Peziza "repanda"

A

B

***Peziza "repanda"* Persoon,**[1] *Icones Pictae Rariorum Fungorum* 4:49 (1808)
[Ascomycota, Pezizomycotina, Pezizomycetes, Pezizomycetidae, Pezizales, Pezizaceae]

Derivation: *Peziza*, New Latin, from the Greek *pezis*, referring to a foot; *repanda*, Latin, "folded backward," referring to the wavy margins.

Fruitbody (apothecium): 5–12 cm (2–4¾ in), sessile, irregularly and deeply cup-shaped, margins incurved, scalloped (Fig. A) or deeply lobed and folded inward over the inner (upper) surface, or hymenium (Fig. B), often lacerated; hymenium smooth, deeply concave, typically wrinkled, furrowed, or "blistered" (Figs. A & B); hymenium light to chestnut-brown, sometimes tinged with red; outer surface whitish to ash-grey or ochraceous, covered with pimple-like scales (Fig. A).

Flesh: Thin, pale greyish or greyish-brown; no particular odor or taste.

Stipe: Absent or rudimentary except when very young.

Spore print: White.

Chemical staining: All tissues negative in KOH; hymenium blue-grey in Melzer's reagent, other tissues nonreactive.

Microscopic: Spores (11–)15–16 × (6–)9–10 µm, elliptical, smooth, clear, nonguttulate (no oil droplets); asci 8-spored, apex staining blue with Melzer's reagent.

Occurrence: Saprotrophic. Solitary, in groups or clustered on rotten hardwood, woody debris, or wood chips or on the ground in adjacent soil rich in decayed wood under deciduous trees; widespread and common in North America and Europe; spring and fall, occasionally in summer.

Edibility: Inedible, suspect.

Comments: Similar species include the Bladder Cup, *Peziza vesiculosa* Bulliard, and the Domicile Cup, *P. domiciliana* Cooke. *Peziza vesiculosa* grows on horse and other manure and composted straw and has much larger spores (20–24 × 11–13 µm). *Peziza domiciliana* grows inside homes, cellars, and basements on rotten wood, sheetrock, and plaster. Its spores are similar in size and shape to those of *P. repanda* but usually contain one or two oil droplets. Neither of these species is edible.

Persoon never designated a type specimen for this taxon, and the name *Peziza repanda* has been inconsistently applied by various authors to various taxa growing on varying substrata.[1] The name is thus of uncertain application and is best regarded as a *nomen ambiguum* or *nomen dubium*. Like Gibson (2007),[2] we enclose the specific epithet within quotation marks to indicate that this is not an assignable name. Because the name was validly published (if currently unassignable), we retain the use of italic typeface.

These species have traditionally been distinguished by their occurrence on various substrata, such as horse or other dung; on wood chips; on soil mixed with sawdust or wood chips or enriched with decayed hardwood; on masonry, sheetrock, or other building materials; and on the decayed branches of various hardwood species or those restricted to a specific genus (*Alnus* spp.). Several molecular studies by Karen Hansen and her colleagues[1,3] have revealed that these criteria are not conservative and that much overlap occurs among closely related taxa or even within the same species. These authors found the same to be true of the morphological features commonly used to distinguish between these taxa. *Peziza* appears to be polyphyletic and includes at least six distinct clades (evolutionary lineages), most of which include species in other genera of the family Pezizaceae.

1. K. Hansen et al., 2002, *Mycol. Res.* 106(8):879–902; these authors declared *Peziza repanda* Persoon to be a rejected name.
2. I. Gibson, 2007, Trial field key to the PEZIZACEAE in the Pacific Northwest, http://www.svims.ca/council/Pezizc.htm.
3. K. Hansen et al., 2001, *Mycologia* 93(5):958–990; K. Hansen et al., 2005, *Mol. Phylogenetics Evolut.* 36(1):1–23.

Green Elfcup, Green-Stain Complex

Chlorociboria aeruginascens and *C. aeruginosa*

Chlorociboria cf. *aeruginosa*. (Photos: A, Mike Christison; B & C, Teresa Fortenberry.)

***Chlorociboria aeruginascens* (Nylander) Kanouse *ex* C.S. Ramamurthi, Korf & L.R. Batra**, *Mycologia* 49(6):858 (1958)

Basionym: *Peziza aeruginascens* Nyl., *Obs. Pez.*:42 (1869)

***Chlorociboria aeruginosa* (Oeder) Seaver *ex* C.S. Ramamurthi et al.**, *Mycologia* 49(6):859 (1958)
[Ascomycota, Leotiomycetes, Leotiomycetidae, Helotiales, *incertae sedis*]

Basionym: *Helvella aeruginosa* Oeder, *Fl. Danic.* 3:tab. 534:2 (1770) [see online Appendix A for synonymies]

Derivation: *Chlorociboria*, Greek, from *chloros*, "pale green"; *ciborium*, Latin, "drinking cup"; *aeruginosa*, Latin, "blue-green"; *aeruginascens*, "becoming blue-green."

Fruitbody: Distinctively blue-green disks with central stipe; ***C. aeruginosa***—<0.5 cm (¼ in) across, disk- or cup-shaped (Fig. A), occasionally convex or flat, margins often inrolled, several fruitbodies often arising from a single dark, irregular hyphal mass; ***C. aeruginascens***—to 0.7 cm (2¾ in) across, cup-shaped to spatulate-shaped when young, funnel-shaped when mature, single fruitbody arising from a scarcely differentiated hyphal mass.

Stipe: (Figs. B & C) ***C. aeruginosa***—inner flesh orange-yellow, <0.3 cm (⅛ in) tall by 0.05–0.1 cm wide, central to slightly off-center; ***C. aeruginascens***—inner flesh bluish, <0.6 cm (¼ in) tall by 0.1–0.15 cm wide.

Microscopic: Spores smooth, fusiform, colorless or light green; ***C. aeruginosa***—aseptate or with one septum, 9–15 × 2–4 µm (average 13 × 3 µm), with two prominent and/or several smaller droplets; ***C. aeruginascens***—aseptate, 5–7 (10) × 1–1.5 (2.4) µm (average 6 × 1.5 µm), with droplets at each end.

Occurrence: Saprobic on well-decayed, barkless hardwoods, especially oaks; fruitbodies summer–fall, but green-stained wood appears all year. Both species widespread in North America.

Edibility: Too tiny to be considered.

Comments: These are the only two species of *Chlorociboria* in North America, although a number of other species occur elsewhere. They can scarcely be confused with any other mushroom. *Chlorociboria* species produce xylindein, a penetrating blue-green quinone pigment that gives wood a lovely blue-green color (Fig. C, arrow), making it especially valuable to woodworkers. Though the characteristically stained wood is often seen, the tiny but spectacular fruitbodies are rarely encountered. *Chlorociboria aeruginosa* is generally somewhat smaller than *C. aeruginascens*, and its stipe is typically centrally attached, as shown in the figures above; the stipe in *C. aeruginascens* tends to be more laterally placed. They are reliably distinguished only by the size of their spores.

Orange Eyelash Cup

Scutellinia erinaceus

***Scutellinia erinaceus* (Schweinitz) Kuntze**, *Rev. Gen. Pl.* (Leipzig) 2:869 (1891)
[Ascomycota, Pezizales, Pyronemataceae]

Basionym: *Peziza erinaceus* Schweinitz, *Schr. Naturf. Ges. Leipzig* 1:119 (1822)

Derivation: *Scutellinia*, Latin, from *scutella*, "saucer-like"; *erinaceus*, "hedgehog-like."

Fruitbody: Tiny disk-like or cup-like fungi 1.5–4 (typically 2–3) mm in diameter, sessile; yellowish, pale orange-yellow, to dull orange on upper spore-bearing surface, sterile sides dull brownish-orange, perimeter of cup lined with deeply embedded long (relatively speaking) brown hairs <1 mm in length—0.1–0.7 (usually 0.4–0.6) mm.

Microscopic: Spores long-elliptical, smooth, 14–26 (usually 7–21) × 9–15 µm, length-to-width ratio averaging 1:7; with cytoplasmic granules; paraphyses subclavate or clavate, filled with orange granules in fresh material.

Occurrence: Widespread, highly gregarious; saprobic on well-decayed hardwood, summer–late fall.

Edibility: Inedible.

Comments: The very similar Scarlet Eyelash Cup, *S. scutellata* (Linnaeus) Lambotte (1887), is larger (0.5–1.5 cm in diameter), the marginal hairs are longer (1–2 mm), the inner fertile surface is bright orange-red, and the outer sterile surface is pale brown. The spores are also slightly shorter (18–19 × 10–12 µm), slightly elliptical, with a length-to-width ratio averaging 1:5, and with a rough and warty instead of a smooth surface.

Lemon Fairy Drops, Lemon Disco

Bisporella citrina

A

B

***Bisporella citrina* (Batsch) Korf & S.E. Carpenter**, *Mycotaxon* 1(1):58 (1974)
[Ascomycota, Pezizomycotina, Leotiomycetes, Leotiomycetidae, Helotiales, Helotiaceae]

Basionym: *Peziza citrina* Batsch, *Elench. Fung.*, cont. sec. (Halle):95 (1789)

Derivation: *Bisporella*, Latin, from *bi-*, "double" or "twice," plus Greek, *spora*, "seed" or "sowing," and the Latin suffix *-ella*, "little" (little double-spored), referring to the spores bisected by a septum, which is highly unusual among the Ascomycota; *citrina*, Latin, from *citrin*, "lemon-yellow."

Fruitbody: Tiny shallow cup-like or disk-shaped mushrooms; uniformly pale yellow (Fig. A), bright lemon-yellow, or golden-yellow (Fig. B),[1] this not changing with age; 1–3 mm (up to ~1⁄12 in) in diameter, may be distorted by overcrowding; fertile upper surface and outer sterile surface smooth.

Stipe: Minute (rarely >1 mm tall), rudimentary, tapering, or lacking altogether.

Spores in mass: White.

Microscopic: Spores 9–14 × 3–5 µm, long-elliptical, smooth, hyaline, when mature typically divided into two sections by a transverse septum, with an oil droplet at each end; asci 100–135 × 7–10 µm, with rounded or club-shaped tips; paraphyses clear.

Occurrence: Saprobic; in dense clusters on decayed, decorticated hardwood logs and stumps, especially beech (*Fagus* spp.), summer–fall. Also found growing on the surface of the Blushing Bracket, *Daedaleopsis confragosa* (Bolton) J. Schröter. Widespread and common in North, Central, and South America, Europe, North Africa, Australia, New Zealand.

Edibility: Unknown, but hardly worth the effort.

Comments: Several tiny disk-like mushrooms can be confused with *B. citrina*.

- *Bisporella sulfurina* (Quélet) Carpenter (not shown) looks like an even tinier version of *B. citrina* (0.5–1.0 mm in diameter). It is cushion-shaped to disk-like and usually grows gregariously on hardwoods, usually (but not always) from or with a pyrenomycete, another fungus, recognized by a layer of blackish tissue found associated with the bright yellow disks. Its spores are much longer and more cylindrical (15–24 × 3–4 µm) than those of *B. citrina*; the paraphyses (sterile filamentous hyphal end cells interspersed among the spore-bearing structures) are filled with yellow oil droplets instead of being clear.

C D E

C, *Dacrymyces stillatus*; D, *Dasyscyphus virgineus*; E, *Chlorosplenium chlora*. (Photos: C, Norbert Nagel; D, W. Light; E, John Plischke III.)

- *Dacrymyces stillatus* Nees (Fig. C) has slightly flattened, blob-like fruitbodies 1–5 mm across that are dull orange when young and become somewhat translucent, pale dull brown or grey-brown to olive-brown while retaining just a hint of orange in age. It is found in large clusters on conifer (not deciduous) wood. It has Y-shaped ("tuning fork") basidia that bear only two basidiospores (quite unlike the 8-ascospore-containing asci of *B. citrina*); the spores are also 3-septate (not 1-septate) when mature, and orange (not white) in mass.
- *Dasyscyphus virgineus* S.F. Gray (Fig. D) are tiny white or cream-colored (never bright lemon-yellow or golden-yellow) cups (1–2.5 mm across) atop downwardly tapering stalks up to 3 mm tall. The outer sterile surface and stalk are covered with white hairs. It is saprobic in groups or clusters on hardwood logs, stems, and branches and on beechnut burrs and other debris.
- *Chlorosplenium chlora* (Schweinitz) M.A. Curtis (Fig. E) are tiny cup-like disks the same size as *B. citrina* that are also stalkless or nearly so. The depressed fertile disk is yellow at first but gradually turns greenish when the raised, roughened, yellowish margins contrast strongly with the smooth, greenish, sunken disk. It occurs in dense clusters on decaying hardwood logs and stumps.
- *Hymenoscyphus calyculus* (Sowerby) W. Phillips (not shown) also resembles *B. citrina* but is slightly larger (disks 2–5 mm across) and has a short but well-developed, thick stipe; it is found gregariously on decorticated hardwood twigs and sticks, rather than logs and stumps.
- *Hymenoscyphus fructigenus* (Bulliard *ex* Fries) Gray (not shown) is a uniformly whitish to very pale yellow cup or disk up to 4 mm in diameter perched atop a thin, elongate stalk up to 12 mm or more tall by 1 mm thick. (The stalk of *B. citrina* is rarely taller than 1 mm.) It is smooth or finely downy on, beneath, and over its typically tiny but elongated stem. The spores are longer than those of *B. citrina* (13–25 × 3–5 µm). It resembles *Dasyscyphus virgineus* (see next page) but grows in clusters on beech nuts, hickory shells, and acorns.

1. The yellow color is due to the pigment β-carotene.

Stalked Hairy Fairy Cup

Dasyscyphus virgineus

A B

C

Dasyscyphus virgineus **(Batsch) S.F. Gray** [as *Dasyscypha*], *Nat. Arr. Brit. Pl.* (London) 1:671 (1821)
[Ascomycota, Pezizomycotina, Leotiomycetes, Leotiomycetidae, Helotiales, Hyaloscyphaceae]

Basionym: *Peziza virginea* Batsch, *Elench. Fung.* (Halle):125 (1783)

Derivation: *Dasyscyphus*, Greek, from *dasýs*, a combining form meaning "shaggy" or "hairy," and *scyphus*, "drinking cup"; *virgineus*, Latin, "white," "pure."

Cap: *Tiny* white (Fig. A) to cream-colored (Fig. C) fruitbodies; cup-like when young, maturing to flattened, rimmed disks 0.5–2.5 mm in diameter atop long slender stalks; entire outer, sterile surface of fruitbody covered with minute white hairs. In high humidity, entire outer fruitbody often covered by condensed droplets of water (Fig. A).

Flesh: White.

Inner fertile surface (hymenium): Smooth (Fig. B), concolorous with outer surface, occasionally becoming pale yellowish (Fig. C).

Stipe: Central, 1.5–6 mm (up to ¼ in) tall, concolorous and merging with cap and covered with similar hairs.

Microscopic: Spores (ascospores) 6–10 × 1.5–2.5 µm, spindle-shaped, smooth, clear.

Occurrence: In large clusters on dead hardwood branches, twigs, beech seeds, and birch catkins, April–October; widespread in North America, Europe.

Edibility: Probably inedible, but hardly worth the effort in any case.

Comments: This species is not likely to be mistaken for any other, although several European species resemble it very closely. *Dasyscyphus bicolor* (Bulliard) Fuckel also resembles *D. virgineus* but is readily distinguished by its yellow-orange hymenium. The spores of *D. bicolor* are similar to those of *D. virgineus* (7–9 × 1.5–2 µm) and are similarly spindle-shaped. It is widespread in North America and probably occurs in our area. This group of small cup fungi is still in a state of taxonomic flux, and the limits of both the genera and the species remain to be established.

Purple Jellydisc, Purple Jellydrops

Ascocoryne sarcoides/Coryne dubia/Ascocoryne cyclichnium

A

B

C

D

Photos: A, Emily Johnson; B, Thomas Pruß; C, W. Light; D, Teresa Fortenberry.

***Ascocoryne sarcoides* (Jacquin *ex* S.F. Gray) J.W. Groves & D.E. Wilson**, *Taxon* 16(1):40 (1967)
[Ascomycota, Pezizomycotina, Leotiomycetes, Leotiomycetidae, Helotiales, Helotiaceae]

Basionym: *Lichen sarcoides* Jacquin, *Miscell. Austriac.* 2:20 (1781) *ex* Fries, *Syst. Mycol.* 2(1):217 (1822)

***Ascocoryne cylichnium* (Tulasne) Korf**, *Phytologia* 21(4):202 (1971)

Basionym: *Peziza cylichnium* Tulasne, *Annls Sci. Nat.*, Bot., sér. 3, 19:174 (1853)

Anamorph: *Coryne dubia* (Persoon) Gray, *Nat. Arr. Brit. Pl.* (London) 1:595 (1821)

Basionym: *Acrospermum dubium* Persoon, *Comm. Fung. Clav.* (Lipsiae):92 (1797)

Derivation: *Ascocoryne*, Greek, from *asco*, "sac," and *coryne*, "club-shaped"; *sarcoides*, "fleshy, flesh-like"; *cyclichnium*, meaning not known to us; *dubia*, Latin, feminine form of *dubium*, "doubt, reservation, hesitation."

Fruitbody: ***A. sarcoides***—ascocarp gelatinous, highly variable in form, 0.5–2 cm (¼–⅘ in) across, sessile or with rudimentary stalk; initially club-like (Fig. B), subglobose, or fused into lumpy, irregular, brain-like masses (Fig. A, bottom), becoming disk-like or cushion-like when mature (Figs. A, top, & C); surface smooth to vaguely wrinkled; reddish-purple, greyish-purple, to purplish-pink; fertile upper surface generally smooth or slightly wrinkled, outer surface and/or lower surface often somewhat granular. ***A. cyclichnium*** (Fig. D)—very similar to *A. sarcoides*, but fruitbody generally larger and more disk-like; lacking anamorphic stage.

Flesh: Gelatinous, pinkish-purple; odor and taste not distinctive.

Microscopic: Spores narrowly elliptical to subfusoid, clear, smooth, acutely rounded at each end. ***A. sarcoides***—mature spores 10–19 × 3–5 µm, uniseptate, containing two (rarely three) large oil droplets; ***A. cyclichnium***—mature spores 18–30 × 5–6 µm, multiseptate, with numerous very small oil droplets. Anamorph (form-genus and form-species *Coryne sarcoides*

Tulasne & C. Tulasne) producing asexual short elliptical conidiospores measuring 3–3.5 × 1–2 µm.

Occurrence: Both are saprobic on slightly decaying and living hardwoods (especially beech), less often on conifers; widely distributed in North America, Europe, Asia; late summer–winter.

Edibility: Inedible.

Comments: Although *Ascocoryne cyclichnium* fruitbodies tend to be somewhat larger and more typically disk-shaped, both species are highly variable, with overlapping features, and cannot be separated macroscopically with any confidence. However, the two taxa are quite readily distinguished on the basis of their mature spores. Mature *Ascocoryne sarcoides* spores measure 10–19 × 3–5 µm, have a single septum, and contain two (rarely three) large oil droplets. Those of *A. cyclichnium* are distinctly longer (18–30 × 4–6 µm), are divided into multiple compartments by several septa, and lack large oil droplets.

These fruitbodies somewhat resemble the quite unrelated Wood Ear, *Auricularia americana*, and Brown Witches' Butter, *Tremella foliacea*, and could be confused with them. However, *Ascocoryne* is an ascomycete and bears its spores in nonseptate sac-like asci. *Auricularia* and *Tremella* are basidiomycetes that bear four spores at the tips of basidia that are transversely and longitudinally septate, respectively. See the species accounts of those taxa for other distinguishing features.

The anamorphic (asexual) form of *A. sarcoides, Coryne dubia* (Persoon) Gray, is incorrectly referred to as *Coryne sarcoides* in some online sources. Its anamorph produces small ellipsoid spores (3.0–3.5 × 1.0–2.0 µm) and occurs early in development as fused club-like or brain-like masses (Figs. B & C), with the disk-like or cup-shaped sexual stage (Fig. C) developing later. See online Appendix A for a complete presentation of these nomenclatural issues. No anamorph of *A. cyclichnium* is known.

Class Agaricomycetes: An Overview

The class Agaricomycetes (approximately equivalent to the obsolete class Homobasidiomycetes) of the division Basidiomycota contains one-fifth of all known fungi.[1] These are the mushrooms commonly encountered and are the ones most likely to be collected, purchased, or cultivated for the dinner table or used as fabric dyes, as medicinals, or for their psychedelic properties. They are also the ones involved in mushroom poisoning incidents. They are also being actively explored for their potential use in bioremediation.

This class includes 17 orders, which together include some 105 families containing over 1,000 genera representing some 21,000 species. The order Agaricales ("true agarics") alone contains 33 families with 413 genera, the order Boletales (boletes) has 16 families with more than 95 genera, the order Russulales includes 12 families with over 80 genera, and the order Polyporales (polypores) has 10 families with some 200 genera. Every known fruitbody morphology is exhibited across ordinal lines in this huge class of basidiomycetes: typical gilled agarics with stipes, hydnoid (toothed) forms, puffballs, earthstars, underground truffle-like mushrooms, stinkhorns, bird's nest fungi, crusts, spreads and parchment fungi, club and coral fungi, and shelf-like bracket fungi. And this list is by no means exhaustive. All of these shapes and growth habits have evolved independently from different ancestral forms in more than one order. This is the reason why organizing a book such as this is so difficult: treating related groups as a taxonomic unit works well enough for the systematic specialist, but it separates very similar-looking species and makes comparative identifications difficult. On the other hand, grouping unrelated look-alikes to accommodate telling them apart introduces its own problems. We have tried to steer a middle course through this no-man's-land.

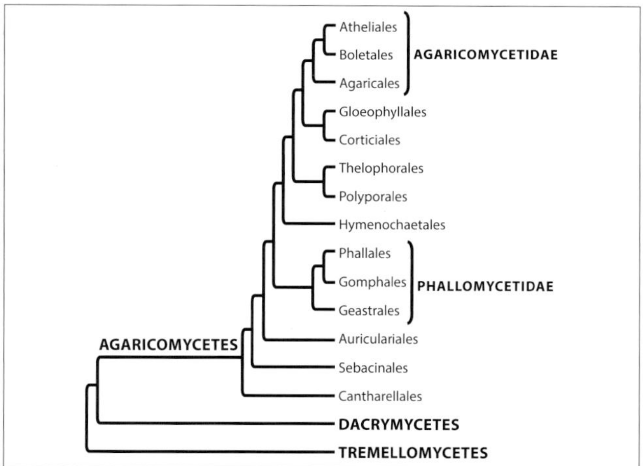

Cladogram of the class Agaricomycetes showing the phylogenetic relationships between its more prominent orders. The Agaricomycetidae and Phallomycetidae represent monophyletic classes. The higher taxonomic affinities of the other orders are not well understood. (Adapted and simplified from Fig. 14.1 in Hibbett et al., 2014.)[2]

The abbreviated cladogram to the left shows the phylogenetic relationships between some of the better-known orders in this class. Three orders each have been grouped into two subclasses: the Agaricomycetidae and the Phallomycetidae. The remaining orders (including those not listed here) are *incertae sedis*, that is, "of uncertain status." Their higher taxonomic affinities are not yet known. The relationship with the classes Dacrymycetes and Tremellomycetes are also shown. The fourth, obscure class Wallemiomycetes, containing a single order with a single family and genus, was omitted.

1. P.M. Kirk, P. Cannon, & J. Stalpers (eds.), 2008, *Dictionary of the Fungi*, 10th ed. (Wallingford: CABI).
2. D.S. Hibbett et al., 2014, Agaricomycetes, *in* D.J. McLaughlin & J.W. Spatafora (eds.), *Systematics and Evolution*, 2nd ed., *The Mycota VII Part A* (Berlin: Springer-Verlag)

Photo: John Plischke III.

Puffballs, Earthstars, Truffles, False Truffles, and Deer Truffles

This diverse and artificial (polyphyletic) group of superficially similar mushrooms vis-à-vis their morphologies and their **hygogeal** (underground), partially hypogeal (**emergent**), or **epigeal** (aboveground) lifestyles actually represent two divisions (phyla), the Basidomycota and Ascomycota, and no fewer than eight orders. These include the true truffles (Ascomycota: Pezizales), deer truffles (Ascomycota: Eurotiales), false truffles (Basidiomycota: Russulales, Agaricales, and Sclerodermatales), true puffballs (Basidiomycota: Agaricales), puffballs-in-aspic (Basidiomycota: Boletales), stalked puffballs (Basidiomycota: Agaricales and Tulostomatales), and earthstars (Basidiomycota: Geastrales and Boletales).

These fungi are either **gasteroid**—their spores developing inside one or more chambers enclosed by a rind-like **peridium**—or **secotioid**—in which the fruitbody has a stipe and a cap-like head that has failed to open and in which the gills are absent or have developed into highly convoluted and anastomosing channels or chambers. As a result, the spores cannot be forcibly discharged and are thus passively released **statismospores**. This gasteroid morphology has evolved independently in each of these separate clades (evolutionary lineages) as an adaptation to adverse environmental circumstances, such as aridity.

Cross-Sections through Various Gasteroid Mushrooms

A B C

A, puffball (*Calvatia*); B, false truffle (*Zelleromyces*); C, true truffle (*Tuber*). (Photos: A, Ron Pastorino; B, Dan Molter; C, Tim Brennemann.)

D

E

F

D, deer truffle (*Elaphomyces*); E, false truffle (*Rhizopogon*); F, earthstar, false truffle (*Scleroderma*). (Photos: D, John Plischke III; E, Alan Rockefeller; F, W. Light.)

Fruitbody: Shapes of Various Gasteroid Mushrooms

G

H

I

G & H, puffballs (*Lycoperdon perlatum*); I, earthstar (*Geastrum*). (Photos: G, Daniel Ullrich; H, W. Light; I, Amanita77.)

J

K

L

M

J, stalked puffballs (*Tulostoma*); K & L, puffballs-in-aspic (*Calostoma*); M, false puffball (*Zelleromyces*). (Photos: J, Mike Christison; K, Emily Johnson; L, Teresa Fortenberry; M, Dan Molter.)

Puffballs

True puffballs are in the order Lycoperdales, representing several families. Puffballs are round or turbinate (top-shaped), sessile, epigeous gasteroid mushrooms (Figs. G & H) with an initially quite undifferentiated, white, soft and pliable, fleshy gleba (Fig. A) that later turns into dark, dry, powdery spores interwoven with long, thread-like, sterile capillitial cells—collectively called the **capillitium**. The spores are released in smoke-like puffs through a pore or tear in the fruitbody when the paper-like endoperidium is touched by an animal or hit by falling raindrops. The peridium consists of two layers, an outer **exoperidium** and an inner **endoperidium**. The exoperidium wears away, exposing the thin, fragile inner endoperidium, which develops an apical pore (**ostiole**) or a tear through which the spores can escape.

The genus *Lycoperdon* is characterized by the presence of (1) a columella—a more or less cylindrical mass of sterile tissue found in the lower portion of the fruitbody and attached to the base, and (2) an interwoven capillitium that is attached at one end to either the endoperidium or the pseudocolumella. The capillitial hyphae are long, typically unbranched, and aseptate. One end of each thread is attached either to the inner wall of the endoperidium or to the columella.

Calvatia has a capillitium like that of *Lycoperdon*, but the hyphae are uniquely septate, with true septa interspersed at regular short intervals. Both layers of the peridium dehisce and flake away over the entire upper portion of the fruitbody to expose the spores.

Stalked Puffballs (*Tulostoma*) and Puffballs-in-Aspic (*Calostoma*)

These curious puffballs belong to two separate orders: *Tulostoma* (*sensu lato*) belongs to the order Agaricales and the family Agaricaceae;[1] *Calostoma* is in the Boletales and the family Sclerodermataceae. These stalked puffballs are epigeous, though some of them begin as underground basidiomata (fruitbodies). Both have small, round gasterocarps (enclosed, sac-like fruitbodies) containing a gleba (spore sac) filled with capillitial threads and spores encased in a peridium. In both genera, the spores become dry and powdery and are released through an apical pore or irregular tear at maturity.

Tulostoma is saprobic on buried roots, wood, and other organic materials and is adapted to xeric (dry) environments. The long stalk fits into a socket at the base of a small, globose to subglobose gasterocarp with a flattened base (Fig. J). Initially hypogeous, the spores mature while the sporocarp is still underground. It becomes epigeous as the stalk elongates. The genus has a cosmopolitan distribution, and the basidiospores are enclosed in a gelatinous, amorphous episporium (envelope).

Calostoma may be stalked (Fig. K) or sessile; when present, the stalk is hygroscopic, absorbing moisture and expanding when wet. The genus is adapted to mesic (humid) environments and is ectomycorrhizal with hardwood trees in the family Fagaceae. The gasterocarp is surrounded and partially to nearly completely covered by a distinctive gelatinous and evanescent exoperidium, which peels away to reveal a brightly colored star-shaped ostiole that splits open to expose and release the spores. The shed exoperidium of the common species in our area, *C. cinnabarinum*, looks for all the world like a slimy mass of amphibian eggs (Fig. L). The endoperidium is suspended from the mesoperidium at the perisotome (thickened tissue around the ostiole). The spores of our local species are elliptical, are densely covered with visible pits (this does not hold elsewhere), and are not enclosed in the gelatinous envelope (**episporium**) that characterizes this genus elsewhere. The genus has a disjunct distribution in eastern North America, Costa Rica, and Brazil, as well as in central-southeastern Asia and Australia–New Zealand.

Earthstars

Geastrales (*Geastrum*) and Boletales (*Astraeus*, *Scleroderma*) are small, puffball-like mushrooms with three peridial layers: an exoperidium and **mesoperidium**, which are usually inseparable, and an inner endoperidium. When wet, the exoperidium and mesoperidium split along radial lines of weakness to form several stellate (star-like) rays (Fig. I), which give these mushrooms their common name. The gleba develops as in the puffballs, and the spores are likewise released through an ostiole or tear in the exposed endoperidium.

True Truffles

In our area, the true truffles include both hypogeous and emergent genera of the order Pezizales [Ascomycota], including the families Tuberaceae, Pezizaeceae, Discinaceae, and Morchellaceae (Fig. C). The true truffle genera *Tuber*, *Imaia*, *Hydnobolites*, and *Hydnotrya* are discussed in detail in the account of the Pecan Truffle, *Tuber lyonii*.

Deer Truffles

The deer truffles, also in the Ascomycota and thus arguably also true truffles, belong to the order Eurotiales and the family Elylaphomycetaceae. The genus *Elaphomyces* is completely hypogeal and is obligately ectomycorrhizal with various hardwoods and conifers. It is characterized by a gleba of powdery, dark ascospores interspersed with greyish, tan, or brown web-like hyphae and a thick peridium (Fig. D). Deer truffles depend largely upon animal mycophagists to disperse their spores. To attract these animal vectors they emit pheromone-like odors as they mature.

False Truffles and Earthballs

This is another polyphyletic catch-all category that includes the orders Boletales, family Sclerodermataceae (*Rhizopogon*, *Scleroderma*, *Pisolithus*), Agaricales (*Bovista*), and Russulales (*Zelleromyces*). False truffles may be initially hypogeous, but they mature aboveground. The exoperidium in some species of *Scleroderma* also dehisces into stellate rays. These hypogeous to emergent mushrooms are more or less globular to irregularly potato-shaped.

Scleroderma

The thick, tough, hard peridium in *Scleroderma* and the initially firm, dark-purple gleba with its interspersed whitish capillitial threads (Fig. F) distinguish this genus from the true puffballs of the order Lyoperdales (soft and pliable peridium with an initially white, soft, and homogeneous gleba). The peridium is attached to the substratum by basal rhizomorphs, which in some species may form a stipe-like "pseudostem." Dehiscence is by an irregular tear at the apex of the sporocarp, by splitting into stellate rays, or, in some species, both. Many species are toxic and can cause severe gastrointestinal distress; deaths have been reported.

Rhizopogon

This genus is hypogeus and obligately mycorrhizal with pines. It is characterized by a thick, leathery peridium or a thin, submembranous one that can be separated from the gleba only with difficulty. The peridium is either simplex or duplex (2-layered). The gleba is loculated (divided into small, irregular, empty cavities) that are initially empty and later become filled with spores (Fig. E). The gleba lacks capillitia, is initially white, and becomes darker with maturity (but never as dark as in *Scleroderma*). The spores are ellipsoid to oblong, smooth, aseptate (but may appear 1-septate due to the equatorial position of the nucleus), and 1- to 2-guttulate (containing one to two oil droplets).

Pisolithus

This genus is represented here (and in North America) by the cosmopolitan Dye Maker's False Puffball, *Pisolithus arrhizus* (Scoploi) Rauschert (1959) [*sensu lato*]. This species is also widely called *Pisolithus tinctorius* (Persoon) Coker & Couch.[2] It is used to dye wool. The species in this genus are ectomycorrhizal with a wide variety of hardwood and conifer trees and shrubs. What was formerly conceived to be a single cosmopolitan and highly polymorphic species in fact appears to be an ill-defined complex of species that cannot be separated on morphological grounds. Some 10 other species have also been described from various parts of the world. Initially globose, this mushroom soon becomes turbinate (top-shaped), with a rudimentary stalk, a thin, fragile peridium, and a blackish, gel-like gleba packed with whitish to yellowish pea-size peridioles (Fig. N)

N

Photo: Vavrin.

that disintegrate from the top downward into a mass of brown powder, which is released into the air. The spores are globose and prominently spinose, with or without reticulations.

Bovista

This genus (Fig. O) resembles the true puffball genus *Lycoperdon*. The fruitbodies are spherical to pear-shaped, with a thin, light-colored, smooth, granular or finely spinose exoperidium that dehisces to expose the firm, membranous endoperidium with its single apical ostiole. Unlike *Lycoperdon*, the gleba lacks a pseudocolumella and has an aseptate capillitium consisting of free (not attached), short, and dichotomously branching threads with no attachment to the endoperidium. The genus *Bovistella* has a delicate peridium that opens via an apical pore and collapses as the spores are released, similar to *Lycoperdon*. Its capillitium is like that of *Bovista*.

Zelleromyces

The genus *Zelleromyces* includes those hypogeous (underground) members of the Russulales without stipes that have anastomosing glebal plates of latex-producing hyphae that bleed profusely when cut (Fig. B). However, molecular studies indicate a close relationship with the milkcap genus *Lactarius*, and several authors suggest that *Zelleromyces* should be synonymized with *Lactarius*. Indeed, Dennis Desjardin described what was essentially a species of *Zelleromyces* with ballistosporic instead of statismosporic spores as a new species in the genus *Lactarius*.[3]

O

Photo: Ron Pastorino.

1. *Fide* E.C. Vellinga, 2009, Nomenclatural overview of the genera in the Agaricaceae Chevall., version 4.1., http://nature.berkeley.edu/brunslab/ev/vellinga_genera_v41_feb2009.pdf. *Tulostoma* was formerly placed in the order Tulostomatales and the family Tulostomataceae. *Calostoma* was formerly considered to be the sole genus in a monotypic family, the Calostomataceae.

2. Many sources refer to this species as *Pisolithus tinctorius* (Persoon) Coker & Couch (1928). However, *Index Fungorum* claims that this combination is an invalid homonym of *Pisolithus tinctorius* (Montagne) E. Fisher (1900). Nevertheless, *Scleroderma tinctorium* Persoon (1801) clearly predates *Polysaccum tinctorium* Montagne (1840). Furthermore, *Pisolithus tinctorius* (Montagne) E. Fischer cannot be prioritized because Persoon's name was sanctioned (by Persoon himself), and this issue remains quite confused to us. And as noted by M.S. Reddy et al., 2005, *Mycologia* 97(4):838–843, this taxon was poorly described, which resulted in numerous synonyms and the epithet *tinctorius* being extensively applied to several different species.

3. *Lactarius rubriviridis* Desjardin, Saylor, & Thiers; D.E. Desjardin, 2003, *Mycologia* 95(1):148–155.

Pear-shaped, Common, and Hedgehog Puffballs

Lycoperdon pyriforme, *L. perlatum*, and *L. echinatum*

A & B, *Lycoperdon pyriforme*; C, *L. perlatum*; D–F, *L. echinatum*. (Photos: A–C, E & F, W. Light; D, William Tanneberger.)

Lycoperdon pyriforme **Schaeffer**, *Fung. Bavar. Palat. Nasc.* (Ratisbonae) 4:128 (1774)[1]

Lycoperdon perlatum **Persoon**, *Observ. Mycol.* (Lipsiae) 1:145 (1796)

Lycoperdon echinatum **Schaeffer**, *Fung. Bavar. Palat. Nasc.* (Ratisbonae) 4:128, table 186 (1774)
[Basidiomycota, Agaricomycotina, Agaricomycetes, Agaricomycetidae, Agaricales, Lycoperdaceae]

Derivation: *Lycoperdon*, Greek, from *lýkos*, "wolf," and *pérdomai*, "fart" (referring to the manner in which members of this genus disperse their spores), "wolf fart"; *pyriformis*, Latin, "pear-shaped"; *perlatum*, from *perlatos*, "circulated," "widespread"; *echinatum*, "covered with spines."

Fruitbody: ***L. pyriforme***—1.5–5 cm wide by 2–5 cm high (⅔–2 × ¾–2 in), typically shaped like an inverted pear or hot-air balloon, but often globose, dry, white to yellowish-brown, darker with age, initially covered with minute white spines (Fig. A) that eventually fall away, leaving no scars and exposing a smooth surface (Fig. B); surface nearly white to yellowish-brown with age; with small, constricted base *attached to woody substratum* or occasionally wood chips by white rhizomorphs. ***L. perlatum***—2.5–7 cm wide by 3–7.5 cm high (1–2¾ × 1¼–3 in), shaped like an inverted pear to subglobose, dry, initially covered with whitish spines, these later falling off, leaving surface scars (Fig. C); surface white at first, later drab yellowish-brown. ***L. echinatum***—2–4 cm wide by 2–3.5 cm high (¾–1⅔ × ⅔–1½ in), somewhat pear-shaped to spherical, becoming flattened, base highly constricted, 6–8 mm across, with white rhizomorphs penetrating substratum; surface densely covered with long spines 3–6 mm (up to ¼ in) long, these fusing distally to form fascicles; spines initially white, gradually becoming dark brown with age (Fig. D), spines eventually falling off and exposing a scurfy, pale brown to dark purple-brown surface (Figs. E & F), the mature fruitbody now vaguely resembling the acorns of a Burr Oak (*Quercus macrocarpa*).

Spore mass (gleba): ***L. pyriforme*** **and** ***L. perlatum***—initially white, subsequently turning yellow, greenish-brown, and then dark olive-brown with age; ***L. echinatum***—white at first, gradually turning yellowish to olivaceous, then greyish-brown, becoming purplish when mature.

Microscopic: ***L. pyriforme***—spores 3.5–4.5 µm, round, smooth,[2] lacking pedicel. ***L. perlatum***—3.5–4.5 µm, round, minutely spiny, lacking pedicel. ***L. echinatum***—spores (4–)4.5–6 µm in diameter, globose, with short warts penetrating a surrounding clear layer, with long, clear pedicels, these often broken.

Occurrence: All three species saprobic. ***L. pyriforme***—cespitose, often forming dense clusters, less often scattered or solitary on decaying hardwood logs, often abundant; later summer–fall, old spore cases may overwinter. ***L. perlatum***—solitary, scattered, or gregarious in clusters on ground in woods beneath hardwoods and conifers, on roadsides, or forming fairy rings on lawns, summer–fall; nearly cosmopolitan. ***L. echinatum***—solitary or gregarious on humus or wood chips (Figs. E & F) or on very decayed hardwood (Fig. D); late summer–fall in periods of ample rainfall, may be locally abundant east of the Rocky Mountains; Europe, East Africa, East Asia.

Edibility: All three species are choice edibles when young and fresh, but make sure that you do not inadvertently collect a button-stage *Amanita*. See the section on mushroom poisoning for a discussion on making this distinction.

Comments: Young stages of *Lycoperdon echinatum*, the Hedgehog Puffball, can be confused with *L. pulcherrimum* Berkeley & M.A. Curtis, which also bears a dense cover of long (3–6 mm), slender, white, terminally fused spines covering the fruitbody and similar-sized spores. However, the spines of *L. pulcherrimum* remain white (rather than turning dark brown) as the sporocarp ages, and the stout, stem-like, sterile base constitutes one-third to one-half of the fruitbody's height, in sharp contrast to the tightly constricted base of *L. echinatum*.

Lycoperdon perlatum, the Common Puffball, and *L. pyriforme*, the Pear-shaped Puffball, are the two puffball species most commonly encountered in eastern North America.

1. On the basis of DNA evidence, the presence of rhizomorphs, and its growth on woody substrata, H. Kreisel & D. Krüger (2002) transferred this species to the genus *Morganella* Zeller—a primarily tropical genus growing on a woody substratum—as a new combination. This has been followed by some but not all subsequent authors. However, *Lycoperdon* species also possess white rhizomorphs, and some (e.g., *L. echinata* and *L. pulcherrimum*) also often grow on woody substrata (see above descriptions). A more recent study involving a wider range of species found that *Lycoperdon* Persoon is a monophyletic (having a single common ancestor) genus and that *L. pyriforme* should be retained within it (E. Larsson & M. Jeppson, 2008, *Mycol. Res.* 112[1]:4–22). These features, plus the presence of a true capillitium of dark-brown sinuous to contorted threads in *L. pyriforme*—typical of *Lycoperdon* but altogether lacking in the genus *Morganella*—and other features that required emending *Morganella* and erecting a new subgenus (*Apioperdon* Kreisel & D. Krüger), support retaining this species within *Lycoperdon*.

2. Scattered warts are visible under scanning electron microscopy but are nearly invisible under brightfield light microscopy.

Orange-staining Puffball

Calvatia rubroflava

A B C D

Calvatia rubroflava (Cragin) Morgan, *J. Cincinnati Soc. Nat. Hist.* 12:171 (1890)
[Basidiomycota, Agaricomycotina, Agaricomycetes, Agaricomycetidae, Agaricales, Lycoperdaceae]

Basionym: *Lycoperdon rubroflavum* Cragin, *Bull. Washburn Coll. Lab. Nat. Hist.* 1(2):30 (1885)

Derivation: *Calvatia*, Latin, from *calvus*, "bald," and *calvaria*, "dome of the skull"; *rubroflava*, "reddish-yellow."

Fruitbody: 2–10 cm wide by 1.5–5 cm tall (⅘–4 × ⅔–2 in), shaped like a top or inverted pear, somewhat flattened above (Fig. A), heavily wrinkled below (Figs. B & D), bluntly conical sterile (lacking spores) base (Fig. B) typically anchored to ground by one or more whitish root-like rhizomorphs (Figs. B & D). Peridium (outer skin) initially white, gradually turning first yellow, then orange to reddish-brown (Figs. A & B) and covered by yellowish patches of bran-like particles (Figs. A & C). Fishy odor when mature.

Spore mass (gleba): Initially white with tiny cavities, becoming a greenish-yellow to dull mustard cottony mass, not becoming powdery; flesh persisting as a very thin layer beneath outer rind, turning quickly yellow or gold when exposed.

Staining: KOH—outer skin (peridium) red, flesh and gleba negative; **Melzer's reagent**—peridium and flesh red, gleba negative.

Microscopic: Spores olive-brown, minute, 3–5 µm, round, with minute warts.

Occurrence: Scattered or clustered on the ground in gardens, grassy areas, and woodlands along streams, often in nutrient-poor sandy soil; widespread but uncommon in eastern North America, August–December. Reported from Pacific Northwest; Brazil; Venezuela, Australia as *Calvatia candida* var. *rubroflava* (Cragin) G. Cunningham (1926).

Edibility: Inedible when mature.

Comments: When mature this orange to reddish-brown puffball with its greenish-yellow or mustard-colored cottony gleba cannot be mistaken for anything else. The fruitbody is soft, velvety, and extremely lightweight—it feels very airy and "insubstantial," for lack of a better description.

Purple-spored Puffball

Calvatia cyathiformis

A

B

C

D

E

F

Photos: A & B, D–F, W. Light; C, Ron Pastorino.

Calvatia cyathiformis (Bosc) Morgan, *J. Cincinnati Soc. Nat. Hist*. 12(4):168 (1890)

Basionym: *Lycoperdon cyathiforme* Bosc, *Mag. Gesell. Naturf. Freunde, Berlin* 5:87 (1811)
[Basidiomycota, Agaricomycotina, Agaricomycetes, Agaricomycetidae, Agaricales, Agariaceae]

Derivation: *Calvatia*, Latin, from *calvus*, "bald," and *calvaria*, "dome of the skull"; *cyathiformis*, Greek, from *cyathus*, "cup," "goblet," "goblet-shaped," referring to the empty spore case.

Fruitbody: Large, 5–20 cm (2–8 in) high and broad, at first broadly rounded or ovoid (Figs. A & B), becoming pear- or top-shaped with a rounded (Fig. D) or flattened (Fig. E) top merging into a narrow, tapering base (Figs. C–E); outer surface initially whitish to pale brown (Fig. F), smooth, sometimes with flat scales (Fig. E), with maturity becoming areolate, with irregular greyish-brown polygons bordered by raised lighter greyish rims (Figs. A & B); periderm 2-layered, both layers dehiscing and peeling away in flakes to expose the spore mass and persisting as a deep, purplish-brown, cup-like mass after spore dispersal (Fig. C).

Spore mass (gleba): At first white and firm, becoming yellowish to yellowish-beige, then deep lilac to dark purple-brown (Fig. C, this color diagnostic) and turning powdery.

Microscopic: Spores 3.5–7.5 µm, globose, smooth to weakly spiny, thick-walled, pale lilac; capillitial threads branched, multiseptate, walls minutely pitted.

Occurrence: Saprobic; scattered to gregarious, exclusively in open grassy areas and forest edges, often forming fairy rings; widespread and common in North America east of the Rocky Mountains; reported from Australia, Galápagos Islands, southern Brazil, West Bengal; fruiting late spring–late fall; empty spore case may persist year-round.

Edibility: Edible, considered good until the flesh turns tan (Fig. E), after which it may cause gastrointestinal upset; absorbs flavors well.

Comments: *Calvatia cyathiformis* (*sensu lato*) may actually represent a complex of cryptic species. In addition to the distinctive areolate surface pattern, this species is distinguished from all other species of *Calvatia* by its purple-brown mature spore mass.

Calvatia craniformis (Schweinitz) Fries is very similar to *C. cyathiformis*. However, its mature spore mass is yellow-brown rather than purple, its surface is smooth or deeply and irregularly wrinkled rather than areolate, it grows in both woodlands and grassy areas, and its spores are smaller (2.5–4 µm). *Calvatia gigantea* (Batsch *ex* Persoon) Lloyd is *huge*—up to 45–80 cm (18–32 in) in diameter or even larger.[1] It has a smooth or tomentose (felt-like) surface and a greenish-brown to yellow-olive spore mass at maturity; it also occurs in both grassy areas and open forests.

Scleroderma species (earthstars) also have dark-brown or purplish spores. However, they have a thick, leathery, 3-layered periderm, the outer two layers of which split apart at maturity to form distinctive star-like rays.

1. The largest giant puffball recorded was 264 cm (8 ft 8 in) in diameter and weighed 48 pounds (A.H. Smith, 1969, *The Mushroom Hunter's Field Guide*, rev. and enl. [Ann Arbor: University of Michigan Press]).

Barometer Earthstar

Astraeus cf. *morganii*

A

Astraeus morganii. (Photo: Bill Blanton.)

***Astraeus* cf. *morganii* Phosri, Watling & M.P. Martín**, in C. Phosri, M.P. Martín, & R. Watling, *IMA Fungus* 4(2):353 (2013) [Basidiomycota, Agaricomycotina, Agaricomycetes, Agaricomycetidae, Boletales, Sclerodermatineae, Diplocystaceae]

Derivation: *Astraeus*, Greek, from Astraios, Titan god of the stars and planets; *morganii*, epithet in honor of the botanist and mycologist Andrew Price Morgan (1836–1907), who described and named the genus *Astraeus*.

Misapplied name (*auct. mult. amer.*): *Astraeus hygrometricus* (Persoon) Morgan, *J. Cincinnati Soc. Nat. Hist.* 12:20 (1889)

Fruitbody: A subglobose spore case with deciduous exoperidium, splitting into 7–20 star-like rays spreading out radially from base.

Rays: Exoperidium forming the rays thick, leathery, hygroscopic (tightly folded over the spore case when dry, expanded when moist); rays 5–7⅔ cm (2–3 in) long, inner (upper when expanded) surface smooth, becoming irregularly cracked and fissured, forming reticular network (Fig. A, *foreground*) or deeply etched into polygonal patterns (Fig. A, background arrow); surface dingy greyish to yellowish-brown, becoming dark brown with age; exterior (downturned) surface fibrous rough, tan to brown.

Spore case: Sessile, subglobose, 2–2.5 cm (¾–1 in) in diameter; covered with thin (1 mm thick), papery endoperidium, surface dry, felt-like or scurfy (covered with a loose scaly crust); whitish, greyish to brownish-grey, with irregular perforation (ostiole) without peristome at apex; columella entirely lacking.

Spore mass (gleba): Initially white and solid, becoming brown and powdery when mature.

Microscopic: Spores 7.5–10 µm, globose, thick-walled, ornamented with minute, rounded warts; brown, nonreactive in Melzer's reagent.

Occurrence: Ectomycorrhizal with oaks and pines. Solitary in groups on the ground in sandy soil along streams and roadsides and in disturbed areas; widespread and occasional to locally common; late summer–fall; central and southern United States south into Mexico, but actual distribution limits uncertain.

Edibility: Inedible.

Comments: The above description is based largely on the original species description for *A. morganii*. This and other North American species of *Astraeus* have long been misidentified as *Astraeus hygrophanus* by mycologists under the mistaken notion that it was a cosmopolitan taxon. The North American representatives were then split into several species, with several species complexes still to be described and named. *Astraeus morganii* is believed to have a north-central and southward distribution; *A. smithii* Watling, M.P. Martín & Phosri occurs in the central and northern United States, and *A. pteridis* (Shear) Zeller is in the Pacific Northwest.

The Barometer Earthstar is hygrophanous, that is, the split outer peridium (exoperidium) expands away as star-like rays when moist, exposing the inner peridium (endoperidium) comprising the subglobular or dome-shaped spore case that encloses the spore mass (gleba). When dry, the rays contract to tightly enclose the spore case. This process is repeated indefinitely as conditions change; ray expansion from a contracted dry state can occur in as little as five minutes.

The combination of distinctive hygroscopic rays with their deeply cracked and geometrically patterned surface, the absence of a peristome, and the lack of a columella distinguish this species from other earthstars in our area.

Species of the genus *Geastrum* typically have a peristome—a raised umbo or distinct disk-like area usually outlined by a circular groove or ridge (as shown in Fig. B, arrow, for *G. saccatum* Fries)—surrounding the ostiole (apical opening) of the spore case, which is not present in *A. morganii*. *Geastrum* species also have smaller spores (<5 µm in diameter) and possess a columella (lacking in *Astraeus*)—a column-like extension of sterile tissue from the stipe into the gleba.

B *Geastrum saccatum*. (Photo: Amanita77.)

Of the *Geastrum* species found in our area, *G. fornicatum* (Hudson) Fries, *G. rufescens*, and *G. fimbriatum* Fries lack a clearly defined peristome. However, the expanded rays of both *G. fornicatum* and *G. rufescens* bend sharply downward to become embedded in the substratum and lift the spore case completely off the ground.[1,2] This contrasts sharply with the flat-lying rays of *Astraeus morganii*, as well as with the other *Geastrum* species discussed here.

Geastrum mirabilis Montagne is slightly hygroscopic, but the fruitbody is only 3 cm (1¼ in) across when fully expanded. The spore case is umbonate and covered with conspicuous adpressed matted hairs (rather than being scurfy); it is light yellowish-brown to reddish-brown instead of dingy greyish to greyish-brown. The upper surface of the expanded rays never exhibits the geometric cracking pattern that characterizes *Astraeus morganii*.

The upper (inside) surfaces of the rays are also often cracked in both *G. rufescens* and *G. fimbriatum*, but they never exhibit the striking reticular or polygonal pattern seen in *A. hygrometricus*; they are also cream or pinkish-cream rather than dingy greyish-brown. Though lacking a clearly defined peristome, the ostioles in both species also open from the tip of a distinct conical beak, which is externally lined with fibers and longitudinally striated.

1. This is done to enhance aerial spore dispersal and is accomplished when the mesoperidium (a middle layer not present in many earthstars) separates from the exoperidium, which becomes attached to or embedded in the substratum, lifting the endoperidium (spore case) and gleba off the ground.

2. This is also true of *G. quadrifidum* Persoon, which also differs from *A. hygrometricus* by its sharply defined silky peristomial disk bordered by a circular groove.

Stalked Puffballs-in-Aspic

Calostoma cinnabarinum, *C. lutescens*, and *C. ravenelii*

A, B, D, F, *C. cinnabarinum*; C, G, *C. lutescens*; E, *C. ravenelii*. (Photos: A–D & G, Emily Johnson; E, Christine Braaten; F, Elizabeth Fortenberry.)

***Calostoma cinnabarinum* Desvaux**, *J. Bot. (Desvaux)* 2:94 (1809)

***Calostoma lutescens* (Schweinitz) Burnap**, *Bot. Gaz.* 23, plate 19, figs. 1 and 2 (1897)

Basionym: *Mitremyces lutescens* Schweinitz, *Schr. Naturf. Ges. Leipzig* 1:60 (1822)

***Calostoma ravenelii* (Berkeley) Massee**, *Ann. Bot.* (Lond.) 2(5):25 (1888)

Basionym: *Mitremyces ravenelii* Berkeley, *Trans. Linn. Soc. London* 22:130 (1857)
[Basidiomycota, Agaricomycotina, Agaricomycetes, Agaricomycetidae, Boletales, Calostomataceae]

Derivation: *Calostoma*, Greek, from *kalos* and *stoma*, "beautiful mouth," referring to the bright-red peristome; *cinnabarinum*, Latin, "cinnabar-red"; *lutescens*, Latin, "of the marshes"; *ravenelii*, Latinized epithet honoring Henry R. Ravenel (1814–1887).

NOTE: This description pertains to *C. cinnabarinum*; *C. lutescens* and *C. ravenelii* are distinguished in the comments section.

Fruitbody: Spore case spherical or egg-shaped mounted atop a short thick stalk covered by a gelatinous mass containing small rounded egg-like inclusions; initially belowground, later emerging as the stipe enlarges.

Spore case (head): Small round or egg-shaped, 1–2 cm (¼–⅘ in) in diameter, consisting of a hard, cinnabar-red mesoperidium (Fig. A) coated with a thick, slimy, translucent, jelly-like exoperidium (Figs. A, B, & F) and an underlying endoperidium. Outer gelatinous coating and mesoperidium gradually breaking down and falling away to produce a gelatinous mass around the base of the fruitbody filled with yellowish, salmon-colored, or reddish fragments of mesoperidium resembling amphibian eggs (Figs. B, D, & F). Surface of newly exposed endoperidium powdery bright red (Fig. A), wearing away to a pale orangish-yellow (Fig. F) or pinkish (Figs. B & D) head, with remnant bright-red apical ridges forming an irregular star-shaped peristome around the ostiole (mouth) (Figs. B, D, & F).

Stipe: Short, stubby (Figs. A & B), 2–4 cm tall and 1–2 cm thick (¾–1⅔ × ⅖–¾ in), with shaggy, coarsely reticulate, or spongy surface also covered with the yellowish gelatinous layer.

Spore mass (gleba): Gleba enclosed within endoperidium, initially white, becoming buff or yellow and powdery with age.

Chemical staining: Mesoperidium clear in KOH, brown in Melzer's reagent; endoperidium clear in KOH and clear to yellow in Melzer's reagent.

Microscopic: Spores clear or pale yellow, 14–20 (up to 24) × 6–11 µm, elliptical to oblong-elliptical, covered with visible pits and reticulations; unlike other *Calostoma* species, spores not nourished by nurse cells.

Occurrence: Ectomycorrhizal with and restricted to oaks (*Quercus* spp.); October–February in wet deciduous woods, generally at higher elevations in the Appalachians; alone or gregarious. Widespread in eastern North America south through the cloud forests of Mexico and Central America to southeastern Brazil; reported from India and East Asia (China, Japan, Korea).

Edibility: Generally considered inedible.

Comments: All three species are commonly referred to as Puffballs-in-Aspic. All occur in our area, and all have a red, star-shaped, ridge-like peristomium around the ostiole (mouth-like opening), which has evoked another common name for members of this genus: "Hotlips." *Calostoma lutescens* and *C. ravenelii* are less common than *C. cinnabarinum* and have a more restricted distribution, but all occur in similar habitats. In *Calostoma*, the endoperidium is suspended from the mesoperidium at the ridge-like, brightly colored peristome.

Calostoma cinnabarinum is immediately distinguished by the jelly-like mass containing egg-like fragments of the mesoperidium and lack of a collar around the base of the bright-red, orange-yellow, or pinkish spore case (Figs. A, B, D, & F).

Calostoma lutescens is easily distinguished from *C. cinnabarinum* in its dry to weakly gelatinous exoperidium, this falling away along with the mesoperidium, leaving the pale to lemon-yellow endoperidium behind as a globose head with a prominent torn collar of tissue around the bottom third (Figs. C & G); *C. cinnabarinum* (Figs. A, B, D, & F) lacks such a collar. The shaggy, open, net-like stipe is much longer and lighter colored (yellow vs. dark brown) than in *C. cinnabarinum*—5–9 cm (2–3½ in) tall by 15–20 mm (⅔–⅘ in) thick: compare Fig. G (*C. lutescens*) with Fig. A (*C. cinnabarinum*). The much smaller spores (5.8–8 × 5.8–8 µm) are spherical (not oblong-elliptical), prominently pitted, each pit surrounded by a polygonal, sharply edged ridge. *Calostoma lutescens* occurs in the United States and Mexico.

Calostoma ravenelii has a smaller (0.7–1 cm; ½–¾ in) grey to grey-brown head with small warts or scales of adhering exoperidial matter (Fig. E). The exoperidium is dry and persists through maturity until old age, when it flakes away, leaving a hard, grey endoperidium as the globose head. Like *C. cinnabarinum*, *C. ravenelii* lacks a prominent collar around the base of the head. The mesoperidium is lacking altogether. Both the exoperidium and endoperidium are clear or stain slightly brown in KOH and yellow in Melzer's reagent. The spores are minutely pitted and appear punctate; otherwise indistinguishable from those of *C. cinnabarinum*.

Calostoma microsporum G.F. Atkinson (1903) differs in no way from *C. ravenelii*, and the size of the spores, the major feature separating the two taxa, also overlaps strongly. *Calostoma microsporum* is considered a variety of *C. ravenelli* by Castro-Mendoza and O.K. Miller in Castro-Mendoza, Miller & Stetler (1983),[1] and we concur with that opinion. This species is also reported from Mexico, Japan, Taiwan, and China.

Despite its common name and superficial resemblance, *Calostoma* is not related to the puffballs, earthstars, stinkhorns, or stalked puffballs but is allied with the boletes. The different layers of the fruitbody (exoperidium, mesoperidium, endoperidium) are described somewhat differently by different authors, which causes some confusion. We have used these terms *sensu* Castro-Mendoza, Miller, and Stetler.[1]

1. E. Castro-Mendoza, O.K. Miller, & D.A. Stetler, 1983, *Mycologia* 75(1):36–45.

Stalked Puffballs

Tulostoma spp.

A B C D

Photos: Mike Christison.

Genus *Tulostoma* Persoon 1794: Persoon 1801, *Neues Mag. Bot.* 1:86 (1749); *Syn. Meth. Fung.* (Göttingen):139, 1801 [Basidiomycota, Agaricomycetes, Agaricomycetidae, Agaricales, Agaricaceae]

Type species: *Tulostoma brumale* Persoon, *Neues Mag. Bot.* 1:86 (1749)

Derivation: *Tulos*, Greek, "knob"; *stoma*, "mouth."

Spore case: Generally globose to subglobose; encased in a peridium consisting of two layers, an outer exoperidium and an inner endoperidium; exoperidium partly deciduous, the remnants forming an acorn-like cup around the base—the peridial sheath; endoperidium enclosing a powdery gleba consisting of capillitia, usually well-developed basidia, and basidiospores; spores released to the exterior through an apical pore.

Stipe: Up to 10 cm (4 in) tall depending on species, woody, brown, scaly, inserted into socket at base of spore case.

Spore deposit: Various shades of cinnamon to rusty-brown when mature.

Microscopic: Spores generally of similar size, (3–)4.2–7(–8) µm, globose, subglobose, slightly ellipsoid to irregular, smooth or ornamented with spines, or fine or coarse warts appearing as tiny plugs on wall of endospore that protrude into gelatinous envelope of epispore; capillitia of two kinds: simple hyaline threads and elators (short cells with thickened spiral bands on inner wall), unique to the genus *Tulostoma*.

Occurrence: Cosmopolitan, most species found on and in sandy soil in arid or semiarid regions; more common in western North America.

Edibility: Inedible, primarily since by the time the fruitbodies appear aboveground they are much too tough.

Comments: The distinctive puffball-like spore case perched atop a long stipe aboveground cannot be confused with any other type of mushroom. The genus *Tulostoma* is distinguished from other stalked puffballs by its stipe (inserted into a socket at the base of the peridium), its septate capillitial hyphae,[1] and the apical pore. Species likely to be found in the Southeast include *Tulostoma brumale* Persoon (1794), *T. campestre* Morgan (1889), *T. beccarianum* Bresàdola (1904) [*T. simulans* Lloyd (1906) is a later synonym], and *T. lloydii* Bresàdola (1904).

However, the 268 or so described species of *Tulostoma*

are notoriously difficult to distinguish either macroscopically or microscopically. The size of the spore caps and shape of the mouth (tubular, fibrillose, mammose, or denticulate) and the presence of a sand case (a thick coating of sand that adheres to the peridium and persists in some species) are important in defining some species. It is important to obtain fresh young specimens before age and weathering have damaged or obliterated important features. Many species fruit in the fall and are often not seen until they have overwintered and weathering has occurred.

Most stalked puffballs are in the cosmopolitan genus *Tulostoma*, but several other genera are also found in semiarid steppes, deserts, and sandy areas in temperate regions.[2] In North America most of these genera are found west of the Rocky Mountains in the arid deserts of the Southwest and/or the sage steppe regions of the Pacific Northwest east of the Cascades. Only a few species occur in the southern Piedmont and Appalachian Mountains.

The genus *Schizostoma* (not found in the Southeast) has a similar stipe, but the capillitial threads are nonseptate, there is no apical pore, and the spore sac peels back at the top to expose the spore mass. Formerly grouped in their own family (Tulostomataceae), these curious mushrooms are now considered to be members of the Agaricaceae. The fruitbody develops underground; it emerges at maturity and is raised above the ground as the stipe elongates.

Calostoma species are brightly colored yellow, orange, or reddish puffballs on a short stalk with a large irregular (denticulate) mouth-like opening outlined in deep red. The spore cap is encased in a slimy jelly-like outer matrix that coats the peridium, but this may wear away in some species. This genus has recently been transferred to the order Boletales and is not closely related to the puffballs, stalked puffballs, earthstars, or stinkhorns. Several *Calostoma* species that are widespread in southeastern North America are discussed in the account of the Stalked Puffball-in-Aspic, *C. cinnabarinum* Desveaux (1809).

Helpful keys and discussions of the stalked puffballs can be found in Arora (1986) and online (L. & M. Bailey 2003).[3]

1. Sterile hyphal threads interspersed within the mass of spores.

2. These include *Battarea* Persoon (1801) and *Podaxis* Desvaux (1809), with some 50 described species, and the monotypic genus *Queletia* Fries (1872) [containing the single species *Q. mirabilis* Fries], all three genera in the Agaricaceae; three monotypic genera in the family Phelloriniaceae: *Phellorinia* Berkeley [with the single species *Phellorina herculeana* (Pers. 1801) Kreisel (1961)], *Dictyocephalos* Underwood [*D. attenuatus* Underwood (1901)], and *Chlamydopus* Spegazzini [*C. meyenianus* (Klotzsch) Lloyd (1903)]; and the genus *Calostoma* Desvaux (1809) [Boletales, Calostomataceae], with some 29 or 30 species worldwide.

3. D. Arora, 1986, *Mushrooms Demystified* (Berkeley: Ten Speed Press), 959 pp.; L. & M. Bailey, 2003, Trial field key to the woody desert fungi found in the Pacific Northwest, http://www.svims.ca/council/Woody.htm

Deer Truffles

Elaphomyces spp.

A B C

Elaphomyces spp. (Photos: John Plischke III.)

***Elaphomyces* Nees *ex* C.D.G. Esenbeck**, *Horae Physicae Berolinenses*:1–20 (1820)

Type species: *Elaphomyces granulatus* Fries, *Syst. Mycol.* (Lundae) 3(1):58 (1829)

Homotypic synonym: *Ceraunium* Wallroth, *Flora Cryptogamica Germaniae* 2:405 (1833)

Heterotypic synonyms: *Phymatium* Chevallier, *Flore Générale des Environs de Paris* 1:361 (1826)
Ceratogaster Corda, *Deutschlands Flora, Abt. III. Die Pilze Deutschlands* 19/20:35 (1841)
Lycoperdastrum Haller *ex* Kuntze, *Revisio Generum Plantarum* 2:858 (1891)
Ascoscleroderma Clémencet, *Botaniste*:14 (1932)
[Ascomycota, Pezizomycotina, Eurotiomycetes, Eurotiomycetidae, Eurotiales, Elaphomycetaceae]

Derivation: *Elaphomyces*, Greek, from *eláfi*, "deer," and *mýkis*, "fungus."

Fruitbody (ascoma): 0.5–4 cm (¼–1⅔ in) across; hypogeous, growing completely underground; round, ovoid, or irregularly potato-shaped; surface of periderm granular to very heavily verrucose, with low, rounded, angular, or elongate warts; ascoma commonly encased in a husk-like mycelial mat incorporating soil, debris, and ectocycorrhizal rhizomes, this and/or interwart hyphal structures often obscuring the warts and making the surface appear smooth; peridium 2- or 3-layered, up to 3 mm thick, surface yellow, brown, reddish-brown, blue, or black; ascoma with a single chamber.

Spore mass (gleba): Initially off-white and cottony, becoming powdery and interspersed with tan, grey, grey-brown, or brown spiderweb-like hyphae at maturity; gleba variously yellow-brown, grey-brown, olive-brown, blue-green, blue, or dark brown to nearly black.

Microscopic: Spores subglobose to globose, ornamented with spinose to reticulate-spinose warts, bars, and spines; with or without oil droplets; clear, pale brown, yellow-brown, to brown.

Occurrence: Ectomycorrhizal with a variety of hardwoods and conifers; worldwide except for Africa and Antarctica.

Edibility: Unknown.

Comments: Deer truffles are an important food source for many mammals; in return, these animal mycophagists are important and necessary agents for spore dispersal. Indeed, the generic name refers to deer feeding on these fungi. As these fungi mature underground they emit potent scent signals that attract a wide variety of animal species. These animals dig up these hypogeous fungi, eat them, and then disperse the spores in their feces at points some distance away from the parent fungus. Spore dispersal by air also occurs.[1] *Elaphomyces* species are also obligately ectomycorrhizal with a variety of hardwood and conifer species.

Elaphomyces spores are not always readily distinguished from those of *Scleroderma* or *Tuber* by light microscopy. The distinction between the "true truffles" and the so-called false truffles is vague. Some authors classify all more or less potato-shaped hypogeous ascomycetes as truffles and relegate all basidiomycetous truffle-like species to the ill-defined bin known collectively as false truffles. By that definition, the deer truffles would be considered true truffles. Most others, ourselves included, restrict the term *truffle* to the genus *Tuber* P. Michelli *ex* F. H. Wiggers and loosely related genera in the family Tuberaceae. On the other hand, some authors consider *all* hypogeous truffle-like fungi—both Ascomycota and Basidiomycota—to be *bona fide* truffles.[2]

Elaphomyces appalachiensis Linder is known from Great Smoky Mountains National Park, Tennessee (the type locality). An uncommon species, it ranges from Iowa south to northern Florida and northeastern Mexico (Nuevo León and Tamaulipas) under a variety of hickories, oaks, and pines. It is immediately distinguished by the sporocarp being completely encased in a lilac to red mycelial mat. The mycelial mats covering other *Elaphomyces* are various shades of pale brown, yellow-brown, or dark brown.

The precise number and identity of *Elaphomyces* species in eastern North America are unclear. Several undescribed species and species complexes in eastern North America have long been assigned the misapplied names of European species that do not occur in North America: *Elaphomyces muricatus* Fries, *E. reticulatus* Vittadini, *E. granulatus* Fries, *E. cyanosporus* Tulasne & C. Tulasne, and *E. variegatus* Vittadini. *Elaphomyces verruculosus* Castellano[3] is a sister taxon to *E. granulatus*, which ranges from Quebec south to Florida and northeastern Mexico along the Coastal Plain. In Georgia it occurs south of the Fall Line, and to our knowledge it has not been reported in our area. However, several undescribed species similar to *E. cyanosporus* and *E. granulatus* are very common throughout the Southeast[4] and are frequently parasitized by *Elaphocordyceps* species.[5] M. A. Castellano and J. M. Trappe have assigned the following unpublished names to several of these complexes: "Elaphomyces americanum" for some of the taxa given the misapplied name *E. muricatus* and "Elaphomyces fallax" for *E. cyanosporus*. A number of other species in these complexes remain undescribed and unnamed.

Scleroderma species have a distinct attachment point at the base that is not present in *Elaphomyces*. In *Tuber* species the gleba remains solid and fleshy.

1. H.T. Reynolds, 2001, Systematics, phylogeography and ecology of Elaphomycetaceae (Ph.D. diss., Duke University), xv + 182 pp.

2. For example, J.M. Trappe et al., 2009, *Diversity, Ecology, and Conservation of Truffle Fungi in Forests of the Pacific Northwest*, U.S. Department of Agriculture, Forest Service Pacific Northwest Research Station, General Technical Report PNW-GTR-772, 194 pp.

3. M.A. Castellano, *in* Castellano et al., 2012, *Revista Mexicana de Micología* 35:17–22.

4. M.W. Beug, A.E. Bessette, & A.R. Bessette, 2014, *Ascomycete Fungi of North America* (Austin: University of Texas Press), 79–83.

5. See our accounts of the Head-like Cordyceps, *Elaphocordyceps capitata* and *E. longisegmentus*, and the Goldenthread Cordyceps, *E. ophioglosoides*.

Pecan Truffle, American Brown Truffle

Tuber lyonii species complex

A B C

Photos: Dr. Timothy B. Brenneman, Department of Plant Pathology, University of Georgia, Tifton campus.

***Tuber lyonii* Butters** [as "*lyoni*"], *Bot. Gaz.* 35:431 (1903) [Ascomycota, Pezizomycotina, Pezizomycetes, Pezizomycetidae, Pezizales, Tuberaceae]

Derivation: *Tuber*, Latin, "tuber," referring to its swollen, fleshy, potato-like underground fruitbody; *lyonii*, epithet in honor of H.L. Lyon, who discovered this species in 1903 near Minneapolis, Minnesota.

Deprecated synonym: *Tuber texense* Heimsch, *Mycologia* 50(5):657 (1959) [1958][1]

Fruitbody (ascocarp): Hypogeous, stipe lacking, subglobose to irregularly lobed with deep furrows (Figs. B & C); 0.75–5.5 cm (⅓–2¼ in) across; peridium 0.5–1 mm thick, surface smooth, furrows scabrous and roughened (Fig. C, inset), yellowish-brown to orangish-brown (Fig. B) or reddish-brown (Fig. C).

Spore mass (gleba): Greyish when young, becoming light pinkish-grey and marbled with narrow white convoluted veins and irregular areas when mature (Fig. A); odor pungent.

Chemical staining: Nonreactive in KOH and Melzer's reagent.

Microscopic: One to four spores (ascospores) per ascus, spores ranging from 19–37 × 15–24 µm excluding ornamentation, the size decreasing with the number of spores in an ascus; oval-elliptical to occasionally subglobose, with a small apiculus; *spores densely covered with minute to prominent spines up to 2–4 µm tall connected by low ridges to form a partial to complete reticulum*; spores yellowish-brown to golden-brown, staining light brown in Melzer's reagent.

Occurrence: Ectomyccorhizal, especially with the Pecan (*Carya illinoinensis*), other hickories (*Carya* spp.), and oaks (*Quercus* spp.), also with hawthorns (*Crataegus* spp.) and American Basswood (*Tilia americana*); fruitbody hypogeous (buried underground) in disturbed soils with limited plant litter and high pH (alkalinity) such as those found in orchards; solitary to gregarious; widespread and common in eastern North America from northeastern Mexico (Nuevo León, Tamaulipas) north to New Mexico east to Florida, north to Ontario and Quebec, and from the Rocky Mountains to the Atlantic Coast.

Edibility: A choice edible when mature. Much sought after, fetching commercial harvesters $160 per pound. **WARNING:** Truffles harvested from pecan orchards treated with the pesticide aldicarb (Temik™) may be extremely toxic!

Comments: Probably representing a complex of two or more species ranging from Ontario south into northeastern Mexico and from the Great Plains to the Atlantic Coast,[2] *Tuber lyonii* was first reported from Georgia in 1989.[3] University of Georgia professor Timothy Brennemann, one of the discoverers of *T. lyonii* in Georgia, publicized it as a possible commercial crop for the state. It has a unique flavor that many restaurants in Savannah and South Georgia find extremely appealing for culinary use. Considered a white truffle like the renowned *Tuber magnatum* of Europe, it has a flavor and texture all its own. Its relatively large size, nutty odor, and light-colored outer tissue make this species relatively easy to identify. These truffles are largely dependent upon mycophagous insects and mammals for spore dispersal. As they mature, many of them release potent odors that mimic mammalian reproductive pheromones to attract these animal vectors.

The genus *Tuber* can be divided into various clades on the basis of the ascospores, which can exhibit various combinations of the following ornamentations: **alveolate** (deep honeycombed cavities), **reticulate** (a network of low ridges), and **spinose** (prominent spiny projections). Some 34 species of *Tuber* have been described for North America, most of them in the Far West.

This is the only species of *Tuber* in the Southeast with spinose ascospores (Fig. D below); all other species have alveolate (hollow honeycombed) spores. The presence of spinose spores and its close association with Pecan trees immediately identify this species and distinguish it from all others.

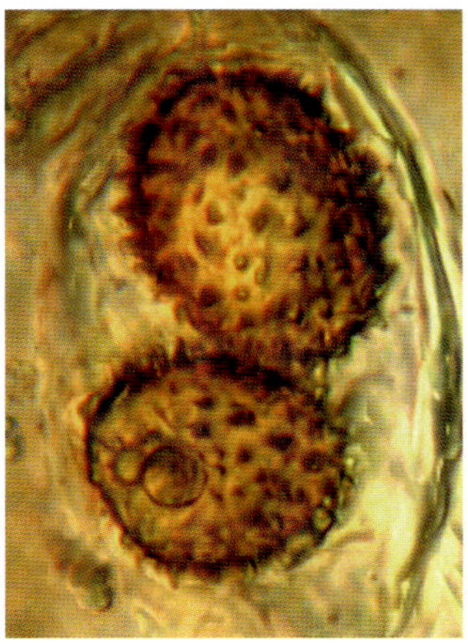

D

Spinose spores of *T. lyonii*. (Photo: Dr. Matthew Smith, Department of Plant Pathology, University of Florida.)

The somewhat smaller (0.5–1 cm [¼–½ in]) Acorn Truffle, *Tuber separans* Gilkey, is associated with both oaks and conifers in the eastern United States and the Pacific Northwest. In our area it can be distinguished from *T. lyonii* by the earthy-brown color of its ascocarp, the inconspicuous veins of its gleba, and its much larger spores (48–58 × 40–56 µm), which lack the spines of *T. lyonii*. The peridium is very thin (0.2–0.26 mm), and the surface of the ascocarp is very minutely verrucose (warty). It is not common but has been reported from Great Smoky Mountains National Park in Tennessee,[4] as well as from an oak forest at an altitude of 1,400 m (4,600 ft) in the Sierra Madre Oriental in northeastern Mexico (state of Nuevo León) associated with the Loquat Leaf Oak (*Quercus rysophylla*).[5]

A Mexican species, *Tuber mexiusanum* Guevara, Bonito & Cázares, is found in the eastern United States (Tennessee, North Carolina, and Georgia) along streams and in bottomlands associated with Eastern Cottonwoods (*Populus deltoides*). Its ascospores are 15–50 × 12–36 µm, and the surface of the ascocarp has 4–7 warts per mm. Though associated with *P. deltoides* in the United States, it is ectomycorrhizal with oaks (*Quercus* spp.) and Texas Madrone (*Arbutus xalapensis*) in north-central and northeastern Mexico (the states of Coahuila and Tamaulipas).[6]

A large truffle, 2–7 cm (¾–2¾ in) across, *Tuber canaliculatum* Gilkey, is found from Quebec and Ontario south to North Carolina and Kentucky associated with Eastern White Pine (*Pinus strobus*); it is also reported from the state of Nuevo León in northeastern Mexico.[4] It is distinguished by its brick-red, reddish-brown, to reddish-orange ascocarp, which is covered by pyramidal warts with scattered angular papillae (1–2 per mm), usually with deep furrows and patches lined with yellow mycelium;[2] thicker peridium (0.5–1.5 mm wide); brownish-grey to dark slaty-grey (when mature) gleba with whitish veins; and larger, globose, dark brown, alveolate (honeycombed) ascospores, which lack the spiny projections of *T. lyonii*, (30–)45–62(–70) × 45–52 µm (excluding the 4–6 µm honeycombed ridges).[7]

Tuber shearii Harkness[8] is also associated with Eastern

White Pine. It is a northern species that has been reported as far south as North Carolina. Its ascocarp is smooth and creamy-buff with white furrows. Its globose to subglobose spores are alveolate and (28–)32–49(–56) µm in diameter (Fig. E at right).

Besides *T. lyonii*, only two other North American species of *Tuber* have spinose ascospores and thus belong to the Rufum Clade: *T. candidum* Harkness and *T. quercicola* J.L. Frank, D. Southworth & Trappe; both are found in western North America. Other members of this clade occur in Europe and Asia (China).

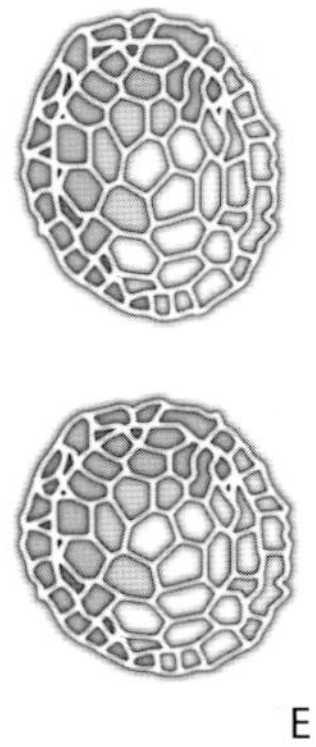

E Alveolate spores of *T. shearii*. (Drawing based on one by Dr. C.L. Shear and published by W.A. Murrill, 1920, *Mycologia* 12[3]:157.)

Key to the Species of Tuber from the Southeastern United States

1. Associated with Pecan; ascospores with spinose projections (Fig. D) connected by low ridges forming a partial to complete reticular network . . . . . *T. lyonii*

1′. Associated with oaks, cottonwoods, or pines; ascospores alveolate, spinose projections lacking (Fig. E) 2

2. Associated with Eastern Cottonwood . . . . . *T. mexiusanum*

2′. Associated with oaks or Eastern White Pine 3

3. 2 cm (¾ in) or less across, ascocarp dull yellowish-brown or earthy-brown, smooth or with very minute scales 4

3′. 2–7 cm (¾–2¾ in) across, ascocarp reddish-brown to reddish-orange, covered by conspicuous polygonal warts . . . . . *T. canaliculatum*

4. Ascocarp 1 cm (½ in) or less in diameter, surface creamy-white to creamy-buff with white furrows; associated with Eastern White Pine . . . . . *T. shearii*

4′. Ascocarp 2 cm (~¾ in) or less in diameter, surface earthy-brown, covered with very minute scales, peridium very thin (~¼ mm); associated with oaks and conifers . . . . . *T. separans*

Other Truffles in Our Area

Besides Tuber, other true truffles are also found in our area; all belong to the order Pezizales, but they represent various families.

1. The Giant Imaia, *Imaia gigantea* (Imai) Trappe & Kovács[9]—the sole species in a monotypic genus in the morel family (Morchellaceae)—also occurs throughout the Appalachian Mountains from Pennsylvania to North Carolina and Tennessee, including in Great Smoky Mountains National Park.[10] Like *Tuber lyonii*, this species is edible and has also been harvested commercially but on a much more limited scale. *Imaia gigantea* is distinguished from *T. lyonii* by its larger (3.5–6 [up to 10 × 15] cm [1½–2½ (up to 4 × 6) in]), warty to papillate (not smooth), and pale yellow to brownish-orange fruitbody. Its spores are larger (35–70 µm, including the epispore), globose to subglobose, and multinucleated; they lack

ornamentation and are encased in a thick, amorphous, mucilaginous epispore.[11]

2. *Hydnobolites californicus* E. Fischer and *Pachyphloeus citrinus* Berkeley & Broome, both in the family Pezizaceae, are widely distributed in North America from the Pacific Coast across the northern tier of states and as far south as Tennessee. Both also occur in Europe.

3. *Hydnotrya cubispora* (E.A. Bessey & B.E. Thompson) Gilkey ranges across northern North America from coast to coast and from Alaska as far south as Tennessee. This truffle belongs to the family Discinaceae and is thus related to the false morels of the genus *Gyromitra*. It is distinguished by its multiple interior, pale grey, and highly convoluted chambers formed by complex infolding of the peridium. Its spores are 30–60 × 25–36 µm (including the epispore), subglobose, broadly ellipsoid, or even keg-shaped or cuboidal (especially when still inside the asci), with a rugose surface, and are initially clear, becoming brown at maturity.

1. J.M. Trappe, A.M. Jumpponen, & E. Cázares, 1996, *Mycotaxon* 60:365–372.

2. J.M. Trappe, 2009, The hunted: Commercially attractive truffles native to North America, *in Les champignons forestiers comestibles à potentiel commercial,* Biopterre, ACCHF, Université Laval, CEF, RNC, 1–5.

3. As *Tuber texense*; *fide* R.T. Hanlin, M. Wu, & T.B. Brenneman, 1989, *Mycotaxon* 34:387–394.

4. D.H. Linder, 1941, *Mycologia* 33:574.

5. E. Cázares, J. García, J. Castillo, & J.M. Trappe, 1992, *Mycologia* 84:347–349, figs. 40–43 (as *Tuber rufidum* var. *nitidum*, misapplied name); G. Guevara, G. Bonito, & E. Cázares, 2013, *Revisión del género* Tuber *(Tuberaceae: Pezizales) de México, Revista Mexicana de Biodiversidad* S39–S49.

6. G. Guevara, G. Bonito, & E. Cázares, 2012, *Mycologia* 105(1):204.

7. 40–48 × 48–56 µm in Gilkey's original species description (see note 4).

8. In W.A. Murrill, 1920, *Mycologia* 12(3):158.

9. The basionym is *Terfezia gigantea* Imai (1933); a deprecated synonym is *Picoa pachyascus* M. Lange (1956); *fide* J.M. Trappe & W.J. Sundberg, 1977, *Mycologia* 69:435; J.M. Trappe & G.M. Kovács, 2008, *in* G.M. Kovács et al., *Mycologia* 100(6):930–939.

10. *Imaia gigantea* has a disjunct distribution in eastern North America and Japan. Such disjunct eastern Asian/eastern North American distributions are seen in several fungal, as well as in various plant and animal groups. This likely represents the remnants of the former continuous mesophytic forest that existed over much of the Northern Hemisphere, including the Arctic Region during the Miocene epoch, and that was subsequently fragmented into its isolated and disjunct remnants in eastern Asia and eastern North America.

11. The structure of this epispore is apparent under TEM (transmission electron microscopy); under the light microscope, the sinuous channels of the epispore can appear as very minute spines in face view (Trappe & Kovács, 2008, op. cit.).

Milky False Truffle

Zelleromyces cinnabarinus

Photo and inset: Dan Molter.

***Zelleromyces cinnabarinus* Singer & A.H. Smith,** *Mem. Torrey Bot. Club* 21(3):19 (1960)
[Basidiomycota, Agaricomycotina, Agaricomycetes, *incertae sedis*, Russulales, Russulaceae]

Derivation: *Zelleromyces*, *Zeller* prefix is in honor of Sanford M. Zeller (1884–1948), a pioneer specialist in the basidiomycetous "false truffles," and *-myces*, Greek, "mushroom," "fungus"; *cinnabarinus*, Latin, from *cinnabaris*, a rich reddish-brown color, referring to the color of the surface and peridium.

Deprecated synonym: *Zelleromyces ravenelii* (Berkeley & M.A. Curtis) Singer & A.H. Smith, *Mem. Torrey Bot. Club* 21(3):20 (1960); *fide* S.L. Miller & Lebel, *Mycotaxon* 72:16 (1999); Desjardin, *Mycologia* 95(1):153 (2003)[1]

Fruitbody (basidiocarp, gastrocarp): 1–3.3 cm in diameter by 1–2.5 cm tall (½–1⅓ × ½–1 in), subglobose or ovoid to somewhat flattened or vaguely irregularly sequestrate (rounded), potato-shaped, sometimes lobed from the fusion of two or more basidiocarps; with prominent, knob-like basal attachment with a tangled knot of rhizomorphs; surface dry, tomentose when young, becoming smooth with age, light brown, brown, cinnamon or cinnamon-brown to various shades of reddish-brown.

Peridium: Very thin, concolorous with surface, merging indistinctly with the glebal flesh.

Spore mass (gleba): Consisting of an anastomosing network of irregular pale buff to pale pinkish-buff plates enclosing irregular, empty chambers, unchanging when injured; basal columella rudimentary to absent; lactiferous hyphae throughout gleba, fresh material exuding copious amounts of white, unchanging latex when cut (inset above); odor pungent; taste not distinctive.

Chemical staining: Surface of fresh gastrocarp very pale olive-green with KOH; intermeshed context (flesh) and glebal plates clear in KOH and pale yellow in Melzer's reagent.

Microscopic: Spores 12–17.5(–19) × (11–)12.5–15.5 µm (including ornamentations), averaging 15–14.6 µm,[2] subglobose to ovoid, surface prominently ridged and reticulate, with scattered spinose projections, with broad, peg-like, symmetrical hilar appendage; ornamentation amyloid (staining blue-black) in Melzer's reagent, clear in KOH; basidia mostly 2-spored; clamp connections lacking.

Occurrence: Thought to be mycorrhizal with pines (*Pinus echinata* Miller, *P. virginiana* Miller, and *P. taeda* Linnaeus); scattered or clustered underground beneath pines in disturbed areas; widespread and common throughout the Southeast, Mississippi Valley from Louisiana to southern Illinois; late fall–early winter.

Edibility: Not known.

Comments: This rounded, underground, truffle-like mushroom is readily recognized by its intricately anastomosing (alternately joining and separating) glebal plates that produce copious amounts of unchanging white latex when cut and its association with pines. It is unlikely to be confused with any other truffle-like species in our area.

The genus *Zelleromyces* Singer & A.H. Smith [type species, *Z. cinnabarinus*] was proposed to include those hypogeous (underground) members of the Russulales without stipes that have dense masses of latex-producing hyphae that bleed profusely when cut. However, the generic arrangement of the gastroid Russulales (*Elasmomyces*, *Gymnomyces*, *Martellia*, *Zelleromyces*) is still very much in flux, and the delimitations of the various genera are unclear.[3] Like most members of the genus *Zelleromyces*, *Z. cinnabarinus* is statismosporic rather than ballistosporic; that is, the spores are not forcibly discharged as they are in *Russula*, *Lactarius*, and related genera.[4]

Although the combination has not yet been officially proposed, several authors have suggested that the genus *Zelleromyces* Singer & A.H. Smith should be synonymized with the genus *Lactarius* Persoon.[5]

The genus *Rhizopogon* is also ectomycorrhizal with pines but produces no latex.

1. *Fide* S.L. Miller & T. Lebel, *Mycotaxon* 72:16 (1999); D.E. Desjardin, 2003, *Mycologia* 95(1):153. Some authors have suggested that because of publication priority this taxon should be called *Z. ravenellii*. However, *Z. cinnabarinus* is the type species for the genus *Zelleromyces*, and *Z. ravenellii* is a heterotypic (subjective) synonym.
2. W.J. Sundberg & J.M. Trappe, 1975, *Mycologia* 67(6):1176–1181. Spores in the type material from Louisiana were reported as somewhat more elliptical (14–17 × 11–13 µm) than in the material from southern Illinois.
3. See D.E. Desjardin, 2003, *Mycologia* 95(1):148–155 for a discussion of this.
4. Only *Z. rogersonii* Fogel & States from the Wassuk Mountains of western Nevada has putatively ballistic (exhibiting an oblique hilar scar) spores (R. Fogel & J. States, 2001, *Mycotaxon* 80:321–326).
5. A. Verbeken & J. Nuytinck, 2013, Not every milkcap is a *Lactarius*, *Scripta Botanica Belgica* 51:162–168.

Earthstar, False Truffle

Scleroderma polyrhizum

A

B

C

***Scleroderma polyrhizum* (J.F. Gmelin) Persoon**, *Syn. Meth. Fung.* (Göttingen) 1:156 (1801)
[Basidiomycota, Agaricomycotina, Agaricomycetes, Agaricomycetidae, Boletales, Sclerodermataceae]

Basionym: *Lycoperdon polyrhizum* J.F. Gmelin, *Systema Naturae*, 13th ed., 2(2):1464 (1792)

Derivation: *Scleroderma*, Latin, "thick, hardened skin"; *polyrhizum*, Latin, "with many roots."

Fruitbody: Potato-shaped, often resembling a round stone (Fig. A); rind-like peridium tough, leathery, subglobose to globose, 5–13 cm (2–5¼ in) in diameter; whitish peridium (rind) 5 mm (¼ in) thick; growing underground before partially emerging as it matures; surface initially whitish, becoming fawn-colored or light brownish; completely sessile base with whitish or cream-colored rhizomorphs (Fig. B). Mature peridium becoming deeply fissured or scaly before splitting at apex to form several star-like rays, with a blackish globular mass of spores at center (Fig. C).

Spore mass (gleba): Black to purplish-black, marbled throughout with whitish to cream-colored threads, at first hard, later becoming powdery, black, or sepia-brown; exposed when peridium splits apart.

Staining: Surface of fruitbody negative or faintly yellowish with KOH.

Microscopic: Spores 6–11 (up to 14) µm (excluding ornamentation), round to egg-shaped, thick-walled, ornamentation subreticulate, with spines up to 0.8 µm high partly connected by low ridges (Fig. D). Clamp connections present.

Occurrence: Solitary to clustered in open areas, along roads, and in open areas in deciduous woodlands; prefers sandy soil; late summer into winter. Generally growing belowground until top of spore case emerges aboveground. Widespread in North America, Europe, Asia.

Edibility: Inedible; many species of *Scleroderma* cause gastrointestinal upset.

Comments: The thick peridium of *S. polyrhizum* distinguishes it from all other *Scleroderma* species in our area except *S. citrinum* Persoon. However, *S. citrinum* is typically ochre-yellow with irregular warts, the surface stains dark red with KOH, and the spores are released through a gaping ragged apical hole; star-shaped rays are not formed, as with *S. polyrhizum*. The spores of *S. citrinum* vary from spiny to mildly or strongly reticulate (Fig. D). *Scleroderma cepa* (Vaillant) Persoon, *S. albidum*,

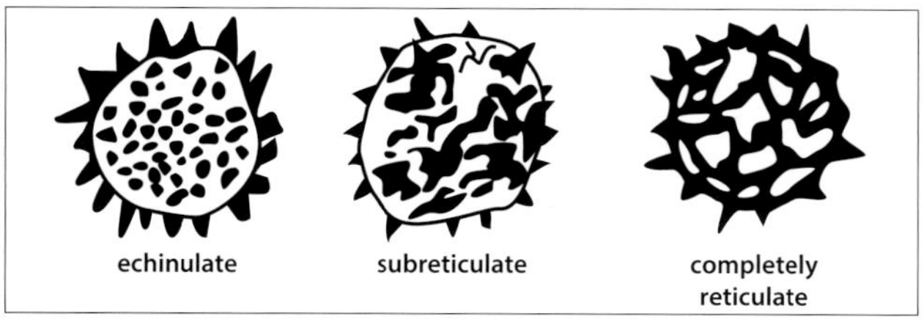

D

and *S. texense* Berkeley have a much thinner peridium, and the spores are echinulate (spiny) rather than subreticulate. The gleba of *S. cepa* is mouse-grey, the peridium is scaly to coarsely cracked without the raised warts seen on *S. citrinum*, and the basal rhizomorphs sometimes form a conical pseudostipe. The whitish to pale olive-brown peridium of *S. texense* is covered with small flat or pyramidal scales and stains yellowish with KOH. The peridium of *Scleroderma bovista* Fries is thin (1–1.5 mm thick), the spores are 11–14 µm in diameter (excluding ornamentation) and are completely reticulated with ornaments 1–2 µm high, and the rhizomorphs form a short pseudostipe. The surface immediately stains red with KOH when fresh and reddish-brown to yellowish-brown when dried. Except for *S. polyrhizum* and *S. cepa*, which open up into star-like rays, all of these species open by rupturing irregularly at the apex.

The prominent basal attachment with its rhizomorphs and purplish-black to black gleba that becomes powdery distinguish the genus *Scleroderma* from the truffle genera *Tuber* and *Imaia*. The dark gleba and the lack of latex-releasing hyphae delineate *S. polyrhizum* from the Milky False Truffle, *Zelleromyces cinnabarinus*.

Stinkhorns and Stinky Squids

(Order Phallales, Family Phallaceae)

The **order Phallales** contains three families:
(1) **Phallaceae** Corda[1]—the stinkhorns, stinky squids, and their relatives and the subject of this section;
(2) **Gomphaceae** Donk—in addition to 12 other genera, this family contains the coral genus *Ramaria* Fries *ex* Bonorden (with a number of species in our area) and the gomphoid-cantharelloid genus *Gomphus* Persoon—*G. clavatus* (Persoon) Gray occurs in our area;
(3) **Claustulaceae** K. M. Curtis, an obscure family with four Australasian genera and very few species.

The Phallaceae are gasteroid fungi; their spores develop within a gelatinous matrix called the **gleba** inside an egg-shaped fruitbody surrounded by a leathery **peridium**. As the fruitbody (**receptacle**) grows it eventually ruptures the peridium and develops into a single column-like stalk with or without an expanded head at the tip, or, depending on the species, it may develop arms that closely resemble squid-like tentacles. The upturned peridium remains as a sac-like volva enclosing the base of the stalk. The gleba is carried upward with the expanding stalk and is discharged through an apical pore and coats the outer surface of the head in columnar genera like *Phallus* and *Mutinus* or the inner surfaces of the arms in the stinky squids like *Clathrus* and *Pseudocolus*.

The gleba is a foul, stinking, olive-greenish-brown slime that attracts flies, carrion beetles, and other insects. These insects are attracted by the smell of decomposing flesh and rotten eggs, which is due in part to compounds like methylmercaptan (CH_3SH) and hydrogen sulfide (H_2S). These insects eat the gleba along with its contained spores, and spores also adhere to the insects' bodies as they move over the slimy mess. These insects then transport these spores over sometimes long distances before depositing them in new locations. Some spores are also released in the insects' feces.

Several types of mushrooms emerge from eggs that look

A, Cross-section of stinkhorn (*Phallus*) egg; B, cross-section of *Amanita* egg; C, cross-section of puffball (*Calvatia*) egg; c = cap, g = gleba, l = lamellae (gills), p = peridium, r = receptacle, s = stipe, uv = universal veil. (Photos: A, Stanislaw Szydlo; B, John Plischke III; C, Ron Pastorino.)

much alike. But a cross-section reveals clear differences between them, as shown in the photos on the previous page. In Figure A the stinkhorn receptacle develops within the greenish gleba inside the egg-shaped peridium. In the *Amanita* egg in Figure B, the developing cap, gills, and stipe of the mushroom-to-be can be seen, all enclosed by a universal veil. In the puffball in Figure C, no internal structures of any sort are visible, only the undifferentiated gleba and the enclosing peridium.

1. The families Clathraceae Chevallier and Lysuraceae Corda are deprecated synonyms of the Phallaceae.

Key to the Stinkhorns Typically Encountered in Our Area

1. Fruitbody columnar, with a single, unbranched tip; when present, gleba covering the outside of the tip 2

1′. Fruitbody topped with tentacle-like arms; when present, gleba contained inside the branches 3

2. Fruitbody topped with a distinct head, resembling an erect human penis (Genus *Phallus*) 4

2′. Fruitbody without a distinct head, resembling a dog's penis (Genus *Mutinus*) 8

3. Arms atop fruitbody arising from a single column-like basal stalk, arms separating at maturity, gleba covering distal inner half of the arms . . . . *Pseudocolus fusiformis*

3′. Arms atop fruitbody, each arising separately from the base, arms remaining fused in maturity, gleba restricted to the distal one-fourth of the arms . . . . *Clathrus columnatus*

4. Mature fruitbody with a prominent net-like or latticework indusium (skirt) hanging from beneath the head . . . . *Phallus duplicatus*

4′. Mature fruitbody lacking a skirt-like indusium 5

5. Stalk white, head not easily separated from stalk 6

5′. Stalk orange to scarlet, head easily separated . . . . *P. rubicundus*

6. Bare head greyish, surface smooth, lacking pits . . . . *P. ravenelii*

6′. Bare head blackish, deeply ridged and pitted 7

7. Fruitbody arising from a white volva . . . . *P. impudicus*

7′. Fruitbody arising from a pinkish or lavender volva . . . . *P. hadriani*

8. Stalk pinkish-rose to rosy-red, conspicuously pitted, gleba restricted to upper one-third or less of fruitbody . . . . *Mutinus ravenelii*

8′. Stalk orange to orangish-red, minutely pitted, gleba occupying upper one-third to one-half of fruitbody. . . . . . . *M. elegans*

Ravenel's, Common, and Devil's Stinkhorns

Phallus ravenelii, P. impudicus, P. hadriani, and *P. rubicundus*

A B C D

A, *Phallus ravenelii*; B, *P. impudicus*; C, *P. hadriani*; D, *P. rubicundus*. (Photos: A, W. Light; B, Jörg Hempel; C, Nathan Wilson; D, John S. Harper.)

Phallus ravenelii **Berkeley & M.A. Curtis**, *Grevillea* 11(57):33 (1882)

Phallus impudicus **Linnaeus**, *Sp. Pl.* 2:1178 (1753)

Phallus hadriani **Ventenat**, *Mém. Inst. Nat. Sci. Arts* 1:517 (1798)

Phallus rubicundus **(Bosc) Fries**, *Syst. Mycol.* (Lundae) 2(2):284 (1823)

Basionym: *Satyrus rubicundus* Bosc, *Mag. Gesell. Naturf. Freunde, Berlin* 5:86 (1811)
[Basidiomycota, Agaricomycotina, Agaricomycetes, Phallomycetidae, Phallales, Phallaceae]

Derivation: *Phallus*, Greek, *phallos*, "penis"; *ravenelii*, Latin, epithet in honor of the American mycologist Henry William Ravenel (1814–1887); *impudicus*, Latin, "shameless, lewd"; *hadriani*, Latin, epithet in honor of the Dutch botanist Hadrianus Junius (1512–1575); *rubicundus*, Latin, "red," "reddish," referring to the color of the stalk.

Fruitbody: All species—initially ovoid, egg-shaped; mature fruitbody upright, cylindrical, conspicuously pitted, hollow, spongy columnar stalk (receptacle) arising from a cup-like volva attached to substratum by rhizomorphs and bearing a swollen or expanded bell-shaped, conical, or ovoid head, typically with a terminal opening (Figs. A & B), strongly resembling the glans of an erect human phallus; head initially coated with a slimy, malodorous, dark olivaceous to blackish-brown spore mass (gleba), this soon eroded away by insect activity. ***P. ravenelii***—fruitbody up to 20 cm (8 in) tall; head bell-shaped, smooth, lacking edges or pits, tipped with open pit surrounded by a white ring-like circle, head initially covered with foul-smelling greenish-brown to olive-brown gleba; stalk white, covered with small round pits, emerging from a whitish to pinkish-lilac volva (Fig. A). ***P. impudicus*** **and** ***P. hadriani***—fruitbody up to 25 cm (10 in) tall; head ovoid to bluntly conical, deeply ridged and pitted (excavated), thimble-shaped (Fig. C), at first covered by slimy, fetid, olive-green gleba (Fig. B), head resembling that of a morel once gleba is removed (Fig. C); stalk white, covered with round pits of varying sizes, emerging from whitish (*P. impudicus*) or rosy-pink to violet volva (*P. hadriani*), the two species otherwise macroscopically indistinguishable. ***P. rubicundus***—fruitbody 15–20 cm (6–8 in) tall; head fragile, easily separated from stalk, wrinkled, conical, fusiform, or narrowly bell-shaped, scarlet red, but initially covered by a fetid, greyish-olive gleba (Fig. D); stalk very coarsely pitted (Fig. D, inset), orange to scarlet above, tapering toward tip, orangish to pale yellow toward base (Fig. D), emerging from whitish volva; a thin, diaphanous, reddish-orange, detachable, membranous veil may or may not be present.

Microscopic: All species—spores cylindrical, smooth, clear, 3–4.5 × 1–2 µm.

Occurrence: Saprobic, solitary, scattered, or in large groups. ***P. ravenelii***—on or near decaying logs, stumps, woody debris, August–April, widespread in the Southeast, by far our most commonly encountered stinkhorn; ***P. impudicus*** and ***P. hadriani***—associated with decomposing tree stumps and roots in sandy soil and cultivated areas, late summer–winter, widespread, occasional; ***P. rubicundus***—in sandy soils, gardens, lawns, and wood mulch, widespread but occasional in Southeast, spring–winter. All four species widely distributed on various continents.

Edibility: Edible when immature (in the "egg" stage), considered a delicacy in some cultures, but the offensive odor discourages most folks from trying these.

Comments: The foul-smelling spore mass coating the tips of all stinkhorns attracts carrion- and offal-loving flies and other insects. The slimy mass adheres to their bodies and is dispersed when the insects fly away. Some of these insects also feed directly on the gleba and pass the spores out in their feces.

Phallus impudicus and *P. hadriani* are morphologically almost indistinguishable, but *P. hadriani* arises from a pinkish to lavender volva; that of *P. impudicus* is white.

Stinkhorns in the genus *Phallus* Junius *ex* Linnaeus are distinguished from those in *Mutinus* Fries by having a clearly differentiated head or cap, which is initially covered by the gleba. In *Mutinus* species (the dog stinkhorns), the gleba simply coats the upper, tapered portion of the stalk, although the expanded spore mass sometimes superficially resembles a head in *M. ravenelii*. *Phallus rubicundus* is often confused with *Mutinus* species, but the fragile head is easily separated from the stalk. The skirt-like veil occasionally seen in *P. rubicundus* and even less often in *P. ravenelii* never occurs in any *Mutinus* species.

The Netted Stinkhorn, *Phallus duplicatus* Bosc, is immediately distinguished by its prominent whitish to yellowish net-like or lace-like indusium (skirt) that covers the top 2½ or so inches of the stalk, although this may become concolorous with and adhere to the stalk in older specimens. A vestigial veil is rarely seen in *P. rubicundus* or *P. ravenelii*. See the following account.

Netted Stinkhorn, Veiled Stinkhorn

Phallus duplicatus

A

B

Photos: A, Walt Sturgeon; B, John Plischke III.

Phallus duplicatus **Bosc**, *Mag. Gesell. Naturf. Freunde, Berlin* 5:86 (1811)
[Basidiomycota, Agaricomycotina, Agaricomycetes, Phallomycetidae, Phallales, Phallaceae]

Derivation: *Phallus*, Greek, "penis"; *duplicata*, Latin, "duplicate."

Homotypic synonyms: *Hymenophallus duplicatus* (Bosc) Nees, *Syst. Pilze* (Würzburg):251 (1816)
Dictyophora duplicata (Bosc) E. Fischer, in Berlese, De Toni & Fischer, *Syll. Fung.* (Abellini) 7:6 (1888)

Fruitbody: Initially egg-shaped; mature fruitbody consisting of an upright, hollow, spongy, columnar stalk (receptacle) arising from a cup-like, flesh to pinkish-brown to faintly purplish–light brown, somewhat wrinkled volva (Fig. B, 2); stalk 10–24 cm tall by 6 cm in diameter at base, tapering to 2.5–3.5 cm at apex (4–9⅔ × 1–1½ in); stalk whitish, surmounted by a distinct, deeply pitted and ridged, irregularly reticulate to alveolate (honeycombed), ovoid, conical (Fig. B), to campanulate (bell-shaped, Fig. A), pinkish-tan head with a whitish, thick-rimmed apical opening (Figs. A & B); entire fruitbody resembling an erect human phallus (Fig. B).

Indusium: A conspicuous whitish to yellowish, net-like skirt (indusium, Fig. A) 3–6 cm (1¼–2½ in) long suspended from beneath base of head (Fig. A), this often collapsing back against the stalk (Fig. B, 1).

Spore mass (gleba): A thick, slimy, greenish-brown to olive-brown, stinking (like rotten carrion) mass (Fig. A, arrow) covering the head; this soon consumed by numerous flies attracted by the smell, leaving behind the bare, pitted head (Fig. B).

Egg: Whitish to flesh-colored, 4–7 cm tall by 4–6 cm in diameter (1⅔–2¾ × 1⅔–2½ in), subglobose to ovoid or slightly flattened.

Microscopic: Spores 3.5–4.5 × 1–2 µm, elliptical to cylindrical, smooth, clear.

Occurrence: Saprobic, solitary or in groups on the ground in hardwood forests, flowerbeds, wood chips, meadows; widespread and occasional to locally fairly common in eastern North America; summer–fall; East Asia and South America.

Edibility: Edible when immature (in the "egg" stage), considered a delicacy in some cultures, but the offensive odor discourages most folks from trying these mushrooms.

Comments: This is the only species of *Phallus* with a prominent indusium found in our area. The closely related and widespread pantropical-subtropical *P. indusiatus* Ventenat does not occur in North America north of about 24° N latitude in Mexico. *Phallus indusiatus* has a much longer indusium of 8–15 cm (3¼–6 in), which reaches to the volva, compared to 2–6 cm (¾–2½ in) for *P. duplicatus*, and which extends no more than halfway down the stalk. *Phallus rubicundus* and even less often *P. ravenelii* very occasionally exhibit a small indusium. But the smooth head of *P. ravenelii* and the deeply pitted bright red stalk of *P. rubicundus* contrast strongly with the deeply honeycombed and ridged head and white stalk of *P. duplicatus*.

Because of the prominent indusium, this species is referred to as *Dictyophora duplicata* (Bosc) E. Fischer in some field guides and online sources. The genus *Dictyophora* Desvaux (1809) was erected for those *Phallus*-like species with an indusium; in 1817 Nees proposed the genus *Hymenophallus* for the same indusium-bearing stinkhorns.[1] However, the wide variability of these veil-bearing stinkhorns, not only the size, color, and form of the veil but even its presence or absence in some cases, has led most mycologists to sink *Dictyophora* and *Hymenophora* into synonymy with *Phallus*.[2]

1. Derivation: *Hymenophallus*, Greek, from humḗn, "skin," "membrane," and *phallus*, "penis"; *Dictyophora*, Greek, from *díktyo*, "net," "network," and *pherōun*, "carry," meaning "carries a net."

2. For example, J.A. Sáen & C.M. Nassar, 1982, *Hongos de Costa Rica: Familias Phallaceae y Clathraceae, Rev. Bio. Trop.* 30(1):41–52.

Little Red and Elegant Stinkhorns

Mutinus ravenelii and *M. elegans*

A B C

A, *Mutinus ravenelii*; B & C, *M. elegans*. (Photos: A, André-Philippe Drapeau Picard; B & C, Teresa Fortenberry.)

***Mutinus ravenelii* (Berkeley & M.A. Curtis) E. Fischer**, in Berlese, De Toni & Fischer, *Syll. Fung.* (Abellini) 7:13 (1888)

Basionym: *Corynites ravenelii* Berkeley & M.A. Curtis [as "*ravenelli*"], *Trans. Linn. Soc. London* 21:151 (1853)

***Mutinus elegans* (Montagne) E. Fischer**, *Syll. Fung.* (Abellini) 7:13 (1888)

Basionym: *Corynites elegans* Montagne, *Syll. Gen. Sp. Crypt.* (Paris):281 (1856)
[Basidiomycota, Agaricomycotina, Agaricomycetes, Phallomycetidae, Phallales, Phallaceae]

Derivation: *Mutinus*, Latin, named after a minor Roman phallic god, Mutinus Titinus; *elegans*, "elegant"; *ravenelii*, epithet in honor of the American mycologist Henry W. Ravenel (1814–1887).

Fruitbody (receptabulum): Both taxa—initially contained within a white ovoid egg with white basal rhizomorphs (Figs. A & B) that rupture and from which a long cylindrical fruitbody emerges that resembles a tumescent canine penis; ***M. ravenelii***—2–9 cm tall by 1.2–1.5 cm thick (¾–3⅔ × ½–⅔ in), more or less straight (Fig. A). ***M. elegans***—10–18 cm tall by 1.2–2.5 cm thick (4–7¼ × ½–1 in), a typically curved fruitbody tapering gradually to a bluntly pointed tip with a terminal opening (Fig. B).

Fertile surface: *M. ravenelii*—fertile surface clearly demarcated from the stalk by a slight constriction, usually somewhat enlarged and restricted to the upper one-third or less of the fruitbody (Fig. A). ***M. elegans***—fertile surface not demarcated from stalk, covering upper one-third to one-half of the fruitbody (Fig. B).

Stalk: Both taxa—cylindrical, hollow, spongy, arising from a white sac-like volva (Fig. A). ***M. ravenelii***—deep rose-red or red above and beneath gleba, paler toward the base (Fig. A); covered with large open pits connecting to the hollow center (Fig. A, inset). ***M. elegans***—orange (Fig. C), pinkish-orange to pinkish-rose above (Fig. B) and beneath gleba, paler below (Fig. B); surface minutely papillated and pitted (Fig. B, inset).

Spore mass (gleba): Both taxa—covering fertile surface, dull greenish, olive-green, or olive-brown, slimy and foul smelling, typically covered with and removed by flies and other insects.

Microscopic: Both taxa—spores ellipsoid to smooth, yellowish-green; ***M. ravenelii***—spores 3.5–5 × 1.6–2.2 µm; ***M. elegans***—spores 4–7 × 2–3 µm.

Occurrence: Both taxa—saprobic; solitary or gregarious on the ground in gardens, lawns, or wood chips or from decayed wood and in roadcuts and hardwood forests; widespread in eastern North America, *M. elegans* being much more common; summer–late fall.

Edibility: Both taxa—like other stinkhorns, edible in the egg stage of development; reported to be very good.

Comments: Of the two species, *M. ravenelii* is distinguished by its smaller size (≤3⅔ in tall), pinkish-rose to rosy-red stalk, larger and more conspicuous open pits covering the surface (Fig. A, inset), and a gleba restricted to the upper one-third or less of the fruitbody. *Mutinus elegans* is larger (¾ in tall), with an orange to orangish-red stalk, its surface is minutely pitted throughout, and the gleba occupies the upper one-third to one-half of the fruitbody.

Stinkhorns in the genus *Phallus* Junius *ex* Linnaeus are distinguished from the genus *Mutinus* Fries by having a clearly differentiated head or cap initially covered by the gleba, which resembles a human glans penis.

Although *M. ravenelii* is widely reported as *M. caninus* in North America by many field guides and online sources, the true Dog Stinkhorn, *Mutinus caninus* (Hudson 1778) Fries (1849), occurs only rarely on this continent, unless *M. ravenelii* proves to be synonymous with *M. caninus*, as some have suggested.[1] In that case, the name *M. caninus* would prevail by reason of priority. A white variety generally known as *M. caninus* var. *albus* Zeller from the Northeast, Upper Midwest, and Pacific Northwest may represent a full species in its own right. The southern extent of its range is unclear.

1. For example, E.A. Burt, 1896, The Phalloideae of the United States. II. Systematic account, *Bot. Gazette* 22(5):381–382. Lloyd strongly disagreed with this idea, noting that the cells are all the same size throughout the fruitbody in *M. ravenelii*; in contrast, in European *M. canina* the cells of the gleba-bearing portion are small and thick-walled, while those of the stalk are larger and thin-walled. On the basis of this distinction, Lloyd authenticated the occurrence of *M. caninus* in eastern North America but noted that it is quite rare here, though quite common in Europe (C.G. Lloyd, 1898, *Mycol. Notes* 1–2:300; C.G. Lloyd, 1902, *Bulletin of the Lloyd Library of Botany, Pharmacy and Materia Medica*, Mycol. Ser. 4, Synopsis of the Known Phalloids [Cincinnati: J.U. & C.G. Lloyd], 2).

Stinky Squid

Pseudocolus fusiformis

A B C

Pseudocolus fusiformis.

***Pseudocolus fusiformis* (E. Fischer) Lloyd**, *Mycol. Writ.* (7):53 (1909)
[Basidiomycota, Agaricomycetes, Phallomycetidae, Phallales, Phallaceae]

Basionym: *Colus fusiformis* E. Fischer, *Neue Denkschr. Allg. Schweiz Ges. Naturwiss.* 32(1):64 (1890)

Derivation: *Pseudocolus*, Latin, from *pseudo*, "false," and *colus*, "distaff" (a staff with a cleft tip), referring to the genus *Colus*, from which *Pseudocolus* was distinguished; *fusiformis*, "spindle-shaped."

Fruitbody: 4–7 cm high (1⅔–2¾ in), consisting of an egg-like volva from which emerges a hollow, columnar basal stalk with 3–4 (usually 3) vertical, wrinkled, tapering arms 2–4 times the length of the stalk; arms initially fused at their tips (Figs. A & B), later separating (Fig. C). The outer face of each arm contains a distinct ridge running the entire length (Fig. B). A malodorous dark green to nearly black slimy mass (the gleba) (Fig. B, arrow) initially fills the space between the arms except near the base, later adhering as distinct patches after the arms separate.

Stipe: 1–3.5 cm tall by 0.5–2.5 cm thick (¼–1 × ¼–1⅖ in), whitish-grey to pale grey, merging into bright orange to red-orange arms (Figs. A–C).

Egg: White to brownish, at first spherical, becoming oblong, arising from white mycelial strands, forming an irregular, tattered volva at base of fruitbody (Figs. A & B). Volva tough, wrinkled, white, whitish-grey, or grey.

Microscopic: Spores variable: 3.5–4.5 × 1.5–2 to 4.5–5.5 × 2–2.5 µm, ovoid to elliptical, smooth, hyaline.

Occurrence: Saprobic. Scattered or in groups on disturbed soil in coniferous or mixed forests, June–October. Widespread in the eastern United States, especially in the Southeast; also reported from Turkey, Japan, Java, the Philippines, Reunion Island, and the Hawaiian Islands. First discovered in Pittsburgh in 1915, thought to be introduced from Southeast Asia.

Edibility: Not recommended; hogs have been reported poisoned by this species.

Comments: The bad smell emanating from this mushroom (and other stinkhorns) from the slimy gleba on the inner surfaces of the arms attracts insects that disperse the spores. This species was first described in 1890 from Reunion Island by Eduard Fischer.

It is distinguished from the similar *Clathrus columnatus* Bosc (Fig. D), which is also common in the Southeast, by the arms arising from a common stem and the gleba covering the distal half or more of the inner surfaces of the arms. In *C. columnatus*, the arms arise separately from the base rather than from a shared stem (see the photo in Fig. D at right of *C. columnatus* taken in South Carolina by Leo Heuser), and the olive-brown gleba is found only in the distal top fourth of the arms under the fused tips. The arms often form a distinct roof-like lattice and, unlike *P. fusiformis*, do not separate in maturity. *Clathrus crispus* (commonly misidentified as *C. ruber*, which does not occur in the East) is fairly common along the Gulf Coast and in Florida. The fruitbody is much larger, 10–15 cm high (4–6 in), and forms an enclosed oval lattice of thick interlaced arms with multiple ovoid to ellipsoid "windows."

D *Clathrus columnatus*. (Photo: Leo Heuser.)

The Boletes

No genus has given me more
trouble than that of the Boleti.
—Elias Magnus Fries

Boletes are grouped together with the gilled mushrooms in the order Agaricales. As such, they are generally characterized by a typical mushroom-shaped fruitbody with (1) a more or less domed to flat or even uplifted and shallowly concave, bilayered pileus (cap); (2) a stout, cylindrical stipe (stalk); and (3) in place of the usual lamellae (gills), a distinct lower spongy layer whose downward-facing surface is covered with pores, each connected to a vertical tube in which the sexual reproductive spores develop. Because of these tubes the boletes are placed in their own family, the Boletaceae. This layer of tubes is known as the hymenophore. One genus represented in our area (*Phylloporus* Quélet) by a single species[1] has secondarily evolved a gill-like hymenium (spore-producing layer) in place of the tubes. Three other genera not represented in the Southeast exhibit misshapen caps and irregularly arranged tubes; they will not concern us here.

The shelf or bracket fungi known as polypores are also characterized by vertically oriented tubes that open to the exterior by pores, but they are typically dry and corky, and the old, hardened, woody fruitbodies often persist for more than a year. In contrast, the boletes are fleshy and moist, and they quickly decompose soon after they reach maturity. Additional distinctions between the boletes and polypores are discussed in the polypore section.

With few exceptions, boletes are ectomycorrhizal with trees and shrubs, generally with a high degree of host specificity; this specificity can be very important in distinguishing between certain bolete taxa. However, the Parasitic Bolete, *Pseudoboletus parasiticus* (Bulliard) Šutara, which ranges southward into our area in the southern Appalachians in North Carolina, is an obligate parasite[2] on the Pigskin Poison Puffball, *Scleroderma citrinum* Persoon.

Depending on the species, the identification of boletes can be either easy and straightforward or difficult to nearly impossible. The color of the cap, pore surface, and stipe can be important features in many boletes, but these can also vary widely between samples and can depend on the age and/or freshness of the specimens. They can range from almost blackish or blackish-grey to varying shades of violet, purple, grey, brown, red, orange, pinkish, or yellow. Some are cream-colored or white, while a great many species are a nondescript brownish or other subdued color. Attempts in the literature and field guides to describe these intergrading variations are rarely helpful. The bruising characteristics of boletes when handled or cut are very useful in some groups, as are the rapidity of onset and duration of the color(s).

Some boletes initially bruise a specific color, which may then fade or transform into another. The same holds true for chemical stains. Some boletes instantly stain a specific color in response to a given chemical, while others produce a more gradual change. Still others exhibit an instantaneous "flash" of one color, which then quickly (or more slowly) changes into another color or a succession of colors. These changes can be very helpful in making a diagnosis. However, many boletes do not exhibit these features; in other cases, a number of close look-alikes all show the same bruising responses.

The dryness, moistness, or sliminess of the cap is an important feature, as is the surface texture, which may be smooth, minutely hairy, velvety, fibrillose, covered with ad-pressed fibers, or scaly. The surface may also be wrinkled or corrugated, cracked, or pitted. The pores can be round, angular, boletinoid (elongated and radially arranged), or meruloid (reticulately ridged like the inner lining of tripe), and their size (how many pores per millimeter) and the thickness of their walls are often important in identification. In most boletes the hymenophore is easily separated from the cap; however, in most species of *Suillus* the tubes can only be torn from the cap with some difficulty. This can be diagnostically important. Taste and odor may not be distinctive, or the taste may be hot and peppery.

Except for *Gastroboletus* (not found in our area), the stipe is central. It may be smooth or tomentose (covered with short, woolly hairs) or exhibit surface grooves and ridges that may be longitudinally striated or form a reticulated network (and may or may not be lacerated and shaggy), or may be covered by raised glandular dots or scabers (elevated scales).

1. See the species account for the Yellow-gilled Bolete, *Phylloporus rhodoxanthus* (Schweinitz) Bresádola.

2. Symbiotic and parasitic relationships can be obligate, meaning one, the other, or both participants can only exist within the relationship, or facultative, meaning that the relationship may be beneficial to one or both organisms but is not obligatory.

Old Man of the Woods

Strobilomyces floccopus, *S. confusus*, and "Strobilomyces dryophilus"

Species in the genus *Strobilomyces*, like most ectomycorrhizal fungi (see the introduction for a discussion of types of mycorrhizal fungi),[1] have historically been reported to associate with a wide variety of hardwood and coniferous host trees. However, this is incorrect. This misunderstanding about their host selectivity is due to taxonomic confusion between very similar species that cannot be readily separated by morphological features.[2] A recent study reported that nearly all *Strobilomyces* species in the western Himalayan Mountains are ectomycorrhizal with fewer than three species of oaks.[3] Indeed, nearly all *Strobilomyces* species are ectomycorrhizal with host trees in the families Fagaceae (beeches and oaks) and Betulaceae (birches) except for one monophyletic lineage that includes *S. confusus*, which also colonizes pines. Reports of *S. floccopus* occurring in association with conifers may thus actually be in error due to confusing *S. floccopus* with *S. confusus* (see the combined account of *S. floccopus* and *S. confusus*).

Strobilomyces confusus is found in Europe, Asia, and North America. It seems likely, therefore, that this taxon exhibits the same broad host selectivity in all three continents. This may be a useful field characteristic to aid in distinguishing between *S. floccopus* and *S. confusus*. That is, if a *Strobilomyces* species is found in association with pines or firs, it is likely to be *S. confusus*. On the other hand, any member of this genus growing with hardwoods in North America could be *S. floccopus*, *S. confusus*, or "Strobilomyces dryophilus." Association with conifers appears to be an evolutionarily derived feature in this genus.[4]

The genus *Strobilomyces* contains three North American species: *Strobilomyces floccopus* (*sensu auct. amer.*),[5] *Strobilomyces confusus*, and "Strobilomyces dryophilus."

1. In contrast to endomycorrhizal fungi, which penetrate the cortical cells of host roots and exhibit high host specificity.

2. H. Sato et al., 2007, *Amer. J. Bot.* 94:1630–1641.

3. V. Pande et al., 2004, *Curr. Sci.* 86:1619–1623.

4. Some isolated populations of *Strobilomyces* are thought to have become mycorrhizal with Pinaceae during Pleistocene glacial regimes (H. Sato et al., 2007, *Amer. J. Bot.* 94:1630–1641).

5. *Strobilomyces floccopus* (Vahl 1797) P. Karsten 1882 is a deprecated synonym of *S. strobilaceus* (Scopoli 1770 *ex* Fries 1828) Berkeley 1851. However, the North American species variously identified as *S. floccopus* and *S. strobilaceus* does not appear to be the same taxon as the European *S. strobilaceus*. It thus represents an unnamed species that is best referred to as *S. floccopus* (*sensu auctore amer.*). For discussions of this, see also R.H. Petersen et al., 2012, *Czech Mycology* 64(2):141–163; M. Kuo, December 2013, *Strobilomyces "floccopus"*: The old man of the woods, retrieved from the MushroomExpert.Com website: http://www.mushroomexpert.com/strobilomyces_floccopus.html.

Key to North American Species of *Strobilomyces*

1. Scales on cap greyish-black to black. 2

1′. Scales on cap dull greyish-pink, pinkish-tan, to various shades of brown; restricted to South; associated with oaks . . . . "Strobilomyces dryophilus"

2. Associated with hardwoods and conifers; spores with cauliflower-like ridges and spines, but no reticulate network. . . . . . . *S. confusus*

2′. Associated only with hardwoods; spores completely covered with a reticulate network of ridges . . . . *S. floccopus*

Strobilomyces floccopus (*sensu auct. amer.*) and *S. confusus*

A

B

C

D

E

F

Photos and micrographs: A–D, William Light; E & F, Dr. Linas Kudzma.

***Strobilomyces floccopus* (*sensu auct. amer.*)** [*non Boletus floccopus* Vahl, *Fl. Danic.* 8:tab. 1252 (1797)]

***Strobilomyces confusus* Singer**, *Farlowia* 2:108 (1945) [Basidiomycota, Agaricomycotina, Agaricomycetes, Agaricomycetidae, Boletales, Boletaceae]

Derivation: *Strobilomyces*, *strobilos*, Greek, "conifer cone," and *myces*, "mushroom," "fungus"; *floccopus*, Latin, from *floccōsus*, "with woolly tufts"; *confusus*, "confusing," reflecting the fact that *S. confusus* is often confused with *S. strobilaceus*.

Cap: Dry, 3–12 (up to 15) cm (1⅛–6 in) across, convex (Fig. A), aging to broadly convex to flat (Fig. B), margin appendiculate with tattered remnants of torn partial veil (Fig. B); whitish to greyish ground color, covered with many flattened or erect grey, purplish-grey, or blackish scales.

Flesh: Whitish, staining reddish or orange-red, gradually turning dark red (Fig. D), then blackish when cut or bruised.

Pore surface: White at first (Fig. C), soon turning grey, then brownish-grey, and finally black. Pores angular (Fig. C), 1–2 mm wide, 1–8 mm deep; unlike most boletes, pores not easily separated from context.

Stipe: Coloration similar to cap, 4–12 cm tall by 1–2.5 cm thick (1⅝–3⅛ × ⅜–¾ in), subequal, occasionally enlarged at base; covered with shaggy remnants of partial veil; ridged above the annulus (Figs. A & C). Flesh bleeding red, then turning black when cut or bruised (Fig. D).

Spore print: Blackish-brown to black.

Staining and bruising: ***S. floccopus***—flesh pale brownish-yellow with NH_4OH, reddish with KOH, and bluish-grey with $FeSO_4$. ***S. confusus***—flesh red, then brown with KOH; bluish-grey with $FeSO_4$. **"S. dryophilus"**—cap surface and flesh pale yellow with NH_4OH, reddish with KOH.

Microscopic: ***S. floccopus***—spores greyish, 9.5–15 × 8.5–12 µm, round to short elliptical, completely covered by reticulum of ridges (Fig. F). ***S. confusus***—spores greyish, 10.5–12 × 9.5–10.5 µm, round to short elliptical, with irregular cauliflower-like ridges and projections but no reticular network (Fig. E).

Occurrence: Both species—June–October, solitary or in groups on the ground, widespread and fairly common in eastern North America; Europe, Asia. ***S. floccopus***—restricted to hardwoods, especially oaks. ***S. confusus***—associated with both hardwoods and conifers.

Edibility: Both species are edible, but not considered choice.

Comments: A number of mycologists have distinguished *S. floccopus* and *S. confusus* on the basis of the very subjective texture and shape of the scales of the cap: "firm" and "erect" vs. "soft" and "downy." However, most current authorities believe that the only definitive distinction between these two species is in the shape and configuration of the spores. However, any *Strobilomyces* found associated with pines has to be *S. confusus*.[1]

On the basis of their genetic studies, Sato et al. (2007) suggested that *S. strobilaceus* in Japan may actually represent as many as seven cryptic species in Asia and that mycorrhizal host specificity is actually much tighter than is currently believed. Different *Strobilomyces* taxa may well prefer different tree species within the family Fagaceae. These authors suggested that these morphologically indistinguishable species might best be separated on the basis of their host tree species.

1. American mushrooms identified as *Strobilomyces floccopus* and *S. strobilaceus* do not appear to be conspecific with European *S. strobilaceus* (Scopoli 1770 *ex* Fries 1828) Berkeley 1851 [*S. floccopus* (Vahl *ex* Fries) P. Karsten 1882 is a deprecated synonym of *S. strobilaceus*], which has had a very confused nomenclatural history. See online Appendix A for a discussion of this. *Strobilomyces confusus* has a clean nomenclatural slate; ironically, its specific epithet reflects the fact that it is easily confused with *S. strobilaceus*.

"Strobilomyces dryophilus"

(No valid scientific name)

A

B

C

D

Photos and micrograph: A & D, Dr. Linas Kudzma; B, Erich Witter; C, Ron Pastorino.

"Strobilomyces dryophilus" *sensu* Cibula & Smith, in Weber & Smith, *A Field Guide to Southern Mushrooms* (Ann Arbor): 76 (1985). **[invalid name]**. See comments in online Appendix A.

Cap, flesh, and stipe: In all respects similar to *S. floccopus* and *S. confusus* except for color. Ground color greyish-pink to pinkish-tan, covered with adpressed or erect darker sepia-brown scales (Figs. A, B, & C). Pore surface initially white (Fig. B), soon becoming grey, then blackish. Bruising and bleeding characteristics as for *S. floccopus* and *S. confusus*. In our experience, this unnamed taxon bleeds faster and brighter red than the other two species.

Chemical staining: Cap surface and flesh stain pale yellow with NH_4OH and reddish with KOH.

Microscopic: Spores greyish, 9–12 × 7–9 µm, subspherical to short elliptical; covered by a complete reticulum; very similar to the spores of *S. floccopus*, but in some spores the reticulation is not complete (Fig. D).

Occurrence: July–December, solitary or in groups on ground under oaks, to which it is restricted; widespread but occasional in eastern North America.

Edibility: Edible, but not choice.

Comments: "Strobilomyces dryophilus" was never validly published and has no status as a scientific name. Nevertheless, most field guides that do include this taxon italicize the binomen and treat it as if the name were valid. It is not. See the comments in online Appendix A.

Some online images misidentified as this species are actually of *Boletellus ananas* (M.A. Curtis) Murrill, which ranges from the Coastal Plain and Gulf Coast of the United States south to Guyana and Brazil. The cap is densely covered with coarsely shaggy and hairy adpressed scales; the young pore surface is yellow, becoming reddish-brown; the spores are larger (17.5–22.2 × 6.4–8 µm) and spindle-shaped, with 12–14 longitudinal ridges connected by minute cross-striae, and the spore print is olive-brown. The cap stains initially yellow and then maroon with KOH and olive-yellow with NH_4OH.

Other Boletes

Yellow-stain Bolete

Boletus curtisii

A

B

C

Boletus curtisii **Berkeley**, *Ann. Mag. Nat. Hist.*, ser. 2, 12:429 (1853)
[Basidiomycota, Agaricomycetes, Boletales, Boletaceae]

Derivation: *Boletus*, Latin, "fleshy-pored fungus"; *curtisii*, specific epithet in honor of M.A. Curtis.

Cap: Bright yellow or orange-yellow (Fig. A), staining paper and fingers yellow; 3–9 cm (1¼–3⅔ in) wide; convex at first (Fig. A), becoming broadly convex to nearly flat; fresh surface smooth, slimy, sticky; margin inrolled, forming yellow band of sterile tissue (Fig. C).

Flesh: Whitish, not bruising when exposed or cut.

Pore surface: Whitish to pale yellowish at first, aging to yellowish-brown, not bruising blue; 2–3 circular to angular pores per mm, tubes 6–12 mm deep.

Stipe: Slimy, 6–15 cm tall by 5–13 mm thick (2⅓–6 × ¼–½ in), nearly equal to expanded base (Fig. B), slimy, somewhat scurfy, whitish near apex, pale yellowish down to base, which is covered with cottony white mycelium (Fig. B); no reticulations, partial veil, or ring.

Bruising/chemical staining: Tissues not bruising blue. **KOH**—pale olive on cap or negative, orangish on flesh; **Fe_2SO_4**—cap negative, greyish on flesh. The slimy cap stains paper and fingers yellow.

Spore print: Olive-brown.

Microscopic: Spores yellowish, smooth, 9.5–17 × 4–6 µm, elliptical to subfusoid-ventricose (subspindle-shaped and somewhat swollen on one side).

Occurrence: Mycorrhizal with hardwoods and conifers; solitary or in scattered groups summer–fall in conifer or mixed woods in eastern and southern North America, often under pines; reported from an unstated locale in Mexico.[1]

Edibility: Edible, but reportedly not appealing.

Comments: The slimy cap and stem of this striking mushroom give it a superficial resemblance to the genus *Suillus*. However, it lacks the radial arrangement of the pores and the glandular dots on the stipe typically found on *Suillus* species. The superficially similar Ornate-stalked Bolete, *Retiboletus ornatipes*, is distinguished from *B. curtisii* by its dry to slightly sticky (never slimy) cap surface that stains reddish-brown (rather than pale olive) with KOH. The stipe is also coarsely reticulate, and the flesh is yellow rather than white.

1. As *Pulveroboletus curtisii*; J.G. Jiménez & F.G Ocañas, 2001, *Conocimiento de los hongos de la familia Boletaceae de México*, *Ciencia UANL* 4(3):336–343, http://eprints.uanl.mx/411/1/conocim_hongos.pdf.

Two-colored Bolete

"Boletus bicolor"

A

B

"Boletus bicolor" Peck, *Ann. Rep. N.Y. St. Mus.* 24:78 (1872) [1871]; *nom. illegit.*, Art. 53.1, homonym [*non Boletus bicolor* Raddi (1806); *non* "Boletus bicolor" Massee (1909), *nom. illegit.*, Art. 53.1, homonym, = *Boletochaete bicolor* Singer (1986)] [Basidiomycota, Agaricomycetes, Boletales, Boletaceae]

Derivation: *Boletus*, Latin, "fleshy-pored fungus"; *bicolor*, "two-colored."

Cap: 4–15 cm across (1⅔–6 in), at first convex, becoming broadly convex to nearly flat; margin entire, but often irregularly notched or incised (Fig. A); surface dry, smooth, often developing cracks; dark red, deep brick-red, purplish-red to rose-red or rose-pink (Fig. A) when young, becoming yellow or reddish-brown with age.

Flesh: Pale yellow, unchanging or slowly and weakly turning blue when injured or cut; odor and taste not distinctive.

Pore surface: Attached to and separated from stipe by a shallow groove, at first yellow, becoming dingy-yellow to olive-yellow with age; pores mostly angular, 1–2 per mm; tube layer shallow, 0.3–1 cm (≤½ in) deep.

Stipe: 4–10 cm tall by up to 3 cm thick (1⅔–4 × 1¼ in), subequal, often swollen at the base (Fig. A); solid, dry, smooth to faintly reticulated toward apex; yellow at apex, rose-red on lower two-thirds (Fig. A).[1]

Spore print: Olive-brown.

Bruising: Pore surface slowly or instantly bruising dark bluish-grey to greenish-blue (Fig. B).

Chemical staining: **$FeSO_4$**—cap blackish, flesh bluish-grey to olive-green; **KOH**—cap negative, flesh pale orange or pale yellow; cap and flesh negative with NH_4OH.

Microscopic: Spores 8–12 × 3.5–5 µm, oblong to somewhat spindle-shaped, smooth, pale brown; nonreactive in Melzer's reagent.

Occurrence: Mycorrhizal with oaks (*Quercus* spp.); solitary or scattered on the ground under oaks; widespread and fairly common throughout eastern North America, summer–late fall; also found in China and Nepal.

Edibility: Edible, but caution advised (see comments).

Comments: The Two-colored Bolete has no legitimate scientific name—an all-too-common conundrum in mycology. The name proposed by Peck and used by nearly all field guides and online sources is a later homonym of *Boletus bicolor* Raddi and is thus illegitimate. *Boletus bicolor* Raddi is restricted to Europe. *Species Fungorum* has synonymized "Boletus bicolor" *sensu* Peck with *Boletus rubellus* Krombholz [as *Xerocomellus rubellus* (Krombholz) Šutara], but this is not accepted by most workers who consider these to be two distinct species.

In our area "Boletus bicolor" (edible) can be confused with several other similar species with reddish caps and bicolored yellow-above and reddish-below stipes, some of which are poisonous. These include *B. campestris* A.H. Smith & Thiers, *B. rubellus* Krombholz, *B. sensibilis* Peck, and *B. subreticulatus* A.H. Smith & Thiers.[2] They can be at least tentatively distinguished as follows:[3]

"Boletus bicolor" (edible): Found on the ground under oaks. The flesh does not change color or bruises blue only slowly and has no distinctive odor. The stipe is often swollen at the base and does not have a yellow basal mycelium. Staining characteristics: $FeSO_4$—cap blackish, flesh bluish-grey to olive-green; KOH—cap negative, flesh pale orange or pale yellow; NH_4OH—cap and flesh negative.

Boletus campestris (edible): The stipe lacks reticulations and is subequal (rarely tapering) with a yellow basal mycelium. It is found fruiting on lawns, in parks and open woods, and along roadsides. The spores are larger, 11–15 × 4.5–7 µm, and yellow-ochre (rather than pale brown). The cap is negative with $FeSO_4$ or pale orangish to dull olive, flesh negative; with KOH—cap yellow, flesh dull orange; with NH_4OH—cap negative, flesh negative (bluish bruising stains removed).

Boletus rubellus (edibility not known): The stipe is subequal or tapered at the base, the flesh is sometimes reddish-orange in the base of the stipe, and the spores are somewhat longer, 10–14 × 3–5 µm. With $FeSO_4$—cap blackish, flesh grey to olive-green; KOH—flesh dull orange to brownish.

Boletus sensibilis (poisonous, causes severe gastrointestinal upset in some individuals): The cap is pinkish-red, often with yellow spots at the margin. The flesh quickly/instantly bruises blue, then fades to whitish and smells like curry. With $FeSO_4$—cap greenish-grey and flesh yellow-orange to yellow-brown; with KOH or NH_4OH—cap and flesh instantly yellow.

Boletus subreticulatus (edible): The upper stipe exhibits more prominent reticulations than the occasional very faint reticulations seen in "Boletus bicolor." With $FeSO_4$—flesh negative; NH_4OH—cap negative.

1. Our samples exhibited a white mold at the base of the stipe that partially obscured the rose-red color (Fig. A).

2. Considered a variety of "Boletus bicolor" by many authors. However, since Peck's name is illegitimate and cannot be used, *Boletus subreticulatus* A.H. Smith & Thiers has to stand on its own.

3. Many of these features are inconsistently and unreliably expressed for these species (e.g., whether they bruise "instantly" or "slowly" and their staining reactions to chemical tests). These characteristics are also inconsistently reported in the literature, making detailed feature-by-feature comparisons impossible. The taxonomic limits of these boletes are not well defined, and identifications can often be problematic.

Red-mouth Bolete

Boletus subvelutipes

A B C D E

Photos: A, B & D, W. Light; C & E, Emily Johnson.

Boletus subvelutipes **Peck**, *Bull. N.Y. State Mus.* 2(8):142 (1889) [Basidiomycota, Agaricomycotina, Agaricomycetes, Agaricomycetidae, Boletales, Boletaceae]

Derivation: *Boletus*, Latin, "fleshy-pored fungus"; *subvelutipes*, "almost velvety."

Cap: 5–13 cm (2–5¼ in) across, convex (Fig. A, young fruitbody lower right), becoming broadly convex (Fig. A) to almost flat (Fig. E) with age; dry, finely velvety to smooth; color variable: reddish-brown to reddish-orange (Fig. A), cinnamon, yellowish-brown to olive-brown (Fig. E).

Flesh: Bright yellow, immediately bruising dark blue to blackish when cut or injured; odor and taste not distinctive.

Pore surface: Red, maroon (Fig. C), reddish-brown or reddish-orange (Fig. B), fading with age, but pores remaining reddish-rimmed (Fig. D); often yellow at the margin (Fig. A), pores round, 1–3 per mm, tubes up to 2 cm deep.

Stipe: 3–10 cm tall by 1–2 cm thick (1¼–4 × ½–¾ in), subequal, solid, dry, smooth, velvety toward base; yellowish, with reddish washes (Fig. A).

Spore print: Olive-brown.

Bruising: All surfaces and flesh immediately bruising dark blue to blackish (Figs. B & C).

Chemical staining: Cap surface orange to dark red and flesh rusty orange with KOH and NH_4OH.

Microscopic: Spores: 13–18(–20) × 5–6.5(–8) µm, more or less spindle-shaped, smooth, pale brown.

Occurrence: Solitary, scattered, or gregarious under hardwoods, especially beeches and oaks, and conifers (Eastern Hemlock in our area); early summer–fall, fairly common; eastern North America.

Edibility: Poisonous, causing gastrointestinal distress.

Comments: In our area *B. subvelutipes* might be confused with either *B. flammans* E.A. Dick & Snell or *B. rubroflammeus* A.H. Smith & Thiers. In *B. flammans* the cap is dark red or brick-red (rather than reddish-brown to reddish-orange), the pore surface is bright red to orange-red with no yellow or orange-yellow marginal zone as in *B. subvelutipes*, the stipe is colored like the cap (usually yellow near the base) and finely reticulated above (not smooth and generally yellowish with reddish washes as in *B. subvelutipes*), and the spores are yellowish, cylindrical to ellipsoid (10–16 × 4–5 µm). It fruits on the ground under pines and hemlocks. The edibility of *B. flammans* is not known, but it should be avoided.

Boletus rubroflammeus is similar to *B. flammans* with a dark-red to deep wine-red cap and concolorous stipe with dark-red reticulations above. However, the stipe is often yellow at the top rather than near the base as in *B. flammans*. Like *B. subvelutipes*, *B. rubroflammeus* is poisonous and causes gastrointestinal distress.

The Firm Bolete, *Boletus firmus* Frost, is another red-pored species that could be confused with *B. subvelutipes*. It differs in the color of its cap (whitish to greyish-tan, as contrasted with reddish-brown or reddish-orange) and in its staining characteristics: cap dingy-yellow to blackish-brown with KOH and pale blue to dull grey with NH_4OH, compared to orange or dark red with both stains in *B. subvelutipes*.

All red-pored, blue-bruising boletes are suspect and should not be consumed, as many of them cause severe gastrointestinal distress. At least one species, *Boletus pulcherrimus* Thiers & Halliing, which is widespread in western North America, has caused a human fatality. This is the only documented case of a fatal human intoxication by a bolete.[1] The nature of the toxin(s) is unknown.

1. D.R. Benjamin, 1995, Red-pored boletes, *in Mushrooms: Poisons and Panaceas—a Handbook for Naturalists, Mycologists and Physicians* (New York: W.H. Freeman and Company), 359–360.

Firm Bolete

Boletus firmus

A B C

Boletus firmus Frost, *Bull. Buffalo Soc. Nat. Sci.* 2:103 (1874) [Basidiomycota, Agaricomycotina, Agaricomycetes, Agaricomycetidae, Boletales, Boletaceae]

Derivation: *Boletus*, Latin, "fleshy-pored fungus"; *firmus*, "firm," "solid."

Cap: 5–18 cm (2–7¼ in) across, initially convex, becoming broadly convex to almost plane with age, margin entire, sterile zone 2–3 mm wide (Fig. B); surface leather-like, finely velvety, becoming smooth with age, whitish, greyish, tan, pinkish-tan (Fig. A), pinkish-cinnamon, or greyish-olive.

Flesh: White to pale yellow, up to 2 cm (¾ in) thick, firm when young; white with yellowish patches in stipe.

Pore surface: Red to reddish-orange (Figs. B & C), pores small, round to subangular, 2–4 per mm, with thick reddish rims (Fig. B, inset), tubes 6–10 mm deep, tube layer depressed around stipe (Fig. B).

Stipe: 5–12.5 cm tall by 1–3.5 cm thick (2–4¾ × ½–1½ in), bulbous at first, becoming subequal with age; solid, dry, smooth or may be finely reticulated (Fig. A) due to clusters of caulocystidia, especially near apex; concolorous with the cap, sometimes with reddish tints.

Spore print: Olive-brown.

Odor and taste: Odor pungent, taste mild to bitter.

Chemical staining: **KOH**—cap dingy yellow, pale brown to blackish-brown, flesh reddish-brown to dull amber; **$FeSO_4$**—cap pale slate-blue, flesh dark grey to greyish-olive; **NH_4OH**—cap pale blue or a pale slate flash, turning dull grey to pale brownish, flesh dull brownish.

Bruising: Cap, flesh, and pores bruising blue when cut or handled (Fig. C); stipe bruising olive to brownish.

Microscopic: Spores (8.4–)9–15 × (2.8–)3.5–5 µm, narrowly ellipsoid, subspindle-shaped to oblong (elongately peanut-shaped), inequilateral, with apiculus, smooth, thin-walled; yellowish in KOH, brownish to clear-ochraceous in Melzer's reagent.

Occurrence: Alone or scattered in broadleaf and mixed broadleaf-conifer forests, associated with oaks (*Quercus* spp.); eastern North America, July–September, uncommon.

Edibility: Inedible, suspected to be poisonous.

Comments: This species can be distinguished from the Red-mouth Bolete, *Boletus subvelutipes* Peck, by the color of its cap and staining characteristics (see the comments in the preceding species account).

Suillus firmus (Frost) Kuntze (1898), *B. satanus* var. *americanus* Coker & Beers (1943), and *B. piedmontensis* Grand & A.H. Smith (1971) are deprecated synonyms.[1,2]

1. *Fide* L.E. Grand & D.J. Lodge, 1978, *Mycologia* 70:1267–1268; A.E. Bessette et al., 2000, *North American Boletes* (Syracuse: Syracuse University Press), 111; D.E. Binion et al., 2008, *Macrofungi Associated with Oaks of Eastern North America* (Morgantown: West Virginia University Press), 51.
2. *Boletus satanas* Lenz is a misapplied name for *Boletus eastwoodiae* (Murrill) Saccardo & Trotter associated with oaks on the West Coast. *Boletus satanas* is restricted to Europe and does not occur in North America. This issue is still confused in most field guides and online sources.

Long Curved-stemmed Bolete

Boletus longicurvipes

A

B

Photos: Emily Johnson.

***Boletus longicurvipes* Snell & A.H. Smith**, *J. Elisha Mitchell Scient. Soc.* 46:325 (1940
[Basidiomycota, Agaricomycotina, Agaricomycetes, Agaricomycetidae, Boletales, Boletaceae]

Derivation: *Boletus*, Latin, "fleshy-pored fungus"; *longicurvipes*, "long curved foot."

Cap: Convex, becoming broadly convex, 1.6–8 cm (⅔–2½ in) across, margin entire; surface smooth, slightly wrinkled, or pitted (Fig. A), olive-green, olive-brown (Fig. A), yellow-brown (Fig. B), to yellow-orange or reddish-orange; slimy and viscid when young (Fig. A).

Flesh: White or pale yellow; odor and taste not distinctive.

Pore surface: Pale yellow, becoming greenish-grey with age, recessed around the stipe when mature (Fig. B); pores circular, 2 per mm.

Stipe: 5–10 cm tall by 1–2 cm thick (2–4 × ½–1⅔ in), solid, subequal, expanded near top (Fig. B) and sometimes toward base, often curved above the base; whitish or yellowish, often with reddish or tan; with whitish scabers, these turning pinkish or reddish-brown with age (Fig. B); with whitish basal mycelium.

Spore print: Olive-brown to dull brown.

Bruising: Flesh unchanging when cut; pore surface may bruise brownish.

Chemical staining: KOH and NH4OH—cap cherry-red to reddish-orange; **$FeSO_4$**—cap and flesh negative.

Microscopic: Spores 13–16 × 4–6.5 µm, narrowly spindle-shaped to oblong, smooth, pale brown; ochraceous-tawny in Melzer's reagent, yellowish in KOH.

Occurrence: Mycorrhizal. Solitary, scattered, or in groups on the ground under oaks (*Quercus* spp.), often in mixed woods; late summer; widespread and fairly common in eastern North America.

Edibility: Edible.

Comments: In our area *Boletus longicurvipes* is likely to be confused only with *B. rubropunctus* Peck, a species of uncertain identity.[1] *Boletus rubropunctus* has a cap that is chestnut-red to reddish-brown and tacky when fresh but is not nearly as viscid as in *B. longicurvipes*, the pore surface is bright (not pale) yellow, the flesh bruises pink, the surface of the stipe is yellow (not white) beneath the scabers, the white scabers on the stipe do not darken with age, the basal mycelium is yellow (not white), and the spores are longer (17–21 × 5.5–7.5 µm). This species is also uncommon, fruiting in the summer through midwinter in hardwood or conifer forests.

1. See the detailed comments about *B. rubropunctus* [as *Leccinum rubropunctum* (Peck) Singer] by Michael Kuo (May 2007, *Leccinum*: Uncertain taxa, MushroomExpert.Com, http://www.mushroomexpert.com/leccinum_uncertain.html#rubropunctum).

Corrugated Bolete

Boletus hortonii

A C B D E

***Boletus hortonii* A.H. Smith & Thiers**, *Boletes of Michigan* (Ann Arbor):319 (1971)
[Basidiomycota, Agaricomycotina, Agaricomycetes, Agaricomycetidae, Boletales, Boletaceae]

Derivation: *Boletus*, Latin, "fleshy-pored fungus"; *hortonii*, epithet in honor of American mycologist Charles Horton Peck (1833–1917).

Cap: 3–14 cm (1¼–5⅔ in) across, broadly umbonate to convex, dry to slightly viscid when wet; deeply sinuously corrugated and wrinkled, tan, ochre-brown, orange-brown, to pale reddish-brown (Figs. A & B).

Flesh: White to pale yellowish, not changing color when cut, or rarely slowly turning weakly bluish.

Pore surface: Pale lemon-yellow when young (Fig. D), aging to dull or olive-yellow (Fig. E). Pores attached to depressed around the stipe (Figs. D & E); round to subangular, small, 2–3 per mm (Fig. D), becoming variably larger, more angular, and irregular with age (Fig. E); tubes 6–10 mm long.

Stipe: Subequal or tapering upward from a slightly expanded base, surface smooth; 6–10 cm tall by 1–2 cm thick, pale yellow to tan, sometimes with reddish flush near base, with minute and inconspicuous, yellowish, tan, or reddish scabers (Fig. C); with white basal mycelium.

Odor and taste: Not distinctive.

Spore print: Olive-brown.

Bruising/chemical staining: Cap nonbruising; pore surface and flesh nonstaining when bruised (Fig. E, arrow) or slowly and weakly bruising blue. **NH_4OH**—cap, a blue-green flash, resolving to olive-brown or grey, flesh negative; **KOH**—cap stains olive-brown, flesh negative; **$FeSO_4$**—cap, pale olive-tan, flesh, bluish-grey.

Microscopic: Spores 12–15 × 3–4.5 µm, elliptical to spindle-shaped, smooth, pale yellow; yellow in KOH and Melzer's reagent.

Occurrence: Solitary, scattered, or gregarious in hardwood and mixed forests associated with oaks (*Quercus*), hickory (*Carya*), beech (*Fagus*), and hemlock (*Tsuga*); mycorrhizal with oaks and possibly other hardwoods; widely distributed and common in eastern North America, June–October.

Edibility: Edible.

Comments: Originally named *Boletus subglabripes* var. *corrugis* by Peck in 1889. The deeply and sinuously corrugated cap of this attractive bolete is distinctive and immediately identifies this species. *Leccinum subglabripes* Peck[1] is similar but is distinguished by its smooth to only slightly wrinkled (never deeply corrugated) cap, pale brown (not yellow) spores, and the following differences in chemical staining: **NH_4OH**—cap, negative reaction (instead of a blue-green flash), flesh, a dirty green (not negative) reaction; **KOH**—a flesh, reddish-orange (rather than negative) reaction; **$FeSO_4$**—a flesh, greyish olive-green (not negative) reaction.

Leccinum rugosiceps (Peck) Singer is distinguished from *B. hortonii* as follows: the stipe has brown scabers that darken with age, the pore tubes are up to 2 (rather than 1) cm long, the cap stains red (rather than olive-brown), and the flesh golden-yellow to rusty or orange (not negative) with KOH, and the flesh bruises pinkish to dull red (instead of not changing color) when cut and then slowly (20–60 minutes) fades to greyish or blackish. The cap is negative or stains reddish (not a green flash fading to brown or grey) and the flesh bruises negative or yellowish with NH_4OH. The spores are also longer and wider, (9.5–)12–21 × (3–)4–6 µm.

1. The genus *Leccinum* S.F. Gray is distinguished from *Boletus* Fries mainly by distinctive scabers on the stipe that darken when mature. The species in this genus are typically mycorrhizal with only a single genus of tree.

Golden-pored Bolete

Aureoboletus auriporus

A

B

C

Photos: A & C, Steven Russell; B, Kristine (JanusGD).

***Aureoboletus auriporus* (Peck) Pouzar**, *Česká Mykol.* 11:49 (1957) [Basidiomycota, Agaricomycotina, Agaricomycetes, Agaricomycetidae, Boletales, Boletaceae]

Basionym: *Boletus auriporus* Peck, *Ann. Rep. Reg. N.Y. St. Mus.* 23:133 (1872)

Deprecated synonym: *Pulveroboletus auriporus* (Peck) Singer, *Am. Midl. Nat.* 37:13 (1947)

Derivation: Latin, from *aureus*, "gold," "golden," and *porus* (Latin from the Greek *poros*), "pore," "a minute opening"; both the generic name and specific epithet mean "golden-pored."

Cap: Initially convex, becoming broadly convex to nearly flat; 3–8 cm (1¼–3¼ in) across; smooth, viscid and tacky when wet; pinkish-brown to pale pinkish-cinnamon (Fig. A).

Flesh: White in the stipe, pale yellow in the cap, not bruising blue when cut or exposed; odor not distinctive, taste somewhat acrid.

Pore surface: Bright golden-yellow to lemon-yellow when young, fading to dull yellow with age, depressed at the stipe (Fig. B); pores angular (Fig. B), initially 1–3 per mm, pores expanding to 2 mm across with age; tubes bright yellow, to 2 cm deep.

Stipe: 4–11.5 cm tall by 0.6–1.7 cm (½–4⅔ × ¼–⅔ in) thick; subequal to enlarged toward the base (Fig. A); nearly smooth, not reticulate; ring lacking; yellow to whitish at apex, pale pinkish-brown below, typically streaked (Fig. A), surface with tiny yellow fibrils when young; base with exposed white mycelium (Fig. C, arrow).

Spore print: Olive-brown.

Chemical staining: **KOH**—cap reddish, orangish-amber, or yellowish, flesh negative; **NH_4OH**—cap stains wine-red to dull blood-red, flesh negative; **$FeSO_4$**—cap and flesh negative to yellowish, pore surface dark green to blue.

Microscopic: Spores (9.2–)11–16(–17.7) × 4–6.5(–8.8) µm, smooth; highly variable, spindle-shaped to elongate-ellipsoid, often with suprahilar depression; pale golden to honey-brown.

Occurrence: Solitary or in small groups under oaks and on shaded lawns, fairly common, July–October; eastern North America south to Costa Rica.

Edibility: Edible, taste somewhat acidic.

Comments: *Aureoboletus auriporus* very closely resembles *A. innixus* (Frost 1874) Halling, A.R. Bessette & A.E. Bessete (2015), and the two species are often confused. *Aureoboletus innixus* is readily distinguished from *A. auriporus* by its chemical staining: the cap flashes green with NH_4OH and then fades to orangish or reddish with a greenish halo; with KOH the cap surface stains dark red to reddish-brown, and the flesh stains pinkish to orangish; with $FeSO_4$ the cap stains pale olive and the flesh negative to greyish.

Pulveroboletus ravenelii (Berkeley & M.A. Curtis) Murrill is distinguished from *A. auriporus* by its prominent powdery partial veil when young and its persistent ring on the upper stipe when mature, the pore surface that bruises greenish-blue then greyish-brown when injured, and its mycorrhizal association with pines, hemlock, and rhododendrons, rather than oaks.

Boletus viridiflavus Coker and Beers (1943) is considered by some to be a later synonym of *A. auriporus*, for example, Singer (1947, *Am. Midl. Nat.* 37:15–16), who stated that it "cannot be anything but the greenish form or stage of *P[ulveroboletus] auriporus*." But this has not yet been fully resolved; the cap of *B. viridiflavus* is distinctively olive-colored and the context of the stipe bruises brick-red.

Ornate-stalked Bolete

Retiboletus ornatipes

A C B

Photos: Emily Johnson.

Retiboletus ornatipes **(Peck) M. Binder & Bresinsky**, *Feddes Repertorium* 113(1–2):37 (2002)

Basionym: *Boletus ornatipes* Peck, *Ann. Rep. N.Y. St. Mus. Nat. Hist.* 29:67 (1878)

[Basidiomycota, Agaricomycotina, Agaricomycetes, Agaricomycetidae, Boletales, Boletaceae]

Derivation: *Retiboletus*, Latin, from *reticulum*, "net-like," and *boletus*, "fleshy-pored fungus"; *ornatipes*, Latin, "ornate foot."

Cap: 4–20 cm (1⅔–8 in) across, convex, becoming nearly flat with maturity, margin even; surface dry, slightly sticky when wet, finely powdery to velvety; color variable: yellow (Fig. A), mustard-yellow, pale grey (Fig. B), olive-yellow, yellowish-brown, often with yellow along margin (Fig. B).

Flesh: Yellow, not bruising blue when injured; odor not remarkable, taste mild to bitter.

Pore surface: Lemon-yellow, aging to olive-yellow, not bruising or bruising brighter yellow, orangish, or brownish; pores circular, 1–2 per mm (larger on expanded caps), tubes up to 15 mm (⅔ in) deep; attached or sunken around stipe (Fig. A).

Stipe: 6–15 cm tall by 1–2.5 cm thick (2½–6 × ½–1 in), subequal, sometimes tapering toward apex or base; coarsely reticulate (Fig. C) over most of its length; pale yellow to deep yellow (Figs. A & C), bruising darker orangish-brown (Fig. B, at base); no partial veil or ring.

Spore print: Olive-brown to yellow-brown.

Chemical staining: **KOH**—surface of cap reddish-brown, flesh orangish; **NH_4OH**—cap orangish to pale brown, flesh negative to pale brownish; **$FeSO_4$**—cap negative, flesh negative to dull white.

Microscopic: Spores 9–13 × 3–4 µm, smooth, elliptical-cylindrical, pale brown.

Occurrence: Isolated or cespitose on the ground beneath hardwood trees, especially oaks and beeches; summer, widespread and common in eastern North America.

Edibility: Inedible, usually very bitter.

Comments: This species characteristically stains the hands yellow and anything else it contacts.

Retiboletus griseus (Frost 1878) M. Binder & Bresinsky (2002) is very similar to *R. ornatipes*, from which it is distinguished by its uniformly pale grey to brownish-grey cap, white flesh, whitish to dusky-grey pore surface, and whitish, coarsely reticulated stipe, which stains yellow from the base upward as it ages. It also stains differently from *R. ornatipes*: **KOH**—cap negative to olive (not reddish-brown); **NH_4OH**—cap negative (not orangish or brownish), flesh negative or pale pinkish-grey (not brownish); **$FeSO_4$**—flesh bluish-grey (not negative or bleaching dull white).

Boletus auriflammeus Berkeley & M.A. Curtis, another look-alike, is a deep golden-orange with white flesh and is generally smaller than *R. ornatipes* (3–9.5 cm). The stipe is much less prominently reticulated. The cap stains reddish-brown or darker orange with a pale blue flash (not pale brown) with NH_4OH, stains yellowish-brown with KOH after dissolving the original yellow, and is negative in $FeSO_4$.

Retiboletus ornatipes and *R. retipes* cannot be reliably separated on morphological grounds. Several recent works and *Index Fungorum* consider *Retiboletus retipes* Berkeley & M.A. Curtis (1872) M. Binder & Bresinksy (2002) to be a synonym of *Retiboletus ornatipes* (Peck 1878) Binder & Bresinsky (2002). It strikes us that since the name of Berkeley & Curtis predates that of Peck by six years, *R. ornatipes* (Peck) should be considered the later synonym of *R. retipes* (Berkeley & Curtis) instead of the other way around. Since DNA sequencing and chemical tests distinguish these two taxa, we follow Smith & Thiers[1] and Binder & Bresinsky[2] in keeping them separate.

1. A.H. Smith & H.D. Thiers, 1971, *The Boletes of Michigan* (Ann Arbor: University of Michigan Press).
2. M. Binder & A. Bresinsky, 2002, *Feddes Repertorium* 113(1–2):30–40.

Russell's Bolete

Boletellus russellii

A

B

Photos: A, Emily Johnson; B, USDA Forest Service.

***Boletellus russellii* (Frost) E.-J. Gilbert**, *Les livres du mycologue tomes I–IV*, vol. 3, *Les bolets*:107 (1931)
[Basidiomycota, Agaricomycotina, Agaricomycetes, Agaricomycetidae, Boletales, Boletaceae]

Basionym: *Boletus russellii* Frost, *Ann. Rep. N.Y. St. Mus. Nat. Hist.* 29:44 (1878)

Derivation: *Boletellus*, Latin, *Boletus* and *-ellus*, "little," "little bolete"; *russellii*, epithet in honor of John Lewis Russell, American mycologist and Unitarian minister (1808–1873).

Cap: 3–13 cm (1¼–5¼ in) across, initially hemispheric to convex (Fig. B), becoming broadly convex (Fig. A), margin incurved, more or less even to slightly wavy, dry or slightly sticky when wet; velvety to smooth when young, becoming cracked or scaly with age; muted yellowish-brown, pinkish-tan (Fig. B), greyish-brown, to dark reddish-brown or vinaceous-brown (Fig. A).

Flesh: Yellow, not discoloring when bruised or exposed; odor and taste not distinctive.

Pore surface: Pale yellow to greenish-yellow, not discoloring or sometimes bruising more yellow; pores large, angular, ≤1 per mm, tubes up to 2 cm deep, usually deeply depressed near stipe (Fig. A).

Stipe: 10–20 cm tall by 1–2 cm thick (4–8 × ⅖–¾ in), subequal, with deep vertical grooves and ridges (Fig. A), these often becoming prominently lacerated, torn, and shaggy (Fig. B); reddish-brown (Fig. A) to pinkish-tan (Fig. B). No partial veil or ring.

Spore print: Olive-brown.

Microscopic: Spores 15–20 × 7–11 µm, elliptical, surface deeply grooved (striated) longitudinally.

Occurrence: Mycorrhizal with oaks, hemlocks, and pines, growing alone or in scattered groups of hardwood or mixed hardwood-conifer forests with oaks in eastern North America;

widespread but rare from Canada south to Central America, possibly also southwestern North America; Asia, Japan, Taiwan, Korea.

Edibility: Edible, but taste bland.

Comments: With its unusually tall, deeply furrowed, and tattered stipe, this species resembles the Shaggy-stalked Bolete, *Heimioporus betula* (Schweinitz) E. Horak. The latter species is distinguished from *B. russellii* by its bright red or bright yellow cap, which is slimy when wet; its bright yellow (not reddish-brown or pinkish-tan), deeply reticulated stipe; and its minutely pitted (not deeply grooved longitudinally) spores. The two species are fairly easily told apart by macroscopic features alone.

Shaggy-stalked Bolete

Heimioporus betula

A B C D

Photos: A, Teresa Fortenberry; B–D, W. Light.

***Heimioporus betula* (Schweinitz) E. Horak**, *Sydowia* 56(2):239 (2004)
[Basidiomycota, Agaricomycotina, Agaricomycetes, Agaricomycetidae, Boletales, Boletaceae]

Basionym: *Boletus betula* Schweinitz, *Schr. Naturf. Ges. Leipzig* 1:90 (1822)

Derivation: *Heimioporus*, Greek, *-póros*, "pore," generic name presumably honors the French mycologist Roger Heim (1900–1979);[1] *betula*, Latin, "birch."

Cap: Reddish to bright red, often with yellow margin (Fig. A), to yellow or orange (Fig. B), or reddish-brown, 3–9 cm (1¼–3¾ in) across; initially convex (Fig. A), becoming broadly convex (Fig. B), viscid when wet, shiny and smooth when dry, may become pitted or wrinkled with age.

Flesh: Yellow, greenish-yellow, or orange-yellow, not discoloring when bruised or injured; odor not distinctive; taste not distinctive or somewhat acidic.

Pore surface: Yellow (Fig. B), becoming greenish-yellow with age, not discoloring when bruised; often deeply sunken around stipe at maturity (Fig. C); pores large (1–2 per mm), circular to angular, tubes up to 1.5 cm (⅔ in) deep.

Stipe: Extremely long, subequal, 10–30 cm long by 1.2–5 cm thick (½–12¾ × ½–2 in), solid; often nearly as wide or even wider than cap in early stage of growth (Fig. A); deeply and coarsely reticulate to shaggy, with raised yellow ridges over reddish base (Figs. A, B, & D), yellow ridges often reddening with age; usually with white mycelium at base attached to white rhizomorphs. No partial veil or ring.

Spore print: Olive to olive-brown.

Microscopic: Spores (13–)16–19(–22) × 6–10 µm, narrowly elliptical, with scattered minute pits.

Occurrence: Ectomycorrhizal with oaks, growing alone or in scattered groups from ground in hardwood and mixed hardwood-conifer forests, often in low areas along streams; summer–fall; widespread in eastern United States, particularly in the southern Appalachians and hills of the northern Piedmont.

Edibility: Edible.

Comments: This striking bolete is not likely to be confused with any other species. The conspicuously tall, deeply reticulated stem and its tidy red or yellow convex cap are distinctive. Russell's Bolete, *Boletellus russellii* (Frost) Gilbert, has a similarly tall and shaggy stalk, but the ridges often appear "winged" and are arranged vertically. The velvety cap, also a muted yellow-brown, greyish-brown, or reddish-brown color, becomes cracked or scaly with age, and the olive-brown spores are longitudinally deeply grooved (striated) rather than finely pitted.

The genus *Boletus* [*sensu lato*] is polyphyletic (i.e., all its species did not arise from a single common ancestor) and thus does not constitute a natural biological group. It is undergoing continual taxonomic revision and has been split into a number of genera, not all of which are accepted by all mycologists. *Heimioporus betula* is still referred to as *Boletus betula*, *Australoboletus betula*, and *Boletellus betula* in many field guides and online sources. See online Appendix A for a full list of synonyms.

1. *Heimioporus* E. Horak (2004, *Sydowia* 56[3–6]:237) is a replacement name for *Heimiella* Boedijn (1951, *Sydowia* 5[3–6]:216)—a later homonym of *Heimiella* Lohmann 1913 (Class Haptophyceae) and thus an illegitimate name. Boedijn did not provide the etymology in his original proposal of his new genus.

Frost's Bolete, Apple Bolete

Exsudoporus frostii

A

B C

Photos: A, Dan Molter; B, John Plischke III; C, Walt Sturgeon.

***Exsudoporus frostii* (J.L. Russell) Vizzini, Simonini & Gelardi**, *Index Fungorum* 183:1 (2014)

Basionym: *Boletus frostii* J.L. Russell, in Frost, *Bull. Buffalo Soc. Nat. Sci.* 2:102 (1874) [1874–1875]
[Basidiomycota, Agaricomycotina, Agaricomycetes, Agaricomycetidae, Boletales, Boletaceae]

Derivation: *Exsudoporus*, Latin, from *exsudare*, "to exude," and *porus*, referring to the pores exuding liquid droplets in young fruitbodies; *frostii*, epithet in honor of the 19th-century mycologist Charles Christopher Frost.

Cap: 5–15 cm (2–6 in) across, initially convex, becoming broadly convex to flat to shallowly depressed; margin at first incurved, entire, sometimes becoming scalloped (Figs. A & B); deep red to crimson, moist, viscid.

Flesh: Up to 2.5 cm (1 in) thick, pale to lemon-yellow (Fig. A, arrow & inset); odor not distinctive.

Pore surface: Dark red at first (Fig. B), becoming paler with age; often exuding pale yellowish to clear liquid droplets (Fig. B, arrow, Fig. C); pores deep red, circular, 2–3 per mm; tubes mustard yellow, 6–15 mm deep.

Stipe: Cylindrical, subequal, sometimes expanded downward; 4–12 cm tall by 1–2.5 cm thick (1⅔–4¾ × ½–1 in); deeply reticulate to alveolate, dark red, yellowish toward the base; when visible, mycelia at the base yellowish to white.

Spore print: Olive-brown.

Bruising: Cap, stipe, flesh, and pore surface bruising deep blue when cut or bruised.

Chemical staining: KOH—cap stains grey to olivaceous, then bright orange, flesh stains bright orange; NH_4OH—cap yellowish to olivaceous, flesh yellowish to amber.

Microscopic: Spores thick-walled, smooth, elliptical to spindle-shaped, 11–17(–18) × 4–5 µm; pleuro-, cheilo-, and caulocystidia present, clamp connections absent.

Occurrence: Eastern North America west to southern Arizona and south into Mexico and Central America; occasional to common; July–October; mycorrhizal with oaks.

Edibility: Edible, but caution advised because of its close resemblance to toxic red-pored species.

Comments: The deep red color of the cap, stipe, and pore surface and the deeply reticulated stipe of this bolete immediately identify it. The other boletes in our area with deeply reticulate stipes have yellow pore surfaces that do not bruise blue. Compare this species with the Ornate-stalked Bolete (*Retiboletus ornatipes*), the Russell's Bolete (*Boletellus russellii*), and the Shaggy-stalked Bolete (*Heimioporus betula*).

Previously known as *Boletus frostii*, this species was transferred to the newly erected genus *Exsudoporus* Vizzini, Simonini & Gelardi (2014). In addition to *E. frostii*, the genus *Exsudoporus* currently contains two other species: *E. permagnificus* (Pöder 1981) Vizzini, Simonini & Gelardi (2014), the type species of the genus, and *E. floridanus* (Singer 1945) Vizzini, Simonini & Gelardi (2014). *Exsudoporus permagnificus* is found in Europe, while *E. floridanus* is restricted to the coastal plain of the Carolinas, Florida, and west to Texas.

Eastern White Leccinum

Leccinellum albellum

A

B

C

D

Photos: A–C, W. Light; D, Dan Molter.

***Leccinellum albellum* (Peck) Bresinsky & Manfred Binder**, in Bresinsky & Besl, *Regensb. Mykol. Schr.* 11:232 (2003)

Basionym: *Boletus albellus* Peck, *Ann. Rep. N.Y. St. Mus.* 41:77 (1888)
[Basidiomycota, Agaricomycotina, Agaricomycetes, Agaricomycetidae, Boletales, Boletaceae]

Derivation: *Leccinellum*, Latinized diminutive form of the genus *Leccinum*, to which the species in *Leccinellum* were formerly assigned: "little *Leccinum*" (*Leccinum* is based on the Italian word *leccino*, a type of rough-stemmed bolete); *albellum*, Latin, from *albus*, "white."

Cap: 2–6.5 cm (¾–2⅔ in) across, convex or broadly convex to almost flat, margin entire; surface dry, smooth, finely velvety, often pitted (Figs. A & B), becoming cracked and areolate with age (Fig. D); color variable, whitish (Fig. A), greyish-buff (Fig. B), pale tan to olive-brown (Fig. D) or brownish-grey.

Flesh: White, not bruising when injured; odor and taste not distinctive.

Pore surface: Initially white (Fig. C), becoming pale grey (Fig. D) to greyish-brown; depressed around stipe; pores small, round to angular (Fig. C), 1–2 per mm.

Stipe: 5–10 cm tall by 0.5–1.2 cm thick (2–4 × ¼–½ in), subequal, with enlarged base (Figs. A & D), often curved or twisted; ring lacking; concolorous with cap; surface densely covered with vertical rows of initially white scabers that darken with age (Figs. A & D).

Spore print: Olive-brown.

Chemical staining: KOH and NH_4OH—cap stains olive, flesh pale yellow; **$FeSO_4$**—cap and flesh stain bluish-grey.

Microscopic: Spores (10–)14–24 × 4–6.5 µm, cylindrical to subspindle-shaped, smooth, pale yellow.

Occurrence: Host specifically mycorrhizal with oaks (*Quercus* spp.); alone or in scattered groups on the ground in deciduous woods; widespread and fairly common in southeastern North America.

Edibility: Edible, but mediocre.

Comments: This species is called *Leccinum albellum* in many field guides and online sources. It is recognized on sight by its distinctively pitted (when young) or cracked and areolate (when mature) cap, its long, thin, and typically twisted stipe with conspicuous dark scabers arranged in vertical rows, and its white to greyish-brown pore surface. No other mushroom in our area resembles it.

Black Velvet Bolete

Tylopilus alboater

A B C

***Tylopilus alboater* (Schweinitz) Murrill**, *Mycologia* 1(1):16 (1909)

Basionym: *Boletus alboater* Schweinitz, *Schr. Naturf. Ges. Leipzig* 1:95 (1822)
[Basidiomycota, Agaricomycotina, Agaricomycetes, Agaricomycetidae, Boletales, Boletaceae]

Derivation: *Tylopilus*, Greek, from *tylos*, "swollen," and *pílos*, "swollen hat"; *alboater*, Latin, "white-black."

Cap: Dark greyish-brown to black (Figs. A & B), often with a deep violet tinge, initially covered with a whitish bloom; velvety when young, smoother with age; 5–15 cm across (2–6 in), sometimes larger; convex to broadly convex (Figs. A & B) to almost flat with age; margin strongly inrolled when young (Fig. C).

Flesh: Thick, firm when young, softer with age; white to greyish-white, bruising pinkish-grey, then blackish when cut, gradually; odor and taste not distinctive.

Pore surface: Initially white (Fig. C), becoming flesh-colored or dull pinkish with age, bruising like the flesh; pores angular to irregular, 1–3 per mm; tubes 5–10 mm deep.

Stipe: Solid, 4–11.5 cm tall by 2–4 cm thick (1⅔–4⅔ × ¾–1⅔ in), subequal or enlarged toward base (Figs. A & B), smooth to faintly reticulated (Fig. B), generally concolorous with the cap but paler when young, especially near the top (Fig. B), darker with age.

Spore print: Dull pink.

Chemical staining: **KOH**—flesh and pore surface pinkish-orange, gradually turning darker brown, cap reportedly amber-orange, but this is not apparent to us against the blackish-purple ground color; **NH_4OH**—flesh olive, then brownish-orange, cap nonreactive; **$FeSO_4$**—instantly greyish-blue or greenish-blue, cap nonreactive.

Microscopic: Spores 7–11 × 3.5–5 µm, oval to ellipsoid, smooth, clear.

Occurrence: Mycorrhizal. Alone or scattered (occasionally cespitose) on the ground in woods under hardwoods, especially oaks (*Quercus* spp.); early summer–fall, fairly common; eastern North America; East Asia.

Edibility: Edible, considered quite good.

Comments: *Tylopilus alboater* is readily recognized on sight and is not likely to be confused with any other species, especially in our area. *Tylopilus atronicotianus* Both (1998) is often included in the differential diagnosis, but this taxon is not reported from the Southeast. The cap is also smooth (never velvety) and generally some shade of olive-brown or bronze-brown, the spore print is reddish-brown to cocoa-brown (not dull pink), the spores are much larger (1.5 µm wide), and the flesh stains pale greyish-yellow with a black halo (not olive to brownish-orange) with NH_4OH.

Violet-grey Bolete

Tylopilus plumbeoviolaceus

A

B

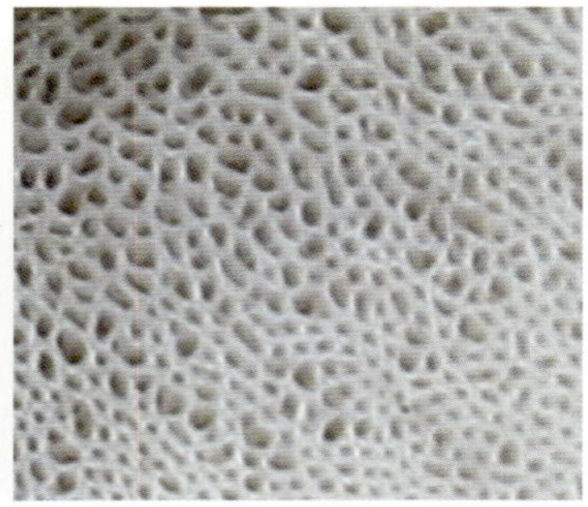
C

D

***Tylopilus plumbeoviolaceus* (Snell & E.A. Dick) Singer**, *Amer. Midl. Nat.* 33:33 (1941)
[Basidiomycota, Agaricomycetes, Boletales, Boletaceae]

Basionym: *Boletus plumbeoviolaceus* Snell & E.A. Dick, *Mycologia* 37:93 (1947)

Derivation: *Tylopilus*, Greek, "swollen cap," from *tylos*, "knob," "protuberance," and *pilos*, "cap"; *plumbeoviolaceus*, Latin, from *plumbeos*, "leaden," "lead-like," and *violaceus*, "violet."

Cap: 4–15 cm (1⅔–6 in) wide, convex at first (Fig. B), becoming flatter (Fig. A) to saddle-shaped when mature; surface finely velvety, becoming dry, smooth; initially purplish, becoming purple-grey to dull brown or tan with age.

Flesh: White, not bruising or staining when cut.

Pore surface: Depressed at and attached to stem; white in young specimens (Fig. C), becoming pinkish or pinkish-tan when mature (Fig. D); reported not to change color when cut or bruised. However, disturbed pores on our specimens exhibited faint bruising. Pores round, angular or elongate (Figs. C & D), occasionally anastomosing (forking and recombining), becoming partly daedeloid; 2–3 per mm, tubes up to 2 cm (0.8 in) deep in mature fruitbodies.

Stipe: 8–12 cm tall by 1–2 cm thick (3¼–4¾ in), mostly equal, enlarging and sometimes bulbous at base, reticulate above, smooth below; initially purple, often mottled with white (Fig. B), becoming purplish-buff to light brown, with white mycelium showing at base (Fig. A).

Chemical staining: 10% KOH—cap surface orange to amber, stipe staining dull yellow, flesh and gills negative; **Melzer's reagent**—not reported in literature, cap and flesh of our specimens immediately staining dark reddish-maroon, flesh, gills, and stipe negative; **NH_4OH**—cap and stipe staining dull yellow or amber, flesh and gills negative.

Spore print: Light pinkish to flesh.

Microscopic: Spores 9–13 × 3.4–4.5 µm, elliptical, smooth.

Occurrence: Mycorrhizal; growing singly, scattered, or clustered in hardwood forests, occasionally in mixed forests with hemlock; June–September east of the Rocky Mountains from Canada to Mexico; also reported from North Korea.

Edibility: Inedible, flesh reported as extremely acrid, leaving bitter aftertaste; our samples were mildly acrid.

Comments: Several species resembling *T. plumbeoviolaceus* have been reported from Florida and Mexico. *Tylopilus violatinctus* Baroni & Roth has a paler, lilac-colored to purplish-pink cap and a reddish-brown spore print, is relatively rare, and is also not reported from our area, but the limits of its distribution are not known.[1] *Tylopilus rubrobrunneus* Mazzer & A.H. Smith, which does occur in our area, has a solid white to brown or olive-brown stipe (which is never lilac or purplish), the spore print is reddish-brown, the whitish flesh slowly turns pinkish when exposed, and the cap bleaches pinkish-brown (not orange or amber) with KOH.

1. A.E. Bessette et al., 2000, *North American Boletes: A Guide to the Fleshy Pored Mushrooms* (Syracuse: Syracuse University Press), 274.

Peppery Bolete

Chalciporus piperatus

Photo: Holger Krisp.

***Chalciporus piperatus* (Bulliard) Bataille**, *Bull. Soc. Hist. Nat. Doubs* 15:39 (1908)
[Basidiomycota, Agaricomycotina, Agaricomycetes, Agaricomycetidae, Boletales, Boletaceae]

Basionym: *Boletus piperatus* Bulliard, *Herb. Fr.* 10:tab. 451, fig. 2 (1790)

Derivation: *Chalciporus*, Greek, "copper-pored"; *piperatus*, Latin, "peppery."

Cap: 2–7 (rarely up to 9 or 10) cm across (¾–2¾ in), convex, becoming broadly convex to almost flat with age, margin even; surface smooth to slightly fibrillose, especially near margin; viscid when wet; cap surface buff to yellow-brown, orange-brown, rusty-brown.

Flesh: Moderately thick, pale yellow or yellowish-buff, bruising cinnamon-brown; odor unremarkable, taste hot and peppery.

Pore surface: Rusty-brown to cinnamon-brown or coppery-reddish, sometimes darkening when injured but never bruising blue; pores angular, large, 1–2 mm in diameter, sometimes appearing to radiate from stipe, attached, sometimes forming vertical ridges at top of stipe.

Stipe: 4–9 cm tall by 6–15 mm thick (1⅔–3⅔ × ¼–⅔ in), subequal, dry, solid, concolorous with cap, base yellow with adhering yellow mycelium.

Spore print: Brown to reddish-brown.

Chemical staining: KOH and NH_4OH—cap dark reddish-brown, flesh grey to purplish-grey; **$FeSO_4$**—flesh negative to pale greyish-green.

Microscopic: Spores 8–12 × 4–5 µm, smooth, narrowly spindle-shaped; pale brown.

Occurrence: Alone or in scattered groups on the ground under pines, late spring–early fall; widely distributed in North America.

Edibility: Uncertain, caution advised.

Comments: The hot, peppery taste of this bolete immediately identifies it in our area. This peppery taste is said to disappear with cooking.

Variegated Suillus

Suillus hirtellus

A

B

***Suillus hirtellus* (Peck) Kuntze**, *Rev. Gen. Pl.* (Leipzig) 3(2):535 (1898)[1]
[Basidiomycota, Agaricomycetes, Boletales, Suillaceae]

Basionym: *Boletus hirtellus* Peck, *Bull. N.Y. St. Mus.* 2:94 (1889)

Derivation: *Suillus*, Latin, from *suis*, "pig-like"; *hirtellus*, "hairy," referring to the adpressed reddish-brown fibrils forming the reticulated pattern on the cap.

Cap: Broadly convex with incurved margin, becoming plane, 5–10 (up to 15) cm across (2–4 in, up to 6 in); ground color yellow, surface initially with greyish to reddish adpressed fibrillose squamules, giving surface a variegated or reticulated pattern (Fig. A); bruising deep vinaceous brown when handled; smooth with age and viscous when fresh; margin with cottony tufts when young.

Flesh: Pale yellow, not bruising blue or other color when cut or handled.

Pore surface: Pores angular and elongate (Figs. B & C), minute, 0.3–0.4 mm in diameter, with thick walls, 3 per mm, irregularly arranged, occasionally coalescing, pale yellow at first (Fig. B), aging to orange-buff (Fig. C); tubes yellowish, 3–8 mm deep. With scattered reddish or reddish-brown glandular spots (Figs. B & C); *typically oozing droplets of white liquid near apex of stipe* (Figs. B & C, arrows).

Stipe: No ring; pale yellow, 3–8 cm tall by 1–2 cm thick (1¼–3¼ × ⅖–¾ in), equal or expanded at base; internal flesh yellowish, becoming olive-brownish-grey in the base; densely covered with yellowish glandular spots, these becoming reddish or reddish-brown with age and/or handling (Figs. B & C) and blackening with age.

Spore print: Cinnamon-brown.

Bruising/chemical staining: Cap surface bruising deep vinaceous brown with handling; **ammonia and NaOH**—all tissues negative; **Melzer's reagent**—cap and pores negative, flesh strongly yellow, stipe context bluish-green or greyish-green; **KOH**—cap wine-colored, vinaceous, flesh negative, pores and tubes yellow-brown, stipe context dark brown or burnt umber.

Microscopic: Spores 7–9 × 3–3.5 µm, smooth, oblong, pale ochraceous to pale tan in KOH and Melzer's reagent.

Occurrence: Scattered or in groups on the ground under conifers, especially Loblolly Pine (*Pinus taeda*); July–September; widespread but occasional in eastern North America.

Edibility: Reportedly edible.

Comments: This species is difficult to distinguish from *Suillus subaureus* (Peck 1887) Snell (1944), and the two are often confused in the literature. *Suillus subaureus* has somewhat larger pores (2 per mm) arranged in a radial (not irregular) pattern. Ammonia stains the cap surface dark olive with a pinkish halo (not negative) and the flesh reddish to pinkish (not negative); in KOH the cap is negative or bluish (not wine-colored), the flesh stains pinkish, greyish, or purplish (not negative), and the pores ooze dark (not white) droplets. *Suillus subaureus* is not restricted to conifers but is also found under hardwoods, especially oaks (*Quercus* spp.), or under Eastern White Pine (*Pinus strobus*), or in mixed forests.

1. *Index Fungorum* incorrectly credits Snell as the authority for this genus and species combination: *Suillus hirtellus* (Peck) Snell, *Lloydia* 7:23 (1944). However, Kunzte's 1898 publication clearly has priority and is the one correctly cited by A.H. Smith & H.D. Thiers, 1971, *The Boletes of Michigan* (Ann Arbor: University of Michigan Press), 34.

Eastern Painted Suillus

Suillus spraguei

A

B

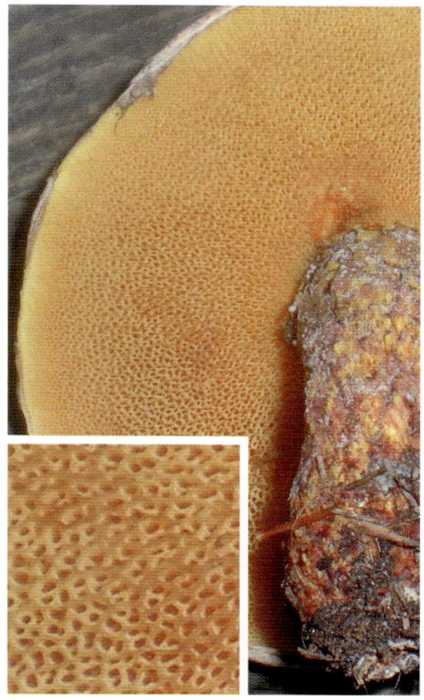

D C

Photos: A & B, John Plischke III; C & D, W. Light.

***Suillus spraguei* (Berkeley & M.A. Curtis) Kunze**, *Rev. Gen. Pl.* (Leipzig) 3(2):536
[Basidiomycota, Agaricomycotina, Agaricomycetes, Agaricomycetidae, Boletales, Suillaceae]

Basionym: *Boletus spraguei* Berkeley & M.A. Curtis, *Grevillea* 1(3):35 (1872).1

Derivation: *Suillus*, Latin, from *suis*, "pig-like"; *spraguei*, epithet in honor of American botanist and lichenologist Charles James Sprague (1823–1903).

Cap: 3–12 cm across (1¼–4¾ in), initially convex with inrolled margin (Fig. A), becoming broadly convex to nearly flat (Fig. B); margin often with whitish remnants of partial veil; surface dry, covered with pinkish, rose-reddish, or dark-red scales (Fig. B), fading with age.

Flesh: Yellow, not bruising blue but sometimes turning slightly reddish when injured; odor and taste unremarkable.

Pore surface: Yellow to dingy yellow, becoming darker with age (Fig. C), bruising reddish or brownish; pores angular, variable in size (Fig. D), 0.5–5 mm across, with thick walls, appearing to radiate from stipe; tubes 4–8 mm deep; covered with white partial veil when young.

Stipe: 4–12 cm tall by 1–2.5 cm thick (1⅔–4¾ × ⅖–1 in), solid, equal, but often expanded somewhat toward the base; lacking glandular dots, but covered with shaggy, scruffy, reddish scales and fibers (Fig. A).

Spore print: Olive-brown.

Bruising/chemical staining: Cap surface blackish with KOH, $FeSO_4$, and NH_4OH; flesh greyish-green, olive, or greenish-black with all three stains.

Microscopic: Spores 8–12 × 3.5–5 µm, subspindle-shaped, slightly asymmetric in profile, smooth, pale brown; nonreactive in Melzer's reagent.

Occurrence: Solitary or in scattered groups on the ground under Eastern White Pine (*Pinus strobus*), with which it is mycorrhizal; late summer–fall, eastern North America from Canada to the Carolinas; fairly common.

Edibility: Edible, turning blackish when cooked.

Comments: The only species in our area with which this species could be confused is *Suillus decipiens* (Berkeley & Curtis) Kuntze (1898). However, *S. decipiens* is smaller, the cap and stipe both being 3.5–7 cm (1¼–2¾ in) across and tall, respectively; the pores are also smaller (0.5–1 mm) across. It also stains differently: **KOH**—cap and flesh greenish-black; **$FeSO_4$**—cap and flesh slowly black; **NH_4OH**—cap rusty-brown, then greenish-black, flesh rusty-brown, then blackish.

This species is called *Suillus pictus* (Peck) Kunze (1898) in nearly all field guides and online sources. However, the basionym, *Boletus pictus* Peck (1870), is a later homonym of *Boletus pictus* Schultz (1806) [= *Polyporus pictus* (Schultz) Fries, a polypore, not a bolete]; it is thus an illegitimate name and cannot be used. In 1986 it was formally replaced by *Boletus spraguei* Berkeley & M.A. Curtis (1872).[1]

1. M.E. Palm & E.L. Stewart, 1986, *Mycologia* 78(3):325.

Southeastern Slippery Jill

Suillus cothurnatus

A–E, *Suillus cothurnatus*; F, *Suillus salmonicolor*. (Photos: A–D, W. Light; E, John Plischke III; F, Bob Zimmer.)

Suillus cothurnatus Singer, *Farlowia* 2:261 (1945) [Basidiomycota, Agaricomycotina, Agaricomycetes, Agaricomycetidae, Boletales, Suillaceae]

Derivation: *Suillus*, Latin, from *suis*, "pig-like"; *cothurnatus*, Latin, from *cothurnus*, from the Greek *kothornos*, a half boot worn by ancient Athenian actors in a tragic drama, and *-atus*, "having the characteristic of."

Deprecated synonym: *Suillus subluteus* (Peck) Snell (1944)

Cap: 1.6–6 cm (⅔–2½ in) across, convex, often with low umbo (Figs. A & C), margin incurved at first (Fig. E).

Flesh: Pale yellow when young (Fig. C), becoming orange-yellow to orange-buff; odor fragrant or not distinctive, taste not distinctive.

Pore surface: Pale yellow (Fig. D) to orange-yellow, darkening to brownish-yellow with age (Fig. E); pores round to ellipsoid to angular, 3–4 per mm, tubes 0.5 cm deep (¼ in).

Stipe: 2.5–6 cm tall by 0.5–1.0 cm thick (1–2½ × ¼–½ in), subequal or slightly tapered toward base, cream-colored to yellowish (Fig. C), becoming brownish with age, covered with tiny pale olive-brown glandular dots (Fig. C), these composed of brownish club-shaped caulocystidia; stipe connected to cap by thick, slimy, whitish, rubbery, baggy partial veil (Figs. A, B, & E), typically lacking a thick fleshy roll where it attaches to the stipe (Figs. A & E).

Spore print: Brown.

Bruising: Not bruising, all tissues unchanging when cut or injured (Fig. B).

Chemical staining: Cap surface brown with KOH.

Microscopic: Spores 8–10 × 2.5–3.5 µm, long-elliptical to subcylindrical, smooth, pale brown.

Occurrence: Mycorrhizal, scattered or in groups beneath pines (*Pinus* spp.); occasional in southeastern North America at least as far north as North Carolina; late spring–fall.

Edibility: Edible.

Comments: In our area this mushroom could be confused with the almost identical *Suillus salmonicolor* (Frost) Halling, a generally more northern species that differs in having a thick fleshy roll at the base of the rubbery, sheet-like partial veil (Fig. F). The cap also stains purplish-red and then black, rather than brown, with KOH.

Ash-tree Bolete

Boletinellus merulioides

A B D

Photos: A, B & D, W. Light; C, James Solomon, USDA Forest Service, Bugwood.org.

***Boletinellus merulioides* (Schweinitz) Murrill**, *Mycologia* 1(1):7 (1909)
[Basidiomycota, Agaricomycetes, Agaricomycetidae, Boletales, Boletinellaceae]

Basionym: *Daedalea merulioides* Schweinitz, *Trans. Am. Phil. Soc.*, ser. 2, 4(2):160 (1832)

Derivation: *Boletinellus*, Latin, "little bolete"; *merulioides*, "like *Merulius*," a genus of polypores with a similarly reticulated or net-like pore surface.

Fruitbody: 5–12 cm (2–4¾ in) wide, irregular, convex when young, becoming flat to concave with age (Fig. A); upper surface various shades of light to darker yellowish-brown or reddish-brown, sometimes bruising pale yellowish-brown; surface dry, slightly viscid when wet, smooth or with minute fibrils; margin even or slightly wavy, with narrow band of sterile tissue.

Flesh: Yellow (Fig. B); taste and odor unremarkable.

Stipe: 2–4 cm long by 0.5–2.5 cm thick (¾–1⅔ × ¼–1 in), not tapering, often curved, eccentrically attached to cap (Fig. B), upper portion yellow, lower portion brownish.

Pore surface: Pale yellow or dull gold to olive, forming net-like, reticular pattern of radiating ribs and cross-veins (Figs. B & D), pores angular to elongate, 1 mm wide and up to 6 mm deep (Fig. D); decurrent for a short distance (Fig. B).

Spore print: Pale yellow to olive-brown.

Staining: Cap surface staining blackish, then reddish or orangish with ammonia and KOH; flesh staining brown.

Microscopic: Spores 7–10 × 6–7.5 µm, oval to short elliptical, smooth.

Occurrence: Widespread in eastern North America; solitary, scattered, or gregarious, associated with ash trees (*Fraxinus* spp.) and occasionally on maples (we have found it on both hosts); summer and fall.

Edibility: Inedible.

Comments: This unusual bolete is not likely to be confused with any other species except *Boletinellus rompelii* (Patouillard & Rick) Watling. However, the stipe of the latter species is dark brown and lacks a yellowish upper portion. *Boletinellus merulioides* forms a mutualistic association with the Woolly Ash Aphid (Leafcurl Ash Aphid), *Meliarhizophagus fraxinifolii* (Riley), on the roots and leaves of ash trees (Fig. C). The fungus forms sclerotia that surround and protect the aphids, which in turn provide nourishment to the fungal mycelia in the form of honeydew. The aphids cause a leafcurl disease and can seriously damage the ash trees.

This bolete is called *Gyrodon merulioides* (Schweinitz) Singer in some field guides and online sources.

Yellow-gilled Bolete, Gilled Bolete

Phylloporus rhodoxanthus

A

B

C

Phylloporus rhodoxanthus (Schweinitz) Bresádola, *Fungi Tridentini* 2(14):95, figs. 15, 30 (1900) [≡ *Phylloporus rhodoxanthus* ssp. *americanus* Singer, *Rev. Mycol.* (Paris) 3:171 (1938)][1] [Basidiomycota, Agaricomycetes, Boletales, Boletaceae]

Basionym: *Agaricus rhodoxanthus* Schweinitz, *Schr. Naturf. Ges. Leipzig* 1:83 (1822)

Derivation: *Phylloporus*, Greek, *phyllos*, "leaf" or "blade," and *poros*, "pore" or "passageway"; *rhodoxanthus*, Greek, from *rhodon*, "rosy," and *xanthos*, "yellow."

Cap: 2.5–8 cm (1–3¼ in) across, initially convex, later becoming depressed or flattening into an even disk; at first tawny olivaceous (Fig. A), becoming cinnamon-brown; surface covered with matted hairs or fibers (Fig. A).

Flesh: Whitish, pale yellow to buff-colored, not bruising when exposed.

Gills: Vivid yellow when young (Figs. B & C), becoming yellowish-brown, not discoloring when bruised; close to subdistant, broad, 8 mm (⅓ in) deep; attached, notched, or decurrent, occasionally forking near stipe (Fig. C), nonanastomosing (not alternately joining and separating); edges even to slightly serrate.

Stipe: 3.5–8 cm tall by 5–15 mm wide (1.5–3¼ × ⅕–⅔ in), subequal, dry, often longitudinally ridged or ribbed at top (Figs. B & C), pale yellowish, with reddish dots and patches (Fig. B). *Basal mycelium yellow* (Fig. D), this feature diagnostic!

D

Spore print: Yellowish.

Bruising/chemical staining: Flesh and gills not bruising when cut. **NH_4OH**—cap surface instantly staining pale blue, this quickly spreading out as a halo and fading; **KOH**—all tissues greenish-blackish-brown, this soon fading to dark yellow-brown; **Melzer's reagent**—cap surface reddish-brown, flesh and gills bluish-greenish-grey.

Microscopic: Spores 8–14.5 × 3.5–5.5 µm, long-elliptical to sausage-shaped, smooth (spinose under SEM), inamyloid (no change) with Melzer's reagent, honey-colored with KOH.

Occurrence: Widespread in North America south to Belize; ectomycorrhizal with oaks, beeches, and pines; solitary, scattered, or clustered; occasional in summer and fall.

Edibility: Edible, but taste not distinctive.

Comments: The pore surface of this bolete and several other closely related species is strongly lamellate; it looks for all the world like the gills of typical gilled mushrooms. The very similar *Phylloporus leucomycelinus* Singer (1978), which also occurs in our area, has a white (rather than yellow) basal mycelium, the yellowish-white flesh in the cap bruises blue and bright yellow (cinnamon just beneath the cuticle), and the spores are dextrinoid (reddish-brown) with Melzer's reagent. It is associated primarily with beeches but also possibly with oaks.

The two species are otherwise very similar, and *P. leucomycelinus* was originally invalidly described as a subspecies of *P. rhodoxanthus*;[2] see M.A. Neves & R.E. Halling (2010, *Mycologia* 102[4]:923–943) for a full discussion of these two species.[3]

Some prominent field guides refer to this taxon as *Phylloporus rhodoxanthus* ssp. *americanus* Singer (1938), but these two names are objective homonyms.

1. *Phylloporus rhodoxanthus* was neotypified (Neves & Halling, 2010, *Mycologia* 102[4]:939) and is now defined by material of *P. rhodoxanthus* subsp. *americanus* collected and identified by Singer, thus making these two names objective identities. See online Appendix A for a discussion of this issue.

2. No Latin description was provided, and no type specimen was designated. Singer subsequently did validly publish this name and elevated the taxon to species rank: *Phylloporus leucomycelinus* Singer, *Persoonia* 9:426 (1978).

3. Another very similar species, *Phylloporus foliiporus* (Murrill) Singer (1978), has pale yellow basal mycelia like *P. rhodoxanthus*, but the gills bruise bluish or greenish, the cap stains blue-green and the flesh does not react with KOH, the spores are amyloid (bluish-black) in Melzer's reagent and stain ochre in KOH; clamp connections are rare but present in the hymenophore (these are lacking in the other two species discussed here). This species is associated with oaks and pines and is restricted to extreme southern Alabama and Florida in North America.

Parasitic Bolete

Pseudoboletus parasiticus

Photo: Hans Hillewaert.

Pseudoboletus parasiticus **(Bulliard) Šutara**, *Česká Mykol.* 45(1–2):2 (1991)

Basionym: *Boletus parasiticus* Bulliard, *Herb. Fr.* 10:tab. 451, fig. 1 (1790)
[Basidiomycota, Agaricomycotina, Agaricomycetes, Agaricomycetidae, Boletales, Boletaceae]

Derivation: *Pseudoboletus*, Latin, from *pseudo*, "false," and *boletus*, "fleshy-pored fungus," false bolete; *parasiticus*, "parasitic."

Cap: 2–8 cm (¾–3¼ in) across, convex, becoming broadly convex; dry, smooth, margin initially incurved; olive to olive-brown.

Flesh: Pale lemon-yellow, not bruising.

Pore surface: Yellowish to olive or olive-brown; pores angular, 1–2 mm wide, tubes up to 6 mm deep (up to 1 cm with age); honey-yellow to greenish-yellow, slowly bruising ochre or reddish-brown.

Stipe: 3–6 cm tall by up to 1.5 cm thick (½–2¼ × ⅔ in), subequal; dry, solid, concolorous with the cap; covered with yellowish-brown matted fibers (visible with a hand lens).

Spore print: Olive-brown.

Chemical staining: **KOH**—cap dark orange-brown, flesh orange-ochre, stipe reddish-orange; **$FeSO_4$**—flesh olive-green.

Microscopic: 12–18.5 × 3.5–5 µm, smooth, elliptical, pale brown; bright yellow in KOH, dull yellowish-tan in Melzer's reagent.

Occurrence: Exclusively parasitic. Alone or in small groups growing on the fruitbodies of the Pigskin Poison Puffball, *Scleroderma citrinum* Persoon; widespread in eastern North America, Canada south to the southern Appalachian Mountains in North Carolina; summer–early fall.

Edibility: Edible.

Comments: This species is immediately identified by only fruiting on the fruitbodies of the Pigskin Poison Puffball, *Scleroderma citrinum* Persoon. It barely enters our area in the southern Appalachians in North Carolina.

Inky Caps

(Coprinoid Mushrooms)

Coprinoid mushrooms—the so-called Inky Caps—are decomposing saprobes. Most leave black or dark brown spore prints and have the odd tendency to autodeliquesce or liquefy from the bottom upward into an inky slime. This transformation takes place rather quickly, making it a real challenge to get them home intact if you are collecting for the table. A very few species do not deliquesce, which is itself an important taxonomic feature.

Certain otherwise edible species in the genus *Coprinopsis* contain coprine, which is metabolized into a compound that causes the disulfiram-ethanol (Antabuse®) reaction when consumed within a few days of drinking beverage alcohol.[1] Drinking even one alcoholic beverage three hours *before* or up to five days *after* eating coprine-containing mushrooms can result in severe headache, flushing, nausea, vomiting, and heart palpitations lasting from a few hours to several days. The symptoms typically arise within 15 minutes and usually subside without complications. But several severe coprine-ethanol poisonings required monitoring for cardiac arrhythmias, and one victim suffered a fatal rupture of the esophagus after extremely severe vomiting.[2]

Coprinoid mushrooms found in the Southeast typically reported to contain coprine include (1) genus *Coprinus*: *C. comatus* and *C. sterquilinus*; (2) genus *Coprinopsis*: *C. atramentaria* and *C. variegatus*; (3) genus *Coprinellus*: *C. disseminatus*, *C. domesticus*, *C. micaceus*, *C. radians*, and *C. truncorum*; and (4) *Parasola plicatilis*.[3]

Despite their similar appearance and tendency to autodeliquesce, coprinoid mushrooms are not monophyletic (i.e., they are not all in the same evolutionary lineage). Once all grouped together in the genus *Coprinus* (*sensu lato*),[4] they are now split up into two distinct families: the Psathyrellaeceae Vilgalys, Moncalvo & Redhead (containing the genera *Coprinellus*, *Coprinopsis*, and *Parasola*) and the Agaricaceae Chevallier (containing *Coprinus* and *Podaxis*). Molecular analysis revealed that the traditional genus *Coprinus* consists of species now placed in six different genera: *Coprinus*, *Coprinopsis*, *Coprinellus*, and *Parasola*[5] plus two monotypic genera not found in North America, *Xerocoprinus* Maire (1906) and *Montagnea* Fries (1836). The former family Coprinaceae became a later synonym of Agaricaceae and thus no longer exists.[6]

Cystidia (sing. cystidium) are large sterile hyphal cells growing on or from the face (pleurocystidia) or edge (cheilocystidia) of a gill or from the surface of the stipe (caulocystidia). The presence or absence of these cystidia is important in distinguishing many coprinoid taxa.

1. Coprine and disulfiram both inhibit aldehyde dehydrogenase, the enzyme that converts acetaldehyde—the first metabolite of ethanol, which is responsible for "hangover" symptoms—into harmless acetic acid (vinegar). This causes acetaldehyde to accumulate to toxic levels in the blood.

2. J.H. Mayer et al., 1971, Esophageal rupture after mushroom and alcohol ingestion, *N. Engl. J. Med.* 285:1323.

3. Although widely reported to contain coprine, this compound does not exist in either *Coprinus comatus* or *Coprinellus micaceus*. See R. & W.Z. Antowiak, 1991, Alkaloids from mushrooms, *in* A. Brossi (ed.), *The Alkaloids: Chemistry and Pharmacology* (San Diego: Academic Press), 298.

4. Some authors still group them all in the genus *Coprinus* (*sensu lato*).

5. S.A. Redhead, R. Vilgalys, J.-M. Moncalvo, J. Johnson, & J.S. Hopple Jr., 2001, *Taxon* 50:203–241.

6. *Coprinus comatus* is the type species of the genus, and *Coprinus* is the type genus of the family Coprinaceae. Since the genus *Coprinus* is now included within the Agaricaceae, the Coprinaceae has disappeared.

Key to Southeastern Coprinoid Mushrooms

1. Mature cap usually taller than 5 cm (2 in); stipe hollow, with a ring and with a string-like strand of fibers running through the cavity (Genus *Coprinus*) 2

1′. Mature cap shorter; stipe lacking both ring and fiber strands 3

2. Cap 5–14 cm (2–5½ in) tall, on disturbed ground or grassy area . . . . *Coprinus comatus*

2′. Cap 3–6 cm (1¼–2½ in) tall, on horse dung or straw . . . . *Coprinus sterquilinus*

3. Stipe arising from a yellowish-brown, orange spiny fibrous carpet or ozonium at the base and/or cap with mica-like granules 4

3′. Not as above 7

4. Stipe arising from a yellowish or orange-colored ozonium . . . . *Coprinellus domesticus*/*C. radians* complex

4′. Stipe not arising from a yellowish or orange ozonium 5

5. Cap up to 5 cm (2 in) wide, radially grooved, studded with mica-like granules, these often washed away by rain; gills deliquescing 6

5′. Cap 2 cm (¾ in) or less in width, lacking mica-like granules at any stage, but covered with minute whitish flecks; gills not deliquescing . . . . *Coprinellus disseminatus*

6. Stipe covered with white powdery granulations, spores mitriform (shaped like a bishop's cap) and with truncated apical pore . . . . *Coprinellus micaceus*

6′. Stipe without white powdery granulations, spores elliptical, with rounded apical pore . . . . *Coprinellus truncorum*

7. Cap tiny, 3.5 cm (1⅖ in) or less across, deeply pleated, lacking scales, granulations or powdery coating; gills not deliquescing . . . . *Parasola plicatilis*

7′. Cap much larger, 3–7.5 mm (1¼–3 in) tall, oval at first, expanding to 7.5–10 cm (3–4 in) across; gills eventually at least partially deliquescing 8

8. Cap covered by tattered remnants of universal veil, these becoming tattered and ragged scales at maturity . . . . *Coprinopsis variegata*

8′. Cap smooth, lacking scales or remnants of universal veil, but usually with fine scales at the center . . . . *Coprinopsis atramentaria*

Tabular Key to Coprinoid Mushrooms

Characteristic	*Coprinus*	*Coprinopsis*	*Coprinellus*	*Parasola plicatilis*
Cap	5–14 cm tall, covered with large shaggy scales at maturity ***C. comatus*** 3–6 cm tall, surface smooth, pale tan at disk, elsewhere whitish, covered with powdery fibrils or delicate scales ***C. sterquilinus***	Smooth to silky, greyish-brown ***C. atramentaria*** Densely covered with tattered remnants of universal veil ***C. variegata***	Tawny, with glistening mica-like granules ***C. micaceus/C. truncorum*** Tawny to yellowish-orange, with brownish scales ***C. radians/C. domesticus***	Cap tiny, 3.5 cm, deeply pleated, lacking scales, granulations, or powdery covering; tawny central eye contrasts with rest of greyish cap
Gills	Initially pinkish, becoming black; always completely deliquescing	Initially white, becoming pinkish-grey, greyish, or purplish grey-brown; completely deliquescing	At first white, soon deliquescing and turning black ***C. radians/C. domesticus*** Initially off-white to grey, becoming brown, then deliquescing to black ***C. micaceus/C. truncorum***	At first white, becoming grey to blackish, not deliquescing
Stipe	Hollow, with central stringlike strand; with inferior ring; pseudovolva present	Central strand lacking; no ring	Central strand lacking; no ring	Central strand lacking; no ring; round, subequal, with subbulbous base
Ozonium	Absent	Absent	Conspicuous, rusty or orange ***C. radians/C. domesticus*** Inconspicuous, unpigmented ***C. micaceus/C. truncorum***	Absent
Microscopic	Purple-brown spores 9–14 × 6–9 µm ***C. comatus*** Dark-brown spores 16–25 × 10–13 µm ***C. sterquilinus***	Spores 7–11 × 4–6 µm, elliptical, smooth, with apical germ pore	Spores 7–11 × 4–7 µm, elliptical ***C. truncorum*** Spores shaped like bishop's mitre ***C. micaceus*** Spores 8.5–12 × 5.5–7 µm ***C. radians*** Spores 6–9 × 3.5–5 µm ***C. domesticus***	Spores smooth, 10–15 × 8–11 µm, angular-ovoid, with prominent eccentric pore; dark reddish-brown in KOH

Shaggy Mane, Shaggy Ink Cap

Coprinus comatus

A B C D

Photos: A & B, W. Light; C, Drahkrub; D, Holger Krisp.

***Coprinus comatus* (O.F. Müller) Persoon**, *Tent. Disp. Meth. Fung.* (Lipsiae):62 (1797)
[Basidiomycota, Agaricomycotina, Agaricomycetes, Agaricomycetidae, Agaricales, Agaricaceae]

Basionym: *Agaricus comatus* O.F. Müller, *Fl. Danic.* 5:tab. 834 (1780)

Derivation: *Coprinus*, Greek (*kóprinos*), "dung," "manure," referring to the substratum in which many of these species grow; *comatus*, Latin, "hair on head," "mane," referring to the shaggy remnants of the universal veil.

Cap: 3–15 cm tall by 2.5–4.5 cm wide (1¼–6 × 1–1¾ in); elongate, columnar, rounded cyclindrical when young (Figs. A & D), becoming bell-shaped with uplifted margin and nearly plane when fully deliquesced (Fig. D at right); mostly white, covered with large shaggy recurved white scales that are pale brownish at the central disk (apex).

Flesh: Soft, thin, white, becoming black as cap deliquesces; taste and odor mild.

Gills: Attached or free, crowded, pink when young (Fig. B), becoming black and liquid as they deliquesce from the margin (Figs. B–D).

Stipe: 5–20 cm (2–8 in) tall by 1–2 cm (0.4–0.8 in) thick; smooth to silky-fibrous, white, hollow, with long string-like cottony strand of fibers, thickened below to bulbous base; ozonium absent; with low thin ring (Figs. A & C).

Spore print: Black.

Microscopic: Spores 9–14 × 6–9 µm, smooth, elliptical, truncate, with slightly eccentric apical pore, smooth, purple-brown.

Occurrence: Saprobic, solitary or in clusters or forming fairy rings in grass, wood chips, or soil, May–November, widespread and common in North America and Europe; introduced in Australia and New Zealand.

Edibility: Edible when young; may cause gastric upset in some people.

Comments: *Coprinus comatus* is one of a number of carnivorous mushrooms that kill and eat soil nematodes (a type of roundworm) as a source of nitrogen. This mushroom uses numerous tiny "spiny balls" no larger than about 50 μm in diameter to inflict mechanical damage to the nematode's cuticle. Because of the worm's high internal turgor, this causes the internal contents to leak out. This mushroom also produces at least seven potent toxins that add to the spiny balls' lethal effect on the worms.

Since this mushroom deliquesces (autodigests its cap and gills) very quickly after it is gathered, it is very difficult to get back home before it dissolves into an inky, sticky mess. Although widely reported to contain coprine, which causes a severe Antabuse®-like reaction when consumed with alcohol, this compound reportedly does not exist in either the Shaggy Mane, *Coprinus comatus*, or the Mica Cap, *Coprinellus micaceus*.[1] Nevertheless, gastrointestinal intoxications resembling the disulfuram-ethanol reaction after eating this species with and without concomitant consumption of alcohol have been reported, and the issue is far from settled.[2]

Coprinus now contains only a dozen or so species, two of which occur in our area: *Coprinus comatus* and *C. sterquilinus* (Fries 1821) Fries (1838). *Coprinus sterquilinus* is smaller and grows on horse dung, manured soil, or straw. The fibrillose to finely scaled cap is 3–6 cm tall by 2.0–2.5 cm wide (1¼–2½ × ¾–1 in) and the stipe is 4–9 cm (1⅔–2⅔ in) tall. The dark-brown smooth spores are much larger (16–25 × 10–13 μm) than those of *C. comatus*.

1. R. & W.Z. Antowiak, 1991, Alkaloids from mushrooms, *in* A. Brossi (ed.), *The Alkaloids: Chemistry and Pharmacology* (San Diego: Academic Press), 298.

2. M.W. Beug, M. Shaw, and K.W. Cochran, 2006, Thirty-plus Years of Mushroom Poisoning: Summary of the Approximately 2,000 Reports in the NAMA Case Registry, *McIlvainea* 16(2):47–68. http://www.cnr.berkeley.edu/brunslab/pmb113/lectures/Beug_poisonings.pdf.

Mica Cap Complex

Coprinellus truncorum and *C. micaceus*

A

B

Photos: A, W. Light; B, Teresa Fortenberry.

Coprinellus truncorum **(Scopoli) Vilgalys, Hopple & Jacq. Johnson**, in Redhead, Vilgalys, Moncalvo, Johnson, & Hopple, *Taxon* 50(1):234 (2001).

Coprinellus micaceus **(Bulliard) Vilgalys, Hopple & Jacq. Johnson**, in Redhead, Vilgalys, Moncalvo, Johnson, & Hopple, *Taxon* 50(1):234 (2001)

Derivation: *Coprinellus*, Greek, *kopros*, "dung," "little *Coprinus*"; *micaceus*, Latin, "micaceous," "mica-like," "glittering." [Basidiomycota, Agaricomycetes, Agaricomycetidae, Agaricales, Psathyrellaceae]

Basionyms: *Agaricus truncorum* Scopoli, *Fl. Carniol.*, 2nd ed. (Wien) 2:426 (1772); *Agaricus micaceus* Bulliard, *Herb. Fr.* 6: plate 246 (1786) [see online Appendix A for synonymy]

Cap: 2.5–5 cm wide; egg-shaped when young, becoming bell-shaped to convex; ochraceous to tawny-brown or amber, becoming paler with age; with radial furrows extending from margin toward center (Figs. A & B); *with glistening mica-like granules* (Fig. A) that are often washed off by rain; margins often tattered or upturned in age.

Gills: Adnate (attached) or free, crowded, off-white to grey, becoming brown and then deliquescing to black from margin to center together with cap trama (flesh).

Stipe: Stipe pruinose (covered with white powdery granulations) due to abundant caulocystidia (*C. micaceus*) (Fig. B) or stipe smooth (*C. truncorum*).[1]

Spore print: Black.

Microscopic: Spores 7–11 × 4–7 µm, elliptical (*C. truncorum*) or mitriform (shaped like a bishop's mitre), smooth, with germ pore at truncated end (*C. micaceus*); pileipellis (cuticle) of cuboidal epithelium; pleurocystidia and cheilocystidia ovoid, elliptical, or subclavate, 150 × 70 µm; stipe with lageniform (flask-shaped, with long necks) caulocystidia up to 100 µm in length (*C. micaceus*), or caulocystidia lacking (*C. truncorum*).

Occurrence: Cespitose; sprouting in dense clusters on and around hardwood stumps and buried wood, spring–fall in forests, parks, and gardens. Cosmopolitan, common and widely distributed in North America up to Alaska, South America, Europe, northern and southern Africa, India, Japan, Australia and New Zealand, Hawaiian Islands.

Edibility: Edible when young.

Comments: The Alcohol Ink Cap, *Coprinopsis atramentaria*, is distinguished from the Mica Cap Complex by the presence of a fleshy annulus near the base of the stipe and very small scattered scales near the center of the cap that do not exhibit a mica-like sheen. The cap is also grey or greyish-brown rather than amber or honey-brown. The spores are also almond-shaped and not truncated or mitre-shaped, as in *C. micaceus*.

The transient presence of mica-like granular remnants of the universal veil on the cap distinguishes *Coprinellus truncorum*/*C. micaceus* from other coprinoid mushrooms. This mushroom is saprotrophic and derives nourishment from well-decayed organic matter. Although widely reported to contain coprine, which causes a severe Antabuse®-like reaction when consumed with alcohol, this compound does not exist in either the Mica Cap or the Shaggy Mane, *Coprinus comatus*.[2]

Evidence suggests that *Coprinellus micaceus* and *C. truncorum* may be genetically identical, in which case *C. micaceus* would become a deprecated synonym of *C. truncorum*.[3] Pending verification of this, these two taxa are treated as distinct for the time being.

1. Alas, as the mushroom ages these caulocystidia are worn away, and the stipe often appears smooth.
2. R. & W.Z. Antowiak, 1991, Alkaloids from mushrooms, *in* A. Brossi (ed.), *The Alkaloids: Chemistry and Pharmacology* (San Diego: Academic Press), 298.
3. M.R. Keirle et al., *Fungal Diversity* 15(4):33–124, see 99.

Alcohol Ink Cap, Common Ink Cap

Coprinopsis atramentaria

A

B

C

Photos: A, Jerzy Opiola; B, Dan Molter; C, Rob Hille.

***Coprinopsis atramentaria* (Bulliard) Redhead, Vilgalys & Moncalvo,** *in* Redhead, Vilgalys, Moncalvo, Johnson, & Hopple, *Taxon* 50(1):226 (2001)

Basionym: *Agaricus atramentarius* Bulliard, *Herb. Fr.* 6: tab. 164 (1786)
[Basidiomycota, Agaricomycetes, Agaricomycetidae, Agaricales, Psathyrellaceae]

Derivation: *Coprinopsis*, Latin, "like *Coprinus*"; *atramentaria*, "inkwell."

Cap: 3–8 cm across (1¼–3¼ in), initially ovoid to egg-shaped (Fig. A), becoming convex (Fig. B) to bell-shaped (Fig. C) with age; grey to greyish-brown, dry, smooth or with small scattered scales over the disk (Figs. A & B), often radially striated or grooved (Figs. A–C); deliquescing from the margin toward the center.

Flesh: Thin, pale grey; odor and taste not distinctive.

Gills: Free, very crowded, initially white, quickly becoming grey, greyish-brown, then black and deliquescing into an inky liquid.

Stipe: 4–15 cm tall by 1–2 cm thick (1⅔–6 × ½–¾ in), equal, with a small fleshy ring near the base (Fig. A, arrow); whitish, smooth to silky-fibrillose.

Spore print: Black.

Microscopic: Spores (6.5–)8–11(–12) × (4–)5–6(–6.5) µm, ellipsoid to almond-shaped, with apical pore, smooth, pale greyish-black; clamp connections present.

Occurrence: Saprobic; in clusters on decaying wood or wood chips or from roots at the base of trees (may appear to be terrestrial if growing on buried wood); widespread and common in grassy areas, meadows, disturbed ground in North America, Europe, and Asia; also in Australia and South Africa; late spring–fall (sometimes in winter).

Edibility: Edible, but like other species in the genus *Coprinopsis*, *C. atramentaria* contains coprine. This compound induces a severe disulfiram (Antabuse®) reaction with beverage alcohol consumed up to five days or more after eating this species (see the comments below).

Comments: The species of the Mica Cap Complex (*Coprinellus micaceus* and *C. truncorum*) are distinguished from *Coprinopsis atramentaria* by the glistening mica-like granular remnants on the universal veil on the cap and the lack of a fleshy annulus at the base of the stipe. However, the mica-like granules are easily washed away by rain and are often absent. The caps of the mica caps are also usually some shade of amber, beige, or honey-brown rather than grey or greyish-brown. *Coprinellus micaceus* also exhibits a dense powdery dust on the stipe due to abundant caulocysticia; unfortunately, these caulocystidia tend to wear away with age, so that the stipe appears smooth.

Coprinopsis atramentaria is the coprinoid mushroom classically associated with the severe, toxic disulfiram-ethanol (Antabuse®) reaction.

Orange-mat Coprinus Species Complex

Coprinellus radians/C. domesticus

A

B

A, fruitbody, arrow indicates orange ozonial mat; B, enlarged view of ozonium.

Coprinellus radians **(Desmazières) Vilgalys, Hopple & J. Johnson**, in Redhead, Vilgalys, Moncalvo, Johnson, & Hopple, *Taxon* 50(1):234 (2001)

Basionym: *Agaricus radians* Desm., *Annls Sci. Nat.*, Bot., sér. 1, 13:214 (1828)

Coprinellus domesticus **(Bolton) Vilgalys, Hopple & J. Johnson**, in Redhead, Vilgalys, Moncalvo, Johnson, & Hopple, *Taxon* 50(1):233 (2001)

Basionym: *Agaricus domesticus* Bolton, *Hist. Fung. Halifax* (Huddersfield) 1:26 (1788)

Deprecated synonym: *Coprinus domesticus* (Bolton) Gray, *Nat. Arr. Brit. Pl.* (London) 1:635 (1821)

Anamorphic form-genus and form-species: *Hormographiella verticillata* Guarro, Gené, de Vroey & Gueho, *Mycotaxon* 45:179–190 (1992)
[Basidiomycota, Agaricomycotina, Agaricomycetes, Agaricomycetidae, Agaricales, Psathyrellaceae]

Derivation: *Coprinellus*, Latin, "little *Coprinus*"; *radians*, "having rays or spokes"; *domesticus*, "belonging to the house," referring to the fact that it often sprouts on carpets inside dwellings.

Cap: Convex to conical, 2–3 cm (⅘–1¼ in) wide, tawny to yellowish-orange, with delicate, radiating, rounded vertical ridges and grooves from near the center to the margin, covered with brownish scale-like remnants of the universal veil reminiscent of many *Amanita* species (Fig. A).

Gills: Attached to nearly free, at first white, soon turning black, deliquescing to inky liquid.

Stipe: 2.5–7.5 cm (1–3 in) tall by up to 5 mm (¼ in) thick. White, hollow, delicate, fragile, often arising from a densely tangled spiny carpet (ozonium) of yellow to bright orange mycelium covering the woody substratum (Fig. B). This feature is unmistakable and diagnostic.

Spore print: Black.

Microscopic: Spores 8.5–12 × 5.5–7 µm, cylindrical-ellipsoid or ellipsoid, with eccentric germ pore, arising from tips of slightly swollen branches of conidiophores. Pleurocystidia 50–120 × 30–65 µm, subglobose, ellipsoid, broadly utriform (flask-shaped) or subcylindrical. Cheilocystidia flask-shaped and 30–60 × 12–20 × 5–10 µm or subglobose, ellipsoid, ovoid, broadly utriform and 40–100 × 25–50 µm. Caulocystidia 35–100 × 10–28 × 7–12 µm, flask-shaped. Anamorphic ozonia characterized by highly septate sterile mycelia lacking clamp connections or with very rare unfused pseudoclamps.

Edibility: Unknown.

Occurrence: Saprobic, a white-rot fungus; growing alone or in groups on wet decaying hardwood logs, May–October. Often invades wet wood in basements and other rooms. The anamorphic ozonium may occur without the fruitbody and vice versa. Widespread in North America, Europe, Asia.

Comments: The almost identical *C. domesticus* can be distinguished from *C. radians* by its smaller spores (6–9 × 3.5–5 µm) and larger cheilocystidia (60 × 100 µm). *Coprinellus radians* ozonial mycelia have occasional unfused pseudoclamp connections; these are lacking in *C. domesticus*. Their microscopic features are otherwise indistinguishable.

The rust-orange spiny ozonia of *Coprinellus radians* and *C. domesticus* are anamorphic mats of mycelia that propagate vegetatively and produce asexual monokaryotic spores. In nature these ozonia can also give rise to teleomorphic fruitbodies, which in turn produce sexual spores on basidia in the gills. *Hormographiella verticillata* Gené & Guarro 1992 is the anamorphic stage of *Coprinellus domesticus*, and *Hormographiella*-type sporulation is also documented in *C. radians*.[1]

1. S.M. Badalyan, M. Navarro-González, & U. Kües, *Proceedings of the 7th International Conference on Mushroom Biology and Mushroom Products* (ICMBMP7), Arcachon, France, October 4–7, 2011, 140–154.

Non-inky Coprinellus, Fairy Bonnet

Coprinellus disseminatus

A

B

C

Photos: A, Jim Edwards; B & C, W. Light.

Coprinellus disseminatus (Persoon) J.E. Lange, *Dansk Bot. Ark.* 9(6):93 (1938)
[Basidiomycota, Agaricomycotina, Agaricomycetes, Agaricomycetidae, Agaricales, Psathyrellaceae]

Basionym: *Agaricus disseminatus* Persoon, *Syn. Meth. Fung.* (Göttingen) 2:403 (1801)

Derivation: *Coprinellus*, Latin, "little Coprinus"; *disseminatus*, Latin, "dispersed," "disseminated."

Cap: Tiny, cap 0.5–1.5 cm across by 0.6–1.0 cm high (¼ × ¼–½ in), convex to broadly parabolic (Fig. A) to bell-shaped (Fig. B); surface dry, minutely hairy, and covered with minute shiny particles when young, becoming smooth at maturity, with darker lines or furrows from the margin almost to the disk (Figs. A & B); disk pale yellow-brown to tawny-brown, greyish-white to grey toward the margin, with darker striae (Fig. B). Extremely fragile, crumbles when handled.

Flesh: Very thin (1 mm), brittle, whitish; odor faintly fungal.

Gills: Attached or free, close to subdistant, 2–4 mm broad, at first white, soon becoming greyish, then blackish; not deliquescing.

Stipe: 1.5–4 cm (⅔–1⅔ in) tall by 1–2 mm thick, equal, straight or slightly bent from the base; hollow, fragile, white to greyish, surface smooth or covered with white bristly hairs (Fig. C). Partial veil lacking.

Spore print: Blackish to blackish-brown.

Microscopic: Spores 6.8–10 × (3.2–)4.4–5.2(–6) µm, elliptical, smooth, truncated, prominent apiculus with central pore; pale to dark brown, occasionally with yellowish to chestnut hues.

Occurrence: Typically in very large, dense troops often numbering hundreds or even thousands of fruitbodies. Saprobic on hardwood debris, at the bases of stumps, and growing from buried and decayed roots and stumps. Widespread, occasional to locally common; North America, South America, Europe, Near East, East Africa, Sri Lanka, Japan, Australia, New Zealand, Hawai'i.

Edibility: Apparently edible, but hardly worth the effort.

Comments: *Parasola plicatilis* (Curtis *ex* Fries) Redhead, Vilgalys & Hopple (2001) is similar and, like *C. disseminatus*, does not deliquesce. It is slightly larger: the cap is 1.0–3.5 cm (⅔–1½ in) across and deeply grooved, and the stipe is up to 10 cm (4 in) tall. The cap is never powdery or covered with minute glistening particles like *C. disseminatus*. The pale greyish-black spores are significantly larger (10–13 × 7–10 µm), broadly and angularly egg-shaped, with a prominent eccentric pore at the angular (pointed) tip.

Variegated Inky Cap, Scaly Inky Cap

Coprinopsis variegata

A

B

Photos: Teresa Fortenberry.

***Coprinopsis variegata* (Peck) Vilgalys, Hopple & Jacq. Johnson**, in Redhead, Vilgalys, Moncalvo, Johnson, & Hopple, *Taxon* 50(1):234 (2001)

Basionym: *Coprinus variegatus* Peck, *Bull. Buffalo Soc. Nat. Sci.* 1:54 (1873) [see online Appendix A for synonymy] [Basidiomycota, Agaricomycetes, Agaricomycetidae, Agaricales, Psathyrellaceae]

Derivation: *Coprinopsis*, Latin, "like *Coprinus*"; *variegata*, "variegated," having patches of various colors.

Cap: 2.5–7.5 cm wide × 5–7.5 cm high, egg-shaped when young (Figs. A & B), whitish, covered with tattered remnants of universal veil (Fig. B), with maturity becoming bell-shaped, grey to grey-brown and covered with large, scattered, loose, white to buff flaky scales (Fig. A).

Gills: Free or attached to stipe; at first white, becoming greyish or purplish grey-brown, later turning black and eventually deliquescing.

Stipe: 5–12.5 cm (2–5 in) tall, 0.5–1 cm (~¼–½ in) in diameter, nontapering, surface white, woolly to scaly, sometimes with wispy transient ring near base; with brownish rhizomorphs running from the base.

Spore print: Dark brown to black.

Microscopic: Spores 7.5–10 × 4–5 µm, elliptical, smooth, appearing truncated due to apical germ pore; pileipellis (cuticle) with hyphae running parallel to surface of cap (pileipellis a cutis); basidia translucent, 14–16 × 6.5–7.5 µm; pleurocystidia numerous, 100–150 × 20–35 µm; cheilocystidia present when young; clamp connections abundant in all tissues.

Occurrence: Spring and early summer through early fall in large clusters on well-decayed surfaces or buried in hardwood or leaf litter; widespread but uncommon east of the Great Plains.

Edibility: Edible, but not very palatable. Like other species in the genus *Coprinopsis*, *C. variegata* contains coprine. This compound induces a severe disulfiram (Antabuse®) reaction with beverage alcohol consumed up to five days or more after eating this species.

Comments: When nutrients are low, this species attacks and consumes certain soil bacteria like *Pseudomonas* and *Agrobacterium*. The hyphae are strongly attracted toward colonies of these bacteria, where they produce haustoria that lyse the bacteria and absorb their contents. The similar Alcohol Ink Cap, *Coprinopis atramentaria*, lacks the scaly remnants of the universal veil. In their *Field Guide to Southern Mushrooms*, Weber and Smith (1985) described a variant of this species under the provisional name "Coprinus americanus." This latter name was never validly published; "Coprinus americanus" is thus a *nomen nudum* with no taxonomic standing. This taxon does not appear to differ in any important respect from *Coprinopsis variegata*.

The Amanitas

(Order Agaricales, Family Amanitaceae R. Heim *ex* Pouzar, Type Genus *Amanita* Persoon)

The classic agaric genus *Amanita* is found worldwide. It is a huge genus with more than 800 species,[1] including some of the most iconically beautiful[2] and deadly poisonous mushrooms known. This genus causes about 90% of all cases of fatal mushroom poisoning, with the Death Caps and Destroying Angels in section *Phalloideae* (see below) being responsible for most of these. The toxin involved is α-amanitin, an amatoxin that destroys the liver and kidneys;[3] its actions and effects are fully described in the section dealing with mushroom poisoning in the introduction.

Although most mycologists (ourselves included) recommend against anyone (even knowledgeable mycophiles) eating any *Amanita* species (even such easily recognized edible species as the Grisette, *A. vaginata*), several species are highly regarded as choice edibles and are regularly consumed in certain regions. Many amanitas are quite delectable, and even the deadly poisonous ones are reported to have been a delicious "last meal."

Two common features uniquely shared by all members of this family are (1) a bilaterally divergent gill tissue (hyphae diverging in opposite directions on either side of a central vertical strand in each gill) and (2) stipes with vertically oriented connective hyphae and numerous, large, terminally inflated, club-shaped cells.

In addition to the genus *Amanita*, this family also contains the monotypic and Neotropical genus *Catarama* containing the sole species *Catarama costaricensis* Franco-Molano, as well as the genus *Limacella* Earle.

Genus *Amanita* Persoon, *Tent. Disp. Meth. Fung.*:65 (1797).

Type species: *Amanita muscaria* (Linnaeus) Persoon, *Tent. Disp. Meth. Fung.*:65 (1797).

Distinguishing features: Gills free or almost free, crowded; cap viscid or dry; lamellar trama (gill hyphae) divergent; universal veil often persisting as a volva and/or patches of tissue on the cap surface or these absent; partial veil (ring) present (persistent or ephemeral) or absent; margin of cap entire or striate; spore print white, spores smooth, clear, subglobose to elliptical, amyloid or not reactive in Melzer's reagent; growing on the ground, rarely on wood. The species in this genus are mostly obligately ectomycorrhizal with oaks and other hardwoods, as well as with conifers, but a few are saprobic.

The genus *Amanita* has been divided up into various **sections**, which are below the level of subgenus but above the level of **stirps**. These sections are briefly characterized as follows:[4]

Section *Amanita*: Type species, *Amanita muscaria* (Linnaeus) Persoon

Features: Stipe typically with a bulbous base; toxins include muscimol and ibotenic acid; a few species also produce muscarine. Species in this section found in our area include *A. muscaria* subsp. *flavivolvata* Singer, *A. muscaria* var. *guessowii* Veselý, *A. muscaria* subsp. *flavivolvata* Singer, *A. muscaria* var. *persicina* D. T. Jenkins, *A. agglutinata* (Berkeley & M.A. Curtis) Lloyd, *A. farinosa* Schweinitz, *A. frostii* (Peck) Saccardo, *A. multisquamosa* Peck, and *A. parcivolvata* (Peck) E.-J. Gilbert.

Section *Caesareae*: Type species, *Amanita caesareae* (Scopoli *ex* Fries) Persoon

Features: Stipe not swollen or with a bulb, saccate volva and annulus (ring) always present. The two species in this section found in our area are *A. jacksonii* Pomerl and *A. spreta* (Peck) Saccardo.

Section *Lepidella*: Type species, *Amanita vittadinii* (Moretti) Vittadini

Features: Margin of cap initially appendiculate (with hanging woolly or powdery tufts); stipe lacking a volva, but a thin flap of the universal veil may remain attached at the top of the bulb; toxic species contain allenic norleucine, which can severely damage the liver and kidneys. Some species in this section are facultatively mycorrhizal or never form mycorrhizal associations; all amanitas in all other sections are obligately mycorrhizal. Certain species in this section—for example, *A. smithiana* Bas, *A. nauseosa* (Wakefield) D.A. Reid, *A. thiersii* Bass, and *A. abrupta* Peck—contain allenic norleucine, a toxin that causes a delayed-onset renal failure syndrome that closely resembles orellanine poisoning except for a faster onset of symptoms (10–12 hours instead of days or weeks).

Species in this section found in our area include *A. abrupta* Peck, *A. cinereopannosa* Bas, *A. cokeri* (E.-J. Gilbert & Kühner) E.-J. Gilbert, *A. daucipes* (Montagne) Lloyd, *A. farinacea* (Cooke & Massee) Cleland & Cheel, *A. hesleri* Bas, *A. longipes* Bas *ex* Tulloss & D.T. Jenkins, *A. pelioma* Bas, *A. ravenelii* (Berkeley & M.A. Curtis) Saccardo, and *A. rhopalopus* Bas.

Section *Vaginatae*: Type species, *Amanita vaginata* (Bulliard *ex* Fries) Lamarck

Features: Stipe lacking a ring, volva present, typically saccate, but readily disintegrates in some species. Many species in this section are edible and are eaten, but as with all amanitas, caution is advised. The two species in this section found in our area are the Grisette, *A. vaginata*, and the American Tawny Grisette, *A.* aff. *fulva* Fries (an undescribed and unnamed species closely allied with the European species, *A. fulva*). Both of these species are edible.

Section *Amidella*: Type species, *Amanita volvata* (Peck) Lloyd

Features: Cap with scanty appendiculate margin, but this often disappears with age; stipe without a bulbous base, but sheathed in a thick, saccate volva, which can be globose, tube-shaped, or large and baggy; fruitbodies often bruising pinkish to brick-red. To the best of our knowledge *Amanita volvata* (Peck) Lloyd is the only described and named species in this section found in our area. It is suspected to be poisonous.

Section *Phalloideae*: Type species, *Amanita phalloides* (Vaillant *ex* Fries) Link

Features: Margin of cap never appendiculate, stipe with persistent ring, a bulbous base, and a flap-like or saccate volva. This section contains the world's deadliest amanitas, which are responsible for some 90% of all fatal mushroom poisonings worldwide. The amatoxins are not inactivated by cooking. The only members of this section in our area are the commonly encountered Eastern Destroying Angel, *A. bisporigera* G.F. Atkinson—often misidentified as *A. virosa* (Fries) Bertillion or *A. verna* (the latter name most likely a *nomen dubium*)—and the rarely encountered Atkinson's Destroying Angel, *A. elliptosperma* G.F. Atkinson.

Section *Validae*: Type species, *Amanita excelsa* (Fries *ex* Fries) Link

Features: Cap margin never appendiculate; stipe always with a persistent but fragile ring and a bulbous base. Toxins include rubescenslysin, which destroys red blood cells, damaging the heart, lungs, and central nervous system; this toxin is destroyed by cooking, and species in the section are marketed and consumed in various regions. Species in this section found in our area include *Amanita* aff. *rubescens* Persoon (*A. rubescens* is a European species that does not occur in North America; the American taxa misidentified as *A. rubescens* represent an undescribed species complex to which Dr. Rodham Tulloss has given the provisional name of "Amanita amerirubescens"), *A rubescens* var. *alba* Coker, *A. flavorubens* (Berkeley & Montagne) Saccardo, *A. flavoconia* G.F. Atkinson, *A. brunnescens* G.F. Atkinson, *A. aestivalis* Singer, and the problematic and poorly understood complex known as *Amanita citrina* (Schaeffer) Persoon (which Tulloss believes likely represents three undescribed species).

The family Plutaceae containing the genera *Pluteus* Fries, *Volvariella* Spegazzini, and *Volvopluteus* Vizzini, Contu & Justo appears to be a sister clade to the Amanitaceae. Both clades exhibit free and crowded gills, and both have gill trama that diverges in opposite directions from a central core running down from the cap. However, the gill trama in the Amanitaceae is divergent, projecting down and outward toward the hymenia from the central core.

In the Plutaceae, the gill trama is convergent (inverse), pointing down and inward toward the central core from the hymenia; the spore print is pinkish to salmon-pink rather than white, and these species more commonly grow on wood. The lepiotoid genera—*Lepiota* (Persoon) S.F. Gray, *Macrolepiota* Singer, *Leucoagaricus*, *Leucocoprinus* Patouillard, and *Chlorophyllum*—exhibit regular, parallel gill trama running straight down from the cap; they never exhibit a volva and have dextrinoid (not amyloid or non-reactive) spores. The three types of gill trama are shown in the following figure.

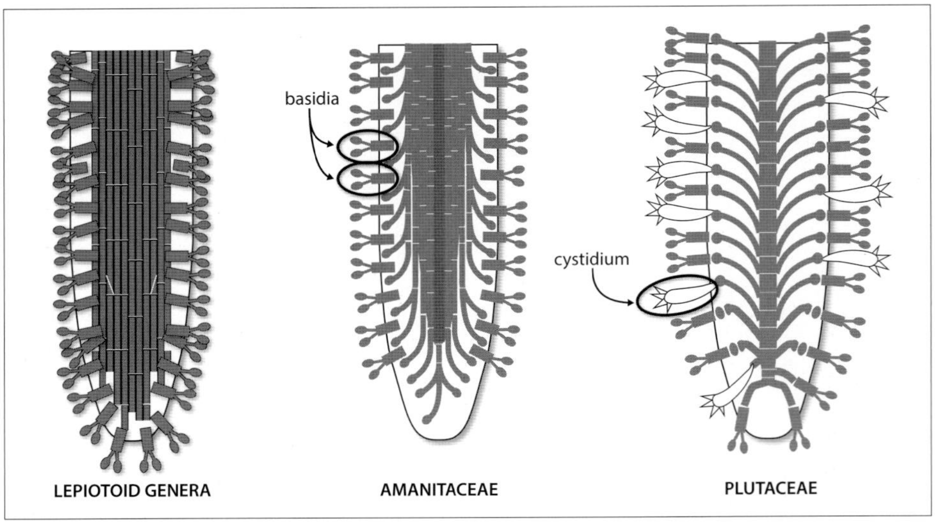

1. One of the leading authorities on this genus, Dr. Rodham Tulloss, lists more than 815 species on his website and believes that many more remain to be described: http://www.amanitaceae.org/?About+Amanitaceae.

2. The classic "Christmas Mushroom" depicted on Christmas cards is the Fly Agaric, *Amanita muscaria* (Linnaeus) Persoon.

3. The unrelated genus *Galerina* Earle has several species with even greater concentrations of this same toxin and are actually deadlier than these notorious amanitas. Some species of *Lepiota* (Family Agaricaceae) also contain amatoxins.

4. The distinguishing features of these sections are largely abstracted from Dr. Tulloss: http://www.amanitaceae.org/?About+Amanitaceae.

Abrupt-bulbed Amanita

Amanita (sect. *Lepidella*) *abrupta*

A

B

Photos: Dan Molter.

***Amanita abrupta* Peck**, *Bull. Torrey Bot. Club* 24:138 (1897) [Basidiomycota, Agaricomycotina, Agaricomycetes, Agaricomycetidae, Agaricales, Amanitaceae]

Derivation: *Amanita*, Latin, presumably named after Mount Amanon in Cilicia; *abrupta*, "abrupt," referring to the prominently swollen bulb at the base of the stipe.

Cap: 4–12 cm (1⅔–4¾ in) across, initially ovoid (Fig. A), becoming convex, broadly convex, to nearly flat (Fig. B); white, shiny, dry, densely covered with white, easily detached conical to pyramidal warts in no particular pattern (Figs. A & B); margin nonstriated, at first with tattered remnants of the universal veil.

Flesh: White, not bruising when injured; up to 7 mm thick over stipe; odor at first unremarkable, smelling like decaying meat with age.

Gills: Free or narrowly attached, with short decurrent lines, close to subdistant, narrow, white to cream, sometimes tinged pale orangish.

Stipe: 6–12 cm tall by ½–1½ cm thick (2½–4¾ × ¼–⅔ in), cylindrical to columnar, with white, skirt-like, persistent, superior ring (Fig. B); abruptly expanded below into a swollen, subglobose to subturnip-shaped bulb (Figs. A & B) often bearing white warts (Fig. B) or low concentric ridges (Fig. A).

Spore print: White.

Chemical staining: Cap negative to faintly pinkish with KOH.

Microscopic: Spores 6–10.7 × 5–9 µm, subglobose to broadly ellipsoid, smooth, clear, amyloid.

Occurrence: Mycorrhizal with hardwoods and conifers. Typically alone (occasionally in groups) on the ground in hardwood, conifer, and mixed forests; widespread and common in eastern North America; fall–winter.

Edibility: Unknown. Avoid; section *Lepidella* contains many highly toxic species.

Comments: The abruptly swollen basal bulb distinguishes this species from all others in our area except the Citron Amanita, *A. citrina* (Schaeffer) Persoon. *Amanita citrina* has a smooth (not verrucose or warty), pale to lemon-yellowish cap, the yellow color of the cap dissolves with KOH and may emit an odor of raw potatoes.

American Caesar

Amanita (sect. *Caesareae*) *jacksonii*

A

B

C

Photos: Teresa Fortenberry.

Amanita jacksonii **Pomerleau**, *Naturaliste Can.* 111(3):329 (1984)

Synonym: *Amanita umbonata* Pomerleau, *Fl. Champ. Québec* (Ottawa):516 (1980)
[Basidiomycota, Agaricomycotina, Agaricomycetes, Agaricomycetidae, Agaricales, Amanitaceae]

Derivation: *Amanita*, Latin, presumably named after Mount Amanon in Cilicia; *jacksonii*, epithet in honor of the amateur Canadian mycologist Henry Jackson (1877–1961).

Cap: 5–15 cm across (2–6 in), initially ovoid (Fig. A), becoming convex to broadly convex or plane with age (Figs. B & C), smooth, lacking patches or remnants of a partial veil, sometimes umbonate; bright red (Fig. A) to orange-red (Figs. B & C), yellowing at margin (Fig. B); margin strongly striated (Figs. A & B).

Flesh: White to yellowish (Fig. C), not bruising; no particular odor.

Gills: Free, crowded, pale yellow; initially covered by a yellowish-orange partial veil.

Stipe: 8–20 cm tall by 1–2 cm thick (3¼–8 × ½–¾ in), slender, subequal to tapering upward; cream to pale yellow, overlain with yellowish-orange scales; with persistent skirt-like ring near apex (Fig. C, arrow); base emerging from a partially buried, white, deep sac-like volva (Figs. A & B, arrows).

Spore print: White.

Chemical staining: KOH turns red cap yellow.

Microscopic: Spores (7.5–)8–10 × 5–6.5(–7) µm, elliptical, smooth, nonreactive in Melzer's reagent.

Occurrence: Mycorrhizal with oaks (*Quercus* spp.) and pines (*Pinus* spp.); solitary or in groups on the ground in mixed woods; widespread in eastern North America, occasional to locally common, summer–fall.

Edibility: Edible.

Comments: This species is identified by its combination of a smooth bright red to reddish-orange cap with striate margins and tall, thin, yellowish stipe with a skirt-like ring near the apex emerging from a deep, partially buried, white, sac-like volva. *Amanita parcivolvata* has prominent cream to yellowish patches on the cap, lacks any trace of a ring, and emerges from a rudimentary bulb-like base. *Amanita muscaria* is covered with whitish to yellowish scale-like remnants of the universal veil, the gills are white (not yellowish), and the ringed stalk arises from a swollen basal bulb (and not a sac-like volva) whose upper portion bears concentric rings of scales.

False Caesar, False Fly Agaric

Amanita (sect. *Amanita*) *parcivolvata*

A B C

Photos: A & C, Teresa Fortenberry; B, Ieponumos.

***Amanita parcivolvata* (Peck) E.J. Gilbert**, *Iconogr. Mycol.* 27 (suppl. 1):226 (1941)

Basionym: *Amanitopsis parcivolvata* Peck, *Bull. Torrey Bot. Club* 27(12):610 (1900)
[Basidiomycota, Agaricomycotina, Agaricomycetes, Agaricomycetidae, Agaricales, Amanitaceae]

Derivation: *Amanita*, Latin, presumably named after Mount Amanon in Cilicia; *parcivolvata*, from *parcere*, "to spare," and *volva*, "a sack," "sparingly volvate," referring to the rudimentary volva.

Cap: 2.5–12.5 cm (1–5 in) across, initially convex, becoming broadly convex with age; sticky when fresh; crimson red (Fig. A) to orange-red (Fig. C), covered with scattered cream or yellow patch-like remnants of universal veil, margin prominently striated (Figs. A & C).

Flesh: White to pale yellow; no distinctive odor or taste.

Gills: Free, broad, crowded; cream to pale yellow, edges powdery.

Stipe: 3–12.5 cm tall by up to 1.7 cm thick (1¼–5 × ⅔ in); ring lacking; pale yellowish (Figs. A & C) to cream (Fig. B), covered with powdery remnants of universal veil (Figs. A & B); tapering gradually upward from small, subglobose base (Figs. A, arrow, & B) typically covered with loose warty remnants of universal veil, these easily detached.

Spore print: White.

Chemical staining: Red cap stains orange, then yellow with KOH.

Microscopic: Spores 9–14 × 6–8 µm, elliptical to cylindrical, smooth, clear, nonreactive in Melzer's reagent.

Occurrence: Mycorrhizal with oaks (*Quercus* spp.); solitary or in scattered groups on the ground under or near oaks in hardwood or mixed hardwood-conifer forests, often in grassy or disturbed areas; widespread and occasional to common in southeastern North America, summer–fall.

Edibility: Unknown, but suspect.

Comments: *Amanita parcivolvata* resembles the Fly Agaric, *A. muscaria* (Linnaeus) Lamarck, which differs in having a cap with faint (not prominent) marginal striations, a stipe bearing a prominent skirt-like ring, and an expanded (not rudimentary), swollen, bulb-like base with several distinct concentric rings of universal veil remnants. The American Caesar, *Amanita jacksonii* Pomerleau, is distinguished from *A. parcivolvata* whose white cap patches have been washed away by its deep and flaring sac-like volva.

American Fly Agaric (Yellow Morph)

Amanita (sect. *Amanita*) *muscaria* var. *guessowii*

A

B

C

D

***Amanita muscaria* var. *guessowii* Veselý** [as "*güssowii*"], *Annls Mycol.* 31(4):254 (1933)
[Basidiomycota, Agaricomycotina, Agaricomycetes, Agaricomycetidae, Agaricales, Amanitaceae]

Derivation: *Amanita*, Latin, presumably named after Mount Amanon in Cilicia; *muscaria*, from *musca*, "fly," because, according to legend, when steeped in milk it attracts flies, which consume the mixture and then become stupefied and drown.

Misapplied name: *Amanita muscaria* var. *formosa* (*sensu auct. amer.*) [*non* Persoon (1800) 1799]

Cap: 4.5–18 cm (1¾–7¼ in) across, initially ovoid (Fig. C), becoming convex, broadly convex (Fig. A), to nearly flat (Fig. D); pale yellow to orangish-yellow, fading with age, covered with numerous cream to yellowish or tan cottony scales (Figs. A–D); dry to viscid, margin often faintly striate (Fig. B).

Flesh: White, not bruising when injured; no particular odor.

Gills: Free or narrowly attached, white, broad, evenly spaced, subcrowded to crowded.

Stipe: 6–15 cm tall by 1–3 cm thick (2½–6 × ½–¾ in), subcylindrical or columnar, expanded below into an enlarged, subglobose to subovoid basal bulb (Figs. A & D); lower stipe and upper portion of bulb with partial or complete rings of universal veil (Figs. A, C, & D); stipe white, cream, or pale yellowish, smooth to somewhat scaly or fibrillose; with superior skirt-like, fragile, evanescent to persistent ring, which may fall away.

Spore print: White.

Chemical staining: Cap negative with KOH.

Microscopic: Spores (7–)9–13(–15) × 6–8(–9.5) µm, ellipsoid to broadly ellipsoid, smooth, clear, nonreactive in Melzer's reagent.

Occurrence: Mycorrhizal with both hardwoods and conifers. Usually gregarious on the ground in woods, under trees, or in grassy areas; summer–fall. Widespread in North America east of the Great Plains as far south in the Appalachians as North Carolina and Tennessee; our specimens were collected in a grassy area under mixed hardwood and conifer trees in the Atlanta area.

Edibility: Edible and widely consumed outside North America when properly prepared, widespread caveats to the contrary notwithstanding (see comments below); hallucinogenic when not detoxified by parboiling.

Comments: *Amanita muscaria* var. *guessowii* is one of three infraspecific American taxa of *Amanita muscaria* found in our area, and our specimens appear to represent a southern extension from its previously reported range. It is immediately identified by the pale yellow to orangish-yellow cap covered with cream to yellowish cottony warts, presence of a ring (which may be lost), and concentric rings of shaggy scales or ridges on the upper portion of an expanded bulb.

The red morph of the American Fly Agaric, *A. muscaria* subsp. *flavivolvata* Singer, is distinguished by its vividly bright red cap (which may fade to pale orange with age) and initially yellow cottony patches that turn white with age. In most other respects the yellow and red morphs are essentially indistinguishable from each other. The False Fly Agaric, *Amanita parcivolvata* (Peck) E.-J. Gilbert, is a virtual look-alike for *A. muscaria* subsp. *flavivolvata*, except that it completely lacks a ring, has deeper marginal striations on the cap, and the unexpanded, rudimentary basal bulb is coated with loose and easily detached warty remnants of the universal veil.

The Peach-colored Fly Agaric, *A. muscaria* var. *persicina* D.T. Jenkins, is distinguished by its soft, pastel rose-peach-colored cap covered by concentric rings of pale yellowish-tan cottony warts and creamy gills with a pale pinkish hue.

Despite widespread and lurid accounts about the allegedly deadly toxic properties of the Fly Agaric in popular field guides, online sources, and the professional literature, not a single, unequivocal, authenticated account exists of a fatality caused by this species. Fatal poisonings have been caused by ingesting a related species, *A. pantherina* (de Candolle *ex* Fries) Krombholz (not found in our area), which contains considerably more muscarine, a potentially lethal neurotoxin; many alleged fatal *A. muscarina* intoxications likely were confused with these fatalities.[1]

Muscimol and ibotenic acid are the main psychoactive agents in the Fly Agaric, which also occur in several other *Amanita* species.[2] These selectively and potently activate $GABA_A$ (γ-aminobutyric acid) and NMDA (*N*-methyl-D-aspartate) glutamate receptors, respectively, in the brain. GABA is the most potent inhibitory neurotransmitter in the central nervous system, and $GABA_A$ agonists have strong anxiolytic, anticonvulsant, and sedative-hypnotic effects. Conversely, glutamate is a potent excitatory neurotransmitter, and it is the interaction of muscimol and ibotenic acid that produces the psychoactive effects sought after by some who consume this mushroom. Muscimol and ibotenic acid are both heat labile and water soluble. Parboiling these mushrooms and discarding the water renders them nontoxic and is widely practiced.

The taxonomic status of the Fly Agarics, *Amanita muscaria* (Linnaeus) Lamarck (*sensu lato*), remains confused, with various authorities employing a wide variety of scientific names involving species, subspecies, varieties, and forms and countless synonyms. Molecular genetic studies have not really cleared things up, and no one really knows what is going on. We follow Rodham Tulloss's approach and consider the basic red-capped North American Fly Agaric to represent a distinct subspecies, *A. muscaria* subsp. *flavovolvata* Singer,[3] with the nominate subspecies restricted to Eurasia and portions of Alaska.

1. W. Rubel & D. Arora, 2008, *Economic Botany* 62(3):223–243. This paper gives an excellent account of this subject and the prevailing misinformation about it.

2. Muscarine, a potentially deadly toxin, is found in only harmless trace amounts in *A. muscaria*, from which it was first isolated in 1869 and after which it was named. It activates the postganglionic muscarinic receptors for acetylcholine in the parasympathetic nervous system.

3. *Amanita muscaria* subsp. *flavivolvata*: http://www.amanitaceae.org/?Amanita muscaria subsp. flavivolvata, accessed December 6, 2013. Many field guides consider this taxon to be a variety: *A. muscaria* var. *flavovolvata* (Singer) D.T. Jenkins.

Peach-colored Fly Agaric

Amanita muscaria (sect. *Amanita*) var. *persicina*

A

B

C

***Amanita muscaria* var. *persicina* D.T. Jenkins**, *Biblthca Mycol.* 57:59 (1977)
[Basidiomycota, Agaricomycotina, Agaricomycetes, Agaricomycetidae, Agaricales, Amanitaceae]

Derivation: *Amanita*, Latin, presumably named after Mount Amanon in Cilicia; *persicina*, feminine adjectival form of *persicum*, "peach," referring to its peach-like color.

Cap: 4–20 cm (1⅔–8 in) across, initially ovoid, becoming convex, broadly convex, to nearly flat with age, often slightly depressed centrally (Figs. A & B); smooth, viscid when fresh; reddish-orange when young (not unlike *A. muscaria* var. *guessowii*), fading with age to a subtle, pastel rose-peach, covered with concentrically arranged cream to pale yellowish cottony scales (Fig. A); margin often faintly striated.

Flesh: White, yellowish underneath cuticle of cap; not bruising when injured.

Gills: Free or narrowly attached, crowded, moderately broad, creamy with faint pinkish tinge (Fig. C).

Stipe: 4–20 cm tall by 1–3 cm thick (1⅔–8 × ½–1¼ in), subequal, columnar; pale yellowish near apex, creamy white below; with fragile, ephemeral ring positioned about midstipe (Fig. B, arrow) that is whitish on upper surface and yellowish below; expanded at base into a globose to ovoid bulb, rarely with cottony remnants of universal veil on upper bulb.

Spore print: White.

Chemical staining: KOH and NH_3—all tissues negative; **Melzer's reagent**—flesh faintly pale yellowish.

Microscopic: (7–)9–13(–18) × (5.5–)6–8.5(–11) µm, ellipsoid to broadly ellipsoid with apiculus, smooth, clear, nonreactive in Melzer's reagent.

Occurrence: Mycorrhizal with oaks (*Quercus* spp.), pines (*Pinus tadea* and *P. virginiana*), and Eastern Hemlock (*Tsuga canadensis*); alone or in scattered groups on the ground; widespread in the Southeast; spring–fall.

Edibility: Unknown; like all Fly Agarics, it contains ibotenic acid and muscimol. See the comments under the American Fly Agaric (Yellow Morph), *Amanita muscaria* var. *guessowii*.

Comments: This is one of the most beautiful mushrooms you will encounter. The subtle peach-like hues of mature fruitbodies almost seem to glow with an inner light and are difficult to capture with a camera. The yellowish-orange to reddish-orange caps of young fruitbodies often resemble the other two infraspecific forms of Fly Agarics that occur in our area, and mature specimens are required for an accurate identification. It is distinguished from the other two infraspecific Fly Agaric taxa by the pastel peach- to melon-colored cap and basal bulb lacking prominent concentric universal veil remnants.

Carrot-footed Amanita

Amanita (sect. *Lepidella*) *daucipes*

A B C D

***Amanita daucipes* (Saccardo) Lloyd**, *Mycol. Writ.* 1(1):8 (1898)

Basionym: *Amanitopsis daucipes* Saccardo, *Syll. Fung.* (Abellini) 5:26 (1887)[1]
[Basidiomycota, Agaricomycotina, Agaricomycetes, Agaricomycetidae, Agaricales, Amanitaceae]

Derivation: *Amanita*, Latin, presumably named after Mount Amanon in Cilicia; *daucipes*, from *daucus*, "carrot," and *pes*, "the foot of."

Cap: 6–30 cm (2½–12 in) across, initially convex (Fig. B), becoming broadly convex (Figs. A & D) to nearly flat (Fig. C) with age; surface dry, cream to pale pinkish-orange or pinkish-beige, covered with concentric rings of dull cream to pinkish-orange or pinkish-brown cake-like (Fig. B), pyramidal, or conical warts (Figs. A & C), which are readily removed without leaving a scar; cap often cracked around the bases of the warts; margin not striate, but often with dangling remnants of universal veil.

Flesh: White to cream, thick (up to 13 mm over stipe, thinning toward margin); odor of chlorine or decayed meat depending on the age of the fruitbody.

Gills: Free, close to almost crowded, broad, with numerous lamellulae (short gills); white, cream, pale yellow, or pale yellowish-cream (Fig. D).

Stipe: 8–20 cm tall by 1–3 cm thick, excluding the basal bulb (3¼–8 × ½–1¼ in), whitish to pale orangish; enlarged below into a deeply rooted, massive, pinkish-orange, spindle-, turnip-, or carrot-shaped basal bulb (Figs. A & D) up to 15–20 cm long by 12 cm wide (6–8 × 4¾ in) that often splits longitudinally; partial veil forming a whitish to pale yellow ephemeral ring toward apex (Fig. A) that typically falls away (Fig. D).

Spore print: White.

Bruising: Not bruising, but old wounds slowly turning pink, then dark brick-red (Figs. A & D).

Chemical staining: **KOH**—all tissues negative; **Melzer's reagent**—cap and flesh faint yellowish-red, then fading away; gills pale grey, turning pale pinkish-grey.

Microscopic: Spores (7.7–)8–12(–13.8) × 5–7(–10) µm, ellipsoid, smooth, clear, amyloid.

Occurrence: Mycorrhizal with oaks (*Quercus* spp.). Alone or in scattered groups on the ground in mixed hardwood and conifer forests; widespread and occasional to common in the Southeast; midsummer–late fall.

Edibility: Unknown, but suspect; many species in section *Lepidella* are highly toxic.

Comments: The massive, deeply rooted fusiform (spindle-shaped) to napiform (carrot-shaped) basal bulb, pale pinkish-orange color of the cap, white to yellowish gills, and odor of chlorine, old ham, or decaying meat distinguish this species. Coker's Amanita, *A. cokeri* (E.-J. Gilbert & Kühner) E.-J. Gilbert, also has an expanded fusiform to napiform basal bulb, which is considerably smaller and more shallowly rooted in the ground. It also emits no distinctive odor, has a shiny white cap with conical warts that are not easily removed and a persistent white superior ring, and the lower stem and upper bulb bear conspicuous, concentrically arranged warts and recurved scales.

1. The designation of the basionym is very confused in the literature, various field guides, and both *Index Fungorum* and *Mycobank*, with various naming authorities being inconsistently listed without explanation. This taxon was originally described as *Agaricus daucipes* by Berkeley & Montagne in 1856, a later homonym (and thus an illegitimate name) of *Agaricus daucipes* Persoon (1828). *Amanitopsis daucipes* was designated as a replacement name by Saccardo in 1887 and is the correct basionym for this taxon. See online Appendix A for a list of synonyms.

Citron Amanita

Amanita (sect. *Validae*) aff. *citrina*

A B C D

***Amanita citrina* (Schaeffer) Persoon**, *Tent. Disp. Meth. Fung.* (Lipsiae):70 (1797)

Basionym: *Agaricus citrinus* Schaeffer, *Fung. Bavar. Palat. Nasc.* (Ratisbonae) 1:tab. 11 (1762)
[Basidiomycota, Agaricomycotina, Agaricomycetes, Agaricomycetidae, Agaricales, Amanitaceae]

Derivation: *Amanita*, Latin, presumably named after Mount Amanon in Cilicia; *citrina*, "yellowish," "citron."

Cap: 2.5–10 cm across (1–4 in), convex (Figs. A & B), broadly convex (Fig. C) to almost flat; surface smooth, margin not striated (or very faintly lined in age), sticky when moist, whitish-grey, pale yellowish, greenish-yellow, or lemon-yellow (Figs. A–C), often with whitish, greyish (Fig. C), brownish (Fig. A), or lavender remnants of the universal veil (Figs. B & C).

Flesh: White; odor of raw potatoes or turnips, or odor and taste unremarkable.

Gills: Narrowly free or adnexed, close to crowded, moderately broad, white to cream (Fig. D), minutely fringed.

Stipe: 5–14 cm tall by 1.5–2.5 cm thick (2–1⅔ × ⅔–1 in), subequal but with massively swollen, bulbous base (Fig. A), with greyish- or yellowish-white ephemeral ring (Figs. A & C), this often disappearing (Fig. D).

Chemical staining: Yellow color of cap dissolving with KOH.

Spore print: White.

Microscopic: Spores 6–9 × 6–9 µm, globose, smooth, clear, amyloid.

Occurrence: Mycorrhizal with hardwoods and conifers; alone or scattered on the ground; widespread and common in eastern North America; late summer–fall.

Edibility: Inedible, suspected to be toxic.

Comments: The leading authority on the genus *Amanita*, Rodham Tulloss, states that the most commonly encountered and widespread Citron Amanita in North America is *Amanita citrina* f. *lavendula* (Coker) Veselý. He also notes that this is a complex that may include up to three distinct and as yet undescribed species.[1] These cryptic taxa apparently all stain purple or lavender under certain circumstances, especially when exposed to near-freezing temperatures. Compare this species with the Abrupt-bulbed Amanita, *Amanita abrupta* Peck.

1. http://www.amanitaceae.org/?Amanita%20citrina%20f.%20lavendula, accessed December 4, 2013.

Cleft-foot Amanita, Brown Blusher

Amanita (sect. *Validae*) *brunnescens*

Photos: A, Joseph O'Brien, USDA Forest Service, Bugwood.org; B–F, W. Light.

***Amanita brunnescens* G.F. Atkinson**, *Proc. Am. Phil. Soc.* 57:354 (1918)
[Basidiomycota, Agaricomycotina, Agaricomycetes, Agaricomycetidae, Agaricales, Amanitaceae]

Derivation: *Amanita*, Latin, presumably named after Mount Amanon in Cilicia; *brunnescens*, from *brunneis*, "brown," referring to its tendency to bruise brown.

Synonym: *Amanita brunnescens* var. *pallida* L. Krieg., *Mycologia* 19(6):308 (1927).

Cap: 5–15 cm (2–6 in), convex at first, becoming broadly convex to nearly flat; sticky when wet; color highly variable, greyish-brown (Fig. A), brownish (Figs. B & E), yellowish-brown, to pale greyish-brown (Fig. C) or greenish-yellow (Fig. F), sometimes with reddish-brown stains (Fig. C, arrow); surface covered with dense (Fig. F) to scattered (Figs. A, B, C, & E) whitish, grey, or greyish-tan cottony patches and scales that may wash away with rain or fall away with age; margin frequently radially striate, entire, no hanging remnants of universal veil.

Flesh: White, sometimes bruising reddish-brown; odor of raw potatoes, especially at the base.

Gills: Free to narrowly attached, white to cream, close to crowded, bruising brown.

Stipe: 5–15 cm tall by 0.5–1.5 cm thick (2–6 × ¼–⅔ in) when expanded, tapering gradually upward from an abruptly expanded, "chiseled" basal bulb with several vertical, wedge-shaped clefts (Figs. A, B, & F, arrows), volva lacking; smooth, creamy white, discoloring brownish to pinkish-brown when bruised (Fig. A, brownish zigzag stripe); with white skirt-like superior ring (Fig. B), upper surface with radial striations (Fig. D, arrow), becoming brown with age, sometimes disappearing.

Spore print: White.

Chemical staining: KOH negative on cap.

Microscopic: Spores (6.5–)7.5–9.5(–10) µm, globose to subglobose, smooth, clear, amyloid.

Occurrence: Mycorrhizal with hardwoods and conifers. Alone or in groups on the ground in mixed hardwood and conifer forests; widespread and common in eastern North America; summer–fall.

Edibility: Inedible; despite contradictory reports, this species is probably poisonous.

Comments: The abruptly swollen and "chiseled" basal bulb with its distinctive vertical clefts and reddish-brown-bruising stipe and fruitbody readily identify this species. However, not all specimens clearly exhibit these features, which can make identification problematic.

A white morph described as *Amanita brunnescens* var. *pallida* is essentially identical, but the cap is white or pale yellowish in color. *Amanita aestivalis* Singer, a northeastern taxon that ranges at least as far south in the Appalachian Mountains as Great Smoky Mountains National Park, is entirely white except occasionally for some pale tan or yellow over the center. The entire fruitbody slowly bruises some shade of brown or reddish-brown. It is currently unclear what its relationship is to *A. brunnescens* var. *pallida* or whether it is a valid species separate from *A. brunnescens*.[1]

1. R.E. Tulloss, *Amanita aestivalis*, http://www.amanitaceae.org/?Amanita%20aestivalis, accessed December 8, 2013.

Coker's Amanita, Coker's Lepidella

Amanita (sect. *Lepidella*) *cokeri*

A

B

C

***Amanita cokeri* (E.-J. Gilbert & Kühner) E.-J. Gilbert**, *Iconogr. Mycol.* 27 (suppl. 1):66 (1940)

Basionym: *Lepidella cokeri* E.-J. Gilbert & Kühner, *Bull. Trimest. Soc. Mycol. Fr.* 44:154 (1928)
[Basidiomycota, Agaricomycotina, Agaricomycetes, Agaricomycetidae, Agaricales, Amanitaceae]

Derivation: *Amanita*, Latin, presumably named after Mount Amanon in Cilicia; *cokeri*, epithet in honor of the American mycologist/botanist William Chambers Coker (1872–1953).

Cap: 7–15 cm (2¾–6 in) across, convex, becoming broadly convex (Fig. B) to nearly flat when mature (Fig. C); white, nonstriated, covered with prominent, concentrically arranged, white to brownish pyramidal warts and scales (Figs. B & C); shiny when dry, tacky when wet.

Flesh: White, thick, firm, unchanging when injured; odor unremarkable.

Gills: Free or narrowly attached, crowded, broad, with numerous short gills (Fig. A); white to cream.

Stipe: 10–20 cm tall by 1–2 cm thick (4–8 × ½–¾ in); solid, white, smooth to fibrillose or with shaggy scales (Figs. A–C), subequal or tapering toward apex; base swollen to a large, spindle- to carrot-shaped bulb (Fig. A) shallowly rooted in the ground; no sac-like volva; lower stem and upper bulb with conspicuous, concentrically arranged warts and recurved scales (Fig. A, arrow); with persistent, white, skirt-like, and often double-edged ring near apex (Fig. A, Figs. B & C, arrows).

Spore print: White.

Bruising: Stem (Fig. C) and especially the bulb (Fig. A) sometimes bruising reddish or rusty.

Chemical staining: Negative with KOH.

Microscopic: Spores (10–)11–14(–15) × 6.5–9 µm, elliptical, smooth, clear, amyloid.

Occurrence: Mycorrhizal with both hardwoods and conifers; alone or in groups on the ground in mixed deciduous and coniferous woods, associated mainly with oaks (*Quercus* spp.) and pines (*Pinus* spp.). Widespread in eastern North America, but centered in Southeast; late summer–late fall.

Edibility: Unknown. Avoid, possibly toxic.

Comments: The combination of a swollen napiform (turnip- or carrot-shaped) basal bulb and shiny white cap with conical warts identifies this species. The only other taxon in our area with a large napiform basal bulb is the Carrot-footed Amanita, *A. daucipes* (Saccardo) Lloyd. *Amanita daucipes* emits a foul odor of chlorine or rotting meat, the cap has a pale orangish hue and is covered with pale orange or reddish-brown warts, the gills range from white to yellowish, the ephemeral ring typically falls away, and the pinkish-orange basal bulb is much more massive: 15–20 cm long by 12 cm wide (6–8 × 4¾ in). Compare this species with the Rag-veil Amanita, *A. cinereopannosa* Bas.

Rag-veil Amanita, Grey Rags Lepidella

Amanita (sect. *Lepidella*) cf. *cinereopannosa*

A B C D

Amanita cinereopannosa **Bas**, *Persoonia* 5(4):522 (1969) [Basidiomycota, Agaricomycotina, Agaricomycetes, Agaricomycetidae, Agaricales, Amanitaceae]

Derivation: *Amanita*, Latin, presumably named after Mount Amanon in Cilicia; *cinereopannosa*, from *cinereo*, "grey," and *pannos*, "rags."

Cap: Large, 7–15 cm (2¾–6 in) across, margin at least initially slightly to massively appendiculate (with dangling, tattered, and pruinose [powdery] remnants of the universal veil [Fig. A]), frequently giving the surrounding substratum a light to heavy coating of white powdery dust; cap studded with distinctive irregular, greyish-brown warts arranged in concentric circles (Figs. B & D).

Flesh: Odor indistinct or reminiscent of fresh dough, sometimes smelling like chlorine with age.

Gills: Free, widely separated from stipe, crowded, blades broad (up to 8 mm wide), white to cream.

Stipe: 10–16 cm tall by 1.5–2 cm thick (4–6½ × ⅔–¾ in), solid, subcylindrical, with expanded, radicating (buried and root-like), bulbous base measuring up to 7.5 cm long by 3 cm thick (3 × 1¼ in) (Fig. C); bulb smooth (Fig. C) or with volval scales (Fig. A) or warts similar to those on the cap; with ephemeral ring near apex, this is easily detached and often missing.

Spore print: White.

Bruising: All tissues nonbruising.

Chemical staining: All tissue nonreactive to KOH, Melzer's reagent, and NH_4OH.

Microscopic: Spores (8–)9–12(–14) × (4.9–)5.5–7(–8.3) μm, elliptical to long ellipsoid, smooth, clear, amyloid; clamps lacking at the bases of the basidia.

Occurrence: New England south to Tennessee and Georgia in mixed forests in the Appalachian Mountains (also found in the Piedmont in the Atlanta area); mid–late summer.

Edibility: Unknown, but to be avoided; the toxic amanitas in this section contain allenic norleucine that severely damages the human liver and kidneys.

Comments: The ecology of this species is largely unreported. *Amanita cinereopannosa* belongs to the *A. cokeri* complex of species, characterized by amyloid spores, a stipe with a radicating bulbous base, and a cap with an appendiculate margin of tattered remnants of the universal veil characterized by woolly tufts and/or a coating of white powdery dust.

Amanita onusta (Howe) Saccardo differs from *A. cinereopannosa* in its smaller size, the cap being 1.5–7.5 cm (⅔–3 in), rarely to 10 cm across, the initially white gills that soon turn yellowish and appear waterlogged, and clamp connections being present at the bases of the basidia.

Amanita cinereopannosa somewhat resembles an unnamed taxon found in the southern Appalachians of Tennessee and North Carolina and the Upper Piedmont of South Carolina, as well as in New York, New Jersey, and Tlaxcala, Mexico.[1] This taxon was unofficially described by Cornelia Bas as "AMANITA TEPHREA, *nom. prov.*," but was never formally described or named. Unfortunately, it is treated as if it were a valid scientific name by most authors.[2] "Amanita tephrea" differs from *A. cinereopannosa* in its smaller size (typically <10 cm, or 4 in, across), in having gills that are narrowly adnate (attached) to adnate (with or without a decurrent line on the upper stipe), a cap with a loosely formed woolly or powdery layer that lacks irregularly surfaced darker polygonal warts and that stains brown with KOH, and having clamp connections at the bases of the basidia.

Compare this species with Coker's Amanita, *A. cokeri* (E.-J. Gilbert & Kühner) E.-J. Gilbert.

1. R.E. Tulloss, http://www.amanitaceae.org/?Amanita%20tephrea, accessed August 28, 2014.

2. C. Bas, 1969, *Persoonia* 5:452, figs. 192–195. In his informal description, Bas noted that "*Amanita tephrea* [*sic*] is a distinct species, but since I have scarcely any information about the fruit-bodies in a fresh state I do not wish to describe it formally" (453).

Eastern American Blusher

Amanita (sect. *Validae*) aff. *rubescens*

Photos: A–D & F: W. Light; E, Teresa Fortenberry.

***Amanita* aff. *rubescens* Persoon,**[1] *Tent. Disp. Meth. Fung.* (Lipsiae):71 (1797)

Sanctioning work: Fries, *Systema Mycologicum* I:18 (1821) [Basidiomycota, Agaricomycotina, Agaricomycetes, Agaricomycetidae, Agaricales, Amanitaceae]

Derivation: *Amanita*, Latin, presumably named after Mount Amanon in Cilicia; *rubescens*, from *rubesco*, "reddened," "blushed," "pink," referring to the characteristic reddening exhibited by injured tissues.

Cap: 5–20 cm (2–8 in) across, initially convex, becoming broadly convex (Figs. A & D) to flat (Figs. B & F); slightly viscid or dry; color highly variable, at first pale yellow, becoming pale tan or brownish, bruising reddish-brown to wine-red (Fig. B, upper white arrow; Figs. D & E), eventually turning various shades of reddish-brown or brown (Fig. F); surface covered with yellowish (initially), pinkish, greyish, or tannish felt-like pyramidal warts, these easily eroded away.

Flesh: White, eventually turning pinkish.

Gills: Free or narrowly attached, with short decurrent ridges; close to crowded, broad; white, often bruising reddish (Fig. B, dotted yellow arrow) to reddish-brown.

Stipe: 5–20 cm tall by 1–4 cm thick (2–8 × ½–1⅔ in), subequal, usually enlarged downward into indistinct to moderately swollen, oval to turnip-like bulbous base lacking a rim and usually with no adhering volval remnants (Fig. A); smooth to somewhat fibrous or scurfy; white to pale flesh, bruising or aging reddish-brown to wine-red (Figs. A & B, white arrows; Fig. E); with a more or less persistent superior ring (Fig. A), this sometimes collapsing, upper surface of ring white, striated (Fig. C), lower surface smooth to velvety, white or sometimes yellow (Fig. C, arrow).

Spore print: White.

Bruising: The characteristic reddish-brown to vinaceous-red bruising of all parts of the fruitbody is the *sine qua non* that defines this species complex.

Microscopic: Spores (7–)8–10 × 5–7 µm, ovoid to short-elliptical, smooth, clear, amyloid.

Occurrence: Mycorrhizal with oaks (*Quercus* spp.). Isolated or in groups on the ground in mixed forests; widespread and common in eastern North America; summer–fall.

Edibility: The European Blusher (and presumably also the American Blusher) is edible when thoroughly cooked to eliminate dangerous hemolytic toxins (see comments). However, caution is advised, as this taxon could be confused with deadly poisonous *Amanita* species.

Comments: Although the colors of this mushroom are highly variable, the characteristic reddish-brown to vinaceous-red bruising together with the pinkish, greyish, or tannish felt-like warts, the persistent superior white ring that is striated above and smooth and sometimes yellow below, and the typically swollen basal bulb that lacks ridges or remnants of a volva identify this species complex. The similar Yellow Blusher, *Amanita flavorubens* (Berkeley & Montagne) Saccardo (Fig. G),[2] has a golden-yellow to bright-yellow cap covered with yellow warts that do not change color. It also has yellow volval remnants on the basal bulb and bruises more slowly and less conspicuously reddish.

The Eastern American White Blusher, *Amanita rubescens* var. *alba* Coker (1917), is essentially an all-white version of the Eastern American Blusher that also bruises and ages reddish. A more northerly taxon, it reaches the southern limits of its distribution in South Carolina.

G Photo: Teresa Fortenberry.

The European Blusher (and presumably also the American Blusher) contains rubescenslysin, a potent hemolytic toxin that destroys red blood cells and causes intravascular hemolysis, massive pulmonary edema resulting in alveolar obstruction, cardiotoxicity, and central nervous system effects. The LD_{50} (the dose that will kill half the test animals) of rubescenslysin in mice is 350 mg per kilogram;[3] humans are known to be more susceptible than mice to this toxin.[4]

1. According to Rodham Tulloss and others, the American Blusher long reported as *Amanita rubescens* by mycologists working in eastern North America represents an undescribed complex of several species and/or varieties. *Amanita rubescens* (*sensu stricto*) does not occur in North America. Tulloss has given this complex the provisional name of "Amanita amerirubescens"; however, the description and name of this taxon have not yet been validly published, and this name has no taxonomic standing. We therefore write this provisional designation in roman typeface and enclose it within quotation marks to show that it is not a valid scientific name. We also formally indicate the fact that this unnamed taxon is closely related to, but is not the same as, the European Blusher, *A. rubescens* Persoon, with the open nomenclatural adjective "aff." (Latin, *affinis*, "related to").

2. *Amanita flavorubescens* G.F. Atkinson (1902), the name used in nearly all the field guides and online sources for this species, is a later synonym of *A. flavorubens*; *fide* R.E. Tulloss, *Amanita flavorubens*, http://www.amanitaceae.org/?Amanita+flavorubens, accessed December 9, 2013.

3. A.L. Harvey, 1990, Cytolytic toxins, *in* W.T. Shier & D. Mebs (eds.), *Handbook of Toxicology* (New York: Marcel Dekker), table 1, 3.

4. R. Seeger, 1980, *Arch. Pharmacol.* 311:95–103.

Eastern Destroying Angel

Amanita (sect. *Phalloideae*) *bisporigera*

A B C

Photos: Dan Molter.

***Amanita bisporigera* G.F. Atkinson**, *Bot. Gaz.* 41(3):348 (1906) [Basidiomycota, Agaricomycotina, Agaricomycetes, Agaricomycetidae, Agaricales, Amanitaceae]

Synonym: *Amanitina bisporigera* G.F. Atkinson, in Bresadola, *Iconogr. Mycol.* 27 (suppl. 1):78 (1941)

Derivation: *Amanita*, Latin, presumably named after Mount Amanon in Cilicia; *bisporigera*, "2-spored" (see comments below).

Cap: 2.5–10 (1–4 in) across, initially deeply convex to oval (Fig. B), gradually becoming nearly plane (Fig. A), smooth, lacking striations or partial veil remnants; dry to slightly sticky when wet, pristine white to ivory, sometimes yellowish or tan tinge with age.

Flesh: White, thin, not bruising, with sweet to somewhat sickening odor.

Gills: White, not attached, close to crowded (Fig. C).

Stipe: 6–14 cm tall by 0.5–2 cm thick (2½–5⅔ × ¼–¾ in); tapering toward apex, smooth (Fig. B) to somewhat scaly (Fig. A), with prominent, persistent, skirt-like ring near apex (Fig. A, arrow), emerging from a prominent sac-like volva (Figs. A & B, arrow) that may be buried or torn.

Spore print: White.

Chemical staining: Cap surface staining yellow with KOH.

Microscopic: Spores 7–10 × 6–9 µm; smooth, thin-walled, clear, subglobose, amyloid. Basidia without clamps; initially 2-spored, becoming 4-spored later in the season (see comments below).

Occurrence: Mycorrhizal, especially with oaks; widely distributed in eastern North America from Canada to Mexico.

Edibility: Deadly poisonous! See the section on amatoxins in "Mushroom Toxins and Mushroom Poisoning" in the introduction.

Comments: *Amanita bisporigera* is by far the most toxic mushroom known in North America. It contains 2.2–5.0 milligrams per gram of dry weight of tissue of both α- and β-amanitin, and this does not appear to vary geographically. The pristine all-white cap, gills, and stipe with its flaring ring and prominent volva, white spore print, and a cap that stains yellow with KOH are the hallmarks of this species. It is not likely to be confused with any other species in our area. However, it has been misidentified as *A. virosa* (Fries) Bertillion and *A. verna* (Bulliard *ex* Fries) Lamarck in many North American field guides.[1] *Amanita virosa*, however, is confined largely to Europe, although a molecularly verified and very likely introduced population of *A. virosa* has been reported from Virginia.[2]

The confusion may arise from the fact that *A. bisporigera* ("two-spored amanita") develops basidia bearing both two and four spores at different stages in the reproductive cycle. As the fruiting season progresses, a change from basidia bearing two spores to four spores takes place.[2] Furthermore, the yellow staining reaction to KOH is intraspecifically variable in several species in this group of amanitas even at the same locality and at the same time. It is also often highly subjective ("weak" vs. "strong") and varies widely within a species at the same place and time.[3] It is thus not a reliable taxonomic feature and should probably be eliminated from species descriptions and keys.

Amanita bisporigera is widely distributed in North America and is frequently encountered. It is one of only two Destroying Angels in the Southeast.[4] The only other amanita with which this species might be confused in our area is the rarely encountered and presumably equally deadly Atkinson's Destroying Angel, *Amanita elliptosperma* G.F. Atkinson. In this species the cap does not stain yellow with KOH, and the spores are long-ellipsoid, (8.0–)8.7–11.2(–11.5) × (5.8–)5.9–7.5(–7.8) µm, instead of subglobose.

1. *Amanita verna* may not be a valid name; no type specimen exists, and it appears to have been described from a misidentified specimen of *A. phalloides*. It should probably be considered a *nomen nudum*.

2. R.E. Tulloss & L. Possiel, 2005, Key to species of *Amanita* section Phaloiideae from North and South America, http://www.eticomm.net/~ret/amanita/key.dir/phallimb.html#decipiens Lamoureux, accessed July 20, 2009.

3. S.A.M. Peterson, 1982, A chemotaxonomic evaluation of selected members of the genus *Amanita* (Ph.D. diss., University of Tennessee), Knoxville.

4. The Death Cap, *Amanita phalloides* (Vaillant *ex* Fries) Link, was introduced into North America from Europe to both the West Coast, where its range is expanding, and the Northeast; it does not occur anywhere close to our area (B.E. Wolfe et al., 2009, Distribution and abundance of the introduced ectomycorrhizal fungus *Amanita phalloides* in North America, *New Phytologist* 1–9, published online in 2009, DOI:10.1111/j.1469-8137.2009.03097.x); 2010, *New Phytologist* 185(3):803–816.

Grisette

Amanita (sect. *Vaginatae*) *vaginata*

A

B

C

D

***Amanita vaginata* (Bulliard) Lamarck**, *Encycl. Méth. Bot.* (Paris) 1(1):109 (1783)

Basionym: *Agaricus vaginatus*, *Herb. Fr.* 3: tab. 98 (1783) [1782–1783]
[Basidiomycota, Agaricomycotina, Agaricomycetes, Agaricomycetidae, Agaricales, Amanitaceae]

Derivation: *Amanita*, Latin, presumably named after Mount Amanon in Cilicia; *vaginata*, "sheathing," referring to the sac-like volva.

Cap: 5–10 cm (2–4 in) across, initially ovoid, becoming convex, bell-shaped, or flat with central umbo; grey (Figs. A & B) to greyish-brown (Fig. C), often with one or more prominent white universal veil patches (Fig. B, arrow), these generally lost with age; margin faintly to prominently radially striated and grooved (Figs. B & C); smooth, tacky when wet.

Flesh: White or pale grey, firm, not changing color when injured; odor and taste mild.

Gills: Free or adnexed, close to moderately crowded (Fig. D); white, sometimes with greyish tint.

Stipe: 10–20 cm tall by 1–2 cm thick (4–8 × ½–¾ in), subcylindrical, tapering gradually toward apex; white, smooth or with white scales (Fig. A); ring lacking, base enclosed in loosely fitting sac-like volva (Fig. D, arrow).

Spore print: White.

Chemical staining: Cap nonreactive or staining yellowish in KOH.

Microscopic: Spores 9–12 µm in diameter, spherical, thin-walled, smooth, clear, nonreactive in Melzer's reagent.

Occurrence: Mycorrhizal with hardwoods and conifers; widespread and common throughout North America in grassy areas, forest edges, and disturbed habitats; Europe; late spring–fall.

Edibility: Edible, but not recommended. Caution advised.

Comments: In our area this species is only likely to be confused with the Mealy Amanita, *Amanita farinosa* Schweinitz, which also lacks a ring and has a radially striated, brownish-grey to brownish cap. *Amanita farinosa* differs in its smaller size (2½ in or less across), the cap is dusted with a fine powdery or mealy coating when young (this often worn away), and the stipe is slightly bulbous at the base and lacks a sheath-like volva. The lack of a volva places it in section *Amanita*, not *Vaginatae*.

The Tawny Grisette, *Amanita fulva* (Schaeffer) Fries, is very similar and was formerly considered a variety of *A. vaginata*. It is distinguished by its yellowish- to orange-brown or reddish-brown cap. See our account of the scientifically unnamed American Tawny Grisette, *Amanita* aff. *fulva*.

American Tawny Grisette

Amanita (sect. *Vaginatae*) aff. *fulva*

A B C D

***Amanita* aff. *fulva* Fries**, *Observ. Mycol.* (Havniae) 1:2 (1815)

Basionym: *Agaricus fulvus* Schaeffer, *Fung. Bavar. Palat. Nasc.* (Ratisbonae) 4:41 (1774)
[Basidiomycota, Agaricomycotina, Agaricomycetes, Agaricomycetidae, Agaricales, Amanitaceae]

Derivation: *Amanita*, Latin, presumably named after Mount Amanon in Cilicia; *fulva*, "tawny."

Cap: 4–10 cm across (1⅔–4 in), initially ovoid, becoming convex (Figs. B & C), flat or even shallowly concave (Fig. A), often with darker central umbo (Fig. B); yellowish-brown (Fig. B), orange-brown, reddish-tan (Fig. A), or tawny (Fig. C); smooth, tacky when moist, margin faintly to deeply grooved and striated (Figs. A–C); rarely with white patches of universal veil.

Flesh: White, soft, not changing color when injured; odor mild.

Gills: Free or adnexed, close to moderately crowded, white, cream, or with faint yellowish tinge (Fig. D).

Stipe: 7–16 cm tall by 0.5–1.5 cm thick (2¾–6½ × ¼–⅔ in), cylindrical, tapering slightly toward apex; smooth or with scattered fibrils or scales; whitish or faintly tinged yellowish or ochraceous (Figs. C & D); ring lacking; base embedded in sac-like volva (Fig. D, arrow).

Spore print: White.

Chemical staining: Cap nonreactive to KOH.

Microscopic: Spores 8–10 µm, spherical to subglobose, smooth, clear, nonreactive in Melzer's reagent.

Occurrence: Mycorrhizal with hardwoods and conifers. Alone or in groups, widespread in eastern North America and Europe; summer–fall.

Edibility: Edible, but caution advised.

Comments: This species is simply a tawny version of the greyish Grisette, *A. vaginata* (Bulliard) Lamarck, and was formerly considered a variety of that species. Like the Grisette, it is identified by its more or less smooth cap with a radially striated and furrowed margin and thin ringless stipe emerging from a sac-like volva. Faded specimens of the American Caesar, *A. jacksonii* Pomerleau, are distinguished from *A. fulva* by a persistent ring on the upper stipe, a deep and flaring sac-like volva, and the reddish cap staining yellow in KOH.

The widespread and common Tawny Grisette found in North America is likely a distinct species from the European *A. fulva*, but it has not been formally described or named. *Amanita* specialist Rod Tulloss has proposed a provisional name for this taxon, which has been cited by some authors. However, this name has no taxonomic standing until and unless it is validly published.

Eastern Mealy Amanita

Amanita (sect. *Amanita*) *farinosa*

Photo: Dan Molter.

***Amanita farinosa* Schweinitz**, *Schr. Naturf. Ges. Leipzig* 1:79 (1822)
[Basidiomycota, Agaricomycotina, Agaricomycetes, Agaricomycetidae, Agaricales, Amanitaceae]

Derivation: *Amanita*, Latin, presumably named after Mount Amanon in Cilicia; *farinosa*, "mealy."

Cap: 2–6.5(–7) cm (¾–2¾ in) across, conical, egg- or dome-shaped when young, becoming shallowly convex to nearly flat with age; surface brownish-grey, dry, dusted with a fine, mealy powder that readily rubs off; margin strongly striated in mature fruitbodies.

Flesh: White, not bruising when injured or exposed; odor and taste not distinctive.

Gills: White, not bruising, free to slightly attached; close to subcrowded, occasionally forked, up to 7 mm broad.

Stipe: 3–9 cm tall by 0.3–1 cm thick (1¼–3⅔ × ³⁄₂₅–½ in), subequal, lacking ring, expanded below into slight but distinct basal bulb (arrow), which may have a powdery dust on upper surface; no trace of a volva.

Spore print: White.

Microscopic: Spores quite variable in size (5–)6.5–8.8(–10.5) × (4.5–)5.5–7.5(–9.0) µm, subglobose to broadly tear- or almond-shaped, smooth, clear, nonreactive in Melzer's reagent; clamps absent at base of basidia.

Occurrence: Occasional in forests with oak (*Quercus* spp.), beech (*Fagus grandifolia*), and hickory (*Carya* spp.), but often overlooked because of its small size; widespread from southeastern Canada to the cloud forests of Costa Rica; summer–late fall.

Edibility: Unknown, but suspect; definitely not recommended.

Comments: Because of its greyish, striated cap and ringless, subequal stipe, this species can be confused with the Grisette, *A. vaginata*. However, the latter species is readily distinguished by its larger size (typically >2½ in across) and the prominent cup-like volva at the base of the stipe. The volva places it in section *Vaginatae* rather than section *Amanita*. *Amanita farinosa* lacks any trace of a volva, and the base forms a small but distinct bulb that is absent in *A. vaginata*. The mealy, dust-like powder covering the cap is also distinctive and is not present in *A. vaginata*.

Spurned Amanita, Hated Amanita

Amanita (sect. *Caesareae*) *spreta*

A B C

***Amanita spreta* (Peck) Saccardo**, *Syll. Fung.* (Abellini) 5:12 (1887)

Basionym: *Agaricus spreta* Peck, *Ann. Rep. N.Y. St. Mus. Nat. Hist.* 32:24 (1880) [1879]
[Basidiomycota, Agaricomycotina, Agaricomycetes, Agaricomycetidae, Agaricales, Amanitaceae]

Derivation: *Amanita*, Latin, presumably named after Mount Amanon in Cilicia; *spreta*, "spurned."

Cap: 5–13 cm (2–5¼ in) across, initially convex or bell-shaped (Fig. B), expanding to broadly convex (Fig. A) to nearly flat, typically umbonate, with radial striations toward the margin (Fig. A); dry, smooth, occasionally with one or more whitish patches; whitish to pallid grey or greyish-brown, usually darker in the center (Fig. A).

Flesh: Thick, firm, white; no particular odor or taste.

Gills: Free, crowded, narrow to rather broad, white to cream.

Stipe: 10–15 cm tall by 1–2 cm thick (4–6 × ½–¾ in), tapering gradually toward the top, dry, smooth, white, with persistent ring near apex (Fig. C, white arrow); base enclosed by a more or less adpressed, whitish, sac-like volva (Fig. C, black arrow).

Spore print: White.

Microscopic: Spores (7.7–)9.4–13.1(–15.5) × (5.2–)5.9–7.8 (–9.0) µm (highly variable dimensions are reported by various authors), long ellipsoid to (less often) cylindrical, smooth, nonreactive in Melzer's reagent.

Occurrence: Mycorrhizal with both hardwoods and conifers. Alone or in scattered groups on sandy soil under hardwoods and pines; widespread and fairly common in eastern North America; summer–fall.

Edibility: Unknown. Variously reported as both edible and poisonous.

Comments: This species has been reported by many authors as deadly poisonous and the cause of numerous fatal poisonings but without documenting any actual cases of poisoning, fatal or otherwise.[1] Coker (1917) made a similar claim but also reported one woman of his acquaintance who ate this species regularly with no ill effects.[2] And a recent monographic review of the subject found no documented cases of poisoning by *Amanita spreta*.[3] Deshpande (2002) lists this species as neurotoxic/muscarinic.[4] Most of these allegations appear to be apocryphal, but given the persistent suspicions, the consumption of this mushroom is definitely recommended against.

1. For example, W.W. Ford, 1909, The distribution of poisons in mushrooms, *Science* 30(760):106–107 (Ford placed *A. spreta* in the same toxic category as *A. phalloides*, *A. virosa*, and *A. verna*); W.W. Ford & E.D. Clark, 1914, A consideration of the properties of poisonous fungi, *Mycologia* 6(4):173; C.H. Kauffman, 1918, *The Agaricaceae of Michigan*, vol. 2, Mich. Geol. Biol. Survey Publ. 26, Biol. ser. 5, 386.

2. W. Coker, 1917, The amanitas of the eastern United States, *J. Elisha Mitchell Sci. Soc.* 33(1&2):22.

3. J.F. Ammirati et al., 1985, *Poisonous Mushrooms of the Northern United States and Canada* (Minneapolis: University of Minnesota Press), 84.

4. S.S. Deshpande, 2002, *Handbook of Food Toxicology* (New York: Mercel Dekker), table 15.1, 756.

The Lepiotas

(Order Agariales, Family Agaricaceae Chevallier)

Some 400 or so species of the lepiotas are known worldwide. Many of these contain amatoxins (including α-amanitin) similar to those in the genus *Amanita* and are deadly poisonous; many fatal intoxications from *Lepiota* species have been reported. Closely related genera include *Macrolepiota* Singer, *Leucoagaricus* Locquin *ex* Singer, *Leucocoprinus* Patouillard, and *Chlorophyllum* Massee, the last three of which are treated in this section.

Genus *Lepiota* (Persoon) S.F. Gray, *Nat. Arr. Brit. Pl.* 1:601 (1821) [*Agaricus* sect. *Lepiota* Persoon, *Tent. Disp. Meth.* 68 (1797)].

Type species: *Agaricus colubrinus* Persoon 1801 [*non* Bulliard 1782; = *Lepiota cypeolaria* (Bulliard *ex* Fries) P. Kummer]; *typus conservandus.*[1]

Distinguishing features: Agaricoid fruitbodies, gills white to cream,[2] free; partial veil often persisting as a ring, sometimes leaving scaly patches on stipe; stipe without volva; cap with concentric rings of scales toward margin or these absent; volva lacking; spore print white, spores smooth, clear, typically dextrinoid; lamellar trama (gill hyphae) regular, parallel, not diverging; all species saprobic, growing on ground in rich, calcareous soils.

1. *Agaricus colubrinus* Persoon is a later homonym of *A. colubrinus* Bulliard (a deprecated synonym of *Lepiota procera*) and is homotypic with the sanctioned name *A. clypeolarius* Bulliard *ex* Fries. Additional confusion surrounded the typification of several taxa, including the genera *Lepiota* and *Macrolepiota*. To maintain nomenclatural stability of these taxa, *Agaricus colubrinus* Persoon (and not *A. colubrinus* Bulliard) was formally conserved under the rules of the *ICBN* (Vienna Code). See W. Gams, 1993, Report of the Committee for Fungi and Lichens: 3, *Taxon* 42(1):112–118.

2. The Green-spored Lepiota, *Chlorophyllum molybdites* (G. Meyer) Massee, has greenish spores. It was formerly assigned to both the genera *Lepiota* and *Macrolepiota* and is still so treated in many field guides.

Green-spored Lepiota, False Parasol

Chlorophyllum molybdites

A

B

C

Photos: A & B, W. Light; C, Nathan Wilson.

***Chlorophyllum molybdites* (G. Meyer) Massee**, *Bull. Misc. Inf.*, Kew:136 (1898)

Basionym: *Agaricus molybdites* G. Meyer, *Prim. Fl. Esseq.*:300 (1818)
[Basidiomycota, Agaricomycotina, Agaricomycetes, Agaricomycetidae, Agaricales, Agaricaceae]

Derivation: *Chlorophyllum*, Greek, from *chloros*, "green," and *phyllon*, "leaf," referring to the greenish gills of mature specimens; *molybdites*, from *molybdos*, "lead," and *-ites*, an adjectival ending denoting "pertaining to."

Cap: (5–)10–30(–40) cm (2–16 in) across, initially egg-shaped or ovoid, becoming convex (Figs. A & B), broadly convex, cone-shaped, to almost plane (Fig. C), with or without an umbo; surface dry, initially smooth, later developing light brown to pinkish-brown or cinnamon scales or patches (Fig. A), these denser near the center.

Flesh: White, close, thick, not changing color when cut or injured or only slowly bruising reddish; odor and taste not distinctive.

Gills: Free, close, broad, white at first (Fig. B), becoming greenish or greyish-green when mature (Fig. C), but this feature not always reliable, as the gills may remain white for an extended period; staining yellowish or brownish when injured.

Stipe: 5–25 cm tall by 1–2.5 cm thick at apex (4–10 × ½–1 in), equal or expanding toward base, volva lacking; smooth, white or bruising brownish; with white, double-edged ring (Fig. B), this becoming movable and brownish on the underside with age (Fig. C).

Spore print: Greenish to olive-greyish, this feature diagnostic.

Microscopic: Spores (8–)10–12(–13) × 6.5–9 µm, subovoid to elliptical, smooth, with small apical pore, dextrinoid.

Occurrence: Alone or in groups, sometimes forming fairy rings, in gardens, lawns, meadows, and other grassy areas. Widespread and occasional to common in North America, summer–midwinter; also found in the Philippines, West Indies, Brazil, Tahiti, East Africa.

Edibility: Widely consumed, but quite poisonous to many people. Definitely not recommended (see comments).

Comments: This species can be confused with the edible Shaggy Parasol, *Chlorophyllum rhacodes* (Vittadini) Vellinga,[1] especially in the button stage, which grows in similar disturbed habitats and often also forms fairy rings. *Chlorophyllum rhacodes* differs from *C. molybdites* in retaining white gills throughout its life, its white (not greenish) spore print, and its stipe flesh that turns pinkish-orange and then reddish-brown when cut. *Leucoagaricus americana* (Peck) Vellinga is also similar and also has a white spore print. Its flesh bruises yellow and then slowly reddish-brown, and it grows near stumps or on wood chips or compost piles.

This is the species most commonly involved in mushroom poisonings in North America. Poisonings are often quite severe and require hospitalization but are not usually life-threatening except in small children, who are susceptible to hypovolemic shock due to loss of fluids. This mushroom is not toxic to all individuals, and some people can consume large amounts without any problems. Symptoms of poisoning typically develop suddenly within one to three hours following ingestion and involve severe gastrointestinal symptoms, including cramps, vomiting, and profuse watery and typically bloody diarrhea. Patients generally recover completely within 24–48 hours with supportive therapy that maintains fluid and electrolyte balance. Some patients present with perspiration, excessive thirst, and dizziness suggestive of muscarine poisoning, but the nature of the toxin(s) remains unknown.[2]

1. Formerly included in the genera *Lepiota* and *Macrolepiota* and still discussed as such in many field guides and online sources.

2. D. Blayney et al., 1980, Mushroom poisoning from *Chlorophyllum molybdites*, *West. J. Med.* 132:74–77; P.F. Lehmann & U. Khazan, 1992, Mushroom poisoning by *Chlorophyllum molybdites* in the Midwest United States: Cases and a review of the syndrome, *Mycopathologia* 118(1):3–13.

Onion-stalked Lepiota

Leucocoprinus cepaestipes

A B C

***Leucocoprinus cepaestipes* (Sowerby) Patouillard**, *J. Bot.*, Paris 3:336 (1889)

Basionym: *Agaricus cepaestipes* Sowerby [as "*cepistipes*"], *Coloured Figs. Engl. Fung. Mushr.* 1:plate 2 (1797) [1795–1797][1] [Basidiomycota, Agaricomycetes, Agaricomycetidae, Agaricales, Agaricaceae]

Derivation: *Leucocoprinus*, Greek, from *leukó*, "white," and *kóprinos*, "dung, manure," "white dung"; *cepaestipes*, Latin, from *cēpa*, "onion," and *stipes*, "stalk," "onion-stalked."

Cap: 2–8 cm (1¾–3¼ in) across, white, occasionally pinkish, discoloring yellowish-grey, profusely covered with white powdery granules (Fig. A); initially bell-shaped, becoming conical to almost flat, umbonate (with brownish, generally smooth central knob); with prominent radial grooves.

Flesh: White, thin, sometimes bruising yellowish or brownish; odor and taste not distinctive.

Gills: Free, crowded, narrow, white (Fig. B), discoloring with age.

Stipe: 5–14 cm tall by up to 6 mm thick (2–5½ × ¼ in), slender, subequal, but swollen here and there and bulbous at the base (resembling a green onion or scallion); smooth to powdery, white or pinkish-grey, with persistent white ring on upper stipe (Figs. B & C).

Spore print: White.

Microscopic: Spores 6–11 × 5–8 µm, broadly elliptical, thick-walled, with germ pore, smooth, clear, weakly dextrinoid in Melzer's reagent.

Occurrence: Saprobic, in compost, wood chips, mulch, rich soil, occasionally in woods, summer; widespread in eastern North America.

Edibility: Probably poisonous.

Comments: The Lemon-yellow Lepiota or Plantpot Dapperling, *Leucocoprinus birnbaumii* (Corda) Singer, is very similar in shape, size, and growth habit, but the cap, flesh, gills, and stipe are bright to pale yellow, and the spores are slightly longer (8–13 × 5.5–8 µm). It also commonly sprouts from the soil in greenhouses and potted plants.

1. Sowerby's original incorrect spelling (*cepistipes*) was emended to *cepaestipes* by Patouillard, and subsequent authors have variously used both spellings. *Cēpae* is the dative, genitive, and vocative declension of *cēpa*, and *cēpaestipes* technically means "stalk of onion." Sowerby's misspelling, *cepistipes* (*cēpistī pes*), roughly translates to "you have seized the foot" (*cēpistī* is the second-person singular perfect active indicative of *capiō*, "to seize, capture"). The *ICBN* (Vienna Code), 2005, Article 60.1 states: "The original spelling of a name or epithet is to be retained, except for the correction of typographical or orthographical errors." Some authors are following the first, independent clause of this sentence instead of the second, adverbial clause. In this case, Patouillard's emendation is correct.

Malodorous Lepiota

Lepiota cristata

A

B

C

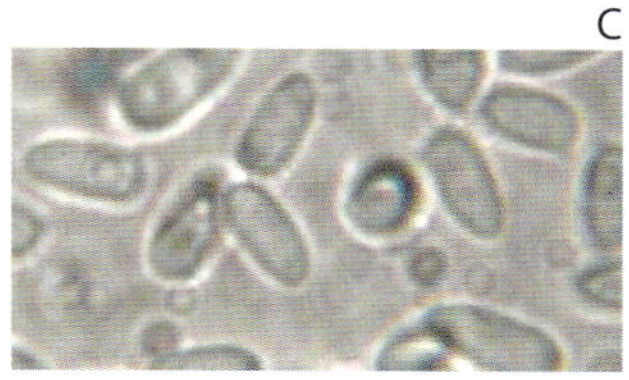
D

***Lepiota cristata* (Bolton) P. Kummer**, *Führ. Pilzk.* (Zerbst):137 (1871)

Basionym: *Agaricus cristatus* Bolton, *Hist. Fung. Halifax* (Huddersfield) 1:7 (1788)
[Basidiomycota, Agaricomycotina, Agaricomycetes, Agaricomycetidae, Agaricales, Agaricaceae]

Derivation: *Lepiota*, Greek, from *lépi*, "scale" or "flake," and *oto*, "ear," "scaly ear"; *cristata*, Latin, "tufted," "crested."

Cap: 2–7 cm (¾–2¾ in) across, initially convex, becoming planoconvex to nearly plane in maturity, dry, initially smooth, developing concentric rings of reddish-brown scales with age (Fig. A); ground color whitish to buff with brownish to reddish-brown center.

Flesh: Thin, white; with unpleasant fishy or pungent smell or odor not distinctive.

Gills: White to buff, free (Figs. B & C), close, narrow, edges entire or finely serrate.

Stipe: 2–7 cm tall by up to 5 mm thick (¾–2¾ × ¼ in), subequal, white to pinkish-buff, often darker toward the base; smooth, fragile, with or without a fragile, often transitory veil or ring on upper portion.

Spore print: White to pale buff.

Microscopic: 5–8 × 3–5 µm, smooth, clear, dextrinoid, distinctively bullet-shaped (Fig. D).

Occurrence: Saprobic. Scattered or in groups on the ground in woods or in grassy areas near trees; summer–late fall; widespread and common in North America and Europe; also reported from Iceland, Japan, and New Zealand.

Edibility: Suspect, possibly poisonous; can be confused with deadly poisonous species.

Comments: *Lepiota cristata* is a variable and widespread complex of species in the temperate regions of the Northern Hemisphere. A number of specific and infraspecific morphological taxa have been variously distinguished and synonymized by different authors, and the limits of these categories are quite unclear (Vellinga 2010).[1]

Lepiota cristata can be confused with several other very similar species found in our area. *Lepiota clypeolaria* (Fries) Quélet typically has tattered remnants of the partial veil hanging from the cap margin, the stipe is covered with shaggy scales from the base to a poorly defined, often transient ring and is smooth above, and the spores are much longer (11–18.5 × 4–6.5 µm) and are fusiform to almond-shaped. *Lepiota acutesquamosa* (Weinmann) P. Kummer is larger,[2] the cap to 11 cm (4½ in) across, the hairy stalk usually has shaggy brown scales similar to those on the cap below a poorly defined but fairly persistent ring, and the spores are long-elliptical (6.5–11 × 2–3.5 µm) and somewhat tapered at each end. All of these lepiotas, including *L. cristata*, likely represent complexes of species that are not at all well delineated.

Lepiota cristata might also be confused with several species in the genus *Leucoagaricus*: *Leucoagaricus americanus* (Peck) Vellinga has a shaggy stipe with a swollen base that bruises yellow, then red, and longer, elliptical spores (8–14 × 5–8 µm). *Leucocoprinus cepaestipes* (Sowerby) Patouillard has a white to pale pinkish cap with white granules or scales and a white, smooth stipe with a swollen base, and the spores are broadly elliptical (6–11 × 5–8 µm) with an apical pore.

1. E.C. Vellinga, 2010, *Mycologia* 102(3):664–674.
2. *Lepiota acutesquamosa* has been synonymized with *L. aspera* (Persoon) Quélet by a number of authors.

Red-tinged Lepiota

Leucoagaricus rubrotinctus

A

B

Photos: A, Teresa Fortenberry; B, Dan Molter.

***Leucoagaricus rubrotinctus* (Peck) Singer**, *Sydowia* 2(1–6):36 (1948)

Basionym: *Lepiota rubrotincta* Peck, *Ann. Rep. N.Y. St. Mus.* 44:179 (1891)
[Basidiomycota, Agaricomycotina, Agaricomycetes, Agaricomycetidae, Agaricales, Agaricaceae]

Derivation: *Leucoagaricus*, Greek, *leukos*, "white," and *agarikon*, "field mushroom"; *rubrotincta*, Latin, "dyed red."

Cap: 3–8 cm (1¼–3¼ in) across, ovoid, becoming broadly convex to nearly plane, with central umbo; initially smooth, becoming radially streaked and scaly, margin often splitting with age; scales cinnamon-buff, pinkish-red, reddish, or reddish-orange, paler toward margin, center darker coral red.

Flesh: White, thin, not changing color when bruised or cut; odor and taste unremarkable.

Gills: Free, white, close, narrow to moderately broad.

Stipe: 5–15 cm tall by ½–1 cm thick (2–6 × ¼–½ in), white, smooth or with scattered white fibrils; equal, solid at first, becoming hollow, base swollen, often deeply embedded in ground; with fairly persistent white ring.

Spore print: White.

Microscopic: Spores 6–10 × 4–6 µm, elliptical, smooth, thick-walled, clear, dextrinoid.

Occurrence: Saprobic. Alone or scattered on the ground in deciduous woods, in leaf litter and compost heaps; summer–fall; widespread in North America.

Edibility: Unknown, but suspect.

Comments: The distinctively colored cap with its dark central umbo, fibrillose radial scales, torn margins, and flesh that does not bruise identify this species. The taxonomic status of the genus *Leucoagaricus*, to which this species currently belongs, is unstable, and this taxon may be reassigned. It is called *Lepiota rubrotincta* Peck in many field guides.

Stropharia-Like Mushrooms

(Order Agaricales, Family Strophariaceae Singer & A.H. Smith)

Genus *Stropharia* (Fries) Quélet

Type species: *Stropharia aeruginosa* (Curtis) Quélet

Genus *Pholiota* (Fries) P. Kummer

Type species: *Pholiota squarrosa* (Bulliard) P. Kummer

Derivations: *Stropharia*, Greek, from *strophos*, "belt"; *Pholiota*, from *pholis*, "scale."

Both *Stropharia* and *Pholiota* are mostly saprobic and grow on woody substrata. Both have smooth spores with apical pores, which may be indistinct in some species, and chrysocystidia on the gills that stain yellowish in KOH.

Stropharia has attached, purplish-grey gills and dark-brown, purplish-brown, to blackish spore prints. The genus is uniquely characterized by vegetative hyphae with numerous stellate spiked acanthocytes (see Fig. C in the account of the Wine-cap Stropharia, *Stropharia rugosoannulata*), used to kill nematode worms, which they digest as a source of nitrogen.

Pholiota species typically have prominently scaly or slimy caps, attached gills that are initially yellowish but turn rusty-brown, rings, and brown to yellowish-brown spore prints. *Pholiota* does not possess acanthocytes or purplish to purplish-brown spore prints.

The genus *Flammulaster* Earle (Family Inocybaceae) has spores that lack an apical pore. The Hedgehog Pholiota, *Flammulaster erinaceellus* (Peck) Watling, occasionally found in our area, is readily distinguished by the staining properties of its cap and flesh (see our species account below).

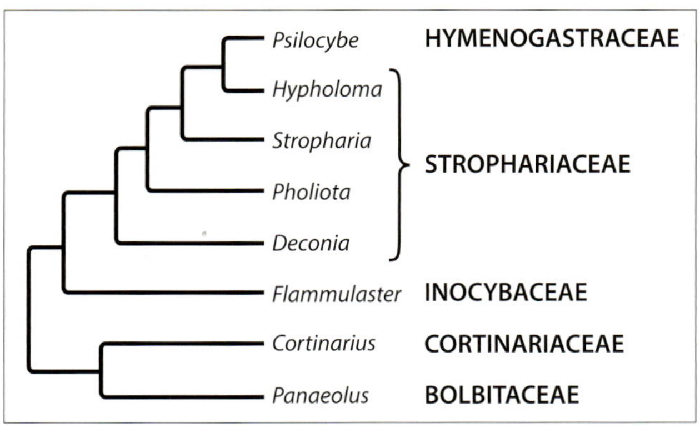

The family Strophariaceae includes the genus *Hypholoma* (Fries) P. Kummer. However, our discussions of the Brick Cap, *H. lateritium* (Schaeffer) P. Kummer, and the poisonous Sulphur Tuft, *H. fasciculare* (Hudson *ex* Fries) P. Kummer, are included under "Miscellaneous Euagarics" in the subsection "Brown Look-Alikes—Some Edible, Some Deadly Poisonous." The family Strophariaceae also contains a number of other genera, including *Agrocybe* Fayod (formerly included in the genus *Pholiota*), some of which are hallucinogenic,[1] as well as *Deconia* (W.G. Smith) P. Kummer, a nonpsychedelic clade that was split off from the psychedelic genus *Psilocybe* (Fries) P. Kummer. *Psilocybe* was formerly in the family Strophariceae but has been transferred to the family Hymenogastraceae Vittadini,[2] which also includes the genus *Galerina* Earle, which has a number of deadly poisonous species. The genus *Panaeolus* (Fries) Quélet was also formerly closely allied with *Psilocybe* in the Strophariaceae but is now placed by various authors in either the Bolbitaceae Singer[3] or the Psathyrellaceae Vilgalys, Moncalvo & Redhead. Species of *Psilocybe* and *Panaeolus* in our area are described in the "Miscellaneous Euagarics" section in "The Magic Mushrooms" subsection. The following simplified cladogram shows our current concepts of the relationships between the Strophariaceae and related families.

1. Several species occur in the southern Appalachians, including *Agrocybe praecox* (Persoon) Fayod, *A. semiorbicularis* (St. Amans) Fayod, and a species complex subsumed under the name *Agrocybe acericola* (Peck) Singer.

2. V. Ramírez-Cruz et al., 2013, *Botany* 91:573–591.

3. *Panaeolus* includes both psychedelic and nonpsychedelic species. The genus *Conocybe* Fayod is also in the Bolbitaceae and has several psychedelic species. As shown in the cladogram, the Bolbitaceae is a sister clade to the Cortinariaceae.

Golden Pholiota Complex

Pholiota aurivella (sensu lato)

A

B

***Pholiota aurivella* (Batsch 1786 *ex* Fries 1821) P. Kummer**, *Führ. Pilz.* (Zwickau):83 (1871)
[Basidiomycota, Agaricomycetes, Agaricales, Strophariaceae]

Basionym: *Agaricus aurivellus* Batsch, *Elench. Fung.*, cont. prim. (Halle):153 (1786)

Derivation: *Pholiota*, Greek, from *pholis*, "scale"; *aurivella*, Latin, "golden fleece."

Cap: Sticky, slimy, 4–16 cm (1⅔–6⅖ in), convex to bell-shaped, often broadly umbonate; uniformly golden-yellow when young, covered with large, flattened to wrinkled reddish-brown scales with age (Fig. A), which may not persist; outer surface viscid to gelatinous.

Gills: Attached, close, margin notched; pale yellow when young, becoming rusty-brown; covered by remnants of a cortina when young (Fig. B).

Flesh: Yellow, firm.

Stipe: 5–10 cm (2–4 in) tall by 5–15 mm (0.2–0.6 in) thick; colored like the cap, finely hairy and increasingly scaly toward the base; rusty ring near top of stipe with off-white zone above.

Spore print: Brown.

Microscopic: Spores 7–11 × 4.5–6.5 µm, smooth, oblong to bean-shaped, with apical pore, but apex not truncate; dull cinnamon in KOH, reddish-tawny in Melzer's reagent.

Occurrence: Saprobic or parasitic on living or dead trunks and logs of both hardwoods and conifers (above specimens photographed on downed Black Willow); widespread in North America; usually in clusters, June–November.

Edibility: Edible, but causes gastric upset in some people.

Comments: *Pholiota aurivella* is not easily distinguished from *P. limonella* (Peck 1878) Saccardo 1887, an extremely variable species. *Pholiota limonella* is supposed to have slightly smaller spores (6–9 × 4.5–5 µm) than *P. aurivella*, but this difference is marginal at best, and there is considerable overlap. The names *P. abietis* A.H. Smith & Hesler, *P. connata* A.H. Smith & Hesler, *P. squarroso-adiposa* J.E. Lange, and *P. subvelutipes* A.H. Smith & Hesler are considered by many to be synonyms of *P. limonella*. It is not yet clear whether *P. limonella* and its synonyms should be placed into synonymy with *P. aurivella*, which would have nomenclatural priority.

Scaly Pholiota Complex

Pholiota squarrosa/*P. squarrosoides*

A

B

A, *Pholiota squarrosoides*; B, *P. squarrosa*. (Photos: A, W. Light; B, Marko Eesti.)

***Pholiota squarrosa* (Vahl) P. Kummer**, *Führ Pilzk*. (Zwickau):22 (1871)

Basionym: *Agaricus squarrosus* Vahl, in Oeder, *Fl. Danic*. 3(9):tab. 491 (1770)

***Pholiota squarrosoides* (Peck) Saccardo**, *Syll. Fung*. (Abellini) 5:750 (1887)

Basionym: *Agaricus squarrosoides* Peck, *Ann. Rep. N.Y. State Mus. Nat. Hist*. 31:33 (1878)
[Basidiomycota, Agaricomycetes, Agaricales, Strophariaceae]

Derivations: *Pholiota*, Greek, from *pholis*, "scale"; *squarrosa*, "tip pointing outward"; *-soides*, "having the likeness of," referring to the earlier named taxon, *P. squarrosa*.

Cap: Both species 2.5–10 cm (1–4 in) wide, blunt to convex, becoming broader, bell-shaped, or almost plane. ***P. squarrosa***—covered with prominent, spiny, pointed, downturned, erect, and/or upturned pale yellowish or yellow, dry (not sticky) scales (Fig. B). ***P. squarrosoides***—with tawny or cinnamon-colored scales on whitish, sticky, viscid surface (Fig. A); margin sometimes with veil remnants.

Gills: ***P. squarrosa***—attached, close, narrow, whitish to pale yellow, then turning greenish before becoming rusty-brown. ***P. squarrosoides***—moderately broad, pale yellow, becoming brown to rust-brown without first turning green.

Flesh: Whitish (both taxa).

Stipe: Up to 1.5 cm (⅔ in) thick, dry, with transient ring or ring zone, covered with spiny scales in both species. ***P. squarrosa***—2.5–10 cm (1–4 in). ***P. squarrosoides***—5–15 cm (2–6 in).

Spore print: Brown (both taxa).

Microscopic: ***P. squarrosa***—spores 5–6 × 3.8–4.5 µm, oval to elliptical, smooth, with apical pore; ***P. squarrosoides***—4–5.5 × 2.5–3.5 µm, oval to broadly elliptical, smooth, apical pore lacking.

Occurrence: ***P. squarrosa***—found on both hardwoods and conifers. ***P. squarrosoides***—saprobic and possibly parasitic, alone or gregarious on hardwoods. Both species widespread in North America, late August–November.

Edibility: Uncertain. *Pholiota squarrosa* reported to be poisonous; *P. squarrosoides* reportedly edible, but some people have experienced gastric upset shortly after consuming this species.

Comments: These two mushrooms are very similar in appearance and habit. *Pholiota squarrosoides* differs from *P. squarrosa* in its slimy cap, pale-yellow gills that become rusty without first turning green, and smaller spores, which lack the apical pore of *P. squarrosa*. Mating and DNA studies have shown that these two taxa are distinct species.

Hedgehog Pholiota, Powder-scale Pholiota

Flammulaster erinaceellus

A B C

Flammulaster erinaceellus (Peck) Watling [as "*erinaceella*"], *Notes R. Bot. Gdn. Edinb*. 28(1):65 (1967)

Basionym: *Agaricus erinaceellus* Peck, *Ann. Rep. N.Y. St. Mus. Nat. Hist*. 30:70 (1878) [1877]
[Basidiomycota, Agaricomycetes, Agaricomycetidae, Agaricales, Inocybaceae]

Derivation: *Flammulaster*, Latin, from *flammula*, "little flame," and *-aster*, suffix implying partial or incomplete resemblance; *erinaceellus*, from *erinaceus*, "hedgehog," and *-ellus*, combining form implying "resemblance" and "small," that is, "resembling a small hedgehog."

Synonym: *Phaeomarasmius erinaceellus* (Peck) Singer (1951).

Cap: 1–4 cm (½–1⅔ in) across, convex (Fig. A), conical (Fig. B), to almost flat, dry, densely covered with erect reddish- to yellowish-brown, easily separated scales that may become powdery; margin frequently with tattered remnants of partial veil (Figs. A–C).

Flesh: Thin, yellowish; no odor, taste metallic or bitter.

Gills: Attached, broad, close, cream to yellowish at first (Fig. C), gradually becoming tawny to cinnamon-brown, margins serrated.

Stipe: 3–6 cm tall by up to 5 mm thick (1¼–2 × ¼ in), thin, equal, solid or hollow, covered with granular scales like the cap up to a fibrous, rudimentary ring near the apex (Fig. B, arrow), stipe smooth and white above that point.

Spore print: Dull cinnamon.

Chemical staining: **KOH**—cap red, then black; flesh reddish-brown. **$FeSO_4$**—flesh olive.

Microscopic: Spores 6–8(–9) × 4–4.5(–5) µm, smooth, elliptical, pale brownish; apical pore lacking.

Occurrence: Saprobic. Alone or in small groups on well-decayed, decorticated, hardwood logs and stumps; widespread, but uncommon in eastern North America; generally considered a more northern species, but we have found it along the Chattahoochee River in the Atlanta area in late summer; early summer–fall.

Edibility: Unknown.

Comments: This small, densely scaly mushroom is not likely to be confused with anything else in our area. The very similar and more northerly distributed *Phaeomarasmius erinaceus* (Fries) Kühner is a darker rusty-brown overall, is smaller (cap not exceeding 1.5 cm across), the gills are distant (not close), and the spores are larger (9–13.5 × 6.5–9 µm) and apiculate. It does not appear to occur in our area.

Wine-cap Stropharia

Stropharia rugosoannulata

Photos and micrograph: A–D, W. Light; E, Ann F. Berger.

***Stropharia rugosoannulata* Farlow ex Murrill**, *Mycologia* 14(3):139 (1922)
[Basidiomycota, Agaricomycotina, Agaricomycetes, Agaricomycetidae, Agaricales, Strophariaceae]

Derivation: *Stropharia*, Greek, from *strophos*, "belt"; *rugoso-annulata*, Latin, "wrinkled ring."

Cap: 4–15 cm (1⅔–6 in) across, initially bell-shaped, becoming broadly convex to flat, dry, smooth; wine-red (Fig. A) or reddish-brown, fading to pale tan; margin often with remnant fragments of partial veil (Figs. A, B, & E).

Flesh: Firm, white.

Gills: Attached, close (Figs. D & E), at first pale grey or whitish, becoming purplish-grey (Fig. E) to purplish-black.

Stipe: 5–15 cm tall by 1–3 cm thick (2–6 × ⅖–1¼ in), base expanded, bulbous (Figs. A & B), with white mycelial threads (Figs. A & B, black arrow) smooth to fibrous, white, staining yellowish with age (Fig. B), with thick, typically persistent split and tattered ring (Fig. A, arrow; Fig. E).

Spore print: Dark purple-brown to blackish.

Chemical staining: Cap surface olive-green with KOH.

Microscopic: Spores 10–14 × 7.5–9 µm, elliptical, smooth, with apical pore; chrysocystidia present on gills that stain yellowish in KOH; vegetative hyphae producing abundant, large, stellate (star-shaped), spiked cells (acanthocytes) (Fig. C), this feature characterizing the genus *Stropharia*.

Occurrence: Saprobic; scattered to gregarious on wood chips, mulch, and grass and in woods.

Edibility: Considered choice.

Comments: The vegetative hyphae of all species of *Stropharia* produce stellate (star-shaped) acanthocytes ("spiny cells") like the one in the micrograph in Fig. C. *Stropharia rugosoannulata* uses them to kill soil nematodes,[1] certain species of which are then digested as a source of nitrogen. The sharp, spiny acanthocytes inflict deep wounds in the cuticles of the worms, whose internal turgor then causes the body contents to leak to the exterior, killing them. The worms are then digested and absorbed by assimilative hyphae as a source of nitrogen and other nutrients for the fungus.[2] The acanthocytes of this species also contain nematocidal toxins.[3] See the section on carnivorous and nematophagous (nematode-eating) fungi in the introduction.

1. Nematodes, so-called roundworms of the phylum Nematoda (Nemathelminthes), are free-living and parasitic worms found in every conceivable habitat on the planet. Most are microscopic, but one sperm whale parasite reaches a length of 13 m (43 ft).
2. H. Luo et al., 2006, *Appl. Environ. Microbiol.* 72(4):2982–2987.
3. J. Yang et al., 2007, *Appl. Microbiol. Biotechnol.* 75:21–31.

The Corts (Webcaps)

(Order Agaricales, Family Cortinariaceae R. Heim *ex* Pouzar, Genus *Cortinarius* Gray)

Derivation: Greek, from *kourtína*, "curtain."

Type species: *Cortinarius violaceus* (Linnaeus) Gray.

Cortinarius is the largest genus of agarics, with over 2,000 species worldwide. Young specimens exhibit a cobwebby veil (**cortina**) between the cap and the stipe. This usually disappears, except for remnants on the stem in some species. All exhibit a rusty-brown spore print, and some of them are dangerously toxic—they contain orellanine, a potent nephrotoxin that can cause a delayed-onset but very severe acute kidney failure that can be fatal and may necessitate a kidney transplant. The renal damage is comparable to that caused by the highly toxic herbicides paraquot and diquat. Many of the species look nearly identical, and identification can be very problematic. Although some edible species exist, as in the case of the amanitas, attempting to eat these mushrooms is definitely not advised. *Cortinarius* can be misidentified as highly prized edible mushrooms such as Blewits (*Clitocybe nuda*) or, in the Pacific Northwest, Matsutake (*Tricholoma magnivelare*).

The two species known to have been involved in fatal cases of orellanine poisoning are the Fool's Webcap, *Cortinarius orellanus* Fries, and the Deadly Webcap, *C. rubellus* Cooke, both found in Europe. Although such poisonings are well known in Europe, the first reported case of orellanine poisoning in the United States occurred in western Michigan in 2008 and involved an unidentified species of *Cortinarius*.[1] This mushroom is closely related to *C. orellanus* from Europe and was subsequently described as a new species: *Cortinarius orellanosus* Ammirati and Matheny sp. nov. (2010).[2] A previous report of delayed nephrotoxicity in the Pacific Northwest resulting from eating a *Coprinus* species was likely due to a misidentification of *Amanita smithiana* Bas,[2,3] a common and widely distributed species in western North America from British Columbia south into Mexico. *Amanita smithiana* appears to be responsible for several cases of delayed-onset renal failure, although the mechanism underlying its nephrotoxicity is unknown. *Cortinarius gentilis* Fries, which occurs in the southern Appalachians,[4] may also cause an orellanine-like poisoning.[1]

Corts grow on the ground and are ectomycorrhizal with a variety of plants; some of these associations are highly specific and can aid in identification. The relative sliminess or dryness of the cap and whether or not it changes color as it is wetted or dried can be important in some species. The shape of the stem and the presence or absence of a ring zone from the disintegration of the cortina is also important, as is the reaction of various tissues to KOH. And finally, the microscopic details of the spores (size, shape, and type and degree of ornamentation) are also important in distinguishing species. And even with all this, identification of many cort species is still very difficult and in some cases almost unachievable.

We will present just a very few representative members of this genus.

1. North American Mycological Association, "Mushroom poisoning syndromes," http://www.namyco.org/toxicology/poison_syndromes.html.
2. In B.S. Judge et al., 2010, *Clin. Toxicol.* 48:545–549.
3. P.L. West et al., 2009, *J. Med. Toxicol.* 5:23–38.
4. R.H. Petersen, 1976, *Checklist of Fungi of the Great Smoky Mountains National Park*, Management Report No. 29, Department of the Interior, National Park Service, Southeast Region.

Viscid Violet Cort, Spotted Cort

Cortinarius iodes

A

B

C

Photos: A & C, W. Light; B, Dan Molter.

Cortinarius iodes Berkeley & M.A. Curtis, *Ann. Mag. Nat. Hist.*, ser. 2, 12:423 (1853)
[Basidiomycota, Agaricomycotina, Agaricomycetes, Agaricomycetidae, Agaricales, Cortinariaceae]

Derivation: *Cortinarius*, Latin, from *cortinas*, "curtains," "sheets"; *iodes*, Greek, "violet."

Cap: Up to 6 cm (2½ in) across, initially bell-shaped (Fig. B), becoming broadly convex to flat (Fig. A), smooth, slimy, dark lilac (Fig. A) to violet, fading with age, developing yellow spots (Fig. B), or becoming yellowish over the disk.

Flesh: Lilac to violet, fading with age.

Gills: Attached, close, at first lilac to violet (Fig. C), becoming rusty-brown to greyish-cinnamon.

Stipe: Up to 7.5 cm tall by 1.5 cm thick (3 × ⅔ in), subequal or enlarged toward base, slimy, smooth, solid, pale violet, with thin ring of purplish or rusty fibers on upper stem left by the cortina (veil) (Fig. C, arrow).

Spore print: Rusty-brown.

Microscopic: Spores 8–10 × 5–6.5 µm, elliptical, finely roughened, brownish.

Occurrence: Scattered or in groups on the ground under broadleaf trees; widespread in eastern North America; summer–late fall.

Edibility: Not poisonous, but, as is true for all corts, not recommended; can be confused with the highly esteemed Wood Blewit, *Clitocybe nuda*.

Comments: *Cortinarius violaceus* (Linnaeus *ex* Fries) Gray has a dark purple to violet to nearly blue-black, scaly, dry cap and stipe and a broadly swollen base (up to 4 cm or 1⅔ in wide), grows on the ground under conifers, has larger spores (12–15 × 7–8.5 µm), and is somewhat rougher. It is much sought after by people who dye wool with mushrooms.

The similar but edible Wood Blewit (*Clitocybe nuda*) is readily distinguished from *C. iodes* by its profuse pale pink to pinkish-cream (rather than rusty-brown) spore print. *Clitocybe nuda* also lacks the purplish or rusty fibers left by the cortina on the upper stipe.

Red-gilled Cort

Cortinarius cf. *semisanguineus*

A

B

Photos: John Plischke III.

***Cortinarius semisanguineus* (Fries) Gillet**, *Hyménomycètes* (Alençon):484 (1876) [1878]

Basionym: *Agaricus cinnamomeus* a *semisanguineus* Fries, *Syst. Mycol.* (Lundae) 1:229 (1821)
[Basidiomycota, Agaricomycotina, Agaricomycetes, Agaricomycetidae, Agaricales, Cortinariaceae]

Derivation: *Cortinarius*, Latin, from *cortinas*, "curtains," "sheets"; *semisanguineus*, "bloody."

Cap: 2.5–6 cm (2–2½ in) across, at first bell-shaped, becoming broadly convex to flat, often with prominent umbo (Fig. B); dry, silky, tawny yellow (Fig. B) to cinnamon-brown (Fig. A), darker in the center with age.

Flesh: Whitish to pale yellowish; odor and taste not distinctive or taste slightly bitter.

Gills: Attached, crowded, narrow; blood-red (Fig. B) at first, becoming rusty-red to cinnabar-red (Fig. A); initially covered by yellowish cortina (Fig. B, lower right).

Stipe: 2–10 cm tall by up to 1.5 cm thick (2–4 × ⅔ in), equal, solid, pale yellow, darker or reddish near base, often with rusty fibrous ring near apex (Fig. B, arrow).

Spore print: Rusty-brown.

Microscopic: Spores 6–9 × 4–5 µm, elliptical, faintly roughened, brownish.

Occurrence: Scattered or clustered on the ground or in moss under hardwoods and conifers; widespread and fairly common in North America; summer–midwinter.

Edibility: Unknown, but suspect; avoid.

Comments: The blood-red gills (when present) distinguish this cort from all others in our area. When present the yellowish cortina distinguishes this species from the similar Maryland Cort, "Cortinarius marylandensis," which has a pinkish to reddish cortina. The cap of the Maryland Cort also initially stains rose before changing to purple with KOH. Pigments from this mushroom are used to dye wool.

Cortinarius semisanguineus in North America appears to comprise a poorly understood species complex. It is unclear whether the taxon found here in the Southeast is the same as that occurring farther north, which is associated with pines and birches.

Maryland Cort

Cortinarius sp. "Cortinarius marylandensis"

A B C

Photos: A, John Plischke III; B & C, Dan Molter.

"Cortinarius marylandensis" [invalid name, Article 28.1, *ICN* (Melbourne Code); never validly published—published only as a provisional name with no valid description or designated type specimen][1]

Basionym: *Dermocybe marylandensis* Ammirati, in Liu, Rogers, Ammirati & Keller, *Beihefte zur Sydowia* 10:144 (1995) [Basidiomycota, Agaricomycotina, Agaricomycetes, Agaricomycetidae, Agaricales, Cortinariaceae]

Cap: 2.5–6 cm (1–2½ in) across, initially bell-shaped (Fig. B), becoming broadly convex to almost plane (Fig. A); dry, smooth to finely hairy; bright brick-red to reddish-brown (Figs. A & B), margin entire.

Flesh: Whitish to pinkish or yellowish; odor and taste not distinctive. Whitish to pale pinkish in cap and upper stem; reddish to brownish in the stem base.

Gills: Attached, close, at first cinnabar-red (Fig. C), becoming cinnamon-brown (Fig. A), initially covered by a pinkish to red cortina.

Stipe: 2.5–7 cm tall by up to 1.5 cm thick (1–2¾ × ⅔ in), subequal; dry, hollow, concolorous with cap near apex, paler cream to yellowish below, with vertical reddish-orange striations (Figs. A–C), sometimes with superior rusty ring zone from the partial veil.

Spore print: Rusty-brown.

Bruising/chemical staining: Cap initially rose, then changing to purple with KOH.

Microscopic: Spores (6.5–)7–9 × 4–5 µm, broadly elliptical, minutely roughened; brown, nonreactive in Melzer's reagent.

Occurrence: Mycorrhizal with oaks (*Quercus* spp.) and American Beech (*Fagus grandifolia*); widespread and occasional in southeastern North America.

Edibility: Unknown.

Comments: This taxon has no valid scientific name. It was published twice by Joseph F. Ammirati and his colleagues, but only as "nom. prov." (a provisional name; see note 1). No type specimen was ever designated nor any formal description published that fulfills the requirements of the *ICBN* or its successor, the *ICN*. Thus, although a number of printed and online mushroom guides treat this as if it were a valid scientific name, it is not, and we present it only within quotation marks and in roman typeface to indicate that fact.

When present, the pinkish to reddish cortina of the Maryland Cort distinguishes this species from the Red-gilled Cort, *Cortinarius semisanguineus*, which has a yellowish cortina. The cap of the Red-gilled Cort also immediately stains purple in KOH without an initial rose phase. Both the Maryland Cort and the Red-gilled Cort are widely used for dyeing wool.

Hesler's Cort, *Cortinarius hesleri* Ammirati et al. (= *C. cinnabarinus* of many American authors),[2] another widespread southeastern species, is very similar and is also associated with oaks and American Beech. Its spores are slightly larger (8–10 × 5–6 μm) and more conspicuously roughened, and the cap stains purple or purplish-black without an initial rose stage.

1. "*Cortinarius marylandensis*, nom. prov." G. Keller & Ammirati, *Mycotaxon* 18(2):364 (1983); "*Cortinarius marylandensis*, nom. prov." Ammirati & A.H. Smith, *McIlvainea* 6(2):58 (1984).
2. *Fide* Ammirati et al., *Botany* 91:91–98 (2013). *Cortinarius cinnabarinus* is restricted to hardwood forests of Europe; *C. hesleri* occurs in mixed hardwood and conifer forests of eastern North America.

Hesler's Cort

Cortinarius hesleri

Photo: John Plischke III.

***Cortinarius hesleri* Ammirati, K.W. Hughes, Liimatainen, Niskanen, and Matheny**, *Botany* 91:96 (2013)[1] [Basidiomycota, Agaricomycotina, Agaricomycetes, Agaricomycetidae, Agaricales, Cortinariaceae]

Derivation: *Cortinarius*, Latin, from *cortinas*, "curtains," "sheets," referring to the cortina; *hesleri*, epithet in honor of American mycologist L.R. Hesler.

Misapplied names: *Cortinarius cinnabarinus*: *auct. mult. amer.*; *Dermocybe cinnabarinus*: *auct. amer.*; [*non Cortinarius cinnabarinus* Fries, *Epicr. Syst. Mycol.* (Upsaliae):287 (1838)].

Cap: 3–9 cm (1¼–3⅔ in) across, initially conical, becoming broadly convex to bell-shaped or almost flat; surface fibrillose (silky) to smooth; cap, gills, and lower stipe initially bright reddish-orange, fading to various shades of brownish-orange to orange-buff.

Flesh: Pale orange, deep orange in the base of the stipe; mild odor of radish.

Gills: Adnexed (attached) to adnate (attached with a notch) or with decurrent tooth, close, broad to moderately broad; concolorous with the cap.

Stipe: 2–10 cm tall by up to 1.5 cm thick near apex (¾–4 × ⅔ in); subequal, dry, silky, pale orangish above, concolorous with cap toward base; sometimes with rusty ring zone.

Spore print: Rusty-brown.

Chemical staining: Cap purple to purplish-black in KOH.

Microscopic: Spores 5–8 × 3–5 μm, broadly spindle-shaped (football-shaped) to broadly elliptical, with prominent warts, pale brown.

Occurrence: Mycorrhizal with oaks (*Quercus* spp.) and possibly also Eastern Beech (*Fagus grandifolia*); scattered to gregarious, widespread in eastern North America; late spring–summer.

Edibility: Unknown.

Comments: This species is misidentified as *Cortinarius cinnabarinus* Fries by most American authors. However, *C. cinnabarinus* is restricted to Europe and does not occur in North America. A closely related species, *C. californicus* A.H. Smith, in western North America is associated with conifers.

The very similar Maryland Cort, "Cortinarius marylandensis" [invalid name], differs from *C. hesleri* in its smaller [(6.5–)7–9 × 4–5 µm], minutely roughened spores and its cap initially staining rose before changing to purple with KOH.

1. The following names are invalid: "Cortinarius hesleri, nom. prov." G. Keller & Ammirati, *Mycotaxon* 18(2):364 (1983); "Cortinarius hesleri, nom. prov." Ammirati & A.H. Smith, *McIlvainea* 6(2):57 (1984); [*nomina invalida*, Art. 39.1 (Melbourne)]. *Cortinarius hesleri* was finally validly published in 2013.

Yellow Webcap

Cortinarius delibutus

A

B

Photos: A, John Plischke III; B, Holger Krisp.

***Cortinarius delibutus* Fries**, *Epicr. Syst. Mycol.* (Upsaliae):276 (1838) [1836–1838]
[Basidiomycota, Agaricomycotina, Agaricomycetes, Agaricomycetidae, Agaricales, Cortinariaceae]

Derivation: *Cortinarius*, Latin, from *cortinas*, "curtains," "sheets"; *delibutus*, Latin, "steeped," "well drenched."

Cap: 2.5–8 cm (1–3¼ in) across, initially hemispherical (Fig. A), becoming convex to almost flat (Fig. B); smooth, sticky when moist (Fig. A), dull and satiny when dry; lemon- to golden-yellow at first (Fig. A), becoming light orange-yellow to yellowish-brown when mature (Fig. B); margin initially inrolled (Fig. A), becoming expanded and thin.

Flesh: Thick over disk, tapering to thin toward margin, white above to greyish-yellow over the gills; odor and taste not distinctive.

Gills: Adnate (attached and notched) with decurrent tooth, close, initially greyish- or bluish-lilac, becoming light rusty-brown with age (Fig. B).

Stipe: 4–9.5 cm tall by 0.5–1.5 cm thick (1⅔–3⅔ × ¼–⅔ in), subequal (Fig. B) or expanded toward the base (Fig. A); viscid or dry, solid at first, becoming hollow; whitish, tinged with pale lilac above and yellowish below (Fig. A), with thin annular zone from partial veil on upper stalk (Fig. B).

Spore print: Rusty-brown.

Microscopic: Spores (6.5–)7–8(–10) × 5.3–6.5(–7) µm, subglobose to broadly ovoid, with apiculus, verrucose (warted); rusty-brown.

Occurrence: Northern North America beneath birches (*Betula* spp.) and Quaking Aspen (*Populus tremuloides*); reported on humus under rhododendrons (*Rhododendron* spp.) in southern Appalachians;[1] occasional, summer–late fall.

Edibility: Reported as edible, but not particularly good; not worth the risk of confusing it with a toxic species.

Comments: This species is identified by its slimy and sticky yellowish cap and white stipe tinged with bluish-lilac above and yellow below. No close look-alikes occur in our area.

1. L.R. Hesler, 1943, Notes on southern Appalachian fungi: V, *J. Tennessee Acad. Sci.* 18(4):294; R.H. Petersen, 1976, *Checklist of Fungi of the Great Smoky Mountains National Park*, Management Report No. 29, Department of the Interior, National Park Service, Southeast Region, p. 68.

Red-banded Cort, Bracelet Cort

Cortinarius armillatus

Photo: John Plischke III.

***Cortinarius armillatus* (Fries) Fries**, *Epicr. Syst. Mycol.* (Upsaliae):295 (1838) [1836–1838]

Basionym: *Agaricus armillatus* Fries, *Observ. Mycol.* (Havniae) 2:61 (1818)
[Basidiomycota, Agaricomycotina, Agaricomycetes, Agaricomycetidae, Agaricales, Cortinariaceae]

Derivation: *Cortinarius*, Latin, from *cortinas*, "curtains," "sheets"; *armillatus*, Latin, "wearing bracelets."

Cap: 5–12 (up to 15) cm (2–4¾(–6) in) across, convex to bell-shaped, becoming nearly flat; margin somewhat inrolled, occasionally with remnants of partial veil; surface dry, smooth to minutely hairy or fibrillose; tawny-brown, brownish-orange to reddish-brown, often darker over the disk.

Flesh: Whitish to pale brownish, not bruising.

Gills: Attached, broad, close to subdistant; pale cinnamon to rusty-brown as spores mature; initially covered by whitish, cobweb-like cortina (above, far left).

Stipe: 7.5–15 cm tall by 1–2.5 cm thick (3–6 × ½–1 in); club-shaped, expanded to swollen or bulbous at the base; solid, dry, smooth to fibrillose; whitish to tawny or pale brownish, with one to four irregular, encircling, reddish-orange bands (remnants of partial veil); base with white mycelium, not bruising when cut.

Spore print: Rusty-brown.

Bruising/chemical staining: Cap surface black with KOH.[1] KOH black on cap surface; purple on orange areas of the stem.

Microscopic: Spores 9–12 × 5.5–7.5 µm; long-elliptical, finely warted; dextrinoid (brownish in Melzer's reagent).

Occurrence: Ectomycorrhizal exclusively with birch (*Betula* spp.). Solitary, scattered, or gregarious on acidic soil under birches in northern mixed hardwood-conifer forests; multiple records from southern Appalachians (Great Smoky Mountains National Park and vicinity)[2] and Southern Inner Piedmont (e.g., Oconee State Park, South Carolina; W. Light & M. Woehrel, personal records; Asheville Mushroom Club Foray); Europe; August and September, fairly common.

Edibility: Reported to be edible; however, caution advised, since many *Cortinarius* species are dangerously toxic.

Comments: This essentially northern species has been reported a number of times from Great Smoky Mountains National Park and other locales in the southern Appalachians. We have encountered it at Oconee State Park in the extreme northwestern corner of South Carolina on the border with Georgia.

The one to four irregular cinnabar-red to orange-red bands encircling the stipe and its association with birches identify this species. In our area it could only be confused with *Cortinarius bolaris* (Persoon) Fries. *Cortinarius bolaris* differs from *C. armillatus* in having a cap densely covered with reddish scales on a white or yellow background (instead of a smooth tawny-brown cap), a stipe with irregular bands and patches of reddish scales and squamules and whose base bruises yellow-orange when scratched or cut, whitish flesh that bruises yellow when injured, smaller (6–8 × 5–6 µm) subglobose to ovoid (rather than long-elliptical) spores, and in being associated with oaks (*Quercus* spp.) and American Beech (*Fagus grandifolia*) rather than birches.

1. *Fide* M. Kuo, 2011, *Cortinarius armillatus*, MushroomExpert.Com, http://www.mushroomexpert.com/cortinarius_armillatus.html, accessed March 16, 2014.

2. For example, see the Great Smoky Mountains National Park All Taxa Biodiversity Inventory, http://www.dlia.org/atbidata/FamilyPage.php?taxon=Family&tname=Cortinariaceae, accessed March 16, 2014.

The Gypsy

Cortinarius caperatus

Photos: A, Teresa Fortenberry; B & D, Strobilomyces; C, John Plischke III.

***Cortinarius caperatus* (Persoon) Fries**, *Epicr. Syst. Mycol.* (Upsaliae):256 (1838) [1836–1838]

Basionym: *Agaricus caperatus* Persoon, *Observ. Mycol.* (Lipsiae) 1:48 (1796)
[Basidiomycota, Agaricomycotina, Agaricomycetes, Agaricomycetidae, Agaricales, Cortinariaceae]

Derivation: *Cortinarius*, Latin, from *cortinas*, "curtains," "sheets"; *caperatus*, Latin, "wrinkled."

Deprecated synonym: *Rozites caperatus* (Persoon) P. Karsten, *Bidr. Känn. Finl. Nat. Folk* 32:290 (1879)[1]

Cap: 5–15 cm (2–6 in) across, broadly convex, becoming irregularly flat, broadly knobbed, and typically somewhat wrinkled (Figs. A & B); dry, silky-scaly when young, with greyish coating of fibers (Fig. B), becoming yellowish-brown when mature (Fig. A).

Flesh: Thick, whitish; odor not distinctive.

Gills: Attached, close, broad, initially covered by a membranous partial veil; creamy white when young (Fig. C), becoming cinnamon-brown with age (Fig. D).

Stipe: 5–13 cm tall by 1–2.5 cm thick (2–5¼ × ½–2 in), equal or somewhat swollen at base, firm, solid; dry, whitish to pale tan, *with thick, persistent white ring at midsection* (Figs. A & C, arrows, & Fig. D); with remnants of partial veil around base.

Spore print: Rust-colored.

Microscopic: Spores 10–15 × 7–10 µm, elliptical to almond-shaped, moderately warted.

Occurrence: Mycorrhizal with conifers and deciduous trees and shrubs; solitary or gregarious on the ground in coniferous and hardwood forests; widespread in eastern North America; uncommon in Europe.

Edibility: Edible, considered choice.

Comments: This mushroom was recently placed in the genus *Cortinarius* on the basis of genetic evidence.[2]

Several species of *Agaricus* resemble *C. caperatus*. They are distinguished by their unattached gills and their chocolate-brown spore print.

An antiviral protein of 10.425 kilodaltons called RC-183 was isolated from this mushroom and shown to be effective against herpes simplex virus types 1 and 2 (HSV-1 and HSV-2), varicella zoster virus (VZV), respiratory syncytial virus (RSV), and influenza A.[3] A second protein of 28 kDA (RC28) also exhibited antiviral activity against HSV-1, HSV-2, VZV, RSV, influenza virus A, and cytomegalovirus.[4]

1. Many field guides call this species *Rozites caperata*, with an incorrect feminine ending.
2. U. Peintner et al., 2002, *Mycotaxon* 83:447–451.
3. F. Pirano & C.R. Brandt, 1999, *Antiviral Res.* 43:67–68.
4. F. Pirano, 2005, *Internat. J. Med. Mushrooms* 7(3):101–114.

Rooted Collybia, Rooted Agaric

(*Hymenopellis* spp. Complex)

***Hymenopellis furfuracea* (Peck) R.H. Petersen**, in Petersen & Hughes, *Nova Hedwigia, Beih*. 137:137 (2010)

Basionym: *Collybia radicata* var. *furfuracea* Peck, *Ann. Rep. Reg. N.Y. St. Mus*. 45:91 (1893) [1891]

***Hymenopellis incognita* (Methven & R.H. Petersen) R.H. Petersen**, in Petersen & Hughes, *Nova Hedwigia, Beih*. 137:156 (2010)

Basionym: *Xerula incognita* Methven & R.H. Petersen, in Petersen & Methven, *Can. J. Bot*. 72(8):1154 (1994)

***Hymenopellis limonispora* (R.H. Petersen) R.H. Petersen**, in Petersen & Hughes, *Nova Hedwigia, Beih*. 137:174 (2010)

Basionym: *Xerula limonispora* R.H. Petersen, in Petersen & Hughes, *Bull. Soc. Mycol. Fr*. 120(1–4):38 (2005) [2004]

***Hymenopellis megalospora* (Clements) R.H. Petersen**, in Petersen & Hughes, *Nova Hedwigia, Beih*. 137:181 (2010)

Basionym: *Clitocybe megalospora* Clements, in Pound & Clements, *Bot. Surv. Nebraska* 4:18 (1896)

***Hymenopellis rubrobrunnescens* (Redhead, Ginns & Shoemaker) R.H. Petersen**, in Petersen & Hughes, *Nova Hedwigia, Beih*. 137:217 (2010)

Basionym: *Xerula rubrobrunnescens* Redhead, Ginns & Shoemaker, *Mycotaxon* 30:384 (1987)

***Hymenopellis rugosoceps* (G.F. Atkinson) R.H. Petersen**, in Petersen & Hughes, *Nova Hedwigia, Beih*. 137:220 (2010)

Basionym: *Collybia rugosoceps* G.F. Atkinson, *J. Mycol*. 8(3):112 (1902)

***Hymenopellis sinapicolor* R.H. Petersen & Justice**, in Petersen & Hughes, *Nova Hedwigia, Beih.* 137:225 (2010) [Basidiomycota, Agaricomycotina, Agaricomycetes, Agaricomycetidae, Agaricales, Physalacriaceae]

Derivations: Latin, *Hymenopellis*, "marriage," "pairing," and *pellis*, "skin," "cuticle";[1] *furfuracea*, from *furfur*, "scurf," and *-aceus*, "of the nature of"; *incognita*, "unknown," "unidentified"; *limonispora*, "lemon-spored"; *megalospora*, "large-spored"; *rubrobrunnescens*, "reddish-brown"; *rugosoceps*, "wrinkled head"; *sinapicolor*, "mustard-colored."

Misapplied names: *Collybia radicata, auct. mult. amer.* [*non Agaricus radicatus* Relhan (1786)]; *Oudemansiella radicata, auct. mult. amer.*; *Xerula radicata, auct. mult. amer.*

NOTE: The following general descriptions are for the eastern North American *Hymenopellis* complex as a whole, except where noted. The individual species details are presented in the dichotomous key following the comments.

Cap: Initially bell-shaped to convex (Figs. A & B), becoming nearly flat (Fig. D) and typically umbonate with age; initially smooth, becoming moderately to deeply radially wrinkled or furrowed (Fig. B); viscid when wet; size highly variable depending on the species.

Gills: White, staining brown in *H. rubrobrunnescens* with age and white to buff to pale olive buff, or sometimes slightly pinkish, in *H. limonispora*; thick, subdistant to distant, broad, often bulging, with numerous short gills (Fig. C); broadly adnate, adnexed, or, more typically, uncinate-sinuate (notched, with a sharp, hooked edge, Fig. C, solid arrow), typically with thin, decurrent ridges (Fig. C, dotted arrow) except in *H. limonispora* and *H. megalospora*.

Flesh: White, thin, not bruising when cut; odor and taste not distinctive.

Stipe: Length highly variable, up to 10 cm (4 in) tall aboveground (up to 20 cm for *H. rubrobrunnescens*, 30 cm for *H. megalospora*, and 8 cm or less for *H. limonispora*, *H. rugosoceps*, and *H. sinapicolor*). Lacking annulus, subcylindrical, expanded to swollen at base and gradually tapering toward apex; white, with or without brownish streaks or chevron-like patterns, or white, becoming brownish-grey, dark olive, cinnamon, or rusty-brown with age.

Pseudorhiza (taproot): A distinctive, prominent pseudorhiza extending deep underground from the base of the stipe, up to 10 cm (4 in) long (shorter in some species), tapering gradually (Fig. D) or abruptly (Fig. E) downward; generally concolorous with the stipe, but often staining yellowish-brown, reddish-brown, or rusty-brown, depending on the species.

Spore print: White to whitish-cream.

Microscopic: Spores broadly ovoid to ellipsoid or lemon- to almond-shaped, with distinct hilar appendage, smooth (finely roughened in *H. megalospora* and *H. rubrobrunnescens*); typically with a large oil droplet (plus several smaller ones in *H. megalospora*); clear, nonreactive in KOH or Melzer's reagent.

Occurrence: Saprobic. Solitary or gregarious in mixed woods, growing from dead and buried hardwood near stumps, but appearing to grow from the ground or occasionally on exposed, well-decayed wood. Widely distributed in eastern North America, but the distributional limits are currently unclear. July–September.

Edibility: All are presumably edible, but not considered worthwhile.

Comments: The long, taproot-like pseudorhiza immediately distinguishes the members of this species complex (Figs. D & E). However, most of these species closely resemble each other, and distinguishing between them can be quite problematic. Even many of the microscopic features overlap. Detailed descriptions of these taxa are available from the sources listed in notes 2–4.

E

Not all of the above species have been recorded in our area, but most have been collected in neighboring areas and will very likely be found here as well. For example, *Hymenopellis sinapicolor* has been collected in Arkansas and Illinois[2,3] and may also occur here. Their true distributions are unclear, because until recently, they were all lumped together and misidentified as *Hymenopellis* (referred to the genera *Xerula*,

Oudemansiella, and *Collybia*) *radicata*, a Eurasian species that does not occur in North America.[5]

In an exhaustive monograph, Petersen & Hughes (2010) provided detailed descriptions of all species in the world formerly referred to *Xerula* and *Oudemansiella*, which they split up into seven separate genera, four of them new. The genus *Oudemansiella* Spegazzini grows directly on woody substrata and is restricted to the tropics. The former genus *Xerula* (*sensu lato*) has been split into *Xerula* R. Maire (*sensu stricto*), with thick-walled setae on the cap and in North America represented by a single species in Mexico; *Paraxerula* R.H. Petersen, with thin-walled pileal setae and in North America represented by a single species in the Rocky Mountains; and *Hymenopellis* R.H. Petersen, the largest genus and characterized by moist to viscid caps.

The following key will not unequivocally distinguish between all of these species. In some cases both the macroscopic and microscopic features overlap to some degree, requiring examination of several specimens and/or several slide preparations.

1. In introducing *Hymenopellis* as a new taxonomic category and new name ["*Hymenopellis* R.H. Petersen, stat. et nom. nov.," a *de facto* new combination and a new genus (and specifically so labeled in a footnote to the table of contents and a section called "Novelties" immediately preceding the list of references)], Petersen did not give its etymology. Hence our derivation is based on the meanings of the combining Latin roots.

2. R.H. Petersen & K.W. Hughes, 2010, The *Xerula/Oudemansiella* complex (Agaricales), *Nova Hedwigia, Beiheft* 137 (Stuttgart: J. Cramer in Gebr. Borntraeger Verlagsbuchhandlung Science Publishers), 625 pp.

3. M. Kuo, 2014, Xeruloid mushrooms, MushroomExpert.Com, http://www.mushroomexpert.com/xeruloid.html.

4. Mycoquébec.org [*Les champignons du Québec*], available at: http://www.mycoquebec.org/bas.php?l=l.

5. *Hymenopellis radicata* appears to be restricted to Europe and extreme western Asia, Asia Minor, and northern Africa.

Dichotomous Key to *Hymenopellis*

1. Gills broadly adnate, not decurrent; spores finely roughened 2

1′. Gills broadly typically uncinate-sinuate, decurrent; spores smooth 3

2. Stipe and pseudorhiza slowly bruising rusty-brown when handled, gills often with rusty edges; spores 13–16 × 7–10 µm. . . . . . . *H. rubrobrunnescens*

2′. Stipe white, unchanging, pseudorhiza staining rusty-brown; gills without rusty edges; spores 15–21 × 8–12 µm . . . . . *H. megalospora*

3. Gills white to creamy 4

3′. Gills white, buff, pale olive buff, or slightly pinkish . . . . . *H. limonispora*

4. Stipe 6–16 cm (2½–6½ in) tall aboveground, pseudorhiza extending 7–10 cm (2¾–4 in) belowground; spores 12–18 × 9–12.5 µm 5

4′. Stipe 3–9 cm (1¼–3⅔) tall aboveground, pseudorhiza extending ≤5 cm (2 in) underground; spores 9–13.5 × 6–9.5 µm 6

5. Cap 8–12 cm across pileocystidia frequent . . . . . *H. furfuracea*

5′. Cap up to 6 cm across; pileocystidia scattered to rare . . . . . *H. incognita*

6. Cap brown to whitish, prominently wrinkled, pitted from center to margin. . . . . *H. rugosoceps*

6′. Cap honey-yellow to mustard-yellow, slightly wrinkled . . . . . *H. sinapicolor*

Oyster-Like (Pleurotoid) Mushrooms

The term "pleurotoid" is a loosely defined adjective referring to gilled mushrooms that are laterally attached to woody substrata either directly by their caps or by a rudimentary to well-developed stipe. However, some of these species have central stipes, and this feature can be quite variable, often within a single species and even a single cespitose cluster. This is often a function of just where on a fallen trunk or log the fruitbody is growing. Most of these mushrooms resemble the oyster mushrooms of the genus *Pleurotus* to a greater or lesser extent, and the term pleurotoid comes from the Greek roots *pleurá*, "side," and *ōtos*, "ear," referring to the ear-like shape they often have. Many of these mushrooms produce shelf-like and often imbricate (overlapping) fruitbodies reminiscent of the polypores, except that they have gills instead of pores. All are saprobic on dead or decaying wood.

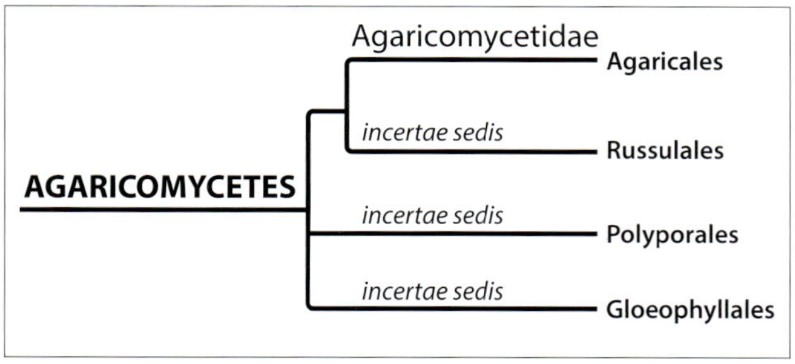

Some of these mushrooms (e.g., *Pleurotus*) are choice edibles, but most are not, and several are poisonous—some, like *Pleurocybella porigens* (Angel Wings) have caused fatalities. The Common Splitgill, *Schizophyllum commune*, has caused several very rare pulmonary infections, some of which have been fatal.

The pleurotoid mushrooms discussed here belong to the class Agaricomycetes and represent three different orders (Agaricales, Russulales, and Gloeophyllales) and seven different families. Two species representing gilled members of the polypore family (Polyporaceae) of the genus *Lentinus* not treated here—*L. strigosus* (Schweinitz) Fries and *L. adhaerens* (Albertini & Schweinitz) Fries—also occur in our area.[1] The pleurotoid morphology and growth habit are thus polyphyletic and have arisen independently several times in different lineages by convergent evolution. This is shown in the following cladogram. The term *incertae sedis* (of uncertain status) means that the indicated orders cannot be assigned to a subclass with any certainty.

1. R.H. Petersen, 1976, *Checklist of Fungi of the Great Smoky Mountains National Park*, Management Report No. 29, Department of the Interior, National Park Service, Southeast Region; A.E. Bessette et al., 2007, *Mushrooms of the Southeastern United States* (Syracuse: Syracuse University Press), 373 pp.

Oyster Mushrooms

Pleurotus ostreatus and *P. pulmonarius*

A–C, *Pleurotus ostreatus*; D–G, *P. pulmonarius*.

***Pleurotus ostreatus* (Jacquin) P. Kummer**, *Führ. Pilzk.* (Zwickau):104 (1871)

Basionym: *Agaricus ostreatus* Jacquin, *Fl. Austriac.* 2:3 (1774)

***Pleurotus pulmonarius* (Fr.) Quélet**, *Mém. Soc. Émul. Montbéliard*, sér. 2, 5:11 (1872)

Basionym: *Agaricus pulmonarius* Fries, *Syst. Mycol.* (Lundae) 1:187 (1821)
[Basidiomycota, Agaricomycotina, Agaricomycetes, Agaricomycetidae, Agaricales, Pleurotaceae]

Derivation: *Pleurotus*, Greek, from *pleurá*, "side," and *ōtos*, "ear," "side ear," referring to the lateral attachment of the stipe; *ostreatus*, Latin, from *ostrea*, "oyster," "oyster-like"; *pulmonarius*, Latin, "pulmonary," associated with the lungs, referring to the shape of the young fruitbodies.

Cap: Both species—margin inrolled at first, becoming wavy or crenulate (Figs. E & G), entire (Fig. B) to scalloped (Fig. F). ***P. ostreatus***—cap typically 4–18 cm (1⅔–7¼ in) across, but may attain widths of 35 cm (~14 in) (Fig. H); convex to flat or

H *Pleurotus ostreatus*. Photo: Alan Rockefeller

depressed, kidney-shaped to fan-shaped or oyster-shaped; smooth, color highly variable: white or cream, pale grey, or more typically yellowish- to greyish-brown (Fig. A), pale brown to dark brown, often with bluish-grey margin (Fig. H). ***P. pulmonarius***—cap typically smaller, 10–13 cm (4–5¼ in); convex to flat or depressed; initially lung-shaped (Fig. F), fan-shaped to semicircular or chanterelle-shaped (Fig. E), often forming overlapping shelves (Fig. D); smooth, white (Figs. D–F) to beige or pale tan, with or without pale dusky margins (Fig. G).

Flesh: ***P. ostreatus***—flesh white, ~2 mm thick, corky; no odor of anise. ***P. pulmonarius***—flesh white, 2–5 mm thick, <1 mm when dry; odor sometimes like anise.

Gills: ***P. ostreatus***—decurrent, close (Fig. B) to subdistant (Fig. A), whitish to greyish, sometimes yellowish in age, 5 mm broad, margin entire. ***P. pulmonarius***—decurrent, close to subdistant, white to cream, 2–3 mm broad, margin entire.

Stipe: ***P. ostreatus***—lacking, rudimentary, or up to 4 cm long by 3.5 cm thick (1⅔ × 1½ in), lateral. ***P. pulmonarius***—usually present, 1–7 cm long by up to 1.5 cm thick (½–2¾ × ⅔ in), eccentric or lateral (rarely central), occasionally lacking.

Spore print: **Both species**—white, cream, or greyish to lavender/lilac.

Chemical staining: ***P. ostreatus***—cap negative with KOH, flesh and gills red in Melzer's reagent. ***P. pulmonarius***—cap orangish with KOH.

Microscopic: ***P. ostreatus***—spores (7–)8–10.5(–12) × 3–3.5 µm, cylindrical, smooth, clear, inamyloid (Fig. C); pileipellis 90–120 µm thick, producing darker fruitbody; hymenophoral trama[1] lacking sclerified hyphae. ***P. pulmonarius***—spores 7–10 × 2.5–5 µm, cylindrical to oblong, smooth, clear, inamyloid; pileipellis 40–50 µm thick, producing lighter-colored fruitbody; hymenophoral trama with sclerified hyphae.

Occurrence: **Both species**—saprobic, causing a white rot in hardwood trees, but especially oaks (*Quercus* spp.); scattered to growing in dense overlapping clusters, especially following heavy rains. ***P. ostreatus***—attacks weakened but living trees, late fall–winter, sometimes into early spring. ***P. pulmonarius***—fruits on dead standing or fallen trees, logs, and stumps, midsummer–early fall, sometimes into early winter.

Edibility: Both species edible and choice.

Comments: Distinguishing between *P. ostreatus* and *P. pulmonarius* is not always easy, although the difference in seasonality (late fall–winter and midsummer–early fall, respectively) usually suffices. However, the fact that these two taxa do overlap in this regard now and then (sometimes considerably, depending on the climate) makes even this distinction problematic at times. This doesn't really matter to the foraging mycophage, since both species are choice edibles (although we personally think the winter oysters have a slight gustatory edge).

One consistent difference is in the width of the pileipellis, or outer cuticle: it is much thicker in *P. ostreatus* (>10 mm) than in *P. pulmonarius* (≤0.5 mm). This 20-fold difference in thickness accounts for the generally darker coloration of the former species.[2] *Pleurotus ostreatus* also attacks weakened but still living hardwood trees, whereas *P. pulmonarius* fruits on dead trees, logs, and stumps. The latter species is also generally smaller.

Like all species in the genus, these oysters are nematophagous, capturing, killing, and digesting nematodes by means of nematostatic toxic microdroplets secreted from short hyphal branches. This appears to be the one unique character that defines, unites, and distinguishes the true oyster genera *Pleurotus* P. Kummer and *Hohenbuehelia* Schulzer.[3]

1. The flesh underlying the hymenium, or spore-bearing tissue on the gills.
2. B.E. Lechner et al., 2004, *Mycologia* 96(4):845–858.
3. R.G. Thorn et al., 2000, *Mycologia* 92(2):241–252.

Smooth Lentinus

Pleurotus laevis

A

B

***Pleurotus laevis* (Berkeley & M.A. Curtis) Singer**, *Lilloa* 22:271 (1951)

Basionym: *Panus laevis* Berkeley & M.A. Curtis, *Ann. Mag. Nat. Hist.*, ser. 2, 12:427 (1853)[1]
[Basidiomycota, Agaricomycotina, Agaricomycetes, Agaricomycetidae, Agaricales, Pleurotaceae]

Derivation: Greek, *Pleurotus*, from *pleurá*, "side," and *ōtos*, "ear," "side ear," referring to the lateral attachment of the stipe; *laevis*, "smooth," "slippery," "glossy."

Cap: 2.5–21 cm (1–8¼ in) wide, convex and centrally depressed (Fig. A) to slightly depressed or more or less flat; at first covered with short hairs, later becoming smooth; white when young, upper surface and flesh becoming yellowish with age (Fig. C, arrow), rarely pastel pink; margin thin, inrolled (Fig. B), more or less entire or with short, erect hairs (Fig. C).

Gills: White, decurrent, crowded to subdistant (Fig. B), broad, margins entire (Fig. C), becoming somewhat irregular or lacerated with age.

Flesh: White, discoloring to yellow with age (Fig. C), up to 2 cm (0.8 in) thick, firm, becoming spongy with age (Fig. C), with or without grapefruit-like odor.

Stipe: White, dry, finely hairy (Figs. C & D), 4–15.5 cm (1⅔–6 in) tall by 1–4 cm (⅖–1⅔ in) thick, with swollen base (Figs. B & D); centrally attached or eccentric, occasionally nearly lateral; no trace of a partial veil.

Spore print: Pale cream.

Microscopic: Spores 9–16 × 4–6 µm, cylindrical, smooth, translucent, inamyloid; context hyphae dimitic.

Occurrence: Alone or in clumps on decayed hardwood logs, trunks, stumps, or occasionally conifers; June–December; widespread but occasional in the Southeast.

Edibility: Edible, but tough.

Comments: This species is morphologically almost identical to *Pleurotus dryinus* (Persoon *ex* Fries) P. Kummer, but the stalk of *P. dryinus* is nearly equal or tapered downward (not swollen, as in *P. laevis*), and the lower portion is enclosed in a whitish, hairy sheath (remnants of the partial veil) topped by a transient ring; remnants of the veil usually remain on the cap margin. *Pleurotus laevis* never has any trace of a ring. These two species cannot be distinguished microscopically. *Pleurotus laevis* has a subtropical to tropical distribution; *P. dryinus* is more temperate and northerly. Both occur throughout the Southeast. Both species have a central rather than lateral stipe, unlike most species of *Pleurotus*. Like all species in the genus, both are nematophagous; they capture, kill, and digest nematodes by means of nematostatic toxic microdroplets secreted from short hyphal branches. This appears to be the one unique character that defines, unites, and distinguishes the true oyster genera *Pleurotus* P. Kummer and *Hohenbuehelia* Schulzer.[2]

C D

1. The specific epithet is incorrectly "modernized" and spelled as "*levis*" in most field guides and online sources. However, the original spelling by Berkeley & Curtis is *lævis* with the ligature intact; both *laevis* and *lævis* are correct spellings for this Latin adjective. There was nothing to "correct," and the naming authors' original spelling should be preserved.

2. R.G. Thorn et al., 2000, *Mycologia* 92(2):241–252.

Veiled Oyster

Pleurotus dryinus

A

B

C

D

***Pleurotus dryinus* (Persoon) P. Kummer**, *Führ. Pilzk.* (Zwickau):101 (1871)

Basionym: *Agaricus dryinus* Persoon, *Comm. Schaeff. Icon. Pict.*:96 (1800)
[Basidiomycota, Agaricomycotina, Agaricomycetes, Agaricomycetidae, Agaricales, Pleurotaceae]

Derivation: *Pleurotus*, Greek, from *pleurá*, "side," and *ōtos*, "ear," "side ear," referring to the lateral attachment of the stipe; *dryinus*, Greek, from *drys*, "oak," and *-inus*, "belonging to," referring to its common association with oaks.

Cap: 4–12 cm (1⅔–4¾ in) across, convex, becoming broadly convex (Fig. B), more or less circular in outline (Fig. C); surface dry, whitish, becoming yellowish with age, covered with minute pale greyish fibrils; margin at first inrolled, expanding and becoming decurved with maturity (Figs. A & C), typically evenly rimmed with remnants of partial veil (Fig. A, arrow; Figs. B & C).

Flesh: Thick, firm, white; odor and taste not distinctive or smell clean and pleasant.

Gills: Decurrent, close (Fig. C) to subdistant (Fig. A), occasionally forking; white, yellowing with age.

Stipe: 4–10 cm tall by 1–3.5 cm thick (1⅔–4 × ½–1½ in), subequal, typically slightly eccentric (Fig. C), occasionally central; concolorous with the cap, tough, solid; lower portion encased in an initially hairy sheath that terminates above in a transient ring (Fig. D, arrow), this feature diagnostic, but ring often lost (Fig. A).

Chemical staining: Cap negative with KOH; yellow with Melzer's reagent, this gradually fading completely.

Spore print: White.

Microscopic: Spores 9–14 × 3.5–5.0 μm, elliptical, smooth, inamyloid; context tissue dimitic and tough.

Occurrence: Saprobic; alone or in clusters on living and dead hardwoods, especially oaks (*Quercus* spp.) and American Beech (*Fagus grandifolia*); widespread and sporadic in southeastern North America, summer–fall.

Edibility: Edible but tough, not choice.

Comments: This is the only *Pleurotus* species with a ring at any stage, which immediately distinguishes it from the otherwise nearly identical Smooth Lentinus, *Pleurotus laevis* (Berkeley & M.A. Curtis) Singer. Both species are widespread but uncommon in our area. The stipe of *P. dryinus* is nearly equal or tapered downward (not swollen, as in *P. laevis*), and the lower portion is enclosed in a whitish, hairy sheath (remnants of the partial veil) topped by a transient ring; remnants of the veil usually remain on the cap margin. *Pleurotus laevis* never has any trace of a ring. These two species cannot be distinguished microscopically but are intersterile and cannot mate.

Pleurotus cystidiosus O.K. Miller is distinguished by longer, subspindle-shaped basidiospores (11–17 × 4.2–5.0 µm), which stain light yellowish in Melzer's reagent. It lacks any trace of a partial veil or ring. It is immediately diagnosed by the presence of anamorphic white coremia (asexual hyphae) topped with black, viscid conidial spores (arthroconidia) covering the stipe, basal mycelium, and surrounding wood. The coremia look like tiny pins with white stems (0.6–1.25 mm tall) and round, black, slimy heads (0.4–0.9 mm in diameter). This co-occurrence of the anamorphic (asexual) state—*Antromycopsis macrocarpa* (Ellis & Everhart) Stalpers, Seifert & Samson—growing on and around the teleomorphic (sexual) fruitbody is highly unusual.

Hypsizygus tessellatus (Bulliard *ex* Fries) Singer is also very similar but lacks a veil and tends to fruit high up on the trunks of living hardwoods, especially elms (*Ulmus* spp.) and Box Elder (*Acer negundo*). The cap is also characteristically covered with round or oval spot-like "watermarks," and the spores are smaller (5–7 µm) and globose rather than elliptical.

Like all species in the genus, *P. dryinus* is nematophagous, capturing, killing, and digesting nematodes by means of nematostatic toxic microdroplets secreted from short hyphal branches. This appears to be the one unique character that defines, unites, and distinguishes the true oyster genera *Pleurotus* P. Kummer and *Hohenbuehelia* Schulzer.[1]

1. R.G. Thorn et al., 2000, *Mycologia* 92(2):241–252.

Orange Mock Oyster

Phyllotopsis nidulans

A C

Phyllotopsis nidulans **(Persoon) Singer**, *Revue Mycol.*, Paris 1:76 (1936)
[Basidiomycota, Agaricomycotina, Agaricomycetes, Agaricomycetidae, Agaricales, Tricholomataceae]

Basionym: *Agaricus nidulans* Persoon, *Icon. Desc. Fung. Min. Cognit.* (Leipzig) 1:19 (1798)

Derivation: *Phyllotopsis*, Greek, *phyllo-*, combining form of *phyllon*, "leaf," and *tops*, "face"; *nidulans*, Latin, from *nidula*, "little nest."

Cap: 2–15 cm (¾–6 in) across, sessile, fan-shaped to shelf-like (Fig. A), often coalescing laterally; flattened to somewhat convex; dry, densely fuzzy, hairy (Figs. B & C); often with whitish coating when young (Fig. A), quickly becoming bright orange, fading to yellowish orange with age; margin entire to undulate, inrolled when young (Fig. C).

Gills: Close to crowded, narrow to broad, straight, forking, radiating from attachment, margins entire; pale to bright orange (Fig. C).

Flesh: Pale orangish, soft, smell often disagreeable or odor lacking.

Stipe: Lacking.

Chemical staining: Negative in KOH.

Spore print: Pale pink.

Microscopic: Spores 5–8 × 2–4 µm, sausage-shaped, smooth, clear, nonreactive in Melzer's reagent.

Occurrence: Saprobic, on dead, fallen hardwoods and conifers, solitary or in overlapping clusters (Fig. A); fall–winter (above photos taken in late December 2012); widespread and fairly common throughout North America and Europe.

Edibility: Inedible, bad-tasting.

Comments: *Phyllotopsis nidulans* somewhat resembles the Stalkless Paxillus, *Tapinella panuoides* (Batsch) E.-J. Gilbert, and the Bear Lentinellus species complex of *Lentinellus ursinus* (Fries) Kühner and *L. vulpinus* (Sowerby) Kühner & Maire. However, the fruitbodies and gills of these latter taxa are never as conspicuously orange as in *P. nidulans*.

Tapinella panuoides differs in having gills that are usually wavy or conspicuously wrinkled and corrugated but sometimes straight. The cap is dull olive- to rusty-brown or yellowish- to orangish-brown, the flesh is whitish to pale cream, the spore print yellowish, tan-brown, to pale ochraceous-brown; the smooth spores are ovoid, smaller

(4–5.5 × 3–3.5 µm), and variably dextrinoid (staining reddish-brown) in Melzer's reagent.

Lentinellus ursinus (see the Bear Lentinellus species account) has a brown to cinnamon-brown cap, the margin is inrolled and often heavily lobed, the flesh is very thin and whitish to pinkish buff, and all tissues stain black in Melzer's reagent and KOH. The gills are glancing (showing an oblique reflective shimmer when rocked back and forth) and whitish to pinkish-brown with coarsely serrated and torn edges, and the spores (4–4.5 × 3–3.5 µm) are warty to spiny and amyloid (staining bluish-black) in Melzer's reagent.

Jack O'Lantern Mushroom

Omphalotus illudens

A C B

Photos: A, W. Light; B & C, Teresa Fortenberry.

***Omphalotus illudens* (Schweinitz) Bresinsky & Besl**, *Beih. Sydowia* 8:106 (1979)
[Basidiomycota, Agaricomycotina, Agaricomycetes, Agaricomycetidae, Agaricales, Marasmiaceae]

Basionym: *Agaricus illudens* Schweinitz, *Schrift. Berlin. Ges. Naturforsch. Freunde* 1:81 (1822)

Derivation: *Omphalotus*, Greek, from *omphalos*, "navel," "central point" (e.g., the navel or center of the earth);[1] *illudens*, Latin, present participle of *illudere*, "to mock," "to ridicule," thus, "mocking," "ridiculing."

Fruitbody: 5–20 cm tall by 5–18 cm wide (2–8 × 2–7¼ in), initially convex, margin inrolled, becoming flattened, depressed, or shallowly funnel-shaped with age (Figs. A & B); surface smooth, bright orange to yellow-orange, turning brownish with age; margin incurved (Figs. B & C), becoming wavy or undulate and often split with age (Fig. A).

Gills: Fertile surface with true nonforking or only occasionally forked,[2] decurrent gills with sharp, thin edges (Fig. C); these are readily detached and are colored like the cap surface. The gills of all *Omphalotus* species are bioluminescent; see comments below.

Flesh: White to yellowish.

Stipe: Central or eccentric, smooth, solid, 5–20 cm (2–8 in) tall by 1–2.2 cm (~½–1 in) thick, more or less equal to slightly spindle-shaped, tapering at the base; colored like the fruitbody or somewhat paler.

Chemical staining: Cap surface green with KOH and ammonium salts.

Spore print: White to pale yellow.

Microscopic: Spores 3.5–5 × 3.5–4.5 µm, spherical to ovoid, smooth, clear, nonstaining in Melzer's reagent.

Occurrence: Saprobic on decaying hardwoods; in dense clusters at the bases of standing dying or dead hardwood trees, especially oaks; may appear to sprout from the ground when growing on buried roots (Fig. A); June–January; widespread and common in eastern North America and northern Europe.

Edibility: Poisonous; nonlethal, causing severe cramps, vomiting, and diarrhea.[3]

Comments: The toxic Jack O'Lantern Mushroom can be distinguished from the edible chanterelles (*Cantharellus* spp. and *Craterellus* spp.), which it resembles, by its sharp, thin, knife-edge gills, which are easily detached. The "gills" of the chanterelles are actually thick, bluntly rounded vertical ridges that cannot be easily detached. The chanterelles also do not form large cespitose clusters, nor do they grow on wood substrata. This usually distinguishes them from the Jack O'Lantern, although the latter can appear to emerge from the ground when growing from submerged roots. The species of *Cantharellus* and *Craterellus* with which the Jack O'Lantern might be confused have spores that are consistently larger (7–12 × 4–6.5 µm) and elliptical rather than spherical to ovoid.

The False Chanterelle, *Hygrophoropsis aurantiaca* (Wulfen) Maire, is distinguished from the Jack O'Lantern Mushroom by being associated with conifers (rather than hardwoods), in having gills that are repeatedly forked (rather than nonforked or only occasionally forked), and in the cap staining red (rather than green) with KOH. The False Chanterelle contains high levels of arabitol, which may account for the gastrointestinal distress experienced by some individuals who consume it.

All Jack O'Lantern Mushrooms have bioluminescent gills and under the right circumstances will emit an eerie greenish glow. This requires placing young specimens in a totally dark room and waiting for a considerable period for your own eyes to become dark adapted. However, trying to see this can be a very frustrating experience, as humorously reported by Michael Kuo, the Mushroom Expert (http://www.mushroomexpert.com/omphalotus_illudens.html). We have not tried this with the Jack O'Lantern, but it mirrors our experience in trying to see the bioluminescence produced by the Luminescent Panellus, *Panellus stipticus* (Bulliard) P. Karsten. We have been unable to see any hint of bioluminescence even in a pitch-black room with fully dark-adapted eyes.

It was first suggested more than a century ago[4] that bioluminescence in fungal fruitbodies functioned to attract arthropods that could help disseminate the spores, and several authors have seconded that notion. A 2015 study[5] provides the strongest support for this hypothesis. These authors report that bioluminescence in a closely related Neotropical mushroom (*Neonothopanus gardneri*) is regulated by a temperature-controlled circadian clock and is only switched on at night when the light is easy to see. They demonstrated experimentally that this bioluminescence was a potent attractor for staphilinid rove beetles, flies, wasps, ants, and hemipterans ("true bugs"), which then carried the fungal spores with them when they left.

Two other species of *Omphalotus* occur in North America: *O. subilludens* (Murrill) H.E. Bigelow, found in Florida, and *O. olivascens* Bigelow, Miller & Thiers, which occurs west of the Rocky Mountains. *Omphalotus mexicanus* Guzmán & Mora is found in the highlands of Mexico and Central America. All species of *Omphalotus* are both poisonous and reported to be bioluminescent.

This species is sometimes called *Omphalotus olearius* (de Candolle) Singer in field guides and online sources, but this European species does not occur in North America.

1. In Greek mythology, the omphalos was a stone in the temple of Apollo at Delphi that was believed to be the center of the earth.
2. Many sources state that the repeatedly forking gills seen in *Hygrophoropsis aurantiaca* is the primary feature distinguishing it from the nonforking gills of *Omphalotus illudens*. However, many specimens of *O. illudens* also have occasional gills that fork.
3. *Omphalotus* species contain two toxic sesquiterpines, illudin S and illudin M, which produce severe gastrointestinal symptoms in those who consume this mushroom: cramps, vomiting, diarrhea, weakness, and exhaustion. These compounds are cytotoxic against a wide variety of cancer cells and have been used to produce an anticancer drug.
4. A.J. Ewart, 1906, Note on the phosphorescence of *Agaricus (Pleurotus) candescens*, *Victorian Naturalist* 13:174.
5. A.G. Oliveira et al., 2015, Circadian control sheds light on fungal bioluminescence, *Current Biology*, doi:10.1016/j.cub.2015.02.021.

Luminescent Panellus

Panellus stipticus

A B C

***Panellus stipticus* (Bulliard) P. Karsten**, *Hattsvampar* 14:fig. 172 (1879)
[Basidiomycota, Agaricomycotina, Agaricomycetes, Agaricomycetidae, Agaricales, Mycenaceae]

Basionym: *Agaricus stypticus* Bulliard, *Herb. Fr.* 3:tab. 140 (1783) [1782–1783]

Derivation: *Panellus*, Latin, diminutive form of *panis*, "bread" or "loaf"; *stipticus*, Latin, "binding, adhesive," referring to its reputed ability to stop bleeding.

Cap: Small, 1–3 cm (~½–1¼ in) across, semicircular to kidney-shaped, convex, with strongly inrolled margin when young, becoming planoconvex with margin slightly inrolled or even; surface cracked or scaly (Fig. A), velvety or woolly; dingy white, tan, or pale yellowish-brown.

Gills: Attached to slightly decurrent, close, narrow, terminating in abrupt line at stipe (Figs. B & C), sometimes forked; pale pinkish- to yellowish-brown.

Flesh: Thin, tough, cream-colored to pale yellow-brown; mildly to severely acrid-tasting in most populations.

Stipe: Rudimentary and stubby, 0.5–1.5 long by 5 mm wide, lateral, velvety, whitish, forming an abrupt white semicircle where it meets the gills (Fig. B).

Spore print: White.

Chemical staining: Cap surface negative in KOH.

Microscopic: Spores 3–6 × 2–3 µm, sausage-shaped, smooth, clear, amyloid (staining bluish-black in Melzer's reagent).

Occurrence: Saprobic, typically in overlapping clusters on decaying hardwood logs and branches (Fig. A), especially beeches, oaks, and birches; May–December, sometimes over-wintering; widespread in North America, especially in the East; Europe, Middle East, Asia, Australia, New Zealand.

Edibility: Inedible, bitter. Taste and subsequent sensations of discomfort highly variable across the species' geographic range.

D Photo: Ylem.

Comments: The distinctive white semicircle at the apex of the stipe where it meets the gills (Fig. B) is diagnostic and immediately identifies this species.

This mushroom is famous for its bioluminescence, but this phenomenon is not easy to observe even in a totally dark room after a prolonged period. The luminescence is reportedly more obvious in some populations than others. The remarkable photo at right taken on August 12, 2009, by "Ylem" required an exposure time of 517 seconds. Despite the fact that the bioluminescent and taste properties of this mushroom vary widely throughout its range, mating studies indicate that they are all conspecific.

Similar-looking species of *Crepidotus* can be distinguished by their brown, rather than white, spore print; likewise, superficially similar *Paxillus* species are separated by their yellow-brown spore prints. The sessile, white to greyish cap of the Common Splitgill, *Schizophyllum commune*, is very densely hairy, and the gill-folds (not gills) are prominently split longitudinally.

Bear Lentinus

Lentinellus ursinus

A B C D

Lentinellus ursinus **(Fries) Kühner**, *Botaniste* 17:99 (1926) [Basidiomycota, Agaricomycetes, *incertae sedis*, Russulales, Auriscalpiaceae]

Basionym: *Agaricus ursinus* Fries, *Syst. Mycol.* (Lundae) 1:185 (1821)

Derivation: *Lentinellus*, Latin, diminutive of *Lentinus*; *ursinus*, Latin, "bear," in reference to the hirsute brown cap.

Cap: Sessile, up to 10 cm (4 in) across, kidney-shaped, semi-circular to irregularly lobed (Fig. A), broadly convex to flat, surface minutely but conspicuously hairy, velvety, especially over inner portion (Figs. A & C); brown to cinnamon-brown toward the center, paler yellowish-brown peripherally; margin inrolled, lobed (Figs. A & B).

Gills: Close to nearly distant, radiating from attachment to substratum, often glancing (Fig. B); whitish to pinkish-brown; edges coarsely serrated and torn (Fig. D).

Flesh: Very thin, whitish to pinkish-buff.

Stipe: Lacking, but a pedicel can resemble a stalk.

Spore print: White.

Staining: All tissues immediately jet black in Melzer's reagent; cap jet black, gills and flesh light shiny brown (like shellac) in KOH.

Microscopic: Spores 4–4.5 × 3–3.5 µm; elliptical; clear, amyloid in Melzer's reagent; finely ornamented with warts and spines.

Occurrence: Saprobic; solitary or gregarious, often overlapping on decaying hardwood, occasionally conifers; summer–fall.

Edibility: Inedible, very bitter, acrid.

Comments: The amyloid spores distinguish the genus *Lentinellus* from *Lentinus* Fries (1825). *Lentinellus ursinus* is very difficult to distinguish from *L. vulpinus* (Sowerby) Kühner & Maire (1934), even based on microscopic features. The basidiospores are nearly identical in both species. The differences given in most field guides often overlap. The short stem supposedly seen in *L. vulpinus* may not be apparent, and the whitish to pinkish-brown color of the "woolly" hairs covering the cap may also intergrade with *L. ursinus*. Moreover, the pedicels of *L. ursinus* sporocarps (fruitbodies) often look like distinct stipes with gills that merge into a common base. To our knowledge, the staining characteristics of *L. vulpinus* have not been reported. *Lentinus ursinus* was even described as a form of *L. vulpinus* (Fr.) by Pilát in 1946. The distinction between these two taxa is problematic. *Lentinellus angustifolius* (Romel) Singer & Digilio (1952) is smaller (cap, 1.0–2.5 by 1.0–3.5 cm), the gills and flesh are yellowish-white to pale yellow, the basidiospores (3.0–4.6 × 2.0–3.5 µm) are subglobose to very short elliptical, and the taste of the flesh is extremely acrid, numbing the tongue for five minutes or longer.

Narrowleafed Cockleshell

Lentinellus angustifolius

A

B

***Lentinellus angustifolius* (Rommel) Singer**, *Lilloa* 25:91 (1952) [Basidiomycota, Agaricomycotina, Agaricomycetes, Russulales, Auriscalpiaceae]

Basionym: *Lentinus angustifolius* Rommel, *Bih. Kungl. Svenska Vetenskapsakad. Handl.* 26(6):7 (1901)

Derivation: *Lentinellus*, Latin, diminutive of *Lentinus*; *angustifolius*, Latin, "narrow-leafed," referring to the thin, crowded gills.

Cap: Very small, sessile, 1–2.5(–5) by 1–3.5 cm (⅖–1 × ⅖–1½ in), convex to plane in profile, kidney-shaped, divided or spathulate from above, margin plane to decurved and inrolled (Fig. B), entire to lobed, occasionally cracked; surface minutely hairy or with matted fibers (Fig. A), sometimes with bristly hairs; yellowish-white, light yellow, greyish-yellow, greyish-orange, or brownish-orange to dark brown.

Gills: Close to crowded, very thin, 2–3 mm broad, radiating from attachment point, margins entire to serrulate (finely toothed) (Fig. B); yellowish-white to pale yellow.

Flesh: Yellowish-white to pale yellow, not staining when exposed; odor fruity, taste acrid, temporarily numbs the tongue.

Stipe: Stipe entirely lacking.

Spore print: White.

Microscopic: Spores (3.1–)3.9–5.0 × (2.5–)2.9–3.9 µm, subglobose to broadly ellipsoid, thin-walled, minutely spiny, amyloid, contents translucent, light yellow in NH_4OH, often with a light yellow oil droplet.

Occurrence: Widely distributed in temperate North America, Europe, South America.

Edibility: Inedible.

Comments: Three species of *Lentinellus* occur in eastern North America: *L. angustifolius*, *L. ursinus*, and *L. vulpinus*.[1] Only *L. ursinus* (see the account of the Bear Lentinus) and *L. angustifolius* occur in our area; *L. vulpinus* is a northern species. *Lentinellus angustifolius* is distinguished by its very small size, tomentose to fibrillose cap, close to crowded and narrow gills, and sessile habit. *Lentinellus ursinus* is also sessile, but its larger size (2.5–14 cm wide), and close to subdistant broad gills distinguish it from *L. angustifolius*. It grows on hardwoods in the East and on conifers in the West. *Lentinellus vulpinus* is even larger, 5–10 up to 25 cm (up to 10 in) wide, and the gills are broad and partly decurrent down a short stipe that fuses with its neighbors to form a large common base.

1. A.N. Miller & A.S. Methven, 2000, *Mycologia* 92(4):792–800.

Train Wrecker, Scaly Sawgill

Neolentinus lepideus

A

B

C

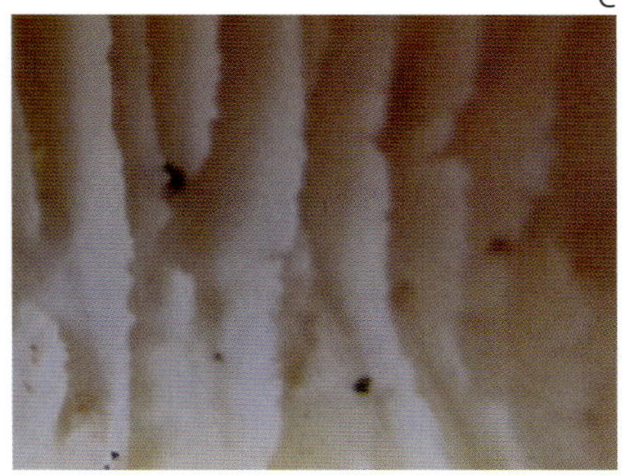

D

***Neolentinus lepideus* (Fries) Redhead & Ginns**, *Trans. Mycol. Soc. Japan* 26(3):357 (1985)
[Basidomycota, Agaricomycotina, Agaricomycetes, Gloeophyllales, Gloeophyllaceae]

Derivation: *Neolentinus*, Latin, "new *Lentinus*," an older genus name plus *lentinus*, "appearing pliable"; *lepideus*, Latin, "scaly."

Basionym: *Agaricus lepideus* Fries, *Observ. Mycol.* (Haviniae) 1:21 (1815)

Cap: 5–12 cm (2–4¾ in) across (usually <10 cm), with inrolled margin (Fig. C), initially convex, later becoming planoconvex; surface cream-white, covered with concentric rings of adpressed cinnamon to yellow- or light-brown scales, these denser toward center (Figs. A & B).

Gills: White, adnate to subdecurrent or decurrent (Fig. C), *margin finely serrate* (Fig. D).

Stipe: *Extremely tough, resists even strong pinching pressure*; 8–10 cm (3¼–4 in) high, often embedded in woody substratum for much of its length, vertical ridges descending from attachment of gills for short distance (Fig. A); partial veil usually leaving a persistent ring near the top, but this sometimes missing; surface cream-white, covered with cinnamon or light-brown scales in the region below the ring (Fig. B), basal portion of stalk light buff.

Spore print: White.

Chemical staining: Nonreactive with KOH or Melzer's reagent; nonbruising.

Microscopic: Spores 7–15 × 4–6.5 µm, long-elliptical, inamyloid in Melzer's reagent, ghostly, barely visible; smooth, thin-walled, with distinct hilar appendage (Fig. E). Sharp, spike-like setae and kinky, narrowly expanded cheilocystidia barely protruding from hymenial layer.

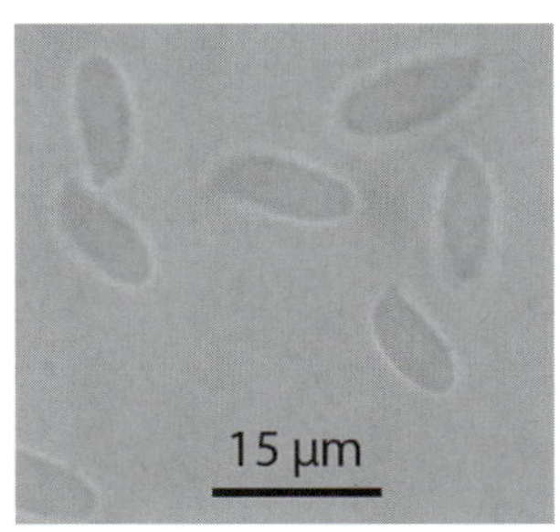

E

Occurrence: Widely distributed in North America and Europe at or around the base of decayed conifer stumps and logs, particularly pines, rarely on hardwoods; fruits from spring to autumn following wet weather.

Edibility: Edible when young, but becoming very tough and woody with age.

Comments: With their inrolled cap margins and adnate to decurrent gills, these mushrooms resemble scaly-capped oyster mushrooms sitting atop a long stout stipe. But the finely serrated edges of the gills immediately give them away as members of the *Lentinus* (*sensu lato*) group. Our specimen agrees in every respect with the macroscopic and microscopic features of *N. lepideus* except for lacking an annulus. In this it agrees with the nearly identical but generally larger *N. ponderosus* found in the western United States, which also lacks a ring. The cap on *N. ponderosus* can reach a width of 30 cm, but the spores are generally smaller (8.5–10 × 3.5–4 µm). However, that taxon has not been reported east of the Rocky Mountains. *Neolentinus lepidus*, *N. ponderosus*, and the seven other members of this genus are brown-rot saprobes, which is partly why they were removed from the white-rot genus *Lentinus* (see Redhead & Ginns, 1985). The name Train Wrecker derives from this tough fungus sprouting from creosote-treated railroad ties.

Common Splitgill

Schizophyllum commune

A

B

C

D

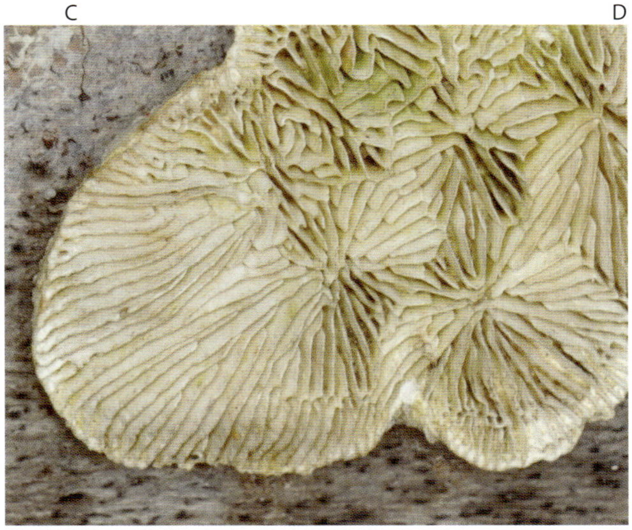

E

Photos: A–D, W. Light; E, Jim Edwards.

Schizophyllum commune Fries, *Observ. Mycol.* (Havniae) 1:103 (1815)
[Basidiomycota, Agaricomycetes, Agaricomycetidae, Agaricales, Schizophyllaceae]

Derivation: *Schizo*, Greek, "split," plus *phyllum*, Greek, "leaf" or "lamella"; *commune*, Latin, "common."

Cap: Very tough, leathery, fan-like to shell-shaped small bracket 1–5 cm (0.4–2 in) across (Fig. C), often lobed or fused with adjacent fruitbodies (Fig. A); sessile, attached at one side (Fig. C); upper surface densely hairy (Figs. A & C), grey to white or greyish-brown.

Flesh: Thin, tough, light grey to brown.

Stipe: Generally lacking, sessile.

Gills: Radiating from point of attachment, well spaced (Figs. B & C), longitudinally split (Fig. D, arrow), white to greyish-brown, aging to yellowish-brown or tan. Resupinate on undersides of branches, forming centrally attached radiating disks or fans (Fig. E).

Spore print: White.

Microscopic: Spores smooth, variously reported as 3–4 × 1–1.5 µm to 5.5–7 × 2–2.5 µm, cylindrical to elliptical.

Occurrence: Saprobic, causing white rot in hardwoods; solitary or clustered on dead logs, branches, and sticks; fruitbodies persisting over winter, shriveling during dry periods, reviving after rain. Very widely distributed on all continents except Antarctica.

Edibility: Generally considered too tough to eat, but widely consumed in Mexico and elsewhere in the tropics.

Comments: *Schizophyllum commune* is the most common and widespread mushroom in the world. The spore-producing gills split longitudinally (the source of its common name) when the mushroom dries out (Fig. D) and close up again when it rehydrates. This mushroom has been estimated to have more than 28,000 different mating strains, the fungal equivalent of sexes (see the introduction for a discussion of sexual reproduction in fungi), thus making it the most successful mushroom in the world. In sexually reproducing fungi, genetically determined differences in mating types ("sexes") exist among individual fungi. Individual hyphae of the same mating type cannot mate with one another. See the section "Mating Types and Sexuality" in the introduction for a discussion of this.

Spores of this cosmopolitan and edible species have been reported as a rare but emerging cause of serious infections of the brain, lungs, mouth (hard and soft palates), and sinuses—at least one of them fatal despite intensive treatment—in both immunocompromised (due to AIDS or corticosteroid treatment) and immunocompetent persons without underlying disease.[1]

1. J.D. Rihs et al., 1996, *J. Clin. Microbiol.* 34(7):1628–1632; L. Sigler et al., 1999, *J. Clin. Microbiol.* 37(10):3395–3398. One bizarre case involved a severely immunocompromised HIV patient who had inhaled Splitgill spores from the mushrooms he had gathered for a meal. He was later found to have a mushroom growing down through the roof of his mouth from his sinus cavity.

Smoked Oysterling, Black Jelly Oyster

Resupinatus alboniger, R. applicatus

A B C D

A & B, *Resupinatus alboniger*; C & D, *R. applicatus*. (Photos: A–C, W. Light; D, Boleslaw Kuznik.)

***Resupinatus alboniger* (Patouillard) Singer**, *Nova Hedwigia* 29(1–2):17 (1978)

Basionym: *Pleurotus alboniger* Pat., *Bull. Soc. Mycol. Fr.* 9:126 (1893)

***Resupinatus applicatus* (Batsch) Gray**, *Nat. Arr. Brit. Pl.* (London) 1:617 (1821)[1]

Basionym: *Agaricus applicatus* Batsch, *Elench. Fung.*, cont. prim. (Halle):171 (1786)
[Basidiomycota, Agaricomycotina, Agaricales, Tricholomataceae]

Derivation: *Resupinatus*, Latin, "resupinate," named for the growth habit of the type species (*R. applicatus*, at that time the only species in the genus); *alboniger*, Latin, "black-and-white" (*albus*, "white," *niger*, "black"), referring to the smoky black fruitbody with its white margin; *applicatus*, Latin, "applied, attached," referring to its close attachment to the substratum.

Cap: ***R. alboniger***—tiny, smooth (occasionally with very fine greyish hairs in spots), sessile, laterally attached, kidney-, shell-, or irregularly fan-shaped (Figs. A & B), strongly reminiscent of an oyster mushroom except for its tiny size and coloration; saucer-shaped and more or less centrally attached when on undersides of logs; 2–10 mm (to ⅖ in) across; dull black to dark grey-brown, often with a whitish margin (Fig. A), smooth, but sometimes with fine greyish fuzz (Fig. B); ***R. applicatus***—7–15 to 20 mm (¼–⅔ to ¾ in) across, covered with fine greyish hairs (Fig. D), centrally attached by rudimentary stem (Fig. C), resupinate, resembling an inverted cup on undersides of logs (Fig. D).

Flesh: Blackish, gelatinous (both).

Gills: ***R. alboniger***—blackish-brown to black, broad, widely spaced, radiating fan-like from lateral point of attachment (Figs. A & B); ***R. applicatus***—broad, widely spaced, pale greyish when wet, dark grey to blackish when dry, radiating radially from a more or less central attachment (Fig. D).

Stipe: ***R. alboniger***—lacking, fruitbody sessile; ***R. applicatus***—centrally attached by rudimentary, minute pseudostem (Fig. C, arrow).

Staining: Negative in KOH (both).

Spore print: White (both).

Microscopic: Smooth, clear, nonreactive in Melzer's reagent. ***R. alboniger***—spores 5.5–8 × 2.5–4 µm, long-elliptical to sausage-shaped; ***R. applicatus***—spores 4–6 × 4–4.8 µm, spherical.

Occurrence: Saprobic, scattered, in groups or dense clusters on dead and decaying hardwoods. ***R. alboniger***—alone or gregarious on dead hardwood trunks and logs, overwinters in warmer regions; ***R. applicatus***—on the undersides of decaying logs. Both widespread in eastern North America, summer–fall.

Edibility: Unknown, hardly worth the effort.

Comments: To our knowledge, *R. alboniger* has never been depicted in a field guide. The few guides that include this genus at all present images and discussions of *R. applicatus* (Batsch) Gray (1821), the type species of the genus. *Resupinatus applicatus* differs from *R. alboniger* in its marginally larger size, smaller and nearly spherical spores, cup-like fruitbody, and inverted sessile (stalkless) habit beneath logs. In addition, the cup-like to conical fruitbodies are more or less centrally attached by an apical "pseudostem" (*R. alboniger* is sessile), the cap is covered with coarse hairs (especially near the pseudostem), the chocolate-brown color when wet fades to ash-grey when dry, and the gills radiate from the central point of attachment. Both species are saprobic on decaying hardwoods and are widely distributed in North America and Europe.

1. Authorship of the genus *Resupinatus* has been variously attributed to some variation of C.G. Nees *ex* Gray and Gray. However, *Resupinatus* Nees was published in violation of Article 33.7 of the *ICBN* and is therefore an invalid name. The genus must be cited with Gray as its correct author. See the discussion under *Resupinatus applicatus* in online Appendix A.

Maroon Ear

Anthracophyllum lateritium

B A

C

***Anthracophyllum lateritium* (Berk. & Curt.) Singer**, *Lilloa* 22:206 (1951) [1949]
[Basidiomycota, Agaricomycetes, Agaricomycetidae, Agaricales, Marasmiaceae]

Basionym: *Xerotus lateritius* Berk. & M.A. Curtis, *J. Linn. Soc., Bot.* 10(45):303 (1868)

Derivation: *Anthracophyllum*, Greek, *anthrakos*, "coal," and *phyllos*, "leaf," presumably based on the sooty-black gills of the type species;[1] *lateritium*, Latin, "brick."

Cap: Laterally attached, 0.5–2.5 cm (⅕–1 in) wide, thin, subcircular, shell-like, to kidney- or fan-shaped; upper surface finely fuzzy to smooth, divided into halves by the indented stipe attachment and a median ridge or groove (Fig. A); surface with radiating folds and grooves from point of attachment; margin at first entire, becoming notched to scalloped; light tan (Fig. A) to reddish-brown (Fig. B) above, becoming dark and iridescent with age (Fig. A, lower right); margin thin, downcurved.

Gills: Strikingly deep reddish-brown to maroon plate-like gills radiating from point of attachment out to the margin; color somewhat lighter and more maroon near margin, which is bordered by a thin lighter line (Fig. C); gills laterally branched and distally forked; black when dried.

Stipe: Absent or very reduced, 0.5–1 mm long.

Flesh: Pale yellow-brown, staining blue-green in alcohol or alkaline solutions, apparently characteristic of the genus.[2]

Spore print: White.

Microscopic: Spores 9.5–12.5 × 5.5–8.0 µm, subovate to oblong-elliptical, tapering to sharp point, smooth, thin-walled, translucent, nonreactive in Melzer's reagent.

Occurrence: Common and widespread on various dead hardwoods in the tropics and subtropics. Year-round in the Gulf Coast and Southeast, often in large clusters; South America, Puerto Rico, Cuba, and the Antilles. Questionably reported from India.

Edibility: Inedible.

Comments: The small size and sharp contrast between the light upper surface of the cap and the deep maroon to dark reddish-black gills make identification of this species straightforward. The genus *Anthracophyllum* is widespread in the tropics and subtropical-temperate regions of the world. It contains 12 described species.

1. *Anthracophyllum beccarinum* Cesati (1879).
2. *Fide* J. Putzke, 2002, *Caderno de Pesquisa Sér. Bio., Santa Cruz do Sul* 14(1):45–66.

Miscellaneous Euagarics

The ill-defined term *agaric* can refer to (1) a species of mushroom in the order Agaricales (more accurately referred to as **euagarics**, or "true agarics") or (2) almost any classically shaped mushroom with a well-defined, umbrella-like pileus (cap) whose underside bears lamellae (gills) and is supported by a pillar-like stipe (stalk). Most species of agaric-like mushrooms do belong to the Agaricales, but this term is also applied to agaric-shaped mushrooms in other orders, including the Boletales, Russulales, and Hymenochaetales. It is also used as part of the common name for certain mushrooms, such as the several species of "fly agarics" (*Amanita* spp.).

Parrot Mushroom

Gliophorus psittacinus

A B C D E

Photos: Emily Johnson.

***Gliophorus psittacinus* (Schaeffer) Herink**, *Sb. Severočeského Musea, Historia Naturalis* 1:82 (1958)
[Basidiomycota, Agaricomycotina, Agaricomycetes, Agaricomycetidae, Agaricales, Hygrophoraceae]

Basionym: *Agaricus psittacinus* Schaeffer, *Fung. Bavar. Palat.* 4:70 (1774)

Derivation: *Gliophorus*, Greek, from *glia*, "glue," "sticky," and *phoros*, "bearer"; *psittacina*, Latin, "parrot-like."

Monotypic synonym: *Hygrocybe psittacina* (Schaeffer *ex* Fries) P. Kummer, *Führ. Pilzk.* (Zwickau):112 (1871)

Cap: Initially conical (Figs. A & C), becoming bell-shaped (Fig. C) to broadly convex or almost flat with age; 1.0–4 cm (½–1⅔ in) across; surface smooth, very slimy (Figs. A & D); color variable, initially dark green, teal green (Fig. A), or olive-green (Fig. D), later becoming pink, yellow, orange (Fig. E), or dingy grey (Fig. C); margin striate (Figs. A, C, D, & E), slightly crenulate (finely scalloped) (Figs. C & D).

Flesh: Thin, colored like the cap, odor and taste not distinctive.

Gills: Attached, distant, thick, waxy; initially light green, becoming yellowish (Fig. B) or pinkish (Fig. D).

Stipe: 3–9 cm tall by 2–5 mm (1¼–3⅔ × ¼ in) thick, hollow; equal to slightly tapering upward, surface viscid and color variable like the cap, but greenish apex often persisting (Figs. A & D).

Spore print: White.

Microscopic: Spores 8–10 × 5–6 µm, smooth, elliptical.

Occurrence: Saprobic, solitary, scattered, or gregarious in damp soil, moss, humus in hardwood and conifer forests; spring–late fall; widespread in North America, western Europe, and Japan.

Edibility: Reportedly edible, but may cause gastric upset.

Comments: Unmistakable when young and slimy green, but when fully mature this mushroom resembles any number of other nondescript little brownish or greyish mushrooms.

Scarlet Waxy Caps

Hygrocybe coccinea and *H. punicea*

A

B

C

A, *Hygrocybe coccinea*; B & C, *H. punicea*. (Photos: A, Emily Johnson; B & C, Rosemary Winnall.)

Hygrocybe coccinea **(Schaeffer) P. Kummer**, *Führ. Pilzk.* (Zwickau):112 (1871)

Basionym: *Agaricus coccineus* Schaeffer, *Fung. Bavar. Palat.* 4:70 (1774)

Hygrocybe punicea **(Fries) P. Kummer**, *Führ. Pilzk.* (Zwickau):112 (1871)

Basionym: *Agaricus puniceus* Fries, *Syst. Mycol.* (Lundae) 1:104 (1821)
[Basidiomycota, Agaricomycotina, Agaricomycetes, Agaricomycetidae, Agaricales, Hygrophoraceae]

Derivation: *Hygrocybe*, Greek, from *hygros*, "moist," and *cybe*, "head"; *coccinea*, Greek, from *kókkinos*, "scarlet"; *punicea*, Latin, from *puniceus*, "pink," "rosy."

Cap: ***H. coccinea***—2–5 cm (¾–2 in) across, at first convex, becoming bell-shaped, conical, with or without central knob, eventually flattened to nearly plane, often centrally depressed; margin initially incurved, later spreading to upraised; surface blood-red when young, slowly turning orangish to yellowish from the margin; viscid, especially when young, becoming dry when mature. ***H. punicea***—3–15 cm (1½–6 in), as above, but dark red fading to brownish- or orangish-red or buff.

Flesh: ***H. coccinea***—yellowish-red, up to 5 mm thick, soft, odor and taste mild; ***H. punicea***—yellowish, whitish near the center, thin.

Gills: ***H. coccinea***—attached to almost free, subdistant to distant, broad, waxy, thick with thin yellowish edges, reddish-orange to yellow-orange, margins entire; ***H. punicea***—narrowly attached, subdistant to distant, thick, waxy, buff to reddish-orange, orangish, peach-colored, or yellow, often with yellowish edges.

Stipe: ***H. coccinea***—2.5–6 cm tall by 5–10 mm thick (1–2½ × ¼–⅖ in), smooth, equal, hollow, ring lacking; red toward the top, becoming progressively orange, then yellow toward the base. ***H. punicea***—3–15 cm tall by up to 15 mm thick (1½–6 × ¼–⅔ in), *vertically strongly fibrillose to striate* (Figs. B & C), equal or tapering toward base, no ring, stringy, readily splitting, various shades of red, fading to orange and yellow, base white.

Spore print: White for both taxa.

Microscopic: Both species—spores 7–11 × 4–5.5 µm, elliptical, smooth; nonreactive in Melzer's reagent.

Occurrence: Both saprobic on hardwoods and conifers; solitary, scattered, or gregarious in woods and moist areas, summer–midwinter; North America, Europe; widespread in the East.

Edibility: Said to be edible, but possibly mildly toxic to some individuals.

Comments: These two species are very similar, but *H. punicea* can be distinguished from *H. coccinea* by its generally larger size and its prominently fibrillose-striated (rather than smooth) stipe with a white (rather than yellow) base.

Goblet Waxy Cap

Hygrocybe cantharellus

A B C D

Hygrocybe cantharellus (Schweinitz) Murrill [as "*Hydrocybe*"], *Mycologia* 3(4):196 (1911)

Basionym: *Agaricus cantharellus* Schweinitz, *Schr. Naturf. Ges. Leipzig* 1:88 (1822)
[Basidiomycota, Agaricomycotina, Agaricomycetes, Agaricomycetidae, Agaricales, Hygrophoraceae][1]

Derivation: *Hygrocybe*, Greek, "moist head"; *cantharellus*, Latin, "small drinking vessel."

Deprecated synonym: *Hygrophorus cantharellus* (Fries) Fries, *Epicr. Syst. Mycol.* (Upsaliae):329 (1838)

Cap: 1–4 cm (½–1⅔ in) across, initially convex, becoming broadly convex to almost plane, often centrally depressed (Figs. A & B), margin scalloped (Figs. A–C); dry, initially scarlet-red, fading to orange-red, orange, orange-yellow, or yellow with age.

Flesh: Thin, white, yellow, or reddish-orange; no particular odor or taste.

Gills: Short decurrent (Figs. C & D), pale yellowish to creamy-orangish; subdistant to distant, broad, with numerous short gills attached to margin (Fig. C).

Stipe: 4–10 (up to 12) cm tall by 1.5–5 mm thick (1⅔–4 or 4¾ × ⅔–¼ in), slender, equal, smooth, concolorous with cap, but pale yellowish to whitish near the base; ring lacking.

Spore print: White.

Microscopic: Spores 7–12(–13) × 4–8 µm, ovoid to long-elliptical, with a small, off-center apiculus; smooth, clear, nonreactive in Melzer's reagent; basidia with four (occasionally two) sterigmata; clamp connections present.

Occurrence: Saprobic. Solitary or clustered in organically rich humus or on well-decayed and rotten wood, often among mosses; widespread and fairly common in eastern North America, summer–late fall; South America, Europe, Africa, Asia.

Edibility: Variously reported as edible or unknown; probably best avoided.

Comments: Statements in several sources that this is one of the few hygrophoroid species that grows on decayed wood are incorrect; many species of *Hygrocybe* occur on rotten wood.[2]

Hygrocybe miniata (Fries) P. Kummer, a northeastern species that is reported from Great Smoky Mountains National Park,[3] is macroscopically and microscopically almost indistinguishable from *H. cantharellus*. *Hygrocybe miniata* tends to have gills that are broadly attached downward and not really decurrent, but this subjective feature is often difficult to distinguish. Their ecological associations are also much the same.

The small Cinnabar-red Chanterelle, *Cantharellus cinnabarinus* (Schweinitz) Schweinitz, looks very similar to *H. cantharellus*. It can be distinguished from *H. cantharellus* by its reddish-orange to pinkish-orange, blunt, and deeply decurrent forking ridges, which are not easily torn away from the cap and which often have cross-veins. *Hygrocybe cantharellus* has true sharp-edged, nonforking, yellowish gills that are readily separated from the cap. See the species account of *C. cinnabarinus*.

1. The Hygrophoraceae are no longer considered to represent a single family but are spread among two separate tribes: tribus Hygrocybeae, which includes *Hygrocybe* and two other genera, and tribus Hygrophoreae, containing the genus *Hygrophorus* (*fide* E. Arnolds, 1995, *Mycotaxon* 53:1–27).

2. Noted throughout Arnolds, 1995.

3. R.H. Petersen, 1976, *Checklist of Fungi of the Great Smoky Mountains National Park*, Management Report No. 29, Department of the Interior, National Park Service, Southeast Region, 77, as "Hygrophorus miniatus (Fr.) Fr. var. miniatus." *Hygrocybe miniata* is a cosmopolitan species with a worldwide distribution.

Honey Mushrooms

Armillaria mellea/A. tabescens/A. gallica

A & B, *Armillaria mellea*; C & D, *A. tabescens*; E, *A. gallica*; F, *Armillaria* sp., exposed rhizomorphs. (Photos: A, Teresa Fortenberry; B, D, & F, W. Light; C, Antonio Abatiello; E, John Plischke III.)

***Armillaria mellea* (Vahl) P. Kummer**, *Führ. Pilzk.* (Zwickau):134 (1871)

Basionym: *Agaricus melleus* Vahl, *Fl. Danic.* 6(17):tab. 1013 (1790)

***Armillaria tabescens* (Scopoli) Emel**, *Le Genre Armillaria* (Strasbourg):50 (1921)

Basionym: *Agaricus tabescens* Scopoli, *Fl. Carniol.*, 2nd ed. (Wien) 2:446 (1772)

***Armillaria gallica* Marxmüller & Romagnesi**, Boidin, Gilles & Lanquetin, *Bull. Trimest. Soc. Mycol. Fr.* 103(2):152 (1987)

[Basidiomycota, Agaricomycotina, Agaricomycetes, Agaricomycetidae, Agaricales, Physalacriaceae]

Derivation: *Armillaria*, Latin, from *armilla*, "bracelet"; *mellea*, "honeyed"; *tabescens*, "melting"; *gallica*, "French."

Cap: All three species, convex when young to broadly convex or flat with age (Figs. A, C, D, & E); ***A. mellea*** (Figs. A & B)—3–15 cm (1¼–3 in) across, surface smooth, lacking squamules, dry or tacky; ***A. tabescens*** (Figs. C & D)—1–4 cm (½–1⅔ in) across, yellow-brown, tawny-brown to cinnamon-brown, dry;

A. gallica (Fig. E)—2–6 cm (¾–6¼ in) across, yellow-brown, covered with small but prominent scales; dry or sticky, tan, pinkish-brown to tawny-brown, finely hairy, often with remnants of partial veil at margin.

Flesh: Whitish to tan.

Gills: Attached to somewhat decurrent, nearly distant, whitish, occasionally discoloring darker when bruised. *Armillaria gallica* initially with a white cortina covering the gills.

Stipe: ***A. mellea***—5–20 cm tall by 5–35 mm thick (2–8 × ¼–1½ in), tapering to base, with persistent, flaring, collar-like ring (Figs. A & B, arrow); ***A. tabescens***—7.5–20 cm tall by 5–15 mm thick (3–8 × ¼–⅔ in), tapering to base; *no ring*, this diagnostic (Figs. C & D); ***A. gallica***—5–10 cm tall by 5–10 mm thick (2–4 × ¼–½ in), base often swollen and bruising yellow, with fleeting cobwebby ring later reduced to a yellow ring zone, this often obscure (Fig. G, arrow).

G

A. gallica. Photo: Dan Molter.

Spore print: White.

Rhizomorphs: Black shoestring-like mycelia (Fig. F) conspicuous and abundant in *A. mellea* and *A. gallica* (fruitbodies of the latter often sprouting from the black mycelia); rarely produced by *A. tabescens*.

Microscopic: **All three species**—spores more or less elliptical, smooth, inamyloid, with prominent apiculus. ***A. mellea***—spores 7–9 × 6–7 µm, *basidia lacking basal clamp connections*, this feature diagnostic; ***A. tabescens***—spores 8–10 × 5–7 µm, basidia with basal clamps; ***A. gallica***—spores 7–9.5 × 4.5–6 µm, basidia with basal clamps.

Occurrence: All three species are largely sympatric and have broadly overlapping distributions; all are widespread in eastern North America, with rare reports from the Pacific Northwest. *Armillaria mellea* and *A. tabescens* are virulent parasitic pathogens of hardwood trees, with *A. gallica* being a fairly harmless saprobe; *A. mellea* and *A. gallica* occasionally occur on conifers. ***Armillaria mellea*** and ***A. tabescens*** grow in dense cespitose clusters of 8–10 (but occasionally up to 30–50) fruitbodies (Figs. A & C) from decaying wood; ***A. gallica*** is solitary to gregarious (never cespitose) and appears to grow from the ground but actually from buried roots, occasionally from wood aboveground. All three species occur in eastern North America (with rare isolates in the Pacific Northwest), Europe, and Asia.

Edibility: Choice edibles; all three species in our area are among the very best. In our opinion, *A. tabescens* is the best of the lot. However, care must be taken not to confuse *Armillaria* species with the Deadly Galerina, *Galerina marginata*, which is truly deadly.

Comments: Culture and mating studies have shown that what was formerly considered as one wide-ranging, polymorphic species in North America referred to as *Armillaria mellea* actually comprises 10 genetically distinct biological species;[1] of these 10, 8 have been formally described, and a ninth may be conspecific with a European species. Only the three discussed here occur in our area.

Armillaria mellea and *A. gallica* reproduce primarily asexually by means of black, shoestring-like rhizomorphs (Fig. F) composed of parallel hyphal strands that spread beneath the bark of a hardwood tree and through the ground to neighboring trees. These characteristic rhizomorphs are revealed in decorticated wood. These mushrooms produce abundant spores, but these are not important in infecting living trees. The fungus can persist in dead roots and stumps for years. *Armillaria tabescens* rarely produces rhizomorphs in nature but sends up fruitbodies from buried roots and rootlets, to which they are attached by white mycelial threads.

Honey mushroom species can be distinguished on the basis of a combination of morphological features, their preferred hosts (hardwoods or conifers), growth habits, and geographical distributions.[2] The color of the fruitbody, presence and type of scales or squamules on the cap, and the presence, structure, or absence of a ring on the upper part of the stipe are all important features. Except for *A. tabescens*, all species have at least a rudimentary ring on the upper stipe; its absence is diagnostic for *A. tabescens*.

Coloration does vary somewhat between the species,

but much intergradation exists, making this feature problematic. The same is true for the microscopic features, except that the total absence of basal clamp connections on the basidia distinguishes *A. mellea* from all other members of the genus.

Both *A. mellea* and *A. tabescens* are virulently pathogenic parasites that kill trees, especially oaks (*Quercus* spp.) weakened by drought or infections by other pathogens. *Armillaria gallica* is for the most part an innocuous saprobe that causes butt rot in hardwoods, especially oaks. However, it can move into more vulnerable tissues in stressed trees and cause major damage and death.[3]

All three species are bioluminescent. However, the saprobic *A. gallica* produces a much greater intensity of bioluminescence in response to both fluorescent and incandescent illumination at night, as well as mechanical disturbance. In sharp contrast, bioluminescence in the two parasitic species (*A. mellea* and *A. tabescens*) is quenched by these stimuli.[4] The ecological role played by these differing responses in these fungi is not known.

1. J.B. Anderson & R.C. Ullrich, 1979, *Mycologia* 71:402–414.
2. The three eastern species are sympatric, but several other species are restricted to the Northeast, Upper Midwest, or Pacific Northwest.
3. *Armillaria gallica* kills many trees in the Northeast that have been defoliated and weakened by the introduced Eurasian Gypsy Moth, *Lymantria dispar*, but it does less damage elsewhere.
4. J.D. Mihail & J.N. Bruhn, 2007, *Mycologia* 99(3):341–350.

Key to Southeastern *Armillaria*

1. Annulus (ring) lacking, cap rarely broader than 4 cm, rhizomorphs uncommon. . . . . *A. tabescens*

1′. With prominent, flaring annulus or rudimentary yellow ring zone, mature cap larger than 4 cm, rhizomorphs common 2

2. With flaring collar, densely cespitose, cap to 15 cm across, smooth, growing from wood, stipe tapering toward base, basidia lacking basal clamp connections . . . . *A. mellea*

2′. Solitary or gregarious but never cespitose, cap to 6 cm across, covered with scales, often appearing to grow from ground, stipe swollen at base, basidia with basal clamps . . . . *A. gallica*

The Magic Mushrooms

(*Psilocybe* and *Panaeolus*)

***Psilocybe* (Fries) P. Kummer**[1]

Type species: *Psilocybe semilanceata* (Fries) P. Kummer [Basidiomycota, Agaricomycotina, Agaricomycetes, Agaricomycetidae, Agaricales, Hymenogastraceae]

***Panaeolus* (Fries) Quélet**

Type species: *Panaeolus papilionaceus* (Bulliard *ex* Fries) Quélet
[Basidiomycota, Agaricomycotina, Agaricomycetes, Agaricomycetidae, Agaricales, Bolbitiaceae]

The blue-bruising species of *Psilocybe* are those containing significant concentrations of the hallucinogenic agent psilocin, which forms bluish and dark black degradation products in the presence of oxygen. Psilocybin is converted into the pharmacologically active psilocin in the body after ingestion. Although *Psilocybe* species are described as "poisonous" in most field guides, this refers to their hallucinogenic properties. There is no known lethal dose for psilocin, no physical dependence associated with chronic recreational use, and no withdrawal syndrome upon cessation. However, the possession and ingestion of the fruitbodies of these mushrooms are illegal in every state except Florida. No federal law prohibits the possession of *Psilocybe* spores, but they are illegal in Georgia, Idaho, and California (except for qualified researchers) because they can be used to grow the psychoactive fruitbodies.

The genus *Psilocybe* was described with the type species *Psilocybe montana* (Persoon 1796 *ex* Fries) P. Kummer in 1871, with some 180 validly described species as of 1996.[2] A number of these species possess potent hallucinogenic properties when consumed due to the presence of the psychoactive compounds psilocybin, psilocin, and baeocystin, all indoleamines closely related structurally to the neurotransmitter serotonin (5-hydroxy tryptamine, or 5-HT). Many of these mushrooms also contain L-tryptophan, the amino acid from which serotonin is derived. The presence of psilocybin and psilocin causes the psychoactive mushrooms to stain bluish-green or dark bluish-black when exposed to oxygen, while their nonpsychoactive congeners do not. All of these *Psilocybe* species were placed together in the family Strophariaceae.

However, in 2000 Montcalvo et al.[1] convincingly showed that the hallucinogenic species of *Psilocybe* centered around *P. semilanceata* (Fries) P. Kummer are not at all closely related to the nonhallucinogenic taxa grouped around *P. montana* and *P. merdaria* (Fries) Ricken. However, this created a problem, since *Agaricus montana* Fries was also the lectotype of the genus *Psilocybe*. This would have necessitated renaming the hallucinogenic species, and the name *Psilocybe*, which was historically associated with the psychoactive mushrooms, would have been lost. A proposal was thus made to conserve the name *Psilocybe* and to assign a new lectotype: *P. semilanceata* (Fries) P. Kummer,[3] which was unanimously accepted by the Nomenclature Committee.

The genus *Psilocybe* now includes the hallucinogenic, blue-staining clade, with *P. semilanceata* as the lectotype; the nonhallucinogenic, nonbluing clade grouped around *P. montana* has been transferred to the genus *Deconica* (W.G. Smith) P. Karsten. The genus *Psilocybe* as now restricted has also been transferred from the family Strophariaceae to the family Hymenogastraceae, which also includes the Deadly Galerina, *Galerina marginata* (Batsch) Kühner. The nonhallucinogenic species in the genus *Deconica* remain in the family Strophariaceae. The genus *Panaeolus* (Fires) Quélet, some of which also produce psilocybin and psilocin, stain blue, and are hallucinogenic and which were previously thought to be closely allied with *Psilocybe*, are now grouped in the Psathyrellaceae or the Bolbitiaceae, depending on the authority. A guide to the psilocybin-containing mushrooms of the world was published by Stamets in 1996.[3]

We treat three hallucinogenic species of *Psilocybe* and two nonhallucinogenic species of *Panaeolus* in this book, but the following dichotomous key includes all six hallucinogenic species of *Psilocybe* recorded from the state of Georgia.

1. J.M. Montcalvo et al., 2000, Phylogenetic relationships of agaric fungi based on nuclear large subunit ribosomal DNA sequences, *Syst. Biol.* 49:278–305.
2. S. Redhead et al., 2007, Proposal to conserve the name *Psilocybe* (Basidiomycota) with a conserved type, *Taxon* 56(1):255–257, https://en.wikipedia.org/wiki/Digital_object_identifier.
3. P. Stamets, 1996, *Psilocybin Mushrooms of the World: An Identification Guide* (Berkeley: Ten Speed Press).

Key to *Psilocybe* and *Panaeolus*

1. Growing on dung 2

1′. Growing on the ground or on wood. 4

2. With membranous ring typically dusted black by spores 3

2′. Ring lacking, with white tooth-like remnants of partial veil hanging from margin . . . . *Panaeolus papilionaceus*

3. Stipe not changing color when injured, ring fragile, spores 18.5–21.0 × 10–11.5 µm, blackish in deposit. . . . . *Panaeolus semiovatus*

3′. Stipe bruising bluish when injured, ring persistent, spores 11.5–17 × 8–11 µm, dark violet-brown in deposit. . . . . *Psilocybe cubensis*

4. Growing on lawns in Bermuda or fescue grass. 5

4′. Not growing on lawns or in grass 6

5. Cheilocystidia irregularly branched. . . . . *Psilocybe atlantis*

5′. Cheilocystidia unbranched . . . . *Psilocybe* cf. *galindoi*

6. Growing in pine needles and woody debris in disturbed ground devoid of shrubs, often under sweetgum trees, stipe with persistent blackish ring. . . . . *Psilocybe caerulescens*

6′. Growing on wood or woody debris along streams and trails with herbaceous plants, ring not black 7

7. Pleurocystidia lacking, cheilocystidia of one type, flask-shaped with slender, neck sometimes forked, rare . . . . *Psilocybe caerulipes*

7′. Pleurocystidia and cheilocystidia of two types: (a) ventricose-rostrate and (b) ovoid- to pear-shaped with a peduculate base, irregularly and locally abundant . . . . *Psilocybe ovoideocystidiata*

Magic Mushroom

Psilocybe cubensis

Photo and micrograph: Alan Rockefeller.

***Psilocybe cubensis* (Earle) Singer**, *Sydowia* 2(1–6):37 (1948)

Basionym: *Stropharia cubensis* Earle, *Inf. An. Estac. Cent. Agr. Cuba* 1:240 (1906)
[Basidiomycota, Agaricomycotina, Agaricomycetes, Agaricomycetidae, Agaricales, Hymenogastraceae]

Derivation: *Psilocybe*, Greek, *psilos* and *kubê*, "bald head"; *cubensis*, Latin, "from Cuba."

Cap: 1.5–9 cm (⅔–3⅔ in) across, at first bluntly conical to bell-shaped, often with a sharp umbo (mushroom at left of figure), becoming convex to almost flat, with or without an umbo; smooth, viscid when wet; whitish to cream, cinnamon-brown over center; margin often with whitish remnants of partial veil.

Flesh: White, firm, bruising blue or bluish-green; odor and taste not distinctive.

Gills: Attached or narrowly attached to seceding (pulling away from stipe), close to crowded, narrow; initially greyish, becoming purple-grey to almost black.

Stipe: 4–15 cm tall by 5–15 mm thick (1⅔–6 × ¼–⅔ in), typically tapering upward from expanded base; whitish, bruising blue or greenish-blue; with thin membranous rim on upper portion, this soon becoming black with deposited spores (middle specimen, arrow).

Spore print: Dark purplish-brown.

Chemical staining: Cap dull orange with KOH.

Microscopic: Spores 11.5–17 × (7–)8–11(–12) µm, broadly elliptical to almond-shaped; smooth, purplish-brown, yellowish-brown in KOH (inset).

Occurrence: Solitary or clustered on cattle, horse, or elephant dung and manured soil in pastures and grasslands;

widespread in Southeast; Mexico, Cuba, Central and northern South America, India, Southeast Asia, Australia; spring–late fall.

Edibility: Edible, but illegal; hallucinogenic, contains psilocin and psilocybin.

Comments: Its size, color, persistent membranous and typically blackish ring, blue-staining stipe and flesh, and occurrence on dung identify this species.

Psilocybe cubensis is moderately potent, with wide variability in the content of psychoactive substances, depending on where the specimens came from and whether they were cultivated or wild, the former generally being more potent.[1]

This is the most widespread and widely consumed psychedelic mushroom in the world. The location where they are found, their age, and the storage methods of the specimens will greatly affect their psilocin and psilocybin content.

1. P. Stamets, 1996, *Psilocybin Mushrooms of the World: An Identification Guide* (Berkeley: Ten Speed Press).

Landslide Mushroom, *derrumbe*

Psilocybe caerulescens

A

B

C

D

Photos and micrograph: A–C, Alan Rockefeller; D, Andrew T. Minnick.

***Psilocybe caerulescens* Murrill**, *Mycologia* 15:20 (1923)

Deprecated synonym: *Psilocybe weilii* Guzmán, Tapia & Stamets, *Mycotaxon* 65:193 (1997)
[Basidiomycota, Agaricomycotina, Agaricomycetes, Agaricomycetidae, Agaricales, Hymenogastraceae]

Derivation: *Psilocybe*, Greek, *psilos* and *kubê*, "bald head"; *caerulescens*, Latin, "greenish-blue," referring to its bruising dark blue to bluish-black when handled.

Cap: 2–7(–10) cm across (~¾–4 in), bluntly conical, convex or bell-shaped, with inrolled margin, becoming broadly convex to nearly flat, with central umbo; margin initially inrolled, often becoming irregularly pleated and eventually entire; smooth, somewhat slimy, hygrophanous—purplish-brown, chestnut-brown (Fig. A), to dark brown when wet, becoming buff or straw-yellow with blackish-brown umbo when dry (Fig. A).

Flesh: Whitish, flesh in the cap, cartilaginous in the stipe; odor described as "farinaceous," whatever that means (variously defined as smelling like flour, cucumbers, or rancid fish oil).

Gills: Attached or sinuate, close, broad, light brown with pale edges (Fig. C), becoming dark chocolate-brown with age (Fig. B).

Stipe: 2.5–7(–13) cm tall by 5–10 mm thick (1–2¾(–5) × ¼–½ in), cylindrical (equal) to slightly swollen at base; whitish to yellowish, becoming bluish-black and drying greyish-yellow with vinaceous tones; hollow, filled with whitish pith; partial veil prominent, but not forming a permanent ring; base sometimes with white rhizomorphs attached.

Spore print: Dark purplish-brown.

Bruising: Cap, flesh, and stipe bruising blue-green to blue-black, stipe often remaining blue or bluish-grey (Fig. D).

Chemical staining: Cap dull orange with KOH.

Microscopic: Spores (5.5–)6–8.5(–9) × 5–6.4(–7.2) µm, subellipsoid to almond-shaped (side view), hexagonal and rhomboid (frontal view); thick-walled, with apical germ pore and short, stubby apiculus (Fig. A, inset); yellowish-brown. Pleurocystidia present, 12–20(–32) × 4.5–9(–10) µm, fusiform, conical, clavate, or utriform, occasionally bifurcate, hyaline, thin-walled. Cheilocystidia 16–27(29) × 4.5–8 µm, flask-shaped, with a flexuous neck 1–2.5 µm broad and sometimes bifurcate.

Occurrence: Saprobic. Gregarious to cespitose on the ground in Loblolly Pine needles (*Pinus taeda*) beneath Sweetgum (*Liquidambar styraciflua*) in red-clay soil often devoid of shrubs; common and widespread throughout northern Georgia; Oconee County, South Carolina; Florida; Tyler County, Texas; south through Mexico and Central America into South America; first described from Montgomery, Alabama, in 1923, but not found there since; September–November.

Edibility: Edible, but illegal; potently hallucinogenic, contains psilocin and psilocybin.

Comments: "Landslide Mushroom" is the common name for this taxon, apparently based on its Spanish name, *derrumbe* (variously meaning "landslide" or "collapse"). This may refer to its typically occurring in disturbed locations but may also be a play on words alluding to the psychedelic "landslide" or "collapse" of ordinary consciousness one obtains by ingesting this species. It is currently used as an entheogen by Mazatec shamans in Oaxaca, Mexico.

This blue-bruising mushroom is highly variable, wide-ranging and common from northern Georgia/South Carolina south through Mexico, Central America, and into South America. It was erroneously described as a new species (*Psilocybe weilii*) found only in north-central Georgia by Guzmán et al. (1997) in having pleurocystidia, which had not been mentioned in earlier descriptions of *P. caerulescens*. However, Alan Rockefeller and his colleagues reexamined the holotype of *P. caerulescens* and found pleuroscystidia to be present not only in that specimen but in other collections of this taxon. Since this was the sole feature separating *P. caerulescens* from *P. weilii*, they consider the latter to be a deprecated synonym. A paper detailing these findings is currently in preparation by Gaston Guzmán and Alan Rockefeller.[1]

The Landslide Mushroom is distinguished from the Blue-foot Psilocybe, *Psilocybe ovoideocystidiata* Guzmán & Gaines, by its occurrence in disturbed, red-clay soils in sunny areas often devoid of shrubs, the lack of a permanent ring, and its subellipsoid to rhomboid spores. *Psilocybe ovoideocystidiata* is found along trails or streams and overflow areas with herbaceous plants, the frequent presence of a thin zone to a persistent annular ring, and its spores are subovoid. The Magic Mushroom, *Psilocybe cubensis* (Earle) Singer, grows on animal dung, and its spores are much larger than those of the other two species (11.5–17 × 8–11 µm compared to about 8–9 × 6–7 µm or less).

It contains the hallucinogens psilocybin (0.61%), psilocin (0.27%), and baeocystin (0.05%) and the indole amino acid tryptophan[2] (0.32%) and is a moderately potent "shroom." Psilocybin is converted into the pharmacologically active psilocin in the body after ingestion.

Although *Psilocybe* species are described as "poisonous" in most field guides, this refers to their hallucinogenic properties. There is no known lethal dose for psilocin, no physical dependence associated with chronic recreational use, and no withdrawal syndrome upon cessation. However, the possession and ingestion of the fruitbodies of these mushrooms are illegal in every state except Florida. No federal law prohibits the possession of *Psilocybe* spores, but they are also illegal in Georgia, Idaho, and California (except for qualified researchers) because they can be used to grow the psychoactive fruitbodies.

A danger exists of confusing this (and other species of *Psilocybe*) with deadly poisonous mushrooms like the related Deadly Galerina, *Galerina marginata* (Batsch)

Kühner [*G. autumnalis* (Peck) A.H. Smith & Singer is a deprecated synonym].[3] *Galerina marginata* has a rusty-brown instead of a violet-black spore print, and the ring is likewise rusty-brown rather than black. It also grows on wood rather than the ground.

1. Alan Rockefeller, personal communication to WHL, 16 November 2015.

2. P. Stamets, 1996, *Psilocybin Mushrooms of the World: An Identification Guide* (Berkeley: Ten Speed Press), 167. L-tryptophan is converted by the body into 5-hydroxy-tryptamine (5-HT, or serotonin), which is involved in inducing sleep and maintaining a stable psychological mood. Psilocybin, psilocin, and baeocystin are structurally very similar to 5-HT. They are potent activators of serotoninergic receptors in the brain, which are largely responsible for their psychedelic effects.

3. *Galerina* and the hallucinogenic genus *Psilocybe* (*sensu stricto*) both belong to the family Hymenogastraceae.

Blue-foot Psilocybe

Psilocybe ovoideocystidiata

A B C D E F

G H I

Photos and micrograph: A, Dan Molter; B–I, W. Light.

Cap: Up to 4.5 cm (1⅘ in) across; initially convex, becoming campanulate (bell-shaped) to plane and umbonate (Figs. B & H) and eventually convex (Figs. A & C), with concentric ridges toward margin (Figs. A & E, arrows; Fig. H); margin entire (Fig. B) to somewhat or deeply wavy (Fig. A); pearly cream to steely-greyish-brown (Fig. C), chestnut to pale brownish (Fig. A), bruising various shades of blue with age and/or handling (Figs. B, E, G, & H).

Flesh: Thick, whitish to pale ochre; odor and taste farinaceous; bruises bluish to greenish.

Gills: Adnate to sinuate-uncinate, thick, margins entire and often sinuous, subdistant, broad, often forked, with numerous short gills (Figs. D & F); uniformly whitish, becoming pale brownish, rusty-brown, lavender, to dark purple-brown.

Stipe: 2.5–6 (up to 9) cm by 2–5 (up to 7) mm thick (1–2½ [to 3⅔] × ¼ in), cylindrical, equal (Figs. B, E, F, & G), smooth above, somewhat scaly below, hollow, base often somewhat expanded, with white mycelium (Fig. G, arrows), may be partially buried; whitish, with various irregular yellowish or brownish tones, bruising various shades of blue with age or handling

***Psilocybe ovoideocystidiata* Guzmán & Gaines**, *Intern. J. Med. Mushrooms* 9:75 (2007)
[Basidiomycota, Agaricomycotina, Agaricomycetes, Agaricomycetidae, Agaricales, Strophariaceae]

Derivation: *Psilocybe*, Greek, *psilos* and *kubê*, "bald head"; *ovoideocystidiata*, Latin, "having ovoid cystidia," referring to the shape of the pleurocystidia and cheilocystidia.

(Fig. A); partial veil varying from a thin, evanescent, and barely visible annular zone (Fig. B, arrows) to a persistent annular ring (Fig. F, arrows).

Spore print: Dark violaceous.

Bruising: Cap (Figs. B, D, E, & H) and stipe (Fig. A) readily bruising blue with age and/or handling, this bruising intensifying with time.

Microscopic: Spores smooth, subovoid to amygdaloid (almond-shaped), thick-walled, with broad apical germ pore at one end (Fig. I, arrow) and short apiculus at the other; yellowish-brown; (7–)8–9(–12) × (5.5–)6–7(–8.5) µm; *two types of pleurocystidia and cheilocystidia*: (1) ventricose-rostrate (swollen in the middle and with a neck-like apical projection) and (2) the more common globose-pyriform (ovoid- to pear-shaped) form with a peduculate base; some cheilocystidia with a short knobby apex, but neck lacking, necks rare on pleurocystidia; basidia 4-spored.

Occurrence: Solitary, gregarious to cespitose, on wood or wood debris in hardwood forests, along trails or places with herbaceous plants, along streams, or in overflow areas; most commonly fruiting April–mid-June through August–September. Widespread in eastern North America with disjunct western populations from Washington to California.

Edibility: Edible, but illegal; moderately hallucinogenic, contains psilocybin, psilocin, and baeocystin.[1]

Comments: This species is easily confused with the very similar but rarely encountered *Psilocybe caerulipes* (Peck) Saccardo 1887, which has the same geographical distribution and occurs in the same habitat. *Psilocybe caerulipes* has a smaller cap (2–3 cm across) and thinner stipe (3–6 cm tall by 1.5–3 mm thick), which is equal, only slightly expanded at the base, and which lacks a ring and at most has a rapidly disappearing fibrillose annular zone. The spores are dark purple-brown, rather than yellowish-brown and terete-elliptical (7–10 × 4–5 µm) instead of subovoid-amygdaloid; pleurocystidia are completely lacking, and the cheilocystidia are lageniform (flask-shaped) with a long and sometimes forked neck. The margins of the gills are whitish and slightly fimbriate.

Five other species of *Psilocybe* also occur in the Atlanta area: (1) *P. cubensis* (Earle) Singer 1948 and (2) *P. caerulescens* Murrill 1923, both of which are discussed above in separate species accounts; (3) *P. atlantis* Guzmán, Hanlin & White 2003;[1] (4) *P.* cf. *galindoi* Guzmán 1978;[2] and (5) *P. caerulipes* (Peck 1885) Saccardo (1887).[3] *Psilocybe atlantis* is known only from central and north-central Georgia; the other two are much more widely distributed; and *P. cubensis* is nearly cosmopolitan, growing on the dung of various mammals (from cattle to elephants). All five species stain blue; all contain psilocybin and psilocin, and all are actively psychedelic.

Both the caps and stems contain moderate amounts of psilocybin, in most cases low amounts of psilocin, and always low concentrations of baeocystin. Smaller, and thus younger, fruitbodies were found to contain higher concentrations of psilocin; the low concentrations of psilocin typically seen in this species are thought to be due to psilocin decomposition with age and upon handling and drying, as evidenced in the progressively blue staining.[4]

1. G. Guzmán, R.T. Hanlin, & C. White, 2003, *Mycotaxon* 86:180.
2. G. Guzmán, 1978, *Nova Hedwigia* 29(3–4):628.
3. P.A. Saccardo, 1887, *Syll. Fung.* (Abellini) 5:1051.
4. J.W. Allen, J. Gartz, D. Molter, F. Castillo Suarez, & P. Sihanonth, December 2009, The occurrence, cultivation, and chemistry of *Psilocybe ovoideocystidiata*, a new bluing species (Agaricales) from Ohio, Pennsylvania and West Virginia, http://www.researchgate.net/profile/John_Allen/publication/235436020_The_Occurrence_Cultivation_and_Chemistry_of_Psilocybe_ovoideocystidiata_a_new_Bluing_Species_%28Agaricales%29_from_Ohio_Pennsylvania_and_West_VirginiaByAllen_John_W._Gartz_Jochen._Molter_Dan_and_Prakitsin_Sihanonth/links/00b7d528d48aea2464000000.pdf.

Petticoat Mottlegill

Panaeolus papilionaceus

A

B

Photos: John Plischke III.

***Panaeolus papilionaceus* (Bulliard *ex* Fries) Quélet**, *Mém. Soc. Émul. Montbéliard*, sér. 2, 5:152 (122 repr.) (1872)

Basionym: *Agaricus papilionaceus* Bulliard, *Herb. Fr.* 1:tab. 58 (1781) [1780–1781]

[Basidiomycota, Agaricomycotina, Agaricomycetes, Agaricomycetidae, Agaricales, Bolbitiaceae]

Derivation: *Panaeolus*, Latin, "all variegated," "all mottled," referring to the gills; *papilionaceus*, Latin, from *papilio*, "butterfly," "butterfly-like."

Cap: 2–5(–6) cm (¾–2½ in) across, bluntly conical to bell-shaped; margin with white, tooth-like remnants of partial veil (Figs. A & B); surface dry, smooth to finely wrinkled to horizontally cracked with age; greyish-brown (Fig. A) to reddish-brown, sometimes with olive tint (Fig. B).

Flesh: Greyish to buff-brown.

Gills: Adnate, soon seceding (attached, soon pulling away from stipe), close to subdistant, moderately broad, initially greyish (Fig. A), becoming mottled dark greyish-black from uneven spore maturation, eventually black overall.

Stipe: 6–14 cm tall by up to 5 mm thick (2½–5⅔ × ¼ in), equal, hollow, fibrous, striated near apex; concolorous with the cap, bruising darker.

Spore print: Blackish.

Chemical staining: Cap dull orange with KOH.

Microscopic: Spores (11–)15–18 × (7.5–)10–12 µm, elliptical to almond-shaped, smooth, with apical pore.

Occurrence: Solitary, scattered, or clustered on cattle and horse dung; widespread and common throughout North America; Mexico, parts of South America, Europe, Iran, Africa, East Asia; spring–fall.

Edibility: Edible, but its habitat should disincline most people. No apparent psychoactive properties.

Comments: This is one of the commonest dung-dwelling species of *Panaeolus* with an almost cosmopolitan distribution. *Panaeolus semiovatus* (Sowerby *ex* Fries) Lundell & Nannfeldt, which also grows on dung, is distinguished from *P. papilionaceus* by its larger size (up to 9 cm, or 3⅔ in, across), viscid cap when wet, margin lacking tooth-like partial veil remnants, and fragile but distinct ring.

Panaeolus campanulatus (Linnaeus) Quélet, *P. retirugus* (Fries) Gillet, and *P. sphinctrinus* (Fries) Quélet are all synonyms of *P. papilionaceus*.

Semiovate Panaeolus

Panaeolus semiovatus

A

B

Photos: A, Jörg Hempel; B, Zonda Grattus.

Panaeolus semiovatus **(Sowerby *ex* Fries) S. Lundell & Nannfeldt**, *Fungi Exsiccati Suecici* 1–12:14 (no. 537) (1938)

Basionym: *Agaricus semiovatus* Sowerby, *Col. Fig. Engl. Fung. Mushr.* (London) 2:plate 131 (1798)
[Basidiomycota, Agaricomycotina, Agaricomycetes, Agaricomycetidae, Agaricales, Bolbitiaceae]

Derivation: *Panaeolus*, Latin, "all variegated," "all mottled"; *semiovatus*, Latin, "semiovate."

Cap: 3–9 cm (1¼–3⅔ in) across, parabolic to semiovoid or semi-egg-shaped (Figs. A & B); viscid when wet, smooth to somewhat wrinkled; initially cinnamon-buff (Fig. B), becoming ivory-white (Fig. A); margin entire, typically smooth.

Flesh: Thin to relatively thick, soft, white; odor and taste not distinctive.

Gills: Adnate, soon seceding (attached and notched, then pulling away from stipe), close to subdistant; initially greyish-white, becoming mottled blackish from uneven spore ripening, then blackish overall.

Stipe: 8–15 cm tall by up to 15 mm thick (3¼–6 × ⅔ in), subequal to slightly expanded at base; solid, becoming hollow and filled with whitish pith; concolorous with the cap; with fragile white ring (Figs. A & B, arrow), this often flattened to stipe and stained blackish by accumulated spores.

Spore print: Blackish.

Chemical staining: Cap dull orange with KOH.

Microscopic: Spores (15–)18.5–21 × (8–)10–11.5 µm, elliptical, smooth, thick-walled, with truncated apical germ pore; dark brown.

Occurrence: Solitary or scattered on cattle and horse dung; widespread and common throughout North America; Europe, Hawai'i; spring–fall.

Edibility: Edible, but its habitat should disincline most people. No apparent psychoactive properties.

Comments: This species is distinguished by its size, viscid cap, entire margin, and membranous ring. This is the only species of *Panaeolus* with a ring, even if a fragile one. *Panaeolus papilionaceus* (Bulliard *ex* Fries) Quélet is smaller (6 cm, or 2½ in), has a dry cap and a margin with prominent, whitish, tooth-like remnants of the partial veil, and lacks any trace of a ring.

Deadly Galerina, Autumn Skullcap

Galerina marginata

A C D B

Galerina marginata (Batsch) Kühner, *Encyclop. Mycol.* 7:225 (1935)
[Basidiomycota, Agaricomycotina, Agaricomycetes, Agaricomycetidae, Agaricales, Hymenogastraceae]

Basionym: *Agaricus marginatus* Batsch, *Elench. Fung.*, cont. sec. (Halle):207 (1789)

Derivation: *Galerina*, Latin, from *galerus*, "cap"; *marginis*, Latin, "margin," "edge."

Cap: 1.7–6.5 cm (~¾–2⅔ in) across, at first convex, bell-shaped (Fig. C), becoming nearly flat with age (Figs. A & B), with or without an umbo; surface smooth, margin inrolled (Fig. D); variously viscid, moist, or dry; various shades of pale to dark brown, tawny, tan, to yellow-ochraceous; hygrophanous (changing colors with changes in humidity).

Gills: Adnate (attached to the stipe) and slightly decurrent, close to subdistant, unforked, broad, in two or more tiers, with shorter lamellulae not reaching stem (Fig. D), initially yellowish, becoming yellowish-orange to pale or rusty-brown.

Flesh: Pale brownish-ochraceous to almost white; thin, pliable; with very faint to strong flour-like odor.

Stipe: 2–10 cm tall by 2–8 mm thick (¾–4 × ¹⁄₁₂–⅓ in), subequal (Fig. B), enlarged toward base, often curved (Fig. A), smooth

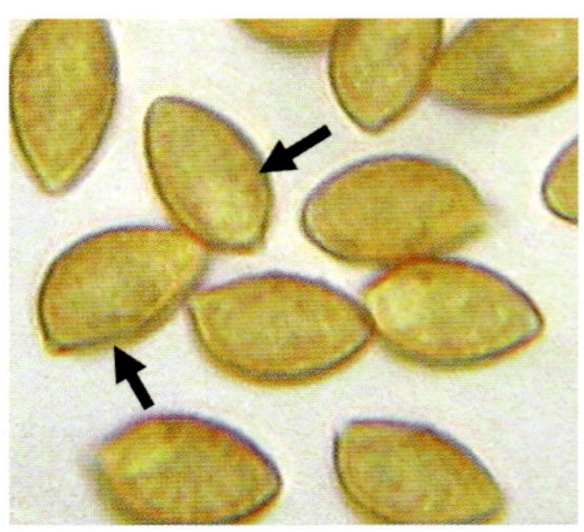

E Photo: Dan Molter.

to powdery at the top, a whitish partial veil over the gills when young, leaving a fragile whitish ring that is soon dusted rusty-brown by spores (Fig. B, arrow), this ring often missing or indistinct with age; pale buff, becoming darker with age; sometimes covered with white mycelial threads; initially solid, becoming hollow from the base upward with age.

Spore print: Rusty-brown.

Microscopic: Spores 8–11 × 5–6.5 µm, asymmetrically elliptical, surface roughened or wrinkled, with a plage (a clear flat area) at the basal end (Fig. E, arrows), this feature not always evident. Fig. E taken from a micrograph by Dan Molter.

Occurrence: Saprobic, solitary or in small clusters mainly on dead and decaying conifer wood, but occasionally also on hardwood; spring–fall. Widespread in the Northern Hemisphere from the tropics to the subarctic. North America, Europe, Asia, Australia.

Edibility: Inedible. ***Deadly poisonous!***

Comments: The Deadly Galerina contains α-amanitin, the same toxin found in the deadly species of *Amanita* that causes catastrophic liver and kidney failure associated with high mortality despite heroic treatment. *Galerina marginata* is as deadly (if not more so) as the Death Cap, *Amanita phalloides*, and the several species of Destroying Angel—one of which (*Amanita bisporigera*) is very common in the Piedmont and southern Appalachians. A 2004 French study showed that the amatoxin content in *G. marginata*, which varied from 78.17 to 243.61 µg/mg of fresh weight, was consistently higher than in *Amanita phalloides*, generally considered to be one of the most toxic of all mushrooms.[1] It also revealed that, contrary to many reports, *Galerina* contains no phallotoxins. See the section on mushroom poisoning for additional information.

Galerina marginata is a classic LBM—a nondescript "little brown mushroom" that looks very much like so many others. Unfortunately, it also closely resembles several much-sought-after choice edible species, including the Honey Mushrooms, *Armillaria mellea*, *A. tabescens*, and *A. gallica*, and the Brick Cap, *Hypholoma sublateritium*, which is eaten in North America and Japan but considered poisonous in Europe. Most poisonings from *G. marginata* occur when "'shroomers" mistake them for very similar looking hallucinogenic species of *Psilocybe*. *Galerina marginata* also sometimes grows alongside dense clusters of edible *Armillaria* species and can be inadvertently collected together with them. *Galerina* has a rusty-brown spore print (Fig. A). Those of the genus *Psilocybe* range from lilac-brown to dark purple-brown, but rusty-brown spore prints are also seen in at least one species, so consuming magic mushrooms can be dangerous in several ways. The spore prints of *Armillaria* species are reassuringly white. See also the comments under Honey Mushrooms.

This species is still referred to as *Galerina autumnalis* in many field guides and online sources. However, molecular and mating studies have combined five former taxa into one morphologically variable species. These formerly separated species, *G. autumnalis* (Peck) A.H. Smith & Singer, *G. oregonensis* A.H. Smith, *G. unicolor* (Vahl) Singer, and *G. venenata* A.H. Smith, are now all placed into synonymy with *G. marginata*.[2]

1. F. Enjalbert, 2004, *Mycologia* 96(4):720–729.
2. G. Gulden et al., 2001, *Mycol. Res.* 105(4):432–440.

Deer Mushroom, Fawn Mushroom

Pluteus cervinus

A

B

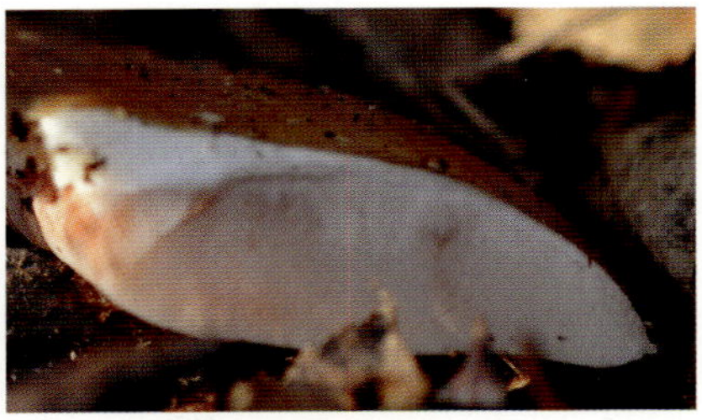
C

D

***Pluteus cervinus* (Schaeffer) P. Kummer**, *Führ. Pilzk.* (Zwickau):99 (1871)
[Basidiomycota, Agaricomycotina, Agaricomycetes, Agaricomycetidae, Agaricales, Pluteaceae]

Basionym: *Agaricus cervinus* Schaeffer, *Fung. Bavar. Palat.* 4:6 (1774)

Derivation: *Pluteus*, Latin, "shelf," "parapet"; *cervinus*, Latin, "deer."

Cap: 3–15 cm (1¼–6 in) across, deeply convex, conical when young (Fig. A), becoming broadly convex to nearly flat in maturity (Fig. C); surface wrinkled or somewhat rugose when young (Fig. A), becoming almost smooth with age (Fig. C); surface smooth, glossy to shiny, sometimes streaked with fibers; evenly dull brown to greyish-brown, pale cinnamon-brown to dark shiny brown; sticky when wet.

Gills: Unattached, separated from stipe by a clear gap, close to crowded (Fig. D), broad (Fig. C), initially white (Fig. B), soon becoming subtly to distinctly pink to salmon (Figs. C & D).

Flesh: Soft, white (Fig. C); odor and taste radish-like or not apparent.

Stipe: 5–13 cm tall by 6–25 mm thick (2–5¼ × ¼–1 in), more or less equal (Fig. A), may be enlarged toward the base (Fig. B), solid; white, sometimes streaked with brown-greyish or brownish fibers. No annulus or volva.

Spore print: Pink, salmon, or pinkish-brown.

Microscopic: Spores 5.5–8 × 4–6 µm, broadly to oblong-elliptical, smooth, clear; pleurocystidia (in the gills) with two or more short apical horn-like projections.

Occurrence: Solitary or in clusters on fallen decaying hardwood or conifer wood, or sprouting on the ground from buried wood or on sawdust piles and other woody debris. Spring–early winter; widespread and common throughout North America; Europe, Africa, Asia, Australia.

Edibility: Edible, but not highly prized.

Comments: The family Plutaceae is distinguished from the similar families Amanitaceae and Lepiotaceae in the introductory section to "The Amanitas."

Similar species in the genera *Volvariella* Spegazzini and *Volvopluteus* Vizzini, Contu & Justo have a distinct *Amanita*-like volva, which is lacking in *Pluteus*. Some

Entoloma species, many of which are toxic, also have pink spores and resemble the Deer Mushroom. However, the gills in *Entoloma* are attached, not free, they grow from the ground rather than on wood, and the spores are distinctively angular, warted or ribbed, and not smooth-elliptical.

Pluteus atromarginatus (Konrad) Kühner can also be confused with and was once considered a variety of *P. cervinus*. It is distinguished by a cap surface that is typically streaked with dark brown fibrils and marginate gills with grey-brown edges.

The Platterful Mushroom, *Megacollybia rodmani* (also called *M. platyphylla* or *Tricholomopsis platyphylla* by many North American authors), has delicate dark-brown, olive, to nearly black streaks radiating to the margin over a silvery-grey ground color. The gills are attached, have wavy, finely serrate margins, and remain pristine white; the spore print is also white.

Velvety Psathyrella

Lacrymaria velutina

Lacrymaria velutina (Persoon) Konrad & Maublanc, *Revisione Hymenomycetes de France*:90 (1925)[1]
[Basidiomycota, Agaricomycotina, Agaricomycetes, Agaricomycetidae, Agaricales, Psathyrellaceae]

Basionym: *Agaricus velutinus* Persoon, *Syn. Meth. Fung.* (Göttingen) 2:409 (1801)

Derivation: *Lacrymaria*, Latin, from *lacrimare*, "to shed tears"; *velutina*, Latin, from *velutinae*, "velvety."

Cap: 3–10 cm (1¼–4 in) across, convex when young, becoming bell-shaped, broadly convex or nearly flat, often with central knob (Figs. A & C), dry, densely hairy when young, becoming almost smooth; margin initially with tattered remnants of veil, becoming furrowed and splitting with age (Figs. A, B, & E).

Flesh: Brownish.

Gills: Attached to stem, often with deep notch, or free; crowded (Fig. B), moderately broad and strikingly mottled, edges faintly lined with white (Fig. D); often exuding droplets of moisture.

Stipe: 5–12 cm tall by 0.5–1.5 cm thick (2–4¾ × ¼–⅔ in); equal or with expanded base, fibrous to woolly/hairy, hollow, with ephemeral ring or ring zone stained dark by spores (Figs. B & E, arrows), white above ring, tawny beige below, with white basal mycelium.

Spore print: Very dark brown to purplish-black.

Chemical staining: Nonreactive or reddish brown with KOH.

Microscopic: Spores 8–12 × 5–8 µm, elliptical, with protruding apical pore, heavily ornamented with unconnected warts.

Occurrence: Saprobic, solitary, gregarious, or in clusters on lawns, pastures, and roadsides and in gravelly soil near recently dead hardwood trees; summer–fall; widespread in North America.

Edibility: Edible, but bland.

Comments: *Lacrymaria velutina* is called *Psathyrella velutina* (Persoon) Singer (1949) in many field guides and online. It is placed in the genus *Lacrymaria* because of its ornamented, warty spores; those of *Psathyrella* (*sensu lato*) are smooth.

1. *Lacrymaria velutina* is considered a later synonym of *Lacrymaria lacrymabunda* (Bulliard) Patouillard (1887) [basionym, *Agaricus lacrymabunda* Bulliard (1785)] by some authors and by *Index Fungorum* and *Species Fungorum*. However, a recent molecular study found that North American specimens identified as *L. velutina* appear to be genetically distinct from each other, as well as from European specimens of *L. lacrymabunda*, and suggested that many species may not, in fact, co-occur in North America and Europe (M. Padamsee et al., 2008, *Mol. Phylogen. Evol.* 46:415–429). The species concepts of these groups are in much need of revision.

Sulphur Tuft

Hypholoma fasciculare

A

B

C

D

E

***Hypholoma fasciculare* (Hudson) P. Kummer**, *Führ. Pilz.* (Zwickau):21 (1871) f. *fasciculare*
[Basidiomycota, Agaricomycetes, Agaricomycetidae, Agaricales, Strophariaceae]

Basionym: *Agaricus fascicularis* Hudson, *Fl. Angl.*, 2nd ed., 2:615 (1778)

Derivation: *Hypholoma*, Greek, from *hyphos*, "web," and *loma*, "edge" or "border"; *fasciculare*, from *fasciculus*, Latin, "small bundle," referring to its tendency to grow in clusters.

Cap: 2–7 cm (~¾–2¾ in) across, initially convex to bell-shaped (Fig. A), often umbonate, becoming broadly convex to nearly flat with age (Fig. B); margin at first incurved (Fig. E), later sometimes rimmed with tattered remnants of partial veil (Fig. D); color highly variable, cream-yellow to sulfur-yellow when young, often tan or darker orange to brownish toward the center (Fig. A), becoming orange-brown to cinnamon-brown with age (Fig. B).

Gills: Attached, close (Figs. C & D), narrow, pale yellow to greenish-yellow when young (Fig. E), later turning grey, then purple-brown with maturity (Figs. C & D).

Flesh: Firm, sulfur-yellow; taste usually bitter.

Stipe: 5–12 cm tall by 0.5–1.5 cm thick (2–4¾ × ¼–⅔ in); equal or slightly expanded in the middle, sometimes tapering toward the base, becoming hollow with age; pale yellow, yellow, or tan, becoming darker with age with vertical fibrillose lines

(Figs. C & D), partial veil sometimes leaving a rusty-brownish fibrillose ring (Fig. E).

Spore print: Purple-brown.

Microscopic: Spores 6–8 × 3.5–5 µm, elliptical, smooth, with apical pore; purplish-brown, nonstaining with Melzer's reagent.

Occurrence: Saprobic, growing in dense clusters on decaying conifer (Fig. A) and hardwood logs and stumps; often appearing to grow in ground litter but actually growing on buried wood (Fig. C), May–November; widespread and common throughout North America; Europe, Asia Minor.

Edibility: Poisonous; generally nonlethal, but causing severe gastric upset: diarrhea, nausea, vomiting, and convulsions.

Comments: Referred to as *Naematoloma fasciculare* (a later synonym) in some field guides and online sources, this mushroom closely resembles two congeners: the Conifer Tuft, *H. capnoides* (Fries) P. Kummer, and the Brick Cap, *H. lateritium* (Schaeffer) P. Kummer, both of which occur in the area covered by this book. *Hypholoma capnoides* is fairly common and grows in dense clusters on conifer logs and stumps (never on hardwood), the young gills are white to greyish-white (never yellow, as in *H. fasciculare*), and it never appears to grow directly on the ground or from conifer litter. *Hypholoma lateritium* (Brick Cap) occurs only on hardwood stumps and logs, the mature cap is up to 10 cm (4 in) across and typically brick-red, especially in the center, fading toward the margin; the gills are whitish to pale greenish-yellow when young but are otherwise as in *H. fasciculare*. The stipe is pale yellow or light ochre at the apex, darkening progressively to a reddish-brown base. It is very uncommon in our area. Both *H. capnoides* and *H. lateritium* are edible.

Brick Cap, Brick Tuft

Hypholoma lateritium

A

B

C

Hypholoma lateritium **(Schaeffer) P. Kummer**, *Führ. Pilzk.* (Zwickau):72 (1871)
[Basidiomycota, Agaricomycetes, Agaricomycetidae, Agaricales, Strophariaceae]

Basionym: *Agaricus lateritius* Schaeffer, *Fung. Bavar. Palat.* 4:22 (1774)

Derivation: *Hypholoma*, Greek, from *hyphos*, "web," and *loma*, "edge" or "border"; *lateritium*, Latin, from *lateres*, "brick," referring to the brick-red color of the cap.

Cap: Initially convex, becoming broadly convex to almost plane, up to 10 cm (4 in) across, brick-red, fading to pale yellow or cream toward the margin (Fig. A); smooth with scattered fibrils.

Gills: Attached, close, narrow, whitish to pale greenish-yellow (Fig. B) when young, later turning greyish (Fig. C), then purplish-grey with maturity.

Flesh: Dull yellow to pale yellowish-brown, thick, firm.

Stipe: 5–10 cm tall by up to 16 mm thick (2–4 × ⅔ in), pale yellow or light ochre at the apex, darkening progressively to a reddish-brown base; partial veil leaving faint zone of fibrils on upper stipe (Figs. B & C).

Spore print: Purple-brown.

Chemical staining: All tissues dark reddish-black in KOH; cap and flesh pale greenish-grey, gills dark grey in Melzer's reagent.

Microscopic: Spores 6–7 × 3.5–4.5 µm, elliptical, smooth, with apical pore, pale brown.

Occurrence: Scattered or in dense clumps on decayed hardwood logs and stumps, never on conifers, summer through winter. Widespread but occasional in eastern North America.

Edibility: Edible, but caution advised (see comments).

Comments: Referred to as *Naematoloma sublateritium* (a later synonym) in some field guides and online sources, *Hypholoma lateritium* closely resembles two congeners: the Conifer Tuft, *H. capnoides* (Fries) P. Kummer, and the Sulphur Tuft, *H. fasciculare* (Hudson) P. Kummer, both of which occur in the area covered by this book. *Hypholoma capnoides* grows in dense clusters on conifer logs and stumps (never on hardwood), and the young gills are white to greyish-white (never greenish-yellow, as in *H. lateritium*). The stipe is pale yellow or light ochre at the apex, darkening progressively to a reddish-brown base. It is fairly common throughout its range but is very uncommon in our area. Both *H. capnoides* and *H. lateritium* are edible.

Hypholoma fasciculare is poisonous and causes severe gastric upset, including diarrhea, nausea, vomiting, and convulsions; it has caused rare fatalities. It grows in dense clusters on both hardwood and conifer stumps and logs, often appearing to grow in ground litter while actually growing on buried wood. The mature cap is smaller, 2–7 cm (~¾–2¾ in) across, and not nearly as red as in *H. lateritium*.

Platterful Mushroom

Megacollybia rodmani

A B

Megacollybia rodmani **R.H. Petersen, K.W. Hughes & E.B. Lickey**, in Hughes et al., *Rep. Tottori Mycol. Inst.* 45:1–57 (2007) [Basidiomycota, Agaricales, Tricholomataceae]

Misapplied name: *Megacollybia platyphylla* (*et synonyma*) *sensu* North American authors; [*non M. platyphylla* (Persoon 1796) Kotlaba & Pouzar 1972]

Derivation: *Megacollybia*, Latin, "large *Collybia*," based on the genus from which it was transferred; *rodmani* is in honor of James Rodman of the U.S. National Science Foundation.

Cap: Up to 20 cm (8 in) wide, smooth, convex with inrolled margin when young, becoming shallowly convex to flat, slightly depressed and more darkly pigmented over the center, with delicate dark-brown, olive, to nearly black streaks radiating to margin over silvery-grey ground color (Fig. A).

Gills: White, narrowly to broadly attached, often with small decurrent notch, edges finely dentate, eroding to unevenly serrate or raggedly undulating (Fig. C); interspersed with incomplete lamellae near margin (Fig. B).

Flesh: Whitish or off-white (Fig. A), not bruising when cut.

Stipe: 5–12 cm (2–4¾ in) tall by 1 cm (⅖ in) in diameter (form *rodmani*, see comments), or somewhat shorter and stouter, 5–9 by 2.5 cm diameter (2–3½ × 1 in, form *murina*); base usually attached to many white rhizomorphs extending for some distance in decayed woody substratum.

Spore print: White.

Microscopic: Spores 6–10 × 5–7.5 µm, smooth, subglobose, ovate or flattened ellipsoid, with offset hilar knob, containing 1 to a few oil droplets; nonstaining in Melzer's reagent.

Occurrence: North America east of the Rocky Mountains, south through Central America (Newfoundland to Costa Rica); saprobic, alone or gregarious on well-decayed logs or buried hardwood; May–July; one of the first conspicuous mushrooms to appear in the spring after the morel season.

Edibility: Edible, taste mild, not distinctive.

Comments: In their original species description, Petersen et al. (2007) distinguished two morphotypes, which they described and named as distinct forms: *Megacollybia rodmani* f. *rodmani* and *M. rodmani* f. *murina*. Form *rodmani* has a taller, thinner stipe (becoming more so on drying) and denser and more conspicuous rhizomorphs, and the color of the cap changes to a neutral tan when dried. Form *murina* has a shorter, stouter stipe, the rhizomorphs are less conspicuous, and the cap remains dark when dried. The type localities of both forms are in Tennessee.

Southeastern Matsutake

Tricholoma cf. *caligatum*

A B C

***Tricholoma caligatum* (Viviani) Ricken**, *Die Blätterpilze*:331 (1914)
[Basidiomycota, Agaricomycotina, Agaricomycetes, Agaricomycetidae, Agaricales, Tricholomataceae]

Basionym: *Agaricus caligatus* Viviani, *I funghi d'Italia* (Genova):40 (1834)

Derivation: *Tricholoma*, Greek, from *trichós*, "hair," and *loma*, "fringe," "border"; *caligatus*, Latin, "wearing heavy boots" (as a private-ranked soldier).

Cap: 4–12 cm (1⅔–4¾ in) across, broadly convex to almost flat, may be shallowly depressed centrally (Fig. A); surface dry, with coarse, flattened, fibrous, reddish-brown to cinnamon-brown scales forming large splotches and reticulations (Fig. A); margin entire, scalloped, or lobed, at first inrolled (Figs. B & C), later becoming uplifted, sometimes with adhering remnants of a partial veil (Fig. C).

Flesh: White to somewhat pinkish, not bruising when cut; variously reported as fragrant, spicy, or odorless; taste mild or strongly bitter, offensive (see comments below).

Gills: Attached, close, moderately broad, white (Fig. B), often with light-brownish stains.

Stipe: 4–10 cm tall by 1–3 cm thick (1⅔–4 × ⅖–1¼ in), subequal, dry, solid, with cottony, membranous, flaring ring (Fig. C) halfway or more up the stalk, ring white above, brown below; stipe white above the ring, covered from the base to the ring by a cinnamon-brown sheath that splits into fibrous, scaly patches, revealing underlying white ground color.

Spore print: White.

Microscopic: Spores 6–8 × 4–5.5 µm, elliptical, smooth, clear, nonreactive in Melzer's reagent; cystidia and clamp connections lacking.

Occurrence: Ectomycorrhizal with hardwoods, especially oaks (*Quercus* spp.). Alone or in groups on the ground associated with oaks in mixed forests, summer–midwinter; widespread but occasional in eastern North America, California, and Mesoamerica.

Edibility: Some mild-tasting forms edible; others far too bitter.[1]

Comments: The North American matsutakes have been variously called False Matsutake, American Matsutake, and Eastern and Western Matsutakes. The situation with regard to the "Matsutake Tricholomas" is still poorly understood despite a number of genetic and molecular studies, and nobody knows just what any of them should be called. We don't know if there is one polymorphic circumboreal species called by different names in Europe, North America, Mesoamerica, and Asia, or if we are dealing with several distinct groups of taxa, each comprising cryptic complexes of several species.

Tricholoma caligatum as it is currently understood (or not understood) includes three main groups ectomycorrhizal with oaks and other hardwood trees in eastern North America, California, and Mesoamerica. Evidence suggests that these three groups may comprise more than one species.

The "True Matsutakes" include *T. matsutake* of Asia and *T. nauseosum* of Europe, now regarded as synonyms, which are mycorrhizal with conifers. *Tricholoma magnivelare* of the Pacific Northwest is equally problematic, apparently being associated with both conifers and hardwoods. The situation is perhaps best described by the German ecological term *Rassenkreis* (circle of races), that is, a suite of one or more (morphologically variable) species or groups of intergrading subspecies linked by gradually changing clines (continuous gradients of gradually changing features in a series of adjacent populations) over a wide geographic distribution. But just when one taxon ends and another begins continues to elude us. For now, we can only refer the Southeastern Matsutake, associated primarily with oaks, to *T. caligatum* and await further developments. We refer you to Chapela & Garbelotto (2004)[2] for a more detailed discussion.

1. This contrasts sharply with the highly prized and much-sought-after "True Matsutake," which may or may not represent a single circumboreal species. The Japanese harvest is much reduced, and most matsutake consumed in Japan is imported from China, the Pacific Northwest of North America, and Scandinavia. In Japan, local matsutake can run as high as $2,000 per kilogram ($900 per pound), compared to $90–$100 per kilogram ($40–$45 per pound) for matsutake imported from elsewhere.

2. A.H. Chapela & M. Garbelotto, 2004, *Mycologia* 96(4): 730–741.

Velvet Paxillus, Velvet-footed Pax

Tapinella atrotomentosa

A B C

Tapinella atrotomentosa **(Batsch) Šutara**, *Česká Mykol.* 46(1–2):50 (1992)
[Basidiomycota, Agaricomycotina, Agaricomycetes, Agaricomycetidae, Agaricales, Tapinellaceae]

Basionym: *Agaricus atrotomentosus* Batsch, *Elench. Fung.*, cont. prim. (Halle):89 (1783)

Derivation: *Tapinella*, derivation not known; *atrotomentosa*, Latin, from *atra*, "black," and *tomentoso*, "hairy," "velvety."

Cap: Semicircular to kidney-shaped, 4–15 cm (1⅔–6 in) across by 1–3 cm (½–1¼ in) thick, convex, later becoming flat (Fig. A) to centrally depressed, vase-like (Fig. C), margin inrolled when young (Fig. A); surface dry, finely hairy, velvety, or smooth; dull olive-brown, rusty-brown, to yellowish-brown, becoming darker with age.

Flesh: Thick, tough, pale whitish to ochre, not bruising when exposed; odor and taste not distinctive.

Gills: Decurrent, close, often forked and with cross-veins near the stipe; dull yellow, yellow-brown, to tan (Fig. A).

Stipe: Typically eccentric to lateral, 2–12 cm tall by 1–3 cm thick (¾–4¾ × ⅖–1¼ in), solid; densely covered with greyish-brown to blackish-brown woolly hairs (Figs. A & B).

Spore print: Dull yellow to pale brown.

Microscopic: Spores 5–7 × 3–4.5 µm, smooth, elliptical, hyaline to pale brown, dextrinoid (staining reddish-brown) in Melzer's reagent.

Occurrence: Solitary, in groups or in clusters; saprobic on decaying conifer logs and stumps, summer–early winter; widespread but occasional in North America; Europe.

Edibility: Suspect, probably poisonous.

Comments: This species is often called *Paxillus atrotomentosus* (Batsch) Fries in many field guides. However, species of the genus *Paxillus* (*sensu stricto*) grow from the soil and are mycorrhizal. The three species of *Tapinella*—*T. atrotomentosa*, *T. panuoides* (Batsch) E.-J. Gilbert (1931), and *T. corrugata* (Batsch) E.-J. Gilbert (1931)—are all saprobic and grow directly from conifer wood. The dark brown to blackish woolly stipe immediately distinguishes *T. atrotomentosa* from the other two species in this genus, which also lack or at most have only a rudimentary stipe. *Tapinella corrugata* also has smaller (3–3.5 × 1.5–2 µm) and more or less oval spores.

White Hardwood Leucopax

Leucopaxillus laterarius

A C B D

***Leucopaxillus laterarius* (Peck) Singer & A.H. Smith**, *Pap. Mich. Acad. Sci.* 28:117 (1943)

Basionym: *Agaricus laterarius* Peck, *Bull. Buffalo Soc. Nat. Sci.* 1(2):43 (1873) [1873–1874]
[Basidiomycota, Agaricomycotina, Agaricomycetes, Agaricomycetidae, Agaricales, Tricholomataceae]

Derivation: *Leucopaxillus*, from the Greek *leukó*, "white," and the Latin *paxillus*, "peg," "small stake," referring to the mushroom genus *Paxillus*; *laterārium*, Latin, "pertaining to the sides."

Cap: 4–20 cm (1⅔–8 in) across, convex when young, margin inrolled, becoming broadly convex (Fig. A) to undulatingly flat (Fig. B) in maturity, smooth; white, sometimes with faint pinkish or buff tinge.

Flesh: White, thick, firm; strong odor of old bread dough, unpleasant.

Gills: Attached to slightly decurrent, crowded, narrow; initially white (Fig. D), becoming pale yellowish-buff.

Stipe: 4–11 cm tall by 6–20 mm thick (1⅔–4.5 × ¼–¾ in), subequal when mature (Fig. B), ring lacking; white, smooth to finely downy; attached to ground and leaf litter by a prominent mycelial mass (Figs. A & B).

Spore print: White.

Chemical staining: KOH—cap negative or pale olive-green, flesh negative, gills pale yellowish-grey.

Microscopic: Spores 3.5–5.5 × 3.5–4.5 µm (not including warts), ovoid to subglobose, thin-walled, with amyloid warts about 0.5 µm high (Fig. C).

Occurrence: Saprobic on litter in hardwood forests under oaks (*Quercus* spp.), beech (*Fagus*), and sweetgum (*Liquidambar*); scattered to gregarious; summer–fall.

Edibility: Inedible, taste very bitter.

Comments: The White Conifer Leucopax, *Leucopaxillus albissimus* (Peck) Singer, is very similar but is associated with conifers, particularly pines (*Pinus* spp.) in our area.

Blewit, Wood Blewit

Clitocybe nuda

A

B

C

D

E

Clitocybe nuda **(Bulliard *ex* Fries) H.E. Bigelow & A.H. Smith**, *Brittonia* 21(1):52 (1969)
[Basidiomycota, Agaricomycotina, Agaricomycetes, Agaricomycetidae, Agaricales, Agaricaceae]

Basionym: *Agaricus nudus* Bulliard, *Herb. Fr.* 10:tab. 439 (1790)

Derivation: *Clitocybe*, Greek, "sloping head"; *nuda*, Latin, "naked."

Cap: 4–14 cm (1⅔–6 in) across, at first convex, becoming nearly plane in maturity (Figs. A–C), margin inrolled, wavy (Figs. D & E); surface smooth, often convoluted (Fig. B), viscid when moist, violet to lilac when young (Fig. A), fading to various shades of lilac-grey (Fig. B), tan, or brown (Fig. C).

Flesh: Thick, soft, pale lilac to cream-colored; odor strong, fragrant to phenolic.

Gills: More or less attached, often with a deep notch (Fig. E), occasionally subdecurrent, close to crowded (Fig. E), often quite sinuous (Fig. D), moderately broad, pale violet or lavender to lilac (Fig. D), becoming tan or light brown with age.

Stipe: With deep ringing notch at apex making the gills falsely appear to be unattached; stipe equal or enlarged or bulbous at base, 3–10 cm tall by 1–2.5 cm thick at apex (1¼–4 × ½–1 in); no partial veil; solid, dry, fibrillose to finely hairy; colored like the cap, bruising darker lavender; base often covered with lilac to buff mycelium.

Macrochemical staining: KOH—pileipellis reddish-brown, other tissues negative; **Melzer's reagent**—pileipellis faint yellowish-green, flesh dark reddish-brown, gills yellow; all tissues negative in NH_3 and NaOH.

Spore print: Pale pinkish-buff.

Microscopic: Spores 5.5–8 × 3.5–5 µm, ellipsoid, minutely rough or smooth, nonstaining in Melzer's reagent.

Occurrence: Fruiting singly or gregariously from ground under hardwoods and conifers, in meadows or from wood chips, sometimes partially buried (as shown) or on decaying organic material; late fall–midwinter. Widespread in North America and Europe; introduced into Australia.

Edibility: Edible when cooked, causes gastrointestinal upset when eaten raw. Considered choice by many (including us), but taste varies widely from one region to another; wild specimens generally better than cultivated ones.

Comments: The Wood Blewit can be confused with several very similar lavender- or violet-colored and potentially poisonous species of *Cortinarius* but is clearly distinguished from the latter genus by its profuse, pale pink to pinkish-cream spore print. *Cortinarius* species have rusty-brown spore prints and typically exhibit a web-like partial veil (cortina) over part or all of the gills when young.

Though still widely used, *Lepista nuda* is a deprecated synonym of *Clitocybe nuda*. The genus *Lepista* W.G. Smith has been synonymized with *Clitocybe* P. Kummer and should no longer be applied. See online Appendix A for a discussion of this issue.

Purple-gilled Laccaria

Laccaria ochropurpurea

A

B

Laccaria ochropurpurea **(Berkeley) Peck**, *Ann. Rept. N.Y. State Mus. Nat. Hist.* 4:275 (1896)
[Basidiomycota, Agaricomycotina, Agaricomycetes, Agaricomycetidae, Agaricales, Hydnangiaceae]

Basionym: *Agaricus ochropurpureus* Berk., *J. Bot.*, Lond. 4:299 (1845)

Derivation: *Laccaria*, from Greek *lakkos*, "pit," and *-aria*, "having"; *ochropurpurea*, Latin, "purple-ochre," referring to the color of the gills.

Cap: 3.5–12.5 (to 15) cm (1¾–5 in) wide, broadly convex to flat, often with central depression, margin entire, upturned or inrolled, finely tomentose to smooth, occasionally slightly scaly or cracked with age; light purplish-brown or pinkish-buff, fading to buff or whitish with age (Fig. A).

Gills: Attached (Fig. B) to slightly decurrent, thick, waxy, close to subdistant, dark reddish-purple or violet, often powdered with white spores.

Flesh: Thick, purplish-brown, similar to cap.

Stipe: 5–15 (to 19) cm (2–6 in) long by 1–2.5 cm (~½–1 in) in diameter, nontapering, but may be swollen at base, tough, fibrous, concolorous with the cap, basal mycelium lilac to reddish-brown.

Spore print: White to pale violet.

Staining: Cap and stipe temporarily orangish in KOH and ammonia.

Microscopic: Spores 6.5–11 × 6.5–9.5 µm, globose to subglobose, with prominent spines 1–2 µm long, translucent, nonreactive in Melzer's reagent or iodine.

Occurrence: Growing from the ground, ectomycorrhizal with hardwoods, especially oaks (*Quercus* spp.) and beech (*Fagus grandifolia*), but also Eastern White Pine (*Pinus albus*); solitary, scattered, or rarely in clusters; widespread and fairly common east of the Great Plains, June–November.

Edibility: Edible.

Comments: The very similar-looking Sandy Laccaria, *Laccaria trullisata* (Ellis) Peck, can be distinguished from *L. ochropurpurea* by its somewhat smaller size and habitat—sand dunes along the Gulf Coast, East Coast, and around the Great Lakes—and its mycorrhizal association with *Pinus* spp. Its basidiospores are also longer (14–21 × 5.5–8 µm), elliptical, and smooth, lacking the spiny projections seen in *L. ochropurpurea* and most other *Laccaria* species. It is often buried up to its cap in sand.

Violet Collybia

Collybia iocephala

A B C D

Photos: A & C, Teresa Fortenberry; B & D, W. Light.

***Collybia iocephala* (Berkeley & M.A. Curtis) Singer**, *Lloydia* 9:116 (1946)

Basionym: *Agaricus iocephalus* Berkeley & M.A. Curtis, *Ann. Mag. Nat. Hist.*, ser. 2, 12:420 (1853)
[Basidiomycota, Agaricomycotina, Agaricomycetes, Agaricomycetidae, Agaricales, Tricholomataceae]

Derivation: *Collybia*, Greek, *kollubos*, "small coin"; *iocephala*, Greek, *ioeidēs* and *cephales*, "violet head."

Cap: Colored various striking shades of deep to pinkish purple or lilac 0.9–4 cm (⅓–1⅔ in) wide; at first convex (Fig. A), becoming flat-convex or flat with age, margin becoming unevenly uplifted, wavy, and scalloped (Fig. B); upper surface prominently grooved and pleated from disk toward margin.

Flesh: Purplish- or pinkish-lilac; odor variously reported as "offensive," smelling like garlic, radish, sauerkraut, or even gunpowder. It is unclear whether this reflects actual differences in the species or merely the subjective experiences of the observers.

Gills: Attached to or free from expanded stipe apex, subdistant to distant, occasionally forked; concolorous with cap, but more pink-hued (Fig. D).

Stipe: Finely hairy, 2–7 cm long by 1–2.5 mm thick (¾–2¾ in), subequal or tapering upward with slightly expanded apex (Fig. D), somewhat enlarged at the base, pliant to tough but not brittle, interior hollow; surface colored various shades of lilac, this often covered up by white pubescence (Fig. C), with matted fibrillose base (Fig. D).

Staining: Cap, gills, and stipe bright greenish-blue or bright blue in KOH.

Microscopic: Spores 6.5–8.6 × 3.2–4.4 µm, teardrop-shaped in profile, ovoid in front view, smooth, clear, nonstaining in Melzer's reagent. Some pinkish-violet pigment interspersed among hyphae when mounted in water. Clamp connections present in all tissues.

Occurrence: August–December; saprobic; scattered, gregarious to subcespitose on leaf litter and humus in mixed woods. Our specimens sprouted from leaf litter in a moist bog.

Widespread in the Southeast; Florida to Texas and north to Massachusetts.

Edibility: Uncertain, probably not advisable. Reportedly used as a garlic substitute.

Comments: This species is placed in the genus *Gymnopus* (Persoon) Roussel by many authors: *G. iocephalus* (Berk. & M.A. Curtis) Halling (1997). The genera *Marasmius* Fries (*sensu lato*) and *Collybia* Fries (*sensu lato*) have been split into 19 and 13 genera more or less, respectively, on the basis of various combinations of morphological characters. However, these distinctions are far from clear, and not all mycologists recognize all of these segregate genera. We are retaining this taxon in the genus *Collybia* (*sensu lato*) for the time being.

Cup-bearing Agaric

Agaricus pocillator

A B C

Agaricus pocillator Murrill, *Mycologia* 33(4):446 (1941) [Basidiomycota, Agaricomycotina, Agaricomycetes, Agaricomycetidae, Agaricales, Agaricaceae]

Derivation: *Agaricus*, Latin, from *agaricum*, "tree fungus," but this meaning was superceded by *Agaricus sensu* Linnaeus (1753), "gilled mushroom"; *pocillator*, Latin, "cup bearer."

Cap: 3–10 cm (1¼–4 in) broad, initially convex, becoming broadly convex to nearly plane with age; whitish (Fig. B) to dingy grey-silver, with grey floccose scales, often with darker greyish-brown umbo in center, with radial striations from center to entire margin (Fig. A); while covering gills when young, partial veil not exuding liquid droplets.

Flesh: White throughout (bruising yellow at base), up to 3 mm deep at disk, thinning to <1 mm at margin. Odor phenolic or not remarkable.

Gills: Free from stipe, close, margins entire, initially whitish, becoming pinkish (Fig. C), then deep chocolate-brown; gills free, up to 3 mm wide (deep).

Stipe: Up to 8 cm tall by 1 cm thick (3½ × ~½ in), often sharply curved near the bulbous, often cup-like base (Fig. B), flesh at base bruising yellow; with a distinct double superior ring, the lower portion becoming a distinct partial veil, the upper portion stretching across the gills when young, afterward remaining as a marginal veil; when covering the gills partial veil not developing dark droplets.

Spore print: Brown.

Macrochemical stains: KOH—cap and flesh yellow.

Microscopic: Spores 4.5–3.0 × 3.0–3.8 µm, ellipsoid to broadly ellipsoid, with thin, dark brown walls and papillate apiculus; hyaline, with brown walls.

Occurrence: Saprobic, widespread in hardwoods and mixed woods throughout the southeastern United States.

Edibility: Questionable, possibly poisonous.

Comments: The double ring, tiny floccose grey scales on the cap, lack of yellow or brown droplets on the partial veil, and the cupulate base distinguish this species. The similar Eastern Flat-top Agaricus, *Agaricus placomyces* Peck, has reddish-brown (rather than grey) scales on the cap, a single (instead of a double) ring on the stipe, and root-like rhizomorphs at the base of a swollen and bulbous stipe.

Eastern Flat-top Agaricus

Agaricus placomyces

A B C

Photos: A & B, W. Light; C, weiliiiiiii.

***Agaricus placomyces* Peck**, *N.Y. State Mus. Bull.* 29:40 (1978) [Basidiomycota, Agaricomycotina, Agaricomycetes, Agaricomycetidae, Agaricales, Agaricaceae]

Derivation: *Agaricus*, Latin, from *agaricum*, "tree fungus," but this meaning was superceded by Linnaeus's use of *Agaricus* to mean "gilled mushrooms"; *placomyces*, Greek, *plak*, "slab-like" or "plate-like," and *mukēs*, "fungus."

Cap: 5–12 cm (2–4¾ in) broad, initially convex, becoming broadly convex to nearly plane with age (Fig. A); whitish to dingy grey-silver (Fig. A), surface covered with reddish-brown adpressed fibrillose scales, often with darker greyish-brown umbo in center (Fig. A); while still covering gills when young, partial veil often exuding yellowish to dark brown droplets.

Flesh: White throughout (bruising yellow at base), 2 mm deep at disk, thinning at margin. Odor phenolic or unremarkable.

Gills: Free from stipe, close, margins entire, initially pale greyish-pink (Fig. B), becoming deep chocolate-brown; gills free, up to 5 mm wide (deep).

Stipe: Up to 15 cm tall by 1.5 cm thick (6 × ⅔ in) more or less equal, often sharply curved near an enlarged, occasionally bulb-like base that lacks a cupule and nearly always bears one to two root-like rhizomorphs (Fig. C); flesh at base bruising yellow; with a distinct single superior ring (Fig. B); when covering gills partial veil often exuding yellowish to dark brown droplets.

Spore print: Brown.

Macrochemical stains: KOH—cap and flesh yellow.

Microscopic: Spores 5–6 (up to 7) × 3.5–4.5 (up to 5) µm, ellipsoid to ovoid, with papillate apiculus; hyaline, with brown walls.

Occurrence: Saprobic, scattered or gregarious east of the Rocky Mountains in hardwood forests and mixed woods; summer–fall; more northern in distribution than *A. pocillator*.

Edibility: Questionable, possibly poisonous.

Comments: This species is distinguished from the very similar *A. pocillator* by the large reddish-brown scales on the cap extending beyond the central disk toward the margin, the yellowish to brown droplets often exuded by the partial veil, the single ring, and the one to two root-like rhizomorphs on the bulbous base of the stipe.

Almond Mushroom, Royal Sun Agaricus, "Agaricus blazei"

Agaricus subrufescens

Photo: Nathan Wilson.

***Agaricus subrufescens* Peck**, *Ann. Rep. Reg. N.Y. St. Mus.* 46:25 (1894) [1893]
[Basidiomycota, Agaricomycotina, Agaricomycetes, Agaricomycetidae, Agaricales, Agaricaceae]

Derivation: *Agaricus*, Latin, from *agaricum*, "tree fungus," but this meaning was superceded by *Agaricus sensu* Linnaeus (1753), "gilled mushroom"; *subrufescens*, Latin, from *rufescens*, present participle of *rufescere*, "to become red," and the prefix *sub*, "below," "underneath."

Deprecated synonym: *Agaricus rufotegulis* Nauta, *Persoonia* 17(2):230 (1999)

Illegitimate name: *Agaricus brasiliensis* Wasser, Didukh & Stamets, *in* da Eira, Didukh, Stamets, Wasser, & de Amazonas, *Internat. J. Med. Mushrooms* 4(4):274 (2002)[1]

Misapplied name: *Agaricus blazei* Murrill *sensu* Heinemann, *Bull. Jard. Bot. Belgium* 62:365–368 (1993)[2]

Cap: 7–25 cm (2¾–10 in) across, convex, marshmallow-shaped when young, becoming nearly flat with age; whitish to pinkish-buff, yellowing with age; dry, initially covered with matted, pale brown fibrils, these breaking up into small brownish scales.

Flesh: White, thick, firm, not bruising or discoloring very slightly yellow or yellowish-orange in the cap, but flesh in the stipe unchanging when cut; strong odor and taste of almonds.

Gills: Free, close, initially whitish, then pinkish, reddish-brown, and finally dark chocolate-brown. At first covered by a thick, membranous partial veil with cottony patches on the undersurface.

Stipe: 5–15 cm tall by 1–4 cm thick (2–6 × ½–1⅔ in), equal to swollen at the base; partial veil leaving a skirt-like ring, stipe white to pinkish above the ring, smooth or scaly below; base often with attached mycelial threads, surface (but not the flesh) yellowing with age or when bruised.

Chemical staining: Cap yellow with KOH.

Spore print: Dark brown.

Microscopic: Spores 5.5–7.5 × 4–5 µm, elliptical, smooth, pale brown, nonreactive in Melzer's reagent; basidia variously bi-, tri-, and tetrasporic (2-, 3-, and 4-spored) in both wild and cultivated specimens.

Occurrence: Saprotrophic. Solitary or clustered in leaf litter, rich humus, compost heaps, and disturbed areas; widespread but rare in eastern North America, California; common in the Atlantic region of Brazil; western Europe, Hawaiian Islands; cultivated in Japan from isolates imported from Brazil;[2] also in the Philippines and Taiwan, and recently reported from Thailand.[3]

Edibility: Edible, considered choice, with aroma and taste of almonds; widely cultivated and sold as an edible and medicinal mushroom.

Comments: This mushroom can be identified by its strong and unmistakable almond aroma and taste together with its thick, membranous partial veil with cottony patches on the lower side that initially cover the gills, gills that start out white and then turn pinkish, then reddish-brown and finally dark chocolate-brown, stipe that bruises yellowish near the base, and its dark-brown spore print.

The poisonous (to some people) Yellow-staining Mushroom, *Agaricus xanthodermus* Genevier, is similar but is distinguished by its strong, unpleasant odor of phenol (carbolic acid), iodine, or disinfectant (that of *A. subrufescens* is like almonds) and its white flesh in the stipe immediately turning intensely bright yellow when cut. Although the surface of the stipe of *A. subrufescens* often bruises yellowish at its base, the flesh does not change color when the stipe is sliced open.

The Almond Agaricus was extensively cultivated in the northeastern United States from the late 19th to the early part of the 20th century but was soon replaced by *A. bisporus* (Lange) Imbach—the widely sold edible mushroom known as the Button Mushroom when young and Portobello when fully mature. It has been widely cultivated in Brazil, Japan, Korea, Denmark, and the Netherlands, where it is sold as a gourmet food and medicinal specialty.

Under the name "Agaricus blazei" or some variant thereof,[4] the Almond Mushroom is widely published and marketed as one of the most potent and effective of all medicinal mushrooms. It has been used in traditional medicine to treat diabetes, hyperlipidemia, atherosclerosis, heart and circulatory disease, digestive disorders, and cancer and to prevent osteoporosis, peptic ulcer, and other diseases. In recent decades this mushroom has been intensively studied for its bioactive and medical properties, primarily by Asian but increasingly also by Western investigators. This is discussed more fully in the introductory section on medicinal mushrooms.

Much nomenclatural confusion has surrounded this species. It is usually misidentified as *Agaricus blazei* Murrill[5] or occasionally as *Agaricus sylvaticus* Schaeffer. Neither of these two misapplied names are synonyms of *A. subrufescens* but are misidentifications.[2] It is also referred to as *Agaricus brasiliensis*, but this is a homonym of a preexisting validly published name and is thus an illegitimate name.[6] Molecular genetic and breeding studies finally revealed that the North American, Brazilian, European, Hawaiian, and Asian populations are all later synonyms of a North American species, *Agaricus subrufescens* Peck, originally described from New England. A list of deprecated synonyms and the misapplied names is given above and in the online Appendix A.

This mushroom has long been referred to as *Agaricus blazei* Murrill (a very different species). This stems from a misidentification by a Belgian mycologist to whom the Brazilian specimens were sent by the Japanese collector who brought samples of this species to Japan.[5,6] This identification error was later recognized, and the species was described as a new species, *Agaricus brasiliensis*, by Wasser et al.[7] However, this is a homonym of another species entirely, *Agaricus brasiliensis* Fries (1830), and is therefore an illegitimate name.[8] Molecular genetic and breeding studies finally revealed that the North American, Brazilian, European, Hawaiian, and Asian populations are conspecific with *Agaricus subrufescens* Peck (1893).[9] Nevertheless, this species and the medicinal extracts and pharmaceutical products prepared from it continue to be widely and incorrectly called *A. blazei* (and to a lesser extent also *A. brasiliensis*) in both the medical research and popular literature and in commercial packaging. Bioactive extracts from this mushroom are still typically referred to as "ABM [for *Agaricus blazei* Murrill] glucans" in the biomedical literature, and a steroid recently isolated from this mushroom in Japan (where it is popularly called *himematsutake*) was recently named "blazein" on the basis of this continuing misidentification.[10]

1. *Nom. illegit.*, Art. 53.1, homonym; *non Agaricus brasiliensis* Fries, *Linnaea* 5:509 (1830); *non Agaricus brasiliensis* Montagne, *Annls Sci. Nat.*, Bot., sér. 4, 5:361 (1857) [1856]; [*nom. illegit.*, Art. 53.1, homonym, *non* Fries (1830)]. See online Appendix A for further details.

2. R.W. Kerrigan, 2005, *Mycologia* 97(1):12–24. *Agaricus sylvaticus* belongs to genus *Agaricus* section *Sanguinolenti*, whereas *A. subrufescens* is in *Agaricus* section *Arvenses*. K. Wisitrassameewong et al., 2012, *Chiang Mai J. Sci.* 39(2):281–291.

3. K. Wisitrassameewong et al., 2012, *Chiang Mai J. Sci.* 39(2): 281–291.

4. Including the unfortunate abbreviation "ABM," standing for *Agaricus blazei* Murrill.

5. P. Heinemann, 1993, *Bull. Jard. Bot. Belgium* 62:365–368. *Agaricusyus blazei* is apparently known only from Florida.

6. R. W. Kerrigan, 2005, *Mycologia* 97(1):12–24.

7. Wasser et al., 2005, *in* A. F. da Eira et al., 2002, *Internat. J. Med. Mushrooms* 4(4):274 (2002).

8. Despite this, *A. subrufescens* is often still erroneously called *A. brasiliensis*.

9. *Agaricus subrufescens* Ellis & Everhart (1894) is an illegitimate name; it is a homonym of *A. subrufescens* Peck (1893). This species is actually *Tricholoma subrufescens* Saccardo (1895), which belongs to an entirely different and unrelated family (Tricholomatacea) than the genus *Agaricus* (Agaricaceae).

10. H. Ito et al., 2009, *Oncol. Rep.* 20(6):1359–1361.

Yellow Trumpet Mushroom

Gerronema strombodes

A

B

C

Photos: A & B, W. Light; C, Teresa Fortenberry.

***Gerronema strombodes* (Berkeley & Montagne) Singer**, Sydowia 15(1–6):50 (1962)

Basionym: *Agaricus strombodes* Berkeley & Montagne, *Syll. Gen. Sp. Crypt.* (Paris):112 (1856)
[Basidiomycota, Agaricomycotina, Agaricomycetes, Agaricomycetidae, Agaricales, Hygrophoraceae]

Derivation: *Gerronema*, etymology not provided in original description[1] but presumably Latin, from *gerro*, "trifler," and *nemus*, "grove" or "woods with open glades"; *strombodes*, derivation not known.

Cap: Typically 5–10 cm (2–4 in) wide, planoconvex when young, becoming convex to funnel-shaped with age (Figs. A & C), pale yellowish with darker brown fibrils radiating from depressed center (Fig. C), margin more or less entire (Fig. C) to wavy or dissected (Figs. A & B).

Flesh: Thin, whitish to yellowish; odor not distinctive to sweetish, taste not distinctive or somewhat bitter.

Gills: Decurrent, broad, distant, initially whitish or cream-colored, becoming pale yellow, often forked, short gills numerous (Fig. B).

Stipe: Subequal, up to 6.5 cm tall by 1.2 cm wide (2⅔–½ in), solid, dry, smooth, white to greyish.

Spore print: White.

Microscopic: Spores 7–9 × 4–5.5 µm, elliptical, smooth, clear, nonreactive in Melzer's reagent.

Occurrence: Saprobic on decaying hardwood; alone to more typically in dense clusters; very occasional to locally and sporadically abundant, July–August; widely distributed in and restricted to eastern North America.

Edibility: Unknown.

Comments: *Gerronema strombodes*, formerly known as *Chrysomphalina strombodes* (Berk. & Mont.) Clémençon, is restricted to eastern North America.[2]

Chrysomphalina chrysophylla (see note 2), an unrelated look-alike, is distinguished from *G. strombodes* by its cap (brownish-orange at the center, with a yellowish-brown edge), golden-yellow to orange-yellow (not cream-colored) *unforked* gills, yellow (not white) spore print, and larger spores (8.5–15.5 × 4.5–7 µm) and by growing on decayed conifer (not broadleaf) wood. It is widespread but uncommon in North America.

1. R. Singer, 1951, *Mycologia* 43(5):598–604.

2. The European taxon misidentified as *Chrysomphalina strombodes* by Clémençon and other European authors is a distinct but related species now known as *Gerronema xanthophyllum* (Bresádola) Norvell, Redhead, & Ammirati. *Gerronema strombodes*, *G. xanthophyllum*, and the closely related *G. subclavatum* (Peck) Singer *ex* Redhead are not at all closely related to the type species of *Chrysomphalina*, *C. chrysophylla* (Fries) Clémençon. Despite a superficial resemblance, the two genera thus represent two distinct and quite unrelated evolutionary lineages (S. Redhead et al., 2002, *Mycotaxon* 83:19–57).

Yellow Unicorn Entoloma

Entoloma murrayi

A

B

Photos: A, Dan Molter; B, W. Light.

Entoloma murrayi **(Berkeley & M.A. Curtis) Saccardo**, *Syll. Fung*. (Abellini) 14:127 (1889)
[Basidiomycota, Agaricomycetes, Hymenomycetes, Agaricales, Entolomataceae]

Basionym: *Agaricus murrayi* Berkeley & M.A. Curtis (1859)

Derivation: *Entoloma*, Greek, from *entos* and *lóma*, "inner fringe," referring to the ragged margin; *murrayi* is the Latinized form of the surname of the original collector of this species, Dennis Murray.

Cap: 1–3 cm (⅖–1¼ in) wide, conical to bell-shaped, with prominent acute, often whitish umbo (Figs. A & B), more convex with age, pale to bright yellow, fading somewhat with age, smooth, with striate margin.

Gills: Narrowly attached, subdistant, broad, moist, smooth, of differing lengths, yellow, tinged pinkish with age.

Flesh: Yellowish.

Stipe: Central, 5–10 cm (2–4 in) tall by 2–5 mm (0.07–0.2 in) thick, yellow, hollow, fragile, dry, base usually covered with white mycelium.

Spore print: Salmon-pink.

Microscopic: Spores 9–12 × 8–10 µm, angular, 4-sided (characteristic for the genus), smooth, nonreactive in Melzer's reagent/iodine; cystidia club-shaped; clamp connections present.

Occurrence: Common, solitary, scattered, to gregarious in soil in swamps and wet deciduous woods, July–October; Maine to Alabama, west to the Great Lakes; Costa Rica; South America; Japan, China.

Edibility: Unknown, but likely poisonous, as are many (but not all) species in this genus.

Comments: Like many mushrooms, this taxon has a confused nomenclature. It has variously also been called *Inocephalus murrayi* (Berk. & M.A. Curtis) Rutter & Watling, *Nolanea murrayi* (Berk. & M.A. Curtis) Dennis, *Rhodophyllus murrayi* (Berk. & M.A. Curtis) Singer, and *Entoloma murrayi* (Peck) Saccardo. The specific epithet has also been misspelled as "*murraii*."

Charming Marasmius

Marasmius delectans

A B C D E

Marasmius delectans **Morgan**, *J. Mycol.* 11(5):206 (1905) [Basidiomycota, Agaricomycotina, Agaricomycetes, Agaricomycetidae, Agaricales, Marasmiaceae]

Derivation: *Marasmius*, Greek, from *marasmos*, "drying out," referring to the ability of these mushrooms to shrivel up in dry weather and reconstitute in rainy weather; *delectans*, Latin, *dēlectāns*, "charming," "delightful."

Cap: Large for *Marasmius*, 1–4 cm across, initially white, convex (occasionally with an umbo), pleated and parachute-shaped (Fig. C), becoming broadly convex to almost flat (Figs. A & D), margin more or less entire to crenulate (Fig. D); pale yellowish when young, becoming pale buff with age, often with darker center and radiating lines (Fig. A).

Flesh: Very thin, fibrous, whitish to pale buff; odor and taste unremarkable.

Gills: Pale yellowish (Fig. B) to white (Fig. E), attached, but notched at stipe, distant, broad, margin entire, nonforking, with three to four tiers of lamellulae (short gills) in addition to the full-length gills (Fig. E).

Stipe: 2–8 cm (¾–3¼ in) tall by 1–3 mm thick, smooth, hollow, subequal, often very slightly expanded near apex; initially white, yellowish, or reddish (Fig. B), darkening with age from the base upward to dark brown or nearly black (Fig. E), fading to reddish then yellowish at apex; with conspicuous white mycelium at base (Fig. B), this often extensive.

Spore print: White.

Microscopic: Spores 5.5–9 × 3–5 µm, elliptical, smooth, tapering to a point at one end, nonreactive in Melzer's reagent; gills with numerous spine-like cystidia, these staining dark reddish-brown in Melzer's reagent.

Occurrence: Gregarious on leaf litter in hardwood forests, especially under oaks (*Quercus* spp.); widespread in eastern North America, July–September; sporadic, but abundant during periods of sustained rainfall.[1]

Edibility: Unknown.

Comments: The small size, *Marasmius*-like form, thin, wiry, elongate, blackish stipe attached to a white basal mycelium, and gregarious growth in hardwood leaf litter distinguish this mushroom from all other species in its range. It is large for the genus, the caps of most *Marasmius* species being 2 cm across or smaller. And unlike many members of the genus, the noncollariate[2] gills are attached directly to the stipe instead of to a separate collarette ringing the apex of the stipe.

1. C.H. Kauffman, 1918, *The Agaricaceae of Michigan*, vol. 1, Mich. Geol. Biol. Survey Publ. 26, Biol. ser. 5, 30.

2. Lacking a pseudocollarium (depression or gap) at the stipe where the gills have separated or are not quite attached.

Orange Pinwheel Marasmius

Marasmius siccus

A B C D

Photos: A–C, W. Light; D, Karen (Oldmanofthewoods).

***Marasmius siccus* (Schweinitz) Fries**, *Epicr. Syst. Mycol.* (Upsaliae):382 (1838) [1836–1838]

Basionym: *Agaricus siccus* Schweinitz, *Schr. Naturf. Ges.* Leipzig 1:84 (1822)
[Basidiomycota, Agaricomycotina, Agaricomycetes, Agaricomycetidae, Agaricales, Marasmiaceae]

Derivation: *Marasmius*, Greek, from *marasmos*, "drying out," referring to the ability of these mushrooms to shrivel up in dry weather and reconstitute in rainy weather; *siccus*, Latin, "dessicated," "dry."

Cap: Small, 0.5–3 cm (¼–1¼ in) across, bell-shaped to convex, prominently pleated, often with small central pimple-like umbo or depression; dry, velvety, rust-orange to orange (Figs. A–C), fading with age.

Flesh: Very thin, pale, odor not distinctive, taste mild to somewhat bitter; staining reddish-brown with Melzer's reagent.

Gills: White to cream or pale yellowish; attached or free, very distant (Fig. D), moderately broad.

Stipe: 2.5–7 cm tall by 2 mm thick (1–2¾ × 1/12 in), equal, thin, wiry to brittle, hollow, smooth, dry; deep brownish below, becoming pale yellowish toward the apex (Figs. A & B).

Spore print: White.

Microscopic: Spores (13–)16–23 × 3–5 µm, spindle- or club-shaped to cylindrical, smooth, clear, nonreactive in Melzer's reagent; large refractive pleuro- and cheilocystidia present.

Occurrence: Saprobic. Typically gregarious on hardwood leaves and woody debris, occasionally on needles of White Pine (*Pinus strobus*); widespread and fairly common in eastern North America; Europe; summer–fall.

Edibility: Unlikely.

Comments: This species is not likely to be mistaken for any other in our area. Its small size, growth on leaf litter, deeply pleated, umbrella-like cap, and long, thin, wiry stipe are distinctive.

Marasmius pulcherripes Peck is very similar in size and shape, but the cap is pink instead of orange, and it typically grows on conifer needles, especially those of White Pine. *Marasmius fulvoferrugineus* Gilliam has a hazel- to reddish-brown cap and a faint flour-like or mealy odor and taste. Cystidia are lacking.

Bell Omphalina and Cross-veined Troop Mushroom

Xeromphalina campanella and *X. kauffmanii*

A, E, *Xeromphalina campanella*; B–D, F, *X. kauffmanii*.

Xeromphalina campanella **(Batsch *ex* Fries) Kühn & Maire**, *Icones Selectae Fungorum*, 6 Texte général 6:236 (1934)

Xeromphalina kauffmanii **A.H. Smith**, *Pap. Mich. Acad. Sci.* 38:81 (1953)

Basionym: *Agaricus campanella* Batsch, *Elench. Fung.*, cont. prim. (Halle):73 (1783)
[Basidiomycota, Agaricomycotina, Agaricomycetes, Agaricomycetidae, Agaricales, Mycenaceae]

Derivation: *Xeromphalina*, Greek, "dry little navel"; *campanella*, Latin, "little bell"; *kauffmanii*, epithet in honor of the American mycologist Calvin Henry Kauffman (1904–1931).

Cap: Both species—5–25 mm (¼–1 in) across, initially convex, becoming broadly convex or flat with deep central depression (Figs. A & B); smooth to minutely hairy, dry to moist but never viscid; margin initially somewhat inrolled, faintly striated, often pleated (Figs. A & B); yellowish, brownish-yellow, rusty, or orange, paler toward margin, fading with age.

Flesh: Both species—tough, rubbery.

Gills: Both species—decurrent, distant, with numerous cross-veins (Fig. C). ***X. campanella***—pale yellow to orangish; ***X. kauffmanii***—pale yellow.

Stipe: Both species—1–5 cm (½–2 in) tall by 1–3 mm thick; subequal, slightly expanded at apex (Figs. C & F); yellowish above, brownish to reddish-brown below (Figs. E & F), wiry and tough; ***X. campanella***—smooth above, finely hairy toward base, typically sharply curved (Fig. E); ***X. kauffmanii***—smooth or finely hairy, more or less straight (Fig. F).

Spore print: Both species—white to buff.

Chemical staining: Both species—flesh and cap yellow-brown in 3% KOH; flesh of cap red-brown in Melzer's reagent.

Microscopic: Both species—spores elliptical, smooth, faintly amyloid (light blue or blue) in Melzer's reagent; clamp connections present; ***X. campanella***—5–8 × 3–4 µm; ***X. kauffmanii***—4–6 × 2.5–3 µm.

Occurrence: Both species—saprobic, typically growing in huge, dense, cespitose clusters, often numbering in the hundreds (Fig. D); spring–fall in cool moist weather; ***X. campanella***—on dead and decayed conifer stumps and logs; widespread in North America, Europe, Asia, North Africa;[1] ***X. kauffmanii***—in erect cespitose clusters (Fig. F) on dead and decayed hardwood stumps and logs; central and eastern North America; Costa Rica, Japan.[1] Substratum (hardwood or conifer) diagnostic.

Edibility: Inedible but not thought to be toxic.

Comments: The Bell Omphalina, *Xeromphalina campanella*, and the Cross-veined Troop Mushroom, *X. kauffmanii*, are very similar, and both are quite widespread and common. They are readily and reliably distinguished from each other by the woody substratum on which they grow in dense cespitose clusters. *Xeromphalina campanella* always occurs on dead and usually well-decayed conifer logs and stumps, in our area mainly pines (*Pinus* spp.) and Hemlock (*Tsuga canadensis*), while *X. kauffmanii* is restricted to hardwoods, primarily oaks in our area (*Quercus* spp.). This distinction has been confirmed by mating studies. In addition, *X. campanella* typically exhibits mostly sharply curved stipes (Fig. E); those of *X. kauffmanii* are mostly straight (Fig. F).

The common names Bell Omphalina and Golden Trumpet apply to both species. The genus *Xeromphalina* includes those small (≤5 cm across) mycenoid yellowish-brown to reddish-brown mushrooms with a convex cap and a deep central depression, thin, wiry stipes, pale yellow or pale orange, decurrent, distant gills with numerous cross-veins (a defining characteristic of the genus), and amyloid spores. A good monographic review of the North American species of *Xeromphalina* can be found in Miller (1968).[2]

The only other species of *Xeromphalina* in our area is the nearly cosmopolitan *X. tenuipes* (Schweinitz) A.H. Smith (1953), which appears to be found on hardwood substrata almost everywhere except in Europe, northern Asia, and most of Polynesia.[1] *Xeromphalina tenuipes* can be distinguished from the other two taxa by its somewhat larger size (cap 2–7 cm [¾–2¾ in] across, stipe 5–8 cm [2–3¼ in] tall), orange-brown and olive-tinged cap when young and fresh, the disk later becoming ochraceous and the margin tawny and translucently striate when old and wet. The velvety to minutely hairy stipe is colored like the cap, and the gills are adnate (not decurrent) and initially white before turning pale yellow. Its spores are also larger (7–9 × 4–5 µm) than those of *X. kauffmanii* (4–6 × 2.5–3 µm), and it fruits in April–July.

1. *Fide* S.A. Redhead, 1989, *Can. J. Bot.* 67:3003–3062.
2. O.K. Miller Jr., 1968, *Mycologia* 60(1):156–188.

Fairy Ring Mushroom, Scotch Bonnet

Marasmius oreades

A B C

***Marasmius oreades* (Bolton) Fries**, *Anteckn. Sver. Ätl. Svamp.*:52 (1836)

Basionym: *Agaricus oreades* Bolton, *Hist. Fung. Halifax*, App. (Huddersfield):151 (1792) [1791]
[Basidiomycota, Agaricomycotina, Agaricomycetes, Agaricomycetidae, Agaricales, Marasmiaceae]

Derivation: *Marasmius*, Greek, from *marasmos*, "drying out"; *oreades*, Greek, named after the Oreiades, Greek woodland nymphs of the hills and valleys.

Cap: 1–5 cm (½–2 in) across, initially bell-shaped with central bump (Fig. A), with slightly inrolled margin, becoming plano-convex (Fig. B), often retaining the central umbo; dry, smooth, pale tan to buff, sometimes whitish to reddish-tan.

Flesh: Tough, white; odor and taste not distinctive.

Gills: Attached or free; broad, distant, with interspersed shorter lamellae; white, pale tan to greyish-buff (Fig. C).

Stipe: Up to 8 cm tall by 4–6 mm thick (3¼ × ¼ in), tough, pliant, subequal whitish to concolorous with the cap; ring lacking.

Spore print: White.

Microscopic: Spores 7–10 × 4–6 µm, smooth, unevenly fusiform-elliptical, with knob-like apiculus; nonreactive in Melzer's reagent.

Occurrence: Saprobic, in grassy areas, typically in gregarious troops or forming arcs or fairy rings; widespread in North America; spring–fall.

Edibility: Edible, but the stems are tough; caution indicated, as several toxic look-alikes exist.

Comments: The pale tan to buff cap, distant gills, thin, tough stipe, and white spore print distinguish this mushroom from other species that form fairy rings. The Meadow Mushroom, *Agaricus campestris* Linnaeus *ex* Fries, has a blackish-brown spore print; that of the lilac-tinged Blewit, *Clitocybe nuda* (Fries) Bigelow & Smith, is pale pinkish-buff. The toxic[1] Sweating Mushroom, *Clitocybe dealbata* (Sowerby) P. Kummer, which commonly co-occurs with *M. oreades*, has a paler whitish cap with no umbo and narrowly spaced gills.

Marasmius oreades readily revives with water following complete desiccation.

1. It contains the neurotoxin muscarine.

White Marasmius

Marasmiellus candidus

A

B

C

D

***Marasmiellus candidus* (Bolton *ex* Fries) Singer**, *Pap. Mich. Acad. Sci.* 32:129 (1948)

Basionym: *Agaricus candidus* Bolton, *Hist. Fung. Halifax* (Huddersfield) 1:39 (1788)
[Basidiomycota, Agaricomycotina, Agaricomycetes, Agaricomycetidae, Agaricales, Omphalotaceae]

Derivation: *Marasmiellus*, Greek, diminutive form of *Marasmius* (from *marasmos*, "drying out"), "little *Marasmius*"; *candidus*, Latin, "white," "bright," "lustrous."

Cap: 1–4 cm (½–1⅔ in) across, convex (Fig. A), becoming nearly flat with depressed center (Fig. B), surface pleated, radially grooved, margin wavy, grooved (Figs. B & C); surface dry, smooth to minutely hairy, shiny white, becoming buffish-white and often staining pinkish with age.

Flesh: Whitish, very thin, soft; odor and taste not distinctive.

Gills: Attached to decurrent, distant, moderately broad, with prominent cross-veins (Fig. D); white to pinkish.

Stipe: 1–2 cm tall by 1–4 mm thick (¼–¾ × 1⁄32–2⁄16 in), equal, with slightly expanded base, central or eccentric solid, dry; white, blackish toward base (Fig. D).

Spore print: White.

Microscopic: Spores 10–15 × 5–6 µm, teardrop-shaped, smooth, clear, nonreactive in Melzer's reagent.

Occurrence: Scattered or in dense clusters and rows on twigs and stems of shrubs, bushes, and brambles; widespread and occasional in Southeast and throughout North America; summer–fall.

Edibility: Unknown, but too small to be of interest.

Comments: This tiny but conspicuous mushroom is often called *Marasmiellus albuscorticis* in field guides. However, *Marasmiellus albuscorticis* Secretan *ex* Singer (1951) is a later and deprecated synonym of *M. candidus*.

Marasmius rotula (Scopoli) Fries is very similar to *Marasmiellus candidus*, but it has a deeply pleated and scalloped cap and margin (like a deployed parachute), and the gills lack cross-veins and are attached to a free or collapsed collar encircling the stipe. The central disk also typically has a dark spot that is not present in *Marasmiellus candidus*.

Magnolia Cone Mushroom

Strobilurus conigenoides

A B C

Strobilurus conigenoides growing on *Magnolia grandiflora* cones. (Photos: A, W. Light; B, Ron Pastorino; C, Christine Braaten.)

***Strobilurus conigenoides* (Ellis) Singer**, *Persoonia* 2(3):409 (1962)

Basionym: *Agaricus conigenoides* Ellis, *Bull. Torrey Bot. Club* 6:76 (1876)
[Basidiomycota, Agaricomycotina, Agaricomycetes, Agaricomycetidae, Agaricales, Physalacriaceae]

Derivation: *Strobilurus*, Latin, from *strobilus*, "cone," and *urus*, "arising from," "arising from a cone"; *conigenoides*, Latin, "resembling *conigenus*," referring to *Strobilurus conigenus* (Persoon) Gulden, which grows on conifer cones.

Cap: 5–20 mm (¼–¾ in) across, initially conic or bell-shaped (Fig. A), becoming convex to broadly convex (Figs. B & C); white, becoming pale yellowish-grey to tan over disk, cream-colored toward margin; surface dry, minutely downy.

Flesh: Very thin, white; odor and taste not distinctive.

Gills: Adnexed to almost free, well separated to nearly distant, white (Fig. C).

Stipe: 2–5 cm tall by 0.75–2 mm thick (¾–2 × ~1⁄12 in), very thin, subequal, tapering gradually toward the apex; dry, densely covered with tiny hairs (Fig. A, inset); initially whitish (Fig. A), becoming yellowish-grey to pinkish-beige (Figs. B & C), remaining white near the apex (Fig. C), cottony white at base.

Spore print: White.

Microscopic: Spores (5.6–)6–7(–7.5) × 2.8–3.3(–3.7) µm, elliptical, smooth, thin-walled, clear, nonreactive in Melzer's reagent; pleurocystidia and cheilocystidia fusoid-ventricose with subacute to acute apices, often with thickened walls and a sort of pedicel.

Occurrence: Saprobic, growing in clusters on the decaying cones of Southern Magnolia (*Magnolia grandiflora* Linnaeus), Mountain Magnolia (*M. fraseria* Walter), and Sweetgum fruits (*Liquidambar styraciflua* Linnaeus). Widespread and common in southeastern North America; fall–midwinter.

Edibility: Unknown, but too tiny to consider.

Comments: This is the only species of *Strobilurus* in our area. It is immediately identified by its substratum (magnolia cones), thin, wire-like stalks, and whitish to greyish-beige color.

In our area the very similar Conifer Cone Cap (*Baeospora myosura* Singer) grows on fallen Eastern White Pine cones (*Pinus strobus* Linnaeus) and on other conifer cones elsewhere; it has close to very close (rather than nearly distant) gills and slightly smaller ellipsoid to ovoid spores (3.3–4.1 × 2–3.1 µm) that are somewhat reactive (rather than inamyloid) in Melzer's reagent.

The Russulas and Milkcaps

The order Russulales includes 12 families exhibiting highly diverse fruitbodies that include agaricoid (gilled mushrooms with central stipes), pleurotoid (oyster-like), hydnoid (toothed, both sessile and stiped), pore-bearing to daedeloid crusts, parchment-like resupinate spreads, coral fungi, and hypogeous (underground) gasteroid (truffle-like) morphologies. Both the agaricoid and pleurotoid types of fruitbodies in the Russulales evolved independently from their morphological counterparts in the Agaricales. Seven of these 12 families are represented by one or more species accounts in various sections of this book.[1] In this section we will be concerned only with the closely related genera *Russula*, *Lactarius*, and *Lactifluus*.

Both *Russula* and *Lactarius* are distinguished by their white to dark-yellow spore print, attached to decurrent, white to yellow gills, brittle cap, and hollow, white, and brittle stipe (both of which readily shatter), and a stipe that lacks an annulus or volva. *Lactarius* is distinguished from *Russula* by the white to yellowish (reddish in some species) milky latex that is released when the gills are cut or scratched,[2] its typically dull-colored caps and stipes, and the short lamellulae that are regularly inserted among the gills. In contrast, *Russula* species do not exude milky latex when injured and typically have brightly colored caps that contrast sharply with their white gills and stipes, and most species do not have short lamellulae regularly inserted among the gills.

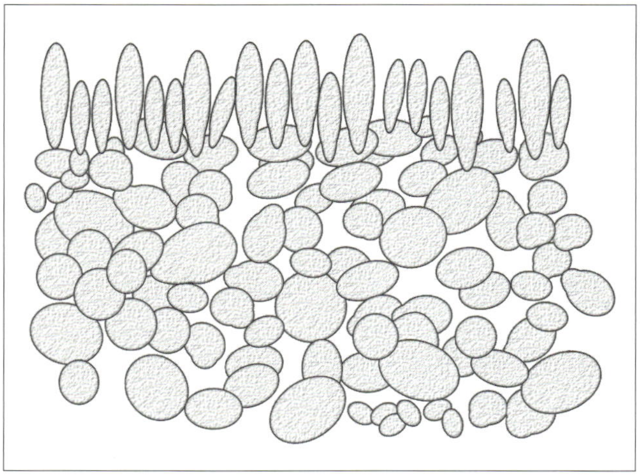

The flesh in the cap and stipe of these mushrooms is composed of loose globular cells called **sphaerocysts** (shown at left) instead of the filamentous and intertwined hyphae typical of other agaricoid mushrooms. This makes them characteristically brittle and causes them to readily shatter or crumble. In *Russula*, sphaerocysts are present in both the cap and the gill trama, making the gills also very brittle. Indeed, another common name for the russulas is "brittle gills." In *Lactarius*, sphaerocysts are present in the cap but not in the gill trama. Sphaerocysts are not present in the resupinate, hydnoid, polyporoid, or coral-like members of the Russulales.

Important characters for distinguishing between the species in both genera include the taste and smell of the flesh, spore ornamentation, types of cystidia, color of the fruitbody, and the spacing and degree of forking of the gills. In *Russula* the reaction to iron salts ($FeSO_4$) and other chemicals, color of the spore print, and bruising of the flesh are also important, as is the "peeling percentage"—the distance from the center the cuticle of the cap can be peeled away. Species complexes are fairly easily determined, but distinguishing between the species is often highly problematic. In *Lactarius*, the color of the freshly exuded latex and whether or not it subsequently changes color can be crucial.

Both genera have edible and toxic species. Many russulas cause mild to severe gastrointestinal poisonings, which can vary idiosyncratically between individuals. This is particularly true of the red-capped complex known as the "Sickeners." An Asian species can induce rhabdomyolysis and has caused several fatalities. Some russulas also concentrate extremely high levels of heavy metals from their environment. *Lactarius* also includes both edible species—several of which are considered choice—and toxic ones, but none are deadly poisonous.

Both *Russula* and *Lactarius* and the several genera recently split off from them (see comments below) are obligately ectomycorrhizal with a variety of hardwoods and conifers. Indian Pipe (*Monotropa uniflora*) forms mycoheterotrophic parasitic associations with several *Russula* species, as described in the introduction.

Molecular studies revealed that several species formerly assigned to *Russula* and others assigned to *Lactarius* actually form a monophyletic clade for which the genus *Multifurca* has been erected. Another group of morphologically typical *Lactarius* species was shown to comprise a third clade for which a previously synonymized genus (*Lactifluus*) has been resurrected. The 20% or so of *Lactarius* species that need to be transferred to *Lactifluus* (a genus proposed by Roussel back in 1806) have been shown molecularly to be in a clade farther away from *Lactarius* than *Lactarius* is from *Russula*. As a result, *Lactarius* was split up into *Multifurca*, its close sister clade with only a very small number of species, and *Lactifluus*. *Russula* is more closely related to *Lactarius* and *Multifurca* than the latter are to *Lactifluus*. This was a better solution than the only other option: expanding the already huge genus *Russula* to also include the 650 or so species presently in *Lactarius* (*sensu lato*).

The genus *Lactifluus* is primarily tropical; however, widely distributed temperate species include *Lactifluus volemus*, *L. hygrophoroides*, *L. corrugis*, *L. glaucenscens*, and *L. piperatus*, all of which occur in our area, and all formerly included in *Lactarius*. The false truffle genus *Zelleromyces* (discussed in the section "Puffballs, Earthstars, Truffles, False Truffles, and Deer Truffles") also belongs in the genus *Lactarius*, but this combination has not yet been formally proposed.[3]

1. Russulaceae, Albatrellaceae, Auriscalpaceae, Bondarzewiaceae, Hericiaceae, Peniophoraceae, and Stereaceae.

2. This feature works well in temperate regions, but tropical *Lactarius* species often do not produce latex, making this trait problematic.

3. A. Verbeken & J. Nuytinck, 2013, Not every milkcap is a *Lactarius*, *Scripta Botanica Belgica* 51:162–168.

Russulas

Emetic Russulas, the "Sickeners": Red-capped and White-stemmed Russulas

Russula silvicola/R. nobilis/R. cessans

A–C, *Russula nobilis*; D–F, *R. silvicola*. (Photos: A, B, D & E, W.H. Light; C, Emily Johnson; F, Teresa Fortenberry.)

***Russula silvicola* Shaffer**, *Beih. Nova Hedwigia* 51:229, Figs. D–F (1975)

***Russula nobilis* Velenovský**, *České Houby* 1:138, Figs. A–C (1920)[1]

***Russula cessans* A. Pearson**, *Naturalist*:101 (1950) [Basidiomycota, Agaricomycotina, Agaricomycetes, *incertae sedis*, Russulales, Russulaceae]

Derivation: *Russula*, Latin, feminine form of *russulus*, "reddish"; *silvicola*, Latin, "sylvan, forest dweller"; *nobilis*, Latin, "noble, high-born"; *cessans*, Latin, "delaying, resting."

Cap: All three species—rose-colored, deep pinkish-red to bright red; 3–11 cm (1¼–4⅓ in) across; convex when young (Figs. B, C, & E), becoming broadly convex to flat, usually with a central depression (Figs. A & D) when mature; surface subviscid or viscid to dry, usually smooth, but occasionally wrinkled, margin smooth to striate (Figs. A, D, & E); all parts very brittle, crumbling when handled (Fig. F); bright red to scarlet; center often darker (Figs. A, C, & D). ***R. silvicola***—fading in age or wet weather to deep pink, reddish-orange, or pale yellowish-pink, often with white or yellowish blotches (Figs. D & E), 3–10 cm (1¼–4 in) across, peeling to three-quarters of its radius or even almost to the center; ***R. nobilis***—3–9 cm (1¼–3⅔ in) across, weakly viscid, peeling to one-third of its radius, often damaged by slugs; ***R. cessans***—crimson to purplish-red, often with darker center, 3–8 cm (1¼–3¼ in), peeling easily.

Flesh: *R. silvicola* and *R. nobilis*—white except for thin layer immediately beneath the cuticle, unchanging on exposure; odor fruity, taste very hot and peppery; ***R. cessans***—as above, but odor not distinctive, and taste mild.

Gills: *R. silvicola*—attached, may be notched or slightly decurrent; white, becoming pale cream, close or crowded; ***R. nobilis***—attached (Fig. C), white (may have greenish tinge), close to crowded, extremely brittle, crumbling when handled; ***R. cessans***—unchanging pale yellow from the outset.

Stipe: *R. silvicola*—white, not bruising, smooth; 4–10 cm tall by 1–3 cm thick (1⅔–4 × ⅕–1¼ in), equal or tapering toward the apex or base; ***R. nobilis* and *R. cessans***—3–5 cm tall by 1–2 cm thick (1¼–2 × ⅕–¾ in), swollen toward base (Fig. C).

Spore print: *R. silvicola* and *R. nobilis*—white; ***R. cessans***—yellow.

Microscopic: Spores of all three species nearly spherical to broadly elliptical, ornamented with warts, with ridges often forming a partial reticulum; amyloid (staining bluish-grey to blackish in Melzer's reagent). Spore sizes broadly overlapping and not distinctive: ***R. silvicola***—6.5–10.7 × 5.7–8.6 µm; ***R. nobilis***—8–11 × 7.5–8.5 µm; ***R. cessans***—8–9 × 7–8 µm.

Occurrence: *R. silvicola*—early summer–midwinter in mixed forests, mycorrhizal with Eastern White Pine (*Pinus strobus*), Virginia Pine (*P. virginiana*), and American Beech (*Fagus grandifolia*); ***R. nobilis***—summer–winter, restricted to American Beech; ***R. cessans***—late fall–early winter, associated primarily with *Pinus strobus* and possibly other pines.

Edibility: *R. silvicola* (Sickener) and *R. nobilis* (Beechwood Sickener)—inedible, toxic but not life-threatening; sesquiterpenes induce severe gastrointestinal distress, with nausea, vomiting, and abdominal cramps (see comments); ***Russula cessans***—reportedly edible, but caution is advised.

Comments: We have often observed Eastern Grey Squirrels (*Sciurus carolinensis*) eating these red-capped, white-stemmed mushrooms, apparently without ill effects, and it is rare to find undamaged caps that have not been nibbled on by these rodents.

A number of other highly variable species that sometimes exhibit red caps and white stems also cannot reliably be distinguished from the species discussed here on the basis of morphological features. The limits of these species are still vague, and it is uncertain how many cryptic (hidden) species are involved. Some of these species are also poisonous, while others are not. And some seem to affect some people adversely, while others can eat them with apparent impunity. Although the toxins that cause gastrointestinal distress reportedly can be inactivated by cooking, consuming these mushrooms is not recommended.

The species in this complex have historically been misidentified as *Russula emetica* (Schaeffer) Persoon (1796), and they are still called that in many field guides and online sources. However, *R. emetica* is a European species found only in sphagnum bogs; it may or may not actually occur in North America.

1. *Russula nobilis* is called *R. mairei* Singer (1929) in several sources. However, *R. mairei* Singer is both a later synonym of *R. nobilis* and an illegitimate name: it is also a homonym of *R. mairei* Peck (1872)—a completely different and valid species in its own right. See online Appendix A and M. Kuo, March 2009, *Russula mariae*, MushroomExpert.Com, http://www.mushroomexpert.com/russula_mariae.html.

Red-capped and Rosy-stemmed Russulas

Russula tenuiceps, *R. pulchra*, and *R. cystidiosa*

A B C

Russula tenuiceps **Kauffman**, *Report Mich. Acad. Sci.* 11:81 (1909)

Russula pulchra **Burlingham**, *Mycologia* 10(2):95 (1918)

Russula cystidiosa **Murrill**, *Bull. Torrey Bot. Club* 67:145 (1940) [Basidiomycota, Agaricomycotina, Agaricomycetes, *incertae sedis*, Russulales, Russulaceae]

Derivation: *Russula*, Latin, feminine form of *russulus*, "reddish"; *tenuiceps*, Latin, "fragile head," referring to the tendency of the cap and stipe to disintegrate when handled; *pulchra*, Latin, "beautiful"; *cystidiosa*, Latin, "replete with cystidia."

Cap: All three species—rose-colored, deep pinkish-red (Fig. C) to bright red; 5–11 cm (2–4⅓ in) across; convex when young, becoming broadly convex to flat, usually with a central depression (Fig. C); dry or sticky when wet.

Flesh: All three species—white, not bruising when cut or handled; *R. tenuiceps* is extremely friable and brittle when handled even for a russula.

Gills: All three species—white when young (Fig. B), aging to cream or pale yellow. ***R. cystidiosa***—attached to barely decurrent, equal, not forked, broad, distant; ***R. tenuiceps***—close, crowded, narrow, mostly unforked, but occasionally forked adjacent to the stipe; ***R. pulchra***—all forked or mostly forked, subdistant, thin.

Stipe: All three species—5–7 cm tall by 1–2 cm thick (2–2¾ × ⅖–⅘ in); white with varying amounts of pink (Fig. A); subequal, smooth or covered with minute hairs formed by caulocysticia.

Spore print: *R. cystidiosa* **and** ***R. pulchra***—creamy; ***R. tenuiceps***—pale yellow, buff, or orangish.

Staining: KOH—cap surface negative, yellowish, or orange, not definitive to species.

Microscopic: *R. cystidiosa*—spores with prominent isolated warts or sometimes also with connecting ridges and reticulated zones; spores 7–12 × 6–10 µm, subspherical to broadly elliptical, with abundant club-shaped pileocystidia and pleurocystidia; ***R. pulchra***—spores 8.5–10 × 7–7.5 µm, egg-shaped to ellipsoid, pileipellis lacking pileocystidia, but with multiseptate end cells; ***R. tenuiceps***—spores 6–9 × 5–7 µm, spherical to egg-shaped, with cylindrical or sub-club-shaped pileocystidia.

Occurrence: Summer–fall, solitary or scattered in hardwoods or mixed woods, widespread and common in southeastern North America; mycorrhizal with oaks (*R. pulchra* and *R. cystidiosa*); in mixed oak and maple woods (*R. tenuiceps*).

Edibility: Unknown.

Comments: The red-capped russulas with varying degrees of rosy-colored stems associated with oaks and other hardwoods are an extremely difficult group to identify. Species descriptions and keys are typically vague and rarely compare the same morphological features or chemical staining characteristics even within a couplet. The microscopic features are also quite variable, making identification to species problematic. It is not at all clear where one taxon ends and another begins.

Russula sanguinea Fries (1838) is a very similar southeastern species that has a bright-red to purplish-red cap that is viscid when wet, a creamy to yellowish spore print, and cream to yellow gills; it is associated with pines (*Pinus* spp.) rather than oaks. It has been misidentified as *R. rosacea* in the literature. See online Appendix A for nomenclatural notes on this taxon. *Russula rubescens* Beardslee (1915) has an initially white (later greyish) stem that bruises quickly or slowly reddish, then black, when injured.

Russula xerampelina (Schaeffer) Fries (1838) has a deep red to wine-colored cap and emits a distinctive "fishy" smell of raw shrimp or crab; the gills, flesh, and stipe slowly turn yellow-brown to brown when bruised or cut and stain green to grey-green with $FeSO_4$. The stipe ranges from pristine white to distinctly rose-tinted.

Blackish-red Russula

Russula atropurpurea

A

B

C

***Russula atropurpurea* (Krombholz) Britzelmayr**, *Bot. Zbl.* 54:99 (1893)

Basionym: *Agaricus atropurpureus* Krombholz, *Naturgetr. Abbild. Beschr. Schwämme* (Prague) 9:6 (1845) [Basidiomycota, Agaricomycotina, Agaricomycetes, *incertae sedis*, Russulales, Russulaceae]

Derivation: *Russula*, Latin, feminine form of *russulus*, "reddish"; *atropurpurea*, Latin, from *atro*, "black-colored," and *purpureus*, "purple," referring to the blackish-purple cap.

Deprecated synonyms: *Russula vinacea* Burlingham; *Russula krombholzii* Shaffer

Cap: 5–10 cm (2–4 in) across, initially convex to broadly convex, becoming nearly flat; smooth, tacky to slimy and shiny when moist, dull when dry; red to purplish-red toward the margins and blackish-red over the disk (Fig. A), this feature distinctive and fairly consistent; cap readily peeling for ½–⅔ of the radius.

Flesh: White, firm, brittle; odor not distinctive or mildly fruity, taste variable, sweet or sour.

Gills: Attached, occasionally notched, close to subdistant, moderately broad (Fig. C); white to yellowish-white; developing rusty-brown to dark brown spots with age (Fig. C, inset).

Stipe: Robust, 5–10 cm tall by 1.5–4.5 cm thick (2–4 × ⅔–1¾ in), subequal, but typically enlarged or swollen at the base (Fig. B); white, becoming ashy-grey at the base and developing brown spots with age; solid, dry, smooth to minutely velvety, usually with faint longitudinal wrinkles.

Spore print: Yellowish-white.

Chemical staining: All tissues nonreactive in KOH, NH_4OH, NaOH, and Melzer's reagent.

Microscopic: Spores 7–10 × 5–8 µm, subglobose to broadly oval; clear, with prominent warts (up to 1.4 µm high) and interconnecting ridges forming a partial reticulum; amyloid (staining blackish-brown in Melzer's reagent).

Occurrence: Ectomycorrhizal with oaks (*Quercus* spp.); also found with beeches (*Fagus* spp.) and pines (*Pinus* spp.). Solitary or in scattered groups on the ground in deciduous and mixed woods; widespread and common throughout eastern North America; in the Southeast this is the first russula to appear in the late spring (late April/early May through November); reported but not confirmed from the Pacific Northwest; widespread in Europe, British Isles, Iceland.

Edibility: Edible.

Comments: The purplish-red cap with its distinctive blackish-red center and early appearance under deciduous trees distinguish this species from all other russulas.

Russula cessans A. Pearson of the "Sickener" complex also has a reddish cap with a dark center. However, it appears under pines, especially Eastern White Pine (*Pinus strobus*), from late fall through midwinter, and its cap stains yellow with KOH (that of *R. atropurpurea* is nonreactive).

White Russula

Russula albidula complex

A–C, *Russula albidula*; D, *R. brevipes*. (Photos: A–C, W. Light; D, Joseph O'Brien, USDA Forest Service, Bugwood.org.)

***Russula albidula* Peck**, *Bull. Torrey Bot. Club* 25:370 (1898) [Basidiomycota, Agaricomycotina, Agaricomycetes, *incertae sedis*, Russulales, Russulaceae]

Derivation: *Russula*, Latin, feminine form of *russulus*, "reddish"; *albidula*, Latin, feminine adjectival ending of *albidulus*, "whitish."

Cap: 3–8 cm (1¼–3¼ in) across, convex at first, becoming flat or shallowly depressed (Fig. A); white, often with tinge of yellowish or buff at center; smooth, somewhat sticky when wet, otherwise dry, cuticle easily peeled off; typically dirty and covered with debris in the field (Fig. A).

Flesh: White, fragile, very bitter-tasting.

Gills: Variously adnate, adnexed (Fig. C), or free, close, brittle; white or cream-colored, yellowing with age, often forking adjacent to stipe.

Stipe: 2.5–7 cm tall by 1–2.5 cm thick (1–2¾ × ½–1 in), smooth, white, dry, subequal or expanded downward (Fig. B), becoming spongy or hollow with age.

Spore print: Pale yellow.

Microscopic: Spores 7–10 × 5.8–7 µm, subglobose to ovoid; thin-walled, warted, with partially reticulate, amyloid ridges.

Occurrence: Alone or in small groups in mixed deciduous and conifer woods in poor soil along trails and in roadcuts, never in large numbers; often associated with oaks (*Quercus* spp.); widespread in eastern North America, summer–early winter.

Edibility: Edible, but very bitter.

Comments: Another all-white or mostly white russula is the Short-stemmed Russula, *R. brevipes* Peck (Fig. D), which is distinguished by its larger size (7–30 cm [3–12 in] across) and, in maturity, concave, shallowly vase-shaped cap, relatively short, thick, stubby stipe, and white to cream-colored spore prints. Unlike *R. albidula*, the gills of *R. brevipes* are unforked, have intervenose ridges (having vein-like ridges between them), and develop a faint yellowish or tan tinge with age (see Fig. D). The cuticle does not easily peel off the cap. *Russula brevipes* is not conspecific with *R. delica* Fries, the latter being a misapplied European name erroneously used by some American authors.

Quilted Green Russula Complex

Russula virescens/R. crustosa/R. parvovirescens

A

B

C

D

E

Photos: A, Teresa Fortenberry; B–E, W. Light.

***Russula virescens* (Schaeffer) Fries**, *Anteckn. Sver. Ätl. Svamp.*:50 (1836)

Basionym: *Agaricus virescens* Schaeffer, *Fung. Bavar. Palat.* 4:40 (1774)

***Russula crustosa* Peck**, *Ann. Rep. Reg. N.Y. St. Mus.* 39:41 (1886)

***Russula parvovirescens* Buyck, Mitchell & Parrent**, *Mycologia* 98(4):612–615 (2006)
[Basidiomycota, Agaricomycotina, Agaricomycetes, *incertae sedis*, Russulales, Russulaceae]

Derivation: *Russula*, Latin, from *russus*, "red," because many species have a red cap; *virescens*, Latin, "becoming green"; *crustosa*, Latin, "inlaid," "encrusted"; *parvovirescens*, Latin, from *parvo*, "small," and *virescens*, "a little virescens."

Cap: *R. virescens* and *R. crustosa*—5–12.5 cm (2–5 in) across, occasionally up to 15 cm (6 in); ***R. parvovirescens***—4–8 cm (1½–3¼ in), sometimes to 10 cm (4 in). Cap convex, broadly convex to flat, with central depression (Figs. A–C & E); surface dry and dull, never viscid (except in *R. crustosa* when wet), soon cracking up into crustose patches, outer cuticle readily separable for up to half the radius of the cap, margin slightly to distinctly striate. Color quite variable, but all three species developing large blotches (areolae) of darker color and rougher texture (Figs. A, B, & E). ***R. virescens***—usually a soft, pale olive-green to grass-green, margin smooth or striate; ***R. crustosa***—usually yellowish-brown, but sometimes dull yellowish-green, faintly olive, or a blend of these colors; ***R. parvovirescens***—bluish-green to greyish-green or dark olive-green, soon becoming disrupted into large crustose patches that expose the underlying greyish- to bluish-green or pale brownish-grey surface, becoming rust colored with exposure, often entirely pale brown at the center.

Flesh: All species—white, thick, firm when young, brittle when mature; taste mild to acrid, odorless.

Gills: All species—white to pale cream, attached to nearly free, close to subdistant, some forking, no interveins (Fig. D).

Stipe: *R. virescens* and *R. crustosa*—creamy white (Figs. A & D), soon becoming hollow and brittle; 3–9 cm tall by 1.5–2 cm thick (1¼–3⅔ × ⅔–¾ in); ***R. parvovirescens***—3–6 cm tall by 0.8–1.3 mm thick (1¼–2½ × ⅓–½ in).

Spore print: *R. virescens* and *R. parvovirescens*—white to pale cream; ***R. crustosa***—pale yellow or buff.[1]

Chemical staining: *R. virescens*—cap negative in KOH, stipe pink in $FeSO_4$; ***R. crustosa***—cap negative in KOH, stipe negative or pink in $FeSO_4$ (often faint and delayed); ***R. parvovirescens***—negative in KOH; stipe flesh brownish-orange, but stipe surface nonreactive in $FESO_4$.

Microscopic: Spores: 6–9(–10) × 5–7 µm, elliptical, clear, surface with warts, elliptical to subglobose; warts nearly absent to moderately dense to dense, about 0.5 µm high, forming occasional reticulations, amyloid; the three species nearly indistinguishable in this regard. ***R. virescens* and *R. crustosa***—pileipellis a cutis (uppermost hyphal layer in the cap) overlaid with crustose patches of pseudoparenchyma (interwoven hyphae resembling a tissue) consisting of chain-like cells that diminish sharply in width from base to tip, with an elongate, tapered, awl-like terminal cell; ***R. parvovirescens***—chained cells in pileipellis a cutis not diminishing noticeably in width and not awl-shaped.

Occurrence: Widespread in eastern North America, summer–fall. ***R. virescens***—mycorrhizal with hardwoods, solitary or gregarious in both deciduous and mixed forests; ***R. crustosa***—mycorrhizal with oaks and possibly other hardwoods; ***R. parvovirescens***—mycorrhizal with oaks, other hardwoods, and conifers.

Edibility: Edible, all three species considered choice. Elsewhere *R. virescens* and greenish *R. crustosa* could possibly be confused with the deadly poisonous *Amanita phalloides*, but that lethal mushroom has thus far not been reported in the Southeast.[2]

Comments: *Russula crustosa* typically looks like a yellowish-brown version of *R. virescens* and is usually not likely to be mistaken for it. However, greenish specimens are regularly encountered, which can cause confusion, but the sticky cap when moist and the buff (not white) spore print distinguish this morph from *R. virescens*. *Russula parvovirescens* is significantly smaller than the other two species.

Several similar greenish-hued species are more uniformly colored and lack the distinctive large patches seen in the foregoing three taxa (although they may be obscurely or minutely areolate). *Russula subgraminicolor* Murrill is the size of *R. parvovirescens* (6.5–8 cm [2½–3¼ in]) and is uniformly bluish grass-green; its spore print is pale straw-yellow.

Russula aeruginea Fries is 5–8 cm (up to 12 cm) across (2–3¼, up to 4¾ in), the smooth, noncracking cap is greyish olive-green to moderately yellow-green and somewhat sticky, the gills are pale cream-white initially, later becoming pale yellow and often forking near the base. The spore print is dark cream. The cap stains orange in KOH, and flesh and stem surface become slowly pink in iron salts.

Russula variata Banning is 5–15 cm (2–6 in) across and is typically dull olivaceous with a central purple tinge that fades toward the margins. The conspicuously and repeatedly forking gills are flexible and not brittle. The cap is negative or orangish in KOH; the flesh and stipe are negative in $FESO_4$.

Russula cyanoxantha (Schaeffer) Fries is very similar to *R. variata*. The cap color ranges from violet-blue to deep green to even yellow and often has fine radial veins and greasy, flexible gills that fork only near the stipe. The cap stains delayed orange in KOH, and the stipe is negative or very slightly greenish in $FeSO_4$. The flexible, rather than brittle, gills immediately separate *R. variata* and *R. cyanoxantha* from the other species treated here (and from most other *Russula* species as well).

This complex of greenish *Russula* species with crustose (bran-like) patches (areolae) on the cap includes more than a dozen species (several undescribed) in eastern North America that are not easily distinguished from each other and that have often been confused in the literature.[3] Except for the form of the chained cells in the areolae, the commonly used microscopic features such as spore size and ornamentation and the form of hyphal extremities are much too variable to separate out the species in the *Russula virescens-crustosa* complex with any confidence.[4] Likewise, one cannot fully rely on macroscopic characters like the color and size of surface areolations (patches of color and/or texture) and marginal striations on the cap or even the color of the spore print.[5,6]

1. This is problematic. The color of the spore print for *R. crustosa* is variously reported as (1) both white and pale yellow and (2) exclusively either white–pale cream or pale yellow; in one work it was reported as white–pale cream in one place and as pale yellow in another. In several of these species the color of the spore print is described as varying from white to pale yellow.

2. *Amanita phalloides*, a European species, was separately introduced into North America in two different regions: (1) the West Coast from California to British Columbia and (2) the East Coast from Maryland to Maine (B.E. Wolfe et al., 2010, *New Phytol.* 185[3]:803–816).

3. B. Buyck, D. Mitchell, & J. Parrent, 2006, *Mycologia* 98(4): 612–615.

4. B. Buyck, J. Parrent, & R. Vilgalys, 2005, *Inoculum* 56(4):11.

5. Michael Kuo's humorous (but accurate!) description of the current state of *Russula* taxonomy cannot be equaled, let alone surpassed; see M. Kuo, March 2009, The genus *Russula*, MushroomExpert.Com, http://www.mushroomexpert.com/russula.html.

6. The macroscopic and microscopic characteristics given in Bart Buyck's provisional key to the *Russula* subsection *Virescentinae* in the United States are equally variable and nonspecific.

Milkcaps

Celandine Milky

Lactarius chelidonium

A B C D E F G H

Lactarius chelidonium Peck, *Ann. Rep. N.Y. St. Mus.* 24:74 (1872) [Basidiomycota, Agaricomycetes, Russulales, Russulaceae]

Derivation: *Lactarius*, Latin, "milky"; *chelidonium*, Latin, adjectival form of *chelidonia*,[1] "celandine," referring to the pale yellow latex exuded by this species.

Cap: 3–12 cm (1¼–4¾ in) across, initially convex, becoming planoconvex, often centrally depressed (Fig. A); initially viscid, soon becoming dry; color highly variable, azure-blue when young, but more commonly various shades of yellowish-grey (Figs. A & E), yellowish-brown, or dull orange, often mottled (Fig. A), often with bluish-green tints with age; generally azonate, but may have faint darker zones near margin (Fig. A).

Flesh: Yellowish, becoming blue-green with age; cut flesh bruising reddish, slowly turning to blue (Fig. B, arrow).

Gills: Attached to slightly decurrent (Figs. D & E), crowded, often forked near the stipe and occasionally also near the margin (Figs. D & H); color highly variable, at first dingy yellow (Figs. B, C, & E), becoming dull orange to coppery-brown (Fig. D) or vinaceous-lavender (Fig. H), bruising green or bluish-green in old specimens (Fig. G).

Stipe: 2–6 cm tall by 1–3 cm thick (¾–2½ × ½–1¼ in), subequal, smooth, tough, hollow, concolorous with the cap, sometimes stained bluish-green with age.

Latex: Scant, often not visible in older specimens; pale yellow when first exuded, quickly turning yellowish-brown (Fig. C, arrow).

Microscopic: Spores 7–9 × 5–7 µm, oval to subellipsoid, with warts and ridges up to 1 µm high forming partial reticulum (Fig. F), clear, amyloid in Melzer's reagent.

Occurrence: Mycorrhizal with pines (*Pinus* spp.), scattered or in groups on the ground, September–November.

Edibility: Considered edible.[2,3]

Comments: The scanty, pale yellow latex that quickly turns yellowish-brown is diagnostic for this species. Its closest relatives in the "Lactarius deliciosus" complex (*sensu* American authors) are distinguished by carrot-orange caps and gills and moderately abundant orange-red latex, which does not change color. However, as one major recent DNA study has concluded, "None of the North American specimens identified as *L. deliciosus* are conspecific with the European *L. deliciosus.*"[4]

Many online sources and field guides separate *Lactarius chelidonium* into *L. chelidonium* var. *chelidonium* Peck and *L. chelidonium* var. *chelidonioides* (A.H. Smith) Hesler & A.H. Smith. Both varieties are reported to be widespread in eastern North America, and both are mycorrhizal with pines. However, the literature descriptions are confused, and the boundaries between these two taxa remain obscure. *Index Fungorum*, *Species Fungorum*, and *MycoBank* have synonymized these two varieties, and we think it best to adopt that view.

1. Originally from the Greek word *chelidóni*, "swallow," because of an ancient belief that this bird used the sap of the yellow-flowered Greater Celandine (a poppy) (*Chelidonium majus*) to give sight to its young.
2. *Fide* C. H. Peck, 1900, Edible fungi of New York, *Mem. N.Y. State Museum* 3(4):151.
3. L.R. Hesler & A.H. Smith, 1979, *North American Species of Lactarius* (Ann Arbor: University of Michigan Press), 19.
4. J. Nuytinck & A. Verbeken, 2007:828. Worldwide phylogeny of *Lactarius* section *Deliciosi* inferred from ITS and glyceraldehyde-3-phosphate dehydrogenase gene sequences. *Mycologia* 99(6):820–832. *Lactarius deliciosus* (Linnaeus *ex* Fries) Gray is a European species that does not occur in North America; "Lactarius deliciosus" has traditionally been and continues to be misapplied to a clade of several closely related and poorly understood taxa found in the United States and Canada. Unfortunately, as is too often the case with North American mushrooms, these widespread and fairly common species have no validly published scientific names, and the names currently applied to them are not valid.

Dispersed Milkcaps

Lactarius dispersus and *L. speciosus*

A

B

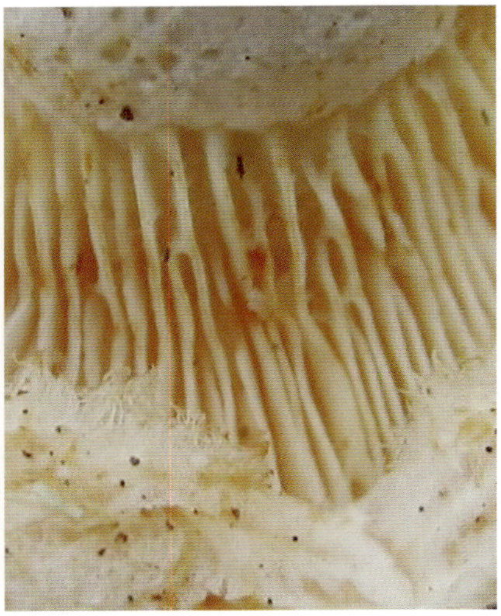

C

Lactarius dispersus **Hesler & A.H. Smith**, *North American Species of Lactarius* (Ann Arbor):225–226 (1979)

Lactarius speciosus **Burlingham**, *Mem. Torrey Bot. Club* 14:34 (1908)
[Basidiomycota, Agaricomycetes, Russulales, Russulaceae]

Derivation: *Lactarius*, Latin, "milky"; *dispersus*, Latin, "scattered," "dispersed."

Cap: 3–10 cm (1¼–4 in) wide, convex at first, becoming broadly flattened and funnel-shaped (Fig. B), with strongly inrolled margin; margin with conspicuous cottony edge when young (Fig. C); cap conspicuously zonate, with various shades of yellowish-tan and/or caramel alternating with creamy-white or pinkish-tan and cinnamon-brown (Figs. A & B), becoming darker and more sharply zonate with age; viscid, often with adhering leaves and other debris.

Gills: Decurrent, cream to pinkish-buff, close to subdistant, moderately broad, frequently forked to anastomosing, with wavy edges. Reported as bruising dingy to dull lilac when cut, but our specimens did not bruise at all.

Flesh: Thick, cream-colored, not bruising when cut.

Latex: Sparse, white, unchanging. Reported as staining flesh dull lilac; this not seen in our specimens.

Stipe: 4–6 cm (1⅔–2½ in) tall by 1–1.7 cm (½–⅔ in) thick (*L. dispersus*); 3–5 cm (1¼–4 in) tall by 1–2.5 cm (½–1 in) thick (*L. speciosus*).

Staining: Flesh staining strongly and immediately reddish-brown with Melzer's reagent; negative in KOH, which reverses Melzer's reaction.

Spore print: Yellowish (*L. dispersus*) or white (*L. speciosus*).

Microscopic: Spores 7.5–10 × 6.0–7.5 µm, broadly ellipsoid, surface with small warts and ridges, not reticulated (*L. dispersus*) or 10.0–13.5 × 9–11 µm, surface with warts and ridges forming partial reticulum (*L. speciosus*).

Occurrence: Singly or in groups on ground in mixed conifer and hardwood forests; August–October (*L. dispersus*); July–December (*L. speciosus*).

Edibility: Unknown.

Comments: These two very similar species are distinguished primarily by their spore prints and their spores. *Lactarius dispersus* has a yellowish spore print, and the spores are smaller (7.5–10 × 6.0–7.5 µm), with small, unconnected warts and ridges. *Lactarius speciosus* has a white spore print, and the spores are larger (10.0–13.5 ×

9–11 µm), with warts and ridges forming a partial reticular network.

In the nonstaining properties of its gills and latex, our specimens resembled the similar *Lactarius yazooensis* Hesler & A.H. Smith (1979), which is widely distributed in grassy areas beneath oaks along the Gulf Coast. However, the gills in *L. yazooensis* are reported to range from pale vinaceous cinnamon to dull vinaceous or light pinkish-brown and are rarely forked, and the cap margin is not cottony or bearded. The spores are similar to those of *L. dispersus*.

Fleecy Milkcap

Lactifluus subvellereus

A B C

***Lactifluus subvellereus* (Peck) Nuytinck**, in Verbeken, Nuytinck, & Buyck, *Mycotaxon* 118:451 (2011)

Basionym: *Lactarius subvellereus* Peck, *Bull. Torrey Bot. Club* 25:369 (1898)

Synonym: *Lactarius subvellereus* var. *subdistans* Hesler & A.H. Smith, *North American Species of Lactarius* (Ann Arbor):203 (1979)
[Basidiomycota, Agaricomycetes, Russulales, Russulaceae]

Derivation: *Lactifluus*, Latin, from *lactis*, "milk," and *fluere*, "to flow"; *subvellereus*, Latin, from *vellus*, "fleece," "almost fleecy," referring to the velvety texture of the cap.

Cap: 4–15 cm (1¼–6 in) across, broadly convex at first, becoming flat (Fig. C) to centrally depressed to funnel-shaped (Fig. A); margin initially strongly incurved, expanded and uplifted with age, often plicate or corrugated (Figs. A–C); surface dry, finely velvety, white to dingy greyish-white, often with greyish or yellowish tinges, azonate.

Latex: White to cream, slowly turning pale yellowish when exposed (Fig. B); staining gills brownish to pinkish-brown; no odor, taste very bitter.

Flesh: Firm, thin, white, turning dingy-white to yellowish when cut; no odor, taste bitter.

Gills: Initially white to cream (Fig. B), yellowish with age (Fig. C); attached to subdecurrent, subdistant to distant, with numerous lamellulae (truncated gills) of varying lengths, with some forking; often beaded with clear liquid droplets, bruising pinkish to brownish.

Stipe: 2–5 cm tall by 1–3.5 cm thick (¾–2 × ½–1½ in), short, subequal or tapering toward the base (Figs. B & C); dry, finely velvety, concolorous with cap.

Spore print: White.

Chemical staining: KOH—cap pinkish to pinkish-purple; flesh and gills orange.

Microscopic: Spores 7.5–9 × 5–7 µm, ellipsoid to broadly ellipsoid, with low, isolated, often barely visible warts, clear, amyloid.

Occurrence: Mycorrhizal with oaks (*Quercus* spp.), alone or scattered on the ground in deciduous and mixed woods; widespread and fairly common in eastern North America; late summer.

Edibility: Controversial; edible, but some individuals cannot tolerate it.

Comments: A northern form whose distributional range is not fully known has white latex that does not turn yellow when exposed, and the cap surface is more noticeably velvety. It may or may not occur in our area.

Several other all-white *Lactifluus* species with very acrid latex closely resemble *L. subvellereus*. However, they all exhibit densely to extremely crowded (not distant), frequently (not occasionally) forking gills and have stipes with noticeably longer length-to-width ratios. And none of them have caps with crenulate margins.

Lactifluus deceptivus (Peck) Kuntze has an initially smooth cap surface that breaks up into coarse yellowish-brown scales with age, the margin has a distinctive soft, cottony roll of tissue, the stipe is up to twice as tall with the same thickness, and the cap stains reddish-cinnamon with KOH. The latex is unchanging when exposed but may stain the tissues brown.

The gills of both *Lactifluus glaucescens* (Crossland) Verbeken and *Lactarius piperatus* (Linnaeus) Roussel are extremely crowded to the point that they almost appear smooth until examined closely. The cap of *L. glaucescens* reacts either negatively or slowly stains pale orangish with KOH, and the white latex dries pale bluish-green. The latex of *L. piperatus* dries olive to pastel green.

All-white milkcap species are easily distinguished from all-white russulas by their latex. Russulas never exude latex of any kind.

Glaucous Milkcap

Lactifluus glaucescens

A B C D

***Lactifluus glaucescens* (Crossland) Verbeken**, *Mycotaxon* 120:449 (2012)

Basionym: *Lactarius glaucescens* Crossland, *Naturalist*, London:5 (1900)

Synonyms: *Lactarius piperatus* var. *glaucescens* (Crossl.) Hesler & A.H. Smith, *North American Species of Lactarius* (Ann Arbor):186 (1979)
Lactarius pergamenus sensu auct.; *fide* N.W. Legon & A. Henrici, Checklist of Basidiomycota of Great Britain and Ireland (Kew: Royal Botanic Gardens) (2005); http://www.basidiochecklist.info/
[Basidiomycota, Agaricomycetes, Russulales, Russulaceae]

Derivation: *Lactifluus*, Latin, from *lactis*, "milk," and *fluere*, "to flow"; *glaucescens*, Latin, from *glaucus*, "bluish-grey," "greenish-grey," referring to the slowly staining flesh and latex.

Cap: 4–12 cm (1⅔–4¾ in) across, broadly convex to nearly flat (Fig. A), with a central depression (Fig. B); margin entire, initially inrolled, expanding and uplifted with age (Figs. A & D); surface dry, smooth, white to cream, becoming dull yellowish-brown with age, especially in the center (Fig. C).

Flesh: White to pale cream, unchanging when exposed or slowly staining pale bluish-green; thick, hard; odor not noticeable or slightly pungent, taste very bitter.

Gills: Subdecurrent to decurrent, narrow, *extremely crowded* (Figs. B & D), so much so that they appear almost like a smooth surface (Fig. A), margins entire, often forked, with numerous lamellulae (short gills); white to cream.

Stipe: 3–10 cm tall by 1–3 cm thick (1¼–4 × ½–1¼ in), tapering to base (Fig. D), but occasionally subequal (Fig. A), straight or curved; smooth, dry, solid; concolorous with the cap or paler.

Spore print: White to creamy.

Latex: Copious, white, not changing color (Fig. B, arrow) but slowly drying pale bluish-green; slowly stains white paper pale yellow; taste extremely acrid.

Chemical staining: **KOH**—cap surface negative or slowly staining pale orangish; latex staining yellow to orange.

Microscopic: Spores 6–9 × 5–7 μm, broadly ellipsoid, faintly ornamented with inconspicuous warts and short ridges ~0.2 μm high; clear, faintly amyloid.

Occurrence: Mycorrhizal with hardwoods, especially oaks (*Quercus* spp.); alone, scattered, or in groups on the ground; widespread and common in eastern North America, summer–winter.

Edibility: Inedible, toxic: can cause severe vomiting, diarrhea, abdominal cramps, and other gastrointestinal symptoms.[1]

Comments: Several other all-white *Lactifluus* species with very acrid and initially white latex closely resemble *L. glaucescens. Lactifluus piperatus* (Linnaeus) Persoon is nearly identical,[2] but the white latex does not change color when dry, and the cap stains pale magenta (not orangish) with KOH. *Lactarius subvellereus* var. *subdistans* Hesler & A.H. Smith is distinguished from *L. glaucescens* by its subdistant to distant gills, white latex that dries yellowish, and a cap that stains pinkish to pinkish-purple with KOH. *Lactarius deceptivus* Peck has close to subdistant (not extremely crowded) gills, and the cap stains reddish-cinnamon with KOH.

Lactifluus glaucescens was formerly in the genus *Lactarius* Persoon, where many authors continue to place it. However, for the reasons outlined in the introduction to this section, this species and the closely related *L. piperatus* (Linnaeus) Roussel were recombined in the genus *Lactifluus* Roussel, where they had previously been placed by Kuntze in 1891. *Index Fungorum* and *Species Fungorum* retain this species in the genus *Lactarius* Persoon even though it is the type species of the genus *Lactifluus* Roussel, which both indices recognize for other species.

1. C.E. Becker et al., 1976, *Western Journal of Medicine* 125:103, table 2. The nature of this nonlethal but supposedly severe poisoning is unknown and may even be apocryphal. This mushroom is widely cited as poisonous in the mycology and medical toxicology literature, but no definitive citation or specific example is ever given. These events may represent idiosyncratic reactions, which are quite common with mushrooms of all kinds.

2. *Lactifluus glaucescens* is regarded by many authors as a variety of *L. piperatus*.

Tawny, Hygrophorus, and Corrugated Milkcaps

Lactifluus volemus, *L. hygrophoroides*, and *L. corrugis*

A–F, H, *Lactifluus volemus*; G & I, *L. corrugis*; J & K, *L. hygrophoroides*. (Photos: A, Teresa Fortenberry; B–H, J & K, W. Light; I, Dan Molter.)

***Lactifluus volemus* (Fries) Kuntze**, *Rev. Gen. Pl.* (Leipzig) 2:857 (1891)

Basionym: *Agaricus volemus* Fr., *Syst. Mycol.* (Lundae) 1:69 (1821)

***Lactifluus hygrophoroides* (Berkeley & M.A. Curtis) Kuntze**, *Rev. Gen. Pl.* (Leipzig) 2:857 (1891)

Basionym: *Lactarius hygrophoroides* Berkeley & M.A. Curtis, *Ann. Mag. Nat. Hist.*, ser. 3, 4:293 (1859)

Misapplied name: *Lactarius hygrophoroides* var. *rugatus* (Kuehner & Romagnesi) Hesler & A.H. Smith, *North American Species of Lactarius* (Ann Arbor):175 (1979)[1]

Lactifluus corrugis (Peck) Kuntze, *Rev. Gen. Pl.* (Leipzig) 2:856 (1891)

Basionym: *Lactarius corrugis* Peck, *Ann. Rep. N.Y. St. Mus. Nat. Hist.* 32:31 (1880) [1879]
[Basidiomycota, Agaricomycetes, Russulales, Russulaceae]

Derivation: *Lactifluus*, Latin, from *lactis*, "milk," and *fluere*, "to flow"; *volemus*, Latin, from *vola*, "hollow," meaning "hollow of the hand," referring to the copious output of latex, sufficient to fill one's cupped hand; *hygrophoroides*, Greek, from *hugros*, "wet" (*hygro-* is a combining form meaning "moisture"), *-phoros*, "bearer," and *-oides*, "similar to," referring to the resemblance of its distant gills to those of the genus *Hygrophorus*; *corrugis*, Latin, "wrinkled."

Cap: All three species—broadly convex, becoming flat with depressed center to shallowly funnel-shaped with age; margin initially inrolled, uplifted with maturity; surface dry, smooth to somewhat velvety, becoming finely wrinkled with age. ***L. volemus***—5–12 cm (2–4¾ in) across; color highly variable: light yellow-brown (Fig. D), orange-brown (Fig. A), cinnamon-brown, to brownish-red (Figs. B & C), or somewhat lighter or darker; generally azonate or with weak concentric zones of color (Fig. C). ***L. corrugis***—4–20 cm (1⅔–8 in) across; *cap concentrically wrinkled or corrugated toward the margin*; usually dark brownish- to purplish-red (Fig. I); often with a whitish, felt-like bloom when young. ***L. hygrophoroides***—3–10 cm (1¼–4 in) across; dry, finely velvety, margin entire, evenly colored pale yellow-orange (Figs. J & K) to dull orange or cinnamon-orange.

Flesh: *L. volemus* and *L. corrugis*—white, bruising brown when injured, odor fishy when mature, taste mild; ***L. hygrophoroides***—white, unchanging, no distinctive odor, taste mild.

Gills: *L. volemus*—attached to slightly decurrent, close, frequently forked, white to cream (Figs. E & H), bruising brown (Fig. F); ***L. corrugis***—attached to slightly decurrent, close to crowded, occasionally forked, pale ochre, cinnamon-brown to pale golden-brown (Figs. G & I), bruising darker brown; ***L. hygrophoroides***—attached to slightly decurrent, *very widely spaced with cross-veins*, nonforking; cream to yellowish buff (Fig. J).

Stipe: *L. volemus*—5–11.5 cm tall by 0.5–2.5 cm thick (2–4⅔ × ¼–1 in), subequal (Figs. B & H) or tapering toward the base; solid, dry, smooth to finely velvety; dull orange-brown, generally paler than the cap; ***L. corrugis***—4–11 cm tall by 1.5–3 cm thick (1⅔–4½ × ⅔–1¼ in), equal (Fig. I); solid, dry, smooth to finely velvety; concolorous with but paler than the cap, with whitish to greyish felt-like bloom when young; ***L. hygrophoroides***—3–5 cm tall by 0.5–2 cm thick (1¼–2 × ¼–¾ in), equal or tapering toward the base; solid, dry, smooth or finely velvety; concolorous with the cap, but somewhat paler.

Spore print: All three species—white.

Latex: *L. volemus* and *L. corrugis*—copious when fresh, white (Figs. F & I, arrows), staining gills and flesh brown, taste mild (Fig. F); ***L. hygrophoroides***—white, unchanging, not staining tissue, taste mild.

Chemical staining: All three species—cap olive with KOH.

Microscopic: *L. volemus*—spores roughly 7–10 µm, globose to subglobose, clear, prominent warts and ridges up to 1 µm high forming complete reticulum, amyloid; ***L. corrugis***—spores roughly 9–11 µm, globose to subglobose, clear, partially reticulate, ornamentation up to 0.8 µm high, amyloid; ***L. hygrophoroides***—spores (7–)7.5–9(–10.5) × (5.5–)6–7.5 µm, elliptical, clear, with scattered warts and short ridges not forming reticulum, ornamentation <0.5 µm high, amyloid.

Occurrence: *L. volemus*—solitary or scattered on ground under hardwoods and in mixed woods, especially with oaks (*Quercus* spp.), widespread and common in eastern North America, summer; Europe, East Asia; ***L. corrugis***—solitary or scattered on ground under hardwoods and in mixed woods, especially with oaks (*Quercus* spp.), widespread and common in eastern North America, East Asia, summer–late fall; ***L. hygrophoroides***—solitary or scattered on ground under hardwoods, eastern North America, occasional, summer–late fall.

Edibility: Edible.

Comments: The Tawny Milkcap, *Lactifluus volemus*, and Corrugated Milkcap, *L. corrugis*, have close to crowded gills without cross-veins, some of which are forked, and the latex stains the flesh and gills brown. The gills of *L. volemus* are white to cream, are frequently forked, and bruise brown; those of *L. corrugis* are pale ochre to cinnamon-brown, are occasionally forked, and bruise darker brown. *Lactifluus volemus* has a yellow-brown to brownish-red, general azonate cap; that of *L. corrugis* is dark brown to wine-colored and concentrically wrinkled or corrugated. The gills of the Hygrophorus Milkcap, *L. hygrophoroides*, are creamy-white to pale yellow-ochre, *very widely spaced, intervenose (with cross-veins), and nonforking*, and the latex does not discolor the flesh or gills.

These three species were formerly in the genus *Lactarius* Persoon, where many authors continue to place them. However, for the reasons outlined in the introduction to this section, these three taxa were recombined in the genus *Lactifluus* Roussel, where they had previously been placed by Kuntze in 1891.

1. G. Lalli & G. Paccioni (1992, *Mycotaxon* 44[1]:166) elevated *Lactarius hygrophoroides* var. *rugatus* (Kuehner & Romagnesi) Hesler & A.H. Smith to full species status, *Lactarius rugatus* Kuehner & Romagnesi (1953), and eliminated all other varieties, merging them into a single taxon with variable smell, latex, and flesh color change. *Lactarius rugatus* appears to be restricted to the Mediterranean region.

Indigo Milky, Indigo Milkcap

Lactarius indigo

B

A

C

Photos: A & B, Teresa Fortenberry; C, Alan Rockefeller.

***Lactarius indigo* (Schweinitz) Fries**, *Epicr. Syst. Mycol.* (Upsaliae):341 (1838) [1836–1838]

Basionym: *Agaricus indigo* Schweinitz, *Schr. Naturf. Ges. Leipzig* 1:87 (1822)
[Basidiomycota, Agaricomycetes, Russulales, Russulaceae]

Derivation: *Lactarius*, Latin, "milky"; *indigo*, Latin, *indicum* (from Greek *indikon*), "blue dye from India."

Cap: Surface smooth, viscid when fresh; 5–15 cm (2–6 in) across, initially convex, becoming flattened, centrally depressed and broadly funnel-shaped with age (Figs. A & C), margin at first inrolled, later uplifted and flared (Figs. B & C); deep blue when young, becoming greyish- or silvery-blue with age, azonate or zonate, with concentric bands of darker and lighter color, often with brownish blotches in age (Fig. A).

Flesh: Brittle, whitish, bruising indigo blue and then greenish when exposed to air; taste mild to slightly bitter, odor not distinctive.

Gills: Attached (Fig. B) to barely decurrent (Fig. C), close, with three to four tiers of lamellulae (short gills), occasionally forked; dark blue when young, becoming blue-grey with age and tinged yellowish splotches; bruising greenish (Fig. B).

Stipe: 2–8 cm tall by 1–2.5 cm thick (¾–3¼ × ½–1 in), equal or tapered toward the base; initially solid, becoming hollow and brittle when mature; initially dark blue, fading to silver-blue with age, slimy when young, but soon becoming dry.

Latex: Deep indigo blue (Fig. C), slowly turning dark green when exposed to air; production sparse, sometimes nonexistent in older specimens.

Spore print: Cream.

Microscopic: Spores 7–9 × 5.5–7.5 µm, broadly elliptical to subglobose, clear, with warts and ridges forming a partial or complete reticulum, amyloid in Melzer's reagent; hyphae on cap surface gelatinous and lying flat.

Occurrence: Ectomycorrhizal with oaks (*Quercus* spp.) and other hardwoods and conifers (*Pinus* spp.); solitary or scattered on ground; widespread in eastern and central North America, sporadic to common.

Edibility: Edible, considered choice by some, mediocre by others.

Comments: In our area *Lactarius indigo* could be confused with either *Lactarius chelidonium* Peck or *L. paradoxus* Beardslee & Burlingham, both of which have pale blue caps and flesh but never blue gills. *Lactarius chelidonium* has a blue-tinged cap and flesh when young, but the cap becomes muddy orange or dull reddish-brown with age. The gills are dingy yellow to yellowish-brown from the outset, never blue. The scant latex is dirty yellowish to yellowish brown, never deep blue, and the spore print is pale buff, yellowish, or orangish. *Lactarius paradoxicus* has a persistently greyish-blue to greyish-purple cap and stipe that bruise green, and the gills are pinkish-orange. The latex is dark vinaceous-brown, and the spore print is cream to yellow.

Salmon-staining Milkcap

Lactarius subplinthogalus

A

B

C

D

Lactarius subplinthogalus **Coker**, *J. Elisha Mitchell Scient. Soc.* 34(1):50 (1918)
[Basidiomycota, Agaricomycetes, Russulales, Russulaceae]

Derivation: *Lactarius*, Latin, "milky"; *subplinthogalus*, Greek, from *plinthos*, "brick," meaning "more or less brick-colored," referring to the staining of the gills by the latex.

Synonym: *Lactarius marylandicus* A.H. Smith & Hesler, *Brittonia* 14:427 (1962); *fide* Bessette et al., *Milk Mushrooms of North America* (Syracuse, NY: Syracuse University Press):248 (2009)

Cap: 3–7.5 cm (1¼–3 in) across, flat, centrally depressed (Figs. A & B) to slightly funnel-shaped, margin prominently pleated and scalloped (Fig. B), surface smooth to somewhat wrinkled; initially whitish, becoming pale buff, pale tan, to greyish-brown with age.

Flesh: White, slowly staining salmon to brownish-orange when injured or cut; no odor, taste acrid.

Gills: Attached to slightly decurrent, *very* distant (up to 3 mm apart), broad (Fig. C); concolorous with the cap, often stained salmon to brownish-orange with age (Fig. D).

Latex: White, unchanging, staining gills and flesh salmon to brownish-orange (Fig. D); taste bitter.

Stipe: 3–8 cm tall by 1–1.5 cm thick (1¼–3¼ × ⅔ in), subequal or tapering toward the base, at first solid, becoming hollow; smooth, dry, colored like the cap, staining brownish-orange with age.

Spore print: Pinkish-buff.

Microscopic: Spores 7.5–9.5 × 7–8 µm, subglobose to broadly ellipsoid, ornamented with prominent spines and branched ridges up to 2.5 µm high, but not reticulate; clear, ornamentation amyloid.

Occurrence: Mycorrhizal with oaks (*Quercus* spp.); alone or in scattered groups in hardwood and mixed hardwood-conifer forests, especially under oaks; widespread and occasional to common in eastern North America; summer–early fall.

Edibility: Unknown.

Comments: This tiny but distinctive mushroom is readily identified by its flat to shallowly funnel-shaped, whitish-grey, pleated and scalloped cap, extremely widely spaced gills, and latex that slowly stains the flesh and gills salmon to brownish-orange (not always visible). *Lactarius gerardii* Peck is somewhat larger (up to 12.5 cm [5 in]) across, the cap is dark brown to yellow-brown, the gills are subdistant rather than very widely spaced, the spore print is white, the flesh does not discolor when injured, and the latex does not stain the gills or tissues. *Lactarius sumstinei* Peck is a similar species reported from Pennsylvania and Tennessee but is not known to occur in our area. Its cap and gills are pinkish-buff to pale pinkish-cinnamon, the spores are slightly larger (8.5–11 × 9–11 µm) and partially to completely reticulated, and the latex does not stain the flesh or gills.

Chanterelles

The chanterelles differ from the other gilled fungi in having pseudogills, or "false gills," derived from longitudinal folds in the hymenophore rather than the "true gills" of most other gilled fungi. The false gills of the chanterelles are relatively shallow and thick, have bluntly rounded margins, and cannot be readily removed. The true gills of most gilled mushrooms are relatively deep, thin, knife-edged, and easily torn away.

The genus *Craterellus* is largely distinguished from the genus *Cantharellus* in having a hollow stipe with a trumpet-shaped, perforated cap; the stipe in *Cantharellus* is more or less solid, and although the cap can be funnel-shaped, it is not narrowly trumpet-shaped or perforated.

The chanterelles are generally grouped into the so-called yellow chanterelles and the red chanterelles. The former include *Cantharellus cibarius*, *C. lateritius*, and *C. confluens*. A recent study[1] surprisingly found that two small species of yellow chanterelles are actually more closely related to the red forms, which include *C. cinnabarinus*, *C. texensis*, and *C. persicinus*.

The Golden Chanterelle, *Cantharellus cibarius* (*sensu lato*), generally regarded as a highly variable and cosmopolitan species, is turning out on the basis of molecular studies to be a complex of cryptic and much more geographically restricted species than previously thought. Many of these have yet to be formally described and named.

The Jack O'Lantern Mushroom, *Omphalotus illudens* (Schweinitz) Bresinsky & Besl, and the False Chanterelle, *Hygrophoropsis aurantiaca* (Wulfen) Maire, are not at all related to the chanterelles (Order Cantharellales) and belong to very different orders, the Agaricales and Boletales, respectively. They are included in this key because they are nonlethally toxic and can be confused with the edible true chanterelles. The Jack O'Lantern Mushroom species account is included with the pleurotid mushrooms.

Because of their close resemblance, the Scaly Vase Chanterelle, Kauffmann's Scaly Chanterelle, and the Pig's Ear Gomphus—*Turbinellus floccosus*, *T. kauffmannii*, and *Gomphus clavatus*, respectively—all once grouped together in the genus *Gomphus*, were formerly included with the chanterelles in the family Chanterellaceae in the order Cantherellales. They were later placed in their own family (Gomphaceae) and transferred to the order Phallales together with the stinkhorns. But molecular studies have shown that the species formerly assigned to the genus *Gomphus* (*sensu lato*) do not share a common evolutionary lineage. They were accordingly reassigned to four separate genera, and the family Gomphaceae was given its own order, Gomphales. Because of these studies, the family Gomphaceae also now includes *Ramaria*, a huge genus (some 200 or so species) of very different-looking coral fungi, which previously also had its own family.[2]

Although not related, the three species of Gomphaceae that occur in our area are included here because of their close morphological resemblance, potential toxicity, and possible confusion with several *Craterellus* species.

1. J.-M. Moncalvo et al., 2006, *Mycologia* 98(6):937–948.
2. A.J. Giachini & M.A. Castellano, 2011, *Mycotaxon* 115:183–201.

Key to the Chanterelles and Chanterelle Look-alikes

1. Hymenium with true gills that are thin, knife-edged, and readily detached 2

1′. Hymenium smooth or with thick, rounded wrinkles or ridges that are not easily detached 3

Key to the Chanterelles and Chanterelle Look-alikes (*Continued*)

2. Associated with conifers; gills repeatedly forking; cap surface reddish with KOH . . . . *Hygrophoropsis aurantiaca*[a]

2′. Associated with hardwoods; gills nonforking or only occasionally forked; cap surface green with KOH . . . . *Omphalotus illudens*

3. Fruitbody consisting of one or more trumpet- or funnel-shaped, hollow, deeply tubular, often perforated cap 4

3′. Fruitbody not formed of deep, trumpet-shaped hollow caps . . . . (*Cantharellus*) 6

4. Cap dark-colored (dark grey, dark brown, blackish, or bluish-black) when fresh (or if bright yellow or orange, then not associated with conifers), spores smooth. . . . . (*Craterellus*) 5

4′. Cap lighter than above, orange to orange-red, buff, tan, brownish-grey, or initially lilac or lavender (fading to yellowish-brown with age), ectomycorrhizal with conifers, spores ornamented with minute warts . . . (*Turbinellus* and *Gomphus*) 14

5. Fruitbody consisting of a single cap, upper (inner) surface not bright orange or yellow 12

5′. Fruitbody a clustered group of trumpet-like caps arising from a common base; upper (inner) surface bright orange, becoming yellow toward the margin . . . . *Cr. odoratus*

6. Mature cap generally 2 inches or less across, nonreactive or staining dull red with $FeSO_4$ 7

6′. Mature cap typically more than 3 inches across, staining grey with $FeSO_4$. 9

7. Cap brownish overall or with central brown spot; flesh and hymenium dull red with $FeSO_4$ . . . . *C. appalachiensis*

7′. Cap not brown; not reactive with $FeSO_4$ 8

8. Fruitbody vermillion to bright reddish-orange when young; flesh white with reddish tinge near cuticle. . . . . *C. cinnabarinus*

8′. Fruitbody bright yellow to orange-yellow; flesh pale yellow to orange . . . . *C. minor*

9. Hymenium smooth or with only weak, unforked wrinkles and folds. 10

9′. Mature hymenium with prominent forked ridges 11

10. Fruitbodies multipileate—two or more caps arising from a single stipe. . . . . *C. confluens*

10′. Fruitbodies monopileate—only one cap per stipe . . . . *C. lateritius*

11. Mature cap funnel-shaped, up to 6 inches across; pale yellow, yellow, or orangish. . . . . *C. cibarius*

11′. Mature cap broadly convex to planoconvex; salmon- to pinkish-orange to peach-colored. . . . . *C. persicinus*

12. Stipe not distinct from the rest of the fruitbody . . . . *Cr. fallax*

12′. Stipe distinctly set off from a cap-like fruitbody 13

13. Hymenophore grey to violet-grey, yellow stipe often brownish above, flesh brownish-yellow . . . . *Cr. tubaeformis*

13′. Hymenium yellow-orange, stipe yellow throughout its length, flesh orange-yellow . . . . *Cr. ignicolor*

14. Mature cap (upper surface) coarsely scaly at least at the center, fertile surface not lilac or purple (except for occasional bruised spots) . . . . (*Turbinellus*) 15

14′. Cap smooth or with only minute scales, fertile surface usually lilac or purple (or broadly tinged with those colors when mature) . . . . *Gomphus clavatus*

15. Upper surface deeply sunken, pale orange to reddish-orange; outer surface lacks purplish bruises; flesh with no distinctive odor . . . . *T. floccosus*

15′. Upper surface more or less deeply depressed, upper surface whitish (never orange), covered with coarse creamy ochre, pale yellow-brown to buffish scales, stipe sometimes with purplish bruises (but never generally washed with purple); often emits a pungent smell . . . . *T. kauffmannii*

a. The False Chanterelle, *Hygrophoropsis aurantiaca* (Wulfen) Maire, is not commonly encountered in our area.

Scaly Vase Chanterelle

Turbinellus floccosus

A B

Turbinellus floccosus. (Photos: Walter Siegmund.)

***Turbinellus floccosus* (Schweinitz) Earle *ex* Giachini & Castellano**, *Mycotaxon* 115:196 (2011)[1]

Basionym: *Cantharellus floccosus* Schweinitz, *Trans. Am. Phil. Soc.*, ser. 2, 4(2):153 (1832)
[Basidiomycota, Agaricomycotina, Agaricomycetes, Phallomycetidae, Gomphales, Gomphaceae]

Derivation: *Gomphus*, Greek, from *gomphos*, "plug," "large wedge-shaped nail"; *floccosus*, Latin, "full of tufts," "woolly."

Deprecated synonym: *Gomphus floccosus* (Schweinitz) Singer, *Lloydia* 8(3):140 (1945)

Fruitbody: 5–15(–30) cm tall by 3–12.5(–15) cm wide (2–6[–12] × 2–5[–6] in), initially cylindrical, becoming funnel- or vase-shaped (Fig. A); upper (sterile) surface (the "cap") deeply depressed (Figs. A & B), moist, pale orange to reddish-orange, becoming orange-brown with age, covered with coarse, woolly, darker scales; margin thin, wavy, lobed with age.

Flesh: Thick, white, fibrous; odor unremarkable, taste somewhat sour.

Fertile surface (hymenium): Deeply decurrent (Fig. A), surface deeply wrinkled, covered with blunt vertical, forking, and anastomosing folds and connecting interveins; pale yellow to creamy-white or beige.

Stipe: Smooth, concolorous with and not distinctly demarcated from the hymenium, tapering to the base; up to 12 cm tall and 5 cm thick near the top (4¾ × 2 in); hollow with age.

Spore print: Pale ochre to cinnamon.

Microscopic: Spores 11.5–14.5(–20) × 7–8(–10) µm, elliptical, wrinkled to minutely warted, clear, nonreactive in Melzer's reagent.

Occurrence: Ectomycorrhizal with conifers, especially hemlock (*Tsuga* spp.); solitary or in groups on the ground; widespread and fairly common in coniferous and mixed woods throughout northern and montane North America (more common in the Pacific Northwest); occasionally found at higher elevations in the southern Appalachians, spring–fall; also in Korea, China, Tibet, India, and Pakistan.

Edibility: Eaten by some, but this mushroom contains norcaperatic acid, which causes nausea, vomiting, and diarrhea in some individuals.

Comments: This unusual mushroom could only be confused with two other similar species in our area: Kauffmann's Scaly Chanterelle, *Turbinellus kauffmanii* (A.H. Smith) Giachini & Castellano, and the Pig's Ear Gomphus, *Gomphus clavatus* (Persoon) Gray 1821.

Turbinellus kauffmannii occurs in the Pacific Northwest and the Appalachian Mountains and has been reported in Great Smoky Mountains National Park.[2] Its habitat and floral associations are similar to those of *T. floccosus*. Much larger than *T. floccosus*, *T. kauffmannii* can be up to 40 cm (16 in) in height, and the funnel is much shallower. The upper, sterile portion of the fruitbody is covered with much larger and coarser erect or recurved to nearly pyramidal scales, which are creamy-ochre, pale yellow-brown, to buff or pinkish-cinnamon (never orange or reddish-orange) on a whitish ground color and are darker at the center and paler toward the margin. The wrinkled hymenium is similar to that of *T. floccosus* but is cream-colored to pinkish-buff and may have purplish stains. *Turbinellus kauffmannii* often emits a pungent odor not found in *T. floccopus*.

The two species of *Turbinellus* found in our area typically have a single fruitbody on a stipe. In contrast, the Pig's Ear Gomphus, *Gomphus clavatus* (Figs. C & D), almost always exhibits multiple fruitbodies ("caps") arising from a single stipe that often fuses with neighboring cespitose stipes. The upper (sterile) surface is smooth without the prominent scales seen in *Turbinellus*, is not as deeply funnel-shaped, and is dark lilac or violet when young, turning tan or brownish with age. The decurrent, wrinkled hymenium is initially dark lilac to violet (Fig. D), which fades to pale lilac or tan with age (Fig. C). *Gomphus clavatus* is edible and is considered choice by many.

C

D

Gomphus clavatus. (Photos: C, Nathan Wilson; D, Heather Hallan-Adams.)

All three species were formerly placed together in the genus *Gomphus* Persoon. Because of their general cantherelloid appearance they were once included with the chanterelles in the family Cantharellaceae. They were later given their own family and grouped with the stinkhorns in the order Phallales. Molecular studies then revealed that the genus *Gomphus* (*sensu lato*) is polyphyletic (the species do not share a common evolutionary lineage), and they were reassigned to four different genera in their own order (Gomphales, family Gomphaceae) together with *Ramaria*, a huge genus of coral fungi.[3] As the type species of the genus, *Gomphus clavatus* remains in (and thus defines) the genus *Gomphus* (*sensu stricto*), which now contains only the three species discussed above.

1. *Cantharellus floccosus* Schweinitz 1832 is the type species of the genus *Turbinellus* Earle (*Bull. New York Bot. Gard.*:407 [1909]).
2. R.H. Petersen & L.R. Hesler, 1979, *Checklist of Fungi of the Great Smoky Mountains National Park* (Gatlinburg: Department of the Interior, National Park Service, Southeast Region, Uplands Field Research Laboratory, Great Smoky Mountains National Park), 64.
3. A.J. Giachini & M.A. Castellano, 2011, *Mycotaxon* 115:183–201.

Black Trumpet

Craterellus fallax

Photos: A, C & D, W. Light; B, E & F, Teresa Fortenberry.

***Craterellus fallax* A.H. Smith**, *Mich. Bot.* 7:153 (1968) [Basidiomycota, Agaricomycotina, Agaricomycetes, *incertae sedis*, Cantharellales, Cantharellaceae]

Derivation: *Craterellus*, Latin, from *crater*, a vessel for carrying oil or a volcano opening, plus *-ellus*, "little"; *fallax*, Latin, "deceptive," "deceitful."

Fruitbody: 3–14 cm tall by 2–8 cm wide (1¼–5⅔ × ¾–3¼ in), trumpet- or funnel-shaped, hollow, deeply tubular, often perforated (Fig. E), with an expanded, often flaring apex (Figs. A–D); dry, smooth to finely scaly, greyish- to dark brown or blackish (Figs. A–D); margin wavy, decurved (Figs. C & D) or elevated (Figs. A & B), often edged with blue (Figs. C, E, & F).

Outer (fertile) surface: Smooth or with shallow, blunt, vertically oriented ridges, grey, greyish-brown, to blackish (Figs. A & B); some individuals quite yellowish or with salmon tints (Figs. E & F).

Flesh: Thin, fibrous or brittle, greyish or greyish-brown; odor and taste not distinctive or faintly fruity.

Stipe: Not distinct from the rest of the fruitbody, often hollow.

Spore print: Ochre-buff to salmon-colored.

Microscopic: Spores 11–15(–20) × 7–11 µm, broadly elliptical, smooth, clear; clamp connections lacking.

Occurrence: Ectomycorrhizal with both oaks (*Quercus* spp.) and conifers (*Pinus* and *Tsuga*). Scattered or in groups on the ground under deciduous and conifer trees, summer–winter; widespread in eastern North America, occasional to common.

Edibility: Edible and choice, often dried and pulverized as a spice.

Comments: *Craterellus fallax* is not likely to be mistaken for any other species in our area despite its variable shape and color. The somewhat similar but uncommon *Craterellus foetidus* A.H. Smith, also found under oaks, has a spore-bearing surface with more prominent decurrent, gill-like ridges *connected by cross-veins*, a white spore print, and smaller spores (8.5–10[–12] × 5–7 µm).

Craterellus calyculus (Berkeley & M.A. Curtis) D. Reid is considerably smaller than *Cr. fallax*, 3½–4 cm (1½–1⅔ × ½ in) tall by 1 cm across, is flat to shallowly vase-shaped (not deeply trumpet-shaped like the other species discussed here), and has a white (not salmon-colored) spore print. The size of the spores varies considerably depending on the reporting authority and is not a useful characteristic.

Craterellus carolinensis Petersen is known only from the type collection from Macon County, North Carolina. It is also much smaller than *Cr. fallax*, 1.5–4 cm tall by 6–12 mm across (⅔–1⅔ × ¼–½ in), with a distinctly hairy or fimbriated margin; the spores are smaller (8.0–10.5 × 5.2–7.0 µm), are ovoid to broadly ellipsoid, and bear a small apiculus. The color of the spore print is unknown.

Dahlman et al. (2000)[1] synonymized *Cr. fallax* with the European species *Cr. cornucopioides* (Linnaeus) Persoon on the basis of DNA studies involving a single conserved gene region; this has been followed by some but not all subsequent authors. However, using longer DNA sequences, Matheny and his colleagues (2010)[2] subsequently demonstrated that the eastern North American *Cr. fallax* is distinct from the European *Cr. cornucopioides* (which has smaller spores and a white, not salmon, spore print), as well as from samples identified as *Cr. cornucopioides* from California and Oregon. These authors accordingly resurrected *Cr. fallax* as a valid species. See online Appendix A for additional details.

1. M. Dahlman, E. Danell, & J.W. Spatafora, 2000, *Mycol. Res.* 104:388–394.
2. P.B. Matheny et al., 2010, *Mycorrhiza* 20:569–575.

Trumpet Chanterelle, Winter Chanterelle

Craterellus tubaeformis

A

B

Photos: John Plischke III.

Craterellus tubaeformis (Fries) Quélet, *Fl. Mycol. France* (Paris):36 (1888)

Basionym: *Cantharellus tubaeformis* Fries, *Syst. Mycol.* (Lundae) 1:319 (1821)
[Basidiomycota, Agaricomycotina, Agaricomycetes, *incertae sedis*, Cantharellales, Cantharellaceae]

Derivation: *Craterellus*, Latin, from *crater*, a vessel for carrying oil or a volcano opening, plus -*ellus*, "little"; *tubaeformis*, Latin, from *tuba*, "funnel," and -*formis*, "shaped," "funnel-shaped."

Fruitbody: 1–7.5 cm across (½–3 in), initially convex, becoming vase- to funnel-shaped with a deeply depressed center that eventually perforates into a hollow stipe; margin at first inrolled (Fig. B), becoming irregularly wavy (Figs. A & B); surface smooth or with small scales, waxy or somewhat viscid when fresh; dark yellow-brown to smoky-brown (Fig. A).

Outer (fertile) surface: With prominent, blunt, decurrent, forking false-gills with cross-veins, initially yellowish-grey, becoming violet-grey (Fig. B).

Flesh: Very thin, yellowish to brownish; odor and taste not distinctive.

Stipe: 3–10 cm tall by up to 1 cm thick (1¼–4 × ½ in), subequal, smooth, hollow when mature; yellowish to yellowish-orange, sometimes darkening above and/or whitish at the base.

Spore print: Cream to yellowish-buff.

Microscopic: Spores 8–13 × 6–10 µm, elliptical, smooth, clear.

Occurrence: Mycorrhizal with conifers. On the ground, in moss (Fig. B), or on decaying wood (Fig. A) in bogs and wet areas under conifers; widespread and occasional in eastern North America; summer–midwinter.

Edibility: Edible; may cause gastric upset in some individuals.

Comments: *Craterellus tubaeformis* can be distinguished from the similar *Cr. ignicolor* (R.H. Petersen) Dahlman, Danell & Spatafora as follows: *Craterellus tubaeformis* has a grey to violet-grey hymenophore, the flesh is brownish-yellow, and the yellow stipe is often yellowish-brown near the apex. *Craterellus ignicolor* has a yellow-orange hymenium, the flesh is orange-yellow, and the stipe is not brownish above. Genetic evidence indicates that *Cantharellus tubaeformis* Fries and *Cr. infundibuliformis* (Scopoli) Fries are synonyms[1] (the latter taxon is treated as a separate species by some authors).

1. M. Dahlman, E. Danell, & J.W. Spatafora, 2000, *Mycol. Res.* 104:388–394.

Fragrant Chanterelle

Craterellus odoratus

A

B

C

D

Photos: A, Teresa Fortenberry; B–D, Richard Bell.

***Craterellus odoratus* (Schweinitz) Fries**, *Epicr. Syst. Mycol.* (Upsaliae):532 (1838) [1836–1838]
[Basidiomycota, Agaricomycotina, Agaricomycetes, *incertae sedis*, Cantharellales, Cantharellaceae]

Basionym: *Merulius odoratus* Schwein., *Schr. Naturf. Ges. Leipzig* 1:91 (1822)

Derivation: *Craterellus*, Latin, from *crater*, a vessel for carrying oil or a volcano opening, and *-ellus*, "little"; *odoratus*, Latin, "fragrant," "perfumed."

Fruitbody: A clustered group of orange trumpet-like funnels (Figs. A & B), typically 7–15 cm (2¾–6 in) tall and wide, but may be up to 25 cm (10 in) across.

Cap: Single cap: a funnel some 1.3–5 cm (½–2 in) wide, bright orange at the center, becoming yellow toward the margins (Fig. A); margins wavy, lobed; several arising from a common base.

Fertile surface: Pale orange-yellow to creamy-yellow or creamy-white; smooth to wrinkled, vertically rugose or ridged, but not forming deep, sharp gills (Figs. C & D).

Flesh: Thin, orange; with strong fruity odor; taste unremarkable.

Stipe: Short, up to 5 cm tall by 2–5 cm thick (2 × ¾–2 in), smooth, hollow, confluent with fertile surface; one stalk for several trumpet-shaped fruitbodies.

Spore print: Light yellow to pale orange.

Microscopic: Spores 8–12 × 4–6.5 µm, elliptical to narrowly ovate, smooth, clear.

Occurrence: On ground in open woods, associated with hardwoods; August–September. Widespread and fairly common in the Southeast.

Edibility: Edible when young.

Comments: *Craterellus odoratus* forms distinctive structures consisting of several bright orange fruitbodies arising from a single hollow stalk and fused into an irregular composite mass; the flesh is orange and has a strong fruity smell, and the spores in mass are yellow to pale orange. It could be confused with two quite unrelated mushrooms: the poisonous Jack O'Lantern and the False Chanterelle; the latter species is mildly toxic to some individuals who consume it.

The Jack O'Lantern, *Omphalotus illudens* (Schweinitz) Bresinsky & Besl (Order Agaricales), is poisonous but nonlethal, causing severe cramps, vomiting, and diarrhea. It is distinguished from both *Craterellus* and *Cantharellus* by having true nonforking gills with sharp, thin edges rather than bluntly rounded, broad vertical ridges. Unlike the ridges of *Cr. odoratus* and the true chanterelles, the gills of the Jack O'Lantern are easily detached. The stipe is solid, tall, and thin (5–20 by 0.5–2.2 cm [2–8 × ¼–~1 in]), and the spore print is white to pale yellow. It also grows in large, overlapping clusters on decaying wood rather than from the ground, although the wood may be buried. Like *Cr. odoratus*, *O. illudens* has thin orange flesh.

The False Chanterelle, *Hygrophoropsis aurantiaca* (Wulfen) Maire (Order Boletales), also has true gills that are close to crowded and repeatedly forked. The spore print is white, and the spores are ovoid, reddish-brown in Melzer's reagent, and considerably smaller (5–8 × 3–5 μm) than those of *Craterellus odoratus*. The False Chanterelle is considered edible by some authors, but a number of people develop gastrointestinal symptoms from the high levels of arabitol.[1]

Craterellus odoratus can also be confused with several species of quite unrelated yellow chanterelles (Order Cantharellales) in the genus *Cantharellus* Adanson, discussed below. However, except for *C. confluens* (see below), *Cantharellus* species arise as single basidiocarps that do not form a composite mass, the flesh is white (not orange), the stipe is solid (not hollow), and the spore print is usually pinkish-cream to pinkish-yellow.

In *Cantharellus confluens* (Berkeley & Curtis) Petersen two or more fruitbodies may also arise from a common stipe that is yellow above but whitish toward the tapering base. The fertile surface is smooth or has shallow, indistinct ridges quite different from the prominent folds of *Cr. odoratus*, and the stipe bruises rusty-orange when handled.

1. Arabitol is a 5-carbon sugar alcohol that is not well absorbed in the small intestine. Overconsumption of arabitol can produce bloating, diarrhea, and flatulence, and some sensitive persons can become ill on even small amounts.

Chanterelle, Golden Chanterelle

Cantharellus cibarius (sensu lato)

A

B

Photos: A, Andreas Kunze; B, Borch3kawki.

***Cantharellus cibarius* Fries**, *Syst. Mycol.* (Lundae) 1:318 (1821) [Basidiomycota, Agaricomycotina, Agaricomycetes, *incertae sedis*, Cantharellales, Cantharellaceae]

Derivation: *Cantharellus*, Latin, from *cantharus*, a large drinking vessel, and the diminutive ending *-ellus*; *cibarius*, Latin, pertaining to food.

Cap: 1–15 cm (½–6 in) across, at first broadly convex, often with inrolled margin, becoming flat to shallowly depressed (Fig. A) to uplifted and funnel-shaped in maturity; margin initially entire (Fig. A, lower right), becoming wavy and irregular (Figs. A & B); pale yellow (Fig. A), yellow (Fig. B), to orangish.

Flesh: White, solid, not bruising when cut or injured; odor fruity, taste not distinctive or somewhat peppery.

Spore-bearing surface (hymenium): Usually with prominent, decurrent, gill-like ridges (Figs. A & B) and/or rib-like folds (pseudogills), these generally forked and prominently cross-veined (Fig. B); pale yellow, yellow, to pale orange.

Stipe: 2.5–7.5 cm tall by 0.5–2.5 cm thick (1–3 × ¼–1 in); solid, smooth below the false gills, tapering downward, central to eccentric; colored like the hymenium or paler.

Spore print: Pinkish-cream to pale yellow.

Bruising: Stipe sometimes bruising brownish or orangish.

Chemical staining: Flesh pinkish-grey to grey with $FeSO_4$, sometimes with transient reddish flash, hymenium staining dark grey;[1] cap negative to reddish with KOH, negative with NaOH.

Microscopic: Spores (7–)8–11 × 4.5–6.0(–6.5) µm, elliptical, smooth, colorless.

Occurrence: Ectomycorrhizal with hardwoods, especially oaks (*Quercus* spp.); widespread and common throughout North America, summer–early fall in the Southeast; Europe, Asia, Africa (but see comments below).

Edibility: Edible, considered choice.

Comments: *Cantharellus cibarius* is best distinguished from its close look-alike, the Smooth Chanterelle, *C. lateritius* (Berkeley) Singer, on the basis of its prominent gill-like ridges and folds, which typically fork and often exhibit cross-veins (see the account of the Smooth Chanterelle for details). This will distinguish between typical populations of the two species. However, intergrading populations of *C. cibarius* with indistinct pseudogills and *C. lateritius* with more prominent pseudogills are not rare, and identification to species can be problematic. Microscopic and staining characteristics also do not unequivocally separate these two taxa.

The toxic Jack O'Lantern Mushroom, *Omphalotus illudens* (Schweinitz) Bresinsky & Besl, is distinguished from the edible chanterelles (*Cantharellus* spp. and *Craterellus* spp.), which it resembles, by its sharp, thin, knife-edge gills, which are easily detached. The "gills" of the chanterelles are actually thick, bluntly rounded, vertical ridges that cannot be easily detached.

The Fragrant Chanterelle, *Craterellus odoratus*, forms distinctive structures consisting of several bright orange fruitbodies with orange (not white) flesh arising from a single hollow (not solid) stalk and fused into an irregular composite mass. The fruitbodies of *Cantharellus cibarius* all arise from a single stalk.

The Appalachian Chanterelle, *Cantharellus appalachiensis* R.H. Petersen, also has well-developed pseudogills with cross-veins, but the cap and stipe are yellowish-brown to brownish. This species is also unique among the chanterelles in exhibiting a *sustained* dull reddish reaction[2] to iron salts on the flesh and spore-bearing surface.

The general interpretation of *C. cibarius* as a widespread and highly polymorphic species is due to the high variability in both macroscopic and microscopic characteristics of the golden chanterelles within and between regions. Few conserved morphological (including color) characteristics exist that clearly distinguish between populations with intergrading features (e.g., the degree of development of the hymenial ridges, or pseudogills). At least four different species in North America have been referred to as *C. cibarius*, some of which are mycorrhizal with conifers,[3] and European populations likewise probably include several cryptic species.[4] In fact, the Swedish taxon described by Fries in 1821 may not be so cosmopolitan and may not actually occur in North America. But pending the necessary genetic and molecular studies, we have little choice but to refer to these undescribed and unnamed golden chanterelles as *Cantharellus cibarius* (*sensu lato*).

1. M. Kuo, November 2003, Mushroom brief: Reactions to iron salts among the chanterelles, MushroomExpert.Com, http://www.mushroomexpert.com/kuo_06.html.
2. *Cantharellus cibarius* sometimes stains fleetingly reddish, then pinkish-grey or grey with iron salts.
3. *Cantharellus formosus* Corner is associated with Western Hemlock (*Tsuga heterophylla*) and Douglas Fir (*Pseudotsuga menziesii*) in the Pacific Northwest; *C. cascadensis* Dunham, O'Dell, & Molina is found in the *Pseudotsuga* forests of the Cascade Mountains in Oregon; *C. cibarius* var. *roseocanus* Redhead, Norvell, & Danell is associated with spruce (*Picea*) and pines (*Pinus*) in the Pacific Northwest; and *C. californicus* Aurora & Dunham is associated with oaks (*Quercus* spp.) in California.
4. D. Aurora & S.M. Dunham, 2008, *Economic Botany* 62(3):376–391.

Appalachian Chanterelle

Cantharellus appalachiensis

A

B

C

D

Photos: John Plischke III.

Cantharellus appalachiensis **R.H. Petersen**, *Svensk Bot. Tidskr.* 65:402 (1971)
[Basidiomycota, Agaricomycotina, Agaricomycetes, *incertae sedis*, Cantharellales, Cantharellaceae]

Derivation: *Cantharellus*, Latin, from *cantharus*, a large drinking vessel, and the diminutive ending *-ellus*; *appalachiensis*, Latin, "of the Appalachians."

Cap: 1.3–5 cm (½–2 in) across, initially convex (Fig. A), becoming flattened with a central depression (Fig. B) or funnel-shaped (Fig. C); margin inrolled (Fig. A), becoming uplifted and irregularly wavy (Figs. B & C); at first brown (Fig. A), becoming increasingly yellow (Figs. B & C) with a brown spot over the disk (Fig. C).

Flesh: Yellowish-brown, not discoloring when cut; odor fruity, like apricot, taste not distinctive.

Spore-bearing surface (hymenium): With well-developed, ridge-like false gills with prominent cross-veins (Fig. D); cream-colored to pale yellow.

Stipe: 2–5 cm tall by up to 1 cm thick (¾–2 × ½ in), slender, equal but expanded at apex; smooth, concolorous with the cap, whitish at the base (Fig. A), becoming hollow with age.

Spore print: Salmon-buff.

Chemical staining: Flesh and hymenium staining persistent dull red with $FeSO_4$, this feature diagnostic.[1]

Microscopic: Spores (6–)7.5–9(–10) × (4–)4.5–5.5(–6) µm; elliptical, smooth, clear; yellowish in KOH, nonreactive in Melzer's reagent.

Occurrence: Solitary or clustered on the ground under hardwood trees in mixed oak forests; rare to occasional in Appalachian Mountains from Georgia, Tennessee, and North Carolina north to the Great Lakes region and New England; summer–fall.

Edibility: Edible, choice.

Comments: *Cantharellus appalachiensis* is the only chanterelle that stains a persistent red with iron salts. The Golden Chanterelle, *C. cibarius* Fries, sometimes exhibits a transient reddish flash with $FeSO_4$ that then turns pinkish-grey or grey.

1. M. Kuo, November 2003, Mushroom brief: Reactions to iron salts among the chanterelles, MushroomExpert.Com, http://www.mushroomexpert.com/kuo_06.html.

Smooth Chanterelle

Cantharellus lateritius

A B C

Photos: A & B, W. Light; C, Teresa Fortenberry.

***Cantharellus lateritius* (Berkeley) Singer**, *Lilloa* 22:729 (1951) [1949]

Basionym: *Craterellus lateritius* Berkeley, *Grevillea* 1(10):147 (1873)
[Basidiomycota, Agaricomycotina, Agaricomycetes, *incertae sedis*, Cantharellales, Cantharellaceae]

Derivation: *Cantharellus*, Latin, from *cantharus*, a large drinking vessel, and the diminutive ending *-ellus*; *lateritius*, Latin, "brick."

Cap: 2–12 cm across by 12 cm tall (¾–4¾ × 4¾ in), convex, nearly flat, slightly depressed to funnel-shaped (Figs. B & C); dry, upper surface smooth; yellow (Fig. C), yellow-orange, to orange (Fig. B); margin initially incurved (Fig. C, lower right), becoming uplifted and wavy (Fig. C) to deeply scalloped (Fig. B) or lobed (Fig. A).

Flesh: White to yellowish, thick; odor fruity.

Spore-bearing surface (hymenium): Smooth or with very shallow radial creases, ridges, or veins (Fig. A), these often becoming more prominent with age; pale cream (Fig. C) to pale yellowish, pale pinkish (Fig. A), or salmon-buff.

Stipe: Up to 8 cm tall by 2.5 cm (3¼ × 1 in) thick, tapering downward (Figs. A & C), typically single (Fig. A), but often with two to three stipes joined at the base (Fig. C); smooth, solid (never perforated), colored like the hymenium.

Spore print: Pale pinkish-yellow.

Bruising: Cap bruising pale yellow, then reddish-brown.

Chemical staining: Flesh pinkish-grey to grey, and hymenium grey with $FeSO_4$.[1]

Microscopic: Spores 7.5–9.5(–12.0) × 4.5–6.0(–6.5) µm, elliptical, smooth, clear, occasionally with 1 to several oil droplets.

Occurrence: Mycorrhizal with oaks (*Quercus* spp.); solitary, scattered, or in groups on the ground, often along paths, summer–fall; widespread and locally common in eastern North America.

Edibility: Edible, considered choice.

Comments: The smooth to faintly ridged hymenium of this species distinguishes it from the Golden Chanterelle, *Cantharellus cibarius* Fries. The latter species exhibits very prominent, deep, vertical, decurrent, gill-like ridges, which are often forked and may have cross-veins.

The toxic Jack O'Lantern Mushroom, *Omphalotus illudens* (Schweinitz) Bresinsky & Besl, is distinguished from the edible chanterelles (*Cantharellus* spp. and *Craterellus* spp.), which it resembles, by its sharp, thin, knife-edge gills, which are easily detached. The "gills" of the chanterelles are actually thick, bluntly rounded, vertical ridges that cannot be easily detached. The chanterelles also do not form large cespitose clusters, nor do they grow on wood substrata. This usually distinguishes them from the Jack O'Lantern, although the latter can appear to emerge from the ground when growing from submerged roots.

The Fragrant Chanterelle, *Craterellus odoratus*, forms distinctive structures consisting of several bright orange fruitbodies with orange flesh arising from a single hollow (not solid) stalk and fused into an irregular composite mass. The spores in mass are yellow to pale orange. Except for *Cantharellus confluens* (Berkeley & M.A. Curtis) R.H. Petersen, *Cantharellus* species arise as single basidiocarps that do not form a composite mass, the flesh is white (not orange), the stipe is solid (not hollow), and the spore print is usually pinkish-cream to pinkish-yellow.

The Appalachian Chanterelle, *Cantharellus appalachiensis* R.H. Petersen, is unique among the chanterelles in exhibiting a sustained red reaction[2] to iron salts on the flesh and the spore-bearing surface.

The only clear distinction between *C. lateritius* and the Conjoined Chanterelle, *Cantharellus confluens* (Berkeley & M.A. Curtis) R.H. Petersen, is that the latter typically has two or more caps growing from a single stipe. Some authors consider the two taxa to be conspecific, in which case *C. confluens* would have nomenclatural priority.

1. M. Kuo, November 2003, Mushroom brief: Reactions to iron salts among the chanterelles, MushroomExpert.Com, http://www.mushroomexpert.com/kuo_06.html.
2. *Cantharellus cibarius* sometimes stains fleetingly reddish, then pinkish-grey or grey with iron salts.

Conjoined Chanterelle

Cantharellus confluens

A

B

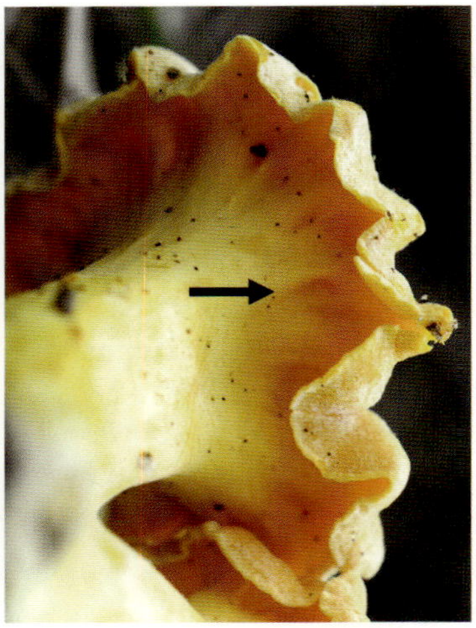
C

***Cantharellus confluens* (Berkeley & M.A. Curtis) R.H. Petersen**, *Sydowia* 32(1–6):201 (1980) [1979]

Basionym: *Craterellus confluens* Berkeley & M.A. Curtis, *J. Linn. Soc., Bot.* 9:423 (1867) [1866]
[Basidiomycota, Agaricomycotina, Agaricomycetes, *incertae sedis*, Cantharellales, Cantharellaceae]

Derivation: *Cantharellus*, Latin, from *cantharus*, a large drinking vessel, and the diminutive ending *-ellus*; *confluens*, Latin, "meeting," "junction."

Cap: 8 cm across by 13 cm tall (3¼ × 5¼ in) when mature; flat (Figs. A & B), becoming centrally depressed, surface smooth, yellow, yellowish-orange, to orange-buff (Fig. B); margin wavy, crenulate, or irregularly lobed, initially incurved (Figs. A–C), uplifted with age; occasionally simple, one cap on a single stipe (Fig. B), but usually multipileate, two or more caps arising from conjoined or united stipes (Figs. A & C), *this feature diagnostic*.

Flesh: Soft, whitish or yellowish.

Spore-bearing surface (hymenium): Smooth or with weak radial wrinkles and folds (Fig. C, arrow); buff-orange to yellowish-orange.

Stipe: 7.5 cm tall by 2.5 cm thick (3 × 1 in), tapering downward; solid, becoming hollow with age; pale yellow to yellow-orange or buffy-orange.

Spore print: Pale to pinkish-yellow.

Bruising: Hymenium bruising darker yellow; stipe bruising rusty-orange.

Chemical staining: Unknown.

Microscopic: Spores (6–)7–10 × (4.5–)5–6(–6.5) µm, ovoid to elliptical, somewhat flattened unilaterally, with off-center apiculus; smooth, clear, nonreactive in Melzer's reagent.

Occurrence: Mycorrhizal with hardwoods; alone or in scattered groups on the ground; occasional in southeastern North America and Mexico (type locality: vicinity of Orizaba, Veracruz, Mexico); summer.

Edibility: Edible.

Comments: Two or more caps emerging from a single stipe is the *sine qua non* that distinguishes *C. confluens* from *C. lateritius*. The two taxa are essentially very similar in all other respects both macroscopically and microscopically, including the smooth to modestly ridged hymenium. *Cantharellus lateritius* may have two to three stipes connected at their very bases, but they are never joined together halfway up, as shown here. *Cantharellus confluens* is much less common and never exhibits the pinkish, apricot, or salmon-orangish tints often seen in *C. lateritius*. Some authors consider the two taxa to be synonymous, in which case *C. confluens* would have nomenclatural priority.

Small Chanterelle

Cantharellus minor

A B C

Photos: A & B, W. Light; C, Nathan Wilson.

Cantharellus minor **Peck**, *Ann. Rep. Reg. N.Y. St. Mus.* 23:122 (1872) [1870]
[Basidiomycota, Agaricomycotina, Agaricomycetes, *incertae sedis*, Cantharellales, Cantharellaceae]

Derivation: *Cantharellus*, Latin, from *cantharus*, a large drinking vessel, and the diminutive ending *-ellus*; *minor*, Latin, "less," "smaller."

Cap: 0.5–3 cm (¼–1¼ in) across; initially convex, becoming flattened and centrally depressed (Fig. A) to shallowly vase-like or funnel-shaped; margin inrolled at first, becoming arched to uplifted and wavy; surface smooth, waxy; bright yellow (Figs. A–C) to orange-yellow.

Flesh: Thin, pale yellow to orange; odor not distinctive.

Spore-bearing surface (hymenium): With prominent, broad, distant, gill-like, decurrent ridges (Fig. C), these often forked (Fig. C, arrow); concolorous with the cap.

Stipe: 1.5–5 cm tall by 3–10 mm thick (⅔–2 × ⅒–ca. ½ in); smooth, subequal (Figs. A & C) to tapering or expanded toward the base (Fig. B), becoming hollow with age; concolorous with the cap.

Spore print: Pale yellow.

Chemical staining: Flesh and hymenium negative with $FeSO_4$.

Microscopic: Spores (6.5–)7–10(–11.5) × (4–)4.5–6.5 µm, ellipsoid to oblong, smooth, clear, with one or more oil droplets; nonreactive in Melzer's reagent.

Occurrence: Mycorrhizal with hardwoods, especially oaks (*Quercus* spp.) and hickories (*Carya* spp.); typically gregarious, common; eastern North America; summer–early fall.

Edibility: Edible, but not worth the effort.

Comments: The smallest of the chanterelles when mature, this tiny, bright-yellow to orange-yellow mushroom is usually recognized by sight. It is distinguished from small immature *C. cibarius* by its generally more slender proportions and common growth habit in moss (Fig. C).

Several species of waxy caps (*Hygrocybe* spp.) superficially resemble *C. minor*. However, all have true gills that are waxy-textured and readily torn away (unlike the nondetachable, fold-like pseudogills of *Cantharellus*) and white (rather than pale yellow) spore prints.

Peach-colored Chanterelle

Cantharellus persicinus

A

B

C

Photos: A, Eddee; B & C, Teresa Fortenberry.

Cantharellus persicinus **R.H. Petersen**, *Nova Hedwigia* 42(1):158 (1986) [1985]
[Basidiomycota, Agaricomycotina, Agaricomycetes, *incertae sedis*, Cantharellales, Cantharellaceae]

Derivation: *Cantharellus*, Latin, from *cantharus*, a large drinking vessel, plus the diminutive ending *-ellus*; *persicinus*, Latin, from *persicum*, "peach," meaning "peach-colored."

Cap: 2–7.5 cm across; initially convex, becoming broadly convex to planoconvex, margin inrolled at first, becoming wavy (Fig. C) or scalloped with age (Figs. A & B); salmon-pinkish-orange to peach-colored, becoming somewhat transparent when wet.

Flesh: White, odor fruity, taste unremarkable.

Spore-bearing surface (hymenium): With blunt, decurrent, forking, and occasionally anastomosing (alternately separating and rejoining) gill-like ridges, with or without cross-veins (Fig. C); concolorous with or paler than the cap.

Stipe: 2–6 cm tall by up to 1 cm thick (¾–2½ × ca. ¾ in), subequal, expanded toward apex; smooth, concolorous with cap, base whitish.

Spore print: Whitish to pinkish.

Chemical staining: Hymenium staining grey with $FeSO_4$.[1]

Bruising: Margin bruising yellowish; stipe bruising brownish-yellow.[1]

Microscopic: Spores 10–12 × 6–7.5 µm; elliptical, smooth, clear.

Occurrence: Mycorrhizal with oaks (*Quercus* spp.) and Eastern Hemlock (*Tsuga canadensis*); occasional in the Appalachian Mountains and vicinity; Pennsylvania, Tennessee, West Virginia, Georgia, possibly Arkansas, distributional limits unknown; summer–early fall.

Edibility: Edible.

Comments: The delicate and distinctive salmon-pink to peach-like color and convex to planoconvex, non-funnel-shaped cap distinguish this taxon from other chanterelles. The Smooth Chanterelle, *C. lateritius* (Berkeley) Singer, also often presents with pinkish tints, but the fruitbody is uplifted, deeply scalloped and lobed, and centrally depressed to funnel-shaped, and the hymenium is smooth to only faintly wrinkled or ridged; it never exhibits the prominent forked and cross-veined gill-like ridges typical of *C. persicinus*.

1. M. Kuo, February 2011, *Cantharellus persicinus*, Mushroom Expert.Com, http://www.mushroomexpert.com/cantharellus_persicinus.html.

Cinnabar-red Chanterelle

Cantharellus cinnabarinus

A

B

C

Photos: A & B, W. Light; C, Bob Zimmer.

***Cantharellus cinnabarinus* (Schweinitz) Schweinitz**, *Trans. Am. Phil. Soc.*, ser. 2, 4(2):153 (1832)

Basionym: *Agaricus cinnabarinus* Schweinitz, *Schr. Naturf. Ges. Leipzig* 1:73 (1822)
[Basidiomycota, Agaricomycotina, Agaricomycetes, *incertae sedis*, Cantharellales, Cantharellaceae]

Derivation: *Cantharellus*, Latin, from *cantharus*, a large drinking vessel, and the diminutive ending *-ellus*; *cinnabarinus*, Greek, *kinnávari*, "cinnabar," "vermillion."

Cap: 1–5 cm (½–2 in) across, initially convex, becoming flat to somewhat depressed centrally; margin incurved at first, becoming uplifted and irregularly wavy with age; surface smooth to matted; reddish-orange (Figs. A & B) to vermillion, fading to pink in maturity (Fig. C).

Flesh: Whitish, with reddish tinge near cuticle, not bruising when injured or cut; odor not distinctive; taste unremarkable or slightly peppery.

Spore-bearing surface (hymenium): With blunt, gill-like, decurrent, forking ridges (Fig. C), often with cross-veins; reddish-orange to pinkish-orange.

Stipe: 1–5 cm tall by 0.5–1.5 cm thick (½–2 × ¼–⅔ in); subequal or tapering downward, often curved; solid, dull reddish.

Spore print: Pinkish-cream.

Chemical staining: Flesh and hymenium negative to very pale grey with $FeSO_4$.

Microscopic: Spores 6–11 × 4–6 µm, elliptical to oblong, smooth, clear; nonreactive in Melzer's reagent.

Occurrence: Mycorrhizal with hardwoods; widespread in eastern North America; alone, scattered, or in dense groups on the ground, often in moss; quite common and abundant in some years, conspicuously absent in others; summer–fall.

Edibility: Edible, but not choice.

Comments: The small size, striking orange-red to cinnamon-red color, and blunt, decurrent ridges that are not easily separated from the cap immediately identify this species. Several superficially very similar waxycap species in the genus *Hygrocybe* are readily distinguished by their sharp-edged, nonforking, and pale yellowish to cream true gills (Fig. D at right), which are easily torn

D

away. Only two other pinkish or red chanterelles are found in North America: *Cantharellus texenis* Buyck, Cruaud, Couloux & Hofstetter and *C. persicinus* R.H. Petersen. The former closely resembles *C. cinnabarinus* but is restricted to southeastern Texas. The latter is found in the Appalachian Mountains but is subdued salmon-orange to pinkish-orange (never vermillion or bright reddish-orange), and the hymenium stains dark grey (not negative or faintly pale grey) with $FeSO_4$.

Corticioid Fungi

(Crusts, Spreads, and Parchment Fungi)

Corticioid is a general, nontaxonomic term referring to fungi that form more or less flat, spreading, or crust-like fruitbodies on the surfaces of living or dead hardwoods and conifers. These may form on the vertical trunks of trees or on the upper or lower surfaces or the sides of downed logs and branches. Many corticioid fungi are saprobic rotters of various trees and shrubs, and some are litter decomposers. Others form ectomycorrhizal relationships with or are parasitic on various plants.

The corticioid fungi include some 15 different orders belonging to the class Agaricomycetes (division Basidiomycota) having superficial macroscopic similarities; four of these orders—Atheliales, Corticiales, Trechisporales, and Jaapiales—have only corticioid forms.[1] These morphologies include effuse (broadly attached), effuse-reflexed (broadly attached and folded up at the margins and often form prominent pileate or cap-like shelves), resupinate (flat over the substratum, with the hymenium [fertile portion] at the edges or facing out over the whole surface), smooth to rugose, wrinkled or cracked, poroid, and even hydnate (toothed) morphologies, several of which can co-occur on the same fruitbody. Two of these corticioid types are shown in the following graphic.

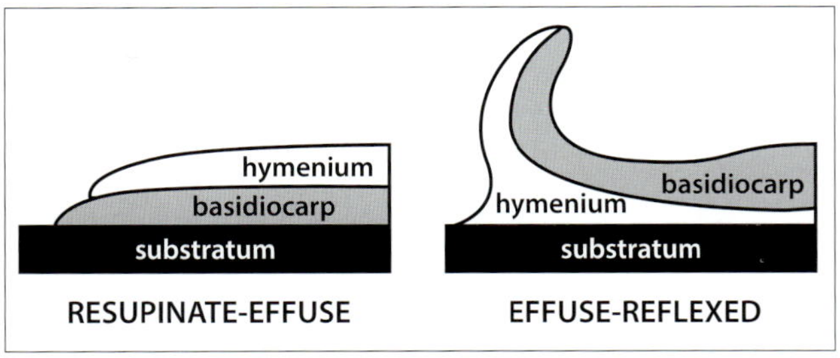

One member of the division Ascomycota, Chestnut Blight, *Cryphonectria parasitica*, and another species belonging to the Basidiomycota, Cedar-apple Rust, *Gymnosporangium juniperi-virginianae*, are also included here primarily because we have no better place to put them. Two other ascomycetes, the Common Tarcrust, *Diatrype stigma* (Hoffman) Fries, and the Hypoxylon Canker, *Biscogniauxia atropunctata* (Schweinitz) Pouzar, are not included here, even though they form distinctive crusts. However, they are presented in detail and keyed out in the section dealing with the carbonaceous fungi in the order Xylariales. Likewise, the Coral-pink and Trembling Phlebias, *Phlebia incarnata* and *P. tremellosa*, often treated as encrusting spreads, are here discussed in the section dealing with the polypores.

Parchment fungi of the genus *Stereum* are also included here, since these polymorphic mushrooms exhibit a combination of effuse, effuse-reflexed, and shelf-like fruitbodies. Thus, the False Turkey Tail, *Stereum ostrea*, is included here, while a close look-alike with which it is often confused by neophyte mushroomers, the Turkey Tail, *Trametes versicolor*, is dealt with in the section on polypores. This reflects a common problem we encountered in putting this book together: many of these taxa can equally well be grouped in various categories, none of which do them full justice.

1. M. Binder et al., 2012, *Mycologia* 102(4):865–880; this number (15) includes two new orders described in this paper.

Chestnut Blight

Cryphonectria parasitica

A

Cryphonectria parasitica. Spore masses on exposed roots of Chestnut Oak (*Quercus montana*) on Pine Mountain, Harris County, Georgia.

Cryphonectria parasitica **(Murrill) M.E. Barr**, *Mycol. Mem.* 7:143 (1978)
[Ascomycota, Pezizomycotina, Sordariomycetes, Sordariomycetidae, Diaporthales, Cryphonectriaceae][1]

Basionym: *Diaporthe parasitica* Murrill, *Torreya* 6:189 (1906) [see online Appendix A for synonymy]

Derivation: *Cryphonectria*, Latin, "hidden swimmer"; *parasitica*, Latin, "parasitic."

Fruitbodies (stromata): Yellowish to orange, pinhead-sized pycnidia sunken within the bark, later appearing on the exterior as orange exudates up to 3 mm wide and 2.5 mm high (Figs. A & B) during moist weather when the perithecia forcibly expel their orange spores. Stem cankers (Fig. B) swollen or sunken, the latter often covered with bark.

Chemical staining: Stromata stain purple in KOH and yellow in lactic acid.

Distribution: Native to Asia; widely introduced in Europe and North America, where it causes a devastating chestnut blight in native American Chestnuts (*Castanea dentata*). Also occurs on other hardwood species, particularly oaks (*Quercus* spp.), with less devastating effects.[2]

Comments: Introduced to North America from Japan or China around 1900, this highly virulent pathogenic fungus virtually eliminated all mature native chestnuts in the country by the 1940s.[3] Nevertheless, a small stand of chestnuts remains in Harris County, Georgia, as well as in California and the Pacific Northwest. Chestnut roots, which produce small trees as shoots, are resistant to the fungus. However, these shoots are usually killed by the

B

Cryphonectria parasitica spore masses on dead downed trunk of Northern Red Oak (*Quercus rubra*) along the Chattahoochee River, Alpharetta, Fulton County, Georgia.

fungus before they can produce seeds, although a sufficient number of seeds are generated to enable ongoing genetic studies to increase resistance. European and Asian chestnuts are also infected, but they have greater resistance to the disease. The fungus is rapidly spread by spores carried by wind and rain splash and on the feet of insects, birds, and other animals. A hypovirulent strain of the blight imported from Italy in the 1960s weakens the virulent form of the fungus and slows the course of infection. This allows the infected tree to develop resistance to the blight. However, inoculation of the hypovirulent strain into lethal cankers on American Chestnuts resulted in remission of disease only in those cankers. It took rigorous treatment of every single canker on all infected trees in a large grove in Connecticut for four years to establish biological remission of the disease. Hypovirulence has been shown to be conferred by a virus whose genome is inserted into the fungal genome, which then produces virons that are incorporated into the fungal spores.

1. The genus *Cryphonectria* was recently removed from the family Valsaceae and placed in its own newly erected family, the Cryphonectriaceae, along with the closely allied genus *Endothia* and certain undescribed related taxa (M. Gryzenhout et al., 2006, *Mycologia* 98[2]:239–249, http://www.mycologia.org/content/98/2/239.full-corresp-1). Indeed, the validity of the family Valsaceae was not supported in a previous study (L.A. Castlebury et al., 2003, *Mycologia* 94[6]:1017–1031). Because of nomenclatural and typification problems (*C. gyrosa*, the type species of the genus, actually belongs in the genus *Amphilogia*), the name *Cryphonectria* was conserved with *C. parasitica* designated as the neotype (H. Myburg, 2004, *Stud. Mycol.* 50:119–142; M. Gryzenhout et al., 2005, *Taxon* 54:539–540).

2. We recently found the characteristic orange spore exudates of this fungus on a dead Northern Red Oak (*Quercus rubra*) along the Chattahoochee River, Fulton County, Georgia (Fig. B). These *Cryphonectria* infections in oaks constitute a persistent wild reservoir for the fungus.

3. Prior to their catastrophic population crash, as many as one in four of all hardwood trees in eastern forests were American Chestnuts.

Cedar-apple Rust

Gymnosporangium juniperi-virginianae

A B C

A, second-year Cedar-apple Rust gall with developing telial horns on Eastern Redcedar before rain; B, gall with gelatinous teliospore-producing horns during spring rains; C, drying telial horns on gall following wet weather. (Photos: A & B, Botond Balogh; C, Jim Edwards.)

***Gymnosporangium juniperi-virginianae* Schweinitz**, *Schr. Naturf. Ges. Leipzig* 1:74 (1822)
[Basidiomycota, Urediniomycetes, *incertae sedis*, Uredinales, Pucciniaceae]

Derivation: *Gymnosporangium*, Greek, from *gumnós*, "naked," *spóros*, "seed," and *angeío*, "vessel," meaning "naked spore carrier"; *juniperi-virginianae*, Latin, genitive declensions of *iūniperus* and *virginia*, meaning "of the Virginia Juniper" (*Juniperus virginiana*).

Description: Infections in junipers first appear a year later as small greenish-brown swellings on twigs in the summer and early fall. These enlarge to a reddish-brown, kidney-shaped to semispherical gall from 2 mm up to 5 cm (~1/10–2 in) in diameter covered with rounded, bumpy swellings, each with a central, shallow, pocket-like pit. During the rainy season in the following spring (second year), each pit produces an elongate, orange-brown, telial horn (Fig. A) up to 4 cm (1⅔ in) long; as they absorb water, these horns transform into light orange to salmon-colored swollen, gelatinous tentacles (Fig. B) up to 10 cm (4 in) in length. During dry periods after the rains, the telial horns shrivel and harden, turn dark orange-brown again (Fig. C), and remain dormant until wet weather resumes, whereupon they swell up and become gelatinous again. This cycle can be repeated several times during a season.

Microscopic: Teliospores 2-celled, 42–65 × 15–21 µm, each cell bluntly conical, thick-walled, brown, borne on a long, thin, thread-like stalk.

Occurrence: Wherever apples or crabapples (*Malus* spp.) and Eastern Juniper (*Juniperus virginiana*) coexist.

Edibility: Inedible, of no interest.

Comments: *Gymnosporangium juniperi-virginianae* is a heteroecious parasitic rust fungus that needs two different hosts—a conifer and a hardwood—to complete its cycle, during which it produces four different kinds of spores. The primary hosts are junipers (*Juniperus* spp.)—in particular the Eastern Juniper, or Eastern Redcedar (*J. virginiana*)—and apples (*Malus domesticus*) or crab-apples (*M. sylvestris*) [Family Malaceae].

The complete life cycle of Cedar-apple Rust requires two years. Diploid ($2n$) teliospores produced by the telial horns of the galls on the juniper undergo meiosis to produce millions of haploid (n) basidiospores. The windborne basidiospores cannot reinfect a juniper but only an apple or crabapple (malacean) host tree up to 9¾ km (6 mi) away. Conversely, the windborne dikaryotic ($n + n$, with two separate haploid nuclei) aeciospores generated by cup-like aecia on the malacean tree can only infect a juniper. These $n + n$ dikaryons then fuse (karyogamy) during the following spring to form diploid ($2n$) teliospores to begin a new cycle. This rust does not seriously damage its juniper host but can cause severe losses in commercial apple and crabapple trees during a particularly wet spring.

Smooth Patch Disease, Hophornbeam Disc

Aleurodiscus oakesii

A

B

C

D

***Aleurodiscus oakesii* (Berkeley & M.A. Curtis) Cooke**, *Grevillea* 3(28):172 (1875)
[Basidiomycota, Agaricomycetes, *incertae sedis*, Russulales, Stereaceae]

Basionym: *Corticium oakesii* Berkeley & M.A. Curtis, *Grevillea* 1(11):166 (1873)

Derivation: *Aleuro-*, Greek, "flour"; *discus*, Greek, "disk"; the epithet *oakesii* is named after the 19th-century botanist William Oakes.

Fruitbody: Tiny cup-like fragments, 1.5–5.5 mm (1⁄25–¼ in) in diameter (up to 10 mm), initially disk- to cup-shaped (Figs. B & C), becoming more or less irregular with age and fusing with adjacent fruitbodies to form a larger mass (Fig. A), broadly but centrally attached, leathery, upper fertile surface pale yellowish-tan (Fig. C) to pale brown or pinkish-buff, fading to greyish-white, smooth; lower sterile surface whitish, minutely fuzzy or cottony, margin reflexed and inrolled (folded upward and curled over upper fertile surface) when dry (Figs. A & B); stipe lacking altogether.

Spore print: Not reported, but probably white.

Microscopic: Spores (16–)18–25(–27) µm by (10–)12–14(–17) µm, egg-shaped to elliptical, with pointed apical knob, mature spores with two germ pores at opposite ends, surface smooth to rough in KOH, spiny and amyloid (blue) in Melzer's reagent. Hymenium with filiform to subclavate acanthophyses bearing scattered acicular (needle-like) prongs on the sides, at the tip, or both.

Occurrence: Saprobic; scattered to clustered on living hardwoods, especially American Hophornbeam (*Ostrya virginiana*) and White Oak (*Quercus alba*); widely distributed in eastern North America, year-round in warm climates; Great Britain, China, Japan.

Edibility: Inedible, tough, too tiny to consider.

Comments: *Aleurodiscus oakesii*[1] consumes the dead outer bark of living trees, leaving behind a smooth, lighter-colored girdle known as "smooth patch disease" that is the telltale mark of this fungus (Fig. D). The mushrooms themselves are very inconspicuous and are not often seen. *Aleurodiscus oakesii* closely resembles *A. amorphus* (Persoon *ex* Fries) J. Schröter (1888). However, *A. amorphus* lacks microscopic spiny acanthophyses, the fertile surface is pinkish-orange, and it is restricted to conifers, especially *Abies*.

1. This species has been misidentified as *Aleurodiscus wakefieldiae* Boidin & Beller and the two species were erroneously synonymized by Binion et al., 2008, *Macrofungi Associated with Oaks of Eastern North America* (Morgantown: West Virginia University Press), 259. The two species are not in fact synonymous, and the fruitbodies are very different in color and form; *Aleurodiscus wakefieldiae* is also confined to Europe and has been transferred to the genus *Gloeosoma* Bresádola (*fide* Wu, Binder, & Hibbett, *Mycologia* 93[4]:720–731 [2001]).

White Patches

Dendrothele spp.

A

B

***Dendrothele* spp.**

Genus *Dendrothele* (*sensu stricto*)

Type species: *Dendrothele papillosa* von Höhnel & Litschauer, *Sber. Akad. Wiss. Wien, Math.-Naturw. Kl.*, Abt. 1, 116:820 (1907) [1906]
[Basidiomycota, Agaricomycotina, Agaricomycetes, *incertae sedis*, Corticiales, Corticiaceae]

Derivation: *Dendrothele*, Greek, from *dendro*, "tree," and *thelé*, "nipple" or "papilla," "tree nipple."

Fruitbody: Resupinate, upper hymenial (spore-bearing) surface facing up, surface smooth (Fig. A) to slightly roughened (Fig. B), forming more or less flat to undulating or rippled, irregularly shaped crusty patches of variable size, up to 2½ cm (1 in) or more by 1 cm (~½ in) thick, that assume the configuration of the underlying bark; white to pale greyish-white.

Spore print: White.

Chemical staining: **KOH**—strongly reactive, light to intermediate brown; **Melzer's reagent**—yellow.

Microscopic: Spores (basidiospores) of various shapes and sizes depending on the species; globose to elliptical, smooth, clear, walls thin or thick; nonamyloid (nonreactive in Melzer's reagent), often cyanophilous (staining red in acetocarmine).

Occurrence: Widespread and common in eastern North America on the bark of living hardwoods, especially oaks (*Quercus* spp.).

Edibility: Inedible.

Comments: These white crusty patches are completely resupinate, and only the hymenial surface is visible. The genus *Dendrothele* is polyphyletic and contains some 80 described species dispersed in 11 different evolutionary lineages.[1] However, by far most of these taxa belong to *Dendrothele sensu stricto* in the corticioid clade.

1. K.K. Nakasone, 2009, Morphological studies of *Dendrothele* species from North America, *North American Fungi* 4(7):1–15.

Milk-white Toothed Polypore

Irpex lacteus

A B C D

***Irpex lacteus* (Fries) Fries**, *Elench. Fung.* (Greifswald) 1:142 (1828)

Basionym: *Sistotrema lacteum* Fries, *Observ. Mycol.* (Havniae) 2:266 (1818)
[Basidiomycota, Agaricomycotina, Agaricomysetes, *incertae sedis*, Polyporales, Merulaceae]

Derivation: *Irpex*, Latin, "rake"; *lacteus*, Latin, "milky."

Fruitbody: A broadly attached (effuse), white to cream-colored resupinate spread (Fig. A), becoming yellowish with age, growing on undersides and sides of fallen logs, often with reflexed margins forming shelf-like to cap-like edges; fertile surface facing outward, initially with 2–3 angular pores per mm (Fig. B); tubes generally 3 mm or less deep, soon splitting and eroding, becoming irregularly tooth-like, then forming a labyrinth of jagged walls (Figs. C & D);[1] when present, caps are whitish, drying yellowish, with a velvety upper surface, 1–4 cm (½–1⅔ in) wide, small relative to extensive spread area.

Flesh: Whitish, thin, tough.

Stipe: Fruitbody effuse, stipe lacking.

Spore print: White.

Chemical staining: All parts brownish-orange with KOH.

Microscopic: Spores 5–7 × 2–3 µm, elliptical to subcylindrical, smooth, clear, thin-walled, nonreactive in KOH and Melzer's reagent; with conspicuous encrusted cystidia, 50–110 × 5–10 µm, clamp connections lacking.

Occurrence: Saprobic on undersides of dead, fallen branches and logs of hardwood trees; widespread and common in most of North America, Europe, Asia; annual, overwintering.

Edibility: Inedible.

Comments: Though a polypore, *I. lacteus* looks for all the world like a hydnate (toothed) spread. Although the pore surface superficially resembles that of the Thick-toothed Polypore, *Spongipellis pachyodon*, the maze-like teeth are much shorter in *I. lacteus* (≤3 mm vs. 6–12 mm), and you usually have to look closely to see them. The effuse fruitbody also grows on the undersides of fallen trees and logs, and the spores are cylindrical (5–7 × 2–3 µm) and thin-walled rather than subglobose ([5–]5.5–7[–7.5] µm) and thick-walled.

1. The older literature often uses the term **iripiciform** to describe these jagged labyrinths.

Snow Dust Fungus

Athelia epiphylla (*sensu lato*)

Athelia epiphylla **Persoon**, *Mycol. Eur.* (Erlanga) 1:84 (1822) [Basidiomycota, Agaricomycotina, Agaricomycetes, Agaricomycetidae, Atheliales, Atheliaceae]

Derivation: *Athelia*, Greek, from *a-*, "not," plus *thele*, "nipples," "lacking nipples" (perithecia); *epiphylla*, Greek, from *epi-*, "upon," and *phullon*, "leaf."

Fruitbody: White to cream-colored, resupinate, sessile, broadly spreading, smooth to slightly bumpy; thin, soft, delicate, and membranous; loosely attached to substratum by diaphanous subiculum, with a gossamer margin (inset).

Spores in mass: Presumably white.

Microscopic: Spores 6–8(–10) × 3–4(–5) µm, ellipsoid, smooth, thin-walled, clear, occasionally guttulate (containing oil droplets); nonreactive in Melzer's reagent; basidia 4-sterigmate; generative hyphae monomitic, with clamp connections; subicular hyphae without clamps.

Occurrence: Saprobic on a variety of hardwood, conifer, and lichen substrata (branches, bark, leaves, thalli); widespread in North America, Europe, the Caribbean.

Edibility: Inedible.

Comments: The limits of the several white snow-dust-like spreads of the genus *Athelia* are quite uncertain. A number of species and several varieties within many species have been described. All are macroscopically nearly identical, and they also cannot easily be distinguished microscopically. *Athelia epiphylla* (*sensu lato*) is currently best considered as representing a widespread species complex.

Ceramic Parchment

Xylobolus frustulatus

A

B

***Xylobolus frustulatus* (Persoon) Boidin**, *Revue de Mycologia* 23(3):318–346 (1958)

Basionym: *Thelephora sinuans* Persoon, *Mycol. Eur.* (Erlanga) 1:128 (1822)
[Basidiomycota, Agaricomycotina, Agaricomycetes, *incertae sedis*, Russulales, Stereaceae]

Derivation: *Xylobolus*, Greek, from *xylon*, "wood," and *-bolus*, "lump"; *frustulatus*, Latin, from *frustum*, "piece," "bit," "flake."

Fruitbody: Resupinate, consisting of large, densely crowded, and closely adpressed aggregations of small, crust-like, hard, flat, irregularly angular, ceramic-like frustules up to 1–2 cm (⅖–¾ in) across and 1–2 mm thick (occasionally thicker) (Fig. A); upper (fertile) surface smooth to minutely pimpled, initially whitish to cream-colored (Fig. A), becoming yellow- to grey-brown (Fig. B) with age; sides (sterile surface) dark brown to black.

Spore print: Pinkish-orange.

Microscopic: Spores 3.5–7 × 2.5–3.5 µm, smooth, elliptical, clear, thin-walled, amyloid in Melzer's reagent; hyphae monomitic, with numerous spine-like acanthocystidia, clamp connections lacking.

Occurrence: Saprobic, causing pocket white rot; perennial, often forming extensive patches of densely crowded frustules on well-decayed logs and branches of oaks (*Quercus* spp.); widespread and common in eastern and south-central North America, Mexico, Central America, Brazil, northern Europe (also Spain and Turkey), Japan, Taiwan, and Malaysia.

Edibility: Inedible.

Comments: This species is not likely to be mistaken for anything else despite suggestions to the contrary in some field guides and online sources. The densely crowded, polygonal, tile-like frustules on dead oak logs and branches immediately identify this fungus in the field. The genus *Xylobolus* P. Karsten, with just four species, is closely related to the genus *Stereum* Hill *ex* Persoon. Spores are released during summer, and most examples are sterile from fall through early spring, when basidia are absent and the hymenium consists mostly of acanthocystidia.

Ceramic Parchment produces frustulosin, an antibiotic that is effective at concentrations of 16 µg/mL against *Staphylococcus aureus* (a human pathogen), as well as *Bacillus mycoides* and *B. subtilis* (common soil bacteria—the last also being a ubiquitous and generally harmless commensal in the human gut).

Ochre Spreading Tooth

Steccherinum ochraceum

A

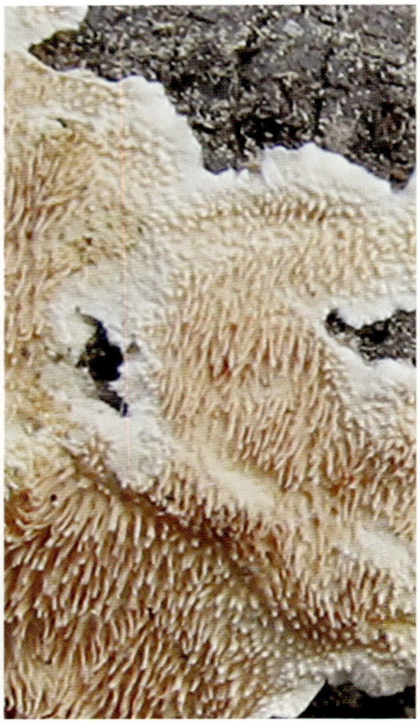

B

***Steccherinum ochraceum* (Persoon) Gray**, *Nat. Arr. Brit. Pl.* (London) 1:651 (1821)

Basionym: *Hydnum ochraceum* Persoon, in J.F. Gmelin, *Systema Naturae*, 13th ed., 2(2):1440 (1792)
[Basidiomycota, Agaricomycotina, Agaricomycetes, Polyporales, Merulaceae]

Derivation: *Steccherinum*, Latin, from *steccherino*, an Italian common name for the prominently toothed Wood Hedgehog mushroom (*Hydnum repandum*); *ochraceum*, Latin, "orange-yellow," "ochre-colored."

Fruitbody: Variable in form, typically a resupinate or effuse-reflexed spread up to 20 cm (8 in) in extent, with or without small, semicircular to fan-shaped shelf-like caps 2–4.5 cm (¾–1¾ in) across (Fig. A, arrow), which may fuse laterally; orange-yellow, orange-tan, pale pinkish-ochre, to pale salmon; margins white, crenate (scalloped) or lobed (Figs. A & B), fimbriate (with a fringed border).

Flesh: Thin, 2–3 mm thick, leathery; white to pale tan; duplex; odor and taste not distinctive.

Upper (adpressed) surface: When present, concentrically furrowed, velvety to woolly, with concentric zones of greyish, brownish, or whitish color.

Spore-bearing surface: Densely covered (3–5 per mm) with prominent, simple (rarely forked), subcylindrical to conical spines 1.5–3 mm tall (Fig. B); these are yellow-orange to ochraceous at first, becoming brownish or reddish with age and salmon-colored when dry.

Stipe: Lacking in sessile basidiomata; when present up to 8 mm tall by 2 mm thick (⅓–1⁄25 in); concolorous with cap.

Spores in mass: White.

Chemical staining: Red with Melzer's reagent; green on Melzer's stain with KOH.

Microscopic: Spores 3–4(–5.5) × (1.8–)2.5–3.5 µm, ovoid to ellipsoid, flattened on one side, with small, oblique appendiculus (hilar appendage); clear, smooth, thin-walled, some spores uniguttulate (containing a single oil droplet); nonreactive in Melzer's reagent. Many large cystidia, these distally expanded and heavily encrusted.

Occurrence: Annual, saprobic on dead hardwood twigs, branches, and stumps: Eastern Beech (*Fagus grandifolia*), maples (*Acer* spp.), oaks (*Quercus* spp.), viburnum (*Viburnum* spp.). Widespread and fairly common throughout central and eastern North America; summer–fall to year-round.

Edibility: Inedible.

Comments: This very attractive and distinctive spread is not likely to be confused with any other species. The outward-facing and prominently toothed, salmon-colored basidiome does not resemble any other taxon in our area.

Job's Brown Crust

Hymenochaete jobii

Hymenochaete jobii **Parmasto**, *Folia Cryptog. Estonica* 37:62 (2001) [2000][1]
[Basidiomycota, Agaricomycotina, Agaricomycetes, *incertae sedis*, Hymenochaetales, Hymenochaetaceae]

Derivation: *Hymenochaete*, Greek, from *hymḗn*, "membrane," and *chaitē*, "hair," "bristle"; *fuliginosa*, Latin, "brown."

Deprecated synonyms: *Hymenochaete ungulata* Burt, *Ann. Missouri Bot. Gard.* 5:338 (1918)

Hymenochaete fuliginosa (*sensu auct. amer.*); [*non Hymenochaete fuliginosa* (Persoon) Léveillé (1846)]

Hymenochaete corticolor sensu Reaves & Welden, *Mycologia* 59(6):1043 (1967); *non* Berkeley & Ravenel (1873) [= *Hymenochaete cervina* Berkeley & M.A. Curtis (1868)]

Fruitbody: Effuse-resupinate (spreading, with fertile surface facing outward), not producing caps, closely attached and not readily separable from substratum, hard when dry; hymenium with scattered or densely aggregated, low, rounded tubercles (warty nodules); surface azonate, either not cracked or with only a few irregular cracks; greyish-brown to café-au-lait, later becoming brown to dark chocolate-brown; margin at first fibrillose and whitish or yellowish-brown, later concolorous with the hymenium.

Flesh: Dark brown.

Spores in mass: White.

Chemical staining: Flesh staining blackish with KOH.[2]

Microscopic: Spores 4.5–6 × 2.2–3.2(–3.5) µm, elliptical, slightly flattened on one side, smooth, clear; hymenium with numerous embedded[3] and projecting (up to 60 µm), stout, rust-colored, awl-like setae (50–)60–110(–120) µm long × 7–12(–15) µm thick, length-to-width ratio of 8.8–11:25 (inset), these naked or with hyphal sheath and lacking encrustations; tramal (flesh) hyphae monomitic, clamp connections lacking.

Occurrence: Saprobic, causing a white rot on various dead hardwoods, especially American Beech (*Fagus grandifolia*) in our area.

Edibility: Inedible.

Comments: The Cinnamon Crust, *Hymenochaete cinnamomea* (Persoon) Bresádola, is very similar and could be mistaken for *H. jobii*. The fruitbody differs in being loosely attached and easily separable from the substratum, smooth when young, and deeply and irregularly cracked with age. It does not exhibit the scattered, low, rounded tubercles (nodules) seen in *H. jobii* and shown in the photograph. It also differs in the proportions of its numerous hymenial setae, which are long, thin, and needle-like (rather than stoutly awl-like, as shown in the inset). They project up to 90 µm from the hymenium, measure 80–100 × 4–5 µm, and have a length-to-width ratio of 16–20:1—twice that of *H. jobii*. This species is solely represented in North America by *Hymenochaete cinnamomea* subsp. *spreta* (Peck) Parmasto, which is also restricted to this continent. It grows on a wide range of dead hardwoods throughout North America as well as on certain conifers (particularly in the West and in Mexico).

Hymenochaete fuliginosa (Persoon) Léveillé is macroscopically indistinguishable from *H. jobii*, and the two taxa have been confused in the literature. It is distinguished from the latter taxon by its cylindrical and somewhat sausage-shaped (not ellipsoid and inequilaterally flattened) spores (4.5–6.5[–7] × 1.8–2.6[–2.8] µm), wider setae (55–110 × 7–15 µm) with a smaller length-to-width ratio of 7:8 (compared to 8:11), and growth on dead conifers instead of hardwoods. *Hymenochaete fuliginosa* does not occur in the Southeast. Reports of *H. fuliginosa* in our area (e.g., Petersen, 1976; Jung, 1985)[4,5] and of this species growing on deciduous wood (e.g., Burt, 1918)[6] and having ellipsoid or ovoid spores may actually be misidentified *H. jobii*.[1]

Hymenochaete cervina Berkeley & M.A. Curtis is similar to *H. jobii* but is generally darker brown. The setae are much like those of *H. jobii* but are encrusted and generally project less (usually 25 µm, rarely to 60 µm) from the hymenium. Crystalline masses are also present in both the hymenium and the context (flesh). Both encrusted setae and crystalline masses in the context are lacking in *H. jobii*. In our area, both species are most often found on American Beech.

Hymenochaete corrugata (Fries) Léveillé has shorter and wider hymenial setae, 35–80(–100) × (6–)7–15(–17) µm (length-to-width ratio 5:6), whose tips are almost always encrusted with small, irregular granules; the spores are 4.5–6.8(–7) × 1.5–2.3 µm, cylindrical and curved (subsausage-shaped).

Other species of *Hymenochaete* all have effuse-reflexed fruitbodies, generally with well-developed caps that are not present in the foregoing taxa.

1. *Fide* E. Parmasto, 2001, Hymenochaetoid fungi (Basidiomycota) of North America, *Mycotaxon* 79:107–176.

2. This is not from a specific extractable pigment but is a xanthochroic reaction to KOH and other alkaline reagents of unstable polymerized polyphenols (E. Parmasto, 1979, *Mycotaxon* 8[1]:201–232).

3. The tramal layer is extremely dense, and the embedded setae are almost impossible to see (see inset in the above image).

4. R.H. Petersen, 1976, *Checklist of Fungi of the Great Smoky Mountains National Park*, Management Report No. 29, Department of the Interior, National Park Service, Southeast Region.

5. H.S. Jung, 1985, A floristic and taxonomic study of the wood-rotting Aphyllophorales of the spruce-fir forest of the Great Smoky Mountains National Park (Ph.D. diss., University of Tennessee, Knoxville), xxiv + 370 pp.

6. E. Burt, 1918, Thelephoraceae of North America: X. Hymenochaete, *Ann. Missouri Bot. Gard.* 5:301–372.

Brown-toothed Crust

Hydnochaete olivacea

A B

***Hydnochaete olivacea* (Schweinitz *ex* Fries) Banker**, *Mycologia* 6(5):234 (1914)

Basionym: *Sistotrema olivaceum* Schwein., *Schr. Naturf. Ges. Leipzig* 1:101 (1822)
[Basidiomycota, Agaricomycotina, Agaricomycetes, *incertae sedis*, Hymenochaetales, Hymenochaetaceae]

Derivation: *Hydnochaete*, Greek, from *hydnum*, "spongy fungus," and *khaitē*, "hair," "bristle"; *olivacea*, Latin, from *olivaceo*, "olive-colored."

Fruitbody: A spreading, closely adnexed, resupinate crust up to 30 cm long by 10 cm wide (12 × 4 in) or even larger; fertile surface densely covered with cylindrical, sharply pointed (Fig. A) to flattened, ragged teeth (Fig. B) up to 6 mm (¼ in) long; dull yellowish- to dull reddish-brown or olive-brown; surface dry, leathery.

Flesh: Up to 3 mm thick, leathery, concolorous with surface; odor and taste not distinctive.

Spore print: White or cream.

Chemical staining: All parts dark brown to black with KOH.

Microscopic: 4.5–7 × 1–1.5 µm, sausage-shaped, smooth, clear; teeth covered with tiny (30–150 µm long), reddish-brown, hair-like setae visible with a 15× hand lens.

Occurrence: Saprobic, covering undersides of dead hardwood branches and twigs, especially oaks (*Quercus* spp.); perennial, spring–winter, often persisting for years; widespread and common in eastern North America. Mexico,[1] China.[2]

Edibility: Inedible.

Comments: This species is not likely to be confused with any other. The characteristic olivaceous to reddish-brown spreading crust densely covered with prominent teeth on the undersides of small hardwood branches and twigs readily identifies this taxon in the field. *Hydnochaete* is an artificial genus erected for spreads with effuse-reflexed fruitbodies and a toothed hymenophore.

1. F. Chanona-Gómez et al., 2007, *Revista Mexicana de Biodiversidad* 78:369–381.

2. Y.-C. Dai et al., 2004, *Ann. Bot. Fennici* 41:233–247.

Cobalt Crust, Velvet Blue Spread

Terana caerulea

A

B

C

***Terana caerulea* (Lamarck) Kuntze**, *Rev. Gen. Pl.* (Leipzig) 2:872 (1891)
[Basidiomycota, Agaricomycotina, Agaricomycetes, *incertae sedis*, Polyporales, Phanerochaetaceae]

Basionym: *Byssus caerulea* Lamarck, *Flore Françoise* 1:103 (1779)

Derivation: *Terana*, derivation unknown; *caerulea*, Latin, "azure," "dark blue."

Fruitbody: A resupinate crust or spread 2–6 mm (up to ¼ in) thick with hymenium (spore-bearing surface) directed away from the substratum, firmly attached to the substratum except at the margins; many fruitbodies broadly coalescing to form extensive patches; outer surface dark blue, indigo, royal-blue, ultramarine, to deep violet, often silvery-white at margins; velvety to waxy when moist; brittle and crusty when dry.

Spore print: White.

Microscopic: Spores (6–)7–10(–12) × 4–6(–7) µm, ellipsoid, smooth, thin-walled, clear; negative in Melzer's reagent.

Occurrence: Saprobic, causing white rot. Widespread but occasional, typically in warm, damp, hardwood forests on the undersides of fallen logs and branches of deciduous trees; annual, August–November; may overwinter in warmer climes. Nearly cosmopolitan: North America (especially in the Southeast), Europe, Asia, Africa.

Edibility: Inedible.

Comments: This beautiful spread is not likely to be confused with any other species. However, it is often overlooked because it tends to grow underneath fallen branches and twigs. This inverted orientation of the hymenium facilitates spore dispersal. The blue color is imparted by derivatives of the terphenylquinone pigment thelephoric acid.

This species is called *Pulcherricium caeruleum* in some field guides and online sources, where it is sometimes incorrectly attributed to Persoon rather than Lamarck. The specific epithet is also sometimes improperly emended as "*coeruleum*" or "*coerulea*." See online Appendix A for the nomenclatural details.

Purple Crust Fungus

Porostereum crassum

A B C

***Porostereum crassum* (Léveillé) Hjortstam & Ryvarden**, *Syn. Fung.* (Oslo) 4:29 (1990)

Basionym: *Thelephora crassa* Léveillé, *Annls Sci. Nat.*, Bot., sér. 3, 2:209 (1844)
[Basidiomycota, Agaricomycotina, Agaricomycetes, *incertae sedis*, Polyporales, Phanerochaetaceae]

Derivation: *Porostereum*, Greek, from *pōros*, "pore," and *stereos*, "thick," "tough"; *crassus*, Latin, "thick."

Fruitbody: Highly variable, resupinate, effuse-reflexed (Figs. A & C), initially disk-like, converging to form a continuous spread up to 20 cm (8 in) or more in extent; no visible pores, surface velvety due to pale brown metuloid (thick-walled) cystidia projecting from the hymenium; exhibiting varying tones of vivid violet, pinkish-purple, to purplish-brown, margins lighter (Figs. A & C) and fringed (Fig. B), becoming brown at maturity.

Flesh: Thin, brown.

Microscopic: Spores (5.5–)6–7 × 3–4 µm, long-elliptical, smooth, thin-walled, clear, nonreactive in Melzer's reagent.

Occurrence: Saprobic on various deciduous trees; fairly common and pantropical; temperate North America, Central and South America, Southeast Asia, Africa, Australia, New Zealand; one record each from Poland and northern Iberia.

Edibility: Inedible.

Comments: The very similar Silverleaf Fungus, *Chondrostereum pupureum* (Persoon) Pouzar, has a smooth (not velvety) surface and tends to develop projecting large lobe-like to fan-shaped caps and overlapping shelves up to 4 cm (~1⅔ in) wide. The thin flesh is white (rather than brown), and it is typically found on *Prunus* spp. (cherries, apples, apricots, and plums). It also has a more northerly distribution than *P. crassum* and is not likely in our area.

This species is also referred to in the literature as *Lopharia crassa* (Léveillé) Boidin, *Phanerochaete crassa* (Léveillé) Burdsall, and *Hjortstamia crassa* (Léveillé) Boidin & Gilles. This species continues to be placed in different genera by various authors who variously describe the fruitbody as monomitic, pseudodimitic, and dimitic and who give different emphasis to such features as the color of the hyphae and presence or absence of encrusted cystidia.

White-and-red Crust

Peniophora albobadia

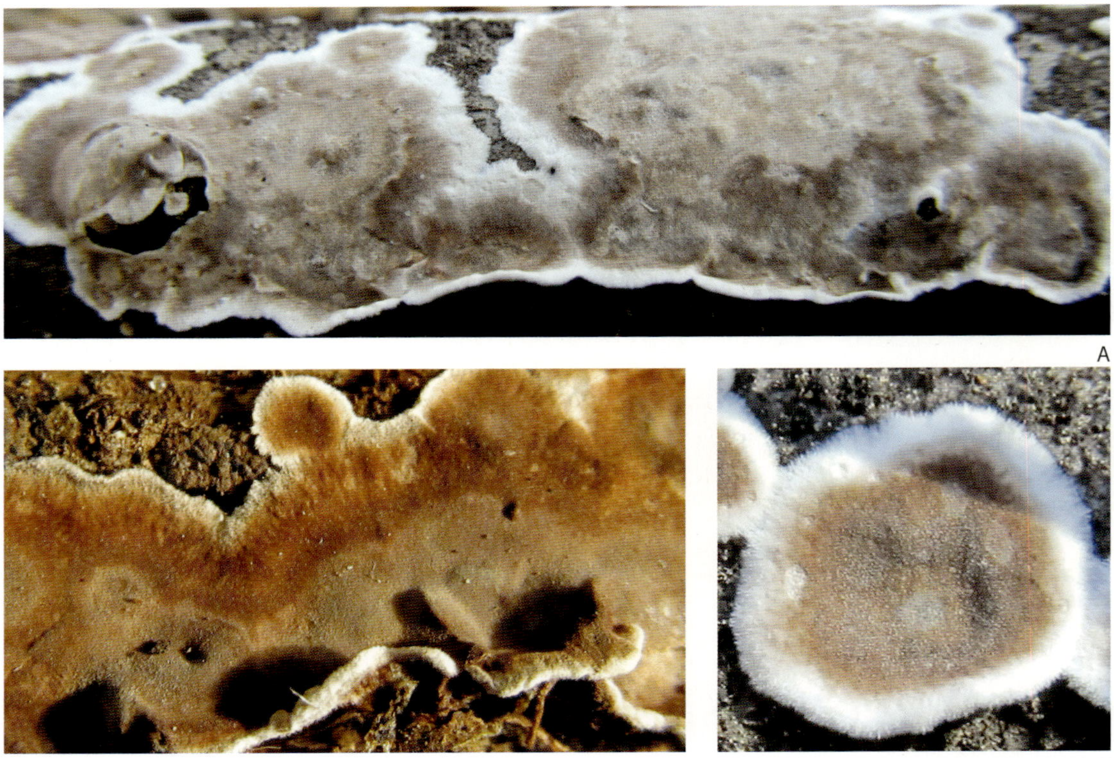

A B C

***Peniophora albobadia* (Schweinitz *ex* Fries) Boidin**, *Revue Mycol.*, Paris 26(3):164 (1961)

Basionym: *Thelephora albobadia* Schweinitz, *Schr. Naturf. Ges. Leipzig* 1:108 (1822)
[Basidiomycota, Agaricomycotina, Agaricomycetes, *incertae sedis*, Russulales, Peniophoraceae]

Derivation: *Peniophora*, Greek, presumably from *penia*, "lack," and *phoros*, "bearing a structure," "lacking a structure"; *albobadia*, Latin, "white-red."

Fruitbody: Leathery, thin, initially resupinate, circular (Fig. C), 5–10 mm (¼–½ in) in diameter, enlarging and coalescing up to 2–5 cm wide and 15–20 cm or more in length (¾–2 × 6–8 in) (Figs. A & B); sometimes narrowly reflexed, upper surface of reflexed portions silky to villose (velvety); hymenial surface velvety, buff-brown to dark brown with areas of purplish-brown to cinnamon-brown, fading to greyish-brown, often somewhat zonate when reflexed (Figs. A & B) about 5 mm (¼ in); margin covered with silky hairs (Fig. C), usually whitish (Figs. A–C) or pale buff.

Spores in mass: White.

Chemical staining: Red in Melzer's reagent; greenish in KOH.

Microscopic: Spores 10.5–13 × 3–4.5(–5) µm, long-ellipsoid to peanut-shaped, flattened on one side, with small off-center apiculus, smooth, clear, nonreactive in Melzer's reagent; 4-sterigmate; clamp connections and encrusted cystidia present.

Occurrence: Saprobic, causing a white rot on dead and fallen deciduous wood of various species; widespread and common in North America year-round; Central and South America, West Indies.

Edibility: Inedible.

Comments: *Peniophora albobadia* is readily recognized by its velvety purplish-brown to rusty-brown resupinate hymenium and distinctive white tomentose border. This species is found only in the New World.

Rosy Crust

Peniophora incarnata

***Peniophora incarnata* (Persoon) P. Karsten**, *Hedwigia* 28:27 (1889)

Basionym: *Thelephora incarnata* Persoon, *Syn. Meth. Fung* (Göttingen) 2:573 (1801)
[Basidiomycota, Agaricomycotina, Agaricomycetes, *incertae sedis*, Russulales, Peniophoraceae]

Derivation: *Peniophora*, Greek, presumably from *penia*, "lack," and *phoros*, "bearing a structure," "lacking a structure"; *incarnata*, Latin, "made flesh," referring to the color.

Fruitbody: Resupinate, forming thin, firmly attached circular patches 2–10 mm (1⁄25–½ in) in diameter or extensive coalesced linear spreads up to 15 cm (6 in) or more in extent on branches; pale pink, salmon-pink, to pale orange, with fringed white margins; surface waxy, initially smooth, becoming papillate, fissured, and irregularly cracked with age.

Flesh: Thin (up to 0.1–0.3 mm thick), white.

Spores in mass: Light pink to pale red.

Microscopic: Spores (7–)8–10(–12) × 3–5 µm, narrowly elliptical to cylindrical to suballantoid (sub-sausage-shaped), with an apiculus, smooth, thin-walled, slightly indented on one side; nonreactive in Melzer's reagent; cystidia encrusted with calcium salt crystals found throughout the hymenium; monomitic, generative hyphae with large clamp connections.

Occurrence: Parasitic and saprobic. On trunks and undersides of branches of a variety of deciduous trees, especially alder (*Alnus* spp.); annual and common throughout temperate, boreal, and arctic North America; Europe, Great Britain, Iceland, Faroe Islands, Svalbard, Northwest Africa, Japan, Australia, New Zealand.[1]

Edibility: Inedible.

Comments: This species is macroscopically nearly identical to two closely related taxa, both of which are restricted to alder branches (*Alnus* spp.) and have a more northern distribution; neither species appears to occur in our area. *Peniophora aurantiaca* (Bresádola) Höhnel & Litschauer is orange-red to pale orange, has flesh up to 1 mm thick, clamp connections, and larger, guttulate (containing oil droplets) spores, (14–)16–17 × (7.5–)8–10 µm, than *P. incarnata*. *Peniophora erikssonii* Boidin is macroscopically identical, but clamp connections are lacking.

1. A report of *P. incarnata* in Taiwan (S.H. Lin & Z.C. Chen, 1990, *Taiwania* 35:69–111) is erroneous. This taxon is actually *Hyphoderma variolosum* Boidin, Lanquetin, & Gilles in the order Polyorales, family Meruliaceae (*fide* S.-H. Wu, 2002, *Bot. Bull. Acad. Sin.* 43:249).

Conifer Parchment

Peniophora gigantea

Peniophora gigantea **(Fries) Massee**, *Brit. Fung.-Fl.* (London) 1:110 (1892)

Basionym: *Thelephora gigantea* Fries, *Observ. Mycol.* (Havniae) 1:152 (1815)
[Basidiomycota, Agaricomycotina, Agaricomycetes, *incertae sedis*, Russulales, Peniophoraceae]

Derivation: *Peniophora*, Greek, presumably from *penia*, "lack," and from *phoros*, "bearing a structure," "lacking a structure"; *gigantea*, Latin, "connected with the giants."

Synonym: *Phlebiopsis gigantea* (Fries) Jülich (1978)

Fruitbody: An effuse-reflexed greyish-white to pinkish-buff crust; when wet, forming an extensive, swollen, waxy sheet 3–30 cm (1¼–12 in) in extent; extensively cracked, fragmented, and tattered; when dry, horny, parchment-like, whitish margins often detached from host and strongly upcurled.

Flesh: White, waxy, becoming horny and parchment-like.

Spores in mass: White.

Microscopic: Spores 4.5–5(–8) × 2.5–3(–4) µm, elliptical, smooth, clear; with encrusted cystidia.

Occurrence: Saprobic, causing a white rot on stumps, fallen branches, and wood of dead conifers, especially on pines (*Pinus* spp.) and Eastern Hemlock (*Tsuga canadensis*) in our area; widespread in North America, summer–midwinter; Central and South America, Europe; East Africa, southern Asia, China, Japan, Australia.

Edibility: Inedible.

Comments: This distinctive species is unlikely to be confused with any other. On contact the hyphae of this decomposing fungus potently antagonize the hyphae of an extremely destructive parasitic root-rot polypore, *Heterobasidion annosum* (Fries) Brefeld (1888)—a process known as hyphal interference. Any *Heterobasidion* hyphae contacting a *P. gigantea* hypha undergo rapid membrane disruption and cellular disorganization. *Peniophora gigantea* is weakly parasitic but poses no threat to healthy trees. Artificial inoculation of pine stumps with *P. gigantea* significantly reduces *Heterobasidion* root-rot infection and was widely used as a biological control agent in the Southeast until 1995, when the U.S. Environmental Protection Agency required its registration as a biological pesticide—a process too expensive for its continued commercial use in this country. It is still available in the United Kingdom.

Crowded Parchment

Stereum complicatum

A B C D

***Stereum complicatum* (Fries) Fries**, *Epicr. Syst. Mycol.* (Upsaliae):548 (1838) [1836–1838]

Basionym: *Thelephora complicata* Fries, *Elench. Fung.* (Greifswald) 1:179 (1828)
[Basidiomycota, Agaricomycotina, Agaricomycetes, *incertae sedis*, Russulales, Stereaceae]

Derivation: *Stereum*, Greek, from *stereós*, "tough," "sturdy," "rigid"; *complicatum*, Latin, "folded back," referring to the reflexed habit of this fungus.

Deprecated synonym: *Stereum rameale* (Schweinitz) Burt, *Ann. Missouri Bot. Gard.* 7(2–3):169 (1920)[1]

Fruitbody: Sessile. Effuse (resupinate, Fig. B) to effuse-reflexed, reflexed (cap-like) portions 0.5–2 cm (¼–¾ in) across, typically laterally fused (Figs. A & B), semicircular (Fig. A) to fan-shaped (Figs. B–D), margins often crimped, crenulate, or lacerate (Figs. C & D); upper surface yellowish-brown, orange-yellow, pinkish-tan, to various shades of reddish-brown, greyish, with paler margin; azonate (Figs. A & B) to zonate and radially furrowed (Fig. D).

Flesh: Thin, leathery, and tough; whitish to buff; odor and taste not distinctive.

Lower surface: Smooth, pale orange, tan to cream- or cinnamon-fawn; may bruise red.

Spore print: White, if you can obtain it.

Chemical staining: Upper surface red, then turning black with KOH.

Microscopic: Spores 5–6.5(–7) × 2–2.5(–3) µm, smooth, clear, cylindrical to slightly curved; amyloid (blue) in Melzer's reagent.

Occurrence: Saprobic, causing a white rot in hardwoods. In effuse mats (Figs. A & B) and/or overlapping and/or laterally fused clusters (Fig. B), commonly fused laterally to form multiple rows covering the twigs, branches, and trunks of fallen hardwoods, especially oaks (*Quercus* spp.); year-round; widespread in North America (except Rocky Mountains) and Europe.

Edibility: Inedible.

Comments: Fresh *Stereum complicatum* may bruise red when injured but does not exude a red liquid. The Bleeding Oak Crust, *Stereum gausapatum* (Fries) Fries, is also found on dead and fallen oaks, and fresh specimens do bleed a reddish liquid, which turns brown when cut;

it first bruises red, then dark violet or blackish. *Stereum sanguinolentum* (Albertini & Schweinitz *ex* Fries) Fries also bleeds a red liquid when cut but is found on conifers.

1. Although *Thelephora ramealis* Schweinitz (1822) predates *Thelephora complicatum* Fries (1838), as a sanctioned name under Article 15, the latter name is automatically conserved against all earlier synonyms or homonyms. Most (but not all) American authors accepting the synonymy between these two names correctly deprecate *Stereum rameale*. Most European authors, on the other hand, fail to understand that Fries's sanctioned name has precedence and incorrectly list *S. complicatum* as a deprecated synonym of *S. rameale*.

False Turkey Tail

Stereum ostrea

A B C D

Stereum ostrea: A & B, fruitbodies; C, pore surface; D, *S. ostrea* with *Phlebia incarnata*, with which it is often asssociated.

***Stereum ostrea* (Blume & T. Nees) Fries**, *Epicr. Syst. Mycol.* (Upsaliae):547 (1838)

Basionym: *Thelephora ostrea* Blume & T. Nees, *Nova Acta Acad. Caes. Leop.-Carol.* 13:13 (1826)
[Basidiomycota, Agaricomycotina, Agaricomycetes, *incertae sedis*, Russulales, Stereaceae]

Derivation: *Stereum*, Greek, from *stereós*, "tough," "sturdy," "rigid"; *ostrea*, Latin, "oyster," "mussel," referring to its shoehorn shape.

Fruitbody: Sessile. Fruitbodies 1–10 cm (½–4 in) in length; semicircular, fan-shaped (Figs. A, B, & D), to spatulate, shoehorn-shaped, or "sliced funnel-shaped" (Figs. A & B); initially smooth or densely felty or tomentose, becoming smooth with age; upper surface grey, light buff, to rusty-brown, with concentric zones of brown, rusty-brown, red, reddish-orange, and buff (Figs. A & B); often greenish with age due to algae.

Flesh: Thin, leathery, tough, up to 1 mm thick; whitish to buff; odor and taste not distinctive.

Spore-bearing surface: Smooth; various shades of whitish-grey to pale yellowish- and/or cinnamon-brown.

Spore print: White, if you can obtain it.

Chemical staining: All surfaces red, then black with KOH.

Microscopic: Spores (5–)5.5–7.5(–8) × 2–3(–3.5) µm, smooth, cylindrical, slightly flattened unilaterally, clear, amyloid (blue) in Melzer's reagent.

Occurrence: Saprobic, causing a white rot on dead hardwood trees; individual fruitbodies clustered or in crowded groups; widespread in North America, Europe, and Asia; annual, overwintering.

Edibility: Inedible.

Comments: This mushroom is called the False Turkey Tail because of its superficial resemblance to the Turkey Tail, *Trametes versicolor* (Linnaeus *ex* Fries) Lloyd. *Stereum ostrea* is immediately distinguished from *T. versicolor* by its smooth lower surface with no sign of any pores, even with a hand lens. In sharp contrast, the lower surface of

T. versicolor is completely covered with pores visible to the naked eye. *Stereum ostrea* also tends to have large and conspicuous shoehorn-shaped fruitbodies; *T. versicolor* often forms characteristic rosette-like clusters.

Stereum ostrea may bleed a yellow or red liquid when injured. Fresh *Stereum complicatum* may bruise red when injured but does not exude a red liquid. The Bleeding Oak Crust, *Stereum gausapatum* (Fries) Fries, is also found on dead and fallen oaks, and fresh specimens do bleed a reddish liquid, which turns brown when cut; it first bruises red, then dark violet or blackish. *Stereum sanguinolentum* (Albertini & Schweinitz) Fries also bleeds a red liquid when cut but is found on conifers.

This mushroom produces several antimicrobial compounds that are effective against both pathogenic bacteria and fungi. Crude culture filtrates of *S. ostrea* inhibited the growth of three Gram-negative and two Gram-positive human pathogens.[1] Extracts from this mushroom were also effective against several phytopathogenic fungi.

1. A. Imtiaj et al., 2007, *Mycobiology* 35(4):210–214; K. Praveen et al., 2012, *J. Biol. Agricult. Healthcare* 2(1):1–5.

Polypores

(Shelf and Bracket Fungi)

The polypores comprise a diverse group of mushrooms, most of which form fruitbodies that are **poroid**, that is, they bear their hymenia inside tubes opening to the exterior through pores instead of on the outside of lamellae (gills). They thus superficially resemble the boletes. They often form shelf-like caps (brackets), which can be sessile or have stubby, rudimentary, or well-developed and typically eccentric or lateral stipes. Others can have stubby to well-developed central stipes and grow on the ground, either attached to buried roots or other woody substrata or in mycorrhizal associations with hardwood and conifer trees. Some polypores form extensive crust-like effuse or resupinate (hymenium facing outward) spreads, with or without reflexed cap-like margins. The **dissepiments**—the walls of the tubes and the rims around the pores—can erode away to become elaborated into quite complex tooth-like structures, which may become elongated and coalesce into daedeloid (maze-like) channels, especially as the fruitbody ages. The shape and size (number per millimeter) of the pores are often important in species determinations, but some taxa exhibit a great deal of variation in these features, sometimes on the same specimen in certain species.

The poroid undersurface of the polypores closely resembles that of the boletes. An important difference is that in the boletes, the hymenium is continuous throughout the lining of the tubes and the underside of the cap. In the polypores, each pore has its own hymenium, which is not connected to that of any other pore. The dissepiment around the pore opening is actually the trama of the pore wall projecting out beyond the hymenium. The hymenial trama also differs between the polypores and the boletes. In the former, the tramal hyphae tend to be either vaguely subparallel or intricately intertwined. In the latter, the hyphae are bilaterally divergent. These features are illustrated in the above figure.

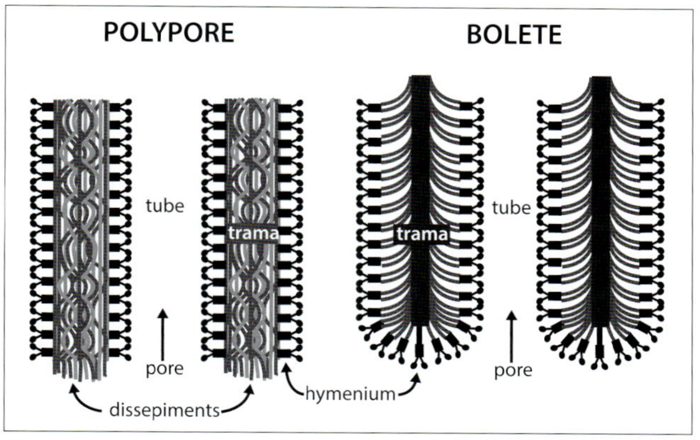

Other differences between the polypores and the boletes are summarized in the following table:

Polypores	Boletes
Fruitbody typically shelf-like, hoof-like, overlapping, effuse-reflexed, or resupinate	Fruitbody typically mushroom-shaped
Often sessile or with rudimentary lateral stipe	Stipitate, never sessile; stipe well developed, typically central
Commonly growing on or from wood	Growing on ground, rarely on wood
Most saprobic or parasitic on wood	Most ectomycorrhizal with trees
Fruitbody typically hard, persistent, nonputrescent, often lasting for years	Fruitbody soft, putrescent, decaying quickly after reaching maturity
Hymenial trama subparallel to intricately intertwined	Hymenial trama bilaterally divergent
Hymenium discontinuous, confined to the inner surfaces of the tubes	Hymenium continuous throughout the tubes and over the pore surface

Most polypores belong to the order Polyporales (formerly called the Aphyllophorales), but some have been transferred to other orders, such as the Russulales, on the basis of genetic studies. Polypores are important agents of wood decay. Many are parasitic and attack living trees and cause considerable damage to forest trees. Some of these are also saprobic and help to decay the trees they have killed. Some polypores degrade lignin and cause white rot in their tree hosts; some white-rotters degrade both lignin and cellulose. Others break down cellulose and produce a brown rot, and most brown-rot fungi are polypores. Their host specificity can be quite wide and some polypores grow equally well on both hardwood and conifer hosts. Others are restricted to hardwoods or conifers; some are restricted to a few genera, while others are highly specific to a single host genus.

Many polypores are involved in highly complex symbiotic associations with plants, other fungi, and animals. An example is the Mossy Maze Polypore (*Cerrena unicolor*), which has a complex ecological relationship with two types of wasps, a siricid woodwasp and four species of giant ichneumon wasps, as well as with a green alga.

Some nesting birds, in particular certain woodpeckers, are dependent upon trees with softer decomposed wood produced by certain polypores. An example is the advanced red ring rot decay of the heartwood in living old-growth pines produced by *Phellinus pini*, which enables endangered Red-cockaded Woodpeckers to hollow out their nesting cavities.

Many polypores produce a variety of highly potent bioactive compounds, and many of these are widely used in traditional medicine. Many species require old growth with a rich supply of rotting wood. As a direct consequence of massive deforestation and the replacement of diverse old-growth forests with plantation monocultures, many species of polypores are becoming increasingly scarce, and a number of them are now listed as threatened in Europe.

"Polypore" comes from the Greek words *polloí* and *póros*, meaning "many pores." It refers to the numerous tubes and their pores opening to the exterior that characterize the hymenia of most of these mushrooms. All are members of the division Basidiomycota. In their typical form, their basidia develop inside these tubes instead of on the surfaces of gills as in the agaricoid mushrooms, and their ballistospores are forcibly discharged just the right distance so that they clear all tissues and fall freely out through the pore-like openings. Most of them form characteristic shelf-like brackets that are either sessile or attached to their woody substrata by a more or less lateral or eccentric and rudimentary to well-developed stipe. Others grow from buried roots but may appear to sprout directly from the ground. All are parasitic or saprobic on trees or woody shrubs, with most causing a white or brown rot in wood.

The pores may be round, ellipsoid, polygonal, or elongate, and in some cases the dissepiments and the walls of the tubes have dissolved away and merged with each other to form **daedeloid** (labyrinthine) or even **lamellate** (gill-like) channels of varying complexity. The shape and size (number of pores per millimeter) of the pores may be constant within a species, or they may vary widely. In some species, this is a function of age, and round pores may exist alongside elongate, daedeloid, or even lamellate ones in the same specimen. This is particularly true of *Trametes elegans* and *Daedalea quercina*, as well as of *Gloeophyllum sepiarium*—a "polypore" (*in sensu tacitae*) that actually belongs to the order Russulales.

However, many polypores have very differently shaped fruitbodies. Some are classically agaricoid and stipitate (conventionally mushroom-shaped with a central stipe), and others (e.g., *Lenzites* and *Gloeophyllum*) have developed gill-like lamellae in place of pores and are known as "gilled polypores." Others are effuse spreads or effuse-resupinate (spreads turned upside down). And in some hydnoid polypores, the dissepiments erode away with age down to the very base, such that they appear like really jagged and fearsome-looking fangs. Examples of this can be seen in *Irpex*, *Spongipellis*, *Trichaptum*, and *Climacodon*.

The term *polypore* has two different and commonly used meanings: (1) as a taxonomic term it can refer to any member of the order Polyporales; and (2) as a nonspecific descriptive term for shelf-like bracket fungi and unrelated look-alikes whose basidia develop inside tubes that open to the exterior through pores, such as described above. In this sense, it includes a number of species in five different and quite unrelated orders. Both meanings of the term will be used here; which connotation is intended should be clear from the context. In both cases, the term applies exclusively to fungi in the division Basidiomycota.

This introductory discussion will focus mainly on the order Polyporales (formerly known as the Aphyllophorales), but species accounts of "polypores" from other orders are also included.

Order Polyporales. All species are saprobic, with most being wood rotters.

The family Polyporaceae comprises the core group, with additional species in the families Fomitopsidaceae, Ganodermataceae, Meripilaceae, and Meruliaceae, plus corticioid (effused spreads) fungi in the Cystostereaceae, Phanerochaetaceae, and Xenasmataceae, as well as the cauliflower fungi in the Sparassidaceae. Other morphologically similar taxa, such as *Bondarzewia* and *Albatrellus* (now placed in the order Russulales), *Gloeophyllum* (now in the Gloeophyllales), *Inonotus* and *Phellinus* (Hymenochaetales), and *Fistulina* (Agaricales), are also informally referred to as "polypores" and are included in this section. Still others, which are in the order Polyporales (*Climacodon*, *Irpex*, *Mycorrhaphium*, *Porostereum*, *Steccherinum*, and *Terrana*), are discussed in the sections on hydnoid (toothed) fungi or together with the crusts, spreads, and parchment fungi because of their superficial resemblances.

White Cheese Polypore

Tyromyces chioneus

A

B

C

D

***Tyromyces chioneus* (Fries) P. Karsten**, *Revue Mycol.*, Toulouse 3(9):17 (1881)

Basionym: *Polyporus chioneus* Fries, *Observ. Mycol.* (Havniae) 1:125 (1815)
[Basidiomycota, Agaricomycotina, Agaricomycetes, Polyporales, Polyporaceae]

Derivation: *Tyromyces*, Greek, from *tyrí*, "cheese," and *mýkos*, "fungus"; *chioneus*, Greek, from *chionátos*, "snow-white."

Cap: Up to 12 cm across, 8 cm deep, and up to 2 cm thick (4¾ × 3¼ × ¾ in), convex, semicircular, kidney-shaped to broadly fan-shaped, broadly attached, finely hairy when young, becoming bald and roughened and wrinkled with age; azonate, initially white or dusky-white (Fig. A), aging to yellowish or greyish.

Flesh: White, watery, and spongy when young, becoming hard, dry, fragile, and easily broken with age (Fig. C), odor pleasantly aromatic, taste mild or indistinct.

Pore surface: White to cream (Fig. B), turning cream- to dusky-yellow (Figs. B & D) with age, pores small, round to angular, with thick walls (Fig. D), 3–4 per mm; tubes up to 8 mm (⅓ in) deep.

Stipe: Sessile, stipe lacking.

Spore print: White.

Bruising: Not bruising.

Chemical staining: Cap and flesh negative in KOH and other chemicals.

Microscopic: Spores (3.5–)4–5 × (1–)1.5–2 µm, cylindrical to sausage-shaped, smooth, clear, nonreactive in KOH or Melzer's reagent.

Occurrence: Saprobic, causing a white rot. Solitary or in small, sometimes fused and/or overlapping clusters on dead and downed hardwood trees, especially birches (*Betula* spp.); widespread and fairly common in eastern North America, summer–fall; Greenland, Europe, Asia.

Edibility: Inedible.

Comments: The similar Smoky Cheese Polypore, *T. fumidiceps* G.F. Atkinson, is generally darker smoky-grey and has somewhat smaller, less-defined pores with thinner walls that turn pale olive with age. The spores are ellipsoid to subglobose rather than cylindrical or sausage-shaped, and the cap stains yellowish and the flesh reddish-orange with KOH.

Occasionally, a specimen whose pore surface is partially or completely stained bright or rusty-orange is encountered. These mushrooms are infected with a parasitic ascomycete, *Protocrea pallida* (Ellis & Everhart) Jaklitsch, K. Põldmaa & Samuels.

The Brown-staining Cheese Polypore, *Oligoporus fragilis* (Fries) Gilbertson & Ryvarden, is found on conifers instead of hardwoods and has somewhat daedeloid pores. Its quick rusty-brown bruising reaction on all surfaces when handled is diagnostic.

Smoky Cheese Polypore

Tyromyces fumidiceps

***Tyromyces fumidiceps* G.F. Atkinson**, *Annls Mycol*. 6(1):61 (1908) [Basidiomycota, Agaricomycotina, Agaricomycetes, Polyporales, Polyporaceae]

Derivation: *Tyromyces*, Greek, from *tyrí*, "cheese," and *mýkos*, "fungus"; *fumidiceps*, Latin, from *fumidus*, "smoky," and *-ceps*, combining form of *caput*, "head," "smoky-head."

Cap: 2–6 cm across by 1–4 cm deep and 1 cm thick where attached (¾–2½ × ½–1⅔ × ½ in); convex, semicircular (Fig. A) to irregularly kidney- or fan-shaped (Fig. D), easily detached; greyish, smoky-grey (Fig. A), creamy-tan (Fig. D), greyish-brown, azonate; soft and watery when fresh, becoming hard and dry, surface initially finely hairy (Fig. B), becoming bald (Fig. D) and rough with age (Fig. C), especially near attachment.

Flesh: White, slightly darker above, brittle, up to 2 cm (¾ in) thick; slightly aromatic, taste sweet.

Pore surface: Dingy-white (Fig. F), becoming creamy-ochraceous, with pale olive tints when mature, nonbruising; pores tiny, round to angular (Fig. G), 4–7 per mm, thin-walled, becoming slightly torn; tubes up to 1 cm (½ in) deep, inner surface often covered with crystals when old.

Stipe: Sessile, stipe lacking.

Spore print: White.

Bruising: Nonbruising.

Chemical staining: Cap yellowish and flesh reddish-orange in KOH.

Microscopic: Spores (2.5–)3–4 × 2–3 µm, ellipsoid to ovoid or subglobose, smooth, clear, nonreactive in KOH or Melzer's reagent.

Occurrence: Saprobic, causing white rot in hardwoods. Alone or in clusters on dead wood of birches (*Betula*), beech (*Fagus*), ash (*Fraxinus*), oaks (*Quercus*), cottonwoods (*Populus*), and willows (*Salix*); always close to the ground along river bottoms and in wetlands, widespread east of the Great Plains; summer–fall, annual; Europe (rare), South America (Brazil).

Edibility: Inedible.

Comments: This species is similar to the more commonly encountered White Cheese Polypore, *T. chioneus* (Fries) P. Karsten. *Tyromyces chioneus* is generally lighter in color, white to creamy- or dusky-white, although it may age to yellowish or greyish. Unlike *T. fumidiceps*, it does not change color with KOH or other chemical stains, the white pore surface becomes dusky-yellow rather than pale olive or greenish with age, and the slightly larger pores have thicker walls and are more conspicuous. The spores are also cylindrical to sausage-shaped rather than ellipsoid to subglobose.

The Brown-staining Cheese Polypore, *Oligoporus fragilis* (Fries) Gilbertson & Ryvarden, is found on conifers instead of hardwoods and has somewhat daedeloid pores. Its quick rusty-brown bruising reaction on all surfaces when handled is diagnostic.

Brown-staining Cheese Polypore

Oligoporus fragilis

B D A C E

***Oligoporus fragilis* (Fries) Gilbertson & Ryvarden**, *Mycotaxon* 22(2):365 (1985)

Basionym: *Polyporus fragilis* Fries, *Elench. Fung*. (Greifswald) 1:86 (1828)
[Basidiomycota, Agaricomycotina, Agaricomycetes, *incertae sedis*, Polyporales, Fomitopsidaceae]

Derivation: *Oligoporus*, Greek, from *oligos*, "few," "scanty," "small," and *póros*, "pore"; *fragilis*, Latin, "fragile," "brittle."

Fruitbody: 1–10 cm wide by up to 5 cm deep (½–4 × 2 in), isolated or in overlapping clusters (Fig. A) and/or laterally fused rows (Fig. B); semicircular, fan-shaped, or shelf-like; upper surface white to yellowish-ochre (Fig. B) or orange-brown, azonate; hairy or fibrillose, heavily wrinkled and rugose, *quickly bruising tawny-brown, then reddish-brown when handled* (Fig. B).

Flesh: Up to 1.5 cm (⅔ in) thick, soft when young, becoming stiff and brittle with age; at first white, discoloring like the cap surface; odor faintly pungent or not distinctive.

Pore surface: White, bruising reddish-brown (Fig. C); pores round to angular to somewhat daedeloid (maze-like) (Fig. D), 3–4(–6) per mm, tubes up to 8 mm deep, walls thin, becoming torn to tooth-like with age (Fig. E).

Stipe: Sessile, stipe lacking.

Spores in mass: White.

Bruising: All surfaces quickly bruising reddish-brown when handled (Figs. A–C).

Microscopic: Spores 4–5 × 1–2 µm, cylindrical to sausage-shaped, smooth, with granular inclusions, nonstaining in Melzer's reagent.

Occurrence: Saprobic, causing brown rot. Annual on decayed conifer trunks, logs, and branches; widespread in North America and Europe, uncommon in our area, late summer–fall.

Edibility: Inedible.

Comments: The reddish-brown bruising reaction of this polypore is distinctive and separates it from all other polypores with which it might be confused. It appears to be quite uncommon this far south, and we have found it only once.

Reishi, Lingzhi, and Hemlock Varnish Shelf

Ganoderma lucidum, *G. curtisii*, and *G. tsugae*

A, B, E & F, *Ganoderma curtisii*; C, D & G, *G. lucidum*; H, *G. tsugae*.

***Ganoderma lucidum* (Curtis) P. Karsten**, *Revue Mycologique Toulouse* 3(9):17 (1881)

Basionym: *Boletus lucidus* Curtis, *Fl. Londin*. 1:72 (1781)

***Ganoderma curtisii* (Berkeley) Murrill**, *N. Amer. Fl.* (New York) 9(2):120 (1908)

Basionym: *Polyporus curtisii* Berk., *Hooker's J. Bot. Kew Gard. Misc*. 1:101 (1849)

***Ganoderma tsugae* Murrill**, *Bull. Torrey Bot. Club* 29:601 (1902) [Basidiomycota, Agaricomycotina, Agaricomycetes, *incertae sedis*, Polyporales, Ganodermataceae]

Derivation: *Ganoderma*, Greek, from *ganos*, "sheen," and *derma*, "skin," "shining skin"; *lucidum*, Latin, "clear, luminous"; *curtisii*, Latin, epithet in honor of American mycologist Moses Ashley Curtis (1808–1872); *tsugae*, Latinized form of the Japanese word 栂 (*tsuga*), meaning "hemlock tree," the preferred host of this species.

Cap: All three species—circular (Fig. E), semicircular (Figs. A & G), kidney-shaped (Figs. B–D) or fan-shaped (Fig. H); 3–35 cm (1¼–14 in) across; ***G. lucidum* and *G. tsugae***—cap typically concentrically rugose, with shiny dark red (Fig. C), brownish-orange (Fig. D), or reddish-black varnish, sometimes with whitish or orange margin (Figs. D & H); both species nearly identical; ***G. curtisii***—bright ochraceous-yellow when young (Figs. A & E), becoming darker with age, especially toward the center (Fig. B), with white margin, typically zonate; varnish much more subdued (or even lacking) than in the other two taxa.

Flesh: Soft when young, corky when mature, up to 3 cm (1¼ in) thick. ***G. lucidum***—pale ochraceous-brown in a layer near the cap, dark brown in a lower layer adjacent to the tubes; **G. curtisii**—white to ochre-yellow above, cinnamon to light sepia-brown adjacent to the tubes, *this layer with several thin, dark-brown interlayers radiating from the stipe parallel to the cap surface* (Fig. I); **G. tsugae**—whitish, often bruising brownish when cut.

Pore surface: Whitish (Fig. F) to dull brown or brownish yellow, pores round to angular; **G. lucidum**—pores 4–7 per mm, tubes up to 2 cm (¾ in) deep, in one to two layers; **G. curtisii**—pores round, 4–5 per mm, tubes ≤1 cm deep; pore surface bruising brown when injured; **G. tsugae**—white to brown, 4–6 per mm, tubes to 2 cm deep.

Stipe: **G. lucidum**—often lacking or rudimentary, when present lateral, concolorous with cap (Figs. C & D), 3–10 cm tall by 2 cm thick (1¼–4 × ca. ¾ in); **G. curtisii**—eccentric to lateral (Figs. A & B), sometimes central (Figs. E & F); ochraceous (Fig. A) to red (Figs. B & F), with dull to shiny varnish, context brown; 3–10 cm tall by 2 cm thick (1¼–4 × ca. ¾ in); **G. tsugae**—3–15 cm tall by 1–4 cm thick (1¼–6 × ½–1⅔ in), lateral or occasionally central, concolorous with and rugose like the cap (Fig. H).

Spore print: **All three species**—brown.

Chemical staining: **G. lucidum**—KOH, immediately jet black on cap, flesh, and tubes, very slowly fading to dark brown; pore surface brownish-grey, slowly becoming darker brown; **G. curtisii**—KOH, yellowish-brown on cap and upper flesh layer, cap not changing, but upper flesh gradually turning reddish-brown, then fading to lighter brown; jet black on tubes and adjacent flesh layer, tubes not changing, adjacent flesh fading to brown; pore surface pale greyish, not changing; **G. tsugae**—unknown to us.

Microscopic: **All three species**—spores smooth, elliptical, double-walled, with embedded spines between the layers (under oil immersion); **G. lucidum**—spores 7–12 × 5.5–8 µm, smooth, elliptical, occasionally with truncated end; **G. curtisii**—9–11 × 5–7 µm, ovoid, echinulate, with truncated end; **G. tsugae**—spores 13–15 × 7.5–8.5 µm, elliptical, with truncated end.

Occurrence: **All three taxa**—annual, May–November, sometimes; **G. curtisii and G. lucidum**—parasitic/saprobic, producing a white butt and root rot at the bases and stumps of deciduous trees, especially oaks (*Quercus* spp.) and maples (*Acer* spp.),[1] occasionally from buried roots; *G. lucidum* cosmopolitan, *G. curtisii* restricted to southeastern North America; **G. tsugae**—found on conifers, particularly Eastern Hemlock (*Tsuga canadensis*).

Edibility: Inedible due to woody consistency, but brewed tea extracts highly valued for multiple medicinal purposes (see comments below).

Comments: *Ganoderma curtisii* (Reishi) has been synonymized with *G. lucidum* (Lingzhi, Reishi) by some but not all authors because they are all saprobic with the same host trees. There is little question that these two taxa are closely related both to each other and to *G. tsugae* (Hemlock Varnish Shelf). However, several consistent geographic, morphological, and chemical staining differences distinguish between these two taxa, and we are among those who consider them to represent two distinct species.

In addition to its whitish to pale yellowish to ochre and much less rugose and laccate (varnished) cap, *G. curtisii* differs from *G. lucidum* as follows: (1) the upper layer of the flesh just below the cap cuticle is whitish to ochre-yellow (not ochraceous-brown); (2) the lower layer of the context adjacent to the tube layer is cinnamon-brown (not dark brown); (3) this lower context layer is also penetrated by several thin, contrasting, dark-brown horizontal layers (Fig. I, arrow) radiating from the stipe that are not present in *G. lucidum*; and (4) the two taxa differ in their staining reactions to 10% KOH: in *G. curtisii* the cap surface is nonreactive, the upper flesh layer gradually turns reddish-brown and then fades to a lighter brown, and the lower layer and the tubes stain jet black, the lower flesh layer fading to brown and the tubes unchanging; in *G. lucidum* the cap surface, flesh, and tube layers all immediately stain jet black, which very slowly fades to dark brown.

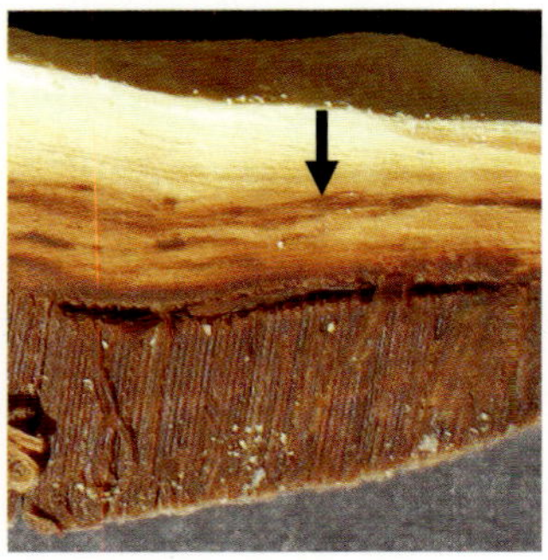

I

Ganoderma curtisii.

Like many polypores, *Ganoderma lucidum* has a tortured and very confused taxonomic history. It was described as a new species in Europe no fewer than 13 times, several times by the same authors. See online Appendix A for a list of its 60-odd synonyms. It is a highly polymorphic species with several color morphs ranging from dark red (Fig. C) to orange-brown (Fig. D) and even blue (Fig. G), with and without a lighter-colored margin. Prominently stipitate and nearly sessile fruitbodies are also commonly encountered in the same cluster.

Reishi (*G. lucidum*) has been used in Asia for over two millennia to promote general health and longevity and as chemotherapy for cancer. Evidence that the bioactive terpenes and polysaccharides in Reishi modulate immune function both *in vitro* and *in vivo* and exert a strong inhibitory effect against invasive cancer cells and metastatic processes is increasingly appearing in the scientific literature. *Ganoderma lucidum* thus has an increasingly acknowledged therapeutic role as a dietary supplement for an alternative treatment of several different cancers. It is not yet known whether these properties are equally present in all populations of *G. lucidum* or if they also are present in *G. curtisii* and *G. tsugae*.

1. The sporocarps of *G. curtisii* shown in Figs. A and F emerging through the pine-needle mulch were actually growing beneath and from the buried roots of maple trees (*Acer* sp.) in an urban park.

Artist's Conk

Ganoderma applanatum

Photos: A, Danny Steaven; B, W. Siegmund; C & D, George Chernilevsky; E, W. Light.

***Ganoderma applanatum* (Persoon) Patouillard**, *Hyménomyc. Eur.* (Paris):143 (1887)

Basionym: *Boletus applanatum* Persoon, *Observ. Mycol.* (Lipsiae) 2:2 (1800) [1799]
[Basidiomycota, Agaricomycotina, Agaricomycetes, *incertae sedis*, Polyporales, Ganodermataceae]

Derivation: *Ganoderma*, Greek, from *ganos*, "sheen," and *derma*, "skin," "shining skin"; *applanatum*, Latin, from *applanatis*, "flattened."

Cap: 5–50(–60) cm across by 5–10 cm thick (2–20[–24] × 2–4 in); sessile, planoconvex, shelf-like (Figs. A, B, & D), semicircular to fan-shaped; alone (Figs. A & E) or in small groups (Figs. B & D); upper surface hard, irregular, often furrowed and zonate (Fig. B); greyish-brown to brown, often covered with a reddish-brown layer of spores.

Flesh: Tough, corky, brown to cinnamon-brown; 3–6 cm (1¼–2½ in) thick above the tube layers.

Pore surface: Initially white (Figs. A, C, & D), becoming greyish to dirty-yellowish to dingy-brownish (Fig. E) with age, bruising brown (Fig. E, arrow); pores minute, circular, nearly invisible to the naked eye (Fig. C), 4–6 per mm; tubes 4–6 mm long, layered.

Stipe: Sessile, stipe lacking.

Spore print: Brown to rusty-brown.

Bruising: Light-colored pore surface immediately bruising brown (Fig. E, arrow); can be used to create fine artistic works.

Chemical staining: Flesh and tubes black with KOH.

Microscopic: Spores 7–11(–12) × (5–)6.5–7.5(–8) µm, broadly elliptical, double-walled, with wart-like ornaments between the two walls; nonreactive in Melzer's reagent.

Occurrence: Parasitic and saprobic on many species of hardwood trees and conifers. Perennial, often persisting for decades. Solitary or in groups on stumps and logs or at the bases of living trees; widespread and common throughout North America; South America, Europe, Asia, Africa.

Edibility: Inedible.

Comments: *Ganoderma applanatum* is immediately distinguished from all other polypores by its dull greyish-brown, typically furrowed, zonate, shelf-like fruitbody and its white to creamy pore surface with invisible pores that *instantly bruise dark brown* when scratched. Very elegant and sophisticated works of art have been etched in the pore surface of the Artist's Conk by skilled artisans using the proper tools.

Like several other *Ganoderma* species, the Artist's Conk produces a number of bioactive chemical compounds with potent medicinal properties. These compounds are discussed in the "Medicinal Properties of Some Southeastern Mushrooms" section in the introduction.

Turkey Tail

Trametes versicolor

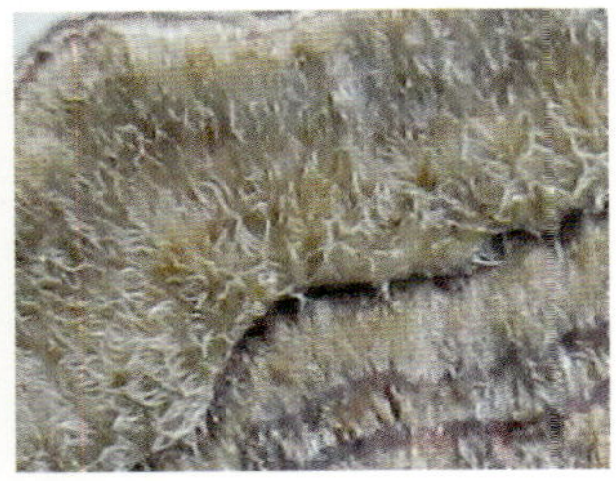

C

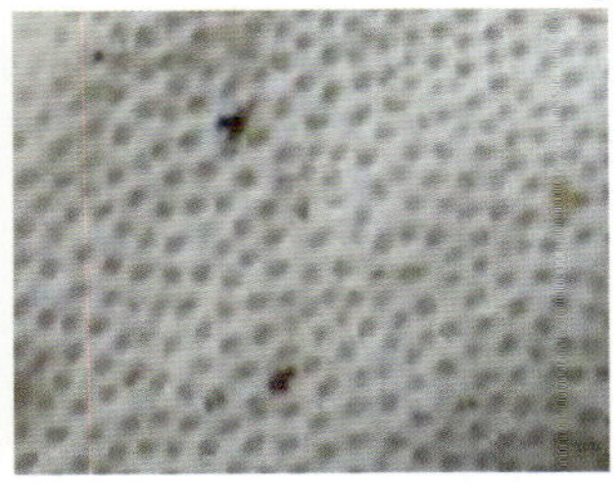

A

B

D

E

***Trametes versicolor* (Linnaeus 1753 *ex* Fries 1821) Lloyd**, *Mycol. Notes* (Cincinnati) 65:1045 (1920)

Basionym: *Boletus versicolor* Linnaeus, *Sp. Pl.* 2:1176 (1753) [Basidiomycota, Agaricomycotina, Agaricomycetes, *incertae sedis*, Polyporales, Polyporaceae]

Derivation: *Trametes*, Greek, from *trama*, "weft," as in a fabric, referring to the weave of hyphae comprising the fruitbody; *versicolor*, Latin, "variably colored."

Fruitbody: Sessile or effuse-reflexed (growing flat with uplifted shelf-like edges), typically forming rows or large clusters of overlapping, tilted tiers, often forming rosettes (Fig. E), occasionally solitary.

Cap: Irregularly fan-shaped to tongue-shaped or forming rosettes; surface plane or wavy, dry; 1.5–6 by 1.5–8 cm (1⅔–2½ × 1⅔–3¼ in) across, occasionally coalescing laterally into larger structures. Surface velvety or silky, covered with woolly hairs (Fig. C), concentrically zonate both in texture and in color; color highly variable, with alternating concentric bands of blue, blue-grey, grey, yellow, orange, reddish, tan, and/or brown in various shades and hues (Figs. A & B); sometimes green due to symbiotic algae. Margin often wavy or irregular, usually with white or creamy edge when young (Fig. A).

Pore surface: Whitish to dull yellowish, pores round when young (Fig. D), 3–5(–8) per mm, becoming somewhat angular with age, walls entire; tubes up to 4 mm deep.

Stipe: Lacking or rudimentary.

Flesh: Tough, corky, cream-colored, up to 5 mm thick, with a black line just below the surface tomentum (woolly layer). Taste not distinctive.

Spore print: White.

Chemical staining: Flesh negative or yellowish in KOH.

Microscopic: Spores 4–6 × 1.5–2 µm, smooth, clear, cylindrical to sausage-shaped; nonreactive in Melzer's reagent.

Occurrence: Saprobic, causing white rot. Solitary or in dense clusters on decaying hardwoods, rarely on conifers; very common and abundant year-round, fruiting mainly winter–spring. Cosmopolitan.

Edibility: Used to make medicinal teas.

Comments: Superficially similar species of *Stereum* can be distinguished from *Trametes* by their lack of visible pores. The very similar and equally variable False Turkey Tail, *Stereum ostrea*, is distinguished by its smooth, white to grey, or reddish-brown undersurface, which lacks any trace of pores even when viewed with a hand lens. The fertile surfaces of *Trichaptum* species like the Violet-tooth Polypore, *T. biforme* (Fries) Ryvarden, are brownish or faintly to vividly tinged with violet and bear prominent teeth instead of pores.

Trametes versicolor can be confused with several look-alikes. They can be distinguished as follows:

Trametes hirsuta (Wulfen) Lloyd: Cap very densely hairy, radially furrowed, with vaguely contrasting zones of whitish, greyish, and brownish hues; margin often brownish, dark brown, or blackish; pore surface white, yellowish, or greyish; 3–4 circular to angular pores per mm; spores 6–9 × 2–2.5 µm.

T. pubescens (Schumacher) Pilát: Cap cream-colored or yellowish, azonate, finely velvety to smooth, pore surface white, yellowish, or umber, 3–5 angular pores per mm, walls thin, often dentate; spores 5–7 × 1.5–2.5 µm.

T. villosa (Swartz) Kreisel: Cap very densely hairy, zonate, with alternating texture and whitish and greyish hues; 1–3 very angular, elongate pores per mm, these becoming daedeloid, tooth-like with age; spores 5.5–8.5 × 2–3.5 µm.

Polysaccharide-K (PSK), a protein-bound polysaccharide isolated from *T. versicolor*, is used as an immunoenhancing agent in treating various cancers in Europe, Japan, and China. PSK appears to be beneficial in treating gastric, esophageal, breast, and small cell and non–small cell lung cancers. It has been used in Asia for more than 30 years and is an approved treatment for cancer in Japan and China.

Shaggy Bracket, Bristly Bracket

Trametes villosa

A

B

C

D

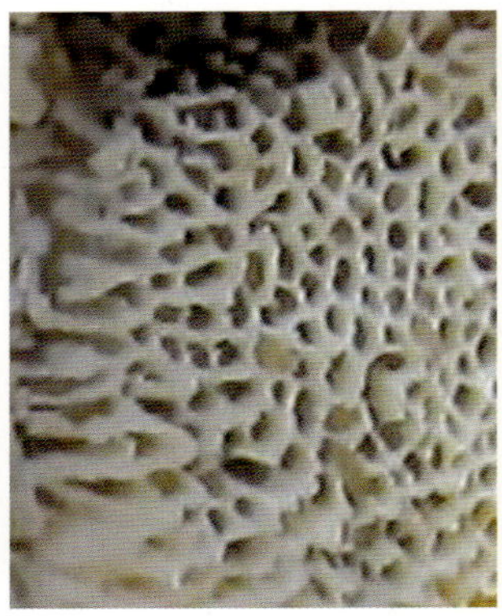

E

F

***Trametes villosa* (Swartz) Kreisel**, *Monografias, Ciencias, Univ. Habana*, ser. 4, 16:83 (1971)

Basionym: *Boletus villosus* Swartz, *Fl. Ind. Occid.*:3 1923 (1806) [Basidiomycota, Agaricomycotina, Agaricomycetes, *incertae sedis*, Polyporales, Polyporaceae]

Derivation: *Trametes*, Greek, from *trama*, "weft," as in a fabric, referring to the weave of hyphae comprising the fruitbody; *villosa*, Latin, "shaggy," "bristly," "hairy," "rough."

Fruitbody: Irregularly bracket-shaped to kidney-shaped (Figs. A & B), sometimes laterally coalescing to form fused fruit-bodies or rosettes (Fig. C); margin typically white, sometimes forming blocky border (Fig. D).

Cap: Very densely hairy (Figs. A & C), with concentric zones of texture and subdued color, with alternating whitish, greyish, and ochre or reddish-brown bands (Figs. A & B); up to 7 cm (2¾ in) across by 3 mm thick.

Pore surface: Whitish at first (Fig. E), becoming darker with age (Fig. F); 1–3 large, angular pores per mm, these elongating, becoming distinctly tooth-like and daedeloid with age (Figs. E & F), resembling the blunt teeth in *Trichaptum* spp.

Flesh: White, tough, corky, homogeneous.

Spore print: White.

Stipe: Fruitbody sessile, stipe lacking.

Chemical staining: Negative or staining pale yellow with KOH.

Microscopic: Spores 5.5–8.5 × 2.5–3.5 µm, smooth, cylindrical.

Occurrence: Saprobic, annual, causing white rot on dead hard-woods and conifers; in clusters on logs and stumps, summer–

fall; widespread in southeastern North America; Central and South America, Europe.

Edibility: Inedible.

Comments: The Hairy Bracket, *Trametes hirsuta* (Wulfen) Lloyd, is a similar species that can be distinguished from *T. villosa* by its larger size (up to 15 cm [6 in] across), its duplex flesh with a black line separating the two layers, and its somewhat smaller pores (3–4 per mm), which are round or only slightly angular and which do not form tooth-like or daedeloid channels. The margin is typically brownish, and it is restricted to hardwoods, particularly beech (*Fagus* spp.).

Elegant Trametes

Trametes elegans

A

B

C

D

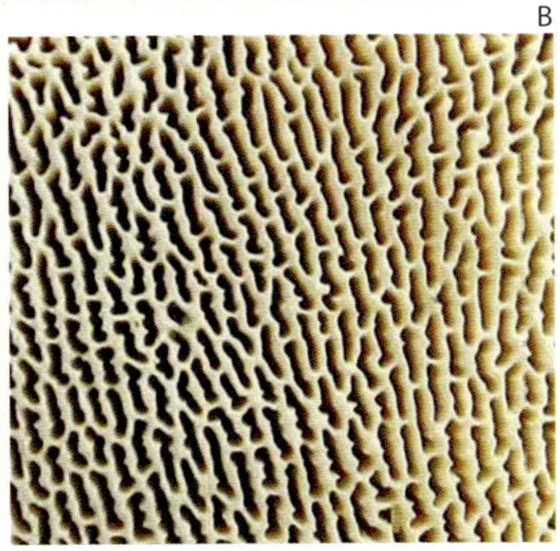

E

***Trametes elegans* (Sprenger 1820) Fries,** *Epicr. Syst. Mycol.* (Upsaliae):492 (1838) [1836–1838]

Basionym: *Daedalea elegans* Sprenger, *K. Svenska Vetensk-Akad. Handl.* 46:51 (1820) [see online Appendix A for synonymy] [Basidiomycota, Polyporales, Polyporaceae]

Derivation: *Trametes*, Greek, from *trama*, "weft," as in a fabric, referring to the weave of hyphae comprising the fruitbody; *elegans*, Latin, "elegant."

Cap: Up to 25 cm (14 in) across, semicircular to kidney-shaped to very irregularly shaped.

Upper surface: Flattened-convex; smooth to lumpy (Fig. A); more or less creamy-white or buff, darker toward the attachment (Fig. B), to darker with concentric zones of texture and/or color (Figs. A & B); sometimes darker with age.

Pore surface: Highly variable. White when young. Poroid, with round (Fig. C) to angular pores (Figs. D & E), often transitioning to a delicate latticework (Fig. D), 1–2 mm wide with very thin walls; all three states often occurring in the same sporocarp. Occasional older specimens may darken considerably.

Context: Flesh white, tough. Does not bruise.

Spore print: White.

Microscopic: Spores 5–7 × 2–3 µm; smooth; cylindrical to long-elliptical; inamyloid, hyaline in KOH. Context hyphae trimitic.

Chemical staining: Flesh stains yellow with KOH.

Occurrence: Saprobic, forming white rot in hardwood trees; annual to perennial, spring–fall. Growing alone or in clusters on logs and stumps (Fig. A); widely distributed in eastern North America south of the Great Lakes.

F

G

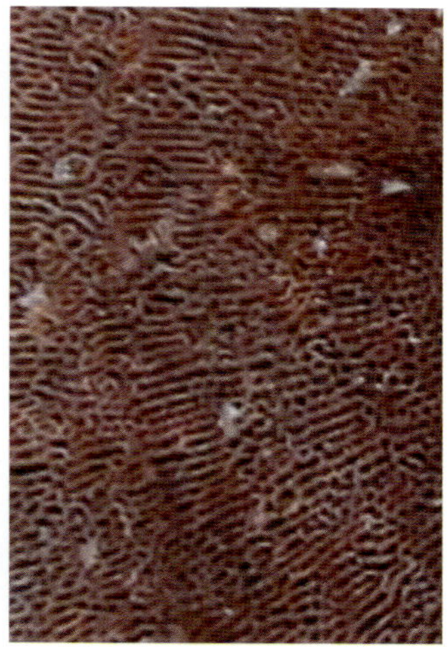

H

Edibility: Tough, inedible.

Comments: Just when you think you have this taxon pegged, you encounter one or more specimens that throw you for a loop. The variability in fruitbody shape (Figs. F & G) and in the variety and color of pore surfaces (Fig. H) encountered in this species is amazing. This variability soon becomes familiar enough, and after several encounters you will be able to identify this species in most of its morphs at a distance.

Little Nest Polypore

Trametes conchifer

A B C

Photos: A–C, W. Light; inset, Walt Sturgeon.

***Trametes conchifer* (Schweinitz) Pilát**, in Kavina & Pilát, *Atlas Champ. l'Europe*, III., Polyporaceae (Praha) 1:264 (1939)

Basionym: *Boletus conchifer* Schweinitz, *Schr. Naturf. Ges. Leipzig* 1:98 (1822)
[Basidiomycota, Agaricomycotina, Agaricomycetes, *incertae sedis*, Polyporales, Polyporaceae]

Derivation: *Trametes*, Greek, from *trama*, "weft," as in a fabric, referring to the weave of hyphae comprising the fruitbody; *conchifer*, Latin, from *concha*, "shell," and the combining form of *-fer*, "producing" (from *ferre*, "to bear," "to carry"), meaning "shell-bearing."

Fruitbody: First forming as a sterile saucer-shaped or cup-like structure (Fig. A), later producing a shelf-like cap with a fertile hymenophore (spore-bearing surface) (Figs. A & B).

Cup: Shallowly or deeply cup-shaped, 1.5 cm (~⅔ in) across, upper (inner) surface smooth, initially brown (Fig. A), then with sharply contrasting circular zones of white and grey-brown, dark brown, or blackish (inset); lower (outer) surface white, with tiny stem-like base; may remain attached to or be overgrown by the base of the cap, with or without a visible remnant (Fig. B, arrow).

Cap: Emerging laterally from the base of the cup, shelf-like, semicircular, fan- to kidney-shaped, 2–5 cm (¾–2 in) across and 1–3 cm (½–1¼ in) deep; margin thin, wavy; white to pale grey or pale brown; smooth to faintly velvety, wrinkled and radially furrowed, with faint zones of contrasting pale brick-red color (Fig. B), especially near the margin. May fuse laterally with other caps or give rise to secondary cups.

Flesh: Thin, tough, white, up to 1 mm (⅖ in) thick, odor and taste not distinctive.

Pore surface: Whitish, becoming yellowish (Fig. C); 2–5(–4) polygonal or irregularly angular pores per mm (Fig. C); tubes thin-walled, up to 2 mm deep.

Stipe: Lateral and knob-like when present.

Spore print: White.

Chemical staining: Cap and pore surface pale yellowish with KOH; cap, flesh, and pore surface strongly orangish-brown with Melzer's reagent, cap and flesh fading to yellowish-brown; all surfaces negative in ammonia.

Microscopic: Spores (sexual spores) 5–7 × 1.5–2.5 µm, smooth, cylindrical, clear, nonreactive in Melzer's reagent; asexual spores produced by cup 3–8 × 3 µm, rod-shaped.

Occurrence: Saprobic, causing white rot on trunks, fallen branches, and dead limbs of elms and other hardwoods. Annual, June–December, cap overwintering; widespread, but occasional in eastern North America.

Edibility: Inedible.

Comments: This species is also called *Poronidulus conchifer* (Schwein.) Murrill (1904) Pilát (1939) in many online sources and some field guides.[1]

This unusual polypore starts out as a small, sterile cup that looks much like a bird's nest fungus, but without the "eggs." The cup produces asexual spores, which, like the sexual spores of its bird's nest counterpart, are also distributed by being ejected out of the cup by falling raindrops. Sexual basidiospores are subsequently produced by the shelf-like cap that emerges from the base of the initial asexual cup and are distributed in the usual way. The cup often persists at the base of the cap but is also frequently overgrown, where it may be barely visible as a lumpy disk with or without an opening or cavity. The cup, with or without the attached shelf, identifies this species on sight; identification is problematic when the cup is not visible.

1. This taxon is placed in both the genus *Trametes* Fries (1835) and the genus *Poronidulus* Murrill (1904) by various authors. *Poronidulus* was erected as a monotypic genus by Murrill, with *P. conchifer* as the type and sole species. This was based on its unique growth pattern of first forming an asexual cup-like stage that later produces a sexual shelf-like fruitbody. However, a recent molecular study indicated that *Poronidulus* falls tightly within a monophyletic *Trametes* clade and should be considered a deprecated synonym of *Trametes* (A. Justo & D.D. Hibbet, 2011, *Taxon* 60:1567–1583).

Violet-tooth Polypore (Hardwoods)

Trichaptum biforme

A

B

C

***Trichaptum biforme* (Fries) Ryvarden,** *Norw. Jl. Bot.* 19:237 (1972)

Basionym: *Polyporus biformis* Fries, in Klotzsch, *Linnaea* 8:486 (1833)
[Basidiomycota, Polyporales, Polyporaceae]

Derivation: *Trichaptum*, Greek, "with clinging hairs"; *biforme*, Latin, "having two forms," referring to its tooth-like and/or daedeloid pore surface (Fig. C).

Cap: 1–8 cm wide, up to 3 mm thick. Semicircular, fan-like, or kidney-shaped; flattened to convex. Often growing in overlapping clusters (Fig. A).

Upper surface: Faintly zonate, finely hairy, velvety, smooth with age; white, greyish, or brownish, margin often pale lilac or purplish (Fig. A), fading with age; often green when covered by algae, when it may be confused with the Mossy Maze Polypore, *Cerrena unicolor*.

Pore surface: Purple to lilac, occasionally stunning (Fig. C), strongest near the margin, soon fading to buff or brown, margin usually retaining some purple tint (Fig. B); 2–5 angular pores per mm, developing spines or teeth with maturity; tubes 1–10 mm deep; not bruising.

Flesh: Monoplex, no gelatinous layer between context (flesh) and tube layer. Flesh 0.5–1.5 cm thick; white to pale buff; tough, leathery.

Spore print: White.

Microscopic: Spores 6–8 × 2–2.5 µm; smooth; cylindrical to slightly curved cylindrical; hyaline in KOH; inamyloid. Cystidia abundant; up to 35 × 5 µm, more or less fusoid, apically encrusted. Hyphae dimitic.

Staining: Not staining, or flesh and cap surface pale yellowish with KOH.

Occurrence: Saprobic and ubiquitous on hardwood logs and stumps throughout North America, often in great numbers; late spring, summer, and fall, often persisting all year.

Edibility: Inedible.

Comments: We often have found anamorphic forms of both *Trichaptum biforme* and its conifer-dwelling congener, *T. abietinum*, growing together with the teleomorphic fruitbodies (see Fig. D at right). These anamorphs appear to be more frequent in *T. abietinum*, but both species exhibit similar morphs. To our knowledge no one has previously commented on this phenomenon.

D

Trichaptum biforme is almost always found on deciduous trees; in contrast, *T. abietinum* occurs almost exclusively on conifers. *Trichaptum biforme* fruitbodies are also generally larger and less brilliantly colored than *T. abietinum*, and the violet hues fade much more quickly. Figure C shows a rare and exceptionally vivid color for this species. The color appeared on old dried sporocarps three days following a good rain; two days later the same fruitbodies had faded to a pale grey and were again tough and leathery. The context is monoplex; that of *T. abietinum* is duplex, and a gelatinous layer separates the context and tube layer. The tiny black club-like fruitbodies of a saprobic ascomycete, the Fairy Pin, *Phaeocalicium polyporaeum* (Nylander) Tibell, are often found growing on the caps of both *Trichaptum biforme* and the Turkey Tail, *Trametes versicolor*. These Fairy Pins are barely 1.5–1.75 mm tall (about 1⁄16 in). See the Fairy Pins species account for details.

Violet-tooth Polypore (Conifers)

Trichaptum abietinum

A

B

Trichaptum abietinum (Dickson) Ryvarden, *Norw. Jl. Bot.* 19:237 (1972)

Basionym: *Boletus abietinus* Dickson, *Fasc. Pl. Crypt. Brit.* (London) 3:21 (1793)
[Basidiomycota, Polyporales, Polyporaceae]

Derivation: *Trichaptum*, Greek, "with clinging hairs"; *abietinum*, Latin, "living in firs," referring to its association with conifers; *fuscoviolaceum*, Latin, "dusky-violet."

Cap: 1–4 (rarely more than 1) cm (½–1⅔ in) wide by up to 0.5 cm thick; flat to slightly convex, similar to *T. biforme*, but generally smaller. Often growing in tiers of overlapping clusters (Fig. A). Fruitbodies may intergrade from completely resupinate to shelf-like forms.

Upper surface: Hairy, zonate, whitish to light grey, usually with prominent purple or lilac margin; Like *T. biforme*, older fruitbodies often green due to colonizing algae.

Pore surface: Bright purple when young, fading to ochraceous or brownish-purple with age, with lilac persisting at the margin. Tubes 1–3 mm deep, pores 4–6 per mm, circular to angular, seldom daedeloid, becoming deeply toothed with age.

Flesh: Usually <1 cm thick, duplex: upper layer whitish, floccose, soft; lower layer white, firm, fibrous; tube layer separated from the context by a thin gelatinous layer, up to 1.5 mm thick.

Spore print: White.

Microscopic: Spores 4–6.5(–7.5) × 2.5–3.0(–3.5) µm, hyaline, smooth, short-cylindrical, slightly curved; inamyloid in Melzer's reagent. Dimitic; skeletal hyphae thin-walled, nonseptate; generative hyphae with clamps.

Occurrence: Saprobic on conifers. Occurs in great abundance on dead trees, stumps, and fallen logs, rarely on living trees; abundant in crevices.

Edibility: Inedible.

Comments: *Trichaptum abietinum* occurs almost exclusively on dead conifer trunks and logs and is very rarely found on hardwoods. The converse is true for *T. biforme*. The fruitbodies of *T. biforme* are also much wider than those of *T. abietinum*, the context is monoplex, and no gelatinous layer separates the pore or tube layer from the context flesh. The pore surface of *T. abietinum* is generally more deeply purple or lilac than in *T. biforme*, and it retains its color longer. A similar species, *Trichaptum fuscoviolaceum* (Ehrenberg) Ryvarden, has rarely been reported from Fraser Fir (*Abies fraseri*) and Loblolly Pine (*Pinus taeda*) in North Carolina. It is distinguished from *T. abietinum* by its effuse-reflexed[1] fruitbody and hydnoid, massively toothed pore surface.

Like *Trichaptum biforme*, to which it is closely related, *T. abietinum* commonly exhibits a chaotic anamorphic stage, which can occur as an isolated morph or be found together with the typical teleomorphic stage (see Fig. D in the comments under *Trichaptum biforme*). Both violet-toothed polypore species are also host to the tiny Fairy Pin, *Phaeocalicium polyporaeum*, which is described in detail in its own species account.

1. Spread over the surface and turned up and back at the margins.

Oak Mazegill Polypore, Thick Oak Maze Polypore

Daedalea quercina

A

B

***Daedalea quercina* (Linnaeus 1753) Persoon**, *Syn. Meth. Fung.* (Göttingen) 2:500 (1801)

Basionym: *Agaricus quercinus* Linnaeus, *Sp. Pl.* 2:1176 1753 [see online Appendix A for synonymy]
[Basidiomycota, Polyporales, Fomitopsidaceae]

Derivation: *Daedalea*, named for Daedalus, the Greek craftsman who built the labyrinth on Crete that housed the Minotaur; *quercina*, Latin, refers to its oak host specificity.

Cap: 4–20 cm (1½–8 in) across, 3–8 cm (1¼–3¼ in) wide, 1.5–7.5 cm (⅔–3 in) thick. Flat to broadly convex; fan-shaped in outline (Fig. A).

Upper surface: Azonate. Smooth or finely fuzzy (generally smoother toward the margin); whitish or ochraceous when young, becoming greyish, brown, or black with age.

Pore surface: Initially porous, but soon producing wide (1–4 mm), deep (up to 4 cm), semilamellate or labyrinthine to partly poroid, maze-like channels (Fig. B) with thick (1–3 mm) walls, often resembling gills; tubes up to 4 cm (1⅔ in) deep. Whitish when fresh, later becoming dingy-yellowish or pale tan; does not bruise like *Daedaleopsis confragosa*.

Flesh: Very tough; whitish, aging to pale brownish; never rusty-brown or dark brown like *Gloeophyllum* spp.

Spore print: White.

Microscopic: Spores 5–6 × 2–3.5 µm; smooth; elliptical to short-cylindrical; context hyphae trimitic.

Staining: All parts stain dark grey to black with KOH.

Occurrence: Widespread in eastern North America (very rare west of the Mississippi), Europe, Asia, North Africa, and Australia. Primarily a brown-rot fungus on oaks (*Quercus* spp.); also reported on Beech (*Fagus grandifolia*), Elm (*Ulmus americana*), White Ash (*Fraxina americana*), Chestnut (*Castanea dentata*), and Black Walnut (*Juglans nigra*).

Edibility: Inedible, tough.

Comments: The hymenophore is elongated-poroid along the margin, the inner portion sinuous-poroid or daedeloid. The whitish to pale brownish pore surface and context distinguish this taxon from *Gloeophyllum trabeum*, which has sepia to umber-brown hymenophore and flesh. The pore surfaces are structurally similar; *G. trabeum* occurs on both hardwoods and conifers. See the discussion under the Blushing Bracket, *Daedaleopsis confragosa*, for further comparisons.

Blushing Bracket, Thin-maze Flat Polypore

Daedaleopsis confragosa

A

B

***Daedaleopsis confragosa* (Bolton 1791 : Fries 1821) J. Schröter**, in Cohn, *Krypt.-Fl. Schlesien* (Breslau) 3.1(25–32):492 (1888) [1889]

Basionym: *Boletus confragosus* Bolton 1791, *Hist. Fung. Halifax*, App. (Huddersfield) 3:160 (1792) [1791] [see online Appendix A for synonymy]
[Basidiomycota, Polyporales, Polyporaceae]

Derivation: *Daedaleopsis*, Greek, *Daedalea* plus *opsis*, "having the appearance of [the genus] *Daedalea*"; *confragosa*, Latin, "rough," "rugged," referring to the coarsely wrinkled and rugose upper surface.

Cap: 2.5–15 cm wide, up to 2 cm thick. Broadly convex to flat, outline fan-shaped to round.

Upper surface: Zonate. Initially minutely hirsute, velvety, becoming glabrous, zonate, with pale greyish, brownish, or reddish- or yellow-brown bands (Fig. A); surface rugose, shallowly grooved (sulcate), radially wrinkled or rugose; margin thin, acute to wavy.

Pore surface: White to greyish, turning brownish with age; *bruises reddish, pinkish, to salmon-brown when fresh* (Fig. B); *central zone of round pores transforming into elongated channels toward the margin*, forming a gill-like maze with channels 0.5–1.5 mm wide, tube walls <1 mm thick, often daedaloid, forked, and slightly anastomosing. May be entirely lamellate in older specimens.

Spore print: White.

Flesh: 0.1–0.2 cm thick, pale buff or brown, not duplex.

Microscopic: Spores cylindrical to elliptical, hyaline, smooth, inamyloid; 7–9 × 1.5–3 μm, variable size not useful.

Staining: All parts stain dark grey to black with KOH.

Occurrence: Saprobic, causes delignifying white rot; solitary or grouped on decaying hardwood logs and stumps, rarely on living trees; June–December. Most common on willow (*Salix* spp.) and birch (*Betula* spp.), but also found on other hardwoods except oaks; very rarely on conifers.

Distribution: Circumboreal; Europe, Asia, and North America. Common in the British Isles, Central Europe, and the Mediterranean. Widespread in North America, especially east of the Rocky Mountains.

C, *Daedaleopsis confragosa*; D, *Daedalea quercina*; E, *Cerrena unicolor*.

Edibility: Tough, inedible.

Comments: This is the type species of the genus *Daedaleopsis* and the only member of the genus occurring in North America. Sometimes confused with the gilled polypores (*Lenzites betula* and *Gloeophyllum sepiarium*), the Thick Oak Maze Polypore (*Daedalea quercina*), and the Mossy Maze Polypore (*Cerrena unicolor*), this polypore is unique in having both pores and gill-like mazes in distinct concentric zones (Figs. C & D) and in bruising red or pink when fresh (thus one of its common names). Unlike *D. quercina*, *D. confragosa* is not (or is only very rarely) found on oaks; the cap margin is thin, sharp; the small pores gradually elongate to form a radiating gill-like maze toward the margin of the sporocarp, which anastomoses only slightly; and the channels are narrow and the walls very thin. In contrast, *D. quercina* has a thick margin and exhibits large, highly anastomosing maze-like channels with very thick walls that are very rarely pore-like (Fig. D). The spores are shorter than 7 µm. *Cerrena unicolor* has a zoned, densely hairy cap usually colored green by symbiotic algae; *Daedaleopsis* and *Daedalea* have much sparser hairs above. The pore surface of *Cerrena* is complexly daedaloid and highly anastomosed (Fig. E).

Mossy Maze Polypore, Canker Rot

Cerrena unicolor

A

B

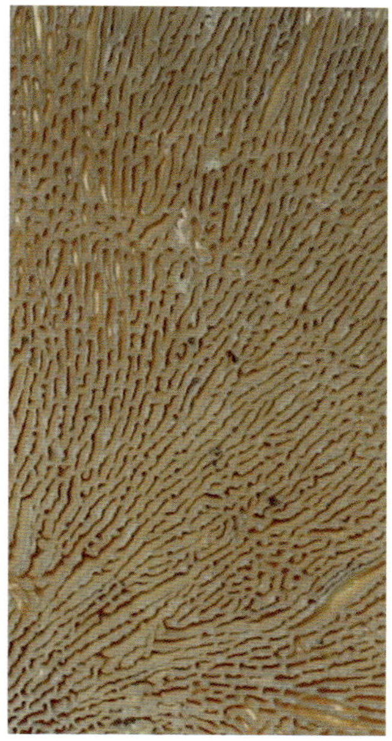

C

Cerrena Gray is a small cosmopolitan genus generally known as the maze polypores found in North and South America, Europe, Asia, Africa, Australia, and New Zealand. Some two dozen species have been described, but most have been transferred to other genera by various authorities. The actual number of species remaining in *Cerrena* is about half a dozen. It is closely related to the genus *Trametes* Ryvarden. *Cerrena unicolor* is the sole species in North America.

***Cerrena unicolor* (Bulliard 1785) Murrill,** *J. Mycol.* 9(2):91 (1903)

Basionym: *Boletus unicolor* Bulliard, *Herb. Fr.* (Paris):tab. 408 (1785) [see online Appendix A for synonymy] [Basidiomycota, Agaricomycotina, Agaricomycetes, *incertae sedis*, Polyporales, Polyporaceae]

Derivation: *Cerrena*, Latin, named after an Italian fungus; *unicolor*, Latin, "of one color."

Cap: 3–10 cm (1¼–4 in) across, typically semicircular to kidney-shaped, often forming overlapping clusters. Stalkless.

Upper surface: Velvety to densely hairy, rugose, concentrically zoned, whitish, greyish, or brownish, but typically covered by a prominent green algal mat (Fig. A). Cap often lacking when growing on the undersides of logs; resembles an isolated pore surface.

Pore surface: Whitish (Fig. B) to greyish (Fig. C), typically daedeloid, forming a distinct maze-like labyrinth of interconnected pores and wall-like passageways up to 2–5 mm deep; this is highly variable. Older fruitbodies may become distinctly toothed.

Flesh: White, duplex. *A thin black line separates the flesh from the upper tomentum (hairy layer).*

Spore print: White.

Microscopic: Spores 4.5–7 × 2.5–4 µm, oval to elliptical, smooth, inamyloid; variable size not useful. Trimitic; generative hyphae with clamp connections.

Staining: Unknown.

Occurrence: Annual, year-round white-rot fungus, parasitic on living hardwoods or saprobic on stumps and logs, often growing in overlapping clusters; rarely on conifers. Widespread in northern, midwestern, and southeastern North America; Central America, Europe, Asia.

Edibility: Inedible.

Comments: The cap surface resembles that of *Trametes versicolor*, but *C. unicolor* is distinguished by its characteristic grey, daedeloid hymenophore (pore surface). The thin black line between the surface tomentum and the underlying flesh is not always obvious, but no other polypore has the combination of a hairy zonate upper surface and a grey daedeloid hymenophore. *Cerrena unicolor* is generally recognized by the green algal mat covering much of its otherwise whitish or greyish upper surface. However, this alone is not definitive, as other polypores, especially the violet-tooth polypores (*Trichaptum biforme* and *T. abietinum*),[1] may also exhibit green algae on the upper surface. To our knowledge the green alga has not been identified. See also the discussion under the Blushing Bracket, *Daedaleopsis confragosa*, for further details.

The most fascinating aspect of this fungus is the 3-way symbiosis it forms with (1) the Pigeon Tremex, *Tremex columba* (Linnaeus) Jurine—a large, nonstinging, siricid woodwasp or horntail, and (2) four species of giant ichneumon wasps of the genus *Megarhyssa* Ashmead. Its association with the green alga is also very likely a mutualistic one, but this relationship remains obscure.

The female woodwasp drills a series of holes with her long ovipositor into several species of weakened or sick deciduous trees. She then deposits a single egg into each of these holes along with asexual spores (oidia) of *Cerrena unicolor*, which she carries in paired sacs at the base of the ovipositor. Some of the oidia are picked up and stored in specialized pouches (mycetangia) of developing female larvae; these are later released into tunnels chewed out of the wood by the larvae. A prolific mycelium develops from the germinating spores that produces several extracellular enzymes that help break down the wood. The wasp larvae then feed on both the wood and the fungus. Woodwasps tend to oviposit in trees weakened by disease, fire, or other causes and avoid healthy trees.

D

The horntail larvae undergo several developmental stages (instars) over a period of up to two years, during which they excavate complex tunnels through the sapwood and into the heartwood. (The tunnels of *T. columba* can attain lengths of 1–3 m [~3–10 ft] but are usually shorter.) Female pupae retain fungal mycelium and oidia in their mycetangia; these are then drawn into specialized pouches at the base of the ovipositor of emerging adult females and deposited in the wood when she lays her eggs. The fungus enables the woodwasps to fully digest wood. In turn, the woodwasp carries the fungus to its specific host trees and injects it beneath the bark past the tree's first line of defense. Although the fungus is able to live independently of the woodwasp, the woodwasp is completely dependent upon the fungus. Eggs laid in the absence of the fungus hatch, but the larvae do not develop beyond the first instar.

In order to keep the woodwasps under control, the Mossy Maze Polypore releases pheromones that attract yet another actor—giant ichneumon wasps of the genus *Megarhyssa*, which parasitize and kill the woodwasp larvae. A female *Megarhyssa* locates a Pigeon Tremex larva or pupa inside its tunnel by means of the vibrations it makes as it chews its way through the wood. She inserts her long ovipositor into the tunnel (Fig. D) and deposits an egg on the larva or pupa of the woodwasp. After hatching, the ichneumon wasp larva feeds on the woodwasp larva, killing it, after which it pupates within the burrow, metamorphoses, and chews its way out of the burrow to emerge as an adult.

These complex interactions between the Mossy Maze Polypore and its algae, the woodwasps, and the giant ichneumon wasps are by no means fully understood. Other actors in this drama may also be waiting in the wings. The Giant Ichneumon Wasp, *Megarhyssa macrura* (Linnaeus, 1763), shown in Fig. D, was photographed in October 2011 depositing her eggs in the burrows of Pigeon Tremex larvae in the Atlanta, Georgia, area.

1. The prominently toothed pore surface of *Trichaptum* is easily distinguished from the daedeloid hymenium of *C. unicolor*.

Cauliflower Mushrooms

Sparassis spathulata and *S. americana* f. *americana*

A

B

A, *Sparassis spathulata*; B, *S. americana* f. *americana*. (Photos: A, W.H. Light; B, Teresa Fortenberry.)

***Sparassis spathulata* (Schweinitz) Fries**, *Elench. Fung.* (Greifswald) 1:227 (1828)

Basionym: *Merisma spathulatum* Schweinitz, *Schr. Naturf. Ges. Leipzig* 1:110 (1822)

Synonym: *Sparassis herbstii* Peck, *Bull. Torrey Bot. Club* 22:207 (1895)

Misapplied names:

Sparassis crispa: *auct. mult.* [*non* (Wulfen 1788) Fries 1821]
Sparassis simplex D.A. Reid, *Trans. Br. Mycol. Soc.* 41(4):439 (1958) [*fide* Burdsall & Miller, *op. cit.*; *nomen dubium*, *fide* Light & Woehrel, *Fungi* 2(4):10 (2009)]

***Sparassis americana* f. *americana* R.H. Petersen**, in K.W. Hughes, A.R. Segovia, & R.H. Petersen, Transatlantic disjunction in fleshy fungi, I: The *Sparassis crispa* complex, *Mycol. Progress* 11 (2013); DOI 10.1007/s11557-013-0927-1

Misapplied names:

Sparassis crispa, *auct. mult. amer.*; [*non* (Wulfen 1788) Fries 1821]
Sparassis radicata, *auct. amer. part.*; [*non* Weir, *Phytopathol.* 7:166 (1917)]
[Basidiomycota, Agaricomycotina, Agaricomycetes, *incertae sedis*, Polyporales, Sparassidaceae]

Derivation: *Sparassis*, New Latin, from the Greek word *sparássein*, "to tear," referring to the spathe-like or fan-like blades (flabellae); *spathulata*, Latin, "spatulate"; *americana*, Latin, "American."

Fruitbody (basidiome): *Sparassis spathulata* (Fig. A)—cauliflower-shaped, basidiome zonate, attaining diameter and height of 35 cm (14–15 in), consisting of densely grouped, flat, undulating, vertical branches or blades (flabellae) emerging from a buried, branching, central base; flabellae whitish, yellowish, or tan, with distinct color zonations, becoming discolored with age, margins entire; ***S. americana* f. *americana***—cauliflower-shaped, basidiome azonate, uniformly white to yellowish, attaining a height of 20 cm and a width of 35 cm (8 × 20 in); typically arising from a single, large, thick (2–4 cm), deeply rooted central stalk; flabellae vertically subdivided, deeply crinkled, and contorted, margins variously incised; somewhat resembling cooked egg noodles.

Spore print: White for both species.

Microscopic: *S. spathulata*—spores smooth, elliptical, 5.9–8.0 × 4.7–5.8 µm; context (flesh) hyphae lacking clamp connections; ***S. americana* f. *americana***—spores smooth, broadly elliptical to subglobose, slightly smaller than in *S. spathulata*,

4.5–6.3(–7.0) × (3.0–)3.5–4.5 µm; context hyphae with clamp connections.

Occurrence: ***S. spathulata***—saprobic/parasitic, emerging singly or in groups from the ground at the base of oaks or pine trees, July–October; widespread and occasional in eastern North America, especially in the South; ***S. americana* f. *americana***—saprobic/parasitic, solitary at the base of pines, August–December; eastern Canada south to the Gulf Coast.

Edibility: Both species have a delicate flavor when young, becoming tough and less flavorful with age.

Comments: *Sparassis americana* Petersen was recently described as a new species with two *formae* for mushrooms in eastern and southwestern North America previously identified as *Sparassis radicata* Weir,[1] a taxon originally described from northern Idaho and the Northwest Coast of North America. Genetic evidence revealed that the eastern and southwestern populations of North American cauliflower fungi are distinct from the western populations of *S. radicata*, and they were described as *S. americana* f. *americana* and *S. americana* f. *arizonica*, respectively.[2]

Both *Sparassis spathulata* and *S. americana* have historically been (and often still are) misidentified as *S. crispa* (von Wulfen) Fries, a European species that does not occur in North America.[1,2] *Sparassis spathulata* has also been (and often still is) called *S. herbstii* Peck (1895), a later, deprecated synonym. See Light & Woehrel (2009),[1] online Appendix A, and Hughes et al. (2013)[2] for a complete discussion of the genus *Sparassis* and its tangled nomenclatural issues.

Sparassis americana superficially resembles the Umbrella Polypore, *Polyporus umbellatus* (Persoon) Fries. However, the many (up to 100 or more) individual shallowly funnel-shaped caps with their centrally attached stalks and prominent decurrent pores distinguish *P. umbellatus* from the poreless, crinkled, lettuce-like, vertical flabellae of *S. americana*.

1. W.H. Light & M. Woehrel, 2009, Clarification of the nomenclatural confusion of the genus *Sparassis* [Polyporales: Sparassidaceae] in North America, *Fungi* 2(4):10–15.
2. K.W. Hughes et al., 2013, *Mycol. Progress* 11; DOI 10.1007/s11557-013-0927-1.

Key to Eastern North American *Sparassis*

1. Flabellae zonate, broadly fan-shaped and gently undulating, arising from a buried, branching central stalk, associated with conifers and hardwoods . . . . *S. spathulata*
2. Flabellae azonate, deeply curled and crinkled, arising from a single deeply rooted central stalk, restricted to conifers . . . . *S. americana* f. *americana*

Umbrella Polypore

Polyporus umbellatus

Photos: Teresa Fortenberry.

***Polyporus umbellatus* (Persoon) Fries**, *Syst. Mycol.* (Lundae) 1:354 (1821)

Basionym: *Boletus umbellatus* Persoon, *Syn. Meth. Fung.* (Göttingen) 2:519 (1801)
[Basidiomycota, Agaricomycotina, Agaricomycetes, *incertae sedis*, Polyporales, Polyporaceae]

Derivation: *Polyporus*, Greek, "many pores"; *umbellatus*, Latin, "umbellate," "umbrella-like."

Fruitbody: A compound structure 30–50 cm (12–20 in) or more in diameter, consisting of up to several hundred small caps arising from a many-branched stock.

Caps: Individual caps small, 1–4 cm (½–1⅔ in) across, deeply umbilicate, shallowly funnel-shaped, arising from branching individual stipes (Fig. B) leading upward from a common fused base; cap surfaces greyish-white, ochre, or dusky-brown (Fig. A), not bruising, often developing scales.

Flesh: White; soft when young, not bruising when exposed, hardening with age.

Pore surface: White, not bruising, pores decurrent far down the individual stems, round to angular, 1–3 per mm (Figs. C & D), tubes to 2 mm deep.

Stipes: White, stout, irregularly sinuous and branching, centrally attached to the caps (Fig. B), fused below into one or more larger masses arising from a blackish, tuber-like, buried nodule (sclerotium).

Spore print: White.

Microscopic: Spores 1.5–10 × 3–4 µm, smooth, cylindrical, non-reactive in Melzer's reagent; hyphae dimitic; cystidia lacking; clamp connections present.

Occurrence: Saprobic on the roots of hardwoods, fruiting at bases of trees, especially oaks and beeches, spring–late fall, often annual, arising repeatedly from same spot; widely distributed, but rare in Southeast and elsewhere. North and Central America, Europe, Asia.

Edibility: Edible, considered choice.

Comments: This mushroom superficially resembles one of the Cauliflower Mushrooms (*Sparassis americana* f. *americana*), Hen-of-the-woods (*Grifola frondosa*), and the Black-staining Polypore (*Meripilus sumstinei*). However, the many individual umbilicate and trumpet-shaped caps with their centrally attached stalks distinguish it from these other species.

Sparassis americana f. *americana*: Fruitbody azonate, uniformly whitish or yellowish, consisting of highly folded and crinkled, lettuce-like flabellae attached to a single, deeply buried, root-like stalk; pores not discernible.

Grifola frondosa: Individual caps larger, 2–10 cm across (¾–4 in), fan-shaped, without central depression; pore surface initially lavender-grey, becoming dirty-whitish or yellowish, 2–4 pores per mm, tubes 5 mm deep; stems attached off center.

Meripilus sumstinei: Many fewer and much larger fan-shaped individual caps, each 5–20 cm (2–8 in) across; pore surface creamy-white, pores tiny, 6–8 per mm, tubes 8 mm deep; stems laterally attached; caps, pore surface, and stems bruising and aging black.

The Umbrella Polypore has a wide variety of medicinal properties, and the sclerotium (*Zhu Ling*) has been used in traditional Chinese medicine as a potent diuretic and to prevent kidney stones.[1] Polysaccharides from the mycelium and *Zhu Ling* have antitumor effects against lung, liver, bladder, gastric, and cervical cancers, as well as leukemia.[2] *Zhu Ling* also has potent antioxidant and free-radical scavenging activity,[3] as well as antibacterial and antiprotozoal (antimalarial) activity.[4]

1. H. Satish, 2006, *Biomed.* 1(2):95–119; L.W. Wang et al., 1964, *Yao Xue Xue Bao* 11:815–818.
2. H. Ito et al., 1976, *Nippon Yakurigaku Zasshi.* 72(1):77–94; T. Ohsawa et al., 1992, *Chem. Pharm. Bull.* (Tokyo), 40(1):143–147; D. Yang et al., 1994, *Zhonghua Wai Ke Za Zhi* 32(7):433–434; 1999, 37(8):464–465; W. Y. Lee et al., 2005, *Bull. Kor. Chem. Soc.* 26(9):1464–1466.
3. F. Liu et al., 1997, *Life Sci.* 60(10):763–771; N. Sekiya et al., 2005, *Biol. Pharm. Bull.* 28(5):817–821.
4. J.J. Li et al., 2000, *Zhongguo Zhong Yao Za Zhi* 25(10):628–630; A. Lovy et al., 1999, *J. Herb Spice Medic. Plant.* 6(4):49–57.

Hen-of-the-woods, Maitake

Grifola frondosa

***Grifola frondosa* (Dickson) Gray**, *Nat. Arr. Brit. Pl.* (London) 1:643 (1821)

Basionym: *Boletus frondosus* Dickson, *Fasc. Pl. Crypt. Brit.* (London) 1:18 (1785)
[Basidiomycota, Agaricomycotina, Agaricomycetes, Polyporales, Meripilaceae]

Derivation: *Grifola*, etymology unknown (anecdotally reported to be based on the name of an Italian fungus); *frondosa*, Latin, "leafy," referring to the multiple caps.

Fruitbody: A complex rosette-like cluster of smaller overlapping caps emerging from a common branched stipe 15–60 cm (6–24 in) across or even larger.

Cap: Individual caps 2–10 cm (¾–4 in) across by 8 mm (⅓ in) thick, fan-shaped or leaf-shaped to spatulate, laterally attached to a thick, branching stalk; greyish- to ochre- or blackish-brown, more or less concentrically zonate; upper surface smooth to finely tomentose, often radially rugose; margins thin, wavy.

Pore surface: Greyish-white to yellowish; pores angular, 2–4 per mm, with thin and tattered rims; decurrent often to the ground; tubes up to 5 mm (¼ in) deep.

Flesh: Firm, white, 2 mm thick in the cap, up to several centimeters thick in the main branches and at the base.

Stipe: Cream-colored, base up to 10 cm (4 in) in diameter, producing many branches, which terminate in individual caps.

Spore print: White.

Bruising: Not bruising when handled or injured.

Microscopic: Spores 5–7 × 3.5–5 µm, ovoid to broadly elliptical, smooth, nonreactive in Melzer's reagent.

Occurrence: Weakly parasitic on living hardwoods, especially oaks (*Quercus* spp.) but also elms (*Ulmus* spp.), maples (*Acer* spp.), and conifers and causing a white rot. Fruiting on the ground from the roots of various hardwoods and conifers; widespread in eastern North America (rare in the Pacific Northwest); Europe, Asia.

Edibility: Edible, considered choice.

Comments: This mushroom is very similar to the Black-staining Polypore, *Meripilus sumstinei* (Murrill) M.J. Larsen & Lombard, and the Umbrella Polypore, *Polyporus umbellatus* (Persoon) Fries. All three species occur at the bases of hardwood trees (particularly oaks) and fruit at similar times of year. *Meripilus sumstinei* has much larger individual caps (5–20 cm [2–8 in] across) and smaller pores (typically 6–8 per mm), and the margins of the caps and pore surfaces bruise black when handled.

Polyporus umbellatus differs in its many deeply umbilicate, trumpet-shaped, and smaller (1–4 cm [½–1⅔ in] across) individual caps with their centrally attached stalks, much larger (1.5–10 × 3–4 µm) cylindrical spores, and shallower tubes (2 mm deep). *Polyporus umbellatus* is also much rarer.

Like the Umbrella Polypore, Hen-of-the-woods (better known in this capacity as Maitake) has potent medicinal properties and is a choice and much-sought-after edible. It is used in traditional Chinese and Japanese medicine to enhance the immune system. Scientific studies have identified a protein-bound β-glucan in this mushroom with immunoenhancing properties and also found that Maitake can stimulate both innate and acquired (adaptive) immune responses, as well as regulate blood pressure, glucose levels, and lipids in the blood and liver. This is discussed in more detail in the section on medicinal mushrooms.

Aborted Polypore

Abortiporus biennis

Anamorph (*Ceriomyces terrestris*): labeled "a" in Figs. A–C; teleomorph (*Abortiporus biennis*): labeled "t" in Figs. A–C.

***Abortiporus biennis* (Bulliard) Singer**, *Mycologia* 36(1):68 (1944)

Basionym: *Boletus biennis* Bull., *Herb. Fr.* 10:tab. 449 (1789) [Basidiomycota, Basidiomyctetes, Meripilaceae]

Derivation: *Abortiporus*, Latin, "miscarried pores"; *biennis*, Latin, "lasting two years."

Anamorph: *Ceriomyces terrestris* Schulzer [anamorphic description], *Verh. Zool.-Bot. Ges. Wien* 24:451 (1874). The most commonly encountered form of this taxon. Surface grotesquely malformed, irregularly and deeply excavated and jumbled (labeled *a* in Figs. A–C, Figs. D & E), when not overgrown by pore surface, similar to upper surface of teleomorphic pileus (cap) in coloration and hairiness (Figs. B–E).

Fruitbody: The rarely seen normal (teleomorphic) morph with central or lateral stipe or sessile (labeled *t* in Figs. A–C, Fig. D); caps solitary, overlapping or fused to form compound fruitbodies, nearly circular and infundibuliform (with central depression) (Fig. A) or irregularly shelf-like (Fig. C), 2–20 cm wide by 1.5–1.5 cm thick (⅘–8 × ⅔ in), often overgrown by pore surface (Figs. B & D); upper surface whitish to tan or pale brown; azonate to faintly zonate; conspicuously tomentose, stiff hairs often matted together (Figs. D & E), never completely smooth. Usually emerging from or fused with larger amorphous, chaotic-looking, anamorphic mass (Figs. A–C).

Pore surface: Whitish, often reported as bruising reddish or pinkish brown. Pores small, round, ovoid, angular to elongate or somewhat daedeloid (maze-like), with thick walls, 2–3 (rarely 4) per mm (Fig. F), often covering stipe.

Spore print: White.

Flesh: Upper context soft and fibrous, light buff to tan, lower portion firm, corky, cream-colored, tubes the same color and continuous with lower context, up to 5 mm long.

Microscopic: Often described as monomitic, but both skeletal and generative hyphae present; skeletal hyphae thick-walled, lacking cross-walls and clamps, 4–8 µm in diameter; generative hyphae thin-walled, with septa and clamps, 3–6 µm in diameter. Basidiospores ovoid to broadly ellipsoid, hyaline, smooth, 5–8 × 3–5 µm. Asexual chlamydiospores abundant, thick-walled, more or less globose, 5–7 µm in diameter or 7–10 µm in diameter, often found developing inside (Fig. G) or forming bead-like buds outside the generative hyphae in the teleomorph and anamorph (Fig. G inset); *the presence of chlamydiospores is diagnostic*. Gloeocystidia present, but usually rare (occasionally abundant), broadly clavate to cylindrical, with swellings and constrictions, not encrusted, 7.5–8.5 µm in diameter × 50–75 µm long, never projecting much beyond the basidia (and have been called cheilocystidia).

Staining/Bruising: Pore surface sometimes bruises red; our samples did not. Young fruitbodies may exude reddish droplets. Flesh negative in KOH; our specimens immediately and distinctly dextrinoid (reddish and brown) in Melzer's reagent.

Occurrence: Widespread but uncommon in North America; saprobic and parasitic; causes white rot in hardwoods, rarely in conifers.

Edibility: Tough, inedible.

Comments: The extremely distorted reef-like shapes of most examples of this fungus have resulted in more than 64 scientific synonyms and very likely a number of others that we have missed. See online Appendix A. This taxon has been described as several different species by more than one author, sometimes in the same paper. The presence of asexual chlamydiospores in both the hymenium and context of this fungus definitively distinguish it from any other species with which it might be confused, for example, *Loweomyces fractipes* (Berkeley & M.A. Curtis) Jülich (formerly in the genus *Abortiporus*) or some species of *Hydnellum* or *Albatrellus*. The massively distorted anamorphic form of this species that is usually encountered does not automatically identify it despite assertions to that effect in the literature or online websites. We have documented similar distorted anamorphic forms associated with other common polypores. Notable examples include *Trichaptum abietinum* (and commonly also *T. biforme*) and *Gloeophyllum sepiarium* (see our descriptions of the first two species), among others. These aberrent anamorphic polypores have become increasingly common in the Atlanta area during the past two drought years. Had we not seen the teleomorphs emerging from and in some cases actually witnessed them transforming into these anamorphic masses as they aged, we would not have made these connections. To our knowledge they have not been reported by anyone in the literature or online.

As it currently stands, *Abortiporus* contains only two or possibly three species: *A. biennis*, *A. roseus* (D.A. Reid) Masuka & Ryvarden (1992), and *A. chocoensis* Læssøe & Ryvarden (2010). *Abortiporus chocoensis* may be a synonym of *A. roseus*.

D

E

F

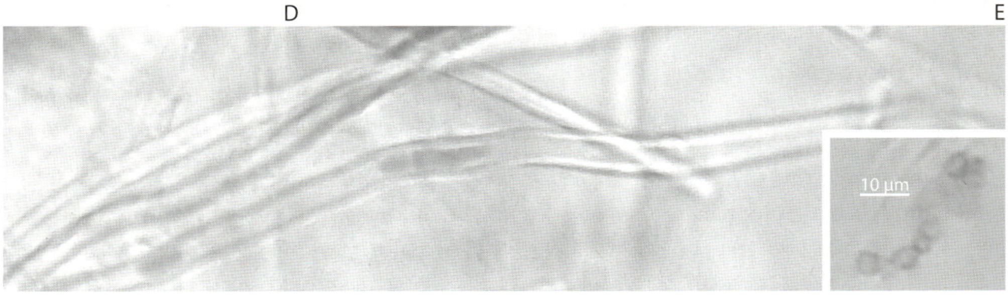

G

Crested Polypore

Albatrellus cristatus

A

B

C

D

***Albatrellus cristatus* (Schaeffer) Kotlaba & Pouzar**, *Česká Mykol.* 11(3):154 (1957)

Basionym: *Boletus cristatus* Schaeffer, *Fung. Bavar. Palat. Nasc.* (Ratisbonae) 4:316 (1774)
[Basidiomycota, Agaricomycotina, Agaricomycetes, *incertae sedis*, Russulales, Albatrellaceae]

Derivation: *Albatrellus*, etymology not known; *cristatus*, Latin, "crested."

Fruitbody: Solitary or in fused compound clusters, with multiple caps arising on branching stalks.

Cap: Individual caps 5–20 cm (4–8 in) across, circular to (more typically) irregularly lobed (Figs. A–C); leathery, surface smooth to minutely velvety, becoming cracked with age; color variable: yellowish, yellowish-brown, greenish-brown to buff; azonate (Fig. B) to zonate (Fig. A).

Flesh: White to cream or buff, bruising yellowish-green when cut; odor not distinctive, taste unremarkable to slightly bitter.

Pore surface: White, eventually becoming yellowish, greenish-yellow, or pinkish; pores usually decurrent, covering stipe (Fig. C), angular, 1–3 per mm (Fig. D), rims becoming thin, somewhat tooth-like and jagged with age; tubes up to 5 mm deep.

Stipe: 2.5–7.5 cm tall by 1–2.5 cm thick (1–3 × ½–2 in), central to eccentric, irregular, often branching and giving rise to several caps; solid, smooth or covered by decurrent pores (Fig. C); white to greenish or concolorous with cap.

Spore print: White.

Chemical staining: Flesh slowly staining pale red with KOH.

Microscopic: Spores 5–7 × 4–5 µm, broadly elliptical to teardrop-shaped, with small apiculus, smooth, clear, non-reactive to weakly amyloid in Melzer's reagent; hyphae simple, septate, lacking clamp connections.

Occurrence: Annual, probably mycorrhizal. Alone or in fused clusters on the ground under hardwood trees, especially associated with oak (*Quercus* spp.), American Beech (*Fraxinus grandifolia*), and maples (*Acer* spp.), occasionally under conifers; widespread and fairly common throughout eastern North America, west to Oregon,[1] summer–fall; Europe, East Asia.[2]

Edibility: Inedible.

Comments: Other species of *Albatrellus* occurring in our area include *A. pes-caprae* (Persoon *ex* Fries) Pouzar, *A. confluens* (Albertini & Schweinitz *ex* Fries) Kotlaba & Pouzar, *A. caeruloporus* (Peck) Pouzar, and *A. peckianus* (Cooke) Niemela.

Albatrellus pes-caprae (Goat's Foot) is widespread but uncommon in the Southeast. It arises from buried wood beneath conifers rather than hardwoods, the cap is grey-brown to orange-brown and scaly (not smooth or finely velvety), and the pores are much larger (each pore 1–2 mm wide, in contrast to 1–3 pores per mm). The spores are slightly larger (8–11 × 6–8 µm), have a more prominent apiculus, and are completely nonreactive (rather than weakly amyloid) in Melzer's reagent; the tramal (in the flesh) hyphae have clamp connections.

Albatrellus confluens is also associated with conifers (not hardwoods) and occurs in many different coniferous ecosystems across North America. It generally ranges from pine forests at lower elevations up into spruce-fir forests at high elevations. Unlike *A. cristatus*, it is rarely found in deciduous woods. It differs from *A. cristatus* in the color (pale orange) of its much smaller (barely visible to the naked eye) pores (3–5 per mm) and somewhat smaller (4–5 × 2.5–3.5[–4] µm) ovoid basidiospores, which, like those of *A. cristatus*, are weakly amyloid. All surfaces of the fruitbody stain purple (instead of pale red) with KOH.

Albatrellus caeruloporus, a mainly northeastern species, is immediately identified by its blue, grey, or greyish-blue cap and a pale blue pore surface (when fresh, both surfaces become brownish with age) and its association with Eastern Hemlock (*Tsuga canadensis*). In our area it has been recorded from Great Smoky Mountains National Park.[3]

Albatrellus peckianus is another northern species that has been reported from Great Smoky Mountains National Park. Its pore surface is bright yellow, and the rest of the fruitbody is tinged with yellow; its spores are smaller (3.5–4.5 × 2.5–3 µm) than those of *A. cristatus* and are nonreactive in Melzer's reagent. It grows on hardwoods or on the ground attached to buried hardwoods, primarily American Beech (*Fagus grandifolia*) and Basswood (*Tilia americana*).[3]

1. R.L. Gilberston & L. Ryvarden, 1986, *North American Polypores*, vol. 1 (Oslo: Fungiflora), 96.
2. H.D. Zheng & P.G. Liu, 2008, *Fungal Diversity* 32:157–170.
3. R.H. Petersen, 1976, *Checklist of Fungi of the Great Smoky Mountains National Park*, Management Report No. 29, Department of the Interior, National Park Service, Southeast Region.

Cinnabar-red Polypore

Pycnoporus cinnabarinus

A

B

Photos: Jim Edwards.

Pycnoporus cinnabarinus **(Jacquin) P. Karsten**, *Revue Mycol.*, Toulouse 3(9):18 (1881)

Basionym: *Boletus cinnabarinus* Jacquin, *Fl. Austriac.* 4:2 (1776) [Basidiomycota, Agaricomycetidae, Polyporales, Polyporaceae]

Derivation: *Pycnoporus*, Greek, "with dense, compact pores"; *cinnabarinus*, Latin, "vermillion."

Cap: Stem lacking, semicircular or kidney-shaped, pore surface plane, while top surface convex, 3–14 cm (2¼–5½ in) across, to 1.5 cm (⅔ in) thick, fruitbodies solitary or grouped, occasionally fused and overlapping.

Upper surface: Bright orange, fading in age to orange; upper surface finely hairy to velvety when young, becoming wrinkled and pockmarked (Fig. A, right) to smooth with age but remaining soft.

Pore surface: Bright orange-red (Figs. A, left, & B), fading less than upper surface; pores round to angular, 2–4 per mm, occasional pores fused to form elongated opening. Thick margin formed by underfolded upper surface.

Flesh: Reddish to pale orange; tough.

Spore print: White.

Staining in KOH: Upper surface immediately black or reddish or purplish, then gradually grey to black; pore surface blue-green or olive-green; flesh slowly reddish or black, or yellowish in older examples.

Microscopic: Spores 5–8 × 2.5–3 µm, smooth, cylindrical to long-elliptical; nonreactive in Melzer's reagent or KOH; trimitic, no cystidia.

Occurrence: Saprobic on dead wood of hardwoods (rarely conifers), causing white rot; widespread, but uncommon in North America.

Edibility: Inedible.

Comments: *Pycnoporus cinnabarinus* closely resembles *P. sanguineus*, which also occurs in our area but is less common. *Pycnoporus sanguineus* has a short stem, 15 by 3–10 mm (6 × 0.12–0.4 in), often flattened, to 25 mm (1 in) wide, with a disk-shaped base, never truly sessile. The cap is smaller and flatter, the cap measuring 1.2–3.5 by 1.5–5.0 cm (0.5–1.4 × 0.6 in) by 0.1–0.45 cm (0.04–0.18 in) thick, and the pores are smaller, 4–6 circular pores per mm. The reddish color in both species is due to cinnabarin, which has some antimicrobial properties and is widely used as a dye.

Rosy Conk

Fomitopsis cajanderi

A

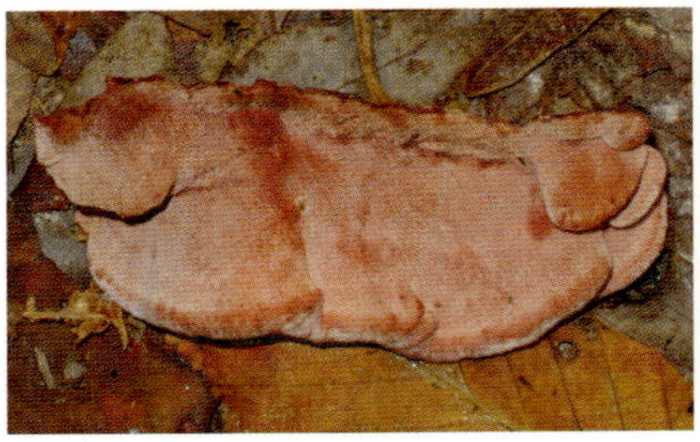

B

C

***Fomitopsis cajanderi* (P. Karsten 1904) Kotl. & Pouzar**, *Česká Mykol.* 11(3):157 (1957)

Basionym: *Fomes cajanderi* P. Karsten, *Finl. Basidsvamp.* 46(11):8 (1904) [1903–1904]
[Basidiomycota, Agaricomycotina, Polyporales, Fomitopsidaceae]

Derivation: *Fomitopsis*, Latin, from *fomes*, "fuel," "tinder," and *opsis*, Greek, "resembling," resembling [the genus] *Fomes*; *cajanderi*, epithet in honor of A.K. Cajander.

Cap: Up to 2.5 cm (1 in) wide, without stalk (Fig. A), or forming laterally fused and overlapping clusters (Fig. B) up to 20 cm (8 in) wide and 10 cm (4 in) deep; with an unmistakable delicate pink to purplish-pink hue when young, fading to grey, brown, or nearly black with age, paler along margin; flat or broadly convex, hirsute at first, becoming smooth with age.

Pore surface: A beautiful deep rose or purplish-pink when young (Fig. C), often fading to reddish-brown with age; pores tiny, round to angular, 4–5 per mm, tube layer up to 2 cm (0.8 in) deep.

Spore print: Off-white.

Flesh: Pinkish-brown, leathery when young, becoming very hard with age.

Microscopic: Spores 4–7 × 1.5–2 µm; smooth, narrow-cylindrical, slightly curved, inamyloid; hyphae dimitic, 2.5–5 µm in diameter, without septa or clamps, long, sinuous; no cystidia.

Staining: Immediately stains black on upper surface and flesh with KOH.

Occurrence: Perennial. Saprobic on dead conifers or parasitic, causing brown rot on living trees, rarely on hardwoods; widespread but uncommon throughout North America.

Edibility: Inedible.

Comments: Young mushrooms may exude drops of amber liquid from the upper surface. It is rare on hardwoods but has been reported on birch (*Betula* spp.), poplar (*Populus* spp.), and fruit trees (*Prunus* spp.). This species closely resembles *Fomitopsis rosea* (Albertini & Schweinitz) P. Karsten (1881), but the latter usually has individually distinct, more hoof-shaped fruitbodies up to 12 cm (4¾ in) across and 6 cm (2.4 in) deep and paler flesh; it has a more northern distribution. It occurs in drier habitats and prefers cut timber in exposed locations, while *F. cajanderi* grows on logs in moist or damp locations. The spores of *F. rosea* are similar to those of *F. cajanderi*: elongate-cylindrical but slightly larger (5.5–7.5 × 2–2.5 µm). Staining and other microscopic features are also similar in both taxa.

Beefsteak Polypore

Fistulina hepatica

A

B

Photos: John Plischke III.

***Fistulina hepatica* (Schaeffer) Withering**, *Bot. Arr. Brit. Pl.*, 2nd ed. (London) 4:405 (1801)

Basionym: *Boletus hepaticus* Schaeffer, *Fung. Bavar. Palat. Nasc.* (Ratisbonae) 4:82 (1774)
[Basidiomycota, Agaricomycotina, Agaricomycetes, Agaricomycetidae, Agaricales, Fistulinaceae]

Derivation: *Fistulina*, Latin, diminutive of *fistula*, "tube," "a small tube"; *hepatica*, Latin, from *hepatis*, "liver-like."

Cap: Up to 30 cm across by 5 cm thick (12 × 2 in), irregular to fan-shaped, spoon-shaped, or tongue-like, margin entire to lobed; surface smooth to velvety, reddish-orange to dark red or liver-colored, aging rusty-brown.

Flesh: Whitish, pinkish, or reddish, marbled with zones of color, gelatinous, watery; exuding a red liquid when squeezed; odor not distinctive, taste sour.

Pore surface: Whitish to pale pinkish (Fig. B), bruising and aging reddish-brown; pores circular, 1–3 per mm (Fig. B, inset), tubes 5–10 mm long.

Stipe: Absent or lateral up to 8 cm (3¼ in) long, concolorous with cap.

Spore print: Pinkish to pinkish-brown.

Microscopic: Spores 3.5–6 × 2.5–4 µm, ovoid, smooth, clear, nonreactive in Melzer's reagent.

Occurrence: Saprobic, causing a brown rot. Solitary or clustered near the base of oaks (*Quercus* spp.) and the American Chestnut (*Castanea dentata*); widespread but uncommon in eastern North America; summer–fall.

Edibility: Edible, but not choice.

Comments: This distinctive mushroom looks much like a slab of raw meat or liver and cannot be mistaken for anything else. Although not a polypore, it is grouped together with them here because of their general resemblance. It was originally placed in the Polyporales, but DNA studies have reassigned it to the order Agaricales (gilled mushrooms). It was formerly much more common before the American Chestnut was decimated.

It is a good source of antioxidants (phenolic compounds and organic acids) and also has potent antibacterial activity. *Fistulina hepatica* is widely believed to be one of the very few mushrooms that can safely be eaten raw because it contains no monomethylhydrazine precursors. However, we have not found a single scientific documentation for that assertion, and we urge anyone to thoroughly cook this species as well.

Coral-pink Phlebia, Coral-pink Merulius

Phlebia incarnata

A

B

***Phlebia incarnata* (Schweinitz 1822) Nakasone & Burdsall**, *Mycotaxon* 21:245 (1984)
[Basidiomycota, Agaricomycotina, Agaricomycetes, Polyporales, Merulaceae]

Basionym: *Merulius incarnatus* Schweinitz 1822 [see online Appendix A for synonymy]

Derivation: *Phlebia*, Greek, "of blood vessels"; *incarnatus*, Latin, "flesh-colored."

Cap: 3–10 cm (1¼–4 in) across; kidney-shaped to irregularly lobate (Fig. A). Bright coral-pink when fresh, fading to pinkish-white; with crenulated (wavy) white margin.

Upper surface: Flat to convex; smooth to finely hairy.

Pore surface: Whitish to pale cream. With distinctive network of folds and ridges and branching, cross-veined reticulations forming a wrinkled pattern of low ridges and hollows (Fig. B).

Flesh: White, may have pinkish tint; soft and spongy when fresh, becoming tough with age.

Spore print: White.

Chemical staining: Cap surface and flesh transiently yellowish-green to olive-yellow, gradually turning pale orange with KOH.[1]

Microscopic: Spores 4–6 × 2–2.5 µm; elliptical, smooth, thin-walled, clear, nonreactive in Melzer's reagent; context hyphae monomitic.

Occurrence: Saprobic, causing a white rot in hardwoods, especially White Oak (*Quercus alba*). Grows in overlapping, irregularly shaped clusters (Fig. A); usually associated with and growing among False Turkey Tail, *Stereum ostrea* (Fig. A). Widespread in eastern North America, Texas, Mexico. Late spring–fall; overwinters along Gulf Coast. Recently reported from southern Brazil.[2]

Edibility: Inedible.

Comments: This mushroom is unlikely to be confused with any other except possibly the Trembling Phlebia, *P. tremellosa* (see below). The bright coral-pink color of the cap surface in fresh specimens stands out against the concentric zones of deep bronzes, tans, and ochres of the

False Turkey Tail, *Stereum ostrea*, with which it is usually found. The network of folds and ridges is not a true pore layer, as in the polypores, in which basidia (spore-bearing cells) are found only lining the tubes. In *Phlebia*, basidia cover the entire lower surface of the fruitbody. The similar congeneric species *Phlebia tremellosa* has a woolly or hairy white upper surface; the pore surface is very similar to that of *P. incarnata*, except for the color, which is orange or pink. The colors of the cap and pore surfaces are exactly reversed in the two species of *Phlebia*. The biggest potential source of confusion is mistaking the upper surface of the cap for the pore surface in *P. tremellosa*. The woolly cap of *P. tremellosa* distinguishes it from the nonhirsute, leathery cap of *P. incarnata*. Arizona specimens reported as *Merulius incarnatus*[3] are likely *Cytidiella albomella* (Bondarzew) Parmasto (*fide* Ginns, 1976).[4]

1. *Fide* M. Kuo, March 2010, *Phlebia incarnata*, at the MushroomExpert.Com website, http://www.mushroomexpert.com/phlebia_incarnata.html.
2. M.C. Westphalen et al., 2010, *Mycotaxon* 114:305–310.
3. R. Gilbertson, 1974, *Fungi That Decay Ponderosa Pine* (Tucson: University of Arizona Press), 45.
4. J. Ginns, 1976, *Canad. J. Bot.* 54(2):134.

Trembling Phlebia, Trembling Merulius

Phlebia tremellosa

A

B

C

D

E

***Phlebia tremellosa* (Schrader 1794 : Fries) Nakasone & Burdsall,** *Mycotaxon* 21:245 (1984)

Basionym: *Merulius tremellosus* Schrader 1794 [Basidiomycota, Agaricomycotina, Agaricomycetes, Polyporales, Merulaceae]

Derivation: *Phlebia*, Greek, "of blood vessels"; *tremellosa*, Latin, "trembling."

Fruitbody: Irregular effuse-reflexed, resupinate, or sessile sheets 3–15 cm (1¼–6 in) across, typically forming projecting stubby, hood-like, often overlapping caps or brackets (Figs. A & B) with fringed margins (Figs. C & D), often fusing laterally into extensive spreads (Fig. B). Solitary (Fig. A), but typically gregarious, with caps in overlapping clusters (Fig. B).

Upper surface: White, woolly upper surface up to 5 mm thick, forming a convex hood (Figs. A & C); stalk lacking.

Pore surface: Translucent, rubbery, gelatinous, pink, pinkish-orange (Fig. B), yellowish- or brownish-orange (Figs. C & D) to reddish; with deep, elaborately folded and wrinkled pockets (Fig. B); network of radiating folds and ridges (Fig. B), often appearing pore-like (Figs. D & E), but true pores lacking.

Spore print: White.

Flesh: White, may have orangish tint; very thin.

Microscopic: Spores 3.5–4.5(–5) × 1–2 µm, smooth, allantoid (sausage-shaped), occasionally biguttulate (containing two oil droplets) in 2% KOH; nonreactive in Melzer's reagent. Hyphae monomitic. Cylindrical to narrowly clavate, scattered leptocystidia occasionally present, barely projecting (20 µm) from hymenium, apices often nodose (with encrusted knobs) with a basal clamp.

Chemical staining: Cap surface dextrinoid (pink to red) in KOH.

Occurrence: Widespread but occasional in North America (except the Southwest) and Europe. Annual, saprobic, causes white rot in stumps and dead branches of hardwoods and decayed wood of living oaks; occasionally on conifers. Annual, late spring–fall, overwinters in warmer climates.

Edibility: Inedible.

Comments: See comments under *Phlebia incarnata* regarding the poroid-appearing pore surface and features distinguishing these two closely related species. The dextrinoid reaction in KOH distinguishes this species from *P. incarnata*.

Warted Oak Polypore

Inonotus dryadeus

A

B

C

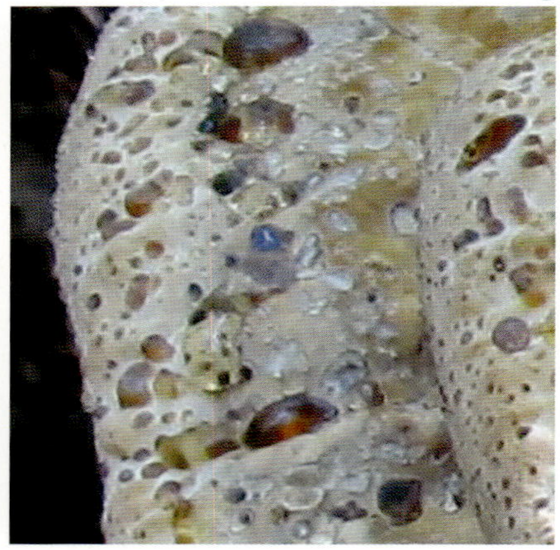

D

***Inonotus dryadeus* (Persoon) Murrill**, *N. Amer. Fl.* (New York) 9(2):86 (1908)

Basionym: *Boletus dryadeus* Persoon, *Observ. Mycol.* (Lipsiae) 2:3 (1800) [1799]
[Basidiomycota, Agaricomycotina, Agaricomycetes, *incertae sedis*, Hymenochaetales, Hymenochaetaceae]

Derivation: *Inonotus*, Greek, from *ino*, "fiber," and *otos*, "ear," "fibrous ear"; *dryadeus*, Greek, from *dryás*, "oak," "of oak."

Fruitbody: Up to 40 cm long, 15 cm wide, and 15 cm thick (15¾ × 10 × 6 in), sessile, semicircular to reniform (kidney-shaped, Fig. A), pulvinate (cushion-shaped, Fig. C), irregular; usually solitary, occasionally in groups that may overlap, forming tiers (Fig. B), or coalesce; surface lumpy, warty, finely velvety at first, later becoming smooth, dingy-grey (Figs. A & D), buff to pale yellowish-brown (Figs. B & C); margin thick, rounded, often paler than the rest of the cap (Fig. A), often exuding clear and/or amber drops when young (Figs. A & D).

Flesh: Thick, soft, and fibrous at first, becoming leathery to corky, yellowish-brown, becoming reddish-brown; somewhat zonate.

Pore surface: Dingy-white to greyish-brown, becoming brown with age; exuding drops of clear and amber liquid when young; tiny circular to angular pores, 4–6 per mm, tubes up to 3 cm deep.

Stipe: Sessile, stipe lacking.

Spore print: White.

Microscopic: Spores 6–8 × 5–7 µm, subglobose, smooth, clear, dextrinoid (staining reddish-brown) in Melzer's reagent, nonreactive in KOH. Hymenium with abundant stout, ventricose, curved and hooked setae projecting into the tubes.

Occurrence: Parasitic and saprobic, causing white butt rot; annual, sometimes forming massive fruitbodies at the base of living oaks (*Quercus* spp.), occasionally on other hardwoods; summer–winter, often persisting for several years; widespread in North America and in northern Europe.

Edibility: Inedible.

Comments: *Inonotus dryadeus* is not likely to be confused with any other polypore. The irregularly lumpy, dingy-grey to buff-colored surface, prominent clear to amber droplets oozing from all surfaces, tiny pores (4–6 per mm), dextrinoid spores, and growth from the very base of living trees are distinctive. Other bleeding polypores ooze liquid droplets primarily from their pore surfaces, have larger pores, and occur higher up on the trunks of living oaks or on dead hardwoods and conifers.

The upper surface of *Inonotus hispidus* (Bulliard *ex* Fries) P. Karsten is covered with coarse, shaggy, reddish-orange hairs that turn reddish-brown to nearly black with age, and the pores are somewhat larger (1–3 per mm). The spores are larger (8–11 × 6–8 µm) and negative (not dextrinoid) in Melzer's reagent, and the spore print is ochre- to chestnut-brown (not white). It also occurs high on the trunks of living hardwoods (mainly oaks), as well as fallen trunks and large branches.

Inonotus dryophilus (Berkeley) Murrill, also found higher up on the trunks of living oaks, has a buff to reddish-brown, hairy or smooth upper surface and is usually solitary, hoof-shaped, and often zonate and cracked. The angular pores are somewhat larger (1–3 per mm), with thin walls that become jagged and lacerate. Setae are lacking.

Inonotus cuticularis (Bulliard *ex* Fries) P. Karsten has a velvety or radially fibrillose (becoming smooth with age), yellowish-brown upper surface, yellowish- to reddish-brown flesh, and a pale brown, glancing pore surface. The spore print is bright yellowish-brown, and the species is saprobic on many species of hardwood trees.

Ischnoderma resinosum (Schrader) P. Karsten (see species account) has distinct concentric zones of color and texture: rusty-brown alternating with dark blackish- or bluish-brown or dull dark blue to vividly bright blue, whitish to pale brown flesh, a white pore surface that quickly bruises brown, and smaller (4.5–7 × 1.5–2.5 µm) cylindrical to sausage-shaped (not subglobose) spores. It also forms overlapping, shelf-like clusters on dead standing and fallen hardwood trees and logs and sometimes on conifers.

Resinous Polypore

Ischnoderma resinosum

A B C D E

Ischnoderma resinosum **(Schrader) P. Karsten**, *Meddn Soc. Fauna Flora Fenn.* 5:38 (1879)

Basionym: *Boletus resinosus* Schrader, *Spicil. Fl. Germ.* 1:171 (1794)
[Basidiomycota, Agaricomycotina, Agaricomycetes, *incertae sedis*, Polyporales, Fomitopsidaceae]

Derivation: *Ischnoderma*, Greek, "withered skin"; *resinosum*, Latin, "with resin."

Fruitbody: Forming semicircular to fan-shaped shelf-like conks.

Cap: Sessile, semicircular to fan-shaped, slightly convex to flat (Figs. A–C), up to 30 cm across by 12 cm deep (12 × 4¾ in) and 2.5–5.0 cm (1–2 in) thick. Margins thick, rounded, somewhat wavy (Figs. A & D); surface velvety to hairy, becoming smoother with age; surface furrowed both radially and concentrically (Fig. C), with distinct concentric zones of color and texture: rusty-brown alternating with dark blackish- or bluish-brown or dull dark blue (Fig. C) to vividly bright blue (Figs. A & B).

Flesh: Whitish, becoming brown to cinnamon-brown; spongy, soft, watery when young, with an anise-like odor; becoming hard, fibrous, and dry with age.

Pore surface: Initially white (Figs. A & D), becoming creamy-white, then pale brownish; bruising brown (Fig. D, arrow), often exuding conspicuous reddish or amber drops of fluid; pores tiny, nearly invisible, 4–6 per mm, round to angular (Fig. E), tubes up to 10 mm long.

Stipe: Lacking.

Spore print: White.

Staining: All parts greyish or blackish with KOH.

Microscopic: Spores 4.5–7 × 1.5–2.5 µm, cylindrical to sausage-shaped, smooth, clear, nonreactive with Melzer's reagent.

Occurrence: Annual, saprobic; forming overlapping shelf-like clusters on dead standing and fallen hardwoods, less commonly on conifers; late summer–early winter, occasionally in spring; widespread and common in eastern North America.

Edibility: Usually erroneously described as inedible.

Comments: We are always astonished that none of the field guides or online sources mention the stunning blue zones of color we frequently see in this species. Even when the cap zonation consists of alternating bands of lighter brown and dark brown or blackish metallic-brown, a distinct bluish tinge is usually also apparent in the latter. This blue does, of course, fade away in older specimens, but it cannot be missed in young, fresh material. This cannot be an artifact of our southeastern geography, since it is also exhibited by specimens photographed in Pennsylvania (Fig. B). Only Arora (1986) mentions this almost in passing as resinous areas being "often darker and more metallic-looking and sometimes slightly bluish."[1]

European authors in particular consider *Ischnoderma benzoinus* (Wahlenberg) P. Karsten (1881) to be a separate species that is restricted to conifers. However, culture studies have not supported that notion, and both major monographs on North American polypores—Overholts (1953) and Gilbertson & Ryvarden (1986)[2]—synonymized these two taxa.

1. D. Arora, 1986, *Mushrooms Demystified* (Berkeley: Ten Speed Press), 573.

2. L.O. Overholts, 1953, *The Polyporaceae of the United States, Alaska and Canada* (Ann Arbor: University of Michigan Press); R.L. Gilbertson & L. Ryvarden, 1986, *North American Polypores*, vol. 1 (Oslo: Fungiflora).

Chicken of the Woods and Sulphur Shelf

Laetiporus sulphureus and *L. cincinnatus*

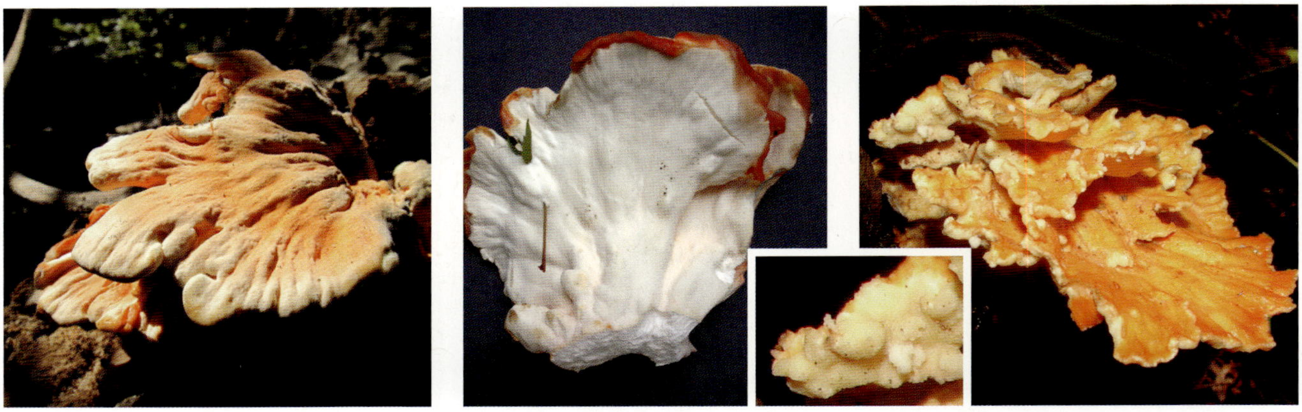

A B C D

A & B, *Laetiporus cincinnatus*; C & D, *L. sulphureus*. (Photos: A & B, W. Light; C & D, Teresa Fortenberry.)

Laetiporus sulphureus sensu stricto **(Bulliard) Murrill**, *Annls Mycol*. 18(1/3):51 (1920)

Basionym: *Boletus sulphureus* Bulliard, *Herb. Fr.* 9:tab. 429 (1789).

Laetiporus cincinnatus **(Morgan) Burdsall, Banik & Volk**, in Banik, Burdsall, & Volk, *Folia Cryptog. Estonica* 33:13 (1998)

Basionym: *Polyporus cincinnatus* Morgan, *J. Cincinnati Soc. Nat. Hist.* 8(3):97 (1885)
[Basidiomycota, Agaricomycotina, Agaricomycetes, *incertae sedis*, Polyporales, Polyporaceae]

Derivation: *Laetiporus*, Latin, from *laetus*, "joyful," "happy," and *porus*, "pore"; *sulphureus*, Latin, "sulfuric," referring to the sulfur-yellow pores; *cincinnatus*, Latin, for Cincinnati, Ohio, near the type locality of this species.

Fruitbody: *L. sulphureus* (Fig. C)—a large, overlapping cluster of flattened, laterally fused, fan-shaped shelves, with lobed margins, occasionally forming rosettes, rarely solitary; bright salmon orange to yellow-orange, usually with a sulfur-yellow, highly crenulated margin, fading to pale orange and turning white in old age; up to 45 cm (18 in) across; upper surface velvety to woolly, radially to irregularly wrinkled; ***L. cincinnatus*** (Fig. A)—typically as a large rosette up to 45 or 60 cm (18–24 in) of flattened to convex, laterally fused, fan-like, lobed caps, each up to 15 cm wide, 10 cm deep, and 3 cm thick (6 × 4 × 1¼ in), attached by a central, solid, branching stipe; rarely as a smaller shelf from a tree trunk; bright salmon orange.

Pore surface: *L. sulphureus*—lemon-yellow to creamy-yellow (Fig. D), pores initially round, becoming angular, 1–5 mm long, 3–5 per mm, decurrent to point of attachment; ***L. cincinnatus***—pale cream, initially round, becoming angular, 1–5 mm long, 2–4 per mm, decurrent on the individual stipes; to white (Fig. B); pores initially round, becoming angular, 1–5 mm long, 2–4 per mm, decurrent on the individual stipes.

Flesh: Soft, fleshy, and pale yellow when young; chalky, friable, and white in old age.

Stipe: *L. sulphureus*—fruitbody sessile or attached to hardwood by a lateral broad, rudimentary, bright salmon-orange stalk; ***L. cincinnatus***—a central, wide, branching stalk arising from a columnar mycelial mass attached to a buried root or exposed stump, 4–9 cm tall by 5 cm wide (1½–3½ × 2 in); whitish to pinkish-yellow.

Spore print: Both species white.

Microscopic: Spores of both species broadly ovoid, clear, thin-walled, smooth, nonreactive in Melzer's reagent; ***L. sulphureus***—spores 5.5–7 × 3.5–5 µm; ***L. cincinnatus***—spores slightly smaller, 4.5–5.5 × 3.5–4 µm.

Occurrence: Causing a brown rot in hardwood trees. ***L. sulphureus***—annual, on living or dead hardwood trunks and logs, primarily oaks (*Quercus* spp.); widespread in eastern North America except along Gulf Coast, where it is replaced by *L. gilberstonii* var. *pallidus* Burdsall; also in Europe; spring–fall; ***L. cincinnatus***—widespread but occasional in East and Midwest, almost exclusively on decaying roots and stumps of old or dead oaks, occasionally at the very base of trees at ground level; May–December. The geographic distributions of the two species are the same.

Edibility: Both are edible and excellent, but, like all species of *Laetiporus*, they cause gastrointestinal upset in some individuals for reasons that are not well understood.[1] Anecdotal reports exist that consuming these mushrooms causes tingling of the lips in some individuals.

Comments: *Laetiporus sulphureus* causes a brown rot in the trunks of living and dead trees ranging from about 1–1¼ m (3–4 ft) to as much as 3 m (10 ft) above the ground; the rot caused by *L. cincinnatus* occurs in the roots to a height of only 1¼–1½ m (4–5 ft) above the ground.

A similar species, *L. persicinus* (Berkeley & M.A. Curtis) Gilbertson, provisionally assigned to the genus *Laetiporus*,[2] is widespread but occasional throughout the Southeast and Puerto Rico. Like *L. cincinnatus*, *L. persicinus* has a centrally or eccentrically (not laterally) stalked fruitbody attached to roots at the base of trees, primarily oaks (*Quercus* spp.) but occasionally pines (*Pinus* spp.). The fruitbody grows as a circular to fan-shaped rosette. The upper surface is brown to pinkish-brown (never bright orange), with a darker brown marginal band, the pore surface and flesh are pale tan, pinkish-tan, or pinkish-cream, and the tubes bruise blackish-brown. The pores are circular, 3–4 per mm. This species does not even look like a typical *Laetiporus*, but much more closely resembles the Dye Polypore, *Phaeolus schweinitzii* (Fries) Patouillard. The fruitbody grows as a basal circular to fan-shaped rosette attached to tree roots by a central (never lateral) stipe.

1. It is not known whether these upsets are due to eating old or improperly cooked specimens, those growing on certain trees (*Eucalyptus* and conifers have been suggested as suspect hosts, but this cannot apply to most species), or some other cause, like an idiosyncratic allergic reaction. We have eaten both *Laetiporus cincinnatus* and *L. sulphureus* with no problems. We find them to be good edibles, but not as delicious as either oyster or honey mushrooms.

2. Although its inclusion in the genus *Laetiporus* is only provisional, we have described it here because it is very rarely included in field guides. Molecular studies indicate that *L. persicinus* is more distant from the *Laetiporus sulphureus* complex than is *Phaeolus schweinitzii*. Indeed, together with *P. schweinitzii* and *Wolfiporia* it forms a separate clade from *Laetiporus (sensu stricto)*. *Laetiporus persicinus* appears to have no close similarity to any known genus, and the erection of a new genus may be needed (D.L. Lindner & M.T. Banik, 2008, *Mycologia* 100[3]:417–430).

Mustard-yellow Polypore

Phellinus gilvus

A

B

C

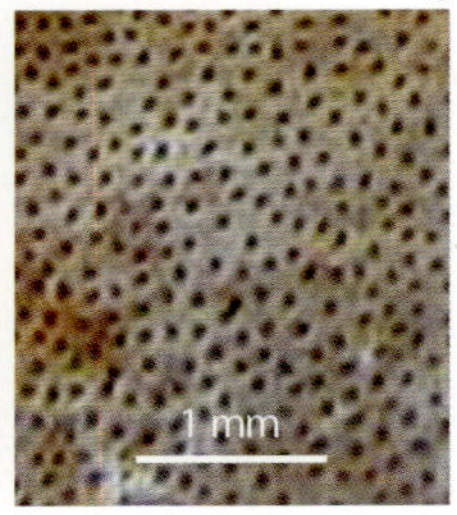

D

***Phellinus gilvus* (Schweinitz) Patouillard**, *Essai Tax. Hyménomyc.* (Lons-le-Saunier):82 (1900)

Basionym: *Boletus gilvus* Schweinitz, *Schr. Naturf. Ges. Leipzig* 1:96 (1822)

Basionym: *Boletus gilvus* Schweinitz 1822 [see online Appendix A for synonymy]
[Basidiomycota, Agaricomycotina, Agaricomycetes, *incertae sedis*, Hymenochaetales, Hymenochaetaceae]

Derivation: *Phellinus*, Greek, "cork-like," referring to the tough flesh; *gilvus*, Latin, "pale yellow," referring to the color of the growing margin.

Cap: 5–15 cm across and 1.5–3 cm thick. Solitary or in overlapping clusters. Lateral stem rudimentary or lacking. Fruitbodies highly polymorphic and variable.

Upper surface: Young fruitbody ochre, bright rust-yellow (Fig. A), or reddish-brown, becoming darker with age (Fig. B); margin usually yellowish, initially velvety, pubescent (Fig. C), becoming smoother with age; azonate or zonate, typically smooth to concentrically furrowed or bumpy.

Pore surface: Dark purplish-brown, greyish-brown, brown, or reddish-brown; *pores tiny*, 6–8 per mm (Fig. D); tubes up to 7 mm deep, multilayered in perennial specimens.

Flesh: Yellowish-brown to orange-brown.

Spore print: White.

Microscopic: Spores 4–5 × 3–3.5 µm; smooth; elliptical; inamyloid. Pointed, awl-like setae abundant in the hymenium, thick-walled. Dimitic.

Staining: All surfaces stain red and then black or simply black with KOH.

Occurrence: Saprobic, causing white rot. Annual or perennial on dead hardwoods, especially oaks (*Quercus* spp.); year-round in warmer climates. Cosmopolitan, especially in the tropics.

Edibility: Tough, inedible.

Comments: This highly variable mushroom has been described over 30 times and has more than 115 synonyms. See online Appendix A. *Phellinus gilvus* and its close look-alike, *P. pini*, differ most strikingly in the color of their spore prints: that of *P. gilvus* is white, and that of *P. pini* is light to dark brown. In addition, *P. gilvus* occurs on hardwood substrata and has *tiny round pores* (6–8 per mm) (Fig. D above). *Phellinus pini* occurs almost exclusively on conifers and has larger angular to slot-like pores (1–4 per mm).

Phaeolus schweinitzii (Fries) Patouillard also resembles

Phelinus gilvus (Thore) A. Ames with its brownish cap and yellow to olive margin. The pore surface is orange to bright yellow when young but fades to brownish or reddish-brown with age. The angular to slot-like pores are much larger (1–2 per mm), and the fruitbody grows from the ground or at the base of living or dead conifers, in the Southeast primarily Eastern White Pine (*Pinus strobus* var. *strobus*) and Loblolly Pine (*P. taeda*). A prominent central stipe is usually present. *Phaeolus schweinitzii* lacks setae, in sharp contrast to the distinctive pointed, awl-like setae seen in *Phellinus gilvus* and *P. pini*.

Pine Bracket, Red Heart Fungus

Phellinus pini

A

B

Phellinus pini **(Thore *ex* Fries) A. Ames**, *Annales Mycologici* 11(3):246 (1913).[1]

Basionym: *Boletus pini* Thore, *Essai Chloris Dép. Landes*:487 (1802)[2]
[Basidiomycota, Agaricomycotina, Agaricomycetes, *incertae sedis*, Hymenochaetales, Hymenochaetaceae]

Derivation: *Phellinus*, Greek, "cork-like," referring to the tough flesh; *pini*, Latin, "of pine trees."

Cap: 2–20 cm across, 1–15 cm thick; fan-shaped, irregularly shelf-like to hoof-shaped, solitary to overlapping clusters (Fig. B). Lateral stem rudimentary or lacking. May assume an effuse-reflexed form on the lower surfaces of branches.

Upper surface: Tawny to dark reddish-brown (Fig. A), brownish-black with age (Fig. B), irregularly curved margin usually brighter; surface hard, crusty, rough, or cracked, finely hairy.

Pore surface: Brownish-yellow to ochre-orange; 1–4 angular to nearly slot-like pores per mm; tube layers up to 6 mm deep.

Flesh: Woody, tough, yellowish-brown to rusty, usually with one or more thin black layers; young fruitbodies often appearing duplex because of the hairy upper layer.

Spore print: Light to dark brown.

Microscopic: Spores 4–6 × 3.5–5.0 µm, globose or subglobose, smooth; inamyloid. Pointed, awl-like setae abundant in the hymenium. Dimitic.

Staining: All parts staining black in KOH.

Occurrence: Widely distributed on living and dead conifers throughout North America. Perennial, all seasons.

Edibility: Tough, inedible.

Comments: *Phellinus pini* is almost indistinguishable from the Mustard-yellow Polypore, *Phellinus gilvus*. However, *P. gilvus* grows on hardwoods rather than conifers, has a white rather than brown spore print, and has much smaller pores: 6–8 rather than 1–4 per mm.

Phellinus pini is one of the most damaging of all heartwood rot pathogens of conifer trees. It is known as "red ring rot," "red heart fungus," and "white speck" to foresters. However, advanced red ring rot decay of the heartwood in old-growth pines (*Pinus* spp.) in the Southeast is essential for the survival of the rare and endangered Red-cockaded Woodpecker (*Picoides borealis*). This is the only woodpecker that excavates its nest cavities exclusively in living pines. The softening of this living heartwood by the fungus—it can require 13–15 years for the fungus to decay a sufficiently large diameter of heartwood to accommodate a nesting cavity[3]—enables these woodpeckers to hollow out their nesting cavities more quickly. Without heartwood degradation by *Phellinus pini*, excavation of a nesting cavity in a living mature pine can take more than 12 years instead of 6 months to 2 years.

The abandoned and usurped/commandeered nesting cavities of Red-cockaded Woodpeckers also provide decades-long nesting and roosting resources for secondary cavity users. More than 30 nonexcavating (secondary cavity-nesting) species also rely on these cavities. The Red-cockaded Woodpecker is thus an important keystone species, particularly in fire-maintained pine forest ecosystems in the southeastern United States.[3,4] The Red-cockaded Woodpeckers' use of living pines for their nesting cavities enables them to exploit the historically fire-maintained upland pine forests with reduced numbers of hardwoods and dead trees.

1. Molecular studies indicate that *Phellinus* as commonly understood is not a monophyletic group (i.e., not derived from the same ancestor). This taxon has therefore been referred to the genus *Porodaedalea* Murrill by several authors. However, due to the incomplete understanding of all these interrelationships, we provisionally retain this species in the genus *Phellinus* [*sensu lato*] for ease of treatment.

2. Most authors list *Boletus pini* Brotero (1804) as the basionym for this taxon. We believe this is incorrect. Brotero's name was preceded by *Boletus pini* Thore (1802). Both Brotero's name and that of Thore appear to have been sanctioned by Fries (1821), leaving the true basionym open to doubt. Pending resolution of this conundrum, we provisionally consider that *Boletus pini* Thore has nomenclatural precedence and should be considered the correct basionym. It is also unclear at this time whether *Boletus pini* Brotero and *B. pini* Thore are homotypic (i.e., based on the same type specimen) or heterotypic (based on different holotypes) homonyms. This is a critical nomenclatural distinction.

3. R.N. Conner et al., 2004, *Texas J. Sci.* 56(4):415–426; S.R. Harding & J.R. Walters, 2004, Dynamics of cavity excavation by red-cockaded woodpeckers, *in* R. Costa and S.J. Daniels (eds.), *Proceedings of the 4th Red-cockaded Woodpecker Symposium: Road to Recovery* (Savannah, GA: Hancock House Publishers), 412–422.

4. L.A. Blanci & J.R. Walters, 2008, Cavity-nest webs in a longleaf pine ecosystem, *Condor* 110(1):80–92.

Cracked Cap Polypore

Phellinus robiniae

Photos: A–D, W. Light; E, Elizabeth Bush, USDA Forest Service, Bugwood.org

***Phellinus robiniae* (Murrill) A. Ames**, *Annls Mycol*. 11(3):246 (1913)

Basionym: *Pyropolyporus robiniae* Murrill, *Bull. Torrey Bot. Club* 30(2):114 (1903)

Misapplied name: *Phellinus rimosus* [*ex auct. amer.*]; [*non Polyporus rimosus* Berkeley (1845)]
[Basidiomycota, Agaricomycotina, Agaricomycetes, *incertae sedis*, Hymenochaetales, Hymenochaetaceae]

Derivation: *Phellinus*, from *phellos*, Greek, "cork," and *-inus*, Latin, "of," "related to," referring to the hard, corky fruitbody; *robiniae*, Latin adjective referring to the genus *Robinia* on which this fungus exclusively grows.

Cap: Semicircular (Fig. E), kidney-shaped (Fig. B), or forming an irregular bracket (Fig. A), initially shallowly convex, becoming more convex to hoof-shaped with age; 40 cm or more across by 20 cm deep (16 × 8 in). Color highly variable, with concentrically furrowed zones of yellowish-brown, brown, and reddish-brown often with areas covered by green algae (Fig. B), becoming darker brown or black and deeply cracked with age (Fig. A).

Flesh: Yellowish cinnamon-brown, rough, woody, with subparallel fibrous layers radiating from the point of attachment, curving sharply downward distally to merge with the tube layer (Fig. D).

Pore surface: Brown, yellow-brown, to reddish-brown (Fig. C), color often varying with viewing angle; pores tiny, circular, 7–8 per mm; tubes up to 9 mm deep, often in several distinct layers, each representing a different growing season.

Chemical staining: KOH—all surfaces and flesh instantly black.

Stipe: Lacking, fruitbody sessile.

Spore print: Brown.

Microscopic: Spores 5–6 × 4.5–5 µm, subglobose to ovoid, smooth, reddish-brown in KOH, nonreactive in Melzer's solution. Setae lacking.

Occurrence: Perennial, parasitic, causing white rot, and later saprobic on dead trees. Solitary or clustered from the tree base (Fig. A) to high up (Fig. E) on dead Black Locust (*Robinia pseudoacacia*) throughout the tree's range in the Appalachian Mountains; also in the Transition Zone on New Mexican Locust (*Robinia neomexicana*) in the American Southwest.

Edibility: Inedible.

Comments: Several other *Phellinus* species in our area develop concentrically furrowed and vertically cracked caps that eventually turn black, *but none are associated with Black Locust.* The species discussed below all have flesh that stains instantly black in KOH and tubes that form stratified annular layers; unlike *P. robiniae*, all have microscopic setae.

Phellinus everhartii (Ellis & Galloway) A. Ames, like *P. robiniae*, has rusty-brown flesh, small pores (4–6 pores per mm), and a reddish-brown spore print, but the older tube layers are filled in with white mycelium.

The Fleck-fleshed Polypore, *Phellinus igniarius* (Linnaeus *ex* Fries) Quélet—*sensu lato*, probably comprising a cryptic species complex—on hardwoods, especially birch and willow, is uniformly rusty-brown to dark grey, eventually becoming greyish-black to black. Like *P. everhartii* but not *P. robiniae*, the older tube layers are stuffed with white mycelium. Unlike *P. robiniae* and *P. everhartii*, it produces a whitish rather than a brownish spore print and has larger pores (4–5 per mm).

The Robust Polypore, *Phellinus robustus* (P. Karsten) Bourdot & Galzin, tends to form deeper, somewhat goblet-shaped conks and occurs on oaks, birches, and other hardwoods (but not on Black Locust), as well as on conifers. Like *P. robiniae*, the pores are small (5–7 per mm); unlike *P. robiniae* and *P. everhartii*, it produces a whitish spore print.

Smoky Polypore

Bjerkandera adusta

A

B

Photos: A & inset, Jim Edwards; B, Josh Milburn.

***Bjerkandera adusta* (Willdenow) P. Karsten**, *Meddn Soc. Fauna Flora Fenn.* 5:38 (1879)

Basionym: *Boletus adustus* Willdenow, *Fl. Berol. Prodr.*:392 (1787) [Basidiomycota, Agaricomycotina, Agaricomycetes, Polyporales, Merulaceae]

Derivation: *Bjerkandera*, named in honor of the Swedish clergyman and naturalist Clas Bjerkander (1735–1795); *adusta*, Latin, "singed," referring to the smoky-grey color of the pore surface.

Fruitbody: 10 cm wide by 6 cm deep by up to 6 mm thick (4 × 2½ × ¼ in), effuse, often resupinate to effuse-reflexed, often forming overlapping sessile shelves (Figs. A & B) with wavy margins; may coalesce laterally (Fig. B) to form large composite fruitbodies up to 20 cm (8 in) or so across or flat sheets on the undersurface of logs; white, cream, or greyish to beige, margin white when young, darkening with age; tomentose when young, becoming smooth with age.

Flesh: Tough, whitish, becoming pale buff or greyish with age, lacking a dark line near the tube layer; leathery, with fungus-like odor; taste sour or not distinctive.

Pore surface: Initially pale grey (inset), becoming light greyish-brown or smoky-grey to blackish-grey; tiny circular to angular pores (inset), 5–7 per mm, tubes 2 mm deep.

Stipe: Sessile, stipe lacking.

Spore print: White.

Microscopic: Spores 4.5–6 × 2.5–3.5 µm, short cylindrical, smooth, thin-walled, clear, nonreactive in Melzer's reagent.

Occurrence: Saprobic and annual, causing a white rot in dead hardwoods, particularly oaks (*Quercus* spp.), rarely on conifers; widespread and common in the Southeast; North America and Europe; summer–midwinter.

Edibility: Inedible.

Comments: The similar Big Smoky Bracket, *Bjerkandera fumosa* (Persoon *ex* Fries) P. Karsten, and the only other species in the genus, is much less common, has generally thicker caps (up to 2 cm [¾ in]), a lighter-colored pore surface, and a thin dark brown line at the base of the flesh next to the tubes that is not present in *B. adusta*. It also emits a disagreeable odor of anise. It has been recorded in Great Smoky Mountains National Park.[1]

1. R.H. Petersen, 1976, *Checklist of Fungi of the Great Smoky Mountains National Park*, Management Report No. 29, Department of the Interior, National Park Service, Southeast Region.

Dye Polypore, Velvet-top Fungus

Phaeolus schweinitzii

A

B

Photos: A, Teresa Fortenberry; B, W. Light.

***Phaeolus schweinitzii* (Fries) Patouillard**, *Essai Tax. Hyménomyc.* (Lons-le-Saunier):86 (1900) [Basidiomycota, Agaricomycotina, Agaricomycetes, *incertae sedis*, Polyporales, Fomitopsidaceae]

Basionym: *Polyporus schweinitzii* Fries, *Syst. Mycol.* (Lundae) 1:351 (1821)

Derivation: *Phaeolus*, from *phae*, Greek, "dusky," and *-olus*, Latin, "little," meaning "little dusky one"; *schweinitzii*, specific epithet in honor of the American mycologist and Moravian Church minister Lewis David von Schweinitz (1780–1834).

Cap: Restricted to conifers, 4–25 cm (1⅔–10 in) or more across, typically large, flattened, fused lobes arising from a single stipe emerging from the ground (Fig. A), often forming a rosette or overlapping clusters, but occasionally forming fused shelf-like brackets at the base of trees or attached to dead, decaying wood (Fig. B).

Upper surface: Faintly to distinctly zonate, with concentric zones of alternating color and texture; woolly, velvety, or hairy, densely matted when young, becoming smooth with age; variously cream-colored, ochraceous, yellow, green-yellow, or orange when young, becoming olive-brown to dark brown from the center toward the margin in maturity; margin yellow, yellow-orange, or olive.

Pore surface: Yellowish, greenish-yellow, or orange when young, becoming brown and then dark reddish-brown with age; quickly bruising dark reddish or brown when handled; 1–2 or 3 angular or slot-like pores per mm, extending onto stipe.

Flesh: Yellowish when young, becoming reddish-brown; soft at first, becoming tough and fibrous.

Stipe: When present, central or lateral, up to 7 cm tall by 5 cm thick (2¾ × 2 in), tapering downward, color and texture like the cap.

Spore print: Whitish.

Microscopic: Spores 5–9 × 3–5 µm, ellipsoid to ovoid, smooth, clear, inamyloid (nonstaining in Melzer's reagent); setae lacking.

Staining: Cap surface and flesh cherry-red (young specimens), and then black or only black with KOH.

Occurrence: Saprobic, solitary or in overlapping clusters or rosettes on roots and at the bases of conifer trees, usually emerging from the ground; or saprobic on dead, decaying conifer wood. Occurring primarily on White Pines (*Pinus strobus*) and Loblolly Pines (*P. taeda*) in southeastern North America, June–November. Widespread on various conifers in North America and Eurasia; introduced into New Zealand, Australia, and South Africa.

Edibility: Inedible and possibly poisonous.

Comments: *Phaeolus schweinitzii* breaks down cellulose, causing a brown rot in the heartwood of the butt and roots and up to 4½ m (15 ft) high in the trunk of living conifers. By the time fruitbodies appear, substantial decay has already occurred in the trunk of the tree. It rarely kills standing trees, but those with advanced infection are susceptible to toppling by wind throw.

This mushroom is highly esteemed for the colors it produces for dyeing wool. When collected, the pore surface often produces a characteristic rusty stain on the hands and everything it is placed on. See the distinguishing remarks under the Mustard-yellow Polypore, *Phellinus gilvus*, which it somewhat resembles.

Onnia tomentosus (Fries) P. Karsten[1] is very similar to *P. schweinitzii* and also grows on the ground or on well-decayed conifer wood. The upper surface of the cap in *O. tomentosus* is velvety or covered with matted hairs and whitish when young, but it soon turns yellowish-brown, ochre, or rusty-brown. Its pore surface is buff when young, becoming greyish-brown to dark brown with age, in contrast to the yellowish, greenish-yellow, or orange pore surface of *P. schweinitzii*. The spore print is pale yellow to pale brown; that of *P. schweinitzii* is white. The hymenium has abundant straight, pointed setae; *P. schweinitzii* lacks setae of any kind. In the Southeast, *P. tomentosus* occurs only in the higher mountains of North Carolina and Tennessee primarily in association with Red Spruce (*Picea rubens*) and Eastern White Pine (*Pinus strobus*). Although *Phaeolus schweinitzii*, like *O. tomentosus*, appears in the mountains, it also descends into the Piedmont, where it is associated with Loblolly Pine (*Pinus taeda*).

Laetiporus persicinus (Berkeley & M.A. Curtis) Gilbertson also resembles *P. schweinitzii*. However, the fruitbody grows as a basal circular to fan-shaped rosette attached to tree roots by a central stipe (never lateral, as often seen in *P. schweinitzii*), the upper surface is brown to pinkish-brown with a darker brown (not yellowish) marginal band, the pore surface and flesh are pale tan, pinkish-tan, or pinkish-cream (not yellowish or greenish-yellow), and the tubes bruise blackish-brown (not reddish-brown). The pores are circular (not slot-like) and 3–4 per mm (not 1 or 2). It is widespread but occasional in the Southeast and is usually associated with oaks (occasionally with pines). See also the comments under Chicken of the Woods and Sulphur Shelf, *Laetiporus sulphureus* and *L. cincinnatus*.

1. *Onnia tomentosa* is referred to as *Inonotus tomentosus* (Fries) Teng in many field guides. However, a recent molecular analysis led to a major revision of the Hymenochaetaceae (T.F.M. Wagner, 2002, *Mycologia* 94[6]:998–1016).

Black-staining Polypore

Meripilus sumstinei

A

C

B

D

Photos: Teresa Fortenberry.

Meripilus sumstinei **(Murrill) M.J. Larsen & Lombard**, *Mycologia* 80(5):615 (1988)
[Basidiomycota, Agaricomycotina, Agaricomycetes, Polyporales, Meripilaceae]

Basionym: *Grifola sumstinei* Murrill, *Bull. Torrey Bot. Club* 31(6):335 (1904)

Derivation: *Meripilus*, Latin, "part of a cap"; *sumstinei*, Latin, epithet named for the American mycologist David Ross Sumstine (1870–1965).

Fruitbody: Multipiliate—with numerous overlapping individual caps radiating from a common stalk (Fig. A); up to 30 cm or more across by 25 cm tall (12 × 8 in).

Individual caps: 5–20 cm (2–8 in) across, surface finely velvety-hairy, more or less spoon-shaped to irregularly fan-shaped; margin initially blunt, becoming thin, sharp, and irregularly undulate to lobed, pale yellowish at first (Figs. B & D), bruising black (Figs. A & C); yellowish- to grey-brown or brownish-grey, darkening with age, often radially streaked and concentrically zonate (Figs. C & D).

Flesh: White to pale tan, 3–16 mm (up to ⅔ in) thick, fleshy-fibrous to stringy, tough.

Pore surface: White, becoming pale tan, bruising dark brown to black; pores minute, (5–)6–8(–9) per mm (Fig. B), round to mostly angular, decurrent to base; tubes up to 8 mm (⅓ in) deep.

Stipe: 1–3 cm long by 11 cm thick (⅖–1¼ × 4½ in).

Bruising: Margins and pore surface bruising black where handled (Figs. A & C).

Spore print: White.

Microscopic: Spore (4.5–)5.0–5.5(–6.5) × (4.0–)4.5–5.5(–6.0) µm, smooth, subspherical to oval, with distinct apiculus, clear, nonreactive in Melzer's reagent.

Occurrence: Annual, causing white rot. Alone or in groups at the base of hardwood trees and stumps, especially oaks (*Quercus* spp.), fruitbodies appearing to grow from the ground away from a tree or stump are actually sprouting from buried roots; summer–midwinter; widespread but occasional in eastern North America.

Edibility: Edible when young.

Comments: This species is often referred to as *Meripilus giganteus* (Persoon *ex* Fries) Karsten (1882) in many North American field guides. However, *M. giganteus*, although very similar, is a distinct European species that does not occur in North America.[1]

Black-staining Polypore closely resembles Hen-of-the-woods, *Grifola frondosa* (Dickson) Gray (1821), which also occurs in our area at the base of oaks and other hardwoods. *Grifola frondosa* is distinguished from *M. sumstinei* by its generally somewhat smaller individual caps (2–7 cm across), larger pores (2–3 per mm), and shorter tubes (2–3 mm) and the fact that the caps and pore surface do not bruise black when handled.

The Kidney-shaped Tooth, *Mycorrhaphium adustum* (Schweinitz) Maas Geesteranus (1962), also has white to tan caps with concentric zones of texture and prominent black margins. However, the pore surface is densely studded with white to purplish or brown teeth 1–3 mm long, which often fuse together. It grows alone or in small groups on sticks and small logs rather than at the base of trees and trunks and never forms massive, multicapped fruitbodies.

Also compare this species with the account of the Umbrella Polypore, *Polyporus umbellatus* (Persoon) Fries (1821).

1. M.J. Larson & F.F. Lombard, 1988, *Mycologia* 80(5):612–621.

Stalked White Polypore

Microporellus obovatus

A B

***Microporellus obovatus* (Junghuhn) Ryvarden**, *Norweg. J. Bot.* 19:232 (1972)
[Basidiomycota, Agaricomycetes, *incertae sedis*, Polyporales, Polyporaceae]

Basionym: *Polyporus obovatus* Jungh., *Praemissa, Flor. Crypt. Javae* (Batavia):65 (1838)

Derivation: *Microporellus*, Latin, "many small pores"; *obovatus*, Latin, "egg-shaped at the tip."

Cap: Plane or somewhat concave above, circular, kidney-shaped, spathulate, or fan-shaped (Figs. A & B), 2–7 cm (⅘–2¾ in) wide, margin entire, usually paper thin; white at first, becoming cream to pale strawberry, often with greyish or umber darker tones; finely hairy or velvety, radially striate at first, becoming smooth with age.

Pore surface: White, colored like the cap; pores minute, irregularly round to angular, 6–8 per mm (inset), decurrent for up to one-third the length of the stipe (Figs. A & B).

Flesh: White, fibrous, tough.

Stipe: 3–7 cm (1¼–2¾ in) tall, laterally attached, even to irregular, may be twisted; dry, rigid; occasionally completely absent.

Spore print: White.

Microscopic: Spores 3.5–5 × 2–4.5 µm, subglobose to ellipsoid, smooth, thin-walled, translucent, nonreactive in Melzer's reagent.

Occurrence: Annual, solitary or in small clusters on dead exposed or buried hardwood, rarely on conifers; widespread and fairly common in North America east of the Rocky Mountains; pantropical: Puerto Rico, South America, India, southeast Asia, subtropical China, Okinawa, Indonesia, Africa.

Edibility: Inedible.

Comments: *Microporellus dealbatus* (Berkeley & M.A. Curtis) Murrill is somewhat larger, 2–10 cm wide by 3–10 cm tall (1¼–4 in), and the upper surface of the cap is sharply zonate with concentric alternating rings of light grey/light brown and dark umber-brown. The basidiospores and context hyphae are dextrinoid in Melzer's reagent; those of *M. obovatus* are nonreactive. *Mycorrhaphium adustum* (Schweinitz) Maas Geesteranus has a thin, flat cap with a very thin wavy margin. The fertile surface bears dense spines up to 3 mm long (¹⁄₁₀ in), and it bruises black with age and/or handling.

Black-footed Polypore

Polyporus varius

A

B

C

D

***Polyporus varius* (Persoon) Fries**, *Syst. Mycol.* (Lundae) 1:352 (1821)
[Basidiomycota, Agaricomycetes, Polyporales, Polyporaceae]

Basionym: *Boletus varius* Persoon, *Observ. Mycol.* (Lipsiae) 1:85 (1796)

Derivation: *Polyporus*, Greek, "many pores"; *varius*, Latin, "variable."

Cap: 2–10 cm (¾–4 in); flat (Fig. A) to broadly convex (Fig. C), round to kidney-shaped, smooth, often with dissected, broadly undulating margin (Figs. B & C); tan or cinnamon-buff to beige (Figs. B & C), with or without radiating striations, fading to greyish white in overwintered fruitbodies.

Pore surface: White when young (Fig. A), later turning pale or dingy-brown, pores usually running down stipe to meet sharply defined black base (Fig. D, arrow); pores round to angular, minute, 4–7 to 7–9 per mm; tubes 3–5 mm deep.

Stipe: Central to lateral; black at base and often for more than half the stipe's length (Figs. A & B), finely tomentose at first, becoming smooth; upper portion pale cream and covered with decurrent tubes.

Flesh: Whitish, dingy-brown, to cinnamon.

Spore print: White.

Microscopic: Spores 7–10 × 2–3.5 (9–12 × 2.5–4) µm, smooth, cylindrical to slightly curved, sausage-shaped.

Occurrence: Annual, saprobic, causing white rot on decaying hardwood sticks and branches; fruitbodies solitary or scattered, summer and fall or year-round in warmer regions. Widespread in temperate zones (to 70° N) of North America, Europe, Asia; reported from India, Malaysia, and Australia.

Edibility: Inedible, tough.

Comments: This species is often described in field guides as *Polyporus elegans* Bulliard, a synonym of *P. varius*. A similar species, *Royoporus badius* (Persoon) A.B. De [also called *Polyporus badius* (Pers.) Schweinitz, a deprecated synonym], has a generally larger cap (4–20 cm [1½–8 in]) and a decurrent pore surface overgrowing a black stipe. However, the cap is dark reddish-brown to dark brown instead of tan, and it generally grows on larger hardwood logs in the fall. A great deal of confusion exists in the literature with regard to this species (*P. varius*), and the synonymy is extensive. See online Appendix A.

Hexagonal-pored Polypore

Polyporus alveolaris

A

B

C

D

Polyporus alveolaris **(de Candolle) Bondartsev & Singer**, *Annls Mycol*. 39(1):58 (1941)
[Basidiomycota, Agaricomycetes, *incertae sedis*, Polyporales, Polyporaceae]

Basionym: *Merulius alveolaris* D.C., in de Candolle & Lamarck, *Fl. Franç*., 3rd ed. (Paris) 5/6:45 (1815)

Derivation: *Polyporus*, Greek, "many pores"; *alveolaris*, Latin, "with little cavities."

Cap: 1–10 cm (0.4–4 in) wide, sessile or stalked, circular to kidney-shaped (Figs. B & D); azonate, upper surface pale reddish-yellow, fibrillose to scaly, with flattened triangular scales (Fig. B), becoming smooth (Fig. C) and pale buff to ivory with age.

Pore surface: White to tan, with large, radially elongate, diamond-shaped pores, 1–2 per mm, pore walls becoming somewhat tattered with age (Fig. A); tube layer up to 5 mm (⅕ in) thick, continuous with flesh.

Stipe: Central to lateral, buff, smooth, 0.5–2 cm long by up to 5 mm thick (¼–⅘ × ⅕ in).

Flesh: Azonate, corky, pale tan to ivory.

Spore print: White.

Microscopic: Spores cylindric, 10–13(–14.5) × 3.5–5 µm, smooth, translucent.

Occurrence: Saprobic, causing white rot; solitary or gregarious on decaying hardwood sticks and logs; annual, early summer through fall. A warm-temperate species widely distributed in North America, Europe, Asia, Australia.

Edibility: Edible but tough, leathery.

Comments: The pores on mature specimens of this species are diamond-shaped rather than hexagonal. In young specimens they are rounder, and the pore walls are much thicker than when mature (Fig. D). Like many mushrooms, this species has a confused nomenclatural history and is called *Favolus alveolaris* in some field guides. However, the most recent DNA evidence supports placing this taxon in the genus *Polyporus*, where it currently resides. Compare this species with the Spring Polypore, *Polyporus arcularius* (Batsch) Fries, and Dryad's Saddle, *P. squamosus* (Hudson) Fries.

Spring Polypore

Polyporus arcularius

B

A C

Photos: Jim Edwards.

***Polyporus arcularius* (Batsch) Fries**, *Syst. Mycol*. (Lundae) 1:342 (1821)
[Basidiomycota, Agaricomycetes, Polyporales, Polyporaceae]

Basionym: *Boletus arcularius* Batsch, *Elench. Fung*., cont. prim. (Halle):97 (1783)

Derivation: *Polyporus*, Greek, "many pores"; *arcularius*, Latin, "with small boxes," referring to the angular pores.

Cap: 1–8 cm (½–3¼ in) across, convex to shallowly vase-shaped (Fig. A); dry, smooth to conspicuously scaly; at first dark brown with concentric zones of mottled color (Fig. A), fading to golden-brown with age; margin densely covered with bristly hairs (Fig. B).

Flesh: Up to 1.5 cm (⅔ in) thick; tough, whitish to buff.

Pore surface: White to pale yellow, pores polygonal to angular, typically radially arranged (Figs. A & C), 0.5–1 per mm, dissepiments (rims) thin, becoming lacerate and jagged (Fig. C).

Stipe: 2–6 cm tall by up to 4.5 mm thick, central, subequal, scaly to almost smooth, concolorous with the cap (Fig. A); base sometimes finely hairy (Fig. A, inset).

Spore print: White.

Microscopic: Spores 7–9 × 2–3 µm, cylindrical, clear, smooth.

Occurrence: Saprobic. Solitary or in groups on dead hardwood or attached to buried wood; causing a white rot; widespread and occasional in North America; spring.

Edibility: Inedible.

Comments: Large, radially elongated pores are also found on the Hexagonal-pored Polypore, *Polyporus alveolaris* (de Candolle) Bondartsev & Singer, the Winter Polypore, *P. brumalis* Persoon *ex* Fries, and Dryad's Saddle, *P. squamosus* (Hudson) Fries.

The Hexagonal-pored Polypore, *Polyporus alveolaris*, has a reddish-yellow to reddish-orange cap, a rudimentary to short, eccentric to lateral, smooth (not scaly) stipe, and larger spores (10–13[–14.5] × 3.5–5 µm). The Winter Polypore, *Polyporus brumalis*, has smaller pores (3–4 per mm) that are not radially aligned and smaller spores (6–7.5 × 2–2.5 µm) than *P. arcularius*. Dryad's Saddle, *Polyporus squamosus*, is much larger, with a pale tan to creamy-yellowish cap 5–50 cm (2–20 in) wide covered with widely spaced and concentrically arranged dark-brown scales, a smooth (not hairy) margin, a lateral stem, and larger spores (10–16 × 4–6 µm). All three species occur in our area.

Dryad's Saddle, Pheasant's Back Polypore

Polyporus squamosus

A B C

Scale in C, inset = 5 mm. The magnificent specimen shown in Figs. B & C measured 24 cm across by 17.5 cm deep (9½ × 7 in) and was collected from a River Birch (*Betula nigra*) in the Atlanta area on April 1, 2014, by Dr. Michael Moribaldi. (Photos: A, Dan Molter; B & C, W. Light.)

***Polyporus squamosus* (Hudson) Fries**, *Syst. Mycol.* (Lundae) 1:343 (1821)
[Basidiomycota, Agaricomycetes, Polyporales, Polyporaceae]

Basionym: *Boletus squamosus* Hudson, *Fl. Angl.*, 2nd ed., 2:626 (1778)

Derivation: *Polyporus*, Greek, "many pores"; *squamosus*, Latin, "scaly."

Cap: Large, 6–30(–50) cm (2½–12[–20] in) across, fleshy, tough; flat, depressed to funnel-shaped (Fig. A); outline kidney-shaped, fan-shaped, circular or dimidiate in outline; margin entire, wavy or sometimes lobed; cap whitish or cream, dingy yellowish or pale buff to dark ochraceous, covered with widely spaced, flattened and overlapping, concentrically arranged, reddish-brown to dark-brown scales; surface smooth, lacking hairs, slightly viscid.

Flesh: White or cream to pale buff, up to 4 cm (1⅔ in) thick; soft and corky when young, becoming tough and fibrous; smells and tastes like watermelon.

Pore surface: White, cream, buff, or yellowish; pores angular, randomly arranged, decurrent to blackish area at base of stipe (Fig. C), large, ¼–1 or 2 per mm (Fig. C, inset), rims entire, becoming lacerate with age; tube layer sooty, tubes up to 1 cm (~½ in) deep.

Stipe: Lateral, eccentric, 1–12 cm in length by 1–6 cm thick (½–4¾ × ½–2½ in), stubby, may be rudimentary; white above, finely hairy and blackish near base (Fig. C, arrow).

Spore print: White.

Microscopic: Spores 10–18 × 4–7 µm, cylindrical, smooth, clear; nonreactive in Melzer's reagent.

Occurrence: Saprotrophic; solitary or in groups or overlapping clusters on many genera of standing or fallen dead hardwood trees; causing a white rot in the heartwood. Widespread and common in temperate and boreal forests of eastern and central North America, west to Alberta and Arizona associated with Aspen (*Populus tremuloides*), spring–late fall; Europe, China, Southeast Asia, East Africa.

Edibility: Edible, considered good when young, becoming tough with age.

Comments: This species somewhat resembles the Hexagonal-pored Polypore, *P. alveolaris*, and the Spring Polypore, *P. arcularius*.

Polyporus alveolaris has a reddish-yellow to reddish-orange cap that lacks the contrasting scale-like pattern of *P. squamosus*. The cap and margin are also fibrillose and hairy (not smooth), the pores are diamond-shaped and more or less radially aligned (not randomly arranged), the flesh is thinner (2 mm or less), and the base of the stipe is slightly swollen and not black.

The Spring Polypore, *Polyporus arcularis* (Batsch) Fries, has a dark brown cap when young, and the margin is densely covered with bristly hairs that are not present in *P. squamosus*. The central stipe is thin and columnar, not lateral and stubby, the pores are radially aligned and not at all decurrent, and the spores are much smaller (7–9 × 2–3 µm) than in *P. squamosus*.

White-rimmed Vinaceous Polypore

Nigroporus vinosus

A

B

C

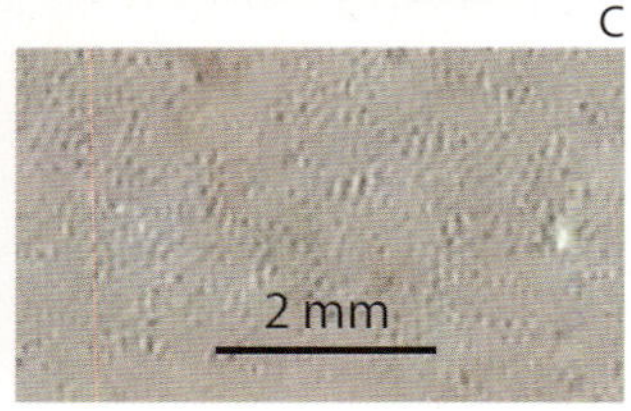

D

***Nigroporus vinosus* (Berkeley) Murrill**, *Bull. Torrey Bot. Club* 32(7):361 (1905)
[Basidiomycota, Agaricomycotina, Agaricomycetes, *incertae sedis*, Polyporales, Polyporaceae]

Basionym: *Polyporus vinosus* Berkeley, *Ann. Mag. Nat. Hist.*, ser. 2, 9:195 (1852)

Derivation: *Nigroporus*, Latin, from *nigro*, "to be black," and *porus*, "having many holes"; *vinosus*, Latin, "tipsy," "drunk with wine."

Cap: 2–12 cm (¾–4¾ in) across, sessile, semicircular, fan-shaped, ear-shaped, to kidney-shaped, flat to depressed (Figs. A & C); initially velvety, becoming smooth with age; often strikingly zonate, with concentric bands of deep maroon-violet, violet-brown, and vinaceous-brown (Fig. A), becoming dark purplish with age; margin white to pinkish-white (Figs. A–C).

Pore surface: Velvety, initially pale pinkish-beige (Figs. B, lower part, & D) to bluish-grey (Fig. B, upper), aging to purplish-brown, violet, or purplish-black; bruising darker brown; pores tiny, round, 7–8 per mm (Fig. D); no odor or taste.

Flesh: Thin, leathery; initially pinkish-brown, maroon-umber to purplish-brown.

Stipe: Sessile, stipe lacking.

Spore print: White.

Chemical staining: All tissues black with KOH.

Microscopic: Spores 3.5–4.5 × 1.0–1.5 µm, cylindrical to sausage-shaped; smooth, clear.

Occurrence: Annual, saprobic. Alone (Fig. B) or in overlapping groups (Fig. A) on well-decayed hardwood trunks (especially Sweetgum, *Liquidambar styraciflua*) and Loblolly Pine (*Pinus tadea*); widespread in the Southeast in the Piedmont and Gulf Coast, summer–midwinter.

Edibility: Inedible, tough.

Comments: This striking polypore is unmistakable and is recognized on sight. It is widespread and occasional in southeastern North America and occurs worldwide in the tropics and subtropics.

Berkeley's Polypore

Bondarzewia berkeleyi

A

B

***Bondarzewia berkeleyi* (Fries) Bondartsev & Singer**, *Annls Mycol*. 39(1):47 (1941)
[Agaricomycotina, Agaricomycetes, *incertae sedis*, Russulales, Bondarzewiaceae]

Basionym: *Polyporus berkeleyi* Fries, *Nova Acta R. Soc. Scient. Upsal*., ser. 3, 1(1):56 (1851)

Derivation: *Bondarzewia*, named in honor of the Russian mycologist Apollinari Semyonovich Bondarzew (1877–1968); *berkeleyi*, named after the British mycologist Miles Joseph Berkeley (1803–1889).

Fruitbody: Typically a very large rosette-like cluster 25–100 cm (10–40 in) or more across, composed of 1 to several fused caps arising from a central stipe.

Cap: Individual caps fan-shaped, 6–30 cm (2½–12 in) across, laterally fused, usually forming a rosette with a central hollow (Fig. A), margins undulate to lobed (Figs. A & B), surface dry, smooth to velvety or densely woolly, surface radially undulate, ground color whitish, cream, to pale tan, faintly to more or less conspicuously zonate (Fig. A).

Flesh: Thick, white, tough, corky to fibrous; odorless or smelling faintly of anisette.

Pore surface: White to creamy; pores large, 0.5–2 mm across, angular to daedeloid (maze-like), decurrent; tubes up to 1 cm (½ in) deep.

Stipe: Central to slightly eccentric, 4–12 cm tall (aboveground) by 3–5 cm thick (1⅔–4¾ × 1¼–2 in), tough, yellowish to pale yellowish-brown; growing from underground sclerotium.

Bruising: Cap and flesh not bruising; pore surface not bruising or faintly bruising brownish.

Spore print: White.

Chemical staining: KOH—surface of cap pale reddish-orange; flesh yellowish.

Microscopic: Spores 7–9 × 6–8 μm, globose, amyloid, with prominent ridges and spines up to 1 μm high. Cystidia lacking.

Occurrence: Parasitic, causing white butt rot; growing from the ground at the base of oaks (primarily) and other living hardwood trees; saprobic, growing around decaying stumps or from buried roots; annual, summer–fall; widespread but occasional east of the Rocky Mountains.

Edibility: Edible when young, bitter and tough when mature.

Comments: The impressive size of the massive fruitbody will grab your attention! In our area this spectacular mushroom occurs both in the mountains and in the Piedmont almost exclusively at the base of living oaks (*Quercus* spp.). The western counterpart to this species is *Bondarzewia mesenterica* (Schaeffer 1774) Kreisel (1984), which occurs west of the Rocky Mountains and is associated exclusively with conifers. The western species is usually called *B. montana* (Quélet 1888) Singer 1940, but this is a later synonym.

Birch Polypore

Piptoporus betulinus

A

B

Photos: A, W. Light; B, Captainpixel, Wikipedia.

***Piptoporus betulinus* (Bulliard) P. Karsten**, *Revue Mycol.*, Toulouse 3(11):9 (1881)
[Basidiomycota, Agaricomycotina, Agaricomycetes, Polyporales, Fomitopsidaceae]

Basionym: *Boletus betulinus* Bulliard, *Herb. Fr.* 7:tab. 312 (1788) [1787–1788]

Derivation: *Piptoporus*, Greek, "falling polypore"; *betulinus*, Latin, "birch dwelling."

Cap: 2.5–25 cm across by 1–7.5 cm thick (1–10 × ⅖–3 in), convex to broadly convex or nearly flat, kidney-shaped to hoof-shaped (Figs. A & B), surface dry, covered by a smooth or velvety, often slightly roughened and cracked crust; whitish, buff, pale brownish, to brownish-grey with darker brownish streaks; margin thickly inrolled around pore surface (Fig. B).

Flesh: White, thick, rubbery at first, becoming corky.

Pore surface: Recessed, initially creamy-white (Fig. B), aging to greyish-brown or dark brown; 3–4 pores per mm, tubes up to 1 cm deep.

Stipe: Lateral, rudimentary and stubby or lacking.

Spore print: White.

Chemical staining: **KOH**—cap surface and flesh dark brownish-black; **Melzer's reagent**—cap, yellowish-green, flesh, greenish-grey.

Microscopic: Spores 3–6 × 1.5–2 µm, smooth, cylindrical to elliptical.

Occurrence: Saprobic and parasitic, solitary or in groups on living and dead birch trees (*Betula* spp.); annual, but fruitbodies persisting more than a year; in our area found in the southern Appalachians; north Georgia.

Edibility: Edible when young, but bitter.

Comments: This mushroom is unlikely to be mistaken for any other. Its restriction to birches, the hoof-like shape of the fruitbody, and the recessed pore surface surrounded by the curb-like margin are distinctive. The cut velvety surface was once used as a razor strop for fine edges on razors, and dried specimens have served as tinder. Several polyporenic acid derivatives from this mushroom have potent anti-inflammatory and biocidal properties, and piptamine ($C_{20}H_{35}N_3$) is strongly antibacterial. This mushroom was carried by Ötzi, the 5,300-year-old mummified Tyrolian Iceman, who is believed to have used it medicinally. We have only found this northern species in the mountains of north Georgia, where it was growing on Yellow Birch, *Betula alleghaniensis* Britton.

Spongy Tooth Polypore

Spongipellis pachyodon

A

B

C

***Spongipellis pachyodon* (Persoon) Kotlaba & Pouzar**, *Česká Mykol.* 19:77 (1965)
[Basidiomycota, Agaricomycotina, Agaricomycetes, *incertae sedis*, Polyporales, Meruliaceae]

Basionym: *Hydnum pachyodon* Persoon, *Mycol. Eur.* (Erlanga) 2:174 (1825)

Derivation: *Spongipellis*, Latin, "spongy-skinned"; *pachyodon*, Greek, "thick-toothed."

Deprecated synonym: *Sarcodontia pachyodon* (Persoon) Spirin, *Mycena* 1(1):64–71 (2001)

Fruitbody: Highly variable, sometimes a spreading pore surface with or without a cap-like edge; when present, caps well developed, tough, leathery, or rubbery (Fig. B), often forming overlapping shelves 2–5 cm (¾–2 in) wide and deep, which may or may not coalesce laterally into extensive rows up to 20 cm (8 in) or more in length (Fig. A); upper surface and teeth white, creamy-white, to pale yellowish or beige.

Flesh: Tough, leathery or rubbery, concolorous with cap, 3–8 mm thick.

Pore surface: Initially white or creamy-white, becoming dull yellowish or light brown; pores angular, forming highly variable daedeloid or labyrinthine channels, tubes up to 1.5 cm (⅔ in) deep, developing into flattened teeth with jagged edges

(Fig. C, this feature diagnostic), eventually becoming hydnoid (awl-like).

Stipe: Lacking.

Chemical staining: Cap surface, flesh, and teeth instantly cinnamon-brown in KOH and deep umber-brown in Melzer's reagent.

Spore print: White.

Microscopic: Spores 5–6.5 µm, smooth, spherical, nonreactive in Melzer's reagent.

Occurrence: Both parasitic and saprobic. Solitary or in overlapping and/or fused clusters on living and dead hardwoods, especially oaks, summer–winter; widespread but occasional in the Southeast.

Edibility: Inedible.

Comments: In order for woodpeckers to be able to excavate their nesting cavities in hardwood snags, the heartwood must first undergo substantial fungal decay. The primary heart-rot decay fungus for most woodpecker species—Downy, Hairy, Red-headed, Red-bellied, and Pileated Woodpeckers and Northern Flickers—is *Spongipellis pachyodon* (Conner, Miller, & Adkisson, 1976; Conner, Dickson, & Williamson, 1983).[1] See also the detailed comments regarding the Red-cockaded Woodpecker in the account for the Pine Bracket, *Phellinus pini.*

1. R.N. Conner, O.K. Miller Jr., & C.S. Adkisson, 1976, Woodpecker dependence on trees infected by fungal heart rots, *Wilson Bull.* 88:575–581; R.N. Conner, J.G. Dickson, & J.H. Williamson, 1983, Potential woodpecker nest trees through artificial inoculation of heart rots, paper given at the Snag Habitat Management Symposium, Flagstaff, Arizona, June 7–9, 1983.

Gilled Polypore

Lenzites betulina

A

B

C

Lenzites betulina **(Linnaeus 1753) Fries**, *Epicr. Syst. Mycol.* (Upsaliae):405 (1838)
[Basidiomycota, Polyporales, Polyporaceae]

Basionym: *Agaricus betulinus* Linnaeus, *Agaricus betulinus* L., *Sp. Pl.* 2:1176 (1753)

Derivation: The genus is named in honor of the German mycologist Harald Othmar Lenz (1798–1870); *betulinus*, Latin, from *betula*, "birch," "associated with birch trees," although this mushroom is associated with many hardwood species.

Cap: Up to 15 cm (6 in) across and 5 cm (2 in) wide; semicircular to kidney-shaped. No stipe.

Upper surface: Flat to convex; densely covered with fine white hairs (Fig. C); concentric zones of texture, ridges, and bumps and alternating bands of whitish, greyish, and brownish color (Fig. A). Flexible when young. As with several other polypores, older fruitbodies may be colored green due to algae.

Pore surface: Deep white or cream-colored gills (not merely gill-like ridges) of various lengths radiating from the point of attachment to the margin (Figs. B & C). Gills may turn a dull pale ochraceous with age.

Context: Flesh white.

Spore print: White.

Microscopic: Spores 5–6 × 2–3 µm; smooth; cylindrical; inamyloid. Hyphal system trimitic, with clamps. No cystidia.

Staining: Flesh nonstaining with KOH.

Occurrence: Solitary or in clumps on dead hardwood, occasionally on conifers; summer through November–December. Causes white rot in the sapwood. Widespread in northwestern, midwestern, and eastern North America and Europe.

Edibility: Tough, inedible.

Comments: *Lenzites betulina* is readily distinguished from the Rusty-gilled Polypore (*Gloeophyllum sepiarium*) by its white gills and flesh and nonstaining properties. The Rusty-gilled Polypore has rusty-brown flesh and golden-brown to brown gills; the flesh stains black in KOH. *Gloeophyllum sepiarium* also grows primarily on conifers rather than hardwoods. The spores of *G. sepiarium* are also much longer (9–13 µm) than those of *L. betulina* (5–6 µm), one of the few cases in which there is no size overlap between compared species.

Rusty-gilled Polypore, Yellow-red Gilled Polypore

Gloeophyllum sepiarium

A

B

C

D

Gloeophyllum sepiarium (von Wulfen 1786) P. Karsten, *Bidr. Känn. Finl. Nat. Folk.* 37:79 (1882)
[Basidiomycota, Gloeophyllales, Gloeophyllaceae]

Basionym: *Agaricus sepiaricus* Wulfen, in Jacquin, *Collnea Bot.* 1:339 (1786)

Derivation: *Gloeophyllum*, Greek, "with sticky leaves"; *sepiarium*, Latin, "sepia-colored."

Cap: Cap 2–3 cm across, 2.5–15 cm wide, 0.5–2 cm thick. Broadly convex to flat, outline fan-shaped to round. No stipe.

Upper surface: Cap with zonate bands of texture and color; yellow to orange when young, later becoming reddish-brown or brown; with yellow or white margin (Fig. A); bruises rust-colored to tawny-yellow. Surface velvety at first, later becoming bristly hairy.

Pore surface: Pore surface mostly gill-like (Fig. B), up to 10 mm deep, 1.5–2 per mm, often mixing with slot-like pores.

Spore print: White.

Context: Dark yellow-brown to dark rust-brown (Fig. C, arrow).

Microscopic: Spores long-cylindrical, smooth, 9–12 × 3–4 µm; hyaline in KOH; trimitic; long-cylindrical cystidia (100 × 10 µm) present.

Staining: Flesh stains black with KOH.

Occurrence: Widespread in North America and Europe. Saprobic, causes brown rot; solitary or grouped or in rosette-like clusters on dead conifers, occasionally on Quaking Aspen; June–November, overwinters in warm climates.

Edibility: Tough, inedible.

Comments: The dark-colored cap, flesh, and rusty-brown gills immediately distinguish this species from the greyish, white-gilled *Lenzites betulina*, as do the much larger and distinctly cylindrical spores. The flesh also stains black with KOH, in sharp contrast to the nonstaining context of *L. betulina*. *Gloeophyllum sepiarium* is distinguished from *G. trabeum* in having a hairy cap and rusty-brown flesh, in being restricted to conifers, and in having a lamellate (1.5–2 per mm) hymenophore and long-cylindrical (9–12 × 3–4 µm) spores; *G. trabeum* has a very finely hairy or hairless cap and paler umber or dull brown (not rusty) flesh, occurs mainly on hardwoods (but also on conifers), has an irregular, semilamellate (2–4 per mm) to labyrinthine or partly poroid hymenophore; its spores are short-cylindrical (6.5–9.5 × 3–4.5 µm). *Gloeophyllum sepiarium* has trimitic hyphae; those of *G. trabeum* are dimitic. This species, like many polypores, is highly polymorphic and may form amorphous reef-like masses (Fig. D).

Hydnoid (Toothed) Fungi

This informal category includes a variety of unrelated fungi in at least eight separate orders (Agaricales, Auriculariales, Gomphales, Hymenochaetales, Polyporales, Russulales, Thelephorales, and Trechisporales) that bear their hymenia on the surfaces of tooth-like or spine-like projections. These fungi have diverse fruitbody morphologies, ranging from normal-looking mushrooms with a cap and stipe, but with teeth instead of gills, pores, or a smooth hymenial surface. The teeth may be long and sharply pointed (*Mycorrhaphium*) or short and bluntly rounded (*Hydnellum*). They may be flat and plate-like (*Climacodon*, *Trichaptum*) or the jagged remnants of the eroding dessipiments of pores. Or they may be intermediate between being poroid and hydnoid. Others are effuse, effuse-reflexed, or resupinate crusts and spreads, while still others (*Hericium*) have dense, dangling clusters of long icicle-like teeth. Great variability exists in the appearance of these structures within a given species as a function of both the fruitbody's age and its environmental circumstances. For example, certain species can present with either a stipitate form or a resupinate form; others (*Irpex lacteus*) can start out resupinate and poroid, then become reflexed (develop upturned edges) and gradually transform into deeply hydnoid caps as the dissepiments erode away.

Stipitate hydnoid fungi (stalked and toothed fungi) produce mushroom-like fruitbodies with a somewhat reduced to well-developed cap atop a prominent (sometimes massive) stipe. Some of them (*Hydnellum*) exhibit indeterminate growth and engulf plant stems and debris as they expand. These mushrooms include the genera *Bankera*, *Hydnellum*, *Hydnum*, *Phellodon*, *Mycorrhaphium*, and *Sarcodon*. These genera represent the orders Cantharellales, Thelephorales, and Polyporales, and all are represented in our area.[1]

Mycorrhaphium is saprobic on dead hardwood logs and sticks. *Hydnum*, *Hydnellum*, *Phellodon*, and *Sarcodon* form

A B C

Hydnoid fruitbody morphologies. A, stipitate; B, shelf-like; C, resupinate.

Hydnoid tooth structures. D, *Hydnochaete olivacea*; E, deeply incised and eroded polypore dissepiments: *Irpex lacteus*; F, long, awl-like teeth: *Hydnellum* cf. *scrobiculatum*; G, round, bluntly rounded teeth: *Hydnum repandum*; H, flat and bluntly rounded teeth: *Spongipellis pachyodon*; I, long, icicle-like teeth: *Hericium erinaceus*. (Photos: D–H, W. Light; I, LeBrac.)

ectomycorrhizal associations with mature trees in old-growth forests and are seriously threatened in Europe and elsewhere where these forests have been largely cut down. In the southern Appalachians these fungi are ectomycorrhizal with Eastern Hemlock (*Tsuga canadensis*) along large major riparian ecosystems. Since 2007, many Eastern Hemlocks have been killed by the introduced (from Japan) Hemlock Woolly Adelgid (*Adelges tsugae*), and the future of these stipitate hydnums appears problematic here as well.[2]

Shelf-like hydnoid fungi typically produce sessile, shelf-like fruitbodies and include the genera *Climacodon*, *Irpex*, *Spongipellis*, and *Trichaptum*. Some, like *Irpex lacteus*, initially produce resupinate, crust-like spreads that later develop shelf-like caps. Indeed, our account of this species is placed in the section on crusts, spreads, and parchment fungi, but it could just as well be included here or with the polypores. The spongy tooth polypores (*Spongipellis*) and the violet-tooth polypores (*Trichaptum*) are treated in the section on polypores.

Resupinate hydnoid fungi, including the genera *Hydnochaete*, *Irpex*, *Sarcodontia*, and *Steccherinum*, form a variety of effuse and effuse-reflexive crusts and spreads, and some of them also produce prominent shelf-like caps when fully mature. *Sarcodontia* is not discussed in this work. The Milk-white Toothed Polypore (*Irpex lacteus*), Ochre Spreading Crust (*Steccherinum ochraceum*), and the Brown-toothed Crust (*Hydnochaete olivacea*). are treated in the section "Corticioid Fungi: Crusts, Spreads, and Parchment Fungi."

1. *Bankera fuligineoalba* (J.C. Schmidt) Coker & Beers is recorded from Great Smoky Mountains National Park (R.H. Petersen, 1976, *Checklist of Fungi of the Great Smoky Mountains National Park*, Management Report No. 29, Department of the Interior, National Park Service, Southeast Region).

2. R. Baird et al., 2013, *Fungal Diversity* 62(1):41–114.

Wood Hedgehog

Hydnum repandum

A B C

Hydnum repandum Linnaeus, *Sp. Pl.* 2:1178 (1753) [Basidiomycota, Agaricomycotina, Agaricomycetes, *incertae sedis*, Cantharellales, Hydnaceae]

Derivation: *Hydnum*, Greek, "spongy plant," "fungus"; *repandum*, Latin, "curved upward."

Cap: 2–15 cm (~¾–6 in) across, convex to nearly plane, sometimes centrally slightly depressed; surface dry, smooth or velvety; margin thick, initially inrolled, becoming wavy to lobed; cream to buff, bruising orange or orange-brown.

Flesh: Whitish or pale buff, thick, fleshy, firm, homogeneous (azonate), up to 2 mm thick; bruising orange-yellow or orange-brown; taste mild to peppery.

Toothed surface: Densely covered with attached or slightly decurrent awl-shaped spines, occasionally completely decurrent to the base of the stipe; these spines cream-colored or dull orange, 2–7 mm long (Figs. B & C).

Stipe: Usually central or slightly off center, 3–10 cm (1¼–4 in) tall by 1–3 cm (⅖–1¼ in) thick; dry, smooth, white, often with orange regions (Figs. A & B) to concolorous with cap, densely covered with decurrent spines.

Spore print: White.

Staining: Yellow with iron sulfate.

Microscopic: Spores 6.5–9 × 5.5–8 µm, nearly spherical or broadly elliptical, with oblique apiculus, smooth, clear.

Occurrence: Summer and fall, solitary or gregarious, widespread; mycorrhizal with conifers or hardwoods in North America, especially oaks in the East; also widespread in Europe, Australia, New Zealand, Japan, Korea.

Edibility: Edible, considered choice.

Comments: *Hydnum repandum* accumulates high levels of heavy metals, especially mercury, in areas of heavy aerial or ground pollution. Some works refer to this taxon as *Dentinum repandum*.

Look-alikes include *Hydnum albidum* Peck (1887) and *H. umbilicatum* Peck (1902), but both of these taxa are considerably smaller than *H. repandum*. *Hydnum albidum* Peck is completely white, with a cap <7–8 cm and a stipe <1 cm thick (2¾–3¼ × <¾ in), and it has smaller spores (6.5–7.3 × 3–4 µm). The teeth are the same length as those of *H. repandum* but appear longer relative to the smaller size of the cap. Some authors consider *H. albidum* to be a variety of *H. repandum*.

Hydnum umbilicatum has a cap that is prominently umbilicate (with a deep, belly-button-like central depression) and generally 5 cm (2 in) or less across. The teeth are either adnate or circumscribed by a depression where they join the stipe; unlike in *H. repandum*, they are not at all decurrent. The spores are also somewhat larger (9–10 × 7.5–8.5 µm). *Hydnum umbilicatum* is also a choice edible; *H. albidum* is typically quite peppery.

The genus *Hydnum* P. Karsten (1879) is distinguished from the other stalked and toothed mushrooms by its combination of determinate growth, a fleshy context, white spore print, and smooth spherical to ovoid spores. The similar but unrelated genus *Hydnellum* Linnaeus (1753)[1] often exhibits indeterminate growth (growing in spurts and engulfing twigs and needles), a leathery context, a brown spore print, and distinctive spherical to oblong spores with warty ornamentation.

1. On the basis of genetic studies, *Hydnum* is now placed in the family Bankeraceae (Order Thelephorales); *Hydnellum* is grouped in the order Cantharellales together with the chanterelles.

Orange Rough-cap Tooth

Hydnellum aurantiacum

A

B

C

D

Hydnellum aurantiacum **(Batsch) P. Karsten**, *Meddn Soc. Fauna Flora Fenn.* 5:41 (1879)
[Basidiomycota, Agaricomycotina, Agaricomycetes, *incertae sedis*, Thelephorales, Bankeraceae]

Basionym: *Hydnum suberosum* var. *aurantiacum* Batsch, *Elench. Fung.*, cont. sec. (Halle):103 (1789)

Derivation: *Hydnum*, Greek, "spongy fungus," *Hydnellum*, "little *Hydnum*"; *aurantiacum*, Latin, from *aurántium*, "orange," pertaining to the cap color.

Cap: 5–18 cm (2–7¼ in) across, solitary, scattered, or in clusters, usually single, but sometimes fused with other caps; broadly convex to irregularly flat (Fig. A) or centrally depressed (Figs. A, right, & D), surface highly rugose and/or pitted (Fig. A); orange to salmon- or rusty-red (Figs. A & D), paler whitish- to pinkish-tan at margin (Figs. A & C).

Flesh: Dimitic (2-layered), upper layer whitish to buff, lower layer brownish- or rusty-orange; odor and taste unpleasant.

Toothed surface: Spines densely crowded, 2–7 mm long, bluntly rounded (Fig. B), initially greyish-white, becoming dingy-brownish.

Stipe: 2–6¾ cm tall by ½–2 cm thick (¾–2¾ × ¼–¾ in), cylindrical, sometimes expanded at base, concolorous with cap.

Spore print: Brown.

Chemical staining: KOH—cap black, flesh olive-green.

Microscopic: Spores 5–9 × 5–7 µm, subglobose, with prominent irregular tubercles; brownish, inamyloid.

Occurrence: Ectomycorrhizal. Alone or clustered, sometimes concrescent; widespread in North America, in the East under or sometimes on the base of Eastern Hemlock (*Tsuga canadensis*); occasional, summer–winter in our area.

Edibility: Inedible.

Comments: Two similar species, the Coalescent Tooth, *H. concrescens* (Persoon) Banker, and the Ridged Tooth, *H. scrobiculatum* (Fries) P. Karsten, are associated with hardwoods, particularly oaks (*Quercus* spp.), or with pines (*Pinus* spp.), respectively, rather than with Eastern Hemlock. The upper surface of the caps in both species is often radially ridged and highly fibrillose, with indistinct to distinct concentric zones of color and texture, although there is much variability in this regard. The fruitbodies form distinctive rosettes in isolation but often fuse together when in groups.

The greyish flesh of *H. concrescens* has a fish-like odor, is not obviously duplex, and in KOH stains yellow-green, which very slowly turns brown with a persistent dark blue-green margin. The brown flesh of *H. scrobiculatum* turns dark carmine in KOH, and the flesh has no distinctive odor. The teeth on the lower surface in both taxa are very short (<3 mm) compared to those of *H. aurantiacum* (2–7 mm).

The three taxa are very similar and are best told apart by their host trees.

Coalescent Tooth, Zoned Tooth

Hydnellum concrescens

***Hydnellum concrescens* (Persoon) Banker**, *Mem. Torrey Bot. Club* 12:157 (1906)
[Basidiomycota, Agaricomycotina, Agaricomycetes, *incertae sedis*, Thelephorales, Bankeraceae]

Basionym: *Hydnum concrescens* Persoon, *Observ. Mycol.* (Lipsiae) 1:74 (1796)

Derivation: *Hydnum*, Greek, "spongy fungus," *Hydnellum*, "little *Hydnum*"; *concrescens*, Latin, from *concrēscere*, "to congeal, coagulate," referring to the tendency of the fruitbodies to merge and coalesce (shared by most *Hydnellum* species).

Cap: 2–10 cm (¾–4 in) across, planoconvex to plane (Fig. A) to shallowly depressed centrally (Fig. B), often coalesced with other caps, finely velvety to smooth, often with pits (Fig. A, arrows); usually developing radial fibrillose ridges and/or rugose growths; white to pale pinkish-buff or pinkish-brown, with concentric zones of color (Fig. B), pale near the margin (Figs. A & B), becoming all buff and turning brown from the center (Fig. B), sometimes fading to white before decaying and turning black; margin bruising brown to blackish (Fig. C, arrows); occasionally exuding a red liquid when young.

Flesh: Greyish to reddish- or purplish-brown, duplex or not, often zonate; with fishy or mealy odor.

Toothed surface: Spines purplish-brown (Figs. C–E), very short (1–3 mm long), bluntly rounded (Fig. D), decurrent (Figs. C & E); initially whitish, becoming purplish-brown (Figs. D & E).

Stipe: 2–4 cm tall by 0.5–2 cm thick (¾–1⅔ × ¾ in) near the top, base often swollen and spongy (Fig. E). Concolorous with the cap.

Spore print: Brown.

Chemical staining: Flesh yellow-green in KOH, very slowly turning brown or black with a persistent dark blue-green margin.

Microscopic: Spores (4–)4.5–6.1 × 3.5–5.4(–5) µm, subglobose to irregular or lobed, covered with conspicuous, angular (i.e., flattened and nodular), and often bifurcated or coalescing wart-like ridges; light brown, inamyloid; clamp connections lacking.

Occurrence: Ectomycorrhizal. Solitary or in groups under hardwoods, especially oaks (*Quercus* spp.); widespread in eastern North America, Europe, China; summer–fall.

Edibility: Inedible, tough.

Comments: *Hydnellum concrescens* is macroscopically indistinguishable from the closely related Ridged Tooth, *H. scrobiculatum* (Fries) P. Karsten. *Hydnellum scrobiculatum* is ectomycorrhizal with pines (*Pinus* spp.)[1] instead of oaks, has spores with rounded instead of angular warts, and the flesh produces a violet or dark carmine flash with KOH, which then turns blue-green to olive-green, rather than simply staining olive-green. It has no fishy or mealy smell. It is nearly impossible to distinguish between these two species with old fruitbodies.

1. Very rarely *H. scrobiculatum* may be associated with hardwoods.

Bleeding Tooth, Devil's Tooth

Hydnellum peckii

A B C D

Photos: A, Teresa Fortenberry; B–D, W. Light.

***Hydnellum peckii* Banker**, *Mycologia* 5:203 (1913) [Basidiomycota, Agaricomycotina, Agaricomycetes, *incertae sedis*, Thelephorales, Bankeraceae]

Derivation: *Hydnum*, Greek, "spongy fungus"; *Hydnellum*, Greek, "little *Hydnum*"; *peckii*, Latin, epithet in honor of mycologist Charles H. Peck (1833–1917).

Cap: Unevenly convex or flattened, sometimes centrally depressed, surface velvety, becoming smooth with age; round or irregular, 2–5 cm (⅘–2 in) across, caps often merging to form larger coalescing fruitbodies (Figs. B & D); initially whitish, becoming brownish, acquiring darker blotches with age; surface and undersides of caps and stipes of young fruitbodies typically oozing a bright red fluid from multiple sites when moist (Fig. A) that leaves dark reddish-black pitted scars (Figs. B, arrows, & D).

Flesh: Tough, fibrous, light brown, cream-brown, or greyish, woody when dry, often with central spongy layer.

Toothed surface: Short, cylindrical teeth tapering to blunt tip (Fig. C), becoming uniformly shorter as they descend the stipe and toward the cap margin, these 3–5 mm or less in length, 7–9 per mm^2.

Stipe: Central, short, up to 5 cm (2 in) tall, but only 1 cm (½ in) or less appearing aboveground, by 1–3 cm (½–1¼ in) thick; often multibranched, upper portion studded with teeth from the underside of the cap, lower portion hairy and usually covered with debris; stipe or stipes merging below into expanded, shapeless, bulbous base (Fig. D).

Spore print: Brown.

Staining: KOH—cap and teeth black; flesh grey to olive-green; **Melzer's reagent**—cap and teeth dark yellowish-brown, flesh pale yellow.

Microscopic: Spores subspherical to ovoid, 4–5 µm in diameter, covered with rounded tubercles, amyloid.

Occurrence: Ectomycorrhizal with conifers; on the ground singly, scattered, or in fused masses. Widespread in North America and Europe; reported in Iran[1] and Korea.[2]

Edibility: Inedible; reported to have a peppery taste; our specimens tasteless.

Comments: *Hydnellum peckii* is sometimes distinguished from *H. diabolus* Banker on the basis of the surface of the cap: pitted and stained in the former; not pitted and covered with fine hairs in the latter. However, these distinctions are vague, and there appears to be considerable overlap between these two confused taxa. The acrid taste test is also widely variable, and no one is sure what it means. They are synonymized by some but not all mycologists.

Hydnellum aurantiacum (Batsch) Karsten has a generally larger (up to 10 cm [4 in] wide) orange to rusty-red cap and an orange stipe. *Hydnellum spongiosipes* (Peck) Pouzar has brown spines running down the stipe to an expanded, bulbous base and grows under hardwoods (especially oaks), not conifers.

1. M.R. Asef, 2008, *Rostaniha* 9(2):115.
2. S.-K. Han, S.-H. Oh, & H.-J. Kim, 2010, *Mycobiol.* 38(2):81–88.

Kidney-shaped Tooth

Mycorrhaphium adustum

A

B

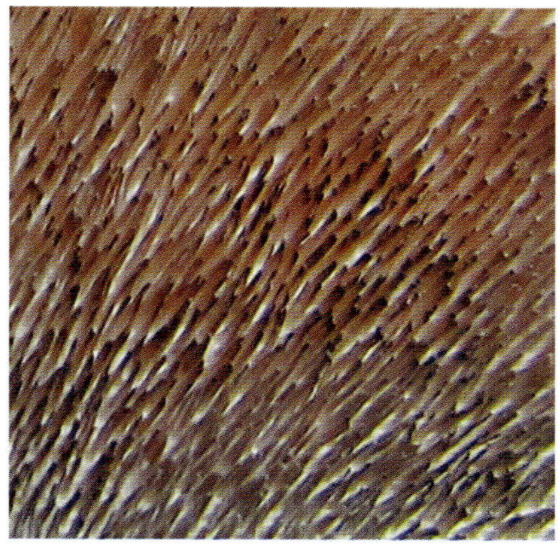

C

***Mycorrhaphium adustum* (Schweinitz) Maas Geesteranus**, *Persoonia* 2(3):394 (1962)
[Basidiomycota, Agaricomycotina, Agaricomycetes, *incertae sedis*, Polyporales, Meruliaceae[1]]

Basionym: *Hydnum adustum* Schweinitz, *Schr. Naturf. Ges. Leipzig* 1:103 (1822)

Derivation: *Mycorraphium*, Greek, combining form of "fungus" and "small needle," referring to the straight, needle-like spines; *adustum*, Latin, from *adustus*, "scorched," "burnt," referring to the blackish margin of the cap.

Cap: 2.5–8 cm (1–3¼ in) wide, kidney-shaped to disk-shaped, upper surface flat to shallowly depressed, especially toward disk, very thin near margin; tough, dry, finely roughened to somewhat velvety, with concentric zones of roughened texture, especially near margin; white to tan, *margin black with age* (Fig. A). Adjacent caps often coalescing.

Flesh: Thin, tough, white, not bruising when cut.

Fertile (spore-bearing) surface: Densely covered with straight needle-like spines up to 3 mm (~⅒ in) long (Fig. B), these often flattened, fused subdistally, and forked at the tips (Fig. C); initially whitish, becoming rose-purple (Fig. B), then dark brown with age.

Stipe: Lacking or central to lateral when present (Fig. B); whitish; 2–3 cm tall by 1–2 cm thick (¾–1¼ × ⅖–¾ in).

Bruising: Bruising greyish with handling.

Spore print: White.

Microscopic: Spores ellipsoid to short cylindrical, 3–5.8 × 1–1.5 μm; smooth, clear, nonreactive in Melzer's reagent.

Occurrence: Saprobic, causing white rot; growing alone or in small, overlapping, and sometimes coalescing clusters on hardwood sticks, branches, and small logs, particularly oaks (*Quercus* spp.); summer–late fall. Widespread and fairly common in eastern North America.

Edibility: Inedible.

Comments: The stiff, hard, circular to kidney-shaped caps *with thin black margins*, hymenial (pore surface) spines, and growing habit on small sticks and logs (rather than at the base of trunks or stumps or on larger logs) immediately identify this small (<8 cm [3¼ in]) mushroom. This species is a polypore and is not closely related to other toothed fungi.

It could be confused with *Climacodon pulcherrimus*, which also can have a broadly convex to flat, fan-shaped or semicircular, and concentrically zoned cap. *Climacodon pulcherrimus* is distinguished from *M. adustum* by its generally larger cap (up to 11 cm [4½ in] across) and its densely matted, woolly (not minutely velvety) surface. In addition, the cap margin is not black, the stipe is lacking, and the hymenial teeth are considerably longer (8 mm). The flesh stains pink or red with KOH.

The Black-staining Polypore, *Meripilus sumstinei* (Murrill) M.J. Larsen & Lombard, also has white to tan caps with concentric zones of texture and prominent black margins. However, the pore surface lacks teeth, has minute pores ([5–]6–8[–9] per mm), and is initially white, becoming pale tan, and bruises dark brown to black. It also grows at the base of trees and trunks and forms massive multicapped fruitbodies up to 30 cm (12 in) across.

1. The family Stecherinaeceae Parmasto (1968) given by some authors is a later synonym of the Meruliaciae Rea (1922).

Charming Tooth

Climacodon pulcherrimus

A

B

C

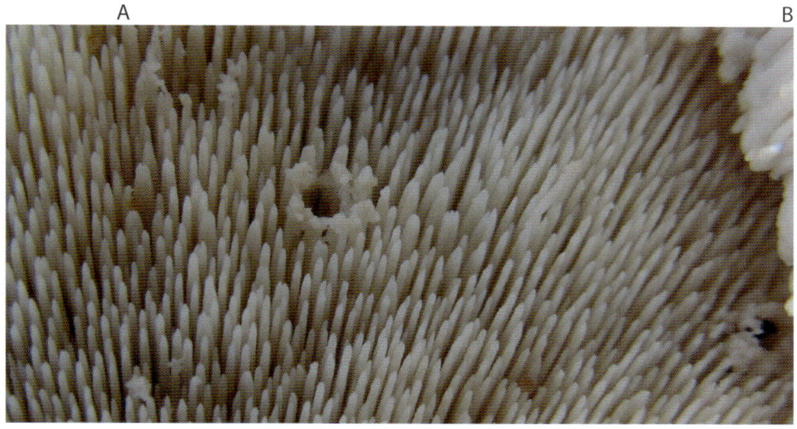

D

***Climacodon pulcherrimus* (Berkeley & M.A. Curtis) Nikolajeva,** *Flora Plantarum Cryptogamarum URSS 6*, Fungi 6(Fungi, 2):194 (1961)
[Basidiomycota, Agaricomycotina, Agaricomycetes, *incertae sedis*, Polyporales, Phanerochaetaceae]

Basionym: *Hydnum pulcherrimum* Berk. & M.A. Curtis, *Hooker's J. Bot. Kew Gard. Misc.* 1:235 (1849)

Derivation: *Climacodon*, Greek, from *klimakos*, "ladder," and *odon*, "tooth," "ladder tooth"; *pulcherrimus*, Latin, "enchanting," "charming."

Cap: 4–12 cm across by 1–2 cm thick (1⅔–4¾ × ½–¾ in), broadly convex to nearly flat, semicircular to fan-shaped (Figs. A & B); surface dry, with dense woolly or matted hairs (Figs. A–C); initially whitish or pale tan (Fig. B), becoming ochre-orangish with age (Fig. A).

Flesh: White, fibrous, not bruising when cut.

Tooth surface: Densely covered with whitish to orangish, bluntly pointed spines up to 8 mm (⅓ in) in length, these often fusing to form hollow columns topped with sharp, tooth-like projections (Fig. D).

Stipe: Sessile, no stipe.

Spore print: White.

Chemical staining: Cap and flesh pink to red with KOH.

Microscopic: Spores 4–6 × 1.5–3 (3.5–4.5 × 1.5–2.0) µm, smooth, elliptical, nonreactive in Melzer's reagent; hyphae of the flesh and cap cuticle often with two to four clamp connections at a single septum.

Occurrence: Saprobic on hardwood stumps and fallen trees, rarely on conifers; solitary, in groups, or imbricate (overlapping) (Figs. A & B); widespread but rare in eastern North America; summer. Also documented in Mexico, Brazil, Europe, Eurasia (Ukraine, Crimea), China, Indonesia.

Edibility: Inedible.

Comments: This is one of two closely related species of *Climacodon* that occurs in North America. *Climacodon septentrionalis* (Fries) P. Karsten is distinguished from *C. pulcherrimus* by its larger size (up to 30 cm across and 5 cm thick [12 × 2 in]), massive overlapping clusters of fruitbodies on the trunks of living hardwoods and stumps of recently dead hardwoods, much longer teeth (up to 2.5 cm [1 in]), and simple septa that may have an occasional single clamp connection.

Genus *Hericium*

(Lion's Mane, Bear's Head Tooth, and Coral Tooth Fungi)

The genus *Hericium* is unmistakable, with its spectacular cushion-like fruitbody from which hangs a dense mass of dangling, icicle-like, white spines. It has aptly been described as looking like a white cheerleader's pom-pom. Indeed, that is one of its common names.

Three species occur in the southeastern United States: *Hericium erinaceus*, *H. americanum*, and *H. coralloides*. All occur on living or dead hardwood trees, stumps, or fallen logs, with one also rarely found on conifers. They are saprobic and very likely also parasitic. The following key to our southeastern *Hericium* species will only work with mature fruitbodies. This is best guaranteed with specimens whose spines are starting to turn yellowish or brownish as they age.[1]

1. Detailed discussions of the systematics of these species are available at Michael Kuo, MushroomExpert.Com, http://www.mushroomexpert.com/hericium.html; N. Hallenburg, 1983, *Mycotaxon* 18(1):181–189; and J. Ginns, 1984, *Mycotaxon* 20(1):43.

Hericium americanum. (Photo: Teresa Fortenberry.)

Key to Southeastern Species of *Hericium*

1. Fruitbody branched 2

1′. Fruitbody unbranched, single, cushion-shaped, with spines longer than 2 inches and often up to 4 inches. . . . *H. erinaceus*

2. Fruitbody consisting of multiple cushion-like masses of hanging, icicle-like spines, these usually 1–2 inches long .... *H. americanum*

2′. Fruitbody consisting of multiple coral-like branches arising from a common base and bearing short (½ in or less) spines resembling the teeth of a comb .... *H. coralloides*

Lion's Mane, Bearded Tooth

Hericium erinaceus

Bear's Head Tooth

H. americanum

A

B

C

A & C, *H. erinaceus*; B, *H. americanum*. (Photos: A, Leblanc; B, John Carl Jacobs; C, W. Light.)

***Hericium erinaceus* (Bulliard) Persoon**, *Comm. Fung. Clav.* (Lipsiae):27 (1797)

Basionym: *Hydnum erinaceus* Bulliard, *Herb. Fr.* 1:tab. 34 (1781) [1780–1781]

***Hericium americanum* Ginns**, *Mycotaxon* 20(1):43 (1984)

Misapplied names: *Hericium coralloides* (*sensu auct. Amer.*), *non Hydnum coralloides* Scopoli (1772); *Hericium erinaceus* (*sensu auct. amer.*), *non Hydnym erinaceus* Bulliard (1781) [Basidiomycota, Agaricomycotina, Agaricomycetes, *incertae sedis*, Russulales, Hericiaceae]

Derivation: *Hericium*, Latin, from *ericium*, "porcupine"; *erinaceus*, Latin, "hedgehog"; *americanum*, Latin, "American."

Mature fruitbody: *H. erinaceus*—up to 40 cm (16 in) across, consisting of a single, unbranched, cushion-shaped mass from which hang dense, crowded masses of long, soft, vertical, beard-like spines, these up to 10 cm (4 in) in length, initially white (Fig. A), becoming yellowish to brownish with age (Fig. C); surface often matted with coarsely clumped mycelial strands (Fig. A, inset); ***H. americanum***—up to 30 cm (12 in) across, forming compact, branching clusters of cushion-shaped, spine-bearing fruitbodies arising from a common base (Fig. B); spines soft, white, densely crowded and beard-like, 0.5–4 cm (¼–1⅔ in) in length, but typically longer than 1 cm (½ in), white when young (Fig. B), becoming yellowish or brownish with age.

Stipe: ***H. erinaceus***—stubby, lateral, unbranched, attached to woody substratum (Fig. C); ***H. americanum***—lateral, branching and/or forked, arising from a common base.

Flesh: Both species—white, soft, not changing when cut; odor and taste not distinctive when young, unpleasant and bitter with age.

Spore print: Both species—white.

Microscopic: ***H. erinaceus***—spores 5–6.5 × 4–5.6 µm, subglobose to elliptical, smooth to minutely roughened, amyolid; ***H. americanum***—spores 5–7 × 4–5.6 µm, subglobose to elliptical, smooth to minutely roughened, amyolid.

Occurrence: Both species—parasitic in wounds of living trees and saprobic on a variety of dead hardwood trees, stumps, and logs, especially oaks (*Quercus* spp.) and American Beech (*Fagus grandifolia*); ***H. erinaceus***—on hardwoods, usually solitary, summer–midwinter; widespread in North America, Europe, and Asia; ***H. americanum***—solitary or gregarious on hardwoods, infrequently on conifers; widespread in eastern North America, late summer–fall.

Edibility: Both species—edible and choice when young and fresh, bitter when old; tastes like lobster or shrimp.

Comments: The only reliable way of separating these two species is with mature fruitbodies, and this is best realized when the spines are beginning to discolor. The two are very similar when young, and then all bets are off. Mature *Hericium erinaceus* consist of a single fruitbody arising from an unbranched base. The spines are also typically longer, up to 4 inches or even longer. *Hericium americanum* forms multiple cushion-shaped structures arising from a branched base. The spines are shorter, generally between ½ and 1⅔ in long. That said, these two taxa cannot possibly be mistaken for anything else. The Coral Tooth Fungus (*H. coralloides*) has coral-like branches, with the spines arranged in rows like the teeth of a comb.[1]

As used in traditional Chinese medicine, these mushrooms have potent anticancer, immunoenhancing, lipid- and glucose-lowering, antioxidant, and neuroprotective properties. They also have antimicrobial and wound-healing effects.

1. *Hericium americanum* was long misidentified by American authors as "*H. coralloides* (Scopoli : Fries) S.F. Gray." Unfortunately, the name *Hericium coralloides* was applied to two different species by European authors. Hallenburg's neotypification of the name (1983, *Mycotaxon* 18[1]:181–189) restricted it to *Hericium coralloides* (Scopoli *ex* Fries) Persoon (1794), the other taxon with which it had been confused being *Hericium alpestre* Persoon (1825). The Bear's Head Tooth, *H. americanum*, does not closely resemble *H. coralloides sensu* (Scopoli) Persoon, but is macroscopically indistinguishable from the European species associated with firs of the genus *Abies*: *Hericium alpestre* Persoon.

Coral Tooth Fungus

Hericium coralloides

A

B

Photos: A, Ken Zinkand; B, W. Light.

***Hericium coralloides* (Scopoli) Persoon**, *Neues Mag. Bot.* 1:109 (1794)
[Basidiomycota, Agaricomycotina, Agaricomycetes, *incertae sedis*, Russulales, Hericiaceae]

Basionym: *Hydnum coralloides* Scopoli, *Fl. Carniol.*, 2nd ed. (Wien) 2:472 (1772)

Derivation: *Hericium*, Latin, from *ericium*, "porcupine"; *coralloides*, Latin, "coral-like."

Fruitbody: 10–40 cm wide by 10–20 cm high (4–16 × 4–8 in), white to cream, consisting of multiple delicate, coral-like branches arising from a common central base (Figs. A & B) attached to woody substratum; main branches branching into smaller collaterals, branches pubescent on upper surface.

Spines: Short, 0.5–1 cm (¼–½ in) long, hanging in fairly even rows along the branches, but arranged in tufts at branch tips (Fig. B); initially white to cream, becoming yellowish to brownish with age.

Flesh: Soft (tough at the base), white, not changing when cut; odor and taste mild, pleasant.

Spore print: White.

Microscopic: Spores 3–5 × 3–4 µm, subspherical to spherical, smooth to minutely roughened; amyloid.

Occurrence: Saprobic. Solitary or gregarious on dead hardwood trees, stumps, and fallen branches; late summer–fall; widespread but uncommon in North America, Europe, Asia (including India), eastern Australia.

Edibility: Like all members of the genus, edible and choice when young and fresh, bitter when old.

Comments: This exquisite coral-like mushroom is the most delicate of the four[1] species of *Hericium* known from North America. It is distinguished by its open arrangement of branches and short (½ in or less), dependent spines arranged in fairly even rows except at the branch tips.

1. *Hericium abietis* (Weir *ex* Hubert) K.A. Harrison is restricted to dead conifers in northern California and the Pacific Northwest.

Bird's Nest Fungi and Artillery Fungi

Bird's Nest Fungi

These fascinating little fungi in the order Agaricales and the family Nidulariaceae with their cup-shaped fruitbodies filled with their spore-containing peridioles really do look for all the world like tiny birds' nests filled with eggs. These are another example of independently derived gasteroid fruitbodies that have evolved from gill-bearing ancestors. These, too, produce their spores within a chamber completely encased by a peridium, and they have converted from forceful-release ballistospory to passive-release statismospory. Within the periderm the spores are produced inside small lens-shaped **peridioles**—the "eggs" that you see in the bottom of the cup-like "nest" formed by the fruitbody.

In these fungi, the peridium has a preexisting pore that either (1) is initially covered by a lid-like **epiphragm** that separates or (2) wears away and uncovers an opening, or the peridium itself simply erodes away to expose the spores. Because of the configuration of the cups, raindrops falling in at an angle of 67°–73° from the perpendicular exert the necessary force at the proper angle to eject the spore packets (peridioles) and send them flying through the air for distances up to 1 m (3⅓ ft)—so-called **splash cup** dispersal. In the two most common genera—*Cyathus* and *Crucibulum*—the peridioles are attached to the inner wall of the peridium by a long elastic **funicular cord**, which is coiled up tightly inside a pouch called the **purse**. The end of the funicular cord consists of a tangled mass of sticky hyphae called a **hapteron**, which serves as an anchor when the peridiole is launched. These structures are shown in Figs. A–C.

When the peridioles are ejected, the funicular cord either (1) unwinds and trails behind so that the hapteron

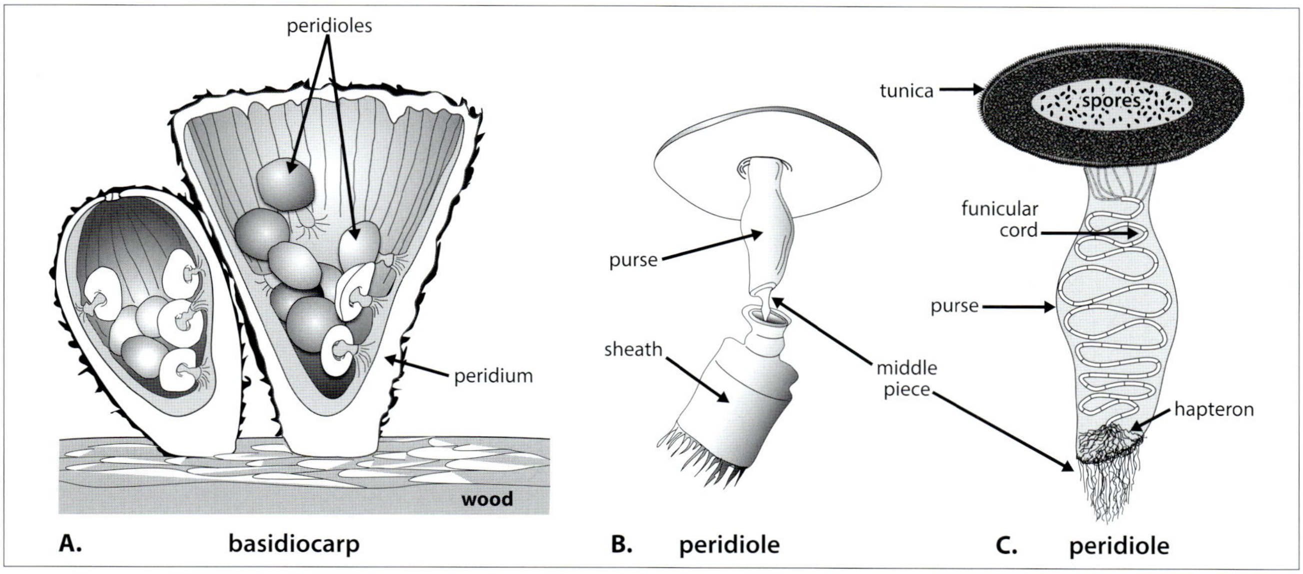

A & B, modified from drawings of *Cyathus* by W.G. Smith in H.C.I. Gymme-Vaughan & B. Barnes, 1927, *The Structure and Development of the Fungi* (Cambridge: Cambridge University Press), 315, fig. 281; C, modified from a drawing by Robert Sasata, redrawn from parts of Figs. 2, 31d, and 36 from H.J. Brodie, 1975, *The Bird's Nest Fungi* (Toronto: University of Toronto Press), 199 pp.

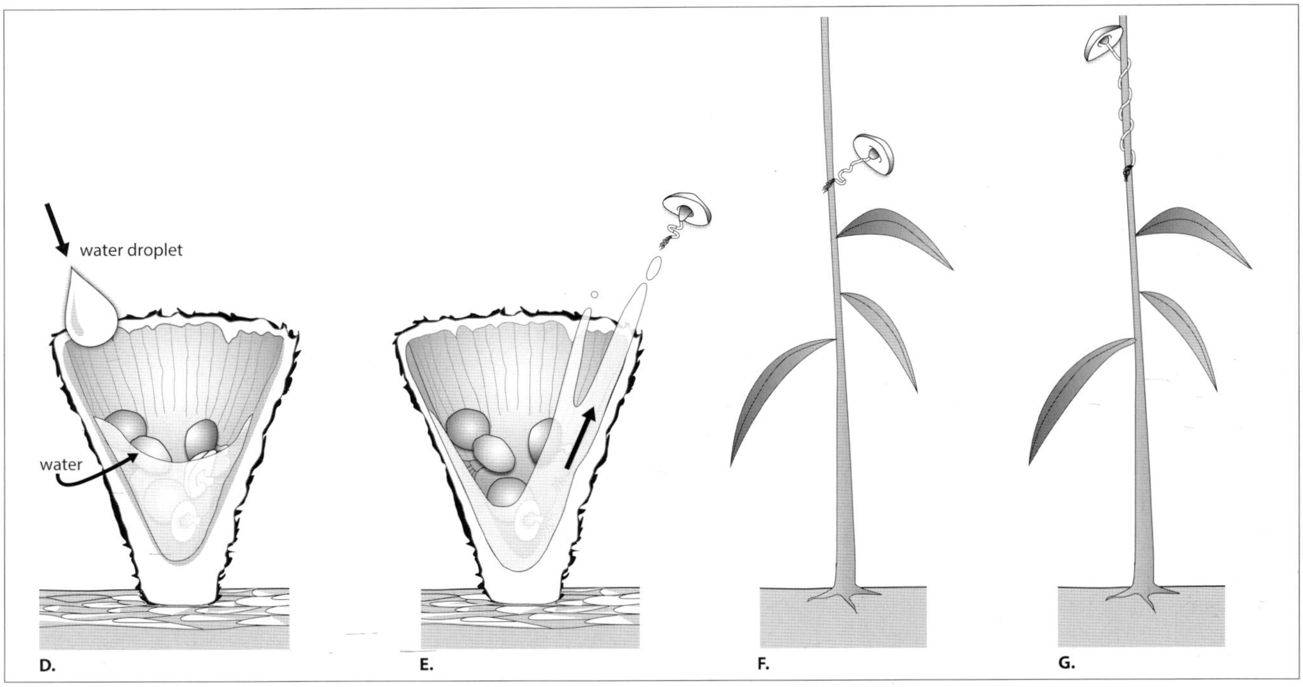

Splash-and-grab peridiole dispersal in bird's nest fungi (*Cyathus* and *Crucibulum*). D, an infalling raindrop provides the force to fracture the purse and break the peridiole free from its attachment to the inner wall of the endoperidium and eject it from the cup (E). When the exposed sticky hapteron encounters a twig, leaf, or other object along the peridiole's in-flight trajectory, it sticks fast to it (F). The peridiole's forward momentum uncoils the funicular cord and provides a centripetal force, causing the cord and attached peridiole to wrap tightly around the stem, twig, or leaf (G) and tether it securely in place. Adapted from a figure in Hassett et al., 2013.[1]

can snag onto twigs, stems, and other objects along the peridiole's trajectory, causing the funicular cord to wrap around it several times from the force of the ejection; or (2) remains coiled inside the purse until the sticky hapteron snags onto an object and adheres to it, causing the funicular cord to then unwind and extend as the peridiole continues its travel under the momentum of the ejection. Recent evidence supports this **splash-and-grab** hypothesis, which is illustrated above in Figs. D–G.[1]

Under either scenario the peridiole is held fast until it eventually disintegrates and releases the spores. The funicular cord in two other bird's nest genera—*Nidula* and *Nidularia*—is much less developed and is not attached to the peridial wall. These peridioles are encased in a mucilaginous matrix, which causes the peridiole to adhere to anything it contacts after it is ejected. In either case, the peridioles are tethered to vegetation in locations where they are likely to be consumed by browsing herbivores. The spores would then be deposited in the animal's feces some distance away, where they would germinate.

Artillery Fungi (*Sphaerobolus*)

However, another bird's nest genus is a notable exception to the above process: *Sphaerobolus*. The so-called Artillery Fungus or Cannonball Fungus produces a single spore-containing spherical peridiole measuring no more than a millimeter in diameter. When mature, all layers of the peridium split radially to produce a jagged tooth-like margin reminiscent of the earthstars, to which they are related. This exposes the peridiole. These peridioles are forcefully shot out of the peridial cup for distances up to 2 m vertically and 6 m horizontally (6½ and 20 ft, respectively)[2]—not too shabby for a tiny projectile less than four-hundredths of an inch in diameter.

Sphaerobolus is not related to the foregoing genera and belongs to a different order: Geastrales, family

Geastraceae, which means it is related to the classic earthstars of the genus *Geastrum*. The two species of *Sphaerobolus* recorded in our area are *S. stellatus* Tode (1790) and *S. iowensis* L.B. Walker (1927).

The peridium consists of six layers: three form the outer exoperidium, two form the inner endoperidium, and the innermost layer dissolves into a liquid in which the single globose to ovoid peridiole is immersed. The innermost layer of the endoperidium is composed of radially oriented palisade cells that swell when they absorb water (the innermost textured grey layer in the graphic to the right). That layer contains the fluid (light grey) in which the peridiole (the oval) is immersed. The next layer out (also in the endoperidium) is more rigid and is composed of laterally arranged hyphae. The thick cross-hatched grey outer layer is the exoperidium. The inner palisade layer is densely packed with glycogen, which exerts little osmotic pressure. But this glycogen is reduced to glucose, mannitol, and trehalose, sugars that do exert powerful osmotic pressure. As these sugars accumulate, water diffuses into the cells osmotically, and the upper portion swells disproportionately relative to the lower regions (Fig. B). This allows considerable pressure to build up in the palisade layer, while the apex of the peridium continues to open up. When a critical point is reached, the accumulation of osmotic pressure in the palisade cells exceeds the containment of the opening peridium, the palisade layer suddenly everts with great force (often with an audible pop), and the peridiole is ejected with great force (Fig. C).

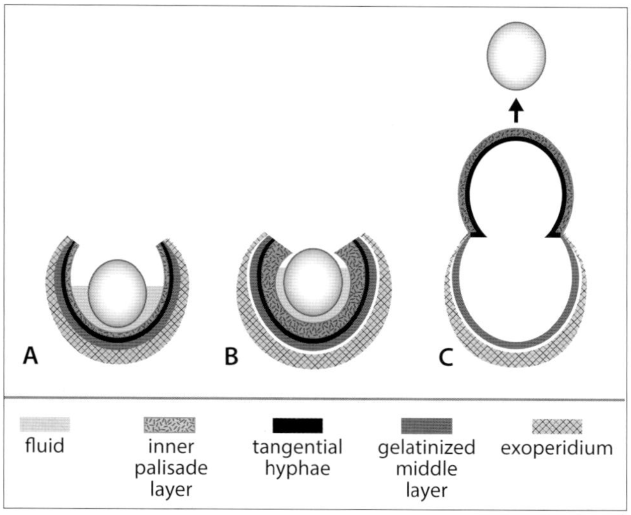

1. M.O. Hassett et al., 2013, *Fungal Biol.* 117(10):708–714. The peridioles of *Cyathus* and *Crucibulum* were ejected at speeds up to 5 m/s (16½ ft/s) for distances of up to 1 m (3.3 ft).
2. J. Geml et al., 2005, *Mycologia* 97(3):680–694.

Key to the Bird's Nest Genera

1. Mature fruitbody oval, 1–2 mm in diameter, exoperidium splitting into six to eight pointed, stellate rays (like a minute earthstar); containing a single ovoid peridiole, which is forcefully projected for up to 6 m (20 ft) .... *Sphaerobolus*

1′. Mature fruitbody cup-like or vase-shaped, margin not split into rays; containing lens-shaped ("egg-like") peridioles ... 2

2. Peridioles attached to inner wall of the peridium by funicular cord, peridioles not embedded in mucilaginous matrix ... 3

2′. Peridioles lacking funicular cord, peridioles embedded in mucilaginous matrix 4

3. Peridium distinctly 3-layered; funiculus complex, with filaments intertwined with spores......................*Cyathus*

3′. Peridium 1-layered, homogeneous; funiculus much simpler, filaments close together, not intertwined with spores........ .... *Crucibulum*

4. Peridium thick, with definite epiphragm that parts at maturity to reveal the urn-shaped concavity and peridioles........ .... *Nidula*

4′. Peridium thin, epiphragm lacking, cup-shaped, tearing open irregularly to reveal peridioles .... *Nidularia*

Fluted Bird's Nest Fungus, Splash Cups

Cyathus striatus

A B C D E

Photos: A, Teresa Fortenberry; B–E, W. Light.

***Cyathus striatus* (Hudson) Willdenow**, *Florae Berolinensis Prodromus*:399 (1787)
[Basidiomycota, Agaricomycotina, Agaricomycetes, Agaricomycetidae, Agaricales, Nidulariaceae]

Basionym: *Peziza striata* Hudson, *Fl. Angl.*, 2nd ed., 2:634 (1778)

Derivation: *Cyathus*, Greek, "cup," "drinking vessel"; *striatus*, Latin, "with furrows or grooves."

Fruitbody (peridium): Mature peridium an inverted, flaring, open, cone-shaped (Fig. B) cup 5–15 mm (¼–⅔ in) tall and wide; inner surface (endoperidium) faintly to deeply grooved (Figs. A–C), but otherwise smooth, brownish to grey-buff to deep brown; outer surface (exoperidium) shaggy or hairy, especially when young (Figs. D & E); young peridium covered by a hairy lid, or epiphragm (Fig. D), which wears away, leaving a thin whitish membrane across the cup opening (Fig. E), this eventually breaking off and exposing the egg-like peridioles at the bottom of the cup (Figs. A & C).

"Eggs" (peridioles): Clustered at the bottom of the cup, more or less disk-shaped or subtriangular, flattened, 1–1.5(–2) mm across, drab grey to dark (Figs. A & C), attached to endoperidium by mycelial cords forming a stalk-like funiculus; spore-bearing basidia are produced inside the flattened upper disk of the peridium.

Microscopic: Spores 15–20 × 8–20 µm, elliptical, smooth, clear, notched at one end.

Occurrence: Saprobic, in clusters on twigs, bark, woody debris, and wood chips in open woods and under shrubs, July–October; widespread in North America.

Edibility: Not considered edible, too tiny to be of interest.

Comments: This species is distinguished from other members of the genus by its hairy outer surface and smooth, vertically grooved inner walls. The walls of the peridium are sloped so that raindrops falling inside will dislodge and eject the spore-bearing peridioles and their funiculi from the cup, thus giving rise to one of its common names: Splash Cups. The sticky, detached end of the funiculus adheres to any surface the peridium may land on. The spores are released when the peridium dissolves or passes through the gut of an herbivore and out in the feces.

White-egg Bird's Nest

Crucibulum laeve

A

B

C

***Crucibulum laeve* (Hudson) Kambly**, *Gast. Iowa*:167 (1936)

Basionym: *Pezizza laevis* Hudson, *Fl. Angl.*, 2nd ed., 2:634 (1778) [Basidiomycota, Agaricomycotina, Agaricomycetes, Agaricomycetidae, Agaricales, Nidulariaceae]

Derivation: *Crucibulum*, Latin, a cup-shaped laboratory container for heating chemicals; *laeve*, Latin, "smooth."

Fruitbody (peridium): Mature peridium an inverted, open, cone-shaped cup (Figs. A–C) 5–10 mm (¼–⅖ in) tall and up to 15 mm (⅔ in) wide; outer surface (exoperidium) velvety (Fig. C), often becoming fairly smooth with age, initially yellowish, becoming brownish; inner surface (endoperidium) smooth, white, not furrowed (Fig. A); young peridium covered by a lid, or epiphragm (Fig. A, on left "nest"), which wears away to expose the "eggs."

"Eggs" (peridioles): Clustered at the bottom of the cup, light buff to whitish (Figs. A & B), more or less disk- or lens-shaped, flattened, 1.5–2 mm across, attached to endoperidium by a stalk-like funiculus.

Microscopic: Spores 4–10 × 4–6 µm, elliptical, clear, smooth.

Occurrence: Saprobic, on dead sticks, wood chips, and other woody debris, widespread in North America (Alaska to Mexico), Europe, South America, Australia, Japan; July–November.

Edibility: Not considered edible, too tiny to be of interest.

Comments: Species of the genera *Crucibulum* and *Cyathus* are distinguished from the other two nidulariacean genera, *Nidula* and *Nidularia*, by the stalk-like funiculus attaching the peridioles to the inner wall of the peridium. The pale buff to whitish peridioles of *C. laeve* and the smooth white endoperidium distinguish it from *Cyathus striatus*, with its dark-colored peridioles and furrowed, brownish to dark-brown inner surface. All other species of *Cyathus* also have dark-colored peridioles.

Raindrops dislodge and eject the peridioles together with their attached funiculi from the cup. The sticky, detached end of the funiculus adheres to any surface the peridium may land on. The spores are released when the peridium dissolves or passes through the gut of an herbivore and out in the feces.

Slime Molds

(Mycetozoa/Myxomycota)

Although no longer considered fungi, the slime molds are fascinating organisms that pique the interest of most mycologists and mycophiles, and they are occasionally presented in mushroom field guides. Of the two major (but unrelated) groups of slime molds, only the plasmodial, or "acellular," slime molds—the **Myxogastria (Myxomycota)**—are large enough to be seen with the naked eye. All of the slime mold species accounts in this book thus fall into this category.[1]

Plasmodial slime molds form a supercell, or **syncytium** (pl. syncytia), consisting of many individual diploid nuclei within a cytoplasmic matrix enclosed within a single cell membrane. In some of the aggregations (e.g., *Physarum polycephalum*), a single plasmodium can contain more than 10,000 nuclei. Such a syncytium is known as a **plasmodium** (pl. plasmodia), or fruiting body, which can range from a few millimeters in size to quite large—up to a few meters across.

Because they were historically considered to be fungi, their nomenclature at the generic and specific levels follows the rules of botanical nomenclature as it pertains to the fungi. However, a dual nomenclature exists at the family level and above, depending on whether the investigator's background is primarily mycological or zoological. We provide both the zoological and botanical names for the suprageneric taxa listed in each species account. But because the plasmodial slime molds are now in the phylum Amoebozoa[2] of the kingdom Protozoa, the zoological names are given priority in the species accounts, with the two names separated by a solidus (forward slash). This is shown as follows:

kingdom Amoebozoa, phylum Mycetozoa/Myxomycota
- class Protostelia/Ceratiomyxomycetidae
  - order Protosteliida/Ceratiomyxales
    - family Ceratiomyxidae/Ceratiomyxaceae
- class Myxogastria/Myxogastromycetidae
  - order Echinostelida/Echinosteliales [not treated here]
  - order Liceida/Liceales
    - family Tubiferidae/Tubiferaceae[3]
  - order Physarida/Physarales
    - family Physaridae/Physaraceae
  - order Stemonitida/Stemonitales
  - order Trichiida/Trichiales
    - family Trichiidae/Trichiaceae
    - family Arcyridae/Arcyriaceae

Not all authors follow the above classification, and you will find other names and other arrangements. But this classification appears to have the widest acceptance at present. Zoologists do not recognize the term *basionym* for the validly published name for a taxon having nomenclatural priority; hence, we have substituted "original combination."

Three basic types of plasmodia are recognized:

PROTOPLASMODIUM—microscopic its entire existence and producing only a single sporangium when fruiting.

APHANEROPLASMODIUM—initially similar to a protoplasmodium but soon elongating and branching into a network of very fine, almost invisible strands. Aphaneroplasmodia are characteristic of the Stemonitida/Stemonitales.

PHANEROPLASMODIUM—likewise starting out like a protoplasmodium but quickly expanding into a large mass of protoplasm, often consolidating into distinct tube-like veins of fluid endoplasm enclosed in a gel-like ectoplasmic membrane. This highly visible multicellular mass exhibits amoeboid movement as it searches for food. Phaneroplasmodia are characteristic of Physarida/Physarales.

A B C D

A, sporangium; B, aethalium; C, plasmodiocarp; D, pseudoaethalium. (Photos: A, B & D, W. Light; C, John Carl Jacobs.)

SCLEROTIUM—under adverse conditions the plasmodium can form a hardened dormant sclerotium that can emerge as a new plasmodium when favorable conditions return.

Plasmodial slime molds form the following types of fruiting bodies:

SPORANGIUM—one to several thousand sessile or stalked, variously shaped sporangia (Fig. A) can arise from a membranous protoplasmic layer, the hypothallus, which usually persists. A tangled network of thread-like fibers called the capillitium typically arises from a central columella or the outer peridiumal layer, which gives rise to the spores. Sporangia are characteristic of most slime molds.

PLASMODIOCARP—produced when the plasmodium becomes concentrated in its main veins and is covered by a peridium (Fig. C). The spores form on the capillitium inside. Plasmodiocarps are seen in *Physarum* and certain species of *Hemitrichia*.

AETHALIUM—a large and conspicuous, typically cushion-shaped sporophore composed of the entire undifferentiated plasmodium enclosed in a peridium (Fig. B). The spores form on the contained capillitium and are released when the peridium breaks open (dehices). Aethalia are best developed in *Fuligo* and *Lycogala*.

PSEUDOAETHALIUM—the individual sporangia retain their identity but are so closely crowded together that the entire structure resembles a true aethalium (Fig. D). Pseudoaethalia are seen in *Tubifera*.

1. See the "Mycetozoa/Myxomycota (slime molds)" section in the introduction to this book for an overview of these organisms. That section provides information on the higher taxonomic conundrums presented by this group and why both the botanical and zoological codes are simultaneously involved in their nomenclature.

2. Some workers treat the Amoebozoa as a kingdom in its own right.

3. This family is called the Reticulariaceae by the Integrated Taxonomic Information System.

Coral Slime

Ceratiomyxa fruticulosa var. *fruticulosa* and *C. fruticulosa* var. *porioides*

A B

A & inset, *Ceratiomyxa fruticulosa* var. *porioides*; B, *C. fruticulosa* var. *fruticulosa*. (Photos: A & inset, W. Light; B, Teresa Fortenberry.)

***Ceratiomyxa fruticulosa* var. *fruticulosa* (O.F. Müller) T. Macbride**, *North American Slime Molds* (New York):18 (1899)

Original combination: *Byssus fruticulosa* O.F. Müller, *Fl. Danica* 4(12):8 (1777)

***Ceratiomyxa fruticulosa* var. *porioides* (Albertini & Schweinitz) Lister**, in Lister, *Monogr. Mycetozoa* (London):26 (1911)

Original combination: *Ceratium porioides* Alb. & Schwein., *Consp. Fung.* (Leipzig):359 (1805)
[Amoebozoa, Eumycetozoa/Myxomycycota, Mycetozoa/Myxomycota, Protostelia/Ceratiomyxomycetidae, Protosteliida (Protosteliales)/Ceratiomyxales,[1] Ceratiomyxidae/Ceratiomyxaceae]

Derivation: *Ceratiomyxa*, Greek, from *kéras*, "horn" (via New Latin *cerata*), and *myxos*, "slime," "mucus"; *fruticulosa*, Latin, from *fruticosus*, "bush" or "shrub"; *porioides*, Greek, from *poros*, "pore," and *-oides*, "similar to."

Description: ***C. f.* var. *fruticulosa***—plasmodium emerging from substratum in the form of erect, simple,[2] slime-covered, pillar-like fruitbodies 1–10 mm tall by 0.5–1 mm thick (up to ~½ in tall) (Fig. B); fruitbodies translucent or yellowish; often forming effusive mats up to 100 cm (~3 ft) or more in extent; ***C. f.* var. *porioides***—fruitbodies poroid and honeycomb-like (Fig. A & inset), otherwise like those of *C. f.* var. *fruticulosa*.

Microscopic: Spores borne externally on the fruitbodies on individual, thread-like stalks. ***C. f.* var. *fruticulosa***—spores 8–15 × 6–10 μm, globose, teardrop-shaped to broadly elliptical, smooth, clear, white in mass; ***C. f.* var. *porioides***—spores 5–7 × 9–10 × 5–7 μm, smooth, clear, ovoid to elliptical, some globose, 6–7 μm in diameter.

Occurrence: Scattered to clustered on dead, rotten wood; plasmodium emerging from substratum in decorticated wood; cosmopolitan, subarctic to the tropics; summer–fall in North America.

Edibility: Inedible.

Comments: The coral slime mold usually appears on wet, rotten wood after a heavy rain in warm weather. The tiny fruiting bodies frequently form massive and readily visible spreads several centimeters across. The plasmodium of this slime mold exists entirely within the woody substratum where it lives. Under the right conditions it spreads out over the surface of the log in a slimy matrix. This matrix then turns white and erects tiny pillar-like fruitbodies, which are densely covered with minute spores[3] borne on individual thread-like stalks.[4]

The externally borne spores and their unusual and complex formation (see note 3) are unique to the Ceratiomyxidae/Ceratiomyxaceae, which is why it is often placed in the Protostelida/Protosteliales (but see note 1). The taxonomic placement of the four known species[5] of *Ceratiomyxa* continues to be difficult and problematic.

1. *Ceratiomyxa* has been considered a protostelid (Class Protostelia) slime mold by many authors because of its external spores and other characteristics discussed above. However, many recent authors have placed it with or near the Myxogastria/Myxomycetes (C. Rojas et al., 2008, *Mycologia* 100[6]:843–850 and references therein). See also L.H. Cavalcanti et al., 2008, *Brazilian Arch. Biol. Technol.* 51(5):971–980.

2. A branched form with up to five antler-like branches connected at the base represents a separate variety, *Ceratiomyxa fruticulosa* var. *flexuosa* (Lister) Lister; this variety is also found in our area in the southern Appalachians.

3. Each spore contains four haploid (*n*) nuclei, derived by meiosis from a diploid (2*n*) protospore, and germinates to produce an elongated **thread stage**, which in turn rounds up into a 4-lobed **tetrad**. This tetrad then divides mitotically to produce an 8-lobed **octet**, and each lobe then becomes a flagellated swarm cell. Two swarm cells fuse to reestablish the diploid (2*n*) condition. The swarm cells lose their flagella to become myxamoeba, which can form a new plasmodium.

4. The stalked spores may actually be sporangia rather than individual spores (R.W. Scheetz, 1964, *Mycologia* 64[1]:38–54).

5. The other three species are much rarer than *C. fruticulosa*, and all are restricted to the tropics.

Dog Vomit Slime, Scrambled Eggs Slime

Fuligo septica

A, plasmodium; B & C, surfaces of aethalia showing interwoven sporangia; D, mature aethalium showing cortex and exposed spore mass; E, mature aethalium with slime trail. (Photos: A–C & E, W. Light; D, Stefanie Hanagud.)

***Fuligo septica* (Linnaeus) F.H. Wiggers**, *Prim. Fl. Holsat.* (Kiliae):112 (1780)

Original combination: *Mucor septicus* Linnaeus, *Sp. Pl.*, 2nd ed., 1656 (1763)
[Amoebozoa, Mycetozoa/Myxomycota, Myxogastria/Myxogastromycetidae, Physarida/Physarales, Physaridae/Physaraceae]

Derivation: *Fuligo*, Latin, "lamp-black," referring to the inner spore mass of the mature aethalium; *septica*, Greek, from *sēptikos*, "rotten."

Plasmodium: Up to 50 cm long by 20 cm wide by up to 2.5 cm thick (20 × 8 × 1 in) or even larger,[1] but most specimens considerably smaller; plasmodium a yellowish, whitish, yellowish-white (Figs. B & C), or yellow-green, irregular, slimy mass (Fig. A) slowly creeping[2] over the ground, logs, tree stumps, and other objects, often leaving a distinct trail of tangled white strands (Fig. E) in its wake.

Aethalium: Pulvinate (cushion-shaped), varying in size from 2 mm to 20 cm (¾–8 in) across; yellowish (Fig. B), greyish-white flesh (Fig. C), pinkish to reddish-brown (Figs. C & D); composed of intricately interwoven and anastamosing (repeatedly branching and reconnecting) sporangia 2–2.5 mm thick with intervening air cavities (Figs. B & C), forming a brittle, calcareous[3] cortex (peridium), which eventually crumbles away to expose the purplish-black spore mass within (Fig. D).

Microscopic: Spores 6–9 µm, spherical, 2-walled, the outer wall minutely spiny; pinkish-brown, dull purplish-black in mass (Fig. D).

Occurrence: Arising on rotten wood, wood chips, litter, or compost, then migrating and consuming a wide variety of microorganisms and organic matter. Cosmopolitan; spring–fall.

Edibility: Edible. Often prepared like scrambled eggs,[4] which, together with its appearance, gives rise to its other common name, Scrambled Eggs Slime.

Comments: The massive cushion-shaped aethalium of *F. septica* with its millions of spores contained inside a crusty calcareous cortex can hardly be mistaken for anything else, with the possible exception of the Many-headed Slime, *Physarum polycephalum* Schweinitz. *Physarum polycephalum* superficially resembles the Dog Vomit Slime, *Fuligo septica* (Linnaeus) F.H. Wiggers, and could be confused with it—especially when it is aggregated in an amorphous mass over a food source. However, it is readily distinguished from *F. septica* by the finger-like exploratory moving fronts with their tube-like veins. *Fuligo septica* typically occurs as a consolidated and vertically rounded cushion-like mass that lacks the finger-like veins so characteristic of *P. polycephalum* (see the following species account).

This slime mold can absorb and withstand levels of zinc (4,000–23,000 parts per million) that would kill any other living organism.[5] This extraordinary resistance to zinc is due to fuligorubin A, a yellow pigment that gives the slime mold its characteristic color and that chelates metals into inactive forms. Fuligorubin A occurs as a stable calcium complex in the aethalium and is thought to function as a photoreceptor in the phototaxis[6,7] and sporulation of this organism. This pigment may also protect the vulnerable plasmodium against microbial attack.[7]

1. One specimen measured 70 by 54 by 3 cm (28 × 21½ × 1¼ in). See H.W. Keller & K.L. Braun, 1999, *Myxomycetes of Ohio: Their Systematics, Biology and Use in Teaching*, Ohio Biological Survey Bulletin, n.s., vol. 13, no. 2, xvi + 182 pp.

2. The maximum speed measured is 2.5 cm (1 in) per hour.

3. The sporangia lay down amorphous and granular deposits of calcium carbonate referred to as "lime knots."

4. In the state of Veracruz, Mexico, it is grilled and is known as *caca de luna* (politely translated as "moon poop").

5. A. Setala & P. Nuorteva, 1989, *Karstenia* 29(1):37–44; D.A. Zhulidov et al., 2002, *J. Environm. Qual.* 31(3):1038–1042.

6. The plasmodium is positively chemotaxic and negatively phototaxic; that is, it moves toward specific chemical signals that indicate food sources and away from light. *Physarum polycephalum* exhibits the same responses to light and chemicals, but the responsible pigment is physarorubinic acid.

7. See A. Gossauer, 2003, Mycropyrrolic natural compounds including tetramic acid derivatives, *in* W. Herz, H. Falk, & G.W. Kirby (eds.), *Fortschritte der Chemie organischer Naturstoffe* (New York: Springer), 88.

Many-headed Slime

Physarum polycephalum

A

B C D

A, actively feeding plasmodial stage; B, plasmodial veins in which oscillatory "shuttle streaming" of protoplasm is taking place (arrow indicates sedentary exploitation phase); C, early development of fruitbodies (sporangia); D, plasmodium showing major vein and fan-shaped exploratory moving front (arrow).

Physarum polycephalum **Schweinitz**, *Schr. Naturf. Ges. Leipzig* 1:62 (1822)
[Amoebozoa, Mycetozoa, Myxogastria/Myxomycota (Myxomycetae), Myxogastromycetidae, Physarida/Physarales, Physaridae/Physaraceae]

Derivation: *Physarum*, Greek, New Latin modification of *physaró*, "bellows"; *polycephalum*, Greek, combining form of *polýs*, "many," and *kefáli*, "head," "many-headed."

Plasmodium: A many-branched, spreading, amorphous, yellowish (Fig. A), greenish-yellow, or yellowish-grey (Fig. B) spreading mass with many branching and anastomosing (alternately separating and recombining), net-like veins (Figs. B & D), often with a fan-shaped moving front with leading pseudopodia (Fig. D); plasmodium producing countless vertical head-like knobs (Fig. C), these lengthening into finger-like columns and eventually forming stalked sporangia tipped with blackish, helvelloid[1] clusters of spores; plasmodium often spread over many square centimeters (Fig. A) or occasionally even square meters.[2]

Microscopic: Spores 9–11 µm, spherical, covered with minute spines, violet-brown; black in mass.

Occurrence: On dead wood, leaves, the fruitbodies of other fungi,[3] and adjacent surfaces in shady, cool, moist areas, typically in forests; widespread in North America; cosmopolitan; summer–fall.

Edibility: Inedible.

Comments: *Physarum polycephalum* superficially resembles the Dog Vomit Slime, *Fuligo septica* (Linnaeus) F.H. Wiggers, and could be confused with it—especially when it is aggregated in an amorphous mass over a food source, as shown in Fig. A above. However, it is readily distinguished from *F. septica* by the finger-like exploratory moving fronts with their tube-like veins (Figs. B & D). *Fuligo septica* typically occurs as a consolidated and vertically rounded cushion-like mass that does not develop the finger-like veins that characterize *P. polycephalum* (see the preceding species account).

The plasmodium of *Physarum polycephalum* is a single gigantic, multinucleate, amoeboid cell (i.e., a syncytium formed by multiple successive mitotic divisions of the nucleus of an original zygote, often coupled with the aggregation of many independent zygotes) that moves and functions as a single organism. This giant macroplasmodium is the dominant stage in the life cycle of *P. polycephalum*, which exhibits coordinated amoeboid movement about various substrata at speeds up to 5 cm (2 in) per hour[4] as it searches for bacteria, other protists and microorganisms, and organic matter, which it engulfs amoeba-style by endocytosis.

Propelled by alternating expansions and contractions of the cortical gel layer (ectoplasm) enclosing the tube-like veins, the more fluid inner protoplasm (endoplasm) flows characteristically back and forth, reversing direction every one and a half to three minutes or so—a process known as "shuttle streaming." This enables the plasmodium to aggregate around discovered food sources (the swollen exploitation phases shown in Figs. B & D) while simultaneously sending out fan-like exploratory moving fronts (Fig. D). Shuttle streaming also circulates nutrients and chemical signals throughout the plasmodium and permits a coordinated response to environmental stimuli.

The plasmodium is positively chemotaxic and negatively phototaxic; that is, it moves toward specific chemical signals that indicate food sources and away from light. When food resources become scarce and in the presence of light, the plasmodium begins to produce specialized spore-generating sporangia. These eventually produce haploid (n) spores by meiosis, which, when released, fuse to form a diploid ($2n$) zygote that then undergoes repeated mitotic nuclear divisions to give rise to a new macroplasmodium.

This slime mold grows easily in culture on oatmeal and is often used in classrooms to teach students as early as middle school. When actively foraging, *P. polycephalum* leaves behind a trail of extracellular slime, which allows it to avoid those areas it has previously explored. The giant supercell thus exhibits a form of spatial memory that enables the slime mold to navigate complex mazes and determine the shortest possible route to a food source located behind impassable barriers.[5] This organism is also able to anticipate repeated cyclic events. And when oat flakes or other food was placed on maps in the same positions as major urban areas in Japan and Europe, *P. polycephalum* created interconnected veins that almost exactly matched the man-made railways and roads between these areas.[6]

1. Multiple convoluted, brain-like lobes faintly reminiscent of fused clusters of grapes.
2. E. Bernitt et al., 2010, Microplasmodium dynamics of *Physarum polycephalum*, *IFMBE Proceedings* 31:1133–1136 (Sixth World Congress of Biomechanics, August 1–5, 2010, Singapore).
3. The plasmodium in Fig. D was covering several dead sporocarps of a Resinous Polypore, *Ischnoderma resinosum*, and was expanding into adjacent territory.
4. D. Kessler, 1982, Plasmodial structure and motility, *in* H.C. Aldrich & J.W. Daniel (eds.), *Cell Biology of* Physarum *and* Didymium, vol. 1 (New York: Academic Press), 145–208.
5. C.R. Reid et al., 2012, *Proc. Nat. Acad. Nat. Sci. USA* 109(43):17490–17494.
6. T. Atsushi et al., 2010, *Science* 327(5,964):439–442; A. Adamatsky & J. Jones, 2010, *Internat. J. Bifurcation and Chaos* 20(10):3065–3084.

Wolf's Milk Slime, Pink Toothpaste Slime

Lycogala epidendrum

A

B

C

***Lycogala epidendrum* (J.C. Buxbaum *ex* Linnaeus) Fries**, *Syst. Mycol.* (Lundae) 3(1):80 (1829)

Original combination: *Lycoperdon epidendrum* J.C. Buxbaum *ex* Linnaeus, *Sp. Pl.* 2:1184 (1753)
[Amoebozoa, Mycetozoa/Myxomycota, Myxogastria/Myxogastromycetidae, Liceida/Liceales, Tubiferidae/Tubiferaceae][1]

Derivation: *Lycogala*, Greek, from *lýkos*, "wolf," and *gála*, "milk," "wolf's milk"; *epidendrum*, Greek, from *epi*, "upon," and *dendron*, "tree," referring to its growth habit.

Fruitbody (aethalium): Plasmodium developing into globose or hemispherical, cushion-shaped aethalia, each 3–15 mm in diameter (~1⁄10–2⁄3 in); cortex (peridium) densely covered with minute papillae, initially pinkish (Fig. A), becoming grey or greyish-brown (Fig. B) with maturity; peridium enclosing a pinkish to pinkish- or reddish-orange semifluid paste, which oozes out when the aethalium is ruptured (Fig. C), this paste later becoming powdery and pinkish-grey or beige.

Microscopic: Spores 6–7.5 µm, spherical, with low reticulated ridges; grey in mass, fading to beige.

Occurrence: Scattered to crowded on well-decayed wood; widespread and common; cosmopolitan; summer–late fall.

Edibility: Inedible.

Comments: This common and distinctive slime mold cannot be mistaken for anything else. One of the common names, Pink Toothpaste, derives from the color and texture of the contents of the pink, immature fruitbodies when ruptured. *Lycogala flavofuscum* (Ehrenberg) Rostafinsky is much less common and is distinguished from *L. epidendrum* by its larger diameter (2–4 cm or even larger [¾–1⅔ in]), its smooth, glossy, and silvery-grey to ochre-colored peridium, and its occurrence on living trees.

1. This genus is placed in the family Reticulariaceae by the Integrated Taxonomic Information System.

Red Raspberry Slime

Tubifera ferruginosa

Photos: A & C, W. Light; B & D, Jim Edwards.

Tubifera ferruginosa (Batsch) J.F. Gmelin, in Linnaeus, *K. Svenska Vetensk-Akad. Handl.* 2:1472 (1791) [1792]

Original combination: *Stemonitis ferruginosa* Batsch, *Elench. Fung.*, cont. prim. (Halle):261 (1786)
[Amoebozoa, Mycetozoa/Myxomycota, Myxogastria/Myxogastromycetidae, Liceida/Liceales, Tubiferidae/Tubiferaceae][1]

Derivation: *Tubifera*, Latin, from *tubus*, "tube," and *-iferus*, "bearing," "tube-bearing"; *ferruginosa*, Latin, from *ferrugo*, "rust," and *-osus*, "abounding in," "rust-colored."

Fruitbody (pseudoaethalium): Fruitbody consisting of a raspberry-like pseudoaethalium up to 15 cm (6 in) in length formed by many columnar sessile sporangia (Fig. D) crowded together into a cushion-shaped mass (Figs. A–C); pseudoaethalia variously colored depending on life stage: initially ochre or rust-colored (Fig. B), becoming pink (Fig. C), red, and finally dark purplish-black (Fig. A) or brown; persistent hypothallus well-developed, spongy, off-white (Fig. D, arrow) to brownish; individual sporangia up to 5 mm tall by 0.5 mm thick, opening at tip to release spores.

Microscopic: Spores 6–8 µm, spherical, double-walled, reticulated, light yellow to brownish; dark brown in mass.

Occurrence: On dead wood, leaves, or litter; widespread in North America; cosmopolitan, all continents except Antarctica; summer–midwinter.

Edibility: Inedible.

Comments: Although the individual sporangia retain their identity, they are so closely compressed that the entire structure resembles a true aethalium like that formed by the Dog Vomit, *Fuligo septica*, or Wolf's Milk, *Lycogala epidendrum*, slimes. In the latter two cases, the fruiting body is a true aethalium formed from most or all of the original plasmodium; individual sporangia are lacking.

1. This genus is placed in the family Reticulariaceae by the Integrated Taxonomic Information System.

Pretzel Slime

Hemitrichia serpula

Photo: John Carl Jacobs.

***Hemitrichia serpula* (Scopoli) Rostafinski**, *Vers. Syst. Mycetozoen* (Strassburg):14 (1873)

Original combination: *Mucor serpula* Scopoli, *Fl. Carniol.*, 2nd ed. (Wien) 2:493 (1772)
[Amoebozoa, Mycetozoa/Myxomycota, Myxogastria/ Myxogastromycetidae, Trichiida/Trichiales, Trichidae/ Trichiaceae]

Derivation: *Hemitrichia*, Greek, from *hemi*, "half," and *trícha*, "hair"; *serpula*, Middle Latin, *serpere*, "to creep."

Fruitbody (plasmodiocarp): Yellowish-white plasmodium forming a mustard-yellow plasmodiocarp (phaneroplasmodium) appearing as a pretzel-like network of vein-like or tube-like channels 0.4–0.6 mm in diameter; individual net meshes up to 5 mm (¼ in) across, the entire plasmodiocarp sometimes covering an area up to 100 cm^2 (about 15 in^2), but usually smaller; darkening to golden-brown with age; plasmodiocarp peridium (cortex) breaking down at maturity to expose a cottony mustard-yellow capillitium containing the spores; underlying hypothallus thin, brown.

Microscopic: Spores 11–16 µm, spherical, coarsely reticulate, pale yellow; mustard-yellow in mass.

Occurrence: On dead wood, leaves, and litter; widespread but uncommon in North America; cosmopolitan; summer.

Edibility: Inedible.

Comments: With its distinctive pretzel-like plasmodiocarp, *Hemitrichia serpula* is unlikely to be confused with any other slime mold. The term *plasmodiocarp* refers to a sessile, branching, ring-shaped or net-like system in which the plasmodium is concentrated in a series of main tube-like veins that do not break up into smaller units when fruiting.

Multigoblet Slime, Wasp Nest Slime

Metatrichia vesparium

Photos: A & C, W. Light; B, Dan Molter.

Metatrichia vesparium **(Batsch) Nannenga-Bremekamp**, *Proc. K. Ned. Akad. Wet.*, ser. C, *Biol. Med. Sci.* 69:348 (1966)

Original combination: *Lycoperdon vesparium* Batsch, *Elench. Fung.*, cont. prim. (Halle):253 (1786)
[Amoebozoa, Mycetozoa/Myxomycota, Myxogastria/ Myxogastromycetidae, Trichiida/Trichiales, Trichidae/ Trichiaceae]

Derivation: *Metatrichia*, Greek, from *metá*, "after," "higher," and *trícha*, "hair," "after [the genus] *Trichia*"; *vesparium*, Latin, "wasp nest," referring to the honeycombed appearance of the dehiced (split-open) sporangia.

Fruitbody (sporangium): Sporangia up to 2.5 cm (1 in) tall, stalked or nearly sessile, arising from a yellowish-white hypothallus, usually clustered or crowded, rarely single; club-shaped to subcylindrical, dark wine-red to blackish-grey, with a dome-shaped lid, which dehices to release the contained capillitium, leaving a resistant cup-like peridium (Figs. A & C); up to a dozen or so sporangia can be connected to a single, united stalk; capillitium brick-red to reddish-brown, consisting of intertwined and entangled elastic threads (elaters) (Fig. B), these spiny, rarely branched. Plasmodium black (Fig. A, arrow) or deep red.

Microscopic: Spores subglobose to broadly elliptical, minutely warted, 9–13 µm; spores brownish-red in mass.

Occurrence: Growing on decaying deciduous wood or bark; widespread and common in Northern Hemisphere; North America, Europe, Asia; less common in Mexico, South America, West Africa.

Edibility: Inedible.

Comments: The persistent, dehiced, cup-shaped peridium remaining after the capillitial contents have been discharged looks for all the world like a minute wasp's nest (Figs. A & B), permitting an immediate identification. This is the only species of *Metatrichia* in our area.

Deceptive Trichia

Trichia decipiens

A

B

***Trichia decipiens* (Persoon) T. Macbride**, *North American Slime Molds* (New York):218 (1899)

Original combination: *Arcyria decipiens* Persoon, *Ann. Bot. Usteri* 15:35 (1795)
[Amoebozoa, Mycetozoa/Myxomycota, Myxogastria/Myxogastromycetidae, Trichiida/Trichiales, Trichiidae/Trichiaceae]

Derivation: *Trichia*, Greek, *trícha*, "hair"; *decipiens*, Latin, from *decipere*, "to deceive."

Fruitbody (sporangium): Stalked, 1.5–3 mm tall by 0.6–0.8(–1.3) mm wide (⅔–1¼ × ¼–½ in), obovate to inversely pear-shaped, distinct calyculus lacking; smooth, but pleated at the stalk (Fig. B), very shiny, initially orange-yellow or orange-red (Fig. A), becoming shiny olive (Fig. B), yellow-olive, or brownish, peridium splitting open at apex to expose capillitium.

Stalk: 0.1–1 mm long, translucent, vertically ridged, at first pallid (Fig. A), becoming darker olive to brown (Fig. B), filled with spore-like cysts (Fig. B, inset).

Capillitium: A dense cottony network of mostly simple (occasionally branched), smooth, olive-yellow to olive-brown, thread-like elaters 4.5–6 µm in diameter, gradually tapering to a long (75–150 µm), slender, pointed tip; elaters with three to five spiral bands.

Hypothallus: Membranous, shiny, colorless or brown.

Plasmodium: White, pink, or orange.

Microscopic: Spores spherical to ovoid, 9–13 µm in diameter, delicately reticulated or minutely warted; olive-yellow in mass.

Occurrence: Growing on dead wood; scattered to densely crowded; widespread and common in North America; cosmopolitan.

Edibility: Inedible.

Comments: *Trichia decipiens* is distinguished by its shiny, membranous, olive-colored peridium and translucent stalk filled with spore-like cells. No other stalked *Trichia* species has this combination of characters.[1]

Trichia decipiens can be confused with both of the yellow-fuzz cone slimes, *Hemitrichia clavata* (Persoon)

Rostafinski and *H. calyculata* (Spegazzini) M.L. Farr, whose sporangia also exhibit various shades of red, orange, or yellow when young and become olive-green or brownish when mature. *Hemitrichia clavata* also has a stalk filled with spore-like cysts, but the emerged capillitial threads are mostly branched (rather than unbranched), and the spores are smaller (7–9 μm) and finely warted (not reticulate). *Hemitrichia calyculata* has similar-size spores to *H. clavata*, the capillitial threads are rounded and untapered (unlike the long tapering tips of *T. decipiens*); macroscopically, the opaque dark reddish-brown to black stalk is expanded into a distinctly demarcated cup-like calyculus with an out-turned petaloid rim that is not present in either *T. decipiens* or *H. clavata*.

1. S.L. Stephenson, 2003, Myxomycetes of New Zealand, *Fungi of New Zealand/Ngā Harore o Aotearoa*, no. 3 (Hong Kong: Fungal Diversity Press), xiv + 238 pp.

Petal-cone Slime

Hemitrichia calyculata

A

B

Photos: A, W. Light; B, Ehecatl (Alonso).

***Hemitrichia calyculata* (Spegazzini) M.L. Farr**, *Mycologia* 66(5):887 (1974)

Original combination: *Hemiarcyria calyculata* Speg., *Anal. Soc. Cient. Argent.* 10(5–6):152 (1880)
[Amoebozoa, Mycetozoa/Myxomycota, Myxogastria/ Myxogastromycetidae, Trichiida/Trichiales, Trichidae/ Trichiaceae]

Derivation: *Hemitrichia*, Greek, from *hemi*, "half," and *trícha*, "hair"; *calyculata*, Latin, from *calyculus*, "calyx," "shell," referring to the cup-like peridium.

Deprecated synonym: *Hemitrichia stipitata* (Massee) T. Macbride

Fruitbody (sporangium): Fruitbody an erect, stalked sporangium 1–3 mm tall; sporotheca top-shaped to subglobose, 0.8–1.4 mm in diameter when fully expanded; various shades of red, orange, or yellow when young, becoming olive-green or brownish when mature (Fig. A, dashed arrow), nearly identical to that of *Trichia decipiens*.

Peridium: Persisting as a thin, membranous, yellowish, funnel-shaped calyx at the top of the stipe (Fig. B, arrow), typically with an out-turned petaloid margin (Figs. A, solid arrow, & B).

Stalk: Thin, reddish-brown to blackish, 0.5–2.0 mm tall, expanded at apex into a cup-like calyx with a flaring petaloid rim (Fig. B, arrow), this feature diagnostic.

Capillitium: A dense network of simple and branching, elastic, yellow to dull orangish filaments 6–7 µm in diameter, each with four to five spiral bands and *untapered* rounded tips.

Hypothallus: Irregular to disk-shaped; pale brown to dark reddish-brown.

Plasmodium: Yellow, becoming pinkish-red when fruiting.

Spores in mass: Dull yellow to ochraceous.

Microscopic: Spores 6.5–8 µm, globose, minutely reticulate; pale yellow.

Occurrence: Decaying wood of hardwood trees; widespread and common in southeastern North America; cosmopolitan.

Edibility: Inedible.

Comments: *Hemitrichia calyculata* can be confused with *H. clavata* (Persoon) Rostafinski and *Trichia decipiens* (Persoon) T. Macbride, both of which also occur in our area and whose sporangia also exhibit various shades of red, orange, or yellow when young and become olive-green or brownish when mature. Unlike *H. calyculata*, *H. clavata* and *T. decipiens* both have stalks filled with spores; in addition, neither species exhibits the flared, petaloid rim on the calyx after the capillitium is exposed that is seen in *H. calyculata*. See also the distinguishing comments in the account of the Deceptive Trichia, *T. decipiens*.

Cotton Candy Slime

Arcyria denudata

A B C

***Arcyria denudata* (Linnaeus) Wettstein**, *Verh. Zool.-Bot. Ges. Wien* 35:353 (1886)

Original combination: *Clathrus denudatus* Linnaeus, *Sp. Pl.* 2:1179 (1753)
[Amoebozoa, Mycetozoa/Myxomycota, Myxogastria/ Myxogastromycetidae, Trichiida/Trichiales, Arcyridae/ Arcyriaceae]

Derivation: *Arcyria*, Greek, "fine network"; *denudata*, Latin, "stripped."

Fruitbody (sporangium): Fruiting body a stalked sporangium arising from a sheet-like, contiguous, silvery hypothallus; 2–6 mm tall by 0.5–1.2 mm wide ($\frac{1}{12}$–$\frac{1}{4} \times \frac{1}{25}$ in), gregarious (Fig. A) to crowded; with pinkish to brick-red cottony capillitium (Figs. A–C) composed of densely intertwined threads (Fig. C, inset), each 3–4 µm in diameter; capillitium fading to brownish or yellow with age (Fig. C, inset, arrow), firmly attached to a shallow, cup-like, peridial calyculus.

Microscopic: Spores spherical, 6–8 µm in diameter, with scattered warts; red to reddish-brown in mass.

Occurrence: On dead deciduous wood; widespread in North America; cosmopolitan, summer–late fall.

Edibility: Inedible.

Comments: This slime mold looks for all the world like a tiny pink ball of cotton candy. It is the pink *Arcyria* species most commonly encountered. However, it can be difficult to distinguish it from several look-alikes: *Arcyria ferruginea* Saut, *A. incarnata* (Persoon *ex* G.F. Gmelin) Persoon, and *A. insignis* Kalchbrenner & Cooke.

In *A. denudata* the capillitium is firmly attached to the entire inside of the calyculus; the spores are 6–8 µm in diameter with a few scattered warts, colorless under the microscope, and red to reddish-brown in mass.

In *A. ferruginea* the capillitium is connected only to the apex of the stalk; the spores are pale rusty and 9–11 µm in diameter.

In *A. incarnata* the capillitium is very loose, expanded, and disheveled in appearance and is loosely connected to the stalk apex and easily detached; the spores are pale pink in transmitted light (i.e., viewed by light microscopy) and salmon-pink in mass.

Arcyria insignis is flesh-colored to pale pink. The capillitium is a close network of delicate threads 2–3 µm in diameter; the spores are similar to those of *A. denudata*.

Japanese Lantern Slime

Cribraria cancellata

A B C

Photos: A & B, Jim Edwards; C, Helen Ginger.

Cribraria cancellata (Batsch) Nann.-Bremek., *Nederlandse Myxomyceten* (Amsterdam):92 (1975) [1974]

Original combination: *Mucor cancellatus* Batsch, *Elench. Fung.*, cont. sec. (Halle):135 (1789)

Derivation: *Cribraria*, Latin, "sifted," "sieved"; *cancellata*, "reticulated," "lattice-like."

Deprecated synonym: *Dictydium cancellatum* (Batsch) T. Macbride, *N. Amer. Slime-Moulds*, 1st ed. (New York):230, table 10, fig. 2 (1899)
[Amoebozoa, Mycetozoa, Myxogastrea, *incertae sedis*, Liceida, Cribrariaceae]

Fruitbody (sporangium): Sporocarp stalked, 5–6 mm (~¼ in) tall. Sporangium subglobose, 0.3–0.7 mm in diameter, often nodding, hollow, consisting of a fragile net-like mesh composed of many rigid longitudinal ribs arising from the top of the stalk and uniting again at the top of the sporangium (Fig. C), these connected by numerous thin transverse filaments; attached to a long thin stalk, generally erect, twisted and sometimes bent at the apex so that the sporangium hangs inverted (Figs. A & B), dark brown to purplish-brown, 2–10 times the height of the sporangium; calyculus typically lacking or rudimentary, but occasionally present as a thin film.

Microscopic: Spores 5–7 µm, smooth, but with one or two adhering plasmodial granules; spores in mass varying shades of brown to purplish, reddish in transmitted light.

Occurrence: Saprotrophic on rotten wood. Often densely gregarious (Figs. A & B). Cosmopolitan, widely distributed throughout the world.

Edibility: Inedible.

Comments: This species is not likely to be confused with any other. The sporangium hanging from its stalk resembles a classic Japanese paper lantern. Most North American workers place this taxon as the sole species in the monotypic genus *Dictydium* (Schrader) Rostafinski, whereas most European authors include it within the genus *Cribraria* (Persoon) Schrader. The major distinction has been based on the presence or absence of a calyx. A recent phylogenetic study of the family Cribrariaceae noted that species with features intermediate between *Cribraria* and *Dictydium*—also seen within various populations of *C. cancellata*—make separating these two genera dubious.[1]

1. J.M. Ramírez-Ortega et al., 2009, *Mycotaxon* 110:331–355.

Chocolate Tube Slimes

Stemonitis spp.

A & C, *Stemonitis splendens*; B, *S. axifera*; D, *S. fusca*. (Photos: A & C, W. Light; B, Dr. Lorne Stobbs; D, River Bissonnette.)

Genus *Stemonitis* Gleditsch, *Method. Fung.*:140 (1753); type species: *Stemonitis fusca* Willdenow, *Fl. Berol. Prodr.*:408 (1787) [Amoebozoa, Mycetozoa/Myxomycota, Myxogastria/Myxogastromycetidae, Stemonitida/Stemonitales, Stemonitidae/Stemonitidaceae]

Derivation: *Stemonitis*, Greek, from *stimónas*, "stamen," referring to the general shape of the sporangium.

Fruitbodies (sporangia): Sporangia hair-like, with tufted, slender, cylindrical sporothecae resembling tiny cattails (Figs. A & B), 0.5–2.5 cm tall by 1 mm thick ($\frac{1}{2}$–1 × $\frac{1}{25}$ in), depending on species, often closely aggregated in a dense cluster (Fig. A); sporangia arising from a sheet-like hypothallus (Fig. D, arrow at bottom), this typically silvery or brown; sporothecae variously brownish-grey to smoky-drab, dark brown, rusty-brown, chocolate-brown, or reddish-brown, depending on the taxon, often becoming paler after shedding spores; stalk typically shiny jet-black, comprising various fractions of the total length of the sporangium, depending on species, stalk extending into sporotheca as the columella; immature sporangia milky-white, resembling a mass of insect eggs (Fig. C) composed of a spore-bearing capillitium, or network, formed from repeated lateral and apical hair-like branchings of a central columella (Fig. E, columella and capillitium after spores were shed).

Microscopic: Spores spherical to subglobose, borne on capillitium (Fig. E), nearly smooth or faintly or strongly warted, warts may be reticulated.

Spore print: Spores in mass dusky- to brownish-violet, depending on species.

Occurrence: On decayed wood and leaves, spring–late fall; most species cosmopolitan, all continents except Antarctica.

Edibility: Inedible.

Comments: In *Stemonitis*, the plasmodium produces a flat, sheet-like hypothallus on the substratum (Fig. D), which gives rise to numerous individual sporangia. The sporangia are initially milky-white (Fig. C) but soon develop a "stalk initial" as protoplasm moves from the plasmodium into the developing sporangium. The stalk then moves upward as the columella and gives rise to the lateral and apical branches that form the fine, net-like capillitium (Fig. E), which produces the variously colored spores characteristic of each species.

The species of *Stemonitis* are not easily distinguished, even using microscopic features.

Stemonitis splendens Rostafinsky and *S. fusca* Willdenow[1] are very similar. Both have sporangia up to 2–2.5 cm (¾–1 in) tall that tend to aggregate (Fig. A) and similar-size spores (7–9 µm).

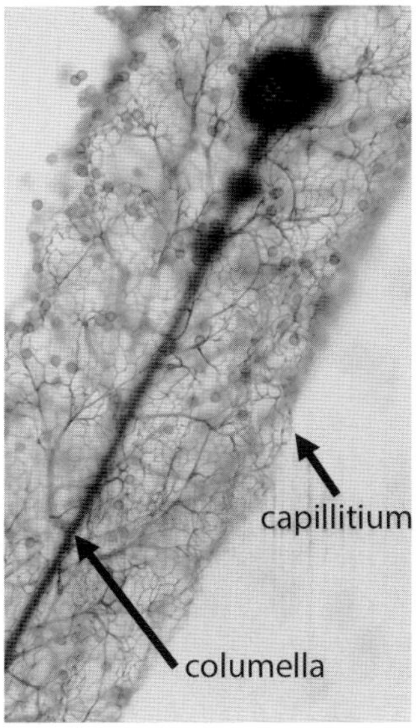

E

Columella and capillitium of *Stemonitis splendens* after spores were shed. (Micrograph: W. LIght)

However, the stalks of *S. fusca* can be nearly half (rather than one-fourth) as long as the entire sporangium, and the spores are prominently to finely warty-reticulate rather than nearly smooth. The sporothecae arise from a brown hypothallus and vary from deep brownish-grey to smoky-drab, which become paler as the spores are shed. In *S. splendens* the spores are nearly smooth, with fine pale warts, and the sporothecae are dark brown to reddish-brown and arise from a shiny, silvery hypothallus.

Stemonitis axifera (Bulliard) T. Macbride has smaller (5.0–7.5 µm) pale rose-brown spores with very fine warts and rusty-brown sporothecae that become pale brown after shedding the spores.

All three species are cosmopolitan.

1. *Stemonitis fusca* Willdenow, *Fl. Berol. Prodr.*:408 (1787); *Stemonitis fusca* Roth, *in* Roemer & Usteri, *Mag. Bot.* 1(2):26 (1787) is a later homonym and is thus an illegitimate name (*ICN* Article 53.1).

Glossary of Some Mycological Terms

abhymenial: Adjective referring to the surface on the opposite side of a fruitbody from the hymenium.

aculeate: A broadly pitted surface.

adnate: Gills attached to the stipe across their total width.

adnexed: Gills tapering toward and narrowly attached to the stipe.

adpressed, appressed: Pressed closely to or flat against a surface.

aereolate: A cake-like pattern similar to dried mud.

aethalium: In a slime mold, a large, conspicuous, and typically cushion-shaped mass composed of the entire undifferentiated plasmodium enclosed in a peridium (cuticle or outer layer of tissue).

aff., *affinis*: Latin, "related to."

agaric, agaricoid: A general term for a gilled mushroom with a central stipe.

allantoid: Sausage-shaped, referring to the spores; long, with bluntly rounded ends, often curved.

alveolate: Pitted and ridged like a honeycomb.

amatoxin: Toxins found in some species of the genera *Amanita*, *Lepiota*, *Conocybe*, and *Galerina*. These potent toxins are selective inhibitors of RNA polymerase II and are responsible for 90% or more of all fatal mushroom poisonings. These toxins are not destroyed by heat and induce a life-threatening hepatorenal syndrome that frequently requires immediate liver and kidney transplants.

Amoebozoa: One of the six major divisions of the domain Eucarya, generally regarded as comprising a kingdom, which includes the phyla (or infraphyla) Mycetozoa (slime molds), Archamoebae (a curious group of degenerate amoebae that lack mitochondria; the Archamoebae includes common human commensals and pathogens of the genus *Entamoeba*), and Lobosa (which includes the familiar "protozoan" amoebae like *Amoeba*, *Arcella*, and *Difflugia*).

amyloid: Staining blue-black with Melzer's reagent or Lugol's solution. Compare with dextrinoid.

analogous: Adjective describing often similar-looking structures or mechanisms that were not derived from a common ancestral form and thus are not biologically related. For example, the very similar-looking Wood Ear, *Auricularia americana*, and Brown Witches' Butter, *Tremella foliacea*, belong to two different classes and independently evolved their convergent appearances and superficially similar lifestyles. Wood Ear is saprobic on decaying hardwood, while Brown Witches' Butter is parasitic on parchment fungi of the genus *Stereum*. In cladistics, an analogous trait is called a homoplasy. Compare with homologous.

anamorph: The asexual (conidial) state or form of a fungus. Compare with holomorph and teleomorph.

anastomosing: Alternately coalescing and separating.

angiocarp: An enclosed (gasteroid) fruitbody within which the spores develop.

annulus: The ring found on the stipe (stem) of many mushrooms; the remnant of the partial veil.

ap., *apud*: Latin, "in the work of." An indirect citation used in scholarly works to secondarily cite a reference on the authority of another reference. Thus, Lindblatt *apud* Fries means that the original source was Lindblatt but that Fries is the author's authority for that reference.

apex: The tip.

aphanoplasmodium: In slime molds, a plasmodium consisting of a network of very fine, transparent, and nearly invisible strands.

Aphyllophorales: An obsolete and polyphyletic order of basidiomycetes that included many diverse groups lacking gills or pseudogills and in which the hymenophore was flattened (Thelephoraceae), club-like (Clavariaceae), tooth-like (Hydnaceae), or poroid (Polyporaceae). No longer used.

apiculus: The part of a basidiospore that attaches to the sterigma, the slender stalk that gives rise to the basidiospore; also called a hilar appendage.

apomorphy: A homologous trait or feature that characterizes an ancestral taxon and all its descendants. Compare with synapomorphy and pleisiomorphy.

apothecium (pl. apothecia): A hymenium covering an exposed flat (disk-like), concave (cup-like), or convex surface of an ascocarp in which the asci are directly exposed to the environment; typical of many ascomycetes and lichens. Compare with perithecium.

appendiculate: Fragments of the universal veil dangling from the margin of the cap.

arbuscular mycorrhizae: Mycorrhizae in which the fungus penetrates the cortical cells of vascular plant roots and produces tree-like structures called arbuscules; arbuscular mycorrhizae belong to the fungal division (phylum) Glomeromycota.

Archaea, Archaebacteria: One of the three domains (empires) of living organisms. It includes organisms at a prokaryotic level of organization; these organisms lack membrane-bound organelles, and their cell walls lack peptidoglycan. Archaea possess unique RNA polymerase and ribosomes, which are closer to those of eukaryotes than to the bacteria. Many Archaea are extremophiles. Compare with Bacteria and Eucarya.

ascocarp: An ascomycete fruiting body containing the asci and ascospores.

ascoma (pl. ascomata): The ascus-bearing structure of an ascomycete.

ascomycete(s): A group of fungi (Ascomycota) that reproduce sexually by the formation of ascospores within a sac-like ascus.

ascomycetous: An adjective referring to the ascomycetes; a member of the division Ascomycota.

Ascomycota: One of the two major divisions (phyla) of the subkingdom Dikarya (higher fungi) characterized by dikaryotic mycelia and sexually reproducing by forming asci and ascospores.

ascospore: A haploid spore produced within an ascus following karyogamy and meiosis.

ascus (pl. asci): A sac-like hyphal chamber arising from a crozier inside which develop typically 8 (sometimes 16) ascospores. Asci are characteristic of the Ascomycota.

aseptate: Lacking septa or partitions; often pertaining to the hyphae seen in zygomycetes (also see coenocytic).

attached: Adjective describing gills that are fully attached to the stipe; a synonym of adnate.

***auct., auctorum*:** Latin, "of authors."

***auct. mult., auctorum multorum*:** Latin, "of many authors."

autonym: The infraspecific name attached to the basionym that is automatically created when a subsequent author creates a new infraspecific taxon (a subspecies, variety, or form). Thus, when *X-us albus* subsp. *ruber* Smith subsp. nov. is proposed as a new subspecies of *X-us albus* Jones, *X-us albus* Jones subsp. *albus* is an automatically created autonym. The fact that Jones's name precedes rather than follows the subspecies name indicates that this is an autonym.

azonate: Not zonate, that is, not exhibiting multiple alternating zones of color and/or texture.

azygosore: A haploid or parthenogenetic zygospore formed by mitosis by some vesicular-arbuscular mycorrhizal fungi and those fungi separated out of the now-defunct division Zygomycota. Compare with zygospore.

Bacteria, Eubacteria: One of the three domains (empires) of living organisms; it includes the Bacteria, Cyanobacteria ("blue-green algae"), and Actinomycetes (Actinobacteria)—all organisms at a prokaryotic level of organization: these organisms lack a nucleus and membrane-bound organelles, have cell walls composed of peptidoglycan, have ribosomes that differ from those of the Archaea and Eucarya, and undergo cell division by binary fission. Compare with Archaea and Eucarya.

ballistospore: A spore that is forcibly released when mature.

ballistosporic, ballistospory: Adjective and noun, respectively, referring to the forcible release of basidiospores from gilled mushrooms and ascomycetes. Compare with statismosporic.

basidiocarp: A fruitbody bearing basidia and basidiospores.

basidioma (pl. basidiomata), basidiome: Synonym of basidiocarp; the fruitbody of a basidiomycete.

basidiomycete, -s: A group of fungi that reproduce sexually by producing basidiospores from a basidium; a member of the division Basidiomycota.

Basidiomycota: One of the two major divisions (phyla) of the subkingdom Dikarya (higher fungi) characterized by dikaryotic mycelia and sexually reproducing by forming basidia and basidiospores.

basidiospore: In the Basidiomycota, a haploid spore produced on a basidium following karyogamy and meiosis.

basidium (pl. basidia): An apical hyphal chamber that gives rise to typically four (sometimes two) prong-like sterigmata, each of which bears a terminal basidiospore at its tip. Basidia are characteristic of the Basidiomycota.

Basionym: In botany the original combination, that is, the genus and epithet (species name) a taxon had when first named and described. The zoological equivalent of this is "original combination."

binary fission: A form of asexual reproduction in which the DNA is replicated and followed by cytokinesis (division of the cell into two parts). In the absence of a mutation, it results in two identical daughter cells. It is the primary mode of reproduction in prokaryotes and also commonly occurs in many unicellular organisms formerly assigned to the defunct kingdom Protista and in fungi formerly placed in the now-defunct division Zygomycota. In eukaryotes, binary fission involves mitosis.

binding hyphae: Long, often nonliving, and frequently branched hyphae in a polypore. Compare with skeletal and generative hyphae.

binomen (pl. binomina): The Latinized scientific name of an organism consisting of a generic name and an epithet (specific name). Binomina should be italicized (or set off in roman typeface when given within an italicized text). The genus is always capitalized, and the specific epithet is always lowercased (including in titles).

boletinoid: Adjective referring to pores that are elongate and somewhat radially arranged to form an intermediate pattern between pores and gills.

boreal: Subarctic ecosystems and climate.

broad: Wide, an adjective referring to the width of the gills in a fruitbody.

brown rot: Wood decay by basidiomycetous fungi in which the cellulose is degraded, leaving the lignin more or less intact.

ca., *circa*: Latin, "around," "about," "approximately."

cambium: The layer of actively dividing cells between the xylem and the phloem in plants.

campanulate: Bell-shaped.

capillitium: Sterile thread-like structures supporting the spores in many puffballs and slime molds.

caulocystidium: A cystidium on the stipe.

central: In the center, referring to the point of attachment of the stipe to the cap of a fruitbody.

cespitose, caespitose: Fruitbodies that are crowded close together but not attached.

cf., *conferre* [*cum*]: Latin, "compare [with]." Often used before the epithet when making an uncertain species identification. Generally not italicized.

cheilocystidium: A cystidium on the edge of a gill.

chlamydospores: Enlarged, thick-walled resting spores seen in certain fungi.

circumboreal: Distributed around the world in northern, subarctic latitudes.

clade: An evolutionary lineage consisting of an ancestral species and all its descendants, both living and extinct. Inherent in this definition is that a clade is monophyletic.

clamp connection: A specialized hyphal bridge arching back over a septum in many basidiomycetous fungi. Clamp connections maintain the dikaryotic state by reestablishing compatible nuclear pairs after synchronous mitotic nuclear divisions.

clavate: Club-shaped.

close: Narrowly spaced, referring to the spacing between the gills of a mushroom.

coenocytic: Adjective referring to a multinucleated cell or compartment; see syncytial, syncytium.

columella (pl. columellae): A sterile structure at the tip of a sporangiophore that extends within and supports a sporangium; commonly extends from the stalk.

comb. nov., *combinatio nova*: Latin, "new combination," referring to the transfer of a specific epithet to a genus to which it had not previously been assigned.

commensal: Adjective describing an organism that obtains benefits (food, shelter, support, mobility, etc.) from another organism without harming the latter; however, the host organism obtains no benefit. Compare with parasitic and mutualistic.

congeneric: Belonging to the same genus.

conidiophore: A specialized hyphal branch on which conidia develop in some fungi.

conidium (pl. conidia): An asexual reproductive propagule—any part of an organism that detaches and forms a new individual or life stage—formed in any manner that does not involve cytoplasmic cleavage. Conidia function as organs of dissemination. *Conidiospore* is an outdated term.

conk: The basidiocarp of a polypore.

conspecific: Belonging to the same species.

context: The inner flesh of a fruiting body.

convergent, inverted: Adjective referring to the gill trama that converges from the hymenia toward a central core running down from the cap. Compare with divergent (bilateral) and regular (parallel).

convergent evolution: The independent evolution of analogous structures or traits with similar appearances and/or functions but that are not derived from a common ancestral feature and thus are not biologically related.

coprophilous: Growing in or on dung.

coremia (sing. coremium): Fruitbodies of some fungi composed of loose bundles of conidiophores.

coriaceous: Having a leathery texture.

corticioid: A general, nontaxonomic term referring to basidiomycetous fungi that form more or less flat, spreading, or crust-like fruitbodies.

cortina: A cobweb-like partial veil.

crenate: Scalloped.

crenulate: Finely scalloped.

crowded: Extremely close, jammed together, referring to the spacing between the gills of a mushroom.

crozier: A recurved, hook-like structure developing at the tip of an ascus-forming hypha, the penultimate cell of which becomes the ascus. In the formation of a crozier, the haploid filaments of two compatible mating types fuse to form a transitory dikaryon. The two nuclei then fuse (karyogamy) and then undergo meiosis to form four haploid ascospores. These then typically undergo mitosis to produce eight ascospores within what has become the ascus. Croziers help maintain the dominant monokaryotic state of most ascomycetes.

cruciate: Cross-like; refers to a basidium in which two vertical or diagonal septa perpendicular to each other form a cross.

cultivar: A cultivated variety of a fungus or plant.

cupule: A small, cup-shaped structure.

cuticle: A surface layer on the cap or stipe that is distinct from the underlying flesh.

Cyanobacteria ("blue-green algae"): A division of nitrogen-fixing bacteria whose photosynthetic pigments include cyanophycin, allo-phycocyanine, and erythro-pycocyanine, which give these algae their blue-greenish hue.

cylindric, cylindrical: Like a cylinder; having parallel walls and a circular cross-section; refers to spores two to three times longer than they are wide.

cystidium (pl. cystidia): Specialized sterile cells in the tissues of some mushrooms; found mainly in the hymenium, but also found in other tissues. Cystidia are given special names according to their location.

daedeloid: Maze-like; refers to a system of hymenial pores that have merged together to form a maze-like labyrinth of channels.

decorticated: Adjective referring to a dead and decayed log whose bark has disintegrated.

decurrent: Gills or gill-like ridges running down the stipe.

dehisce, dehiscent: Verb and adjective referring to a structure that splits along a zone of weakness.

deliquesce, deliquescing: Autodigesting, self-liquefying; characteristic of certain coprinoid mushrooms.

dendroid: Tree-like.

dentate: Toothed.

deprecated synonym (later synonym): A nomenclatural synonym for an organism whose date of publication is later than that of the earliest validly published name for that taxon. Unless it was rejected or formally suppressed for any one of several reasons, and providing that the later synonym was not formally conserved, the earliest validly published name will have priority, and all other synonyms are thus deprecated.

depressed: Adjective referring to the central portion of the cap being lower than the margin.

Deuteromycetes: An artificial, polyphyletic, and "wastebasket" subdivision (Deuteromycota, Deuteromycotina) accommodating those fungi in which only the asexual state is known.

Deuteromycota: See Fungi Imperfecti.

dextrinoid: Staining brick-red or reddish-brown in Melzer's reagent or Lugol's solution.

diagnostic: Adjective describing any feature or suite of features that unequivocally identifies a fungus as belonging to a specific taxon.

Dikarya: The subkingdom of higher fungi that includes the divisions (phyla) Ascomycota and Basidiomycota, which produce dikaryons and whose gametes are never flagellated.

dikaryon: A hyphal compartment formed when the hyphae of two compatible mating types meet and fuse (plasmogamy). Delayed karyogamy (nuclear fusion) results in a dikaryon, that is, two different haploid nuclei sharing the same cell or cytoplasmic compartment. Compare with monokaryon and heterokaryon.

dikaryotic: Adjective describing a dikaryon.

dimidiate: Having a semicircular outline.

dimitic: Adjective describing a polypore whose fruitbodies have generative hyphae plus either skeletal or binding hyphae. Compare with monomitic and trimitic.

diploid: In fungi, a nucleus containing the complete set of chromosomes ($2n$) from the fusion of two nuclei from different but sexually compatible haploid hyphae, each having only half (n) the diploid number of chromosomes. Compare with haploid.

dissepiment(s): The sterile partitions formed by the trama between the pores in a polypore.

distal: Farthest from the point of attachment, origin, or center of the structure of interest. Compare with proximal.

distant: An adjective describing gills that are spaced far apart in a mushroom.

divergent, bilateral: Adjective referring to gill trama that diverges in opposite directions toward the hymenia from a central core running down from the cap. Compare with convergent (inverse) and regular (parallel).

division (phylum): A higher taxonomic category above the rank of class and below the rank of kingdom.

DNA, deoxyribonucleic acid: The genetic material of all prokaryotes and eukaryotes and most viruses.

domain (empire): The highest taxonomic category.

echinate: Covered with stiff, stout spines.

echinulate: Covered with delicate spines.

ectomycorrhizae, ectomycorrhizal: Noun and adjective referring to mycorrhizae in which fungal hyphae grow around the root and between the cells of the epidermis but do not penetrate the cell membranes.

effuse: Spread out, radiate; describes crust-like fungi that are spread out over a substratum.

effuse-reflexed: A spread-out fruitbody whose margins are turned up and may form shelf-like structures.

ellipsoid, elliptical: Oval, with a symmetric curve; broadly football-shaped, but with rounded tips.

emarginate: A gill that becomes abruptly narrower (often notched) just before attaching to the stipe.

***emend., emendatus*:** Latin, "as amended," "emended."

endomycorrhizae: Mycorrhizae in which the fungal hyphae penetrate the root of the plant and the root cell walls and become enclosed by the cell membrane.

entire: A margin or rim that is smooth and lacks notches, indentations, teeth, crenulations, or lobes.

ephemeral: Transient, short-lived.

epigeal, epigean, epigeous: Living above the soil surface. Compare with hypogeous.

epispore, episporium: The outermost and thickest layer of a fungal spore; it gives rise to whatever color and ornamentation the spore may have. Compare with perisporium.

epithet: The specific name part (i.e., the second, lowercased word) of a binomial scientific name; the species name.

epitype: A morphological type specimen chosen by a subsequent reviser when the holotype or lectotype does not exhibit the features needed to distinguish the taxon.

equal: A stipe with an approximately equal diameter from the base to the apex.

equilateral: Having equal sides; the opposite of inequilateral.

erumpent: Abruptly breaking out from beneath a surface or from within a host's tissues.

Eucarya, Eucaryota: One of the three domains (empires) of living organisms, characterized by a true cellular nucleus enclosing DNA complexed with histones, membrane-bound organelles, and which undergo cellular reproduction by mitosis whose germ cells undergo meiosis.

eukaryote: Any organism that has double-stranded DNA complexed with histones within chromosomes that are enclosed within a membrane-bound nucleus, has other membrane-bound organelles, and whose cells divide by mitosis or meiosis. All eukaryotes belong to the domain Eucarya (Eukaryota). Compare with prokaryote.

eukaryotic: Adjective referring to the grade of cellular organization seen in eukaryotes.

***ex*:** Latin, "from," "of," "according to."

excentric: Describes a stipe attached laterally (to one side) of the cap.

f., *forma*: Latin, "form." Not usually italicized when preceding the epithet as part of a scientific name.

facultative: Adjective referring to an optional (not required) biological attribute or way of life, such as a method of feeding, locomotion, deriving energy, reproduction, or association. Thus an organism may be a facultative carnivore, anaerobe, aerobe, parasite, or symbiont; the opposite of obligate.

falcate: Curved like a sickle.

false gills: Folds or ridges (often forked or anastomosing) on the underside of certain mushrooms (e.g., chanterelles) that superficially resemble gills.

farinaceous: Smelling and/or tasting like flour.

$FeSO_4$: "Iron salts," iron sulfate, a chemical stain used to identify certain groups of mushrooms.

fibrillose: Covered by fibrils (fine fibers).

***fide*:** Latin, "on the faith of," "according to" a cited source.

filiform: Thread-like, hair-like.

fimbriate: Covered with hairs.

flagellum (pl. flagella): A whip-like locomotory organelle emerging from many bacteria and unicellular protists and the spores of certain fungi by which these cells move about.

floccose: Fluffy or cottony.

form-taxa: Taxonomic categories erected for anamorphic forms of a fungus; examples include form-genus and form-species.

free: Gills that do not attach directly to the stipe.

fructification: An outdated term for fruitbody, sporocarp.

fruitbody, fruiting body: The sporocarp, the sexual spore-producing multicellular structure that develops from the mycelium of an ascomycete or basidiomycete. Typically the visible mushroom. An outdated term is fructification.

Fungi: One of the several major kingdoms in which organisms are grouped; it is composed of eukaryotic, obligately heterotrophic decomposers and parasites that lack chlorophyll and have cell walls composed largely of chitin.

Fungi Imperfecti (Deuteromycota): An informal, polyphyletic (and thus artificial) grouping of unrelated fungi that are known only by their anamorphic (asexually reproducing) forms. Many of these are the anamorphs of ascomycetes and basidiomycetes, but without sexual fruitbodies their affinities remain obscure.

furcate: Forked.

fusiform, fusoid: Spindle-shaped, tapering toward both ends.

gasteroid, gastraceous: Adjectives referring to Agaricomycetes that produce their spores inside the fruiting body rather than externally on basidia located on gills. The terms mean "stomach-like," "sac-like," or "closed up"; the fruitbodies are often rounded or ball-shaped, but many are erect club-like or coral-like forms. The gasteroid morphology is a homoplasy that has independently evolved several times in unrelated groups. Gasteroid fungi include the puffballs and stinkhorns. Compare with secotioid and sequestrate.

Gastromycetes: A polyphyletic and obsolete class that formerly included fungi that produce spores inside their fruitbodies rather than on a hymenial surface. This defunct category included many quite unrelated fungi.

generative hyphae: Living, spore-bearing hyphae in a polypore. This is the only type of hyphae that can exhibit clamp connections. Compare with skeletal and binding hyphae.

generic: Adjective referring to genus, for example, a generic feature.

gen. nov. et sp. nov., *genus novum et species novum*: Latin, "new genus and new species."

genus: The taxonomic category above the rank of species and below the rank of family; the first and always capitalized word in a binomen (scientific name).

germ pore: A small pore in the outer wall of a fungal spore through which the germ tube exits upon germination. An apical germ pore is located at one end of the spore.

germ slit: A characteristic longitudinal furrow that develops in the spores of certain fungi; it often has taxonomic significance.

germ tube: The initial hypha that develops from a conidium or sexual spore and exits via a germ pore.

gills: The lamellae, or plate-like hymenial structures of agaric mushrooms.

glabrous: Smooth.

gleba: The fleshy, spore-bearing, inner mass of tissue encased within the sporocarp of certain gasteromycetous fungi such as the puffballs and earthstars. The stinkhorns (family Phallaceae) also produce a gelatinous, deliquescent, and foul-smelling gleba on the surface of the cap.

globose: Spherical, globular.

Glomeromycota: A division (phylum) of the kingdom Fungi whose members form arbuscular mycorrhizae with the roots of land plants or the thalli of bryophytes except for one monotypic genus: *Geosiphon pyriformis* forms lichen associations with endosymbiotic cyanobacteria ("blue-green algae") of the genus *Nostoc*.

guttulate: Spores containing one or more oil droplets.

haploid: The number of chromosomes (n) in a gamete, which is half the diploid number ($2n$) in a zygote. The haploid stage predominates in the life cycle of most fungi, which produce gametes (haploid nuclei) by mitosis. During the sexual phase two compatible nuclei fuse (karyogamy) to form a diploid zygote, but meiosis soon follows, resulting in haploid spores that produce new haploid hyphae.

haustorium (pl. haustoria): A specialized appendage of a parasitic fungus that penetrates the host's tissues but does not penetrate the host's cell membranes.

Heterobasidiomycetes: An obsolete term for a polyphyletic class of basidiomycetes with septate basidia: certain smuts, rusts, and jelly fungi of the orders Tremellales, Dacrymycetales, Auriculariales, and Sebacinales; still used informally. Compare with Homobasidiomycetes.

heterokaryon, heterokaryotic: A noun and adjective referring to a multinucleate cell containing genetically different nuclei. Compare with dikaryon and monokaryon.

heterothallic: Adjective describing a fungus that requires mating between two different compatible strains for sexual reproduction to occur. Compare with homothallic.

heterotypic: Adjective referring to a nomenclatural synonym that results when the name of a taxon is reduced to synonymy with a different taxon based on a different type specimen; the equivalent zoological term is *subjective synonym*. Compare with homotypic.

hilar appendage: The part of a basidiospore that attaches to the sterigma, the slender stalk that gives rise to the basidiospore; also called an apiculus.

hirsute: Covered with stiff, bristle-like hairs.

Holarctic: Distributed throughout the Northern Hemisphere.

holobasidium (pl. holobasidia): A basidium consisting of a single cell; an apomorphic feature of the Agaricomycetes. Compare with phragmobasidium.

holomorph: All life stages of a fungus, including its anamorph(s) and teleomorph.

holotype (*holotypus*): The single specimen or element thereof selected by the original author of a validly published new species as the nomenclatural (name-bearing) type specimen. Also simply called the "type" or "*typus*."

Homobasidiomycetes: An obsolete term for a monophyletic class of basidiomycetes with nonseptate basidia; approximately equivalent to the modern Agaricomycetes and still used informally. Compare with Heterobasidiomycetes.

homologous: Adjective describing often dissimilar-looking structures independently derived from a common ancestral form. The sac-like asci of the Ascomycota are believed by many mycologists to be homologous with the pedestal-like basidia of the Basidiomycota. Likewise, the hook-like ascomycetous croziers are thought to be homologous with the arching clamp connections of the Basidiomycota. A homologous trait is also called a synapomorphy. Compare with analogous trait (homoplasy).

homology: Shared ancestry between different structures in various organisms.

homonym: A published binomial name that is identical in spelling to an earlier binomen belonging to a different taxon. A later homonym is a *nomen illegitimatum* (illegitimate name) under Article 53.1 of the *ICN* and is forever suppressed.

homoplasy: In cladistics, an analogous trait that arose independently in two or more clades by convergent evolution. For example, both the loss of ballistospory and the very similar gasteroid fruitbodies of the earthballs (*Scleroderma*, Order Boletales) and the puffballs (*Lycoperdon*, Agaricales) are homoplasies that arose independently from different ballistosporic and gilled ancestral mushrooms. Compare with apomorphy, pleisiomorphy, and synapomorphy.

homothallic: Adjective describing a fungus in which self-fertilization occurs and both fusing nuclei are derived from the same mycelium. Compare with heterothallic.

homotypic: Adjective referring to a nomenclatural synonym that results when two or more different names are based on the same type specimen(s); the equivalent zoological term is *objective synonym*. Compare with heterotypic.

hyaline: Clear, colorless.

hydnate, hydnoid: Bearing teeth or spines on the hymenial surface instead of gills or pores.

hygrophanous: Adjective referring to the change in color and transparency exhibited by certain fungi as they absorb water or undergo desiccation.

hymenial: Adjective referring to the hymenium.

hymenium (pl. hymenia): The layer of fertile tissue giving rise to and bearing the sexual spores, for example, the tissue(s) covering the gills of agarics, the pores of boletes and polypores, or the smooth fertile surfaces of certain corticioid and stereoid fungi and some other groups; it includes the asci of the ascomycetes, the basidia of the basidiomycetes, and any sterile cells such as cystidia or paraphyses.

hymenomycete: An informal name for a basidiomycete whose hymenium is exposed at maturity and whose basidiospores are ballistosporic.

hyperparasite: A parasite that parasitizes another parasite.

hypha (pl. hyphae): A single filament of a fungus.

hypogeal, hypogean, hypogeous: Adjectives describing fruitbodies that remain underground, like truffles and certain earthstars. Compare with epigeal.

hypothallus: The plasmodial remnant in certain slime molds that forms a thin, film-like layer at the base of the sporangia and connects the stalks to the substratum.

***ICBN*:** *International Code of Botanical Nomenclature* (Vienna Code, 2005); superseded by the *ICN* (Melbourne Code).

***ICN*:** *International Code of Nomenclature for algae, fungi, and plants* (Melbourne Code, 2011); the most recent nomenclatural rules for plants, algae, and fungi; it supersedes the *ICBN* (Vienna Code).

***ICNB*:** *International Code of Nomenclature of Bacteria.*

ICTV: International Committee on Taxonomy of Viruses. This committee, established in 1961, is responsible for approving all names of viruses. Unlike all other nomenclatural codes, virus nomenclature does not use a Linnaean system of Latinized binomina but an international English-based system.

ICZN: *International Code of Zoological Nomenclature.*

imbricate, imbricated: Overlapping, like the shingles on a roof.

inamyloid: Nonreactive in Melzer's reagent.

***incert. sed., incertae sedis*:** Latin, "of uncertain placement." Used to indicate that the placement of an organism into one or more higher taxonomic categories (above the rank of genus) is problematic.

incurved: Margin of the cap turning inward toward the stipe.

indusium (pl. indusia): A lacy or net-like skirt.

inequilateral: Having unequal sides; opposite of equilateral.

infrageneric: Any taxonomic category below the rank of genus. In botanical nomenclature the infrageneric categories are ranked in descending order as follows: genus, subgenus, section, subsection, series, subseries, and stirps.

infraspecific: Any taxonomic category below the rank of species. In botanical nomenclature the infraspecific categories are ranked as follows: subspecies, variety (*varietas*), and form (*forma*).

infundibuliform: Funnel-shaped.

inrolled: Margin of the cap rolling up on the side next to the stipe.

intervein: A connecting cross-vein between two gills.

***in vitro*:** Latin, "in a glass"; refers to anything cultured in a laboratory.

***in vivo*:** Latin, "within a living thing."

involute, involuted: Rolled in; refers to the margin of a pileus.

irpicoid: Having lacerate (torn) pores or teeth.

karyogamy: The fusion of two haploid nuclei within a dikaryon to form a diploid zygote or zygospore.

kingdom, regnum: The second highest taxonomic rank; it is below the rank of domain and above division or phylum.

KOH: Potassium hydroxide; a chemical stain used to help identify certain mushroom groups.

lageniform: Flask-shaped.

lamella (pl. lamellae): The gills below the cap in agaric mushrooms.

lamellula (pl. lamellulae): A short lamella or gill that does not extend all the way from the margin of the cap to the stipe.

lanceolate: Lance-shaped.

lateral: To or on the side.

latex: The milky fluid that oozes from certain mushrooms, notably the genus *Lactarius*.

LD50: The dose of a toxin needed to kill 50% of the animals tested. LD_{50}s are commonly expressed as mg/kg, µg/kg, or ng/kg.

lectotype: A type selected as the nomenclatural type by a subsequent reviser (author) from the original type material when no holotype was designated by the original author.

lenticular: Shaped like a biconvex lens, that is, convex on both surfaces.

lichen: A composite organism consisting of a symbiotic association between a fungus (the mycobiont), which forms the thallus of the lichen, and either a photosynthetic alga or a cyanobacterium (the photobiont). The morphology and physiology of a lichen are quite different from those of either symbiont living alone.

lignicolous: Adjective referring to fungi that grow on wood and cause it to decay.

limbate: Edged with a different color, texture, or membranous structure.

loculated: Divided into small cavities or chambers.

Lugol's solution: A mixture of potassium iodide (IKI) and elemental iodine (I_2) in water used by mycologists in lieu of Melzer's reagent. It is much easier to obtain because it does not contain chloral hydrate.

macroscopic: Features visible to the unaided eye. Compare with microscopic.

meiosis: The process by which a diploid (2*n*) set of chromosomes in eukaryotic organisms are first replicated (4*n*), then undergo a reduction division (2 × 2*n*), and then a second reduction to produce four haploid (*n*) gametes or spores. Compare with mitosis.

meiospores: Sexual spores formed by meiosis from the diploid parent protoplast.

Melzer's reagent: A specific mixture of chloral hydrate, potassium iodide (IKI), and elemental iodine (I_2) used by mycologists to identify mushrooms. Because chloral hydrate is a controlled substance, obtaining this reagent requires a physician's prescription and a "tame" compounding pharmacist who is willing and able to obtain the ingredients and put them together. Many pharmacists will not do this.

meruloid: Having shallow, reticulated pores.

mesic: Humid, moist.

metuloid: A thick-walled cystidium (sterile cell) with a blunt apex (tip).

microscopic: Tiny features visible only through a microscope.

mitosis: The process in eukaryotic cells by which the chromosomes contained in a nucleus are first replicated and then separated into two identical copies of the original set, with one of each set going to a daughter nucleus. Compare with meiosis.

mitospores: Asexual spores produced by fission or budding from the haploid parent protoplast and involving mitosis.

monokaryon: A fungal spore or hyphal cell containing only one haploid nucleus. Compare with dikaryon and heterokaryon.

monomitic: Adjective describing a polypore whose fruitbodies have only generative hyphae.

monophyletic: Adjective describing a taxonomic category whose members all share a common ancestor and evolutionary lineage. Compare with polyphyletic and paraphyletic.

monotypic: A taxonomic group that contains only a single immediately subordinate taxon; for example, a monotypic family contains only one genus, and a monotypic genus contains a single species. This term has nothing to do with type specimens.

multiseptate: Having several septa.

mushroom: A generalized term typically used to refer to the visible fruitbodies of an ascomycete or basidiomycete, that is, a higher fungus of the subkingdom Dikarya.

mutualism, mutualistic: Noun and adjective referring to an interaction between two or more different biological species that benefits all parties.

mycangium (pl. mycangia): Specialized structures on or within the body of an animal to transport symbiotic fungi (usually as spores); also called mycetangium (pl. mycetangia).

mycelium (pl. mycelia): The mass of hyphae making up the thallus (vegetative mass) of a fungus.

Mycetozoa: A phylum or infraphylum of the kingdom Amoebozoa that includes the acellular (Myxogastria) and cellular (Dictyostelia) slime molds and certain intermediate forms (Protostelia).

mycobiont: The thallus-producing fungal partner in the symbiotic associations known as lichens. Compare with photobiont.

mycology: The scientific study of fungi.

mycoparasite, mycoparasitic: Noun and adjective, respectively, referring to a fungus that parasitizes another fungus.

mycophagous, mycophagy: Adjective and noun referring to feeding on fungi.

mycorrhiza (pl. mycorrhizae): A symbiotic association between a fungus and the roots of a vascular plant in which the fungi greatly increase the surface area of the plant's root system for the uptake of minerals, nutrients, and water. In return, the plant provides the fungus with carbon-based products of photosynthesis. This association may be ectomycorrhizal, in which fungal hyphae grow around the root and between cells of the epidermis, or endomycorrhizal, in which the fungal hyphae penetrate both the root of the plant and the root cell walls and become enclosed by the cell membrane. Arbuscular mycorrhizae are a type of endomycorrhizae where the fungus penetrates the cortical cells of vascular plant roots and produces tree-like structures called arbuscules.

mycorrhizal: Adjective referring to mycorrhizae.

myxamoeba (pl. myxamoebae): An amoeba-like cell typically lacking flagellae and containing a single haploid nucleus that is released upon the germination of a spore. It moves about, engulfing bacteria and yeast cells, before fusing with other myxamoebae to form a plasmodium. It is a stage in the life cycle of acellular slime molds.

Myxogastria, Myxomycetes: A class within the kingdom (an unranked clade of some authors) Amoebozoa. It includes the plasmodial slime molds.

Myxomycota: See Mycetozoa.

Nearctic: Distributed in North America.

nematode: A nonsegmented worm of the phylum Nematoda, some of which feed upon fungi, and which, in turn, are eaten and/or parasitized by certain carnivorous fungi.

nematophagous, nematophagy: Adjective and noun, respectively, referring to preying upon and eating nematodes. A great many mushrooms—including a number of well-known species widely consumed by humans—practice this form of carnivory as a source of nitrogen.

Neotropical: Distributed in Central and South America.

neotype: A nomenclatural type specimen selected by a subsequent reviser (author) when all of the original type material has been lost.

NH_4OH: Ammonium hydroxide, a chemical stain used to help identify certain groups of mushrooms.

***nom. conserv., nomen conservandum* (pl. *nomina conservanda*):** Latin, "conserved name."

***nom. dub., nomen dubium* (pl. *nomina dubia*):** Latin, "doubtful name," a taxonomic name that cannot be applied with certainty.

nomenclature: In biology, the formal systems of giving scientific names to organisms and degenerate, organically derived, nonliving, parasitic particles such as viruses, prions, and the like. Each major group (or incorporated series of groups) is governed by a specific, internationally recognized nomenclatural code. See *ICBN*, *ICN*, *ICZN*, and *ICTV*.

***nom. illegit., nomen illegitimatum*:** Latin, "illegitimate name," typically a later homonym of a previously validly published name. An illegitimate name may never be legitimately used and is forever suppressed.

***nom. nov., nomen novum* (pl. *nomina nova*):** Latin, "new name," a replacement name for an earlier invalid name.

***nom. nud., nomen nudum* (pl. *nomina nuda*):** Latin, "naked name," a name that has not been validly published and therefore has no taxonomic standing. It is an available name should it ever be validly published.

***nom. provis., nomen provisorium*:** Latin, "provisional name," a name used in a manuscript or a published work that does not meet the requirements of the *ICBN* or the *ICN* for valid publication. *Nomina provisoria* are *nomina nuda* (naked names) and have no taxonomic standing whatsoever until and unless they are eventually validly published together with a formal description that meets the requirements of the Code.

***non*:** Latin, "not [of]," used in taxonomy to mean a name not intended by a specific author or authors. Commonly used when referring to homonyms or misapplied names. Example: *X-us albus* Jones 1940, *nom. illegit.* [homonym, *non* Smith 1840] means that *X-us albus* of Jones is a later homonym of *X-us albus* Smith 1840. Jones's name is illegitimate and can never be used, while that of Smith is valid by reason of publication priority.

nonamyloid: Not changing color in response to Melzer's reagent or iodine.

noncollariate: Lacking a pseudocollarium (depression or gap) at the stipe where the gills have separated or are not quite attached.

nonseptate: Lacking septa.

obligate: "Of necessity," an adjective referring to a biological attribute or way of life, such as a method of feeding, locomotion, deriving energy, reproduction, or association. Thus an organism may be an obligate carnivore, anaerobe, aerobe, parasite, or symbiont; the opposite of facultative.

oblong: Adjective describing a spore that is twice as long (or slightly less) than wide and with flattened ends.

obovate, obovoid: Egg-shaped, with the narrower end pointing outward or inward, respectively.

obtuse: Blunt, rounded.

olivaceous: Olive-grey color.

operculum (pl. opercula): The lid-like cover of an ascus or sporangium.

order: A higher taxonomic category above the rank of family and below the rank of class.

original combination: In zoological nomenclature (as used here in slime molds, for example), the original genus and species by which a new species was described and named; the zoological equivalent of the basionym.

ostiole: An opening or pore in an ascocarp or a pycnidium.

ovoid: Oval or egg-shaped.

Palearctic: Distributed in Europe and Asia.

pantropical: Distributed throughout the tropics.

paraphyletic, paraphyly: Adjective and noun referring to a monophyletic group that does not include all of the descendants of its common ancestor.

paraphysis (pl. paraphyses): Elongate, sterile cells sometimes found in the hymenium along with the asci or basidia.

parasite, parasitic: An organism that lives in or on another organism from which it derives food, protection, and/or other benefits at the host organism's expense.

parthenogenesis: Reproduction by means of an unfertilized egg or spore.

partial veil: A temporary membranous tissue extending from the stipe to the cap margin that covers and protects the hymenium (spore-bearing tissue) until the spores mature. As the partial veil disintegrates, a portion of it may remain on the stipe as an annulus (ring).

pectinate: Resembling the teeth of a comb.

pedicel: A slender stalk.

pellicle: A thin surface skin or film.

peridiole: One of the spore-carrying "eggs" in a bird's nest fungus.

peridium: The outer cortex of certain mushrooms (especially gasteroid forms) and slime molds; the immature, egg-like form of the stinkhorns.

perispore, perisporium: A transient membrane enveloping the developing spores of certain fungi that typically sloughs away. Compare with episporium.

perithecium (pl. perithecia): An enclosed, flask-like ascocarp containing the asci and having an apical ostiole (pore) through which the ascospores are released. Compare with apothecium.

phaneroplasmodium: A large, visible plasmodium in a slime mold.

phloem: The vascular tissue in plants that transports sugars and other photosynthetic products down from the leaves to the rest of the plant.

photobiont: The photosynthesizing algal or cyanobacterial partner in the symbiotic associations known as lichens. Compare with mycobiont.

phragmobasidium (pl. phragmobasidia): A basidium consisting of more than one cell; a primitive (ancestral) feature of the Agaricomycotina restricted to the Tremellomycetes, Dacrymecetes, and the basal Agaricomycetes. Compare with holobasidium.

phylogenetic: Adjective referring to the phylogeny of an organism or group of organisms.

phylogeny: The evolutionary history of an organism or a group of genetically related organisms.

pileipellis: The uppermost hyphal layer in the cap of a fruitbody.

pileocystidium: A cystidium on the pileus (cap) of a mushroom.

pileus: The cap of a fungal fruitbody.

plage, subhilar plage: A bright flattened area near the base of a basidiospore; characteristic of the genus *Galerina*.

plasmodiocarp: Channel-like structures formed when a slime mold plasmodium becomes concentrated in its main veins and is covered by a peridium or cuticle.

plasmodium (pl. plasmodia): In acellular slime molds, a syncytium consisting of up to thousands of individual diploid nuclei within a cytoplasmic matrix enclosed within a single cell membrane.

plasmogamy: The cytoplasmic fusion of two compatible protoplasts or hyphal cells.

pleisiomorphy: A homologous trait that is shared by groups other than just the clade of interest. Compare with apomorphy and synapomorphy.

pleurocystidium: A cystidium on the face of a gill.

pleurotoid: Adjective referring to a gilled mushroom with a laterally attached stipe.

polyphyletic: An artificial taxonomic group consisting of superficially similar members that do not share a common ancestor. Polyphyletic groups are not natural biological categories. Compare with monophyletic and paraphyletic.

***p.p., pro parte*:** Latin, "in part," "partly."

prokaryote: Any single-celled organism (which may aggregate or form biofilms) with single-stranded, naked, circular DNA that is free in the cytoplasm and that lacks a nucleus or any other membrane-enclosed organelles. A member of one of two domains: Bacteria (Eubacteria) or Archaea (Archaebacteria). Compare with eukaryote.

prokaryotic: Adjective referring to the grade or cellular organization of a prokaryote, that is, a bacterium, a cyanobacter, or archaean.

protologue: All of the original published material associated with a new name, including the description of the new taxon, its diagnosis, and any other elements such as illustrations, identification of types, synonymy, and the like.

protoplasmodium: A microscopic plasmodium that never exceeds 1 mm in size and gives rise to a single sporangium.

protoplast: A fungal cell whose cell wall has been removed, such as occurs during the formation of zygospores.

proximal: Nearest to the point of attachment, origin, or structure of interest. Compare with distal.

pruinose: Covered with a powder-like bloom.

pseudoaethalium: In slime molds a structure resembling an aethalium formed by several individual sporangia crowded close together into what appears to be a single cushion-shaped mass.

pseudorhiza: A long, root-like extension of the lower stipe.

pulvinate: Cushion-shaped.

pycnidium: An ovoid or flask-shaped cavity in certain ascomycetes and imperfect fungi that produce conidia.

Pyrenomycetes: A class of the division Ascomycota, subdivision Pezizomycotina, which generally produce asci within flask-shaped perithecia. Also called class Sordariomycetes.

pyriform: Pear-shaped.

receptabulum, receptacle: The elastic column-shaped or arm-like structure containing the gleba that emerges from the egg-like peridium in stinkhorns.

recurved: An object that is curved or bent back on itself.

reflexed: Turned up or backward at the margin.

regular, parallel: Adjective referring to gill trama parallel to a central core running down from the cap. Compare with convergent (inverse) and regular (parallel).

resting spore: A thick-walled, resistant, sexual or asexual spore that germinates following an extended period of dormancy.

resupinate: Upside down, facing upward; refers to an effuse fruitbody whose hymenium (fertile surface) faces away from the substratum (i.e., downward in those spreads growing on the undersides of downed logs).

reticulate: Resembling a net or a network.

reticulum: A network-like surface.

rhizine: A root-like anchoring and/or absorptive filament growing from the thallus of a lichen.

rhizoid: A short, branching, root-like hypha that releases digestive enzymes and absorbs organic matter.

rhizomorph: A mycelial strand of aggregated parallel hyphae attached to the basal portion of some mushrooms.

rostrate: Having an anterior or apical projection or elongation.

rugose: Wrinkled.

saccate: Sac-like, shaped like a bag.

saprobe: A saprobic organism, typically a fungus or bacterium.

saprobic: Adjective describing an organism that feeds on dead organic matter.

saprotrophic: An equivalent term to saprobic, but generally restricted to plants; formerly also applied to fungi.

scaber: Scale-like projections or rough areas.

scabrous: A rough surface with short rigid projections.

sclerotium (pl. sclerotia): A highly condensed mass of undifferentiated sterile (asexual) hyphae typically encased in a hard, woody, thick, dark rind. These structures enable those fungi producing them to survive under adverse environmental conditions.

scurfy: Covered with a loose, scaly crust.

secotioid, secotiaceous: Having an intermediate growth form between a typical mushroom and a closed gasteroid form. Compare with gastroid, gastraceous, sequestrate.

sect., *sectio*: Latin, "section," an infrageneric taxonomic category below the rank of subgenus and above the rank of series.

***sensu*:** Latin, "in the sense of," referring to a taxonomic name as used by a specific author or authors.

***sensu lato* (s. l.):** Latin, "in the broad sense"; in taxonomy referring to a taxon in its widest interpretation. Compare with *sensu stricto*.

***sensu stricto* (s. s.):** Latin, "in the strict sense"; in taxonomy referring to a taxon in its restricted or narrowly circumscribed definition or interpretation. Compare with *sensu lato*.

[*in*] *sensu tacitae*: Latin, "in the 'unspoken' or informal sense"; in taxonomy referring to a polyphyletic group of organisms as loosely understood or defined (e.g., the "polypores," which include species representing five different orders of fungi).

septate: Having one or more septa.

septum (pl. septa): Latin, A "partition," or cross-wall in a hypha, cell, or spore.

sequestrate: Rounded, used in a general way to mean gasteroid or secotioid.

series: A taxonomic category below the rank of section and above the rank of stirps.

serrate: Having a sawtooth-like edge.

serrulate: Very finely serrate.

sessile: A fruitbody lacking a stipe and attached directly to the substratum.

seta (pl. setae): A modified spine-like hyphal structure within or protruding from the tissues of a fruitbody. Typically microscopic but in some cases visible to the unaided eye.

setose: Covered with bristles.

sinuate: A gill that becomes abruptly narrower and forms a notch just before curving back and attaching to the stipe.

skeletal hyphae: Long, often nonliving, thick-walled, unbranched hyphae in a polypore. Compare with generative and binding hyphae.

slime mold: A term loosely used for three polyphyletic groups of unrelated organisms that reproduce by spores: (1) the syncytial or plasmodial slimes (Myxogastria/Myxomycetes), (2) the cellular slimes (Dictyosteliida), and (3) an obscure group of protosteloid amoebae. Historically included with the Fungi, slimes are now placed in the kingdom Amoebozoa; some workers still consider the Amoebozoa to be a phylum in the kingdom Protozoa, but this is no longer generally accepted.

smooth: A surface without any pits, projections, warts, ridges, spines, or other ornamentations.

solitary: Alone, by itself, referring to a single fruitbody.

Sordariomycetes: A class of the division Ascomycota, subdivision Pezizomycotina, which generally produce asci within flask-shaped perithecia; a synonym of Pyrenomycetes.

sp. (pl. spp.): Abbreviation for species; commonly used to refer to an undetermined species whose genus is known (e.g., *X-us* sp.). Not italicized.

species name: A Latinized binomen consisting of a generic name and a specific epithet that refers to a single biological species.

specific epithet: The second part of a scientific binomen, the species name. The specific epithet is always written in lowercase letters, even if it represents a proper noun, and it must always be accompanied by the name of the genus (or its abbreviation when no doubt exists as to which genus is meant) to which it is assigned.

spinulose: Covered with small spines.

sp. nov., *species nova*: Latin, "new species," a designation that follows a newly published name with its description of a species new to science.

sporangiophore: A modified hypha that supports a sporangium.

sporangiospore: An asexual spore produced within a sporangium.

sporangium (pl. sporangia): A sac-like structure that produces asexual sporangiospores.

spore: A general term for a fungal reproductive structure that does not contain an embryo. Spores may have one or several nuclei, be septate or aseptate, and be either asexual or sexual.

sporocarp: The fruitbody of a fungus.

sporodochia: Tight clusters of sporophores produced by the fungal symbionts of ambrosia beetles on which the adult and larval beetles feed.

spp.: Abbreviation for species (plural); commonly used to designate more than one species of a genus. Not italicized.

squamose: Covered with scales.

squamule: A minute scale.

squamulose: Covered with minute scales.

ssp., subsp.: Latin, abbreviation of "subspecies." Not italicized when preceding the epithet as part of a scientific name.

s. s., *sensu stricto*: Latin, in the strict or narrow sense.

statismospore: A spore that is passively released.

statismosporic, statismospory: Adjective and noun, respectively, referring to the passive dispersal of basidiospores from the basidia in secotioid or gastroid (sac-like) basidiomycetes. It is thought to be an adaptation to xeric (dry) conditions that have insufficient moisture for the forcible ejection of spores in gilled forms. Compare with ballistosporic. Gasteroid and secotioid fungi depend on animals to disperse their spores.

stellate: Star-shaped.

sterigma (pl. sterigmata): A small, narrow, stalk-like structure at the apex of a basidium upon which a basidiospore forms; an epibasidium.

stipe: The stalk or pedicel of a fruitbody.

stipitate: Having a stipe, stalked.

stirps: A taxonomic category below the rank of series and above the rank of species.

stolon: A running hypha from which rhizoids and sporangiospores arise.

striate: Having more or less straight lines or minute furrows.

stroma (pl. stromata): A compact mass of fungal tissue on or within which fruitbodies develop.

stuffed: Refers to a stipe filled with a soft, rather cottony center that usually disappears in mature specimens.

subdecurrent: Gills or ridges running down the stipe for a short distance.

subfamily: A taxonomic category below the rank of family and above the rank of tribe.

subgenus: A taxonomic category below the rank of genus and above the rank of section.

subglobose: Not quite round or spherical.

subsp. (pl. subspp.): The abbreviation standing for subspecies; in botanical names, all infrageneric taxa must be preceded by the abbreviation designating which category the infrageneric name refers to.

subspecies: An infraspecific taxon below the rank of species but above the ranks of forma (form) and variety.

substratum (pl. substrata): The surface on or within which a fungus exists; commonly called "substrate" by many authors, but we prefer to restrict that term to a molecule acted upon by an enzyme. The correct plural form of substrate is "substrates," not "substrata."

supra-: Adjective indicating something that is higher or greater (e.g., as in rank) than something else.

swarm cells, swarm spores: The motile asexual spores of some fungi and algae that move by means of flagella; also called zoospores.

symbiosis: Two or more unrelated biological species living together in a complex interrelationship in which they have coevolved over a long evolutionary period. This relationship is often restricted to mutualistic interactions, in which both (or all) partners benefit. Examples include lichens (formed by a fungus and one or more algae) and mycorrhizal associations. Other biologists include any persistent coevolved interaction, including commensalism and parasitism.

symbiote, symbiont: Any one of the partners in a symbiotic association.

sympodial: A type of conidial growth where conidia develop on a sharply bent (knee-like) or zigzag structure. Sympodial conidia form sequentially at the base of an apical blastic cell. The main axis of this cell continues to grow and produce new conidia beyond and to one side of the original conidium.

synapomorphy: A homologous trait or feature shared by two or more taxa that originated in their last common ancestor; synapomorphies are the basis for constructing cladograms (phylogenetic trees) and delineating monophyletic clades. Compare with apomorphy and pleisiomorphy.

syncytium (pl. syncytia): Many individual diploid nuclei within a cytoplasmic matrix enclosed within a single cell membrane. The plasmodia of slime molds consist of syncytia.

synnema (pl. synnemata): A group of stalk-like fused conidiophores that are cemented together and produce conidia at the apex and/or along the sides.

synonyms: Different validly published scientific names for the same species of organism.

heterotypic (subjective) synonyms: Subjective synonyms arise whenever the same taxon is described more than once under different names by the same or different author. In this case, each named and described taxon is based on its own type specimen(s), and the inference that they represent the same species is thus subjective and not absolute. Thus, *Y-us niger* Smith 1949 might be considered to be a subjective (heterotypic) synonym of *X-us albus* Jones 1850. Because *X-us albus* has nomenclatural priority by virtue of being validly published first, *Y-us niger* becomes a deprecated heterotypic synonym. But this synonym is a matter of subjective opinion, since the two names are based upon different types.

homotypic (absolute) synonyms: Absolute (nonsubjective) synonyms based on the same nomenclatural type specimen that has been recombined with one or more different genera. For example, *X-us albus* Jones may be placed in a different genus by Smith: *Y-us albus* (Jones) Smith. But it is still clearly the same biological species. And Doe may subsequently place this taxon in a new genus that he proposes: *Z-us albus* (Jones) Doe. All these combinations refer to the taxon based on the same type specimen(s) and are thus absolute (nonsubjective) synonyms.

syntype: One of two or more specimens in the type series used by the author in the original description of a species or infraspecies taxon when no specimen was designated as the holotype. The zoological equivalent of syntype is paratype. Less often called a cotype.

systematics, systematic biology: The field of study of the evolutionary history, environmental adaptations, phylogenetic relationships, biogeographic distributions, classification, and nomenclature of organisms.

taxon (pl. taxa): Any of several taxonomic categories ranging from kingdom down to the infraspecific level, for example, a division (phylum), class, order, family, genus, or species and any supra- or infralevel category of these groups.

taxonomic: Adjective referring to the classification and/or nomenclature of an organism or group of organisms.

taxonomy: The discipline devoted to the collection, cataloging, classification, and naming of organisms. A subspecialty of systematics.

teleomorph: The sexual stage of a fungus. Compare with anamorph and holomorph.

terete: Round in cross-section.

thallic: A form of conidial development in which a conidium (asexual spore) is formed from a preexisting hyphal segment or cell.

thallus: The fruitbody structure of a mushroom or lichen.

toadstool: A general and vague term for a mushroom usually regarded as inedible or poisonous.

tomentose, tomentum: Adjective and noun referring to being woolly, velvety, covered with thick, soft hairs; a mat or pubescence of fine hairs.

topotype: A specimen collected from the species' type locality.

trabecular, trabeculate: Characterized by short beams, struts, or rods.

trama: Latin, *trāma*, "warp," "woof"; the context (flesh) of the cap, gills, or stem of a fruitbody.

tremelloid basidium: A basidium divided into compartments by vertical or diagonal septa; a longitudinally septate basidium.

tribe: A taxonomic category below the rank of subfamily and above the rank of genus.

trimitic: Adjective describing a polypore whose fruitbodies have all three major types of hyphae: generative, skeletal, and binding. Compare with monomitic and dimitic.

truncate: Cut off sharply.

tuberculate: Having small wart-like structures.

turbinate: Shaped like a top.

type: Specifically, the holotype or *holotypus*; the single specimen specified by the naming author on which the description of a new taxon is based and to which its scientific name is permanently attached. Compare with paratype, syntype, neotype, epitype, and lectotype.

type genus: An unofficial term of convenience in botanical nomenclature that has no official status. Type specimens are used in botanical nomenclature for all taxa up to the rank of family. In botany, the type genus is unofficially used just as it is in zoology; it is the genus after which the family is named and to which the type species belongs. In zoology, the name of a family is formally and permanently affixed to a designated type genus.

type locality: The specific geographic location where the type specimen(s) of a taxon was/were collected.

type species: The species that contains the type specimens and to which the name of a genus or subgenus is permanently linked.

type specimen (*typus, holotypus*): The single specimen, sample, or illustration of an organism designated or used by the author of a new species or infraspecific taxon (subspecies, variety, form, etc.) in the original description of that taxon and to which the name is permanently attached.

umbilicate: Centrally depressed, especially pertaining to the cap of a mushroom.

umbo: A rounded, convex, knob-like swelling in the center of a mushroom's cap (pileus).

umbonate: Having a rounded, humped, or dome-like shape, especially pertaining to the central disk of the cap of a mushroom.

universal veil: The membrane initially surrounding the entire fruitbody of a mushroom.

utriform: Having a slight constriction below an expanded and rounded head; bladder-shaped.

var., *varietas*: Latin, "variety." Not usually italicized when preceding the epithet as part of a scientific name.

ventricose: Expanded or inflated in the middle (may be on one side), urn-shaped.

verrucose: Warty, ornamented with warts.

vinaceous: A deep wine-red color.

viscid: Extremely slimy and sticky.

volva: A cup-like remnant of the universal veil at the base of the stipe in some mushrooms; particularly characteristic of the genus *Amanita*.

white rot: Wood decay by basidiomycetous fungi in which both the cellulose and lignin are degraded but at different rates, with some cellulose remaining until very late stages.

xeric: Dry.

xylem: The dead woody tissue in plants that transports water and nutrients from the soil.

zonate: Having multiple, often concentric zones of alternating color and/or texture.

Zygomycota, zygomycete: A former (and now defunct) polyphyletic division (phylum) of unrelated lower fungi that has been split up into several other divisions/phyla and unranked clades. The only feature zygomycetes had in common was their form of sexual reproduction by means of conjugation and the generation of zygospores, a feature they also share with the filamentous green algae (Chlorophyta).

zygospore: A thick-walled sexual spore formed by the conjugation (fusion) of two similar gametangia followed by karyogamy; characteristic of the most primitive of the terrestrial fungi, including the food molds.

zygote: A diploid cell in which two compatible haploid nuclei have undergone karyogamy (fusion).

Index of Common Names

Aborted Entoloma, 167
Aborted Polypore, 529–530
Abrupt-bulbed Amanita, 283
"Agaricus blazei", 404–406
Alcohol Ink Cap, 272–273
Amanita Eater, 165
Amanita Mold, 165
Amanitas
  Abrupt-bulbed Amanita, 283
  American Caesar, 284
  American Fly Agaric (Yellow Morph), 287–288
  American Tawny Grisette, 305–306
  Brown Blusher, 293–294
  Carrot-footed Amanita, 290–291
  Citron Amanita, 292
  Cleft-foot Amanita, 293–294
  Coker's Amanita, 295–296
  Coker's Lepidella, 295–296
  Eastern American Blusher, 299–300
  Eastern Mealy Amanita, 307
  False Caesar, 285–286
  False Fly Agaric, 285–286
  Grey Rags Lepidella, 297–298
  Grisette, 303–304
  Hated Amanita, 308–309
  Peach-colored Fly Agaric, 289
  Rag-veil Amanita, 297–298
  Spurned Amanita, 308–309
American Brown Truffle, 204–207
American Caesar, 284
American Fly Agaric (Yellow Morph), 287–288
American Tawny Grisette, 305–306
American Wood Ear, 105–106
Appalachian Chanterelle, 459
Apple Bolete, 247–248
Artist's Conk, 504–505
Ash-tree Bolete, 261
Autumn Skullcap, 382–383

Barometer Earthstar, 196–197
Bear Lentinus, 352–353
Bearded Tooth, 588–589
Bear's Head Tooth, 588–589
Beautiful Coral, 140–141
Beefsteak Polypore, 535
Beetle Cordyceps, 149–150
Bell Omphalina, 412–413
Berkeley's Polypore, 565–566
Birch Polypore, 567
Bird's Nest Fungi
  Fluted Bird's Nest Fungus, 594–595
  Splash Cups, 594–595
  White-egg Bird's Nest, 596
Black Hairy Earth Tongue, 120–121
Black Jelly Oyster, 359–360
Black Knot, 97–98
Black Morel, 64
Black Raspberry Hypoxylon, 86
Black Trumpet, 451–452
Black Velvet Bolete, 250–251
Black Witches' Butter, 107
Black-footed Polypore, 558
Blackish-red Russula, 424
Black-staining Polypore, 555–556
Bleeding Tooth, 581–582
Blewit, 397–398
Blue-foot Psilocybe, 378–379
Blushing Bracket, 519–520
Bolete-eating Hypomyces, 158–159
Boletes
  Apple Bolete, 247–248
  Ash-tree Bolete, 261
  Black Velvet Bolete, 250–251
  Corrugated Bolete, 237–238
  Eastern Painted Suillus, 257–260
  Eastern White Leccinium, 249
  Firm Bolete, 233–234
  Frost's Bolete, 247–248
  Golden-pored Bolete, 239–240
  Long-curved Stem Bolete, 235–236
  Old Man of the Woods, 224–227
  Ornate-stalked Bolete, 241–242

Boletes (*continued*)
Parasitic Bolete, 264
Peppery Bolete, 254
Red-mouth Bolete, 231–232
Russell's Bolete, 243–244
Shaggy-stalked Bolete, 245–246
Two-colored Bolete, 229–230
Variegated Suillus, 255–256
Violet-grey Bolete, 252–253
Yellow-gilled Bolete, 262–263
Yellow-stain Bolete, 228
Bracelet Cort, 332–333
Brick Cap, 390–391
Brick Tuft, 390–391
Bristly Bracket, 508–509
Brittle Cinder, 78–79
Brown Blusher, 293–294
Brown Witches' Butter, 104
Brown-staining Cheese Polypore, 500
Brown-toothed Crust, 482
Bubblegum Fungus/Cuban Club, 89–90

Candlesnuff Fungus, 87
Canker Rot, 521–522
Carbon Antlers, 87
Carbon Balls, 80–82
Carbon Cushion, 84–85
Carbon-like Fungi
Black Knot, 97
Black Raspberry Hypoxylon, 86
Brittle Cinder, 78–79
Bubblegum Fungus/Cuban Club, 89–90
Candlesnuff Fungus, 87
Carbon Antlers, 87
Carbon Balls, 80–82
Carbon Cushion, 84–85
Common Tarcrust, 76
Cramp Balls, 80–82
Cuban Club/Bubblegum Fungus, 89–90
Dead Man's Fingers, 88
Fairy Sparklers, 91–92
Hickorynut Club, 95–96
Hypoxylon canker, 75
Magnolia Cone Fungus, 94
Red Carbon Cushion, 83
Sweetgum Fruit Fungus, 93
Woodwort, 84–85
Yellow Sack Fungus, 77
Carrot-footed Amanita, 290–291
Cauliflower Mushroom, 523–524
Cedar Apple Rust, 470–471
Celandine Milky, 430–431
Ceramic Parchment, 477
Chanterelle, 457–458
Chanterelles
Appalachian Chanterelle, 459
Black Trumpet, 451–452
Chanterelle, 457–458
Cinnabar-red Chanterelle, 465–466
Conjoined Chanterelle, 462
False Chanterelle, 456
Fragrant Chanterelle, 455–456
Golden Chanterelle, 457–458
Peach-colored Chanterelle, 464
Scaly Vase Chanterelle, 448–450
Small Chanterelle, 463
Smooth Chanterelle, 460–461
Trumpet Chanterelle, 453–454
Winter Chanterelle, 453–454
Charming Marasmius, 409–410
Charming Tooth, 585–586
Chestnut Blight, 468–469
Chicken of the Woods, 544–545
Chocolate Tube Slimes, 615–616
Cinnabar-red Chanterelle, 465–466
Cinnabar-red Polypore, 533
Citron Amanita, 292
Cleft-foot Amanita, 293–294
Club-like Tuning Fork, 126
Clubs. *See* Earth Tongues and Clubs
Coalescent Tooth, 579–580
Coalescing White Jellyfungi, 111–112
Cobalt Crust, 483–484
Coker's Amanita, 295–296
Coker's Lepidella, 295–296
Common Ink Cap, 272–273
Common Puffball, 191–192
Common Splitgill, 357–358
Common Stinkhorn, 214–215
Common Tarcrust, 76
Conifer Parchment, 488
Conjoined Chanterelle, 462
Coral Slime, 599–600
Coral Tooth Fungus, 590
Coral-pink Merulius, 536–537
Coral-pink Phlebia, 536–537

Corals
 Beautiful Coral, 140–141
 Crested Coral Fungus, 134–136
 Crown-tipped Coral Fungus, 143–144
 False Coral Fungus, 113–114
 Grey Coral, 134–136
 Magenta Coral, 130
 Straight-branched Coral, 142
 White Coral, 133
 White Spindles, 131–132
 White Worm Coral, 131–132
 Yellow Antler Coral, 139
 Yellow-tipped Coral, 140–141
Corrugated Bolete, 237–238
Corrugated Milkcap, 438–440
Corticioid Fungi. *See* Crusts, Spreads, and Parchment Fungi
Corts (Webcaps)
 Bracelet Cort, 332–333
 Gypsy, The, 334–335
 Hesler's Cort, 329–330
 Maryland Cort, 327
 Red-banded Cort, 332–333
 Red-gilled Cort, 326
 Spotted Cort, 325
 Viscid Violet Cort, 325
 Yellow Webcap, 331
Cotton Candy Slime, 613
Cracked Cap Polypore, 550–551
Cramp Balls, 80–82
Crested Coral Fungus, 134–136
Crested Polypore, 531–532
Cross-veined Troop Mushroom, 412–413
Crowded Parchment, 489–490
Crown-tipped Coral Fungus, 143–144
Crusts, Spreads, and Parchment Fungi
 Brown-toothed Crust, 482
 Ceramic Parchment, 477
 Cobalt Crust, 483–484
 Common Tarcrust, 76
 Conifer Parchment, 488
 Crowded Parchment, 489–490
 False Turkey Tail, 491–492
 Hophornbeam Disc, 472–473
 Job's Brown Crust, 480–481
 Purple Crust Fungus, 485
 Rosy Crust, 487
 Snow Dust Fungus, 476
 Trembling Phlebia, 538–539
 Velvet Blue Spread, 483–484
 White-and-red Crust, 486
Crystal Brain Fungus, 110
Cuban Club/Bugglegum Fungus, 89–90
Cup-bearing Agaric, 402
Cups and Disks
 Devil's Urn, 170–171
 Green Elfcup, 177
 Green-stain Complex, 177
 Hairy Rubber Cup, 172–173
 Lemon Fairy Drops, 179–180
 Orange Eyelash Cup, 178
 Orange Peel Fungus, 174
 Palomino Cup, 175–176
 Peanut Butter Cup, 172–173
 Purple Jellydisc, 182–183
 Purple Jellydrops, 182–183
 Recurved Cup, 175–176
 Smooth Patch Disease, 472–473
 Stalked Hairy Fairy Cup, 180–181

Dead Man's Fingers, 88
Dead Man's Rubber Gloves, 110
Deadly Galerina, 382–383
Deceptive Trichia, 609–610
Deer Mushroom, 384–385
Deer Truffles, 202–203
derrumbe/Landslide Mushroom, 375–377
Devil's Stinkhorn, 214–215
Devil's Tooth, 581–582
Devil's Urn, 170–171
Disks. *See* Cups and Disks
Dispersed Milkcaps, 432–433
Dog Vomit Slime, 601–602
Dryad's Saddle, 562–563
Dye Polypore, 553–554

Earth Tongues and Clubs
 Black Hairy Earth Tongue, 120–121
 Club-like Tuning Fork, 126
 Flat-topped Club Coral, 124–125
 Golden Spindle Club Fungus, 137–138
 Green-capped Jelly Club, 128–129
 Hickory Nut Club, 95–96
 Jelly Babies, 128–129
 Little Periscope, 127
 Ochre Jelly Club, 128–129
 Orange Earth Tongue, 123

Earth Tongues and Clubs (*continued*)
- Pestle-shaped Club Coral, 124–125
- Vase Thelephore, 145–146
- Velvety Fairy Fan, 122–123
- Yellow Spindle Club Fungus, 137–138

Earthstar, 210–211
Eastern American Blusher, 299–300
Eastern Destroying Angel, 301–302
Eastern Flat-top Agaricus, 403
Eastern Mealy Amanita, 307
Eastern Painted Suillus, 257–260
Eastern White Leccinum, 249
Elegant Stinkhorn, 218–219
Elegant Trametes, 510–511
Elfin Saddles, 68–69
Emetic Russulas, 420–421

Fairy Bonnet, 276–277
Fairy Pins, 166
Fairy Ring Mushroom, 414
Fairy Sparklers, 91–92
False Caesar, 285–286
False Chanterelle, 456
False Coral Fungus, 113–114
False Fly Agaric, 285–286
False Morels, 66–67
False Parasol, 311–312
False Truffle, 210–211
False Turkey Tail, 491–492
Fawn Mushroom, 384–385
Firm Bolete, 233–234
Flat-topped Club Coral, 124–125
Fleecy Milkcap, 434–435
Fluted Bird's Nest Fungus, 594–595
Fragrant Chanterelle, 455–456
Frost's Bolete, 247–248

Gilled Polypore, 570
Glaucous Milkcap, 436–437
Goblet Waxy Cap, 366–367
Golden Ear, 103
Golden Hypomyces, 163–164
Golden Pholiota Complex, 319
Golden Spindle Club Fungus, 137–138
Golden-pored Bolete, 239–240
Goldenthread Cordyceps, 157
Green Elfcup, 177
Green-capped Jelly Club, 128–129
Green-quilted Russula, 427–429
Green-spored Lepiota, 311–312
Green-stain Complex, 177
Grey Coral, 134–136
Grey Rags Lepidella, 297–298
Grisette, 303–304
Gypsy, The, 334–335

Hairy Rubber Cup, 172–173
Half-free Morel, 64
Hated Amanita, 308–309
Head-like Cordyceps, 156
Hedgehog Pholiota, 322
Hedgehog Puffball, 191–192
Helvella, 68–69
Hemlock Varnish Shelf, 501–503
Hen-of-the-woods, 527–528
Hesler's Cort, 329–330
Hexagonal-pored Polypore, 559–560
Hickory Nut Club, 95–96
Honey Mushrooms, 368–370
Hophornbeam Disc, 472–473
Hypoxylon Canker, 75

Indigo Milkcap, 441–442
Indigo Milky, 441–442
Ink Caps
- Alcohol Ink Cap, 272–273
- Common Ink Cap, 272–273
- Fairy Bonnet, 276–277
- Mica Cap Complex, 270–271
- Non-inky Coprinellus, 276–277
- Orange-mat Coprinus Species Complex, 274–275
- Scaly Inky Cap, 278–279
- Shaggy Ink Cap, 268–269
- Shaggy Mane, 268–269
- Variegated Inky Cap, 278–279

Insect Eaters
- Beetle Cordyceps, 149–150
- Goldenthread Cordyceps, 157
- Head-like Cordyceps, 156
- Long-spored Cordyceps, 156
- Military Cordyceps, 151
- Snake-tongued Cordyceps, 157

Jack O'Lantern Mushroom, 348–349
Japanese Lantern Slime, 614

Jelly-like Fungi
- American Wood Ear, 105–106
- Black Jelly Oyster, 359–360
- Black Witches' Butter, 107
- Brown Witches' Butter, 104
- Coalescing White Jellyfungi, 111–112
- Crystal Brain Fungus, 110
- Dead Man's Rubber Gloves, 110
- Golden Ear, 103
- Orange Jelly, 102
- Orange Witches' Butter, 102
- Purple Jellydisc, 182–183
- Purple Jellydrops, 182–183
- Silver Ear Fungus, 108–109
- Snow Fungus, 108–109
- Trembling Phlebia, 538–539
- Yellow Witches' Butter, 101

Job's Brown Crust, 480–481

Kidney-shaped Tooth, 583–584

Landslide Mushroom/*derrumbe*, 375–377
Lemon Fairy Drops, 179, 180
Lepiotas
- False Parasol, 311–312
- Green-spored Lepiota, 311–312
- Malodorous Lepiota, 315–316
- Onion-stalked Lepiota, 313–314
- Red-tinged Lepiota, 317

Lingzhi, 501–503
Lion's Mane, 588–589
Little Nest Polypore, 512–513
Little Periscope, 127
Little Red Stinkhorn, 218–219
Lobster Mushroom, 160–161
Long-curved Stem Bolete, 235–236
Long-spored Cordyceps, 156
Lorchels, 66–67
Luminescent Panellus, 350–351

Magenta Coral, 130
Magic Mushroom, 373–374
Magic Mushrooms
- Blue-foot Psilocybe, 378–379
- *derrumbe*, 375–377
- Landslide Mushroom, 375–377
- *Magic Mushroom*, 373–374

Magnolia Cone Fungus, 94
Magnolia Cone Mushroom, 416–417
Maitake, 527–528
Malodorous Lepiota, 315–316
Many-headed Slime, 603–604
Maroon Ear, 361
Maryland Cort, 327
Mica Cap Complex, 270–271
Military Cordyceps, 151
Milkcaps
- Celandine Milky, 430–431
- Corrugated Milkcap, 438–440
- Dispersed Milkcaps, 432–433
- Fleecy Milkcap, 434–435
- Glaucous Milkcap, 436–437
- Indigo Milkcap, 441–442
- Indigo Milky, 441–442
- Salmon-staining Milkcap, 443–444
- Tawny Milkcap, 438–440

Milk-white Toothed Polypore, 475
Milky False Truffle, 208–209
Molds and Rusts. *See also* Crusts, Spreads, and Parchment Fungi
- Aborted Entoloma, 167
- Amanita Eater, 165
- Amanita Mold, 165
- Black Knot, 97–98
- Bolete-eating Hypomyces, 158–159
- Canker Rot, 521–522
- Cedar Apple Rust, 470–471
- Chestnut Blight, 468–469
- Fairy Pins, 166
- Golden Hypomyces, 163–164
- Lobster Mushroom, 160–161
- Yellow-green Russula Mold, 162

Morels, False Morels, and Elfin Saddles
- Black Morel, 64
- Elfin Saddle, 68–69
- False Morel, 66–67
- Half-free Morel, 64
- Helvella, 68–69
- Lorchels, 66–67
- North American Yellow Morel, 63–65

Mossy Maze Polypore, 521–522
Multigoblet Slime, 608
Mustard-yellow Polypore, 546–547

Narrowleafed Cockleshell, 354
Netted Stinkhorn, 216–217

Non-inky Coprinellus, 276–277
North American Yellow Morel, 63–65

Oak Mazegill Polypore, 518
Ochre Jelly Club, 128–129
Ochre Spreading Tooth, 478–479
Old Man of the Woods, 224–227
Onion-stalked Lepiota, 313–314
Orange Earth Tongue, 123
Orange Eyelash Cup, 178
Orange Jelly, 102
Orange Mock Oyster, 346–347
Orange Peel Fungus, 174
Orange Pinwheel Marasmius, 411
Orange Rough-cap Tooth, 577–578
Orange Witches' Butter, 102
Orange-mat Coprinus Species Complex, 274–275
Orange-staining Puffball, 193
Ornate-stalked Bolete, 241–242
Oyster Mushroom, 340–341
Oyster-like Mushrooms
  Bear Lentinus, 352–353
  Black Jelly Oyster, 359–360
  Common Splitgill, 357–358
  Jack O'Lantern Mushroom, 348–349
  Luminescent Panellus, 350–351
  Maroon Ear, 361
  Narrowleafed Cockleshell, 354
  Orange Mock Oyster, 346–347
  Oyster Mushroom, 340–341
  Scaly Sawgill, 355–356
  Smoked Oysterling, 359–360
  Smooth Lentinus, 342–343
  Train Wrecker, 355–356
  Veiled Oyster, 344–345

Palomino Cup, 175–176
Parasitic Bolete, 264
Parchment-like Fungi. *See also* Crusts, Spreads, and Parchment Fungi
  Ceramic Parchment, 477
  Conifer Parchment, 488
  Crowded Parchment, 489–490
  False Turkey Tail, 491–492
Parrot Mushroom, 363
Peach-colored Chanterelle, 464
Peach-colored Fly Agaric, 289
Peanut Butter Cup, 172–173
Pear-shaped Puffball, 191–192
Pecan Truffle, 204–207
Peppery Bolete, 254
Pestle-shaped Club Coral, 124–125
Petal Cone Slime, 611–612
Petticoat Mottlegill, 380
Pheasant's Back Polypore, 562–563
Pine Bracket, 548–549
Pink Toothpaste Slime, 605
Platterful Mushroom, 392
Polypores
  Aborted Polypore, 529–530
  Artist's Conk, 504–505
  Beefsteak Polypore, 535
  Berkeley's Polypore, 565–566
  Birch Polypore, 567
  Black-footed Polypore, 558
  Black-staining Polypore, 555–556
  Blushing Bracket, 519–520
  Bristly Bracket, 508–509
  Brown-staining Cheese Polypore, 500
  Cauliflower Mushroom, 523–524
  Chicken of the Woods, 544–545
  Cinnabar-red Polypore, 533
  Coral-pink Merulius, 536–537
  Coral-pink Phlebia, 536–537
  Cracked Cap Polypore, 550–551
  Crested Polypore, 531–532
  Dryad's Saddle, 562–563
  Dye Polypore, 553–554
  Elegant Trametes, 510–511
  Gilled Polypore, 570
  Hemlock Varnish Shelf, 501–503
  Hen-of-the-woods, 527–528
  Hexagonal-pored Polypore, 559–560
  Lingzhi, 501–503
  Little Nest Polypore, 512–513
  Maitake, 527–528
  Milk-white Toothed Polypore, 475
  Mossy Maze Polypore, 521–522
  Mustard-yellow Polypore, 546–547
  Oak Mazegill Polypore, 518
  Pheasant's Back Polypore, 562–563
  Pine Bracket, 548–549
  Red Heart Fungus, 548–549
  Reishi, 501–503

- Resinous Polypore, 542–543
- Rosy Conk, 534
- Rusty-gilled Polypore, 571–572
- Shaggy Bracket, 508–509
- Smoky Cheese Polypore, 498–499
- Smoky Polypore, 552
- Spongy Tooth Polypore, 568–569
- Spring Polypore, 561
- Stalked White Polypore, 557
- Sulphur Shelf, 544–545
- Thick Oak Maze Polypore, 518
- Thin-maze Flat Polypore, 519–520
- Trembling Merulius, 538–539
- Turkey Tail, 506–507
- Umbrella Polypore, 525–526
- Violet-tooth Polypore (Conifers), 516–517
- Violet-tooth Polypore (Hardwoods), 514–515
- Warted Oak Polypore, 540–541
- White Cheese Polypore, 496–497
- White-rimmed Vinaceous Polypore, 564
- Yellow-red Gilled Polypore, 571–572

Powder-scale Pholiota, 322
Pretzel Slime, 607
Puffballs
- Barometer Earthstar, 196–197
- Common Puffball, 191–192
- Earthstar, 210–211
- Hedgehog Puffball, 191–192
- Orange-staining Puffball, 193
- Pear-shaped Puffball, 191–192
- Purple-spored Puffball, 194–195
- Stalked Puffballs, 200–201
- Stalked Puffballs-in-Aspic, 198–199

Purple Crust Fungus, 485
Purple Jellydisc, 182–183
Purple Jellydrops, 182–183
Purple-gilled Laccaria, 399
Purple-spored Puffball, 194–195

Rag-veil Amanita, 297–298
Ravenel's Stinkhorn, 214–215
Recurved Cup, 175–176
Red Carbon Cushion, 83
Red Heart Fungus, 548–549
Red Raspberry Slime, 606
Red-banded Cort, 332–333
Red-capped and Rosy-stemmed Russulas, 422–423
Red-capped and White-stemmed Russulas, 420–421
Red-gilled Cort, 326
Red-mouth Bolete, 231–232
Red-tinged Lepiota, 317
Reishi, 501–503
Resinous Polypore, 542–543
Rooted Agaric, 336–338
Rooted Collybia, 336–338
Rosy Conk, 534
Rosy Crust, 487
Russell's Bolete, 243–244
Russulas
- Blackish-red Russula, 424
- Emetic Russulas, 420–421
- Green-quilted Russula, 427–429
- Red-capped and Rosy-stemmed Russulas, 422–423
- Red-capped and White-stemmed Russulas, 420–421
- Short-stemmed Russula, 425–426
- "Sickeners, The," 420–421
- White Russula, 424

Rusts. *See* Molds and Rusts
Rusty-gilled Polypore, 571–572

Salmon-staining Milkcap, 443–444
Scaly Inky Cap, 278–279
Scaly Pholiota Complex, 320–321
Scaly Sawgill, 355–356
Scaly Vase Chanterelle, 448–450
Scarlet Waxy Caps, 364–365
Scotch Bonnet, 414
Scrambled Eggs Slime, 601–602
Semiovate Panaeolus, 381
Shaggy Bracket, 508–509
Shaggy Ink Cap, 268–269
Shaggy Mane, 268–269
Shaggy-stalked Bolete, 245–246
Short-stemmed Russula, 425–426
"Sickeners, The," 420–421
Silver Ear Fungus, 108–109
Slime Molds, 597–616
- Chocolate Tube Slimes, 615–616
- Coral Slime, 599–600
- Cotton Candy Slime, 613
- Deceptive Trichia, 609–610
- Dog Vomit Slime, 601–602
- Japanese Lantern Slime, 614
- Many-headed Slime, 603–604

Slime Molds (*continued*)
- Multigoblet Slime, 608
- Petal Cone Slime, 611–612
- Pink Toothpaste Slime, 605
- Pretzel Slime, 607
- Red Raspberry Slime, 606
- Scrambled Eggs Slime, 601–602
- Wasp Nest Slime, 608
- Wolf's Milk Slime, 605

Small Chanterelle, 463
Smoked Oysterling, 359–360
Smoky Cheese Polypore, 498–499
Smoky Polypore, 552
Smooth Chanterelle, 460–461
Smooth Lentinus, 342–343
Smooth Patch Disease, 472–473
Snake-tongued Cordyceps, 157
Snow Dust Fungus, 476
Snow Fungus, 108–109
Southeastern Matsutake, 393–394
Splash Cups, 594–595
Spongy Tooth Polypore, 568–569
Spotted Cort, 325
Spreads. *See* Crusts, Spreads, and Parchment Fungi
Spring Polypore, 561
Spurned Amanita, 308–309
Stalked Hairy Fairy Cup, 180–181
Stalked Puffballs, 200–201
Stalked Puffballs-in-Aspic, 198–199
Stalked White Polypore, 557
Stinkhorns
- Common Stinkhorn, 214–215
- Devil's Stinkhorn, 214–215
- Elegant Stinkhorn, 218–219
- Little Red Stinkhorn, 218–219
- Netted Stinkhorn, 216–217
- Ravenel's Stinkhorn, 214–215
- Stinky Squid, 220–221
- Veiled Stinkhorn, 216–217

Stinky Squid, 220–221
Straight-branched Coral, 142
Stropharia-like Mushrooms
- Golden Pholiota Complex, 319
- Hedgehog Pholiota, 322
- Powder-scale Pholiota, 322
- Scaly Pholiota Complex, 320–321
- Wine-cap Stropharia, 323

Sulphur Shelf, 544–545
Sulphur Tuft, 388–389
Sweetgum Fruit Fungus, 93

Tawny Milkcap, 438–440
Thick Oak Maze Polypore, 518
Thin-maze Flat Polypore, 519–520
Toothed Fungi
- Bearded Tooth, 588–589
- Bear's Head Tooth, 588–589
- Bleeding Tooth, 581–582
- Brown-toothed Crust, 482
- Charming Tooth, 585–586
- Coalescent Tooth, 579–580
- Coral Tooth Fungus, 590
- Devil's Tooth, 581–582
- Kidney-shaped Tooth, 583–584
- Lion's Mane, 588–589
- Ochre Spreading Tooth, 478–479
- Orange Rough-cap Tooth, 577–578
- Wood Hedgehog, 575–576
- Zoned Tooth, 579–580

Train Wrecker, 355–356
Trembling Merulius, 538–539
Trembling Phlebia, 538–539
Trooping Club, 151
Truffles
- American Brown Truffle, 204–207
- Deer Truffles, 202–203
- False Truffle, 210–211
- Milky False Truffle, 208–209
- Pecan Truffle, 204–207

Trumpet Chanterelle, 453–454
Turkey Tail, 506–507
Two-colored Bolete, 229–230

Umbrella Polypore, 525–526

Variegated Inky Cap, 278–279
Variegated Suillus, 255–256
Vase Thelephore, 145–146
Veiled Oyster, 344–345
Veiled Stinkhorn, 216–217
Velvet Blue Spread, 483–484
Velvet Paxillus, 395
Velvet-footed Pax, 395
Velvety Fairy Fan, 122–123

Velvety Psathyrella, 386–387
Violet Collybia, 400–401
Violet-grey Bolete, 252–253
Violet-tooth Polypore (Conifers), 516–517
Violet-tooth Polypore (Hardwoods), 514–515
Viscid Violet Cort, 325

Warted Oak Polypore, 540–541
Wasp Nest Slime, 608
Webcaps. *See* Corts (Webcaps)
White Cheese Polypore, 496–497
White Coral, 133
White Hardwood Leucopax, 396
White Marasmius, 415
White Patches, 474
White Russula, 424
White Spindles, 131–132
White Worm Coral, 131–132
White-and-red Crust, 486
White-egg Bird's Nest, 596
White-rimmed Vinaceous Polypore, 564
Wine-cap Stropharia, 323
Winter Chanterelle, 453–454
Wolf's Milk Slime, 605
Wood Blewit, 397–398
Wood Hedgehog, 575–576
Woodwort, 84–85

Yellow Antler Coral, 139
Yellow Morel, North American, 63–65
Yellow Sack Fungus, 77
Yellow Spindle Club Fungus, 137–138
Yellow Trumpet Mushroom, 407
Yellow Unicorn Entoloma, 408
Yellow Webcap, 331
Yellow Witches' Butter, 101
Yellow-gilled Bolete, 262–263
Yellow-green Russula Mold, 162
Yellow-red Gilled Polypore, 571–572
Yellow-stain Bolete, 228
Yellow-tipped Coral, 140–141

Zoned Tooth, 579–580

Index of Scientific Names

Abortiporus biennis, 529–530
Agaricus placomyces, 403
Agaricus pocillator, 402
Agaricus subrufescens, 404–406
Albatrellus cristatus, 531–532
Aleuria aurantia, 174
Aleurodiscus oakesii, 472–473
Amanita abrupta, 283
Amanita bisporigera, 301–302
Amanita brunnescens, 293–294
Amanita cf. *cinereopannosa*, 297–298
Amanita aff. *citrina*, 292
Amanita cokeri, 295–296
Amanita daucipes, 290–291
Amanita farinosa, 307
Amanita aff. *fulva*, 305–306
Amanita jacksonii, 284
Amanita muscaria var. *guessowii*, 287–288
Amanita muscaria var. *persicina*, 289
Amanita parcivolvata, 285–286
Amanita aff. *rubescens*, 299–300
Amanita spreta, 308–309
Amanita vaginata, 303–304
Annulohypoxylon cohaerens, 84–86
Annulohypoxylon multiforme, 84–85
Annulohypoxylon truncatum, 86
Anthracophyllum lateritium, 361
Arcyria denudata, 613
Armillaria gallica, 368–370
Armillaria mellea, 368–370
Armillaria tabescens, 368–370
Artomyces pyxidatus, 143–144
Ascocoryne cylichnium, 182–183
Ascocoryne sarcoides, 182–183
Astraeus morganii, 196–197
Athelia epiphylla, 476
Aureoboletus auriporus, 239–240
Auricularia americana, 105–106

Biscogniauxia atropunctata, 75
Bisporella citrina, 179–180
Bisporella sulfurina, 179–180
Bjerkandera adusta, 552
Boletellus russellii, 243–244
Boletinellus merulioides, 261
Boletus bicolor, 229–230
Boletus curtisii, 228
Boletus firmus, 233–234
Boletus hortonii, 237–238
Boletus longicurvipes, 235–236
Boletus subvelutipes, 231–232
Bondarzewia berkeleyi, 565–566

Calocera cornea, 126
Calostoma cinnabarinum, 198–199
Calostoma lutescens, 198–199
Calostoma ravenelii, 198–199
Calvatia cyathiformis, 194–195
Calvatia rubroflava, 193
Cantharellus appalachiensis, 459
Cantharellus cibarius, 457–458
Cantharellus cinnabarinus, 465–466
Cantharellus confluens, 462
Cantharellus lateritius, 460–461
Cantharellus minor, 463
Cantharellus persicinus, 464
Ceratiomyxa fruticulosa var. *fruticulosa*, 599–600
Ceratiomyxa fruticulosa var. *poroides*, 599–600
Cerrena unicolor, 521–522
Chalciporus piperatus, 254
Chlorociboria aeruginascens, 177
Chlorociboria aeruginosa, 177
Chlorophyllum molybdites, 311–312
Chlorosplenium chlora, 180
Clavaria fragilis, 131–132
Clavaria zollingeri, 130
Clavariadelphus americanus, 124–125
Clavariadelphus pistillaris, 124–125
Clavariadelphus truncatus, 124–125
Clavulina cinerea, 134–136
Clavulina coralloides, 134–136
Clavulina cristata, 134*n*, 136*n*

Clavulinopsis aurantiocinnabarina, 137–138
Clavulinopsis corniculata, 139
Clavulinopsis fusiformis, 137–138
Climacodon pulcherrimus, 585–586
Clitocybe nuda, 397–398
Collybia iocephala, 400–401
Coprinellus disseminatus, 276–277
Coprinellus domesticus, 274–275
Coprinellus micaceus, 270–271
Coprinellus radians, 274–275
Coprinellus truncorum, 270–271
Coprinopsis atramentaria, 272–273
Coprinopsis variegata, 278–279
Coprinus comatus, 268–269
Cordyceps militaris, 151
Cortinarius armillatus, 332–333
Cortinarius caperatus, 334–335
Cortinarius delibutus, 331
Cortinarius hesleri, 329–330
Cortinarius iodes, 325
"Cortinarius marylandensis," 327
Cortinarius cf. *semisanguineus*, 326
Coryne dubia, 182–183
Craterellus fallax, 451–452
Craterellus odoratus, 455–456
Craterellus tubaeformis, 453–454
Cribraria cancellata, 614
Crucibulum laeve, 596
Cryphonectria parasitica, 468–469
Cyathus striatus, 594–595

Dacrymyces palmatus, 102
Dacrymyces stillatus, 180
Dacryopinax elegans, 127
Dacryopinax spathularia, 127
Daedalea quercina, 518
Daedaleopsis confragosa, 519–520
Daldinia childiae, 80–82
Daldinia petriniae, 80–82
Daldinia vernicosa, 80–82
Dasyscyphus virgineus, 180–181
Diatrype stigma, 76

Elaphocordyceps capitata, 156
Elaphocordyceps longisementis, 156
Elaphocordyceps ophioglossoides, 157
Elaphomyces spp., 202–203
Entoloma abortivum, 167
Entoloma murrayi, 408
Entonaema liquescens, 77
Exidia glandulosa, 107
Exsudoporus frostii, 247–248

Fistulina hepatica, 535
Flammulaster erinaceellus, 322
Fomitopsis cajanderi, 534
Fuligo septica, 601–602

Galerina marginata, 382–383
Ganoderma applanatum, 504–505
Ganoderma curtisii, 501–503
Ganoderma lucidum, 501–503
Ganoderma tsugae, 501–503
Gerronema strombodes, 407
Gliophorus psittacinus, 363
Gloeophyllum sepiarium, 571–572
Grifola frondosa, 527–528
Gymnosporangium juniperi-virginianae, 470–471
Gyromitra carolineana, 66–67
Gyromitra esculenta, 66–67
Gyromitra infula, 66–67

Heimioporus betula, 245–246
Helvella albella, 68–69
Helvella crispa, 68–69
Helvella elastica, 68–69
Helvella macropus, 68–69
Hemitrichia calyculata, 611–612
Hemitrichia serpula, 607
Hericium americanum, 588–589
Hericium coralloides, 590
Hericium erinaceus, 588–589
Hydnellum auranticum, 577–578
Hydnellum concrescens, 579–580
Hydnellum peckii, 581–582
Hydnochaete olivacea, 482
Hydnum repandum, 575–576
Hygrocybe cantharellus, 366–367
Hygrocybe coccinea, 364–365
Hygrocybe punicea, 364–365
Hygrophoropsis aurantiaca, 456

Hymenochaete jobii, 480–481
Hymenopellis spp. Complex, 336–338
Hymenoscyphus calyculus, 180
Hymenoscyphus fructigenus, 180
Hypholoma fasciculare, 388–389
Hypholoma lateritium, 390–391
Hypomyces aurantius, 163–164
Hypomyces chrysospermus, 158–159
Hypomyces hyalinus, 165
Hypomyces lactifluorum, 160–161
Hypomyces luteovirens, 162
Hypoxylon fragiforme, 83
Hypoxylon howeianum, 83

Inonotus dryadeus, 540–541
Irpex lactea, 475
Ischnoderma resinosum, 542–543

Kretzschmaria deusta, 78–79

Laccaria ochropurpurea, 399
Lacrymaria velutina, 386–387
Lactarius chelidonium, 430–431
Lactarius dispersus, 432–433
Lactarius indigo, 441–442
Lactarius speciosus, 432–433
Lactarius subplithogalus, 443–444
Lactifluus corrugis, 438–440
Lactifluus glaucescens, 436–437
Lactifluus hygrophoroides, 438–440
Lactifluus subvellereus, 434–435
Lactifluus volemus, 438–440
Laetiporus sulphureus, 544–545
Leccinellum albellum, 249
Lentinellus angustifolius, 354
Lentinellus ursinus, 352–353
Lenzites betulina, 570
Leotia atrovirens, 128–129
Leotia lubrica, 128–129
Leotia viscosa, 128–129
Lepiota cristata, 315–316
Leucoagaricus rubrotinctus, 317
Leucocoprinus cepaestipes, 313–314
Leucopaxillus laterarius, 396
Lycogala epidendrum, 605
Lycoperdon echinatum, 191–192
Lycoperdon perlatum, 191–192
Lycoperdon pyriforme, 191–192

Marasmiellus candidus, 415
Marasmius delectans, 409–410
Marasmius oreades, 414
Marasmius siccus, 411
Megacollybia rodmani, 392
Meripilus sumstinei, 555–556
Metatrichia vesparium, 608
Microglossum rufum, 123
Microporellus obovatus, 557
Morchella angusticeps, 63–65
Morchella diminutiva, 63–65
Morchella esculentoides, 63–65
Morchella punctipes, 63–65
Morchella virginiana, 63–65
Multiclavula mucinda, 131–132
Mutinus elegans, 218–219
Mutinus ravenelii, 218–219
Mycorrhaphium adustum, 583–584
Myxarium nucleatum, 110

Neolectra irregularis, 123
Neolentinus lepideus, 355–356
Nigroporus vinosus, 564

Oligoporus fragilis, 500
Omphalotus illudens, 348–349
Ophiocordyceps melolonthae, 149–150

Panaeolus papilionaceus, 380
Panaeolus semiovatus, 381
Panellus stipticus, 350–351
Peniophora albobadia, 486
Peniophora gigantia, 488
Peniophora incarnata, 487
Peziza repanda, 175–176
Phaeocalicium polyporaeum, 166
Phaeolus schweinitzii, 553–554
Phallus duplicatus, 216–217
Phallus hadriani, 214–215
Phallus impudicus, 214–215
Phallus ravenelii, 214–215
Phallus rubicundus, 214–215
Phellinus gilvus, 546–547

Phellinus pini, 548–549
Phellinus robiniae, 550–551
Phlebia incarnata, 536–537
Phlebia tremullosa, 538–539
Pholiota aurivella, 319
Pholiota squarrosa, 320–321
Pholiota squarrosoides, 320–321
Phylloporus rhodoxanthus, 262–263
Phyllotopsis nidulans, 346–347
Physarum polycephalum, 603–604
Piptoporus betulinus, 567
Pleurotus dryinus, 344–345
Pleurotus laevis, 342–343
Pleurotus ostreatus, 340–341
Pleurotus pulmonarius, 340–341
Pluteus cervinus, 384–385
Polyporus alveolaris, 559–560
Polyporus arcularius, 561
Polyporus squamosus, 562–563
Polyporus umbellatus, 525–526
Polyporus varius, 558
Porostereum crassum, 485
Pseudoboletus parasiticus, 264
Pseudocolus fusiformis, 220–221
Psilocybe caerulescens, 375–377
Psilocybe cubensis, 373–374
Psilocybe ovoideocystidiata, 378–379

Ramaria formosa, 140–141
Ramaria striata, 142
Ramariopsis kunzei, 133
Ramariopsis lentofragilis, 133
Resupinatus alboniger, 359–360
Resupinatus applicatus, 359–360
Retiboletus ornatipes, 241–242
Russula albidula, 425–426
Russula atropurpurea, 424
Russula brevipes, 425–426
Russula cessans, 420–421
Russula crustosa, 427–429
Russula cystidiosa, 422–423
Russula nobilis, 420–421
Russula parvovirescens, 427–429
Russula pulchra, 422–423
Russula silvicola, 420–421
Russula tennuiceps, 422–423
Russula virescens, 427–429

Schizophyllum commune, 357–358
Scleroderma polyrhizum, 210–211
Scutellinia erinaceus, 178
Sebacina concrescens, 111–112
Sebacina sparassoidea, 111–112
Sparassis americana f. *americana*, 523–524
Sparassis spathulata, 523–524
Spathularia flavida, 123
Spathularia velutipes, 122–123
Spongipellis pachyodon, 568–569
Steccherinum ochraceum, 478–479
Stemonitis spp., 615–616
Stereum complicatum, 489–490
Stereum ostrea, 491–492
Strobilomyces confusus, 224–227
"Strobilomyces dryophilus," 224–227
Strobilomyces floccopus, 224–227
Strobilurus conigenoides, 416–417
Stropharia rugosannulata, 323
Suillus cothurnatus, 259–260
Suillus hirtellus, 255–256
Suillus spraguei, 257–258

Tapinella atrotomentosa, 395
Terana caerula, 483–484
Thelephora vialis, 145–146
Trametes conchifer, 512–513
Trametes elegans, 510–511
Trametes versicolor, 506–507
Trametes villosa, 508–509
Tremella aurantia, 103
Tremella foliacea, 104
Tremella fuciformis, 108–109
Tremella mesenterica, 101
Tremellodendron pallidum, 113–114
Trichaptum abietinum, 516–517
Trichaptum biforme, 514–515
Tricholoma cf. *caligatum*, 393–394
Tricia decipiens, 609–610
Triglossum hirsutum, 120–121
Tuber lyonii species complex, 204–207
Tubifera ferruginosa, 606
Tulostoma spp., 200–201
Tylopilus alboater, 250–251

Tylopilus plumbeoviolaceus, 252–253
Tyromyces chioneus, 496–497
Tyromyces fumidiceps, 498–499

Urnula craterium, 170–171

Xeromphalina companella, 412–413
Xeromphalina kauffmanii, 412–413
Xylaria cubensis, 89–90
Xylaria hypoxylon, 87
Xylaria liquidambar, 93
Xylaria magnoliae, 94
Xylaria oxyacanthae, 95–96
Xylaria polymorpha, 88
Xylaria tentaculata, 91–92
Xylocoremium flabelliforme, 89–90